Optical Lithography

Here Is Why

SECOND EDITION

Optical Lithography

Here Is Why

SECOND EDITION

Burn J. Lin

SPIE PRESS
Bellingham, Washington USA

Library of Congress Cataloging-in-Publication Data

Names: Lin, Burn Jeng, 1942– author.
Title: Optical lithography : here is why / Burn J. Lin.
Description: Second edition. | Bellingham, Washington : SPIE, [2021] | Includes bibliographical references and index.
Identifiers: LCCN 2020049077 (print) | LCCN 2020049078 (ebook) | ISBN 9781510639959 (hardcover) | ISBN 9781510639966 (pdf)
Subjects: LCSH: Microlithography. | Semiconductors–Etching. | Lasers–Industrial applications.
Classification: LCC TK7872.M3 .L56 2021 (print) | LCC TK7872.M3 (ebook) | DDC 621.3815/31–dc22
LC record available at https://lccn.loc.gov/2020049077
LC ebook record available at https://lccn.loc.gov/2020049078

Published by
SPIE
P.O. Box 10
Bellingham, Washington 98227-0010 USA
Phone: +1 360.676.3290
Fax: +1 360.647.1445
Email: books@spie.org
Web: www.spie.org

Printed in the United States of America.
First Printing.
For updates to this book, visit http://spie.org and type "PM329" in the search field.

SPIE.

Contents

Preface to the Second Edition

The first edition of this book was published in 2010. Eleven years is a long time in the field of lithography, even though much of the original content still withstands the passage of time and the advances in the technology nodes. Here are the reasons I decided to update this book:

1. Over these years, I am happy to have fulfilled my original goal of making this book suitable for (1) newcomers to the field who are interesting in growth into a technology career, (2) seasoned professionals who seek more depth into the technology, and (3) managers and executives who seek more breadth. People from each of these groups have told me that the book helped them and that they would recommend it to new readers. With this new edition, I hope to make the book even more useful.
2. For more than three years since my retirement from TSMC in Nov. 2015, I have taught the course Innovative Lithography, using the materials in this book. My students have given me the inspiration and enthusiasm to improve and update the book.
3. Lithography technology is fun to learn, fun to practice, and fun to teach. It is gratifying for me, once again, to preserve this knowledge in a book so that the torch can be passed.

This second edition features the following updates:

Chapter 2 Proximity Printing. This is a completely new chapter. In universities and many research labs, people need to make patterns with less-expensive equipment. Therefore, proximity printing is still very popular, even though there are not many publications on proximity imaging. I show rigorous and approximate ways to simulate the diffracted image at contact and in near, medium, and far fields, as well as the region of validity of these methods. The unintuitive positive- and negative-resist images from the proximity image are plotted. The E-G diagrams (from various authors) for quantifying proximity imaging are also covered. An appendix is provided to document our (two of my graduate students' and my) extensive research on methods to determine the region of validity of the approximation methods studied.

Chapter 3 Exposure Systems. To complement the coverage of historical and current exposure systems for replicating patterns described in the first edition, a carefully prepared step-by-step illustration of the mask and wafer movements in a step-and-scan system has been added. I also clarify that the projection-printed image is a mirror image, despite the common impression that only proximity printing produces a mirror image.

Chapter 4 Image Formation. In addition to adding derivations of the resolution-scaling and DOF-scaling equations and analyses of spatial frequency, light–resist interaction, and resist image development, I almost completely rewrote the section on Zernike polynomials, making their concepts easier for lithography engineers to grasp. The simulated partially coherent images in this chapter have been updated.

Chapter 5 E-D Methodology. This chapter is timeless. I provide a detailed explanation of how to construct E-D trees and emphasize why the log scale is preferred for intensity and exposure. The term apparent exposure —the reciprocal of intensity—is introduced, and its use in E-D diagrams is explained.

Chapter 6 Hardware Components. In this long chapter on lithography components, I added examples of some enlightening resist development phenomena to help people visualize the resist development process. The coverage of chemically amplified resists is expanded. The coverage of wafers, the wafer stage, and alignment systems is also enriched.

Chapter 7 Processing and Optimization. This is another long chapter. I added new insightful derivations of off-axis illumination and a demonstration of extracting the overlay error components with a redundant number of data points to improve the accuracy. Multiple patterning is also extensively discussed, and the G-rule for double patterning is introduced.

Chapter 8 Immersion Lithography. This chapter continues the thorough coverage of this technology from the first edition with an outlook of its extendibility and its impact on the semiconductor technology. The best scaling equations for resolution and DOF are given, and the numerical aperture of the reduction immersion system is clarified.

Chapter 9 EUV Lithography. I almost completely rewrote this chapter, which is quite understandable because of EUVL's rapid growth during the last decade. Given that there are already other books on EUVL, I made sure that my contribution provides a valuable, unique perspective on the technology.

I omitted the material on MEB direct write to leave open the possibility of producing a single volume on this important topic for future development. Finally, I upgraded the figures with color for the eBook. Indeed, all of the figures in the second edition have been modernized.

I credited my wife Sue for her support of the first edition and of my life in general. Before starting my work on this revision, we celebrated our 50th anniversary in 2018. Sue continues to be an indispensable partner for her support during my writing of this second edition and her support in my professional life, family life, and spiritual life.

Burn J. Lin
July 2021

Chapter 1
Introduction

Lithography creates a resist image on the wafer. The subsequent etching, lift off, or ion implantation process is masked by the resist image at the areas dictated by the lithography mask. Hence, the thin-film material on the wafer is selectively removed or built up, or its characteristics are selectively altered. The resist image is produced by replicating the mask pattern, except in the cases of mask making or direct writing on the wafer. Figure 1.1 depicts the mask replication process with an imaging lens. Light from the source is collected by the condenser to illuminate the mask pattern. It passes through the imaging lens to form an aerial image that selectively exposes the resist. After development, the resist image as shown is produced. Figure 1.2 illustrates various forms of image transfer from the resist to the underlying thin film. The film can be isotropically or anisotropically etched, lifted off, plated, or implanted, using the patterned resist as the mask. Detailed descriptions of these transfer processes are given in Chapter 4.

The image formation process is facilitated with an information-carrying beam consisting of photons, electrons, or ions. Optical lithography uses photons to carry out this process. Photons with energy ranging from visible to x-ray wavelengths can be used. However, in this book, the wavelengths of consideration are between 157 and 436 nm, as well as 13.5 nm. These wavelengths are already fully employed in manufacturing semiconductor integrated circuits, except for 157 nm, which was forsaken after a lot of heavily invested learning in research and development, and was replaced by 193-nm water immersion, resulting in an equivalent wavelength of 134 nm. More details on 193-nm immersion lithography are given in Chapter 8. Chapter 9 covers the 13.5 nm extreme-UV (EUV) light that has started to be used for manufacturing 7-nm circuits.

The purpose of this book is to present the working principle of each area in optical lithography by providing examples instead of recipes on *how* to do things. This way, readers understand *why* some techniques are used under certain circumstance and why other techniques are not used. The aim is for

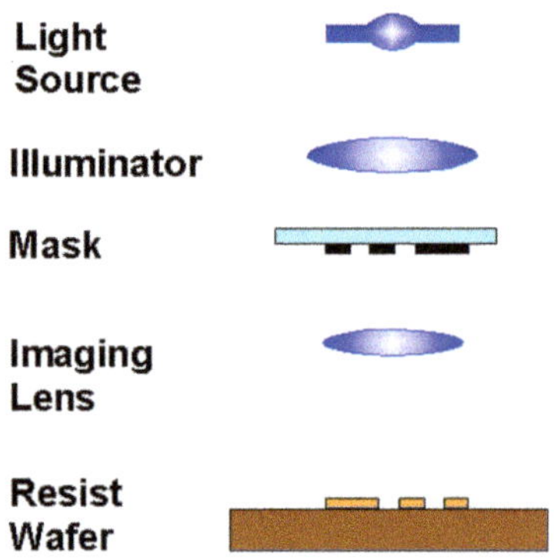

Figure 1.1 Optical lithography replicates the mask pattern through an imaging lens.

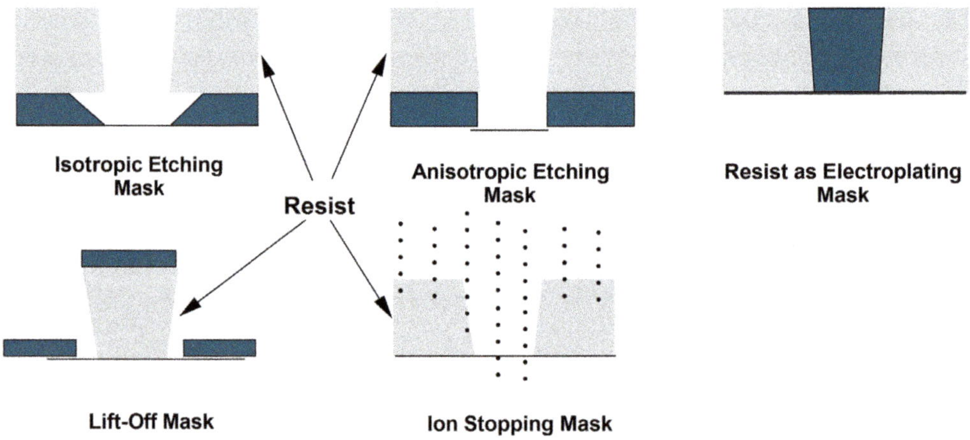

Figure 1.2 Exposed resist image transferred to the underlying thin film by isotropic etching, anisotropic etching, electroplating, lift-off, or ion implant.

readers to become better equipped with the innovation needed to improve the technology and solve problems.

1.1 The Role of Lithography in Integrated Circuit Fabrication

Lithography is an important part of semiconductor manufacturing technology because it is needed for every masking level. In a typical 0.13-μm CMOS integrated circuit fabrication run having 4 metal layers, there are more than 30 masking levels using 474 processing steps, in which 212 steps are related to lithographic exposure and 105 steps are related to pattern transfer using a resist image. For 7-nm CMOS technology, 8 technology nodes later, the number of masking layers is even larger. Relevant steps for a masking levels are listed as follows.

Lithography-exposure-related steps:

- Wafer cleaning and priming to improve adhesion
- Application of anti-reflection coating before and/or after resist coating
- Resist coating
- Post-application bakes
- Exposure/alignment
- Post-exposure bakes
- Resist development
- Resist hardening
- Critical dimension (CD) metrology
- Alignment metrology
- Selective removal of anti-reflection coating
- Stripping of the anti-reflection coating and the resist

Pattern-transfer-related steps:

- Etching
- Ion implantation
- CD metrology
- Surface preparation
- Electroplating

Note that the actual sequence and number of processing steps may vary depending on the particular masking level. For example, the bottom anti-reflection coating (BARC) at the wafer–resist interface is deposited on the wafer prior to priming and resist coating if the BARC is inorganic. Otherwise, it is applied between the priming and resist coating steps. In the latter case, an extra baking step is required for the BARC. A top anti-reflection coating (TARC) may be applied after the resist coating and baking with or without the BARC. Another baking step follows the application of the TARC. A block diagram showing one possible sequence and the relation of these processing steps is shown in Fig. 1.3.

Lithography is important not only because it is needed for all masking levels, but also because it is often the limiting factor for entering the next technology node. For every node, the minimum feature sizes and their separations are reduced by a factor of $\sqrt{2}$. This way, the circuit density is reduced by a factor of 2. Therefore, the generation following 1-μm lithography is 0.7 μm, followed by 0.5 μm, 0.35 μm, 0.25 μm, 0.18 μm, 0.13 μm, etc. As the technology continues to advance, the node names are no longer being defined by the minimum feature size. Additionally, the names do not strictly follow the 70% multiplication factor. Nevertheless, the trend is

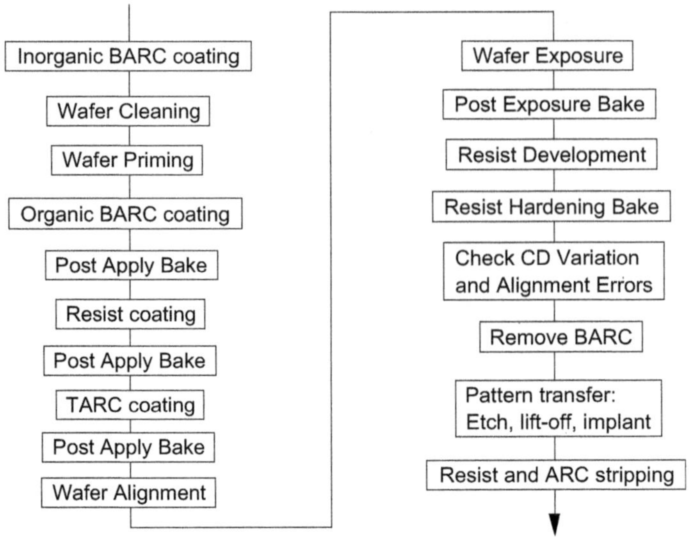

Figure 1.3 Block diagram of lithography processing steps.

followed. Therefore, there are 90-nm, 65-nm, 40-nm, 28-nm, 20-nm, 16/14-nm, 10-nm, and 7-nm technology nodes. Scaling, regardless of whether it is still a fixed percentage of the previous node, necessitates improvements in resolution and overlay accuracy, which call for many of the following improvements: numerical aperture (NA) increase, wavelength reduction, suppression of reflections, better resists, better masks, higher-accuracy stepping, higher-precision alignment, less lens distortion, better wafer flatness, and many others. These will be discussed in detail in the chapters that follow.

1.2 The Goal of Lithography

To make a circuit usable, the features fabricated on it must meet certain criteria. The most important criterion is edge position control. The location of any given edge of a feature on the chip must be within a given tolerance from a nominal position. When the edge meets this requirement, both the linewidth and the overlay controls have been maintained. This is illustrated in Fig. 1.4, where each of the six edges of the L-shaped feature must fit within the six windows surrounding the edges. The nominal position, represented by the centerline, is where the edge ideally should be. Either feature size or feature placement contributes to deviation from the ideal edge positions. Although it is convenient to separately deal with feature size control and feature placement, their combination must meet the ultimate edge placement requirement.

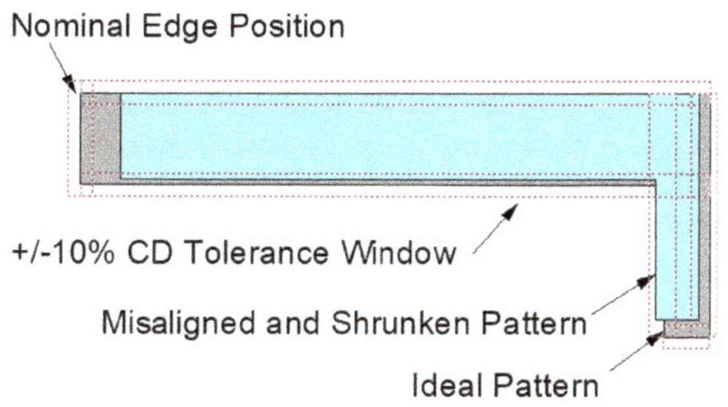

Figure 1.4 Edge placement tolerance.

1.3 The Metrics of Lithography

Following the goal presented in the previous section, a measurement of success lies in the size of the process window in the feature-size and placement controls that keep a given feature edge within its tolerance. The feature-size-control parameters are focus and exposure dosage, whereas the feature placement parameters are alignment accuracy, magnification, and rotations. All of these parameters are mutually dependent and can be set during wafer exposure. Among the five parameters, the exposure-defocus (E-D) window shown in Fig. 1.5 and the mutual tradeoff of exposure latitude and depth of focus (DOF) depicted in Fig. 1.6 are the most frequently used. Incorporating alignment, magnification, and rotations generalizes the metrics to include overlay-related parameters. These metrics will be fully described and extensively used in this book.

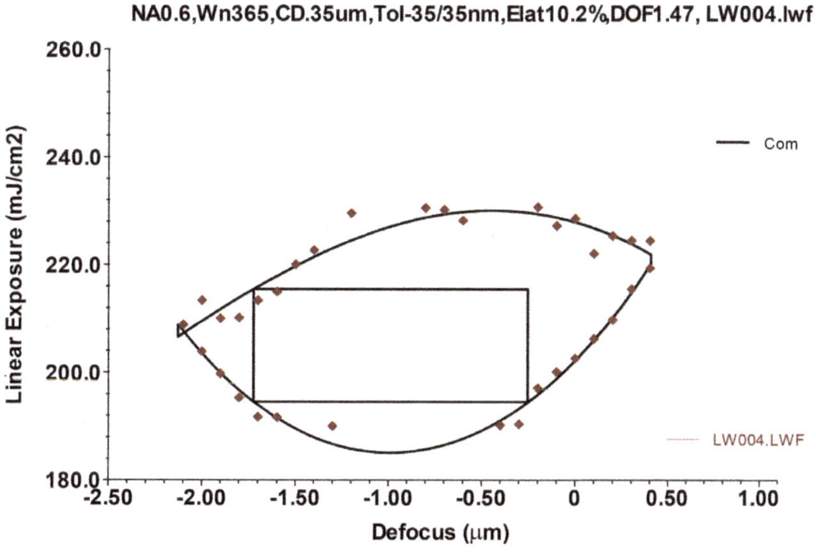

Figure 1.5 A typical E-D window in an E-D tree.

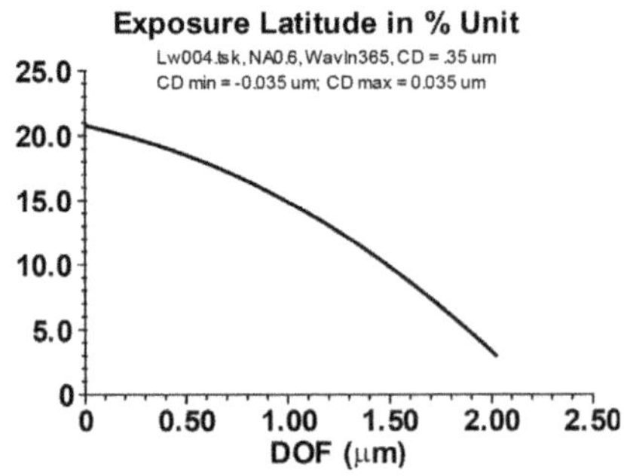

Figure 1.6 Exposure latitude versus DOF tradeoff curve.

1.4 Introduction to the Contents of this Book

In the following chapters, exposure systems, image formation, and the metrics of lithography for proximity printing and projection printing will be covered, followed by components of optical lithography, including processing and optimization, then immersion lithography and EUV lithography.

Chapter 2 covers the proximity printing system of proximity imaging, the region of validity for various approximations of diffraction, proximity images, and the exposure-gap (E-G) diagram. Chapter 3 covers the projection printing systems: step-and-repeat, step-and-scan, and 1X and reduction systems.

Chapter 4 deals with image formation, starting with the formation of aerial images diffracted from the mask and imaging through the lens with its aberrations, followed by creation of the latent image, the resist image, and the transferred image.

The metrics of lithography provided in Chapter 5 include the basic principles of E-D trees, forests, and windows. Their applications include defining the common and individual exposure latitudes and DOFs of different feature types and sizes, at different cuts in a feature, using a mixture of CD tolerances, resist processing tolerances, mask errors, and proximity and projection printings. These metrics help to check experiment against theory, characterize lens aberrations, quantify the performance of phase-shifting masks, and guide proximity corrections.

Chapter 6 covers the components of optical lithography, which consist of light sources, illuminators, masks, the imaging lens, photoresists, wafers, wafer stages, and alignment systems. Chapter 7 deals with processing and optimization. Processing includes exposure, overlay, resist processing, k_1 reduction, optical proximity correction, and improving the CD uniformity. Optimization involves optimizing the NA and illumination coherence, wavelength, bias, off-axis illumination angle, and mask tolerance.

Chapter 8 explains the theory and practice of immersion lithography. Chapter 9 presents EUV lithography using the 13.5-nm wavelength and discusses how long this technology will carry the torch of optical lithography.

Chapter 2
Proximity Printing

2.1 Introduction

Soon after the integrated circuit (IC) was invented[1] in 1958, lithography was used to replicate the designed pattern on the mask to the wafer. The most intuitive way to do so is to copy the mask pattern by direct contact of the mask and the wafer. No lens is needed, the field size can be as large as the mask, and the wafer does not even need to be precisely focused. This replication technique was called contact printing, but this is a misnomer. Unless a deliberate effort is made to achieve intimate contact, the mask and the wafer contact each other only at a few points over their entire area. Even if intimate contact is achieved, because there is a photoresist layer between the mask and the wafer, the mask is still in proximity to the wafer by the thickness of the photoresist. In reality, intimate contact is not desirable for the following three reasons. (1) When the mask and the wafer are in intimate contact, high-contrast interference fringes will form wherever there are slight imperfections of contact, as shown in Fig. 2.1. These fringes make the contact exposure nonuniform. (2) The mask and the wafer become difficult to separate after intimate contact. (3) Damage to the mask and the wafer is inevitable. Therefore, the proper lensless way to replicate a mask pattern is proximity printing, not contact printing.

The proximity printing system consists of a light source, an illuminator, a mask, a wafer, and a wafer stage that holds the wafer at a fixed distance from the mask, usually by putting spacers on the mask. Either the mask or the wafer is pressed hard against the other, with the spacers in between. The stage also has the means to move the wafer with respect to the mask for alignment purposes; there are alignment-viewing devices to guide lateral positioning on the wafer. A schematic of a proximity printing system is shown in Fig. 2.2. The system consists of illumination from the light source and the condenser. The mask and wafer are held in their respective holders to maintain a distance on the order of 20 μm between the absorber surface of the mask and the resist surface of the wafer. A larger distance prevents damage to the mask and the wafer; however, resolution is reduced rapidly as the mask-to-wafer gap

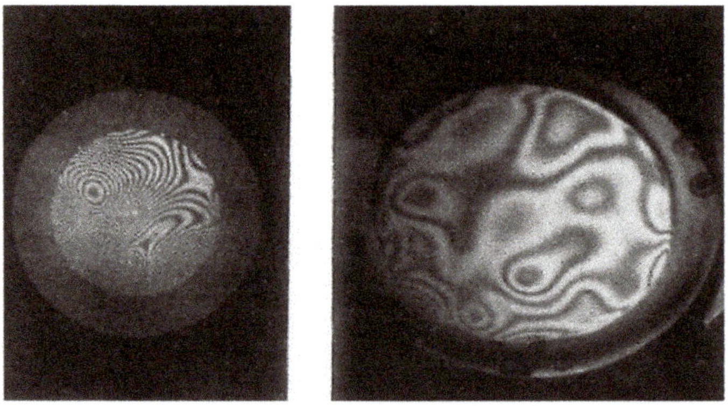

Figure 2.1 Contact fringes are formed when the mask and the wafer are in intimate but imperfect contact. These fringes make the contact exposure nonuniform.

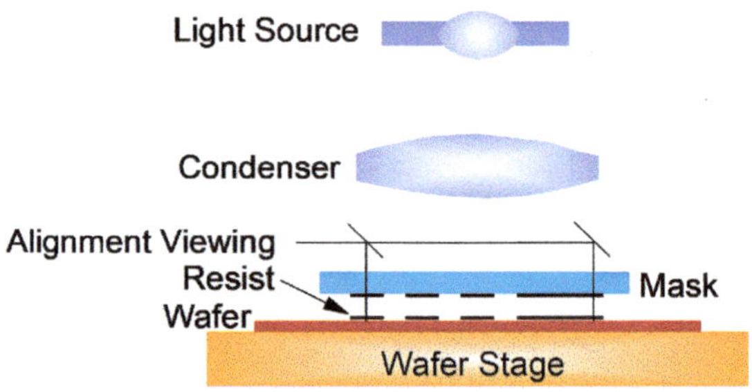

Figure 2.2 Schematic of a proximity printing system.

increases. In addition, alignment viewing must accommodate two focal positions. To achieve this alignment, either the mask or the wafer must move laterally with respect to the other. Detailed coverage of proximity printing has been given by this author elsewhere[2] in 1980. Forty years later, a lot more is known about proximity printing, and it is still practiced in many laboratories. Realizing its importance in research laboratories, in this chapter I have provided up-to-date coverage of the topics entailed in this technology.

Because of the absence of an imaging lens, it is relatively easy to change the wavelength of the imaging system. Therefore, proximity printing is often the forerunner of wavelength-reduction schemes used to screen potential resist candidates.[3,4,5,6] It also enables printing at a spectral region where the refractive lens material is absolutely impossible, e.g., in the vicinity of a 1-nm wavelength for x-ray proximity printing. Using a Fresnel number threshold of 3, 0.25-μm features can be delineated at a 20-μm gap. This is the basis on which x-ray proximity printing claims a very high-resolution potential.[7,8] The Fresnel number ν is an indicator of the proximity of the wafer and is defined in Eq. (2.1).

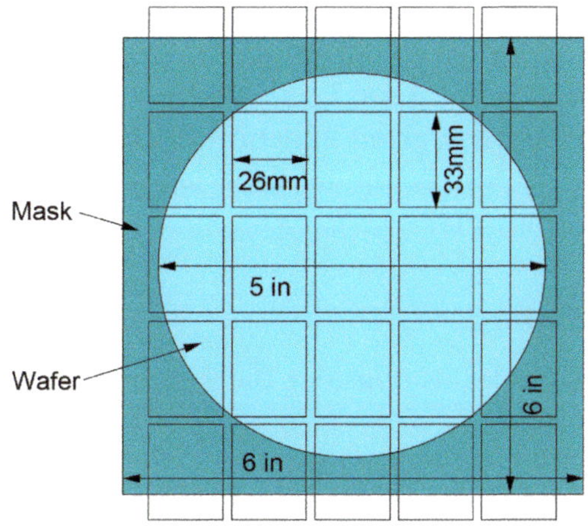

Figure 2.3 Field size in proximity printing and projection printing. For proximity printing, the 6 × 6-inch mask can cover the entire wafer even after allowing a 0.5-inch border on all sides. For projection printing, nine full scanner fields plus eight sizable partial fields are needed to cover the wafer.

Another advantage of proximity printing is the field size. In proximity printing, the size of a semiconductor chip can be as large as the mask allows. The entire printable area on the mask can be its field. Whereas, in projection printing using 4X reduction, the largest field size possible is limited in width by the image field of the projection lens; in length it is limited to 1/4 of the mask size. Figure 2.3 shows a 6 × 6-inch mask and a 5-inch-diameter wafer to illustrate the field size in these two types of printing. For proximity printing, the 6 × 6-inch mask can cover the entire wafer even after allowing a 0.5-inch border on all sides. For projection printing, nine full scanner fields plus eight sizable partial fields are needed to cover the wafer. The chip size and field size of projection printing systems are extensively discussed in Chapter 3.

A unique characteristic of proximity printing is that its depth of focus (DOF) is "identical" to the working distance. The mask-to-wafer gap is designed such that an acceptable proximity image on the wafer can be held within this gap. Hence, the DOF is between the absorber side of the mask and the top surface of the wafer. The working distance is the DOF with the thickness of the photoresist removed.

The smallest feature produced in manufacturing with proximity printing is 2 μm in the near UV (NUV) and mid-UV, and 1.5 μm in the deep UV. Their spectral bands are 450–350 nm, 350–260 nm, and 260–200 nm, respectively. The lower-bound wavelength of 350 nm for UV is chosen so that the popular mercury i-line wavelength of 365 nm is included. Similarly, the upper-bound wavelength of 260 nm for the deep UV includes the 254-nm mercury line.

Until the 1970s, UV proximity printing was widely used in production. Commercial deep-UV proximity printers were available in the late 1970s until the early 1980s. By about that time, lens-based projection printing systems started to replace proximity printing in high-volume manufacturing calling for low defect counts, high resolution, large working distance, and high overlay accuracy. This progression is given in detail in Chapter 3. Today commercial proximity printers are still being used in laboratories and in less-critical manufacturing environments.

2.2 Proximity Imaging

For proximity imaging, let us consider the Fresnel number ν defined by the following equation:

$$\nu = \frac{W^2}{\lambda G}, \tag{2.1}$$

where ν is an indicator of the optical closeness of the wafer to the mask. A large ν indicates a close optical proximity to the mask and thus supports a higher resolution at the expense of mask/wafer damage. The diffraction pattern from the mask is supposed to be identical with the same ν, under a given condition, according to Section 2.3. The mutual dependence of minimum feature size W, wavelength λ, and mask-to-wafer gap G is governed by trading off W, λ, and G, with ν being held constant. Equation (2.1) is derived from the near-field diffraction equation using the Fresnel approximation.[9] When ν is above a certain threshold, the square of the feature size is deemed large with respect to λG, indicating that the image is sufficiently close to the mask that it acceptably resembles the pattern on the mask, and the system is considered capable of the resolution derived from it. For example, at $\nu = 1$, with a wavelength of 250 nm and a gap of 10 μm, the system can support printing 1.6-μm features. To be rigorous, one should determine the ν threshold using the exposure-gap (E-G) methodology given in Section 2.5. This method is similar to determining the usable value of k_1 using the exposure-defocus (E-D) methodology in Chapter 5.

A previous work[10] investigated diffraction from an infinitely long slit extending from the mask plane to any distance from it. The transverse electric (TE) field diffraction pattern from this study shows the high-contrast periodicity of a wavelength in the mask plane, as shown in Fig. 2.4(a), where a 3λ slit has a three-peak fringe pattern in the slit before propagating out into two peaks, then has a single peak that gradually spreads out. Figure 2.4(b) shows the transverse magnetic (TM) aperture field being identical to the plane-wave incident magnetic field, then it quickly turns into a three-ripple shape that gradually becomes two ripples, then turns into a single-peak shape as in the TE case. The TE and TM cases become very similar at a distance as short as 5λ. The respective boundary conditions can be observed in this figure.

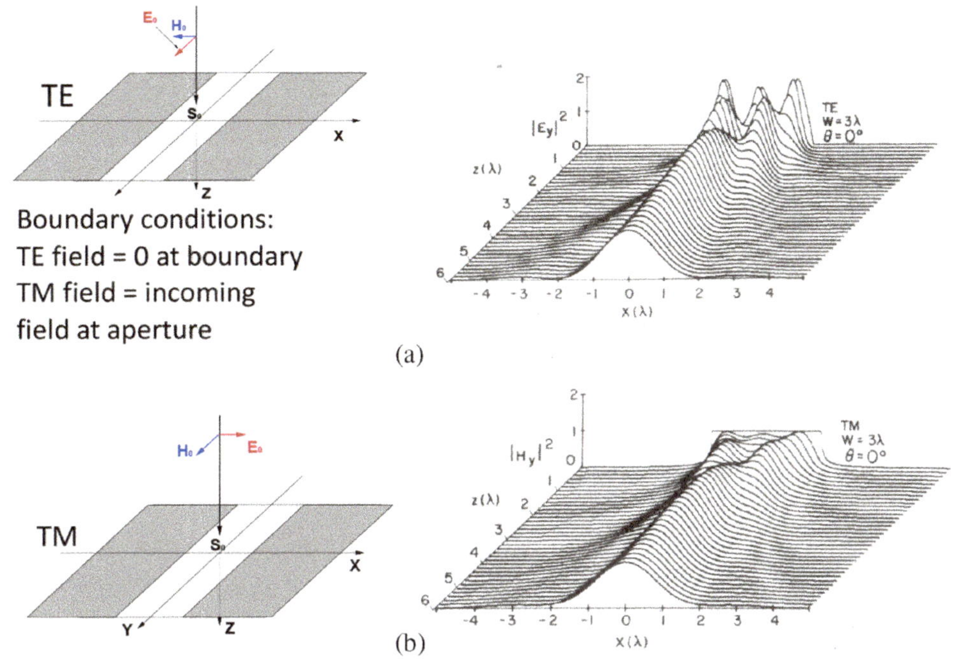

Figure 2.4 Near-field diffraction at normal incidence in the (a) TE and (b) TM modes. The aperture in the mask is a 3λ-wide, infinitely long slit.

To evaluate the diffraction pattern exactly requires solving Maxwell's equations with a set of boundary conditions defined by the mask pattern. Solving Maxwell's equations in close form is not always possible. Fortunately, for proximity printing, good approximations can be made for arbitrary patterns. Figure 2.5 shows the coordinate of the object plane $z = 0$, which is where the mask is located. There is also an arbitrary image plane where (x, y) resides.

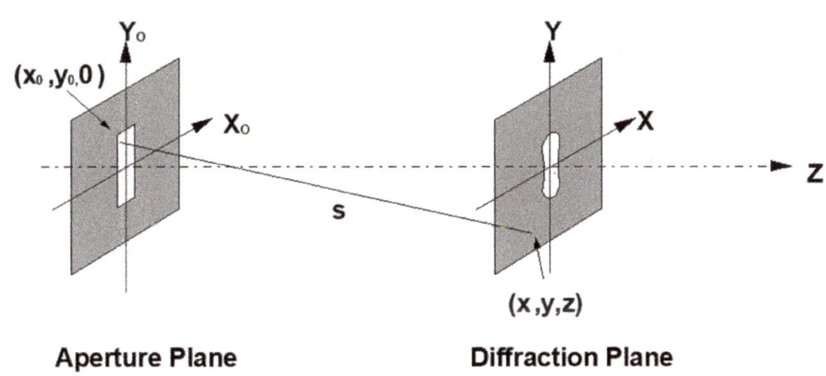

Figure 2.5 Aperture and diffraction coordinates.

The propagating waves can be represented by

$$E(x,\,t) = A\cos(kx - \omega t) = A\cos 2\pi\left(\frac{x}{\lambda} - \frac{t}{T}\right), \tag{2.2}$$

where A is the amplitude of the wave, λ is the wavelength, and T is the periodicity.

In Fig. 2.6, the wave started at $t = 0$ as shown. After $t = T/4$, the wave has moved by $\pi/2$. Usually the time-dependent part is not written out explicitly, with the understanding that it is always there. The wave is simply written as $A\cos(kx) = A \cdot \mathrm{Re}e^{ikx}$.

For a wave radiating from a point source,

$$E(r) = A\frac{e^{ikr}}{kr}, \tag{2.3}$$

where r is the radius from the source to a spherical wavefront. Once again, the time-dependent part is not written out explicitly. We are now ready to introduce Huygens' principle, which states that the field at an arbitrary point (x, y, z) can be evaluated from the field in a given area of radiation (such as the aperture area in the $z = 0$ plane in the light-transmitting area shown in Fig. 2.4):

$$U(x,\,y,\,z) = \frac{2\pi}{\lambda^2}\int_{y_0m}^{y_0M}\int_{x_0m}^{x_0M} U(x_0,\,y_0,\,0)\frac{e^{iks}}{ks}dx_0dy_0, \quad \text{in the } z = 0\,\text{plane}, \tag{2.4}$$

where $s^2 = (x - x_0)^2 + (y - y_0)^2 + z^2$, the wafer number $k = 2\pi/\lambda$, and x_{0m} and x_{0M} are the minimum and maximum of x_0, respectively. The case is similar for y_{0m} and y_{0M}.

Hence, one can substitute the electric field distribution $E_y(x_0,\,y_0,\,0)$ into Eq. (2.4) to evaluate $E_y(x,\,y,\,z)$. In fact, when the amplitude and phase of the field at any plane are known, the field at any arbitrary plane can be evaluated with Eq. (2.4).

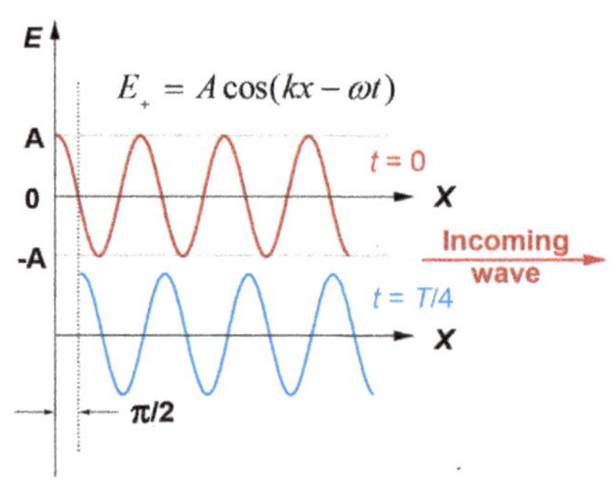

Figure 2.6 A propagating plane wave.

When the field at any one given plane is difficult to calculate, an intuitive approximation can be used to evaluate the diffracted field to an arbitrary z. This method involves setting the field to be the incident field in the aperture plane, and 0 elsewhere, and is called the physical optics approximation (POA). Referring to Fig. 2.4, the TE aperture field is quite different from that of the POA. Only the opaque part where $E_y = 0$ is true. On the other hand, the TM aperture field fits the POA in the aperture of the opening, but it is not 0 in the opaque part. Fortunately, the nonzero field quickly approaches zero away from the edge of the aperture. Comparing the TE and TM fields, in the case of a 3λ infinite slit opening, the diffracted fields converge at $z \geq 5\lambda$. Therefore, the POA is valid in this zone.

The region of validation will be defined more rigorously in the following sections. For now, we introduce more approximations that take advantage of the fact that the z of interest is usually much larger than the size of the aperture. From Eq. (2.4),

$$s^2 = z^2 + (x - x_0)^2 + (y - y_0)^2 = z^2\left[1 + \frac{(x - x_0)^2}{z^2} + \frac{(y - y_0)^2}{z^2}\right],$$

$$s = z\sqrt{1 + \frac{w_x^2}{z^2} + \frac{w_y^2}{z^2}} = z\left[1 + \frac{w_x^2}{2z^2} + \frac{w_y^2}{2z^2} + \frac{w_x^4 + w_y^4}{8z^2} + \ldots\right], \tag{2.5}$$

where $w_x^2 \equiv (x - x_0)^2$ and $w_y^2 \equiv (y - y_0)^2$ truncate the quadratic terms and beyond in s in the exponential ($\because z \gg x$ and y) and make $s \rightarrow z$ in the denominator:

$$U(x, y, z) \cong \frac{2\pi}{\lambda^2} \frac{e^{ikz}}{kz} \iint U(x_0, y_0, 0) e^{i\pi\left(\frac{w_x^2}{\lambda z} + \frac{w_y^2}{\lambda z}\right)} dx_0 dy_0. \tag{2.6}$$

This is the Fresnel approximation and is the basis of the Fresnel number:

$$\frac{w^2}{\lambda z} \equiv \text{Fresnel number}. \tag{2.7}$$

When z is taken to be G, the gap between the mask and the wafer, Eq. (2.7) becomes Eq. (2.1). The Fresnel approximation is useful because it can be evaluated from the well-tabulated Fresnel integrals, $C(X)$ and $S(X)$:

$$\begin{aligned}\int_0^Y \int_0^X e^{i(x^2+y^2)} dx dy &= \int_0^Y e^{iy^2} dy \int_0^X e^{ix^2} dx \\ &= \int_0^Y e^{iy^2} dy \int_0^X (\cos x^2 + i \sin x^2) dx \\ &= \int_0^Y e^{iy^2} [C(X) + i S(X)] dy.\end{aligned}$$

As z becomes even larger, another approximation can be made, resulting in the Fraunhofer approximation:

$$U(x,y,z) \cong \frac{1}{\lambda z}\iint U(x_0,y_0,0)e^{ikz\left[1+\frac{w_x^2}{2z^2}+\frac{w_y^2}{2z^2}\right]}dx_0dy_0$$

$$= \frac{e^{ikz(1+\frac{x^2}{2z^2}+\frac{y^2}{2z^2})}}{\lambda z}\iint U(x_0,y_0,0)e^{ikz\left(\frac{x_0^2}{2z^2}+\frac{y_0^2}{2z^2}-\frac{xx_0}{z^2}-\frac{yy_0}{z^2}\right)}dx_0dy_0.$$

Dropping $\frac{x_0^2}{2z^2}+\frac{y_0^2}{2z^2}$ when z becomes even larger yields

$$U(x,y,z) \cong \frac{e^{ikz}}{\lambda z}\iint U(x_0,y_0,0)e^{\frac{-ik}{z}(xx_0+yy_0)}dx_0dy_0. \tag{2.8}$$

This is the Fraunhofer approximation. $U(x, y, z)$ is now the Fourier transform of $U(x_0, y_0, 0)$. The Fraunhofer approximation is a far-field approximation. Fresnel approximation is considered near field, despite the fact that the POA can handle an even closer field.

Figure 2.7 shows the diffraction pattern as a function of the Fresnel numbers 2 through 30 and 50. A small ν indicates a longer distance z between the mask and the wafer or a smaller feature size W. As the distance becomes shorter, the number of peaks increases, just as shown in Fig. 2.3. However, the resemblance is closer for larger physical dimensions of W. That is, even at the same ν, when $W = 50\lambda$, the Fresnel prediction is better than when $W = 1\lambda$. This becomes obvious after the arguments in Section 2.3.

The case of $\nu = 1 \sim 3$ is often used for proximity printing. At $\nu = 51$, the wafer is too close to the mask. Take deep-UV proximity printing at $\lambda = 254$ nm as an example. To print 0.5-μm features for $\nu = 1$, a gap of only 1 μm is allowed. This gap must include the resist thickness. Therefore, there is intimate contact between the wafer and the mask, with the resist in between. With the propagation medium completely in the resist, the wavelength in the Fresnel number should be corrected to match the refractive index of the resist. In this case, the resist is polymethyl methacrylate (PMMA). Its refractive index is 1.56 at $\lambda = 254$ nm. λ now becomes 162.8 nm. The z distance for the same ν is now 1.535 μm. Figure 2.8 shows 0.5-μm features and 0.5-μm spaces in 1.78 μm of PMMA resist. Hence, we have demonstrated contact printing at $\nu = 0.86$. However, this was printed at a specific optimum condition. For high-volume manufacturing, ν should be kept between 1 and 3.

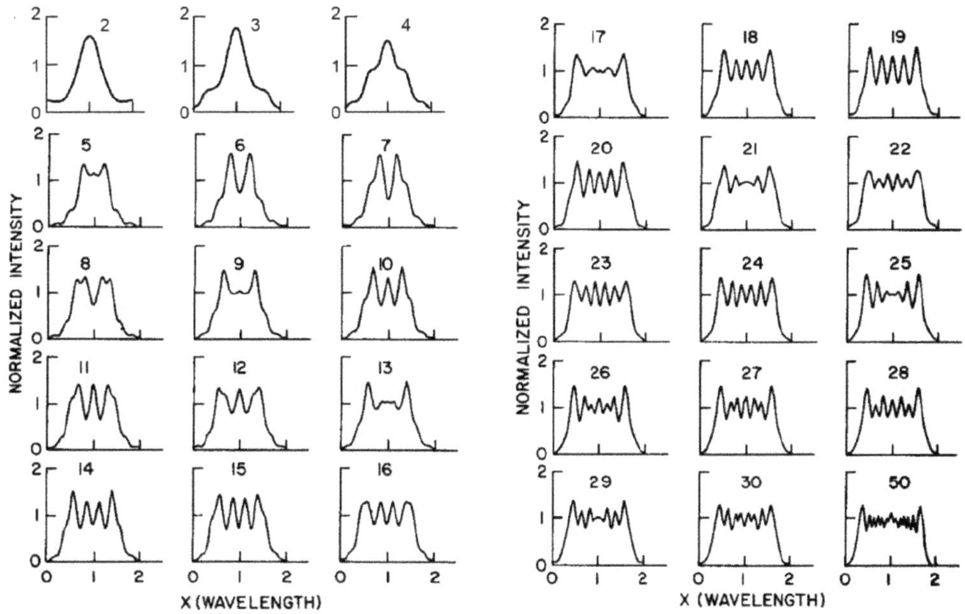

Figure 2.7 Fresnel diffraction patterns. The number above the diffraction patterns is the Fresnel number of the combination of *W* and *z* according to Eq. (2.1). Note that the *x* coordinate is normalized to the wavelength λ.

Figure 2.8 0.5-μm chevrons and the same-size space in 1.78 μm of PMMA.

2.3 Region of Validity for Various Approximations of Diffraction

How accurate are the POA, the Fresnel approximation, and the Fraunhofer approximation? The conditions of ν = 51, 11, and 1.126 are plotted in Figs. 2.9 to 2.11 to compare the Fresnel approximation and the POA to exact solutions of the TE and the TM modes using the method in Ref. 10. It application on a 3λ slit is shown in Fig. 2.4.

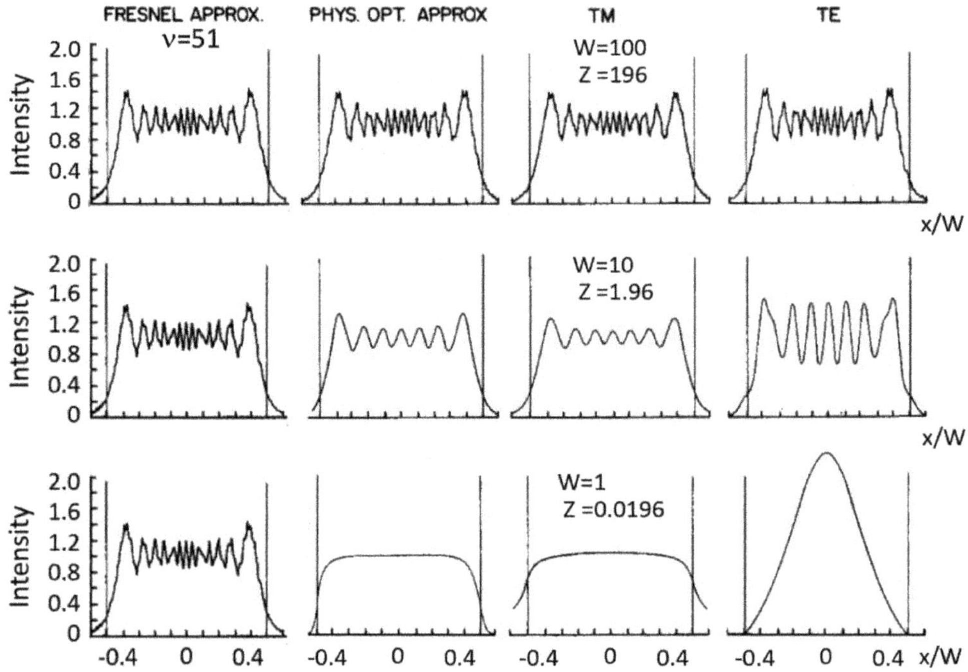

Figure 2.9 Comparison of diffraction approximations at ν = 51.

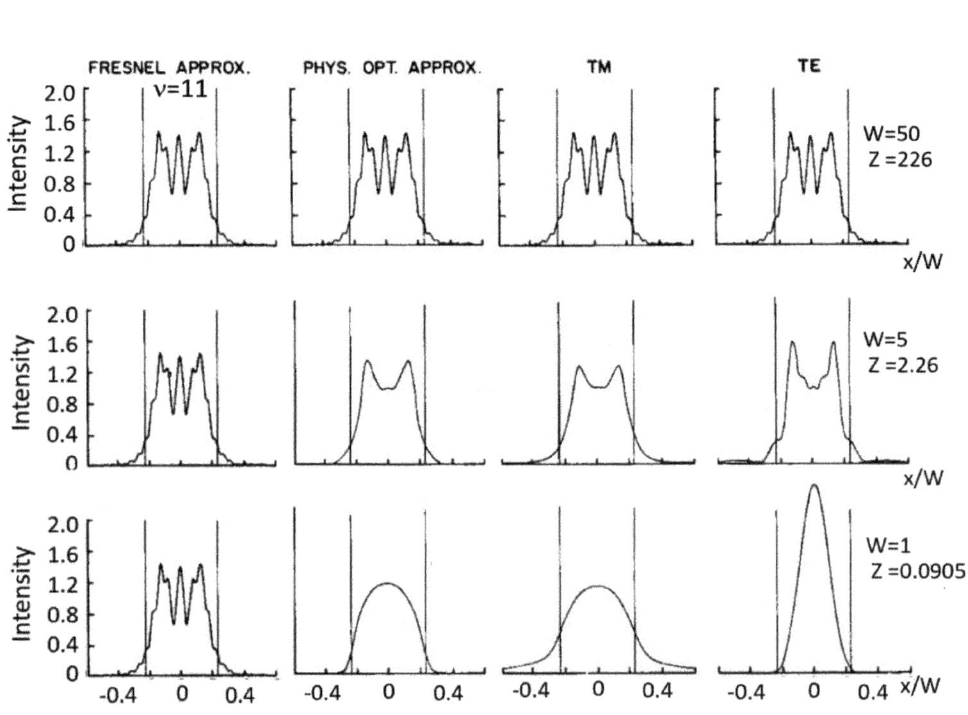

Figure 2.10 Comparison of diffraction approximations at ν = 11.

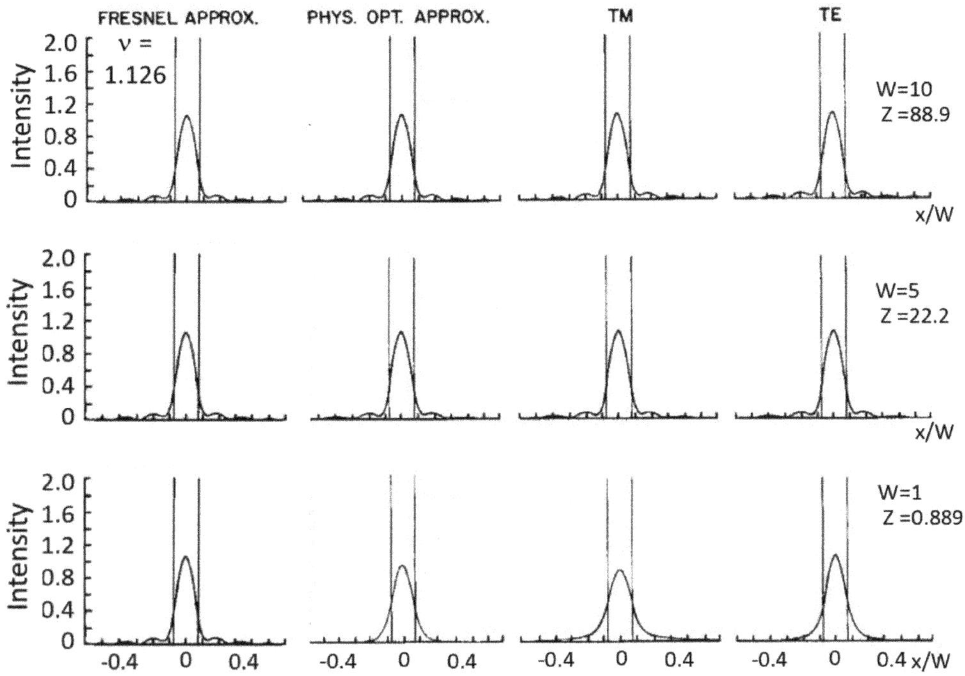

Figure 2.11 Comparison of diffraction approximations at ν = 1.126.

Figure 2.9 consists of plots of the TE, TM, and scalar fields diffracted from infinite slits of widths 100λ, 10λ, and 1λ, all at ν = 51. With the Fresnel approximation, there is only one diffraction pattern at each ν value, no matter the slit width, but the diffraction pattern is different for TE, TM, and POA. At W = 100λ, the diffraction patterns are visually identical for TE, TM, and POA. The Fresnel diffraction looks like the others, except for the number of peaks and their distribution. At W = 10λ, the Fresnel diffraction pattern is completely different from the other types of diffraction patterns. POA and TM are similar, but TM is slightly wider and has slightly less contrast; TE has the highest contrast at the fringes. For W = 1λ, POA and TM are still similar, with TM being wider, but TE exhibiting a strong focusing effect. TM has a larger spread outside of the mask boundary compared to POA because POA assumes a zero field outside the boundary, while the boundary condition of TM allows fading out of the magnetic field from the aperture field that is identical to the incident magnetic field.

In Fig. 2.10, ν = 11. At W = 50λ, all diffraction patterns look alike. At W = 5λ, they still look alike, except for a slightly higher contrast for TE. At W = 1λ, TM spreads out more than is shown in Fig. 2.9. The focusing effect of TE is even stronger.

In Fig. 2.11, ν = 1.126. The diffraction field is far enough that for $W = 10\lambda$, 5λ, and 1λ there is not much difference in these four cases. Only a slightly sharper peak is seen for TE at $W = 1\lambda$.

Qualitatively, the approximations are better for small values of ν. At the same ν, the approximation is better for larger W and z. For a quantitative comparison of the region of validity, the root mean square (RMS) error is used:

$$\delta = \sqrt{\frac{\int_{x_1}^{x_2} (F_{\text{approx}} - F_{\text{exact}})^2 dx}{x_2 - x_1}}, \text{ where } x_2 > x_1. \tag{2.9}$$

Values of δ are plotted in the W-z space in Figs. 2.12 to 2.17 for $2\lambda \leq W \leq 200\lambda$ and $2\lambda \leq z \leq 1000\lambda$. Each parameter is normalized in wavelength. The RMS error distribution is divided into zones logarithmically, ascending from 0 to 19. Each pixel in the W-z space is represented by a color. The zones are 2.5 decibels apart. Hence, a separation of 2 zones corresponds to approximately a factor of 2 difference in RMS error. The percentage of RMS error is related to the zone number (level) shown in Table 2.1.

Figure 2.12 shows the region of validity for Fraunhofer approximation of an infinite slit, referring to the TE solution. The latter is evaluated with

Table 2.1 Zone level and its corresponding RMS error.

Level	0	1	2	3	4	5	6	7	8	9	10
RMS (%)	0.01	0.02	0.04	0.08	0.16	0.31	0.63	1.25	2.50	5.00	10.00

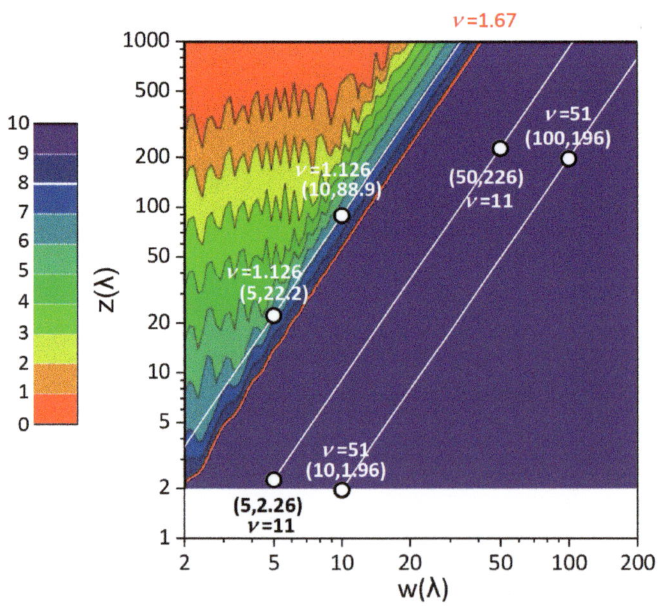

Figure 2.12 Region of validity for Fraunhofer–TE.

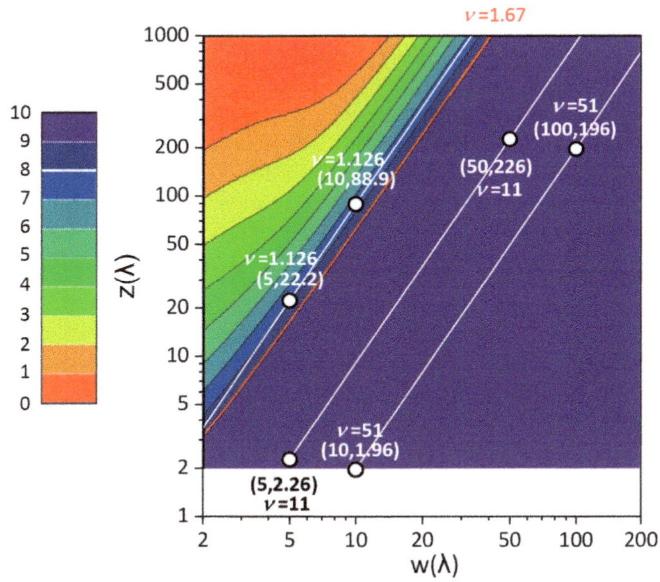

Figure 2.13 Region of validity for Fraunhofer–TM.

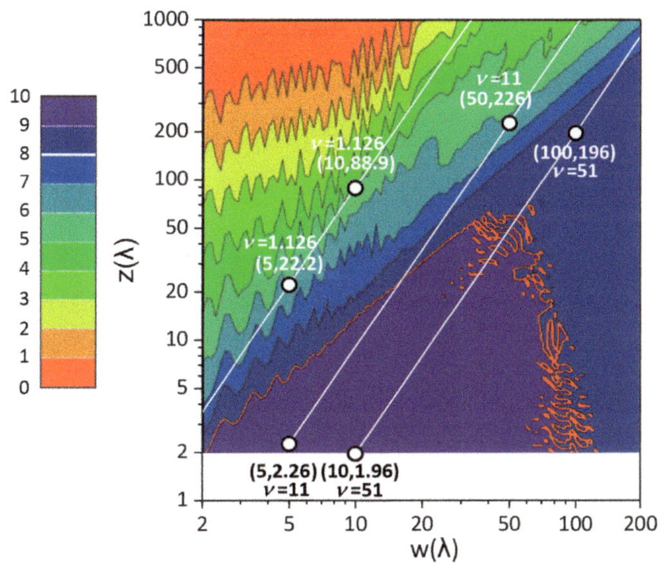

Figure 2.14 Region of validity for Fresnel–TE.

synthesis of the Sommerfeld half-planes.[10] We set 5% as the threshold for acceptable RMS accuracy. Here the ν = 1.126, 11, and 51 lines are plotted with the *W*-*z* coordinates from Figs. 2.12 to 2.17. A red line corresponding to $\nu = 1.67$ to the right of the $\nu = 1.126$ line is the $\delta = 5\%$ border. At the right-hand side of this red line, the RMS error is over 5%. Hence, the Fraunhofer approximation fails RMS accuracy at $\nu > 1.67$. The $\nu < 1.67$ region is

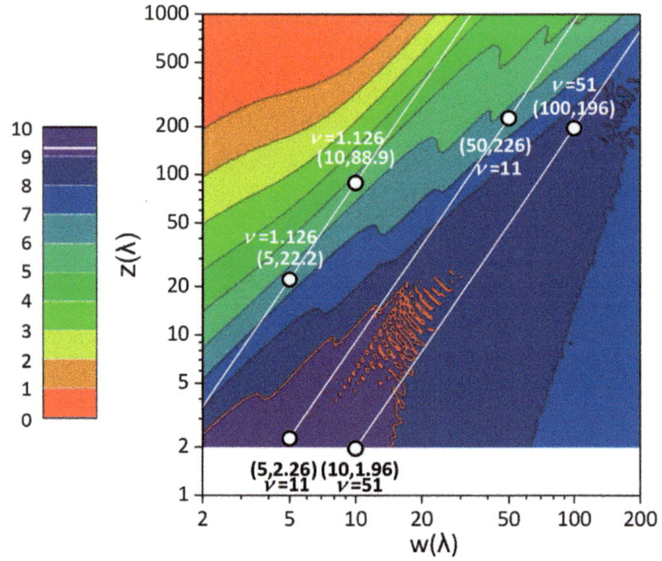

Figure 2.15 Region of validity for Fresnel–TM.

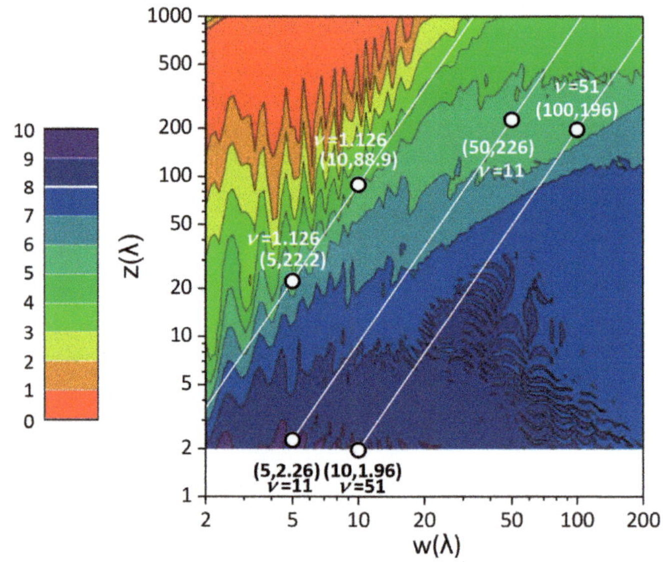

Figure 2.16 Region of validity for POA–TE.

acceptable. Confirming intuition, the RMS error decreases at larger z values along constant W lines, and similarly decreases at smaller W values along constant z lines. The intensity contours fluctuate, indicating that the RMS error is not a smooth function of W because the boundary condition in the TE case is quite different from the POA assumption that leads to the Fraunhofer

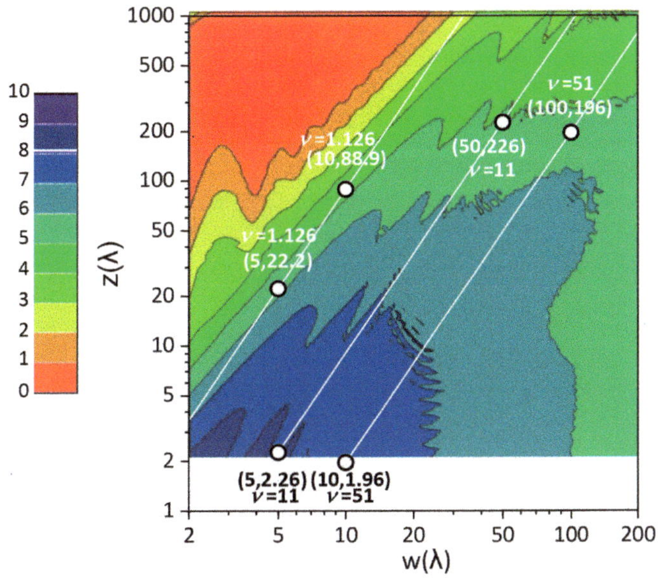

Figure 2.17 Region of validity for POA–TM.

approximation. This concept is further discussed after all of the RMS errors are shown—through Fig. 2.17.

Figure 2.13 shows the region of validity for Fraunhofer–TM. The $\delta = 5\%$ zone boundary is similar to $\nu = 1.67$, with the lower part of the red line moving up slightly. Figure 2.14 shows the region of validity for Fresnel–TE. There is now a larger valid W-z region because Fresnel is a better approximation than Fraunhofer. The validity boundary at zone $\delta = 5\%$ is indicated by the red line. The zone boundaries are again oscillating due to the different boundary conditions of the two methods, as explained in the discussion of Fig. 2.12. However, unlike Fig. 2.12, Fig. 2.14 has a usable region up to $\nu = 51$, at $W > 50\lambda$.

Figure 2.15 shows the region of validity for Fresnel–TM. Like the Fraunhofer case, compared to the TE mode, the TM mode enjoys better validity in the low-W region. In addition, the boundaries are smoother than those in the previous figure.

Figure 2.16 shows the region of validity for POA–TE, and Fig. 2.17 shows the same for POA–TM. The region boundary is near the very bottom of the figure. This means that the POA is valid even at very short mask-to-wafer distances. It is quite flat; i.e., δ is insensitive to changes to W. This effect is even stronger for the TM mode, consistent with the TM boundary condition. In this case, there is no area where δ is larger than 5% within the plotted area.

We have been taking the RMS error as an indicator of the validity of the approximation methods. For lithography, one need not be too concerned with the correctness of the shape of the diffraction patterns. The slope of the

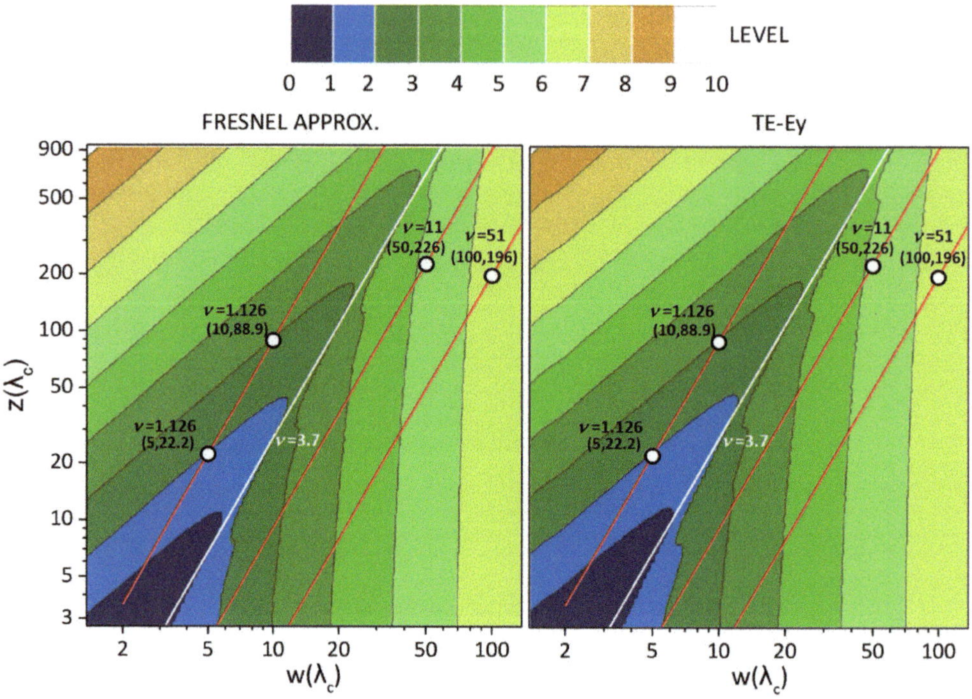

(a) (b)

Level	0	1	2	3	4	5	6	7	8	9	10
CD (μm)	0.4	0.8	1.6	3.1	6.3	12.5	25.0	50.0	100.0	200.0	400.0

Figure 2.18 CD contour plots for (a) the Fresnel approximation and (b) the TE exact solution (w = 0.5~50 μm, z = 1~50 μm, λ_c = 365 nm).

diffraction patterns at the edge of concern and the CD that results from the intersection of the diffraction pattern and a given exposure level are more relevant. In Figs. 2.18 and 2.19, the CD is taken as the intersection of the diffraction pattern and half of the peak intensity. Figure 2.18(a) is the CD plot for the Fresnel approximation, and Fig. 2.18(b) is the CD plot for E_y of the TE case; Fig 2.19(a) is the CD plot for E_x, and Fig. 2.19(b) is the CD plot for H_y, the electric and magnetic field distributions of the TM case, respectively. H_y is the field distribution that is transverse, whereas E_x is the electric field distribution that the photoresist reacts to. In all cases, the contour plots look alike. They all have a near-vertical section until the z/W^2 ratio becomes very large. This occurs when the field distribution is so far that the peak of the diffraction pattern is very small, and the log slope of the diffraction pattern is too low to be practical. The near-vertical contours mean that the CD is insensitive to the distance from the mask. Regardless of whether the case is for Fresnel, TE, or TM, the ν = 3.7 line is the turning boundary. Hence, this line

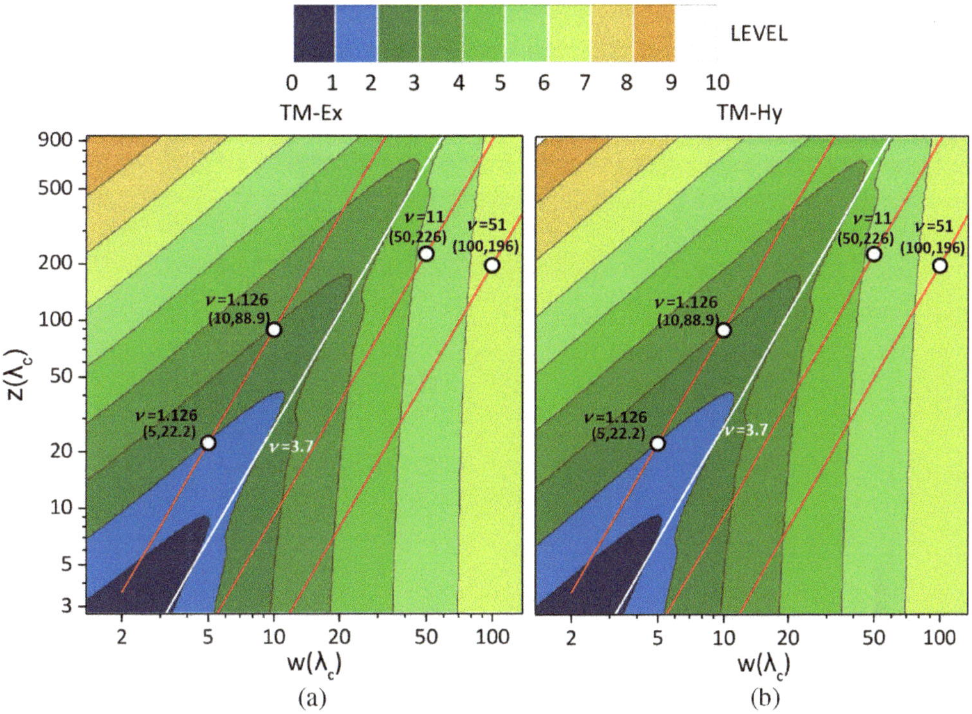

Figure 2.19 CD contour plots for the TM exact solution with (a) E_x and (b) H_y (w = 0.5~50 μm, z = 1~350 μm, λ_c = 365 nm).

determines the usefulness of the approximation. So, even when the required RMS accuracy is met, the $\nu = 3.7$ boundary must be checked.

Now, we check the CD error of the Fresnel approximation against the TE and TM exact solutions in Figs. 2.20 and 2.21, respectively. The scale of the color levels is identical to that of Figs. 2.12 to Fig. 2.17. Unlike Figs. 2.12 to 2.17, the largest error occurs in the vicinity between $W = 3\lambda$ and 5λ and at small z values. This can be understood by looking at the diffraction patterns at $\nu = 11$ in Fig. 2.10. The Fresnel–TE plot is less smooth due to different boundary conditions being used in the other figures.

2.4 Proximity Images

We now show some simulated and experimental proximity images. Figure 2.22 shows the intensity distribution evaluated with POA using 365-nm normally incident illumination on two 2.5 μm × 12.7 μm bars separated by 2.5 μm. The proximity image is 12.7 μm from the mask. The intensity levels are 1.5 dB (decibels) apart. The incident intensity is 1. Level 19 equals an intensity of 4. Level 0 is an intensity smaller than

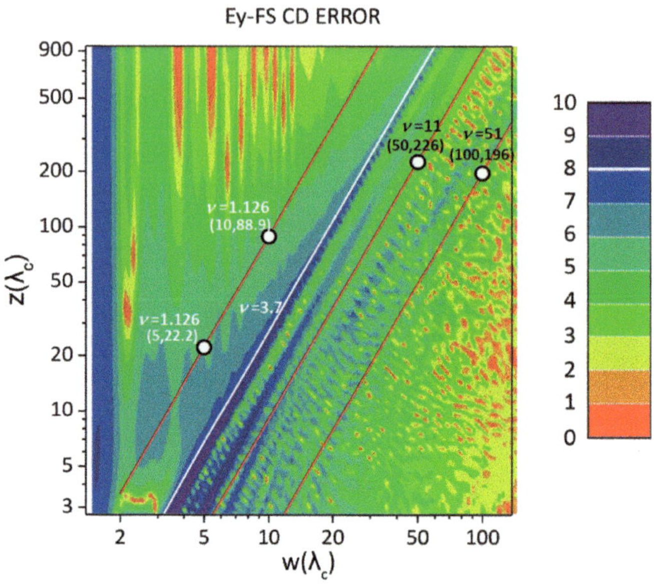

Figure 2.20 CD error contour plots for the transverse electric field distribution of the TE exact solution with respect to the Fresnel (FS) approximation [w = 0.5~50 μm, z = 1~350 μm, λ_c = 365 nm; error (%) = ($CD_{FS} - CD_{TE}$) × 100/CD_{FS}].

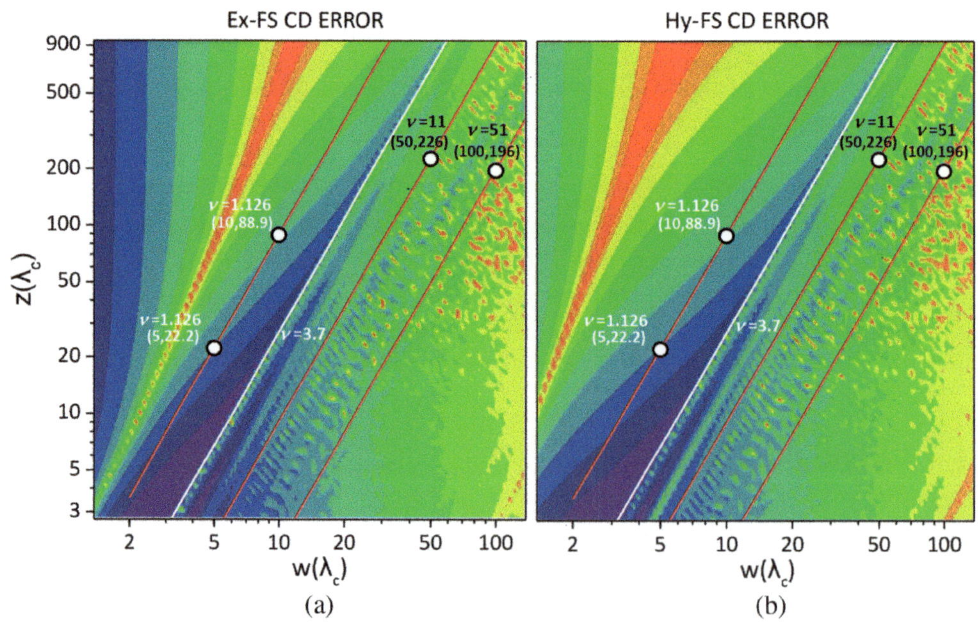

Figure 2.21 CD error contour plots for the transverse magnetic fields of the TM exact solution with respect to the Fresnel approximation [w = 0.5~50 μm, z = 1~350 μm, λ_c = 365 nm; error (%) = ($CD_{FS} - CD_{TE}$) × 100/CD_{FS}].

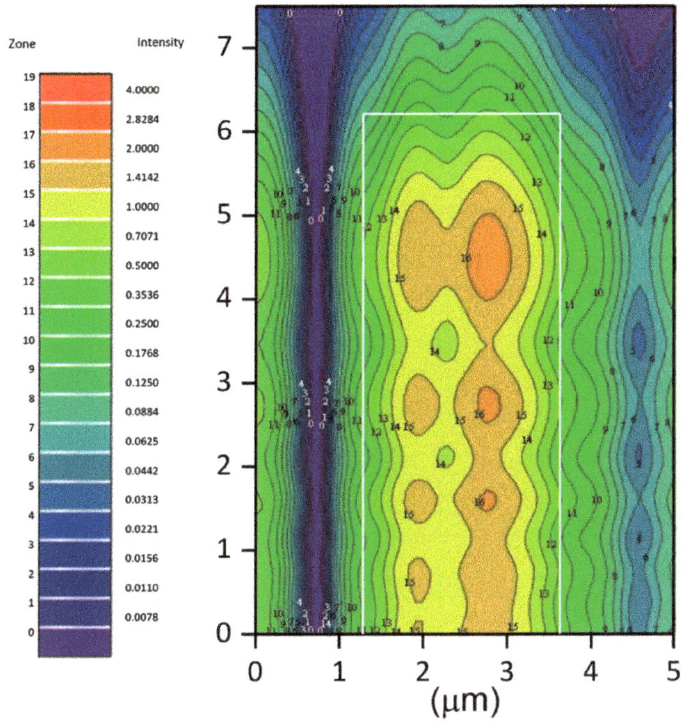

Figure 2.22 Simulated proximity image using the POA. The image plane is 12.7 μm from the mask where there are two 2.5 × 12 μm bar openings 2.5 μm apart. The illumination is a λ = 365 nm normally incident plane wave. Due to symmetry, only a quadrant of the image is simulated.

0.0078. Taking advantage of symmetry, only a quadrant of the image is shown. The Fresnel number is $2.5^2/0.365 \times 12.7 = 1.35$.

The image according to the contour of level 12 best resembles the mask pattern. The corners of the bars are rounded because of a loss of the higher spatial frequencies in the diffracted image. The contour is not straight due to interference from the edges. In the middle of the two bars, there is a level-12 bright spot at $x = 0$ and $4 < y < 5$, as well as smaller bright spots near $y = 2.7$ and 1.6. To prevent the spot from being exposed, the level-12 exposure cannot be used unless an extra bright line in the middle of the two bars is allowed. This is, of course, not possible. The extra line is called a ghost line and must be suppressed. One can suppress the ghost line by using levels 13 or higher. Unfortunately, this would narrow the main bars. The desired bar width can be maintained by manipulating the illumination to break the coherence in the illumination.

Figure 2.23 shows the simulated proximity image located 25 μm from a rectangle with dimensions of 3.8 × 8.6 μm. Normally incident illumination of 405 nm is used. The exposure level for the closest resemblance to the mask pattern is again level 12. The image takes the shape of a "figure 8"

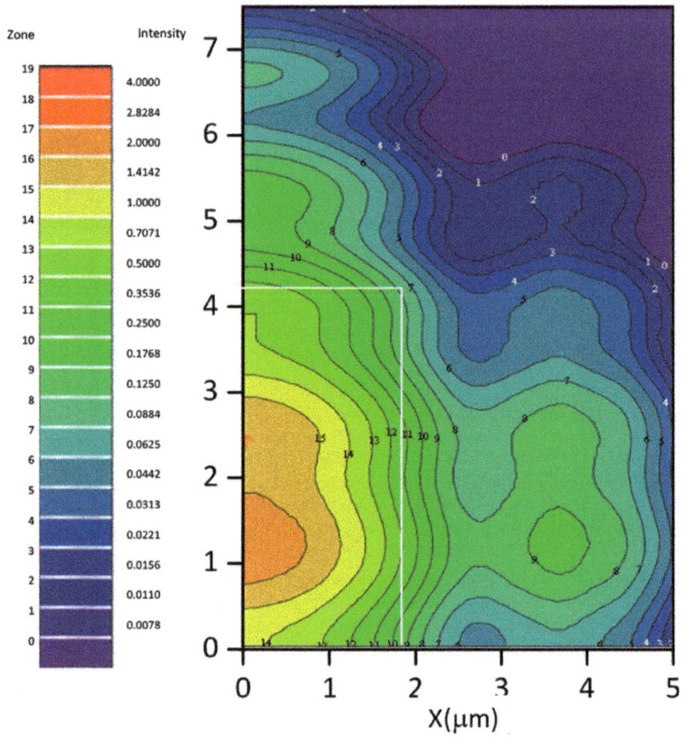

Figure 2.23 Simulated proximity image located 25 μm from a rectangular opening of 3.8 × 8.6 μm. The illumination is normally incident with λ = 405 nm. Only a quadrant of the image is simulated.

with its corners rounded and waist narrowed. The Fresnel number is $3.8^2/(0.405 \times 25) = 1.43$.

Figure 2.24 shows the simulated proximity image 25 μm from a cross opening that is 2.5 μm in width and has a span of 25 × 51 μm. The illumination wavelength is 405 nm. This type of image is characterized by the bright spot at a 45-deg angle from the center of the cross. The Fresnel number is $2.5^2/(0.405 \times 25) = 0.617$.[i]

Figure 2.25 shows experimental proximity images of various sizes and with varying gaps between each image and the mask. The gaps are 65, 80, 100, 140, 180, 250, 400, 500, 550, 650, and 800 micro-inches, or 1.65, 2.03, 2.54, 3.56, 4.57, 6.35, 10.2, 12.7, 14.0, 16.5, 20.3 μm, respectively. For the case of the largest rectangle, its shape remains consistent. Only the corners become

[i]Figures 2.12 to 2.24 were updated from Ref. 2 by Yen Hui Hsieh and Ming Xiang Hsieh, my teaching assistant and graduate student for the course Essence of Lithography at the National Chiao Tung University, respectively. Both were PhD candidates. Yen Hui has graduated, and Ming Xiang is in his third year. It required much work to reprogram, re-integrate, and replot the figures. Many new insights have been gained through this effort and are captured in the Appendix of this second edition.

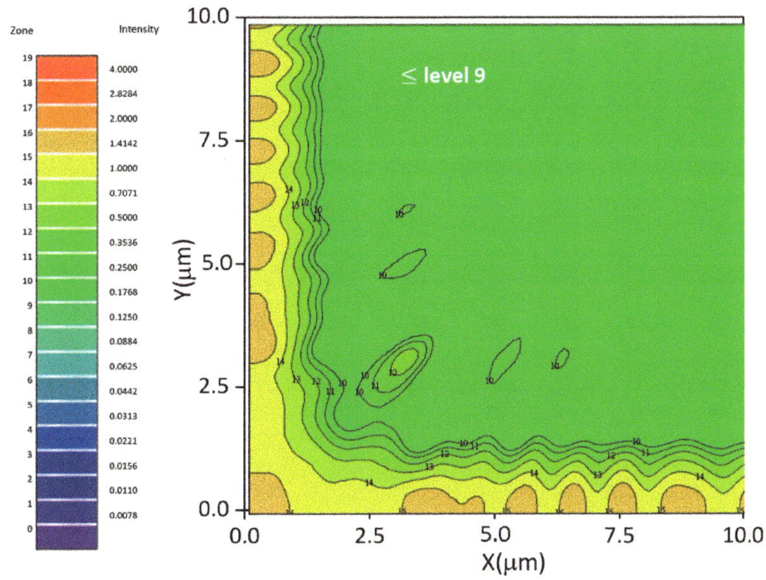

Figure 2.24 Simulated proximity image located 25 μm from a cross opening that is 2.5 μm in width and has a span of 25 × 51 μm. The illumination wavelength is 405 nm.

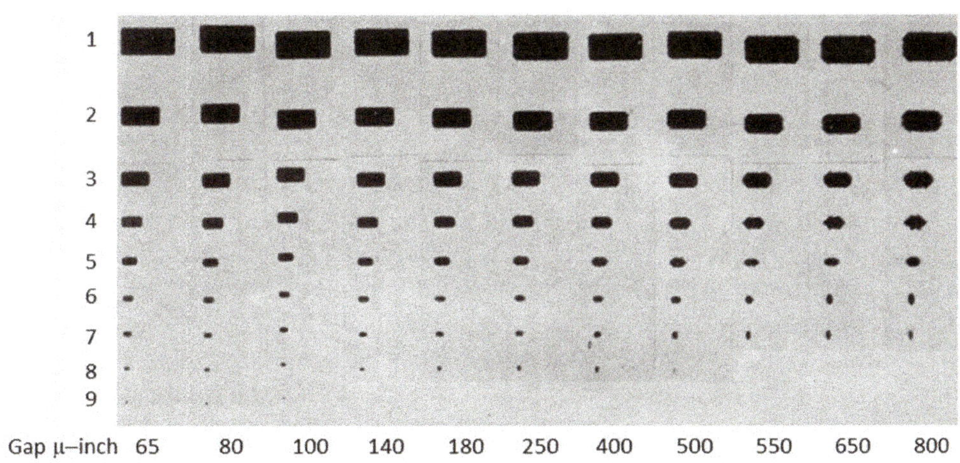

Figure 2.25 Experimental proximity image in various sizes and with varying gaps (from the mask) measured in micro-inches (adapted from Ref. 11).

more rounded at larger gaps. For the Group 3 rectangles, the figure-8 shape shown in Fig. 2.18 starts to appear. The effect is more pronounced for Group 4. The Group 5 images go through the figure-8 shape then approach the shape of an ellipse. For Group 6, the ellipse becomes a circle at larger gaps then turns into an ellipse again, but with the long and short axes exchanged. The image is entering the Fraunhofer regime, i.e., the Fourier transform

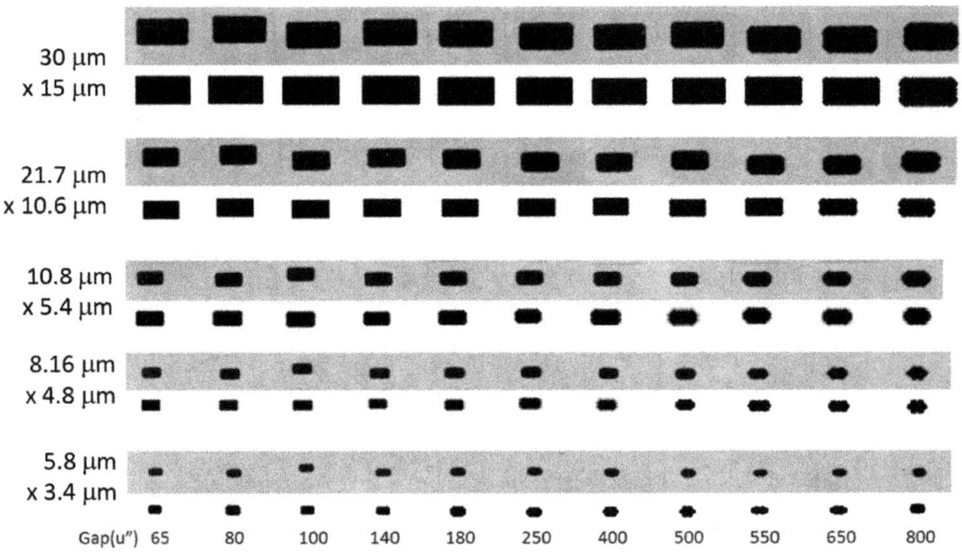

Figure 2.26 Comparison of simulation and experimental results.

regime. The Group 8 patterns are barely printed and appear to be more vertical than horizontal compared to the Group 7 images at larger gaps.

The rectangle size at the mask in Fig. 2.25 is no longer available. However, my students in the 2016 Innovative Lithography class used a 400-nm wavelength to simulate the proximity images at the given gaps to match the experimental results. Good matching is shown in Fig. 2.26. From it, the students estimated the mask patterns to be 30 × 15, 21.7 × 10.6, 10.8 × 5.4, 8.16 × 4.8, and 5.8 × 3.4 μm^2 for Groups 1, 2, 3, 4, and 5, respectively.

Proximity printing is unique due to the high-resolution fringes at the contact plane. As shown in Fig. 2.4, these fringes are about one wavelength in width at the mask. The number of fringes decreases to two as the diffracted wave propagates from the mask to the wafer. The diffractive wave eventually becomes a single-peak image. To capture these fringes in the resist, the mask is coated with resist as shown in Fig. 2.27. The mask is illuminated according to the diffraction phenomenon being investigated. Here a normally incident plane wave is used to illuminate the mask from the glass side. The resist is partially developed to the plane of interest. Figure 2.27 shows the diffraction fringes at $z = 0.16$ μm and 0.35 μm from a $W = 1.1$-μm slit opening in 2-μm pitch. The illumination wavelength is 254/1.56 = 162.8 nm, where 1.56 is the refractive index of the PMMA resist.

The developed resist image can be quite different from the aerial image. As an example, Fig. 2.28 shows the diffracted aerial image from three 0.5-μm slits spaced 0.5 μm apart at distances of 0, 0.6, 1.5, 2.4, and 3.3 μm from the mask. The illumination consists of six spectral lines whose spectral characteristics are shown in Table 2.2. At $z = 0$, i.e., at the mask, the

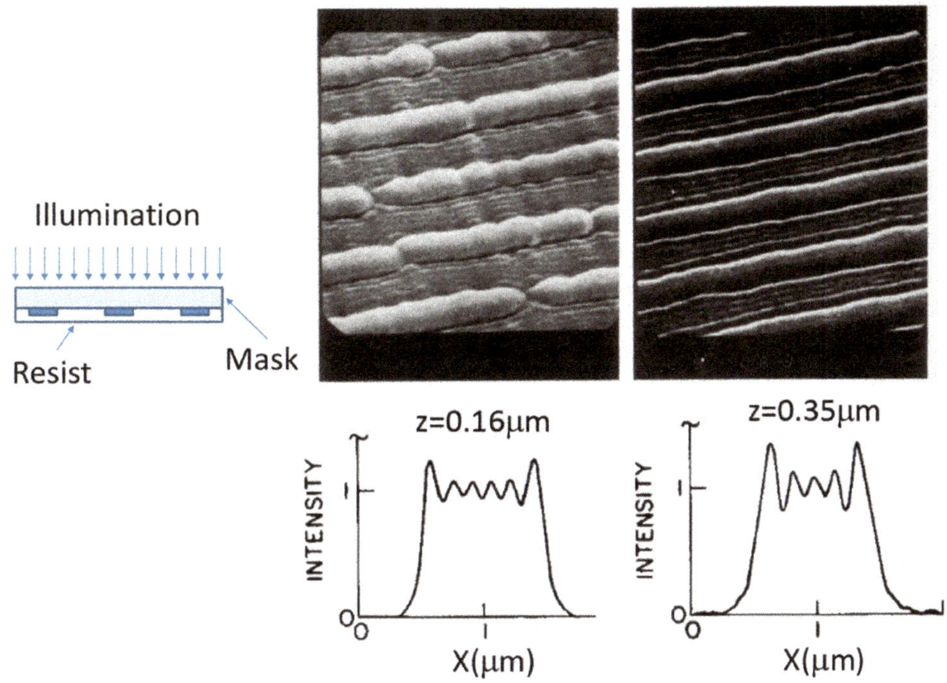

Figure 2.27 Experimental results of near-field diffraction.

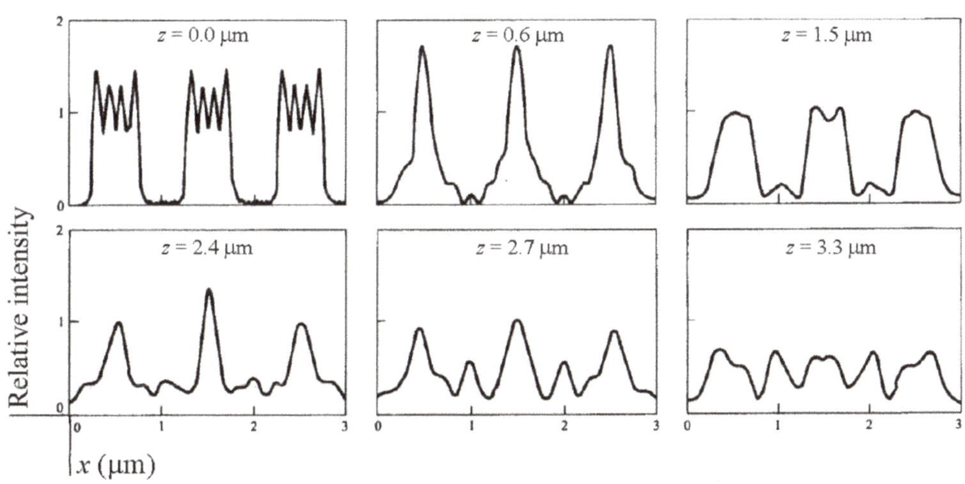

Figure 2.28 Proximity image in 0-, 0.6-, 1.5-, 2.4-, 2.7-, and 3.3-μm gaps.

intensity distribution has fringes as expected, even though polychromatic illumination is used. At z = 0.6 μm, the image sharpens as it is in focus. It broadens at 1.5 μm and sharpens again at 2.4 μm. At 2.7 μm, the three main

Table 2.2 Spectral characteristics of polychromatic illumination and the resist.

Spectral line	1	2	3	4	5	6
Wavelength (nm)	205	215	225	235	245	255
Refractive index	1.596	1.588	1.579	1.571	1.562	1.534
Relative spectrum	2	8	13	22	30	29
Relative spectral photosensitivity	254.6	131.2	62.5	15.46	3.22	0.214

peaks become much lower, while two growing peaks appear between the major peaks. At 3.3 μm, there are five peaks of similar peak intensity.

We now look at the latent and developed resist images shown in Fig. 2.29. The latent image is the side view of the exposed and unexposed areas in normal exposure, overexposure, and underexposure. The positive-resist image shows that the three-line image extends much deeper than predicted by the aerial image. The reason is that the two exposed areas in the middle of the main peaks are surrounded by unexposed areas and are shielded from the developer. In the overexposure case, the protected, unexposed resist is too thin on one side. The developer penetrates the exposed area. In the underexposure case, the exposed areas are developed shallowly. Two cylindrical latent images

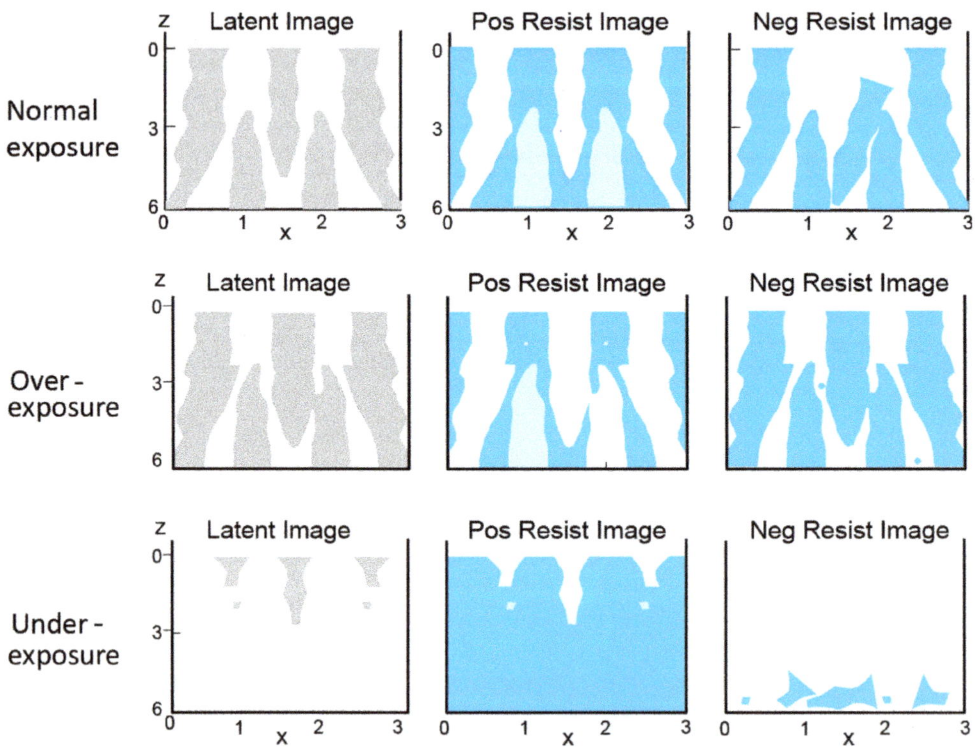

Figure 2.29 Latent, positive-resist, and negative-resist images from the aerial image of Fig. 2.28.

are protected by the surrounding unexposed areas. The negative-resist image is more problematic. With normal exposure, the middle resist island falls to the wafer for lack of support. In the overexposure case, the middle piece is held at one side by the resist on top of the wafer. In the underexposure case, all resist pieces collapse for lack of support. In addition to being unsuccessful at the local site, these fallen resist pieces are a defect source.

2.5 E-G Diagram

To quantitatively assess the process window of an image resulting from proximity printing, an exposure-gap (E-G) diagram is used. This is like the exposure-defocus diagram presented in Chapter 5. Constant-feature-width contours are plotted in the $E_{\log}$–gap space. Each contour is an E-G branch. A combination of these contours is called an E-G tree. The branches of the upper and lower limits of the feature width define the usable region in the E-G space. The 0% contour is the value of the critical dimension (CD). Each feature has its usable E-G region. An E-G common window is a rectangle fitted to the common region. This is the process window in which a combination of features can be printed. The concept of a common window and a common region is explained in detail in Chapter 5.

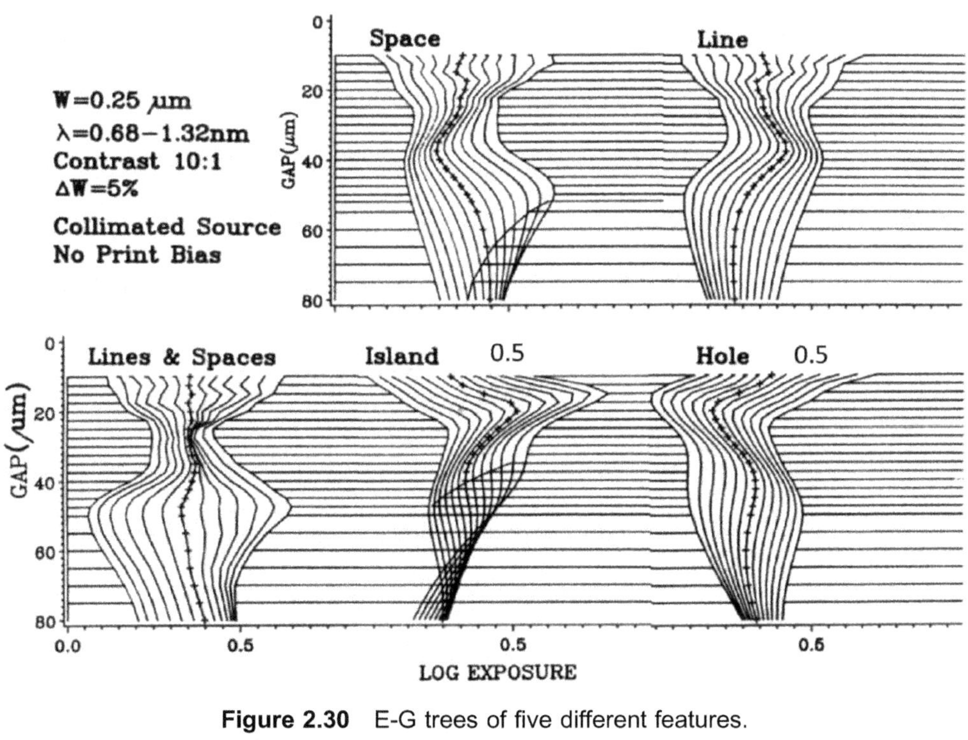

Figure 2.30 E-G trees of five different features.

Figure 2.30 shows the E-G trees of an isolated opaque space, an isolated line opening, packed lines and spaces, an isolated island, and an isolated hole.[11,12,13] The 1D features have a width of 250 nm, and the 2D features are 250 nm × 250 nm. Five spectral lines at 0.68, 0.84, 1, 1.16, and 1.32 nm with equal intensity and nominal incidence are used to illuminate the mask, which provides a contrast of 10:1 with the x-ray absorber. The E-G branches are spaced 5% apart from –30% to 30%. The branches of these trees exhibit narrowing and broadening before finally converging to the unusable point. Take the tree of lines and spaces as an example—the largest exposure latitude (EL) is found near z = 50 μm, contrary to the intuitive thought that a smaller gap is always better.

This non-intuitive behavior is explained by Fig. 2.31(a). At z = 25 μm, the intensity log slope is shallow at the exposure for the nominal feature size. Thus, the EL is small. At z = 50 μm, the log slope becomes steep, making the spacing between the contours much larger. For opaque objects such as islands and isolated spaces, the usable region is further reduced by ghost lines, as illustrated in Fig. 2.31(b). For example, at z = 50 μm, an extra bright band will be printed in the nominally dark line if the exposure threshold is set to produce the nominal linewidth. The onset of these ghost lines is marked by the tic marks on the contours in Fig. 2.30.

The CD ±10% E-G trees of the lines and spaces, isolated line opening, and isolated opaque space 1D features are plotted together to find their common window. Figure 2.32(a) shows the individual and common E-G windows for the 500-nm 1D features. Figure 2.32(b) is the same as Fig. 2.32(a), except for ±5% CD tolerance. Figure 2.32(c) changes the feature size to 350 nm at ±10% CD tolerance. Case (a) supports a large process window with a usable gap of up to 70 μm at an EL of 0.045/0.4 = 11.25%. If a larger EL is needed, the usable gap would reduce to 50 μm. Case (b) has a usable gap of 50 μm for EL = 10%. Case (c) can support a large EL, but the usable gap is limited to less than 35 μm.

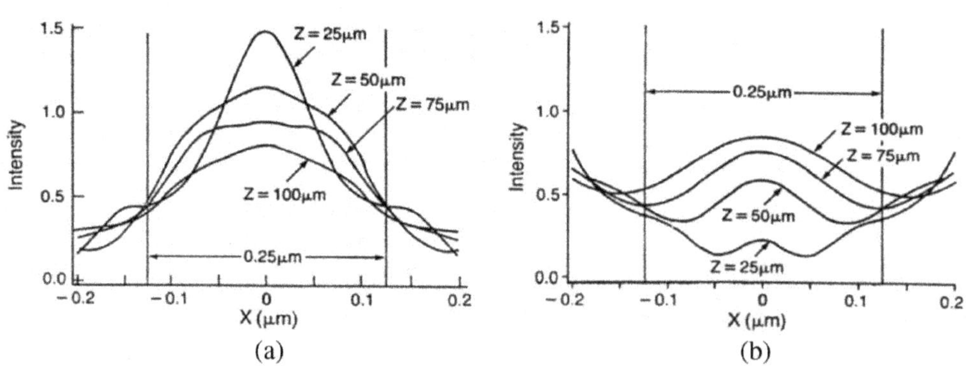

Figure 2.31 Intensity distributions of (a) the 250-nm lines and spaces and (b) the 250-nm island in Fig. 2.30.

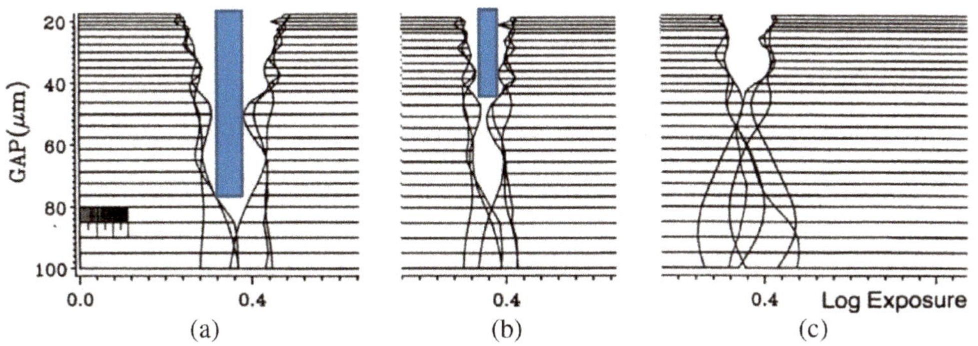

Figure 2.32 Common E-G region of 500-nm features with (a) ±50-nm tolerance and (b) ±25-nm tolerance. (c) Common E-G region of 350-nm features.

The meaning of the so-called E-G window should be expounded further. Does it indicate the DOF or the working distance (WD)? The former means that at any z and W within the rectangle, the image supports the CD tolerance. This is not the case for our E-G plots because the range from $z = 0$ μm (i.e., the contact plane) to $z = 20$ μm is not plotted. Even if it were plotted, Fig. 2.32 shows that, near the contact plane, the proximity image can oscillate between narrow and wide, and is not very useful. Moreover, this is a restricted zone for proximity printing. In this zone, the wafer and the mask can be damaged by accidentally touching each other. Hence, this rectangle definitely indicates a WD for which the wafer can be separated from the mask; however, it is not very safe to perform proximity printing at the short end of this WD. To obtain the DOF, approximately 10 to 20 μm must be deducted from the WD. The bottom of this rectangular window is more useful; it sets the maximum gap size at which the given set of features can be printed.

Let us explore the printability of 250-nm features. Figure 2.33 shows the E-G trees for three 1D features with $CD = 250$ nm. The CD tolerance remains at ±10%. The maximum gap is 30 μm. When 350- and 500-nm features are to be printed together, the WD drops to 15 μm, as shown in Fig. 7 of Ref. 13. In the same article, many ways to extend this WD, such as using a larger source

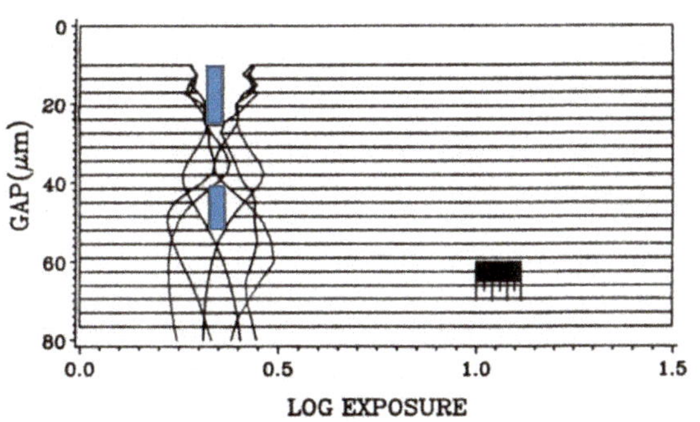

Figure 2.33 E-G windows of three 250-nm 1D features.

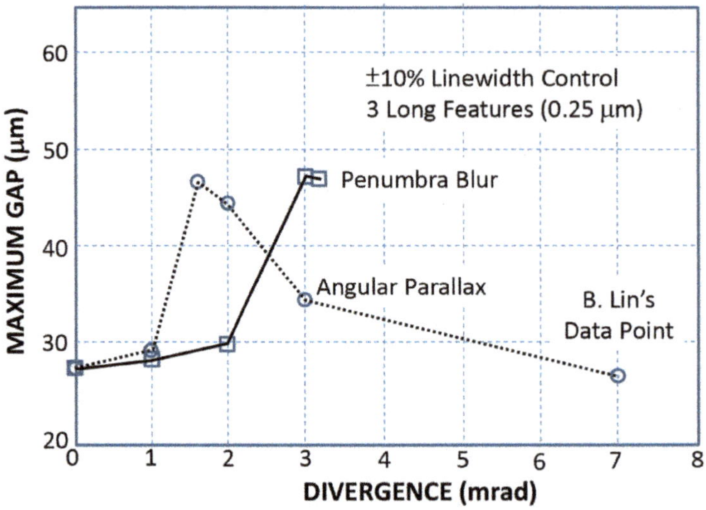

Figure 2.34 Maximum gap for three 1D 250-nm features with ±10% EL and ±10% CD tolerance as a function of beam divergence (reprinted from Ref. 15).

size, physical perturbation, and biasing, have been explored. The WD can be extended up to 23 μm, but the methods to achieve this are not very promising.

Fortunately, Oertel et al.[14] found other avenues to extend the WD. Figure 2.34 shows that by optimizing either the penumbra blur of the x-ray source or the angular parallax, the WD can be extended to about 47 μm. Further exploring physical disturbances, such as using double exposure with the mask or wafer slightly shifted between the exposures, the WD can be

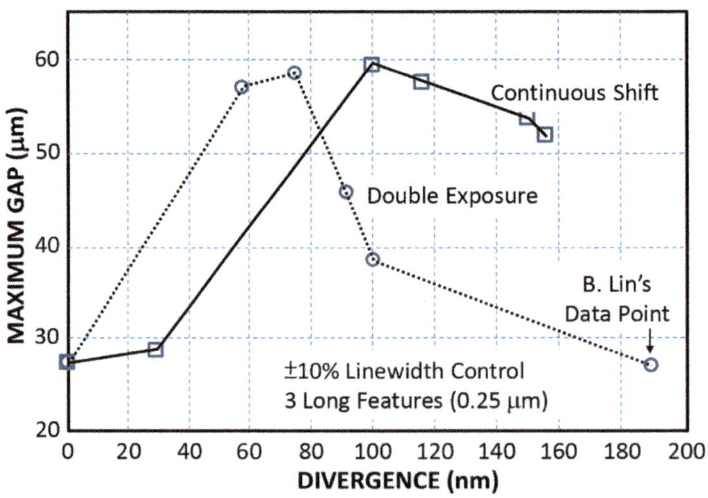

Figure 2.35 Maximum gap for three 1D 250-nm features with ±10% EL and ±10% CD tolerance as function of a continuous shift between the mask and wafer and double exposure (reprinted from Ref. 15).

extended to 59 μm. Shifting the mask or wafer continuously while exposing can extend the WD to 60 μm, as shown in Fig. 2.35. These last two techniques are less practical than the first two.

Oertel et al. plotted the E-G windows in the W-G space as in Lin's articles, as shown in Fig. 2.36. They also explored plotting E-G windows in the W-CD space, as shown in Fig. 2.37.

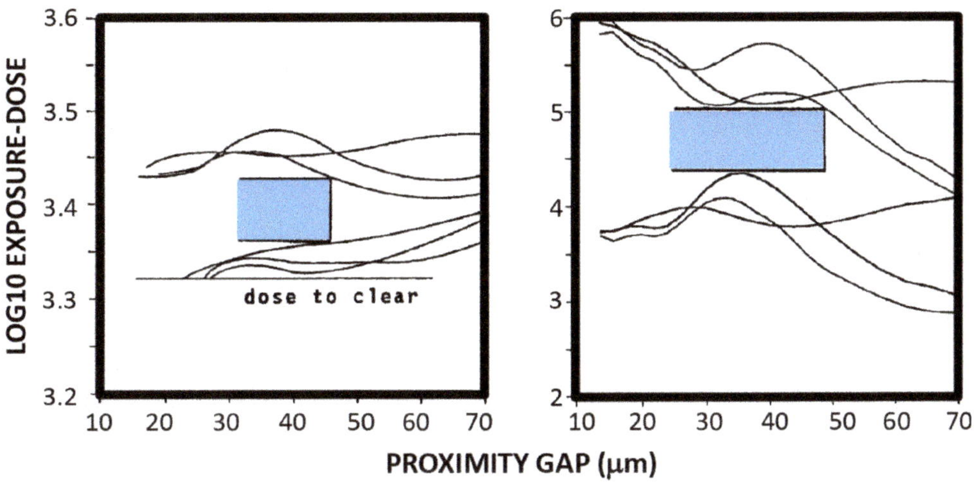

Figure 2.36 E-G window for three 1D features with 250-nm design rule and ±10% CD tolerance: (left) 1CS resist and (right) aerial image (reprinted from Ref. 15).

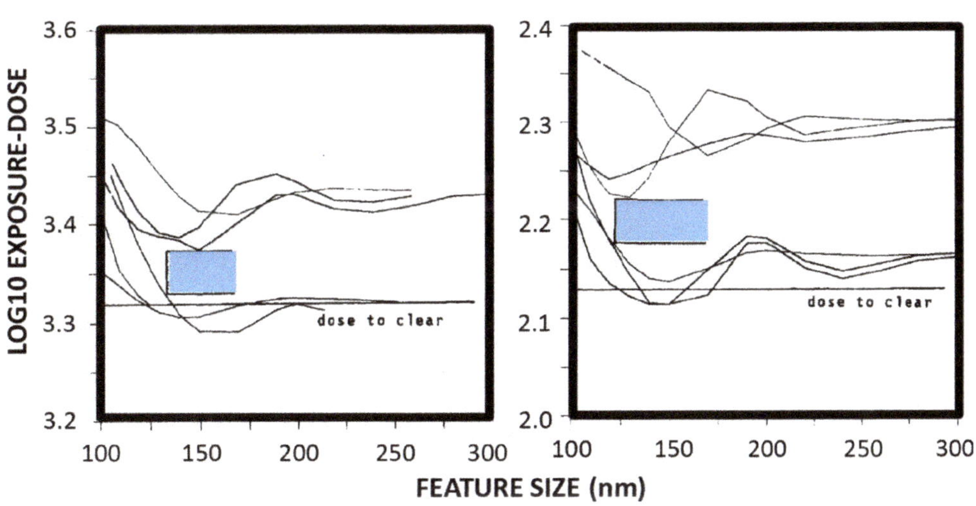

Figure 2.37 Process window in the exposure-feature-size space for three 1D features with ±20- nm CD tolerance and 10% exposure latitude: (left) with 1CS resist and (right) with aerial 3CS resist (reprinted from Ref. 15).

2.6 Conclusion

Proximity printing is attractive because of its simplicity, large field size, and polychromatic nature. Wavelengths from the visible to single-digit nanometers can be used as long as mask and resist materials can be developed. This technology attracted a lot of corporate and governmental funding to extend lithography to the submicron regime.

However, the scalability of x-ray proximity printing is limited because $W \propto z^{1/2}$. By reducing the wavelength from 254 to 1 nm, the resolution can be improved by only 16X. There is not much room to reduce z; it would be somewhere between 20 and 40 μm. The 1X mask is another problem. It is not only difficult to make, but the contribution to CD tolerance is direct, without reduction as from a reduction mask. There is not much possibility for resolution enhancement. Proximity correction is possible, but it requires much more accuracy because of the 1X mask. So, while x-ray proximity printing is struggling with 250-nm resolution, optical projection lithography has sailed past 250 nm to 180, 130, 90, 65, 40, 28, 20, 16, 10, and 7 nm. Even 1X projection lithography stopped at no less than 700 nm in resolution.

Despite all of its drawbacks, proximity printing in the UV and deep UV is still being used in universities and laboratories due to both the lower initial cost of the tools and the tolerance of the defect density of the product. These are the reasons that it persists, with the understanding that the resolution would stop at 500 nm or slightly below.

References

1. J. S. Kilby, "Miniaturized Electronic Circuits," U.S. Patent 3,138,7431 (1964).
2. B. J. Lin, "Optical methods for fine line lithography," Chapter 2 in *Fine Line Lithography*, R. Newman, Ed., American Elsevier Publishing Co. (1980).
3. B. J. Lin, "Deep UV lithography," *J. Vac. Sci. Technol.* **12**, p. 1317 (1975).
4. M. Feldman, D. L. White, E. A. Chandross, M. J. Bowden, and J. Appelbaum, "A demonstration of photolithography at 1850 Å," paper given to B. J. Lin in 1975 without specific citation.
5. K. Jain, C. G. Wilson, and B. J. Lin, "Ultrafast high resolution contact lithography using excimer lasers," *Proc. SPIE* **334**, pp. 259–262 (1982) [doi: 10.1117/12.933585].
6. J. C. White, H. G. Craighead, R. E. Howard, L. D. Jackel, R. E. Behringer, R. W. Epworth, D. Henderson, and J. E. Sweeney, "Submicron, vacuum ultraviolet contact lithography with an F_2 excimer laser," *Appl. Phys. Lett.* **44**(1), p. 22 (1984).

7. D. L. Spears and H. I. Smith, "Prospects for x-ray fabrication of Si IC devices," *Electron. Lett.* **8**, p. 102 (1972).
8. M. Hasegawa, Y. Nakayama, K. Yamaguchi, T. Terasawa, and Y. Matsui, "Printing characteristics of proximity x-ray lithography and comparison with optical lithography for 100-nm node and below," *Proc. SPIE* **3997**, pp. 96–104 (2000) [doi: 10.1117/12/390116].
9. Equation 2.22 of Ref. 2.
10. B. J. Lin, "Electromagnetic near-field diffraction of a medium slit," *J. Opt. Soc. Am.* **62**, pp. 977–981 (1972).
11. B. J. Lin, "A comparison of projection and proximity printings—from UV to x-ray," *Microelectronic Engineering* **11**(1–4), pp. 137–145 (1990).
12. B. J. Lin, "A new perspective on proximity printing—from UV to x-ray," *J. Vac. Sci. Technol. B* **8**(6), p. 1539 (1990).
13. J. Z. Y. Guo, G. Chen, V. White, P. Anderson, and F. Cerrina, "Aerial image formation in synchrotron-radiation-based x-ray lithography: the whole picture," *J. Vac. Sci. Technol. B* **8**(6), p. 1551 (1990).
14. H. K. Oertel, M. Weiss, H. L. Huber, Y. Vladimirsky, and J. R. Maldonado, "Modeling of illumination effects on resist profiles in x-ray lithography," *Proc. SPIE* **1465**, pp. 244–253 (1991) [doi: 10.1117/12.47361].

Chapter 3
Exposure Systems

Before getting into imaging theory and practice, I would like to present exposure systems so that readers have the right perspective to approach Chapters 4 and 5.

An exposure system reproduces the mask image at the wafer side to expose the photoresist layer on a wafer. The reproduced image can be the same size as the mask image in a 1X system. Reproduction on a modern exposure system usually reduces the reproduced image by a factor of 5 or 4, making it a 5X or 4X system, respectively. After development, this resist image is used as an etch, implant, plating, or lift-off mask for pattern transfer to a thin-film layer on the wafer.

There are different systems for reproduction. When the feature size was in the 2- to 5-μm regime, and the budget for semiconductor manufacturing was low, the aerial image in the 20- to 40-μm proximity of the mask sufficiently produced a useable resist image, maintaining feature size and placement control as covered in Chapter 2.

Below 2 μm, infidelity of the aerial image, defect generation, and alignment viewing difficulties drove the bulk of exposure systems to 1X full-wafer projection printing. The 1X full-wafer coverage fills the requirement of a smooth transition from proximity printing to projection printing.

When the minimum feature size of integrated circuits reached 1 μm, the field size of full-wafer systems grew with the wafer size from 50 to 125 mm in diameter. This ever-increasing wafer size and reduced feature size—together with the associated requirements of feature-size control and overlay accuracy—drove exposure systems to reduction step-and-repeat. Reduction is used to addresses the need to resolve smaller features, tight feature-size control, and overlay accuracy. The step-and-repeat feature is used to accommodate the ever-increasing wafer size in order to overcome the limited image field size of a projection system.

This chapter describes projection-printing systems, 1X and *n*X systems, and full-wafer, step-and-repeat, and step-and-scan systems. The significance of reduction imaging is also analyzed.

3.1 Projection Printing and a Comparison to Proximity Printing

Two problems with proximity printing led to the use of an imaging lens for lithography pattern transfer.

The first problem is related to defects. As the design rule is reduced to 2 μm and below while the wavelength is kept constant in the UV region, the maximum printable mask-to-wafer gap is squeezed to below 20 μm, making it extremely likely that the wafer will accidentally scratch the mask, and vice versa. Mask damage issues and edge acuity requirements drove the replacement of emulsion masks by chrome masks. Wafer damage issues drove manufacturers to clean their masks after every 100 exposures. These preventative measures still could not stop the tide of projection printing, which can print at a much larger working distance with no fear of the mask touching the wafer.

The second problem with proximity printing is related to the absence of a reduction capability, which became apparent only after the feature size dropped below 1 μm, and mask making at 1X became too demanding. With these persistent problems, lithography has needed to rely solely on projection, regardless of the wavelength. Even if the wavelength is reduced to 1 nm, as in the case of x-ray proximity printing, the 1X mask requirement is simply too stringent. More discussions on 1X and reduction systems can be found in Section 3.6.

A schematic of a projection-printing system is shown in Fig. 3.1. The system looks very similar to the proximity-printing system, except for the

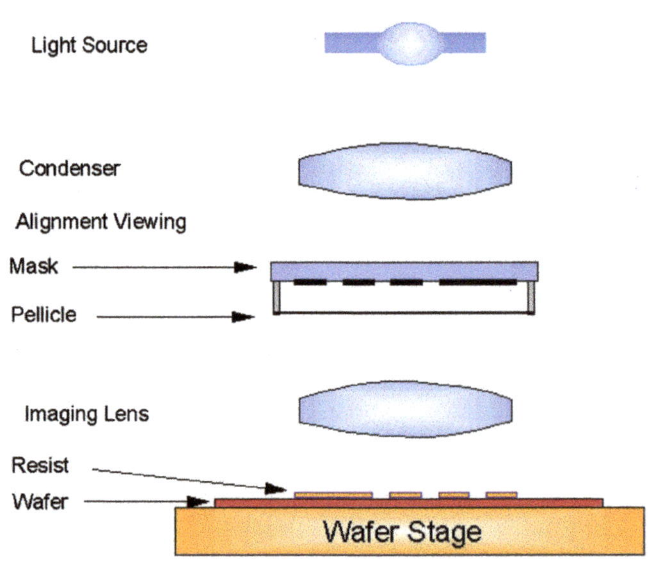

Figure 3.1 Schematic of a projection-printing system.

addition of the imaging lens and the large separation between the mask and the wafer.

The alignment system is not shown because there are many different ways to view alignment. The alignment marks may be viewed through the imaging lens or in an entirely separate system. The mask and the wafer marks may be viewed together or separately.

In a projection system, the mask-to-wafer distance is typically 80 cm or larger. The working distance between the front lens element and the wafer surface is on the order of millimeters. Even more forgiving, the working distance between the mask and the first physical element is on the order of centimeters. This facilitates high-speed movements between the mask and the wafer without any danger of damaging either of them. Projection printing also allows placement of mask-protecting pellicles at both sides of the mask. The pellicle is a thin transparent membrane of negligible optical thickness, stretched over a frame attached to the mask. In Fig. 3.1, only the pellicle at the absorber side of the mask is shown. If any foreign particle falls on the pellicle, it is safely kept out of focus by remaining outside the imaging plane. The system is therefore immune to particulate-induced defects on the mask or defects caused by the wafer touching the mask.

There is a practical difference between proximity printing and projection printing. The proximity image is always a mirror image of the mask when the latter is viewed from the patterned side, whereas the projected image is usually not considered as a mirror image if no mirror or an even number of mirrors is used in the projection lens system.

Figure 3.2 shows the wafer images from proximity printing and from projection printing with the mask viewed from the glass side. The proximity image stays unchanged, and the projected image is the mask image rotated by

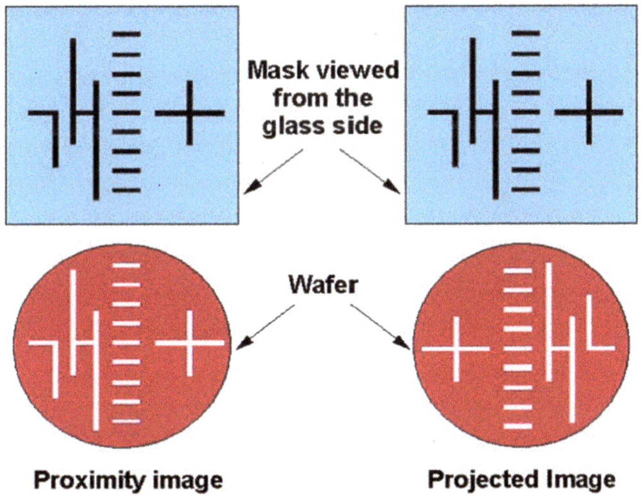

Figure 3.2 Proximity and projected images.

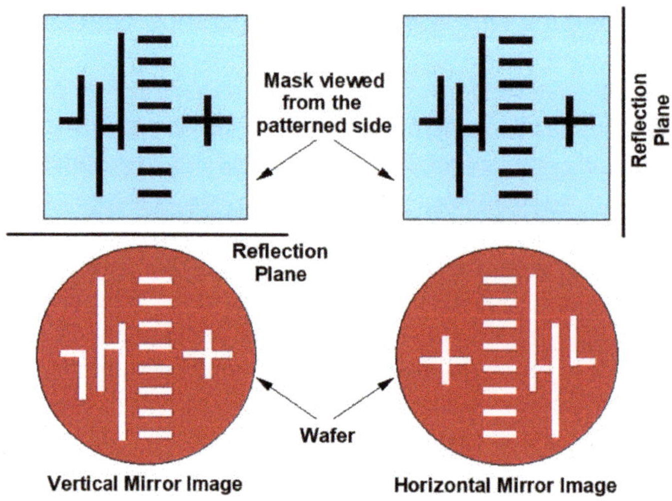

Figure 3.3 Vertical and horizontal mirror image.

180 deg. However, this rotated image can also be considered as a mirror image. Figure 3.3 shows the mask viewed from the patterned side. The proximity wafer image is a vertical mirror image with the reflection plane at the bottom of the mask. The projected wafer image is a horizontal mirror image with the reflection plane at the side of the mask.

During the introduction of full-wafer 1X projection printing using mirrors, an extra mirroring step was introduced to create an image that is identical to that of the masks used in proximity printing so that the the two systems are compatible. This full-wafer-field 1X system is discussed in Section 3.2.

The resolution and DOF of a projection-printing system subscribe to the following relationships when the NA is small:

$$W \propto \frac{\lambda}{NA} \tag{3.1}$$

and

$$DOF \propto \frac{\lambda}{NA^2}, \tag{3.2}$$

where W is the minimum half pitch in the pattern that can be resolved, λ is the wavelength, and NA is the numerical aperture of the lens defined in Eq. (4.2). The proportional constants for Eq. (3.1) and Eq. (3.2) are k_1 and k_2, respectively. These constants will be fully discussed in Chapter 4. There, Eq. (3.2) is shown to be only true for dry and paraxial systems, whose NA is relatively small. Here, Eqs. (3.1) and (3.2) are simply used to point out the difference between projection and proximity printings.

Note that, like proximity printing, the resolution improves when the wavelength is reduced. However, the improvement is much more efficient with projection printing. It is linearly related to wavelength, whereas it is only related to the square root of the wavelength in proximity printing.

In projection printing, there is a penalty in DOF when the wavelength is reduced, whereas in proximity printing, the imaging impact of a shorter wavelength does not reduce the DOF. In projection printing, the DOF is independent of the working distance. The former is usually on the order of a few micrometers to a few decades of double-digit nanometers; the latter is given in millimeters. In proximity printing, the DOF and the working distance are identical. This value is usually larger than that of projection printing in terms of DOF, but insufficient in terms of working distance.

One important aspect of projection printing is the exposure field size that dictates the size of the chips that can be produced and the level of defects. If many identical chips can be printed in the same field, a defect on one of the chips has a small impact on the yield, even though the same field is repeatedly stepped to cover the entire wafer. On the other hand, if the field is so small that it can only accommodate a single chip, the defect is repeated on every chip on the wafer, and the yield drops to zero. Multiple identical chips make possible the use of chip-to-chip inspection (commonly known as die-to-die inspection), rather than depending on die-to-database inspection. In proximity printing, the field size is limited only by the size of the mask substrate and the illumination system. There is no other fundamental limit. Exposing a 100-mm-diameter wafer in one shot is not formidable. In projection printing, the field size is dictated by the imaging lens. The field diameter is only given in double-digit millimeters, leading to the following discussion of methods for covering the wafer.

3.2 Full-Wafer Field

A full-wafer-field exposure system covers the entire wafer in one exposure step. The mask covers a field as large as the wafer. In terms of accommodating as many chips as possible in one exposure, full-wafer field exposure is desirable. It also helps to improve throughput because there is no need to wait for the wafer stage to move though all of the exposure positions, as opposed to when a small field must be stepped through the entire wafer. In the early days, the full-wafer requirement kept projection printing out of manufacturing until a clever scheme enabled it to cover the full-wafer field, as proximity printing is inherently capable of. This is the popular all-reflective 1X projection-printing system[1,2] shown in Fig. 3.4.

The inventors took advantage of the low-aberration ring-field (slit-field) zone of two concentric mirrors and scanned a segment of the ring field to cover the entire wafer. A key component of this system is the three-face mirror

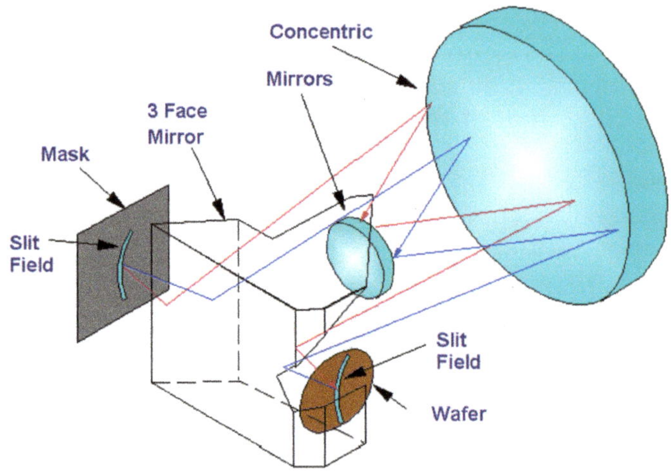

Figure 3.4 A 1X all-reflective, full-wafer field-projection-printing system (reprinted from Ref. 1).

assembly. Without this mirror assembly, the mask object ring section and the wafer image ring section are coplanar and must be scanned in opposite directions to cover the entire mask, as shown in Fig. 3.5.

Scanning with a precision of much less than 1 μm between the mask and wafer should obviously be avoided due to difficulty and cost. One mirror at the mask side and another one at the wafer side changes the direction of the

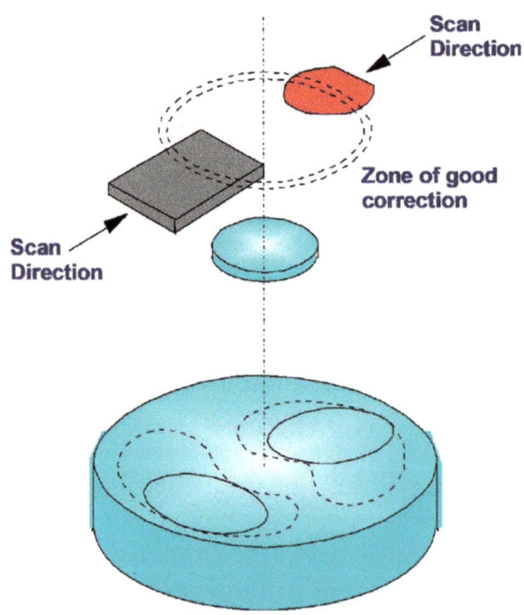

Figure 3.5 Ring zone of good correction, and mask and wafer scanning directions (reprinted from Ref. 1).

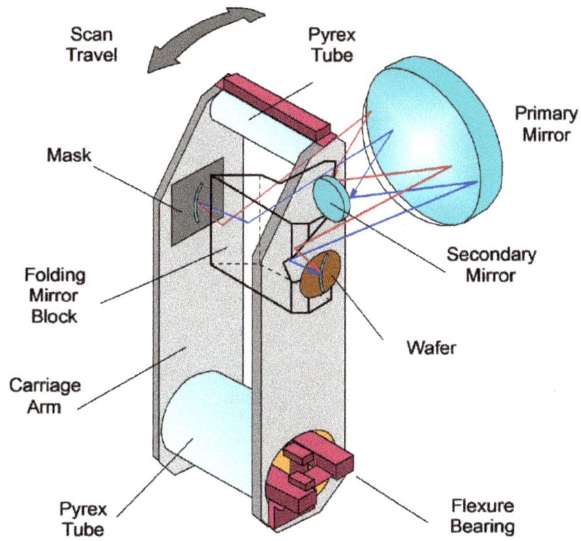

Figure 3.6 The 1X all-reflective, full-wafer-field projection printing system with the mask-wafer carriage rotated around the flexure bearing (reprinted from Ref. 2 with author permission).

light beams so that the mask and wafer can now be scanned together in one direction on a single carriage, as shown in Fig. 3.6. Note that there are two mirrors at the wafer side instead of just one at the mask side. The additional mirror is inserted to produce a mirror image at the wafer to match that from the proximity printers. The mask and the wafer are mounted on the mask-wafer carriage to pivot against the flexure bearing during scanning.

The hard-to-build flexure bearing can be eliminated using the design of the Colbilt system,[3] as shown in Fig. 3.7, which rearranges the three mirrors so that the mask and wafer are coplanar and can be scanned together on one rigid platform. The tradeoff is the lack of a single prism piece that can mount the three mirrors precisely and permantly in the factory, as in the Hemstreet–Markle system. The three separately mounted mirrors can cause the need for tedious alignment on the manufacturing floor when the mirrors lose their preferred positions.

Because the law of reflection is independent of wavelength, these all-reflective systems are broadband and can be used for new, shorter wavelengths just as conveniently as in a proximity-printing system. An all-reflective system has all the advantages of a proximity system without its drawbacks. In fact, this system inherently replicates the correct image of the mask, which means that the circuit designer can design the mask without worrying that the resultant image on the wafer is a mirror image.

Starting in 1974, the 1X projection system quickly replaced proximity printing to become the industry's workhorse. As wafer size continued to

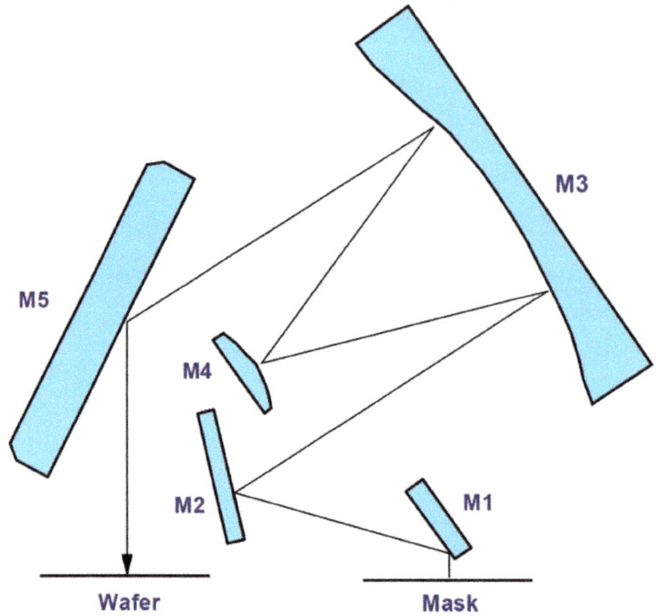

Figure 3.7 Imaging lens in the Colbilt 3000 1:1 all-reflective mask aligner (reprinted from Ref. 3).

increase, the original 1X all-reflective projection-printing system was expanded in 1984 to include two sets of concentric mirrors[4,5] and some refractive elements, making it a catadioptric (a mixture of refractive and reflective optical elements) system. A schematic drawing of the system is shown in Fig. 3.8. In addition to expanding to a larger field size, the mask and wafer can now be scanned together via an airbearing on a common rectangular column. The refractive shells increase the ring-field slit width from 1 mm to 3 mm. The strong refractive shell enables fine tuning of

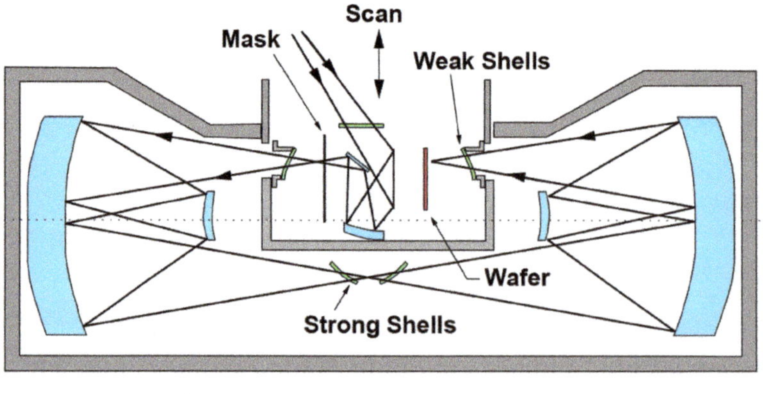

Figure 3.8 The 1X all-reflective full-wafer-field projection-printing system with a mask-wafer carriage that can be moved linearly and a flexure bearing (reprinted from Ref. 2).

magnification during the scan by moving in the axial direction, coupled with the movement of the mask with respect to the wafer. The three-faced roof prism shown in Figs. 3.4 and 3.6 is eliminated, saving the effort to make such an expensive element and the slight distortion in the middle of the field, as well as some loss of resolution that it caused. When the illumination intensity is increased to raise the wafer throughput, the secondary mirrors in the mirror blocks become warm and generate a warm air plume that passes through the center of the ring field. As a result, the optical path is filled with He, which provides better heat conduction and a lower refractive index.[6]

This system is capable of covering a 125-mm wafer in full field. Due to scanning, the field is not limited by the lens in the scan direction. For 150-nm wafers, a $125 \times 150\ \text{mm}^2$ field is used. The loss of exposed wafer area reduces profit because many more chips would be produced otherwise. So, further growth of the wafer size requires growing the mask and the lens field. Neither endeavor is easy. As feature sizes approach 1 μm, making a 1X mask to specification has become difficult. In addition, the concentric mirror scheme is usable only up to about 0.18 NA. [NA stands for numerical aperture, which is defined in Eq. (4.2).] The resolution of the imaging lens is inversely proportional to the NA, as indicated in Eq. (3.1). A detailed discussion on the NA is given in Chapter 4. Above 0.18 NA, it is difficult to avoid aberrations to achieve the resolution expected from a higher NA. It became inevitable that the full-wafer-field concept could not be sustained. Step-and-repeat systems succeeded the full-wafer-field systems. The smallest feature piloted for manufacturing with the full-wafer-field system is 1 μm, using deep-UV light.[7]

The all-reflective system is regaining popularity by optical necessity at the 13.5-nm EUV wavelength, where there is no transparent material. With four to six extremely high-precision aspherical mirror surfaces in the imaging lens, the NA can be 0.35 in production. Even 0.55 NA has been demonstrated. More on this topic is covered in Chapter 9.

3.3 Step and Repeat

When the imaging-lens-dictated field size is smaller than the wafer size, the only way to cover the entire wafer is to step identical exposures through the wafer. Although, in principle, one can change the mask between exposures, this is rarely done due to throughput and overlay considerations. A step-and-repeat system steps a mask repeatedly to cover the wafer. Figure 3.9 depicts a circular lens field and an exposure field with the unused part of the lens field coarsely blocked out with mechanical blades and finely blocked by the mask absorber. As many chips as possible are fitted into the exposure field for defect, throughput, and mask inspection considerations. The chip is also called a die. In general, the former is used by people concerned with the

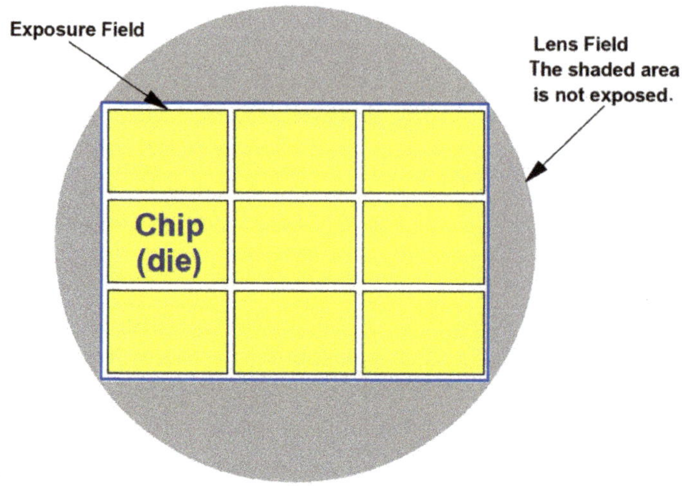

Figure 3.9 Schematic showing chips (dies), exposure field, and lens field.

electrical properties of the piece, while the latter is used by those who are interested in its physical structure, such as those who cut the wafer into dies.

The exposure field is stepped repeatedly to cover the entire wafer, as shown in Fig. 3.10. Note that the out-of-bound fields are exposed. The wafer area is very valuable; even though the chips that are partially or completely out of bounds are lost, the chips that are in the same field but are exposed in full can still be used as product chips. In the nanometer era, another reason to expose the partial fields is to maintain even loading both physically and chemically.

A typical step-and-repeat projection-printing system using a refractive lens is shown schematically in Fig. 3.11. This is the most popular configuration for production of integrated circuits with feature sizes of 1.5 μm and below. Although not required by projection printing but essential to improving mask accuracy, most step-and-repeat projection-printing systems[8,9,10] use a reduction ratio of five. Initially, there were 10X reduction systems for easier mask making at the expense of the field size.[11,12] After the 5X system, 4X systems started appearing.[13] The tradeoffs of reduction ratio in imaging are discussed in Section 4.5.

The step-and-repeat projection-printing system is not limited to dioptric (refractive) reduction systems. Figure 3.12 depicts a 1X catadioptric lens[14] that is used in a step-and-repeat system, which has proven to be cost effective and remains widely used for technology nodes in the longer submicrometers, as well as for noncritical levels in the shorter-submicrometer technology nodes. Short submicrometer is the regime below 0.3 μm; mid-submicrometer is 0.3 to 0.7 μm; and long submicrometer is 0.7 to 1 μm. Short submicrometer is sometimes referred to as deep submicrometer.

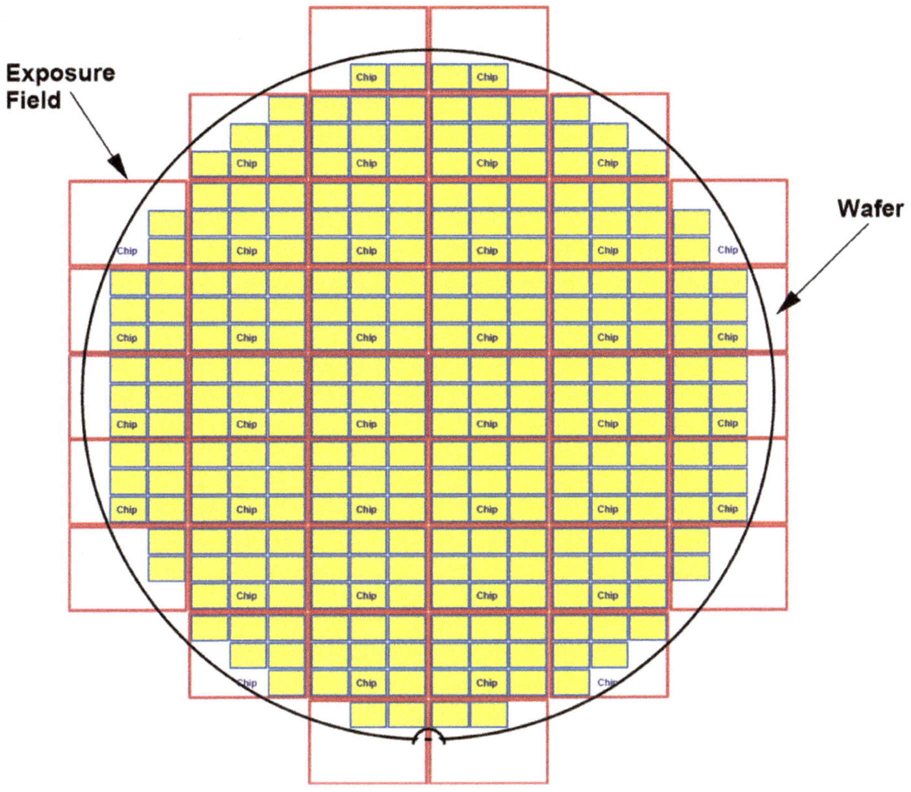

Figure 3.10 Exposure field stepped repeatedly to cover the entire wafer. The chips that have spilled over at the wafer edge are not plotted.

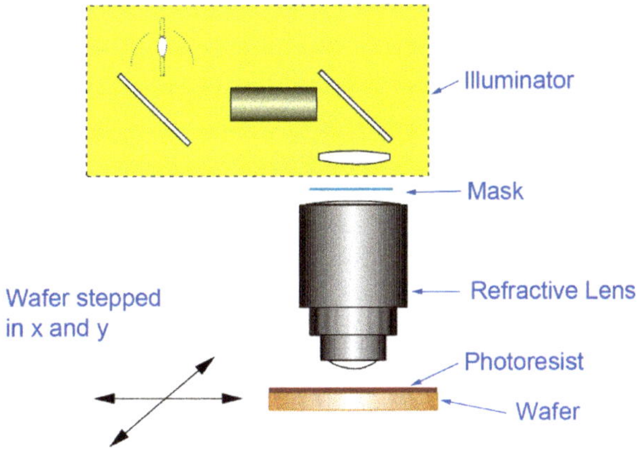

Figure 3.11 A step-and-repeat projection-printing system using a dioptric reduction imaging lens.

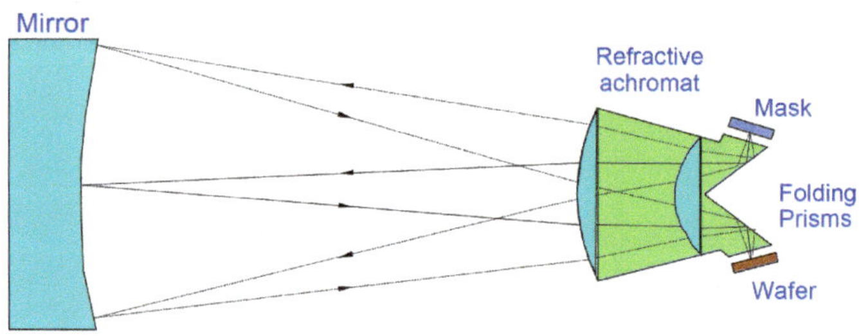

Figure 3.12 1X catadioptric projection lens used for step-and-repeat (reprinted from Ref. 13).

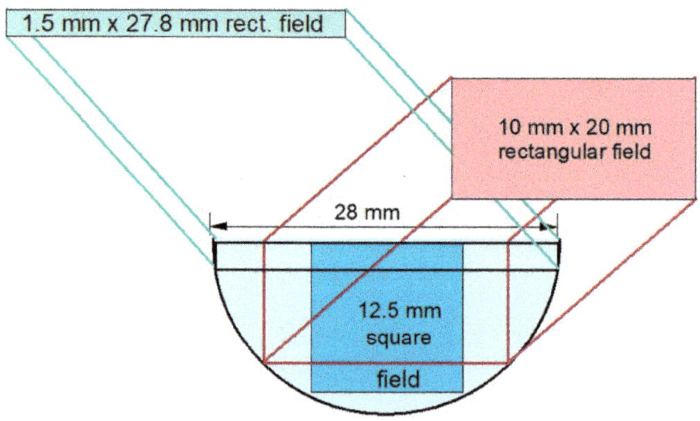

Figure 3.13 Semicircular lens field of the system in Fig. 3.12 and the three possible rectangular fields to fit the lens field.

This catadioptric system does not have a full circular lens field but a semicircular field instead because the incident and reflected lights need to give way to each other. Three possible retangular fields that fit the semicircular lens field are shown in Fig. 3.13.

Initially, masks used in a step-and-repeat system were called reticles. This term was borrowed from the usage of the photorepeater,[15] which reduces a reticle and steps it repeatedly on a mask blank to make the mask. The mask can be 1X or nX. The patterns on the reticle must be bigger than those on the mask, so that the reticle patterns can be generated by stepping a focused optical spot on a photosensitive substrate. The tool that performs this task is called a pattern generator.[16] E-beam mask writing eliminated the need for an optical pattern generator and a photorepeater. Today, the terms mask and reticle are almost interchangeable.

3.4 Step and Scan

The increase in chip size drives the increase in field size, without which wafer throughput, yield, and productivity cannot be maintained. On the other hand, high packing density demands higher resolution, which most often calls for increasing the lens NA. High packing density also demands higher overlay accuracy, which tightens both the lens distortion specifications and the mask fabrication specifications. These competing requirements compound the complexity and cost of the imaging lens. The limit of these three lens parameters has become difficult to exceed. The step-and-scan system was developed to meet these requirements.

The step-and-scan principle[5] was introduced by Markle. Markle's system uses a stationary vertical slot and scans the vertically oriented mask and wafer in the same horizontal direction. Modern scanners use a stationary horizontal slot and scan the horizontally oriented mask and wafer in opposite horizontal directions. Figure 3.14 shows a step-and-scan sequence of a modern scanner.

The starting position of the mask and the wafer is shown in Step 1, which is at the lower left corner of the wafer. The mask is below the slot, and the wafer is above the slot. In Step 2, the mask scans upwards, and the wafer scans downwards. The situation at the end of the scan is shown in Step 3. Then, the wafer steps to the left by one field position to prepare for scanning the next field. This sequence is shown in Steps 4 and 5. At Step 6, the next field is being scanned. The end of this scan is shown in Step 7. We then skip to the end of the scanning of the next field, in Step 8. The wafer is stepped downwards to prepare for scanning the field in the 2nd row. We then skip to the ready position for scanning the 1st field in the upper row and continue to skip to Step 11 at the position for scanning the last field. Scanning is carried out at Steps 12 and 13. By Step 14, the wafer is unloaded, leaving the stationary slot and the mask to wait for the next wafer.

Figure 3.15 summarizes the step-and-scan movements in a larger wafer with more fields, without depicting the mask movements shown in Fig. 3.14.

With the 1X system described in Section 3.2, the mask and the wafer move at the same speed and in the same direction, making it mechanically simpler to scan them together in sync. However, 1X systems require unattainable specifications for the mask in short-submicrometer applications; a reduction system must be used, requiring the mask to move at a speed that is higher than that of the wafer by the reduction ratio. For example, with a 4X system, the mask must move exactly four times faster than the wafer. At the end of the travel, the wafer is moved to a new location for the next scanning. This procedure repeats until the entire wafer is exposed. Note that the scanning direction must be reversed from one step to the other because the slot is at the other end of the mask. The reverse scan effectively produces the wafer image without wasting time moving the slot to the first end of the mask.

A schematic representation of a step-and-scan projection system is shown in Fig. 3.16. The system shares many basic components of a reduction step-and-repeat system, except for the field-defining slot and the scanning

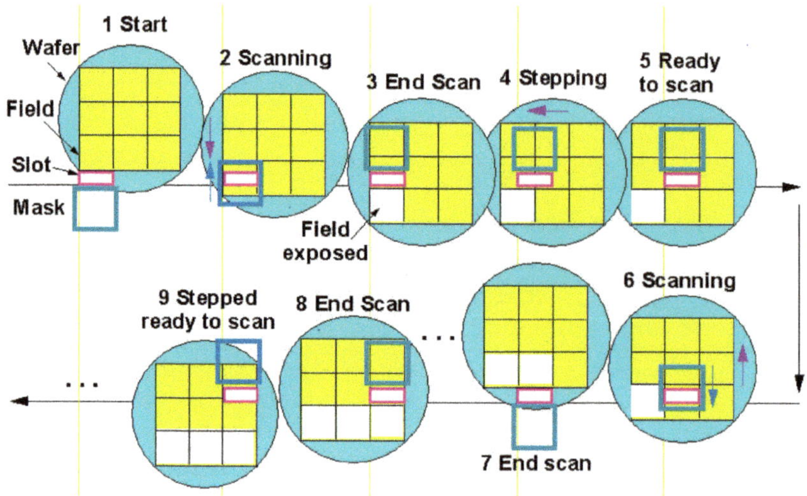

(a)

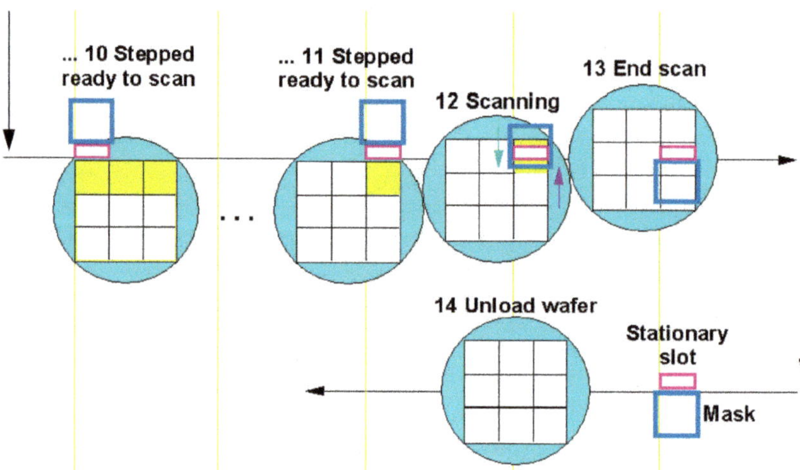

(b)

Figure 3.14 The step-and-scan sequence.

mechanism for the mask and the wafer. The slot remains stationary with the imaging lens, while the mask is scanned four times faster than the wafer for this 4X reduction system. Most step-and-scan systems[17,18,19] employ a dioptric (refractive) reduction lens. However, an early step-and-scan system uses a catadioptric lens with a beam splitter,[20,21,22] as shown in Fig. 3.17. This imaging system produces a mirror image so that the mask and the wafer can be scanned in the same direction. It is also arranged such that the mask and wafer are vertical. Other step-and-scan systems need to scan the mask and the wafer in opposite directions, making them slightly preferable in terms of mechanical balancing, but slightly less desirable due to the increased difficulty

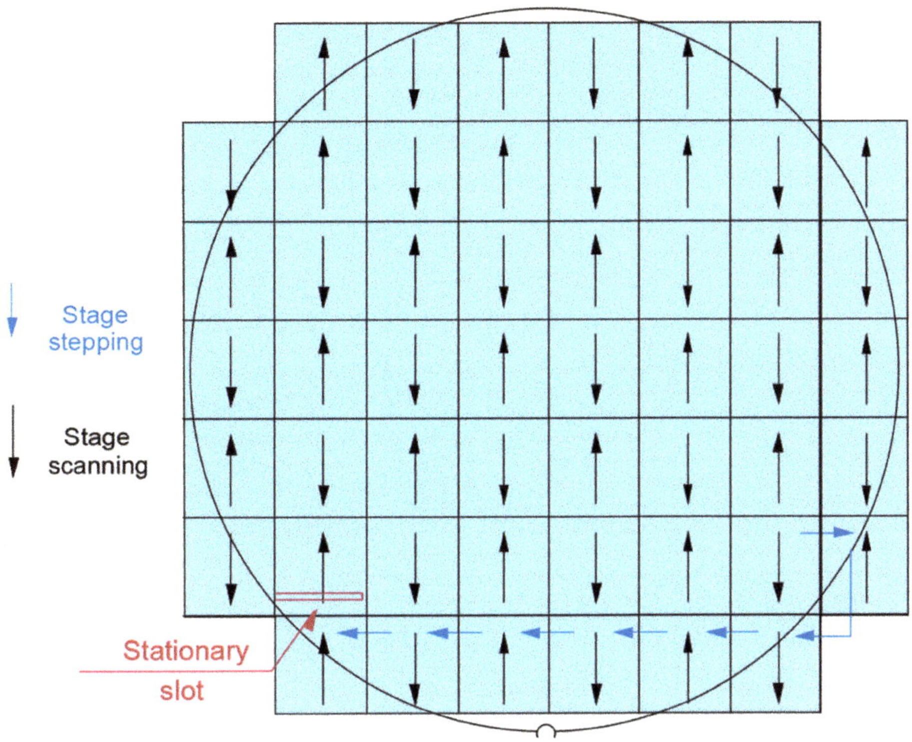

Figure 3.15 Summary of the step-and-scan movements.

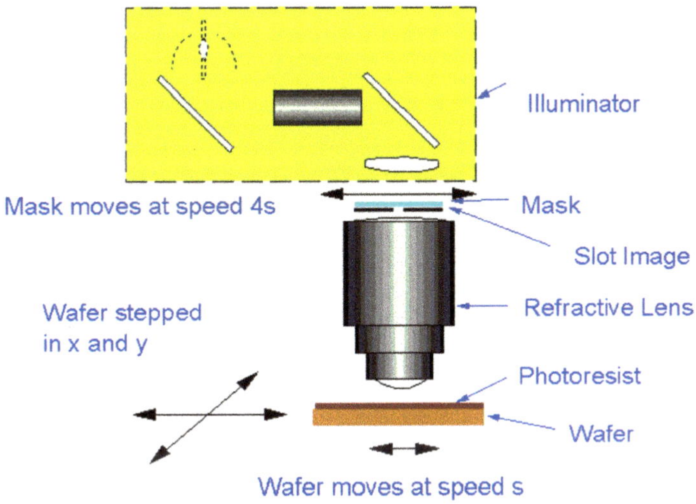

Figure 3.16 A step-and-scan projection-printing system using a dioptric reduction lens.

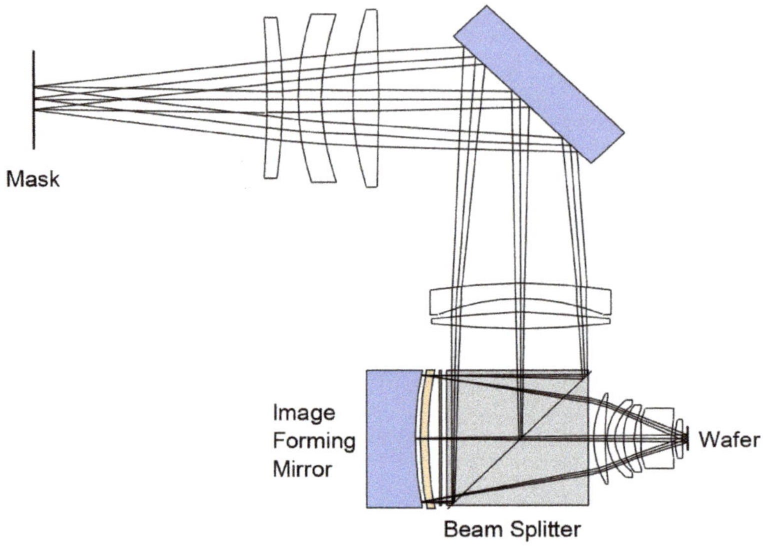

Figure 3.17 A catadioptric lens system with a beam splitter, as used in a step-and-scan system (reprinted from Ref. 20).

in controlling the movement of the mask with respect to the movement of the wafer. The mask and the wafer in these systems are placed horizontally. In either case, there can be mask sagging, but the sagging in the horizontal system is more symmetrical than that in the vertical system. The dioptric, catadioptric, and reflective lens systems will be discussed in Section 6.4.2.

In the 1X full-wafer-field-exposure system, a circular scanning slit is used, as was shown in Fig. 3.4. The slit has a curvature that follows the contour of the imaging ring field. Good imaging cannot be maintained outside of the curved, narrow imaging field. On the other hand, the illuminated slot area used for limiting the field of a lens that is actually capable of a full circular field is straight.[23] The scanning distance traveled using a slit is longer than that using a slot because of the curvature, which requires an extra lead time and trailing time. As a result, slightly more scanning time is wasted with the slit field. The scanning field of the extreme ultraviolet (EUV) lithography system uses a slit field due to the all-reflective system required of EUV light. Producing a uniform field to support a straight slot is too much of a luxury.

A major advantage of step-and-scan systems is the capability to extend the field size in one direction without adding any burden on the imaging lens. Figure 3.18 shows a typical square stepper field size of 22 $\times$ 22 mm^2 that shares the same circle with a 1:2 stepper field size of 14 $\times$ 28 mm^2. When the same lens is used for scanning, an 8 $\times$ 30 mm^2 slot supports a 30 $\times$ 33 mm^2 field for 150-mm reticles; the field can be extended to 50 mm in the scan direction when 225-mm reticles become available.

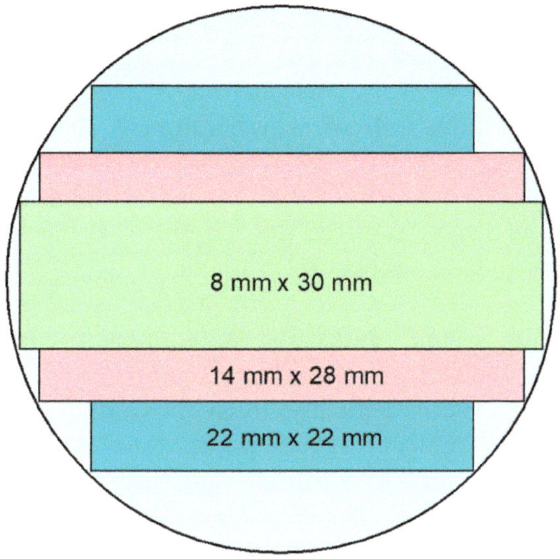

Figure 3.18 Stepper and scanner fields from the same circular field area.

Another advantage of scanners is the unification of lens distortion in the scan direction. Figure 3.19(a) shows a simulated dynamic distortion map[24] of the static distortion in the slot area of a stepper lens. When the scanning movement is perfect, this is what is expected. An actual distortion map is shown in Fig. 3.19(b). The difference between them is induced by scanning

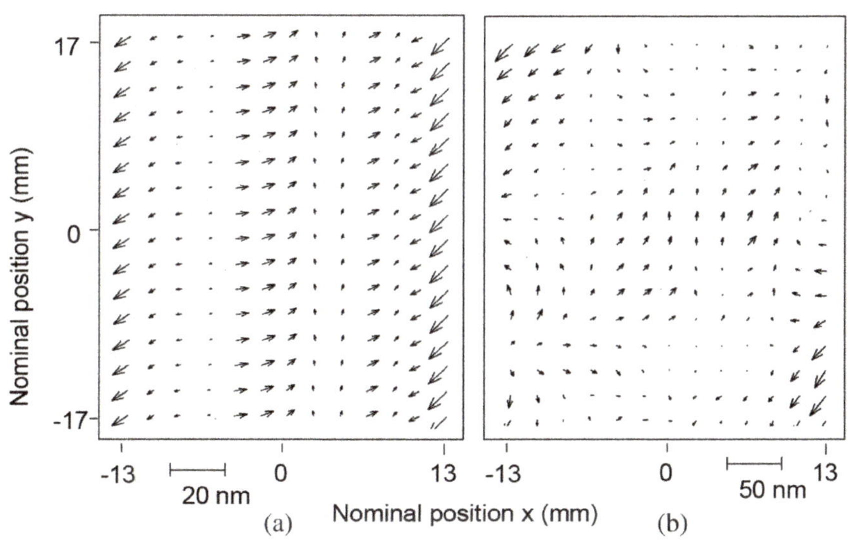

Figure 3.19 (a) Simulated dynamic distortion from measured static distortion. (b) Actual dynamic distortion from the same lens as in part (a) (reprinted from Ref. 23).

imperfections and measurement inconsistencies. Note that the error vectors in parts (a) and (b) of the figure differ by a factor of 2.5X.

Further advantages of a scanner include *in situ* focusing, leveling, and alignment during scanning, all of which improve the lithographic performance. There is no reason that the width of the slot should be uniform. Perturbation of the slot width may be used to compensate for illumination nonuniformity in the direction of the slot length. Also, the scanning speed can be modified to produce an exposure gradient or exposure variations in the scanning direction.

A step-and-scan system gains its advantages by trading off optical complexity for mechanical precision and complexity. For example, to maintain throughput, the wafer stage must move at 250 mm/sec or higher. Modern scanners have achieved speeds higher than 500 mm/sec. Needless to say, the mask must move at 1 and 2 m/sec for wafer speeds of 250 mm/sec and 500 mm/sec, respectively, in a 4X system. Maintaining the positioning accuracy to a few nanometers at such speeds cannot be taken for granted by any means.

3.5 Reduction and 1X Systems

Lithography started with 1X replication because of its simplicity and versatility. As discussed above, 1X systems were popular until 1X-mask specifications could no longer be maintained due to the ever-reducing feature size and the overlay accuracy that followed. The switchover point was 1.5 μm for most manufacturers. From this point on, mask specification was of key importance in selecting the reduction ratio of the replication system.

Table 3.1 is an update of an earlier analysis[32] of the CD tolerance contributions from mask fabrication and wafer processing. We assume that the wafer resist feature size is controlled to 10% of the critical dimension. The wafer etch control is given at 6%, and the hot-process-induced CD error is 5%. The resist control and etch control of the mask CD are 8% and 4.8%, respectively—better than those of the wafer. When the same control is applied to a 4X system, the mask control is improved by a factor of 4 to the first order,

Table 3.1 CD tolerance contributions from mask fabrication and wafer processing with mask error enhancement factor *MEEF* = 1.

CD tolerance	4X	1X	1X, Improve wafer	1X, Improve mask & wafer
Mask resist	8%/4	8%	8%	7%
Mask etch	4.8%/4	4.8%	4.8%	4.8%
Wafer resist	10%	10%	4.3%	5.8%
Wafer etch	6%	6%	6%	6%
Hot process	5%	5%	5%	5%
RSS sum	12.9%	15.75%	12.9%	12.9%

while the wafer tolerances are identical. The root-sum-square (RSS) of the CD error is 15.75% for 1X systems, compared to 12.9% for 4X systems—a 22% disadvantage. To make the CD control performance identical for both systems, the wafer resist image error of the 1X system must be reduced to 4.3%, which is quite impossible. Improving wafer and mask resist errors to 5.8% and 7%, respectively, makes the RSS of the CD error identical for 1X and 4X systems. In either case, there are few reasons that the improvement cannot be applied to 4X systems as well, making it self-defeating for 1X systems to catch up.

Normally, any dimension change is linearly transferred to the wafer; i.e., a 10-nm change on a 1X mask becomes a 10-nm change on the wafer. Similarly, a 40-nm change on a 4X mask translates to a 10-nm change on the wafer. However, when the wafer image size is pushed to well below λ/NA, the mask error enhancement factor $MEEF \equiv \Delta CD_{wafer}/\Delta CD_{mask}$ is no longer unity. It can exceed 4 in extreme cases. The contributions from the mask must be multiplied by the MEEF in both the 1X and 4X cases. The situation is shown in Table 3.2. The increase in CD tolerance in the 4X case is undesirable, whereas the situation of the 1X case is disastrous. The entire CD tolerance budget is easily taken by the mask tolerance. Making the wafer contribution zero is still insufficient.

Table 3.3 is a similar update of an earlier analysis of the overlay budget. The mask placement tolerance is taken from an e-beam placement tolerance of 30 nm. When one masking layer is aligned with another masking layer, the overlay tolerance is the RSS sum of the two placement tolerances:

Table 3.2 Same as Table 3.1, except with $MEEF$ = 4.

CD tolerance	4X	1X	1X, Improve wafer	1X, Improve mask & wafer
Mask resist	32%/4	32%	32%	0%
Mask etch	19.2%/4	19.2%	19.2%	19.2%
Wafer resist	10%	10%	0%	0%
Wafer etch	6%	6%	0%	0%
Hot process	5%	5%	0%	0%
RSS sum	15.7%	39.4%	37.3%	19.2%

Table 3.3 Overlay budget contributions from mask fabrication and wafer processing.

	4X	1X	E-beam direct write
Mask placement tolerance	42/4 nm	42 nm	42 nm
Alignment tolerance	15 nm	15 nm	12 nm
Magnification tolerance	15 nm	12 nm	12 nm
Lens distortion	10 nm	10 nm	0
Laser table tolerance	10 nm	10 nm	10 nm
RSS sum	27.6 nm	48.7 nm	46.8 nm

$\sqrt{2} \times 30$ nm. The overlay error of a 1X system can be 76% more than that of a similar 4X system, as seen in columns 2 and 3 of Table 3.3. In fact, even when an e-beam direct-write system is used, the overlay accuracy is still on the order of 70% worse than that of a 4X reduction system. The original is worse than the copy! This shows the leverage of a reduction system.

Hence, from the point of view of CD and overlay control, a high reduction ratio is always desirable. However, a larger reduction ratio requires a larger mask substrate, as well as a longer and faster scanning to maintain the same field size on the wafer. These specifications are expensive to implement in commercial systems. Conversely, one can use the same substrate size and reduce the wafer field size. As will be discussed in Section 7.1.5, a longer stepper time is required to support a smaller field size. In addition, the productivity loss is prohibitive. When the mask contribution is no longer dominating, further increasing the reduction ratio has a diminishing return.

3.6 1X Mask Fabricated with a Reduction System

To leverage the advantage of reduction, optically reducing a 4X mask made with e-beam direct write to make 1X masks has been considered. The tolerance analysis must include the extra step of mask making. For CD control, the tolerance components are shown in Table 3.4. The CD tolerance on the 4X mask indeed provides the 4X-reduction leverage. However, the CD control needed to fabricate the 1X mask is like that needed when fabricating the wafer; nevertheless, because the mask substrate is planar, it is possible to control the resist and etch images on the 1X mask to 8% and 4.8% instead of 10% and 6%. The result is still worse than 1X lithography using conventional mask making.

The overlay budget using a reduction-fabricated mask is shown in Table 3.5. The contributions from magnification error, lens distortion, and laser table error in reduction mask making are added to the components shown in Table 3.3. The lens distortion component is smaller because an

Table 3.4 CD tolerance analysis of the reduction-fabricated mask situation.

	CD tolerance
4X mask resist	2%
4X mask etch	1.2%
1X mask resist	8%
1X mask etch	4.8%
Wafer resist	10%
Wafer etch	6%
Hot process	5%
RSS sum	16%

Table 3.5 Overlay budget analysis of the reduction-fabricated mask situation.

Overlay error component	Tolerance
4X mask placement	42/4 nm
Reduction-mask-making magnification	15 nm
Reduction-mask-making lens distortion	6 nm
Reduction-mask-making laser table	10 nm
Wafer-printing alignment	15 nm
Wafer-printing magnification	10 nm
Wafer-printing lens distortion	10 nm
Wafer-printing laser table	10 nm
RSS sum	32 nm

extremely low-distortion lens can be selected for reduction mask making. The resultant overlay tolerance of 32 nm is much better than that of conventional 1X lithography, but it is still 14% worse than in conventional 4X-reduction lithography. An inherent disadvantage of the 1X system is that the DOF of the mask is the same as that of the wafer, instead of being larger by the square of the reduction ratio.

3.7 Summary

The progression of optical exposure tools from proximity printing to projection printing, 1X full-wafer to reduction step-and-repeat, and then to step-and-scan, is given. The interplay of resolution, working distance, field size, CD/overlay control, and productivity has changed the configuration of the optical exposure system to the ultimate reduction step-and-scan system in the span of half of a century. Equipped with a general perspective of the optical exposure system, we are poised to move to the next chapter on optical imaging.

References

1. H. S. Hemstreet, D. A. Markle, W. H. Newell, and A. Offner, "Optical projection apparatus," U.S. Patent 4,011,011 (1977).
2. D. A. Markle "A new projection printer," *Sol. State Tech.* **17**, p. 50 (June 1974).
3. T. W. Novak, "A new VLSI printer," *Proc. SPIE* **0135**, pp. 36–43 (1978) [doi:10.1117/12.956111].
4. J. D. Buckley, "Expanding the horizons of optical projection lithography," *Sol. State Tech.* **25**, p. 77 (May 1982).
5. D. A. Markle, "The future and potential of optical scanning systems," *Sol. State Tech.* **27**, p. 159 (September 1984).
6. D. A. Markle private communications.
7. J. G. Maltabes, S. J. Holmes, J. R. Morrow, R. L. Barr, M. C. Hakey, G. Reynolds, W. R. Brunsvold, C. G. Willson, N. J. Clecak, S. A.

MacDonald, and H. Ito, "1X deep-UV lithography with chemical amplification for 1-micron DRAM production, *Proc. SPIE* **1262**, pp. 2–7 (1990) [doi:10.1117/12.20090].

8. K. Ushid, M. Kameyama, and S. Anzai, "New projection lenses for optical stepper," *Proc. SPIE* **633**, pp. 17–23 (1986) [doi:10.1117/12.963698].
9. A. Suzuki, "Double telecentric wafer stepper using laser scanning method," *Proc. SPIE* **538**, pp. 2–8 (1985) [doi:10.1117/12.947740].
10. S. Wittekoek, "Step and repeat wafer imaging," *Proc. SPIE* **221**, pp. 2–8 (1980) [doi:10.1117/12.958617].
11. J. M. Roussel, "Step and repeat wafer imaging," *Proc. SPIE* **135**, pp. 30–35 (1978) [doi:10.1117/12.956110].
12. K. Suwa, K. Nakazawa, and S. Yoshida, "10:1 step-and-repeat projection system," Kodak Micro-Electronics Seminar (1981).
13. R. DeJule, "Cover story: mix-and-match: a necessary choice," *Semicond. Int.* **23**(2), pp. 66–67 (2000).
14. R. Hershel, "Optics in the Model 900 Projection Stepper," *Proc. SPIE* **221**, pp. 39–45 (1980) [doi:10.1117/12.958622].
15. D. J. Elliot, *Integrated Circuit Mask Technology*, McGraw-Hill Book Co., p. 161 (1985).
16. ibid, p. 146.
17. S. Murakami, "Optical exposure system—today and future," *Proc. SEMI Technology Symposium* **94**, p. 397 (1994).
18. R. Ebinuma, K. Iwamoto, H. Takeishi, H. Itoh, M. Inoue, K. Takahashi, and M. Ohta, "Imaging performance of scanning exposure systems," *Proc. SPIE* **3334**, pp. 437–447 (1998) [doi:10.1117/12.310772].
19. M. van den Brink, H. Jasper, S. Slonaker, P. Wijnhover, and F. Klaassen, "Step-and-scan and step-and repeat, a technology comparison," *Proc. SPIE* **2726**, pp. 734–753 (1996) [doi:10.1117/12/240936].
20. D. M. Williamson, "Optical reduction system," U.S. Patent 4,953,960 (1990).
21. D. M. Williamson, "Catadioptric microlithographic reduction lenses," *Proc. Int. Optical Design Conference* **22**, OSA (1994).
22. D. M. Williamson, J. A. McClay, K. W. Andresen, G. M. Gallatin, M. D. Himel, J. Ivaldi, C. J. Mason, A. W. McCullough, C. Otis, J. J. Shamaly, and C. Tomczyk, "Micrascan III: 0.25-um resolution step-and-scan system," *Proc. SPIE* **2726**, pp. 780–786 (1996) [doi:10.1117/12.240939].
23. B. J. Lin, "The paths to subhalf-micrometer optical lithography," *Proc. SPIE* **922**, pp. 256–269 (1988) [doi:10.1117/12.968423].
24. G. de Zwart, M. A. van den Brink, R. A. George, D. Satriasaputra, J. Baselmans, H. Butler, J. B. P. van Schoot, and J. de Klerk, "Performance of a step-and-scan system for DUV lithography," *Proc. SPIE* **3051**, pp. 817–835 (1997) [doi:10.1117/12.276002].

Chapter 4
Image Formation

As discussed in Chapter 1, the goal of lithography is to place the edges of a given mask pattern on the wafer within a specified tolerance. Doing so with effective controls requires a good understanding of the image formation process—from the mask image to the final transferred image. The aerial image is formed from the light that illuminates the mask and carries the information of the mask pattern through the lens to focus on the image plane. The aerial image becomes reflected and refracted as it propagates into the resist and is reflected and refracted numerous times by the resist, as well as by the multiple film layers on the wafer substrate. The superposition of reflected and refracted images in the resist exposes this medium, producing chemical and physical changes. The distribution of these changes is the latent image. The exposed resist has a distribution of dissolution rate according to the latent image. The developer removes the resist according to the dissolution rate distribution and the topography of the resist surface. The developed resist is now the resist image. The transferred image is dictated by the resist image and the characteristics of the transfer process. These images are discussed below.

4.1 The Aerial Image

We define the aerial image as light distribution at the vicinity of the image plane without any resist or multiply reflecting surface. The image must be treated as light waves in physical optics as opposed to light rays in geometric optics. When the light wave has a perfect wavefront to form the image, there is no aberration to deteriorate the image. The only limit on resolution comes from the finite extent of the wavefront dictated by the aperture of the imaging lens. A larger aperture produces a better diffracted image. Such an imaging system is called a diffraction-limited system. We start our investigation using a perfect spherical wavefront with a finite lens aperture to form a point image, which leads to the relationship of diffraction-limited resolution to wavelength and aperture size, followed by the relationship of depth of focus (DOF) to the same set of parameters. Deviation from the spherical wavefront is treated in

Section 4.1.4. The formation of a diffraction-limited aerial image from a mask pattern and three different types of illumination are given in Sections 4.1.5 and 4.1.7. The concept of spatial frequency and angular spectrum is further treated in Section 4.1.6.

4.1.1 Effects of a spherical wavefront and deviations from it

Lens designers and lens makers take great pains to make the imaging lens diffraction limited. Their goal is to make every image point in the lens field converge from spherical wavefronts of identical aperture. Therefore, the spherical wavefront plays a key role in understanding diffraction-limited image formation. This diffraction-limited image is a function of the size of the aperture defining the extent of the sphere. A smaller aperture makes the diffracted image wider and thus lowers the resolution. Deviation from the spherical wavefront restricted by a finite aperture further deteriorates the diffraction-limited image. In reality, even though the deviation is suppressed to tenths of a wavelength, aberrations induced from the deviation cannot be ignored when optical lithography is pushed to its ultimate limit. Therefore, Section 4.1.4 covers analytic treatments of deviation from the spherical wavefront.

4.1.2 Spherical wavefront

Let us consider the imaging of a point source. Given a point object, a perfect point image is to be reproduced in the image plane. The perfect wavefront in this case is a spherical wavefront (as shown in Fig. 4.1, where the largest range of θ is from $-\pi/2$ to $\pi/2$):

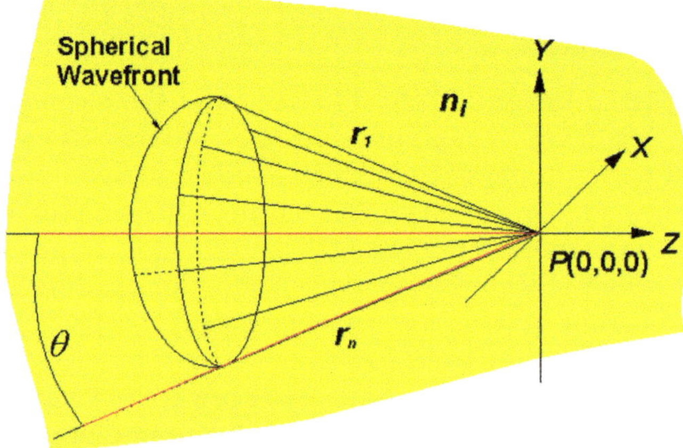

Figure 4.1 A spherical wavefront with imaging beams converging to P. The finite wavefront sustains an angle θ.

$$E = Ae^{ikr}, \tag{4.1}$$

where E is the electric field distribution of the light wave, A is the magnitude of the electric field, and r is the distance between the image point P and a point at the wavefront. $k \equiv 2n\pi/\lambda_0$ is the wave number, where λ_0 is the wavelength in vacuum. n_i in Fig. 4.1 is the refractive index of the medium between the spherical wavefront and the image plane.

In Fig. 4.1, $P(0,0,0)$ is the origin of the coordinate system at the image plane. All rays normal to the spherical wavefront converge to P. θ defines the size of the sphere.

All points at the spherical wavefront are equidistant to $P(0,0,0)$ and thus are in phase with each other. The electric field (instead of the magnetic field in the electromagnetic wave of the propagating light) is used here because the photoresist records light intensity that is proportional to $|E|^2$. This phenomenon is routinely observed in the practice of optical lithography and is easily verified with standing waves in the resist, as is discussed in Section 6.6.2.5. This spherical wavefront produces the strongest E field distribution (therefore, the sharpest image) at $P(0,0,0)$, but it is not infinitesimally small due to diffraction, even when the angle θ sustains the full range from $-\pi/2$ to $\pi/2$. In Fig. 4.1, a simple lens is used for illustration. The lens pupil coincides with the diameter of the simple lens. For an actual compound lens, the pupil plane is determined in the lens design process. The size of the exit pupil is represented by the numerical aperture (NA) of the imaging lens, which is defined as the product of $\sin\theta$ and the refractive index n_i of the medium at the image side (between the spherical wavefront and the point P), as shown in Fig. 4.2 and defined in Eq. (4.2):

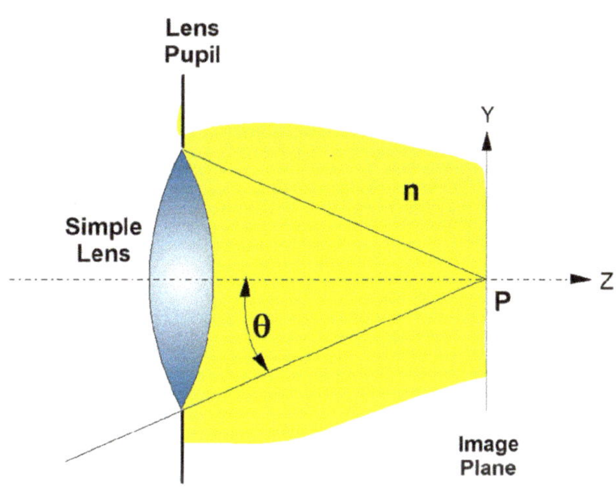

Figure 4.2 Numerical aperture (NA) is defined by $n_i \sin\theta$, where n_i is the refractive index of the imaging medium.

$$NA = n_i \sin\theta. \tag{4.2}$$

Note that the NA of the lens is different at the mask and the wafer sides in a reduction system, as shown in Fig. 4.3(a), with a double-telecentric system used in modern projection printers. The object A_oB_o is reduced to the image A_iB_i, while the aperture angle θ_o increases to θ_i. By lens design,

$$\frac{NA_i}{NA_o} = \frac{n_i \sin\theta_i}{n_o \sin\theta_o} = \frac{A_oB_o}{A_iB_i} = m, \tag{4.3}$$

where m is the NA ratio, defined as the NA on the image side with respect to that on the object side. The NA ratio is identical to the reduction ratio M, which is defined as the object size with respect to the image size.

A compound lens makes it possible to maintain this relationship. The situation is different when a single-element lens is used [as shown in Fig. 4.3(b)]:

$$M = \frac{C_o}{C_i} = \frac{\tan\theta_i}{\tan\theta_o} = \frac{\sin\theta_i \cos\theta_o}{\sin\theta_o \cos\theta_i},$$

so (4.4)

$$m \equiv \frac{NA_i}{NA_o} = \frac{n_i \sin\theta_i}{n_o \sin\theta_o} = M\frac{\cos\theta_i}{\cos\theta_o}, \text{ when } n_i = n_o.$$

The NA ratio m is no longer identical to the reduction ratio M, except for small angles. Take $NA_i = 0.93$, which is a typical value of modern projection scanners with a 4X reduction system. The NA ratio is 10.58 instead of 4.

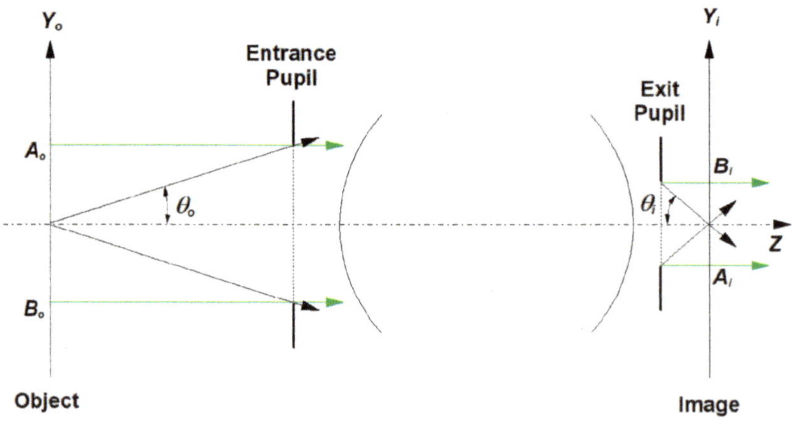

(a)

Figure 4.3(a) A 2X reduction double-telecentric system depicting the NA at the mask and the wafer sides. The reduction ratio M and the NA ratio m are identical.

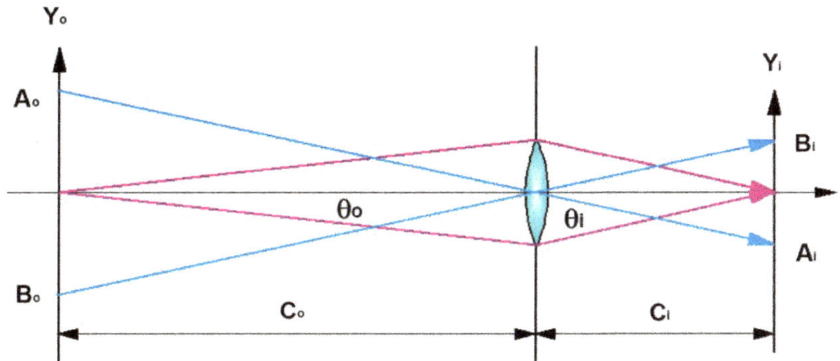

(b)

Figure 4.3(b) An *m*X-reduction single-element lens system cannot keep the NA ratio identical to the reduction ratio, except for small angles.

4.1.3 The effect of a finite numerical aperture on the spherical wavefront

Let us start with the imaging of a point object as shown in Fig. 4.4. The point object diffracts the illuminating light into a wide spectrum of spatial frequencies covering the entire hemisphere behind the object. Only the spatial frequencies within the acceptance angle of the imaging lens are collected, and their spherical fronts are reversed by the imaging lens to focus on the image plane.

We now show the effect of a finite NA on $E[P(x,0,0)]$. To simplify the calculation, the spherical wavefront is flattened to a cylindrical shape, and the y axis is omitted from integration, as shown in Fig. 4.5.

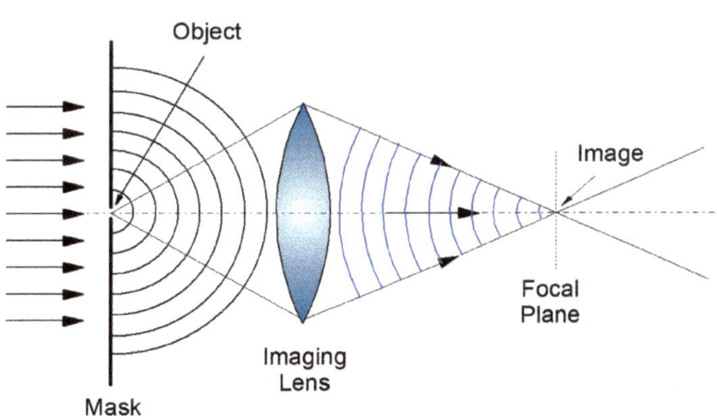

Figure 4.4 The spherical wavefronts from a point object. The object diffracts the illuminating light into a wide spectrum of spatial frequencies covering the entire hemisphere behind the object. The lens collects only the spatial frequencies within its aperture and reverses these wavefronts to focus on the image plane.

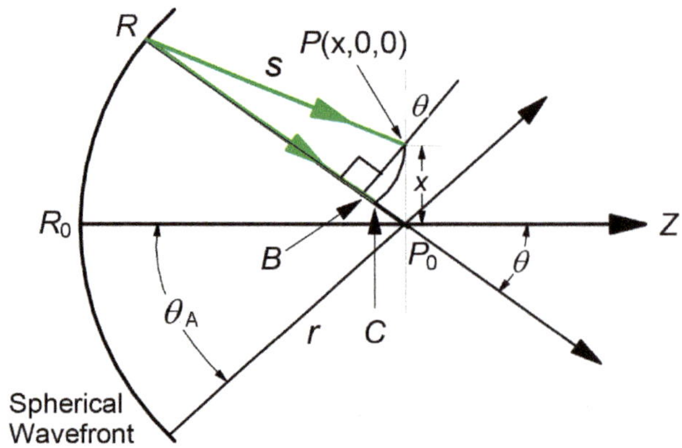

Figure 4.5 Path length difference ($\delta = RP_0 - RP$) at an off-axis point $P(x,0,0)$.

Let the NA-defining angle be θ_A, and let an arbitrary point R on the wavefront make an angle θ with the optical axis z. At $P(x,0,0)$ the electric field distribution of the image is determined by the accumulation of light on all points of the sphere within the angle θ_A. The amplitude of the electric field vector E_x propagating in the z-direction is

$$E_x[P(x,0,0)] = \frac{1}{2\theta_A}\int_{-\theta_A}^{\theta_A} Ae^{iks}\cos\theta d\theta, \tag{4.5}$$

where s is the distance between R and P, and r (see Fig. 4.5) is the distance between R_0 and P_0. $RP_0 = R_0P_0 = r$ by the definition of a spherical wavefront. We introduce points B and C as seen in Fig. 4.5. Point B is the perpendicular intersection from P to RP_0. Point C is an equal distance to R; i.e., $RC = RP$. Therefore, distance s is defined as

$$\begin{aligned} s = RP &= \sqrt{(r - x\sin\theta)^2 + (x\cos\theta)^2} \\ &= \sqrt{r^2 + x^2(\sin^2\theta + \cos^2\theta) - 2rx\sin\theta} \\ &= r\sqrt{1 - 2\frac{x}{r}\sin\theta + \frac{x^2}{r^2}} \approx r\sqrt{1 - 2\frac{x}{r}\sin\theta} \approx r - x\sin\theta \\ &\because\ x/r \ll 1, \end{aligned} \tag{4.6}$$

where x is usually given in nanometers, while r is generally in centimeters.

The difference δ between optical paths RP and RP_0 is

$$\delta = r - s = x\sin\theta, \tag{4.7}$$

where the first null intensity takes place at $\delta = \lambda/2$,

and

$$x = 0.5\frac{\lambda}{\sin\theta} = 0.5\frac{\lambda_0}{NA}. \tag{4.8}$$

Continuing from Eq. (4.5), a sinθ term would be there if the x-component of propagation were considered. Here only the z-component is of interest. The x-component is zero due to symmetry with respect to the z-axis:

$$E[P(x,\,0,\,0)] = \frac{1}{2\theta_A}\int_{-\theta_A}^{\theta_A} Ae^{ik(r-x\sin\theta)}\cos\theta d\theta. \tag{4.9}$$

After integration,

$$E[P(x,\,0,\,0)] = \frac{Ae^{ikr}\sin(kx\sin\theta_A)}{ikx\sin\theta_A}. \tag{4.10}$$

Substituting Eq. (4.2) in Eq. (4.10), the electric field distribution at P is

$$E[P(x,\,0,\,0] = \frac{Ae^{ikr}\sin\left(2\pi\frac{NA}{\lambda_0}x\right)}{2\pi\frac{NA}{\lambda_0}x}. \tag{4.11}$$

Hence, the intensity in the vicinity of the image point is not zero. It becomes zero only when

$$2\pi\frac{NA}{\lambda_0}x = \pi,$$

i.e., when

$$x = 0.5\frac{\lambda_0}{NA}. \tag{4.12}$$

Equation (4.12) is identical to Eq. (4.8). The proportional constant 0.5 may change as a function of the shape of the lens aperture boundary (e.g., square or circular shape). It is designated as k_1 in Eq. (5.1) and is a function of many imaging parameters. Extensive discussions on k_1 will be given in Chapter 5. Here, the main point of Eq. (4.12) is that, even with a perfectly spherical wavefront, the image spot has a finite size dictated by λ and NA. When the wavelength is reduced, the spot size becomes smaller, whereas a smaller NA produces a larger spot size. The smaller the spot size the better the resolution. This dependency of x on NA and λ is the basis of Eq. (5.1).

After studying $E[P(x,0,0)]$, we now examine the effects of a finite NA on the DOF by studying $E[P(0,0,z)]$ in the vicinity of the focal plane.

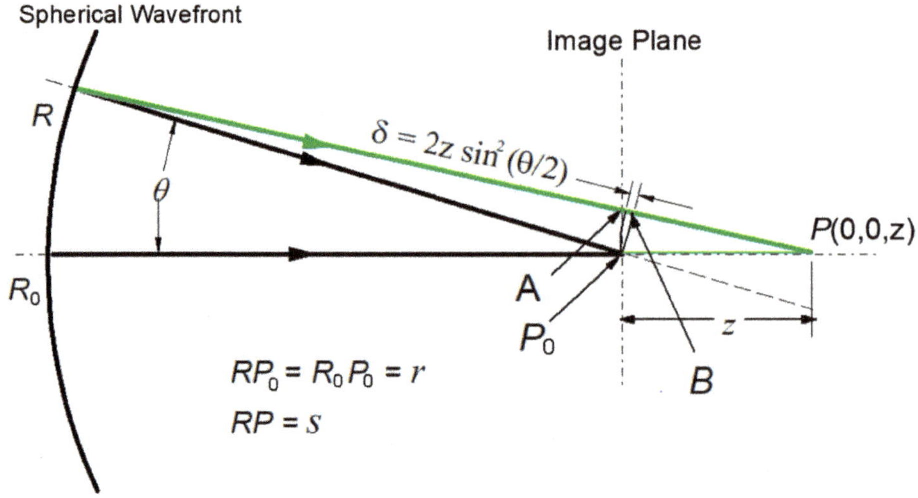

Figure 4.6 Path length difference ($\delta = RP - R_0P$) at an axial off-focus point $P(0,0,z)$.

Equation (4.1) is again used for the derivation. Here, the path difference δ is between R_0P and RP, which is is depicted in Fig. 4.6 and shown to be $2z\sin^2(\theta/2)$:

$$\delta = s - (r+z) = \frac{s^2 - (r+z)^2}{s + (r+z)}, \tag{4.13}$$

where $s = RP$, $r = R_0P_0 = RP_0$, and $z = P_0P$. From the cosine law,

$$s^2 = r^2 + z^2 - 2rz\cos(\pi - \theta). \tag{4.14}$$

Combining Eqs. (4.13) and (4.14) produces

$$\delta = \frac{2rz(\cos\theta - 1)}{s+r+z} = \frac{-4rz\sin^2\left(\frac{\theta}{2}\right)}{s+r+z}. \tag{4.15}$$

Setting $s + r + z \approx 2r$ results in

$$\delta = -2z\sin^2\left(\frac{\theta}{2}\right). \tag{4.16}$$

We are now ready to derive the scaling equation of DOF. We take Eq. (4.5) but evaluate the tangential electric field distribution at $P(0,0,z)$ instead of $P(x,0,0)$:

$$E[P(0, 0, z)] = \frac{1}{2\theta_A} \int_{-\theta_A}^{\theta_A} Ae^{iks} \cos\theta d\theta. \tag{4.17}$$

From Eq. (4.13), we obtain

$$E[P(0, 0, z)] = \frac{1}{2\theta_A} \int_{-\theta_A}^{\theta_A} Ae^{ik(r+z-2z\sin^2\frac{\theta}{2})} \cos\theta d\theta \tag{4.18}$$

and

$$E_z[P(0, 0, z)] = \frac{Ae^{ik(r+z)}}{2\theta_A} \int_{-\theta_A}^{\theta_A} e^{-2kz\sin^2\frac{\theta}{2}} \cos\theta d\theta. \tag{4.19}$$

A closed-form solution cannot be found for this integration. Instead of relying on a closed-form result, one can derive a similar relationship of the finite size of the image point along the z-axis to that along the x-axis, as shown in Eq. (4.12). We then sum up the contribution from different parts of the wavefront by $\Delta\theta$ instead of integrating by $d\theta$. We choose $\Delta\theta$ such that there are 10 wavelets from the wavefront to the image point, as shown in Fig. 4.7. When $z = 0$, all wavelets from the spherical wavefront are in phase and the intensity is the highest. As soon as z increases, the wavelets are out of phase by the increment (1/2) $z \sin^2(\theta_A/2)$ and start to have imaginary components, making the total intensity smaller. The first minimum intensity is reached when

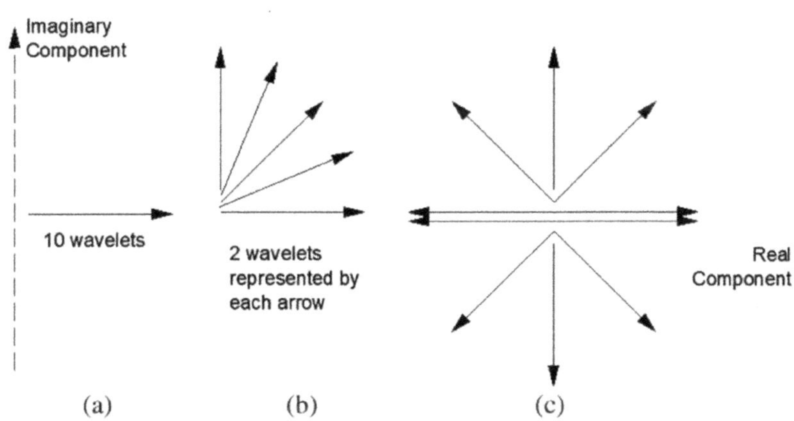

Figure 4.7 Wavelet components moving toward the image point: (a) $z = 0$, (b) $z = 0.125n\lambda/NHA^2$, and (c) $z = 0.5n\lambda/NHA^2$.

$$2kz\sin^2(\theta_A/2) = 2\pi. \tag{4.20}$$

Hence, the intensity reaches the first minimum when

$$z = \frac{0.5\lambda_0}{n\sin^2(\theta_A/2)} = \frac{0.5n\lambda_0}{NHA^2} \approx \frac{2n\lambda_0}{NA^2}; \tag{4.21}$$

i.e., when the NA is small.

The numerical half aperture (NHA) is defined as

$$NHA = n\sin(\theta_A/2). \tag{4.22}$$

Therefore, the DOF of a point source from a perfect spherical wavefront is not zero. It is dictated by λ and θ_A, just as in the case of lateral displacement. One difference is that it is inversely proportional to $\sin^2(\theta_A/2)$ instead of just $\sin\theta_A$. The factor 0.5 is replaced by k_3 in actual situations, just as k_1 is for lateral displacement. By the same token, k_2 is used instead of the factor 2 with small NAs. Equation (4.21) is the basis for Eq. (5.2) in Chapter 5. When θ is small, an approximation can be made as shown in Eq. (4.21).

We have used a simplified analysis to help convey the physical meaning of image formation. The $\sin^2(\theta_A/2)$ dependency of the DOF has been derived by Sheppard and Matthews.[1] For an extensive treatment of imaging a point image from a spherical wavefront, consult Born and Wolf.[2] These authors demonstrate that the normalized intensity distribution in the focal plane from a spherical wavefront with a circular stop is

$$I(0, v) = \left[\frac{2J_1(v)}{v}\right]^2 \tag{4.23}$$

for small NAs, and the normalized intensity distribution along the z-axis is

$$I(u, 0) = \left(\frac{\sin\frac{u}{4}}{\frac{u}{4}}\right)^2, \tag{4.24}$$

where

$$u = \frac{2\pi}{\lambda}\left(\frac{a}{r}\right)^2 z \tag{4.25}$$

and

$$v = \frac{2\pi}{\lambda}\left(\frac{a}{r}\right)x, \tag{4.26}$$

where a is the radius of the circular stop, and r is the distance from the wavefront to the image point. The ratio a/r can be considered as a paraxial approximation of $\sin\theta_A$. Hence, Eqs. (4.23) and (4.24) have the same implications as Eqs. (4.12) and (4.21), respectively. Note that the DOF of a diffraction-limited system of arbitrary NA is not inversely proportional to NA^2 but rather to NHA^2.

4.1.4 Deviation from a spherical wavefront

The diffraction-limited image is produced by a spherical wavefront. When the wavefront deviates from a perfect sphere to reproduce a point source, the image is no longer diffraction limited. Light distribution near the image is much more complicated to analyze. Figure 4.8 shows an aberrated wavefront super-imposed on a reference spherical surface.

With the aberrated wavefront, Eq. (4.1) becomes

$$E = Ae^{ik(r+\psi)}. \tag{4.27}$$

The aberration is characterized by aberration coefficients. These are simply the coefficient of a polynomial that analytically defines the deviation from a perfect sphere. There are two major types of representations—the Seidel and the Zernike aberration coefficients.

4.1.4.1 The Seidel aberration coefficients

The Seidel aberration coefficients are based on a power series expansion of the aberrated wavefront in x_i, y_i, x_p, and y_p, where x_i and y_i are in the image plane, and x_p and y_p are in the exit pupil plane:

$$\psi = \sum_l^{\infty}\sum_q^{\infty}\sum_m^{\infty}\sum_n^{\infty} A_{lqmn} x_i^l y_i^q x_p^m y_p^n, \tag{4.28}$$

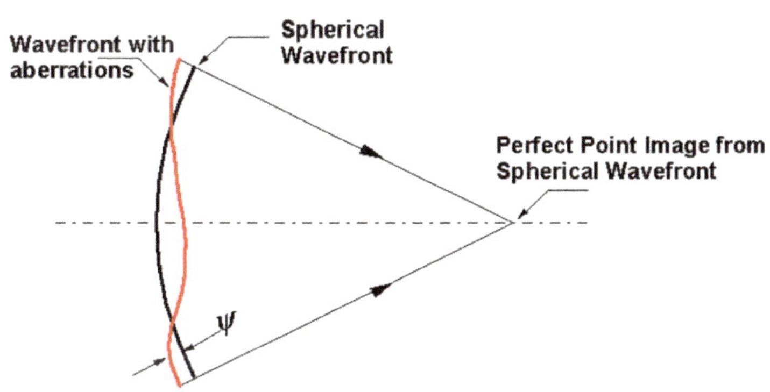

Figure 4.8 Perfect and aberrated wavefronts.

where m, n, l, and q are positive integers. Many simplifications can take place. If one is only interested in an axially symmetrical optical system, then ψ only contains the combination of $x_i^2+y_i^2$, $x_p^2+y_p^2$, and $x_i x_p+y_i y_p$. Furthermore, since only one image point is considered and axial symmetry is present, any image point with an arbitrary coordinate in x_i and y_i can be rotated to x_i, making y_i zero, thus simplifying the combination to x_i^2, $x_p^2 + y_p^2$, and $x_i x_p$. The pupil rectangular coordinates x_p and y_p are replaced by polar coordinates ρ and θ:

$$x_p = \rho\cos\theta. \tag{4.29}$$

$$y_p = \rho\sin\theta. \tag{4.30}$$

Equation (4.28) now takes the form of

$$\psi = \psi^{(0)} + \psi^{(2)} + \psi^{(4)} + \psi^{(6)} + \ldots \tag{4.31}$$

and

$$\psi^{(2)} = A_{200}\rho^2 + A_{111}x_i\rho\cos\theta + A_{002}x_i^2. \tag{4.32}$$

Note that l represents the order of x_i, m represents the order of ρ, and n represents the order of cosθ in A_{lmn}:

$$\begin{aligned}\psi^{(4)} = {} & A_{400}x_i^4 + A_{040}\rho^4 + A_{131}x_i\rho^3\cos\theta + A_{222}x_i\rho^2\cos^2\theta \\ & + A_{220}x_i^2\rho^2 + A_{131}x_i^3\rho\cos\theta.\end{aligned} \tag{4.33}$$

Given the aberrated wavefront ψ, the intersection of the aberrated imaging ray and the image plane determined by the spherical wavefront is[3,4]

$$\Delta x = r\frac{\partial\psi}{\partial x_p}, \tag{4.34}$$

$$\Delta y = r\frac{\partial\psi}{\partial y_p}, \tag{4.35}$$

and

$$\Delta z = \frac{r}{NA^3}\frac{\partial\psi}{\partial\rho}. \tag{4.36}$$

In Eq. (4.33), the A_{400} term disappears after the differentiation in Eqs. (4.33) to (4.35). The remaining five coefficients constitute the five primary Seidel aberrations, namely, A_{040} for spherical aberration, A_{131} for coma, A_{222} for astigmatism, A_{220} for field curvature, and A_{311} for distortion. Detailed descriptions of these aberrations can be found in standard optics textbooks (e.g., Refs. 3 and 4).

4.1.4.2 The Zernike polynomials

The Zernike aberration coefficients are more popular than the Seidel aberration coefficients for optical lithography because the Zernike polynomials[5,6,7,8,9,10] are in cylindrical coordinates that fit the geometry of an imaging lens. Hence, imaging lenses are designed using and characterized by these polynomials. The Zernike polynomials not only form a complete set inside the unit circle, but they are also invariant under rotation. In addition, the Zernike polynomials are orthogonal between orders in the radial and rotational directions. With the Zernike polynomial $Z_n^m(\rho, \theta)$, an arbitrary aberrated wavefront can be expressed by the following equation:

$$\begin{aligned}\psi(\rho, \theta) &= \sum_{n,m} C_n^m Z_n^m(\rho, \theta) \\ &= C_0^0 + \sum_{n=2}^{\infty} C_n^0 R_n^0(\rho) \\ &+ \sum_{n=1}^{\infty}\left[\sum_{m=1,\,2}^{n} R_n^m(\rho, \theta) C_n^m \cos(m\theta) + \sum_{m=-n,\,2}^{-1} R_n^m(\rho, \theta) C_n^m \sin(m\theta)\right],\end{aligned}$$

where

$$R_n^m(r) = \sum_{l=0}^{(n-m)/2} (-1)^l \frac{(n-l)!}{l!\left(\frac{n+m}{2}-l\right)!\left(\frac{n-m}{2}-l\right)!} r^{(n-2l)}. \tag{4.37}$$

The integers n and m are attributed to the radial and circumferential wave numbers, respectively. Also, $n - m$ must be an even number; otherwise, $(\frac{n-m}{2} - 1)!$ and $(\frac{n+m}{2} - 1)!$ would not be valid.

It is helpful to assign a single number j to denote the combination of n and m. The assignment sequence is shown in Table 4.1, according to Lakshminarayanan and Varadharajan.[10] The assignment formulae are

$$j = 1/2[n(n+2)+m], \tag{4.38}$$

$$n = Integer\left[1/2\left(-3+\sqrt{9+8j}\right)\right], \tag{4.39}$$

and

Table 4.1 The single number j according to n and m.

$n \backslash m$	–7	–6	–5	–4	–3	–2	–1	0	1	2	3	4	5	6	7
0								0							
1							1		2						
2						3		4		5					
3					6		7		8		9				
4				10		11		12		13		14			
5			15		16		17		18		19		20		
6		21		22		23		24		25		26		27	
7	28		29		30		31		32		33		34		35

$$m = 2j - n(n+2). \tag{4.40}$$

The Zernike polynomials are orthonormal. In the radial direction,

$$\int_0^1 R_n^m(\rho) R_{n'}^m(\rho)\rho d\rho = \frac{1}{2(n+1)}\delta_{nn'}, \tag{4.41}$$

whereas, in the direction of rotation,

$$\int_0^{2\pi} \cos m\theta \cos m'\theta d\theta = \pi(1+\delta_{m0})\delta_{mm'} \tag{4.42}$$

and

$$\int_0^{2\pi}\int_0^1 Z_j(\rho, \phi) Z_{j'}(\rho, \phi)\rho d\rho d\theta = \frac{\pi}{2(n+1)}(1+\delta_{m,0})\delta_{j,j'}. \tag{4.43}$$

Since Zernike set $R_n^m(1) = 1$, the Zernike polynomials are not normalized.

Zernike polynomials also provide a way to classify lens aberration. Generally, if m is even, the aberrations are called even aberrations; similarly, odd aberrations refer to an odd m. When $m = 0, \pm1, \pm2, \pm3, \pm4$, and ±5, the corresponding aberration is called spherical, coma, astigmatism, 3-foil, 4-foil, and 5-foil, respectively. When $n + |m| = 4$, the aberration is called primary, primary spherical, primary coma, etc.; $n + |m| = 6$ corresponds to secondary aberrations; $n + |m| = 8$ corresponds to tertiary aberrations.

Equation (4.37) can be rewritten as

$$\psi(\rho, \theta) = \sum_j c_j Z_j. \tag{4.44}$$

In this case the mean-square deviation of the wavefront is

$$\overline{(\Delta\psi)^2} = \frac{\int_0^{2\pi}\int_0^1 (\Delta\psi)^2 \rho d\rho d\theta}{\int_0^{2\pi}\int_0^1 \rho d\rho d\theta} = \sum_j \frac{1+\delta_{m,0}}{2(n+1)} c_j^2. \tag{4.45}$$

This is another advantage of the Zernike polynomials. The magnitude of each Zernike term can be independently summed quadratically for the mean-square deviation of the wavefront. Imaging lenses are often specified this way.

The polynomials can readily be written explicitly for each combination of m and n using Eq. (4.37). For example,

$$R_4^2(\rho) = 4\rho^4 - 3\rho^2 \tag{4.46}$$

and

$$R_6^2(\rho) = 15\rho^6 - 20\rho^4 + 6\rho^2. \tag{4.47}$$

We now test the orthogonality of these two terms:

$$\int_0^1 R_4^2(\rho) R_6^2(\rho)\rho d\rho = \int_0^1 \left(\frac{60}{12}\rho^{12} - \frac{125}{10}\rho^{10} + \frac{84}{8}\rho^8 - \frac{18}{6}\rho^6\right) d\rho = 0.$$

.

4.1.4.3 Sample plots of Z_j

This section provides graphic illustrations of Z_j plots taken from Wong[9] using the naming convention of Lakshminarayanan and Varadharajan.[10] Figure 4.9 shows the case where $m = 0$ for $j = 0$, 4, 12, 24, and 40. For $m = 0$, ψ is circularly symmetrical. The shape of the modification of the non-aberrated wavefront when $l = 0$ is often called piston because it is a flat value that does not change the shape of the non-aberrated spherical wavefront. $Z_l = 4$ corresponds to a change of focus without deforming the point image. $Z_l = 12$ represents a primary spherical aberration, $Z_l = 24$ represents a secondary spherical aberration, $Z_l = 40$ represents a tertiary spherical aberrtion, and so on. These spherical aberrations do not change the circular symmetry of the image but do introduce different levels of blur to it.

In Fig. 4.10 with $m = \pm 1$, when $j = 1$ or 2, the modification to a spherical wavefront is a horizontal or vertical tilt, respectively. The image is simply shifted without other deformations. For $j = 7$, the image is no longer circularly

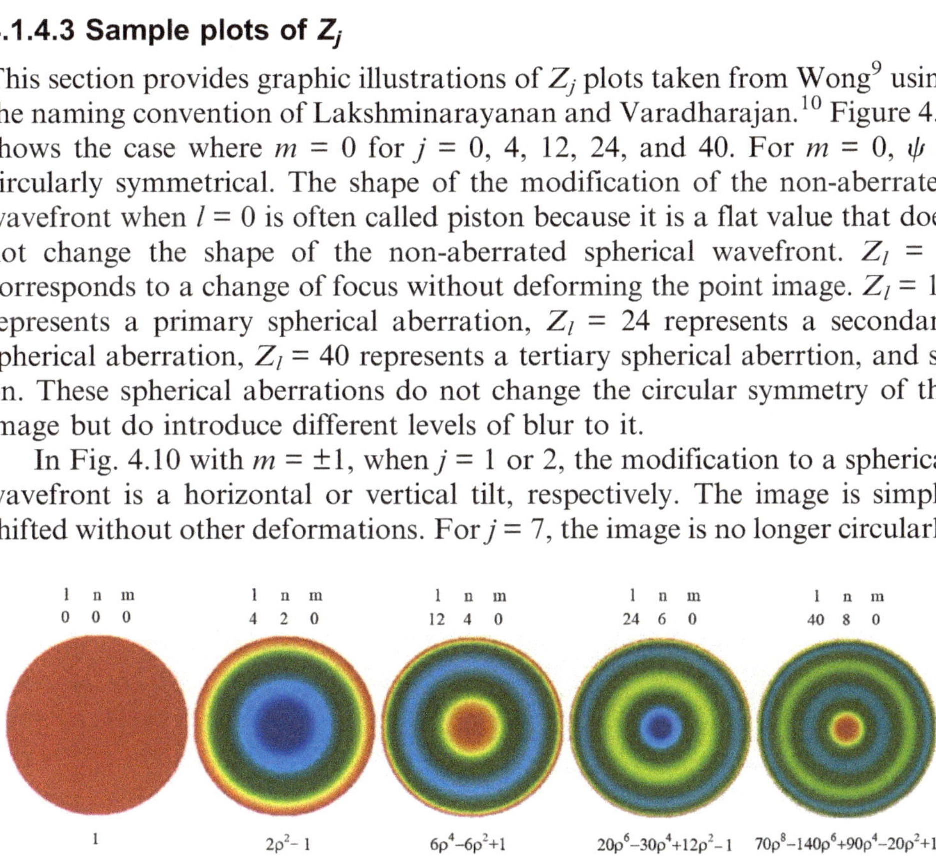

Figure 4.9 Zernike polynomials for $m = 0$, $j = 0, 4, 12, 24$, and 40 (reprinted from Ref. 9).

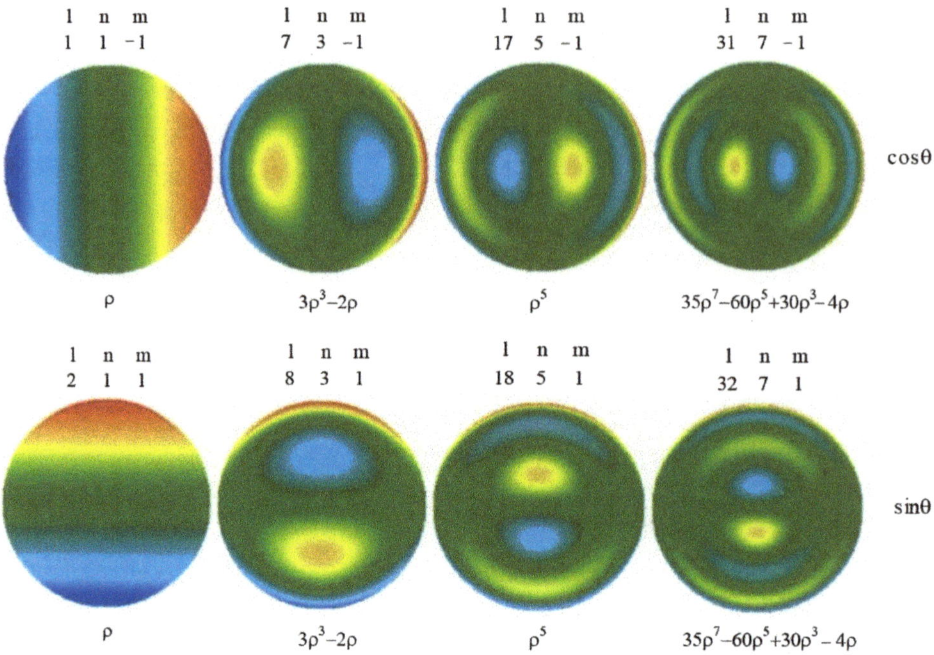

Figure 4.10 Zernike polynomials for $m = \pm 1$, $j = 1, 2, 7, 8, 17, 18, 31$, and 32 (reprinted from Ref. 9).

symmetrical. It has a strong focus on the left but has a blur that increases towards the right. This is called primary coma. $Z_j = 8$ behaves similarly, but the sharp focus is to the top. The higher orders of coma for $j = 17$ and 18 and for $j = 31$ and 32 are called secondary and tertiary coma, respectively. For secondary coma, the sharp focus in now turned 180 deg, and the blurred tail oscillates in brightness twice. For tertiary coma, the bright focus remains in the same direction as in primary coma, but the tail oscillates three times.

We now turn to the situation of Fig. 4.11 with $m = \pm 2$; i.e., the θ dependence has become $\sin\theta$ or $\cos\theta$. For $l = 3$ or 5, the aberration is primary astigmatism. The image alternates its orientation from horizontal to vertical along the optical axis in the case of $\cos\theta$. For $\sin\theta$, the orientation oscillates between 45 deg and 135 deg.

4.1.5 Imaging from a mask pattern

With optical microlithography, one seldom prints point objects unless they are studying the aberrations of the imaging lens by intentionally creating pinhole objects in the mask. Instead of pinholes being printed, long lines, rectangles, squares, elbows, T shapes, or a combination of these basic patterns transformed into complicated circuit patterns are printed. Therefore, the diffraction effects of mask patterns with and without aberrations must be studied. The E-D tools described in Chapter 5 are used to quantitatively study

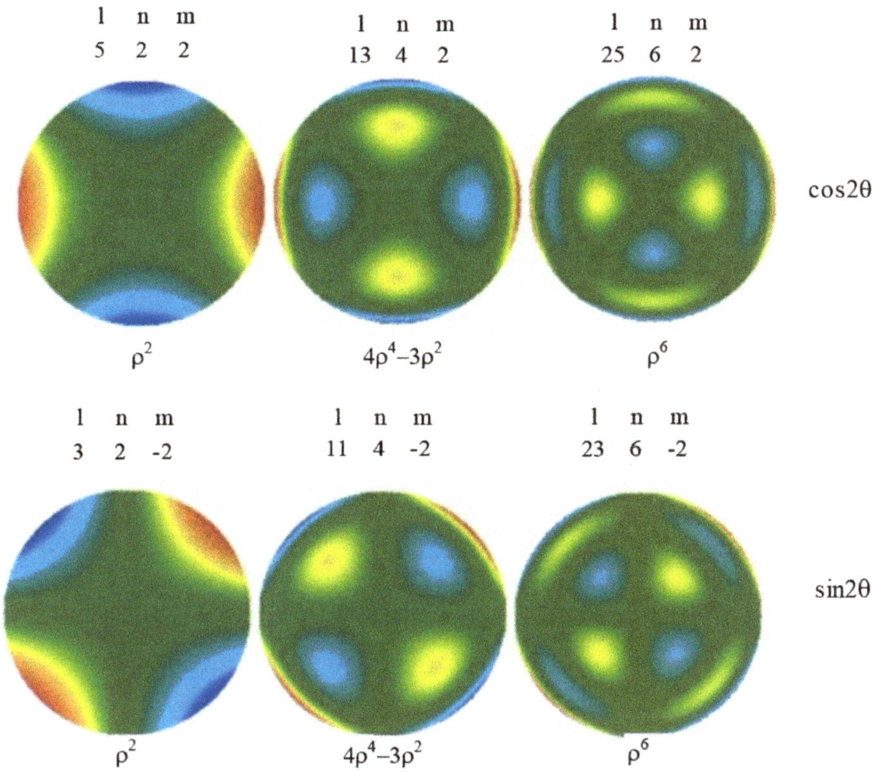

Figure 4.11 Zernike polynomials for $m = \pm 2$, $j = 3, 4, 11, 12, 25$, and 26 (reprinted from Ref. 9).

how the mask pattern shapes and sizes and the placement of these patterns impact the processing window (E-D window). The following sections cover the derivation of the diffracted images from a given mask pattern.

4.1.5.1 Coherent imaging from a mask pattern

Let us consider coherent imaging of an object through a lens, as shown in Fig. 4.12. From geometrical optics considerations, all light rays passing through the objects in the mask plane are bent by the lens in such a way that they converge to the corresponding geometric image positions in the bright part of the image. From a wave-diffraction point of view, light waves from one object point are modified by the lens in such a way that they travel the same optical distance to the image point. In other words, the same number of peaks and valleys, as well as fractions of them, are experienced when the light waves converge on the image point.

The optical distances of all waves are made identical by the lens, regardless of the angles the waves make. Paraxial waves travel through the thicker part of the lens, while waves farther off-axis travel through the thinner

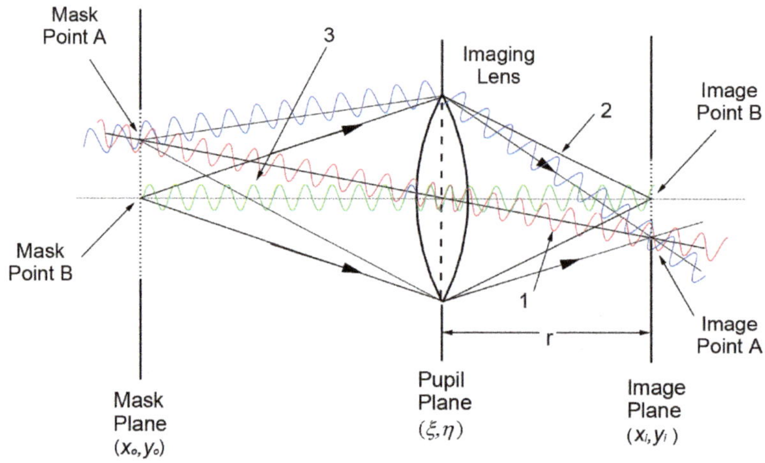

Figure 4.12 Image formation by coherent light. The phase between any imaging beams is deterministic.

part of the lens. Because the wavelength is shortened by the refractive index of the lens media, the larger distances traveled in the lens media for the paraxial waves make up for the shorter physical distance to the image point. With coherent imaging, all waves from a single object point, as well as from any other object points, have a deterministic phase relationship with each other.

For example, point 1 from object point *A* is 180 deg ahead of point 2, which is also from object point *A*, as shown in Fig. 4.12. This relationship holds at any instance. Point 3 from object point *B* also remains in phase with point 1 from point *A* and 180 deg ahead of point 2.

The equation relating the scalar electric field distribution at the mask plane to that of the image plane is[11,12]

$$E_i(x_i, y_i) = \int_{-\infty}^{\infty}\int_{-\infty}^{\infty} E_o(x_o, y_o)K(x_i, x_o; y_i, y_o)dx_o dy_o. \tag{4.48}$$

In a well-corrected system, the transmission function *K* can be taken as

$$K(x_1 - x_o, y_1 - y_o) = \frac{1}{\lambda^2 r^2}\int_{-\infty}^{\infty}\int_{-\infty}^{\infty} G(\xi, \eta)e^{\frac{-ik}{r}[(x_1-x_o)\xi+(y_1-y_o)\eta]}d\xi d\eta. \tag{4.49}$$

The following equation can be dissected into Fourier transform and inverse Fourier transform components with insertion of the pupil function $G(\xi,\eta)$ when performing the inverse transform. In an aberration-free system, $G(\xi,\eta)$ is simply unity in the aperture and zero outside. That is,

$$E_i(x_i, y_i) = \frac{1}{\lambda^2 r^2} \int_{-\infty}^{\infty} \int_{-\infty}^{\infty} \left[\int_{-\infty}^{\infty} \int_{-\infty}^{\infty} E(x_o, y_o) e^{\frac{ik}{r}(x_o \xi + y_o \eta)} dx_o dy_o \right] G(\xi, \eta) e^{\frac{-ik}{r}(x_i \xi + y_i \eta)} d\xi d\eta. \tag{4.50}$$

Strictly speaking, the electric field distribution in the mask plane is very complicated. It must satisfy Maxwell's equations as well as the boundary conditions.[13] For a mask using chromium as the absorber, the transverse electric field distribution is zero because chromium is a conductor and does not support any transverse electric field distribution on its surface. This supports the following assumptions:

$$E(x_o, y_o) = \begin{Bmatrix} 1, \text{ at openings} \\ 0, \text{ at opaque areas} \end{Bmatrix} \tag{4.51}$$

for TE incidence. Bachynski and Bedefi[14] proved that the transverse magnetic field distribution in the opening is identical to the incident transverse magnetic field distribution. Otherwise, Eq. (4.51) does not hold.

Figure 4.13 shows the electric field distribution at the mask plane of a 3λ-wide slit opening and its propagation into the image plane. Here TE incidence refers to the electric field perpendicular to the plane of incidence and parallel to the direction of the slit; TM incidence refers to the magnetic field parallel to the direction of the slit. The fine structure disappears at a distance of a few wavelengths from the mask plane. It is more pronounced for TE incidence. With TM incidence, the magnetic field distribution in the opening of the mask follows conventional assumptions, such as in Eq. (4.51), except that the field distribution in the opaque part of the mask is not zero. Conversely, the electric field distribution in the opaque part is zero with TE incidence, but the field distribution inside the mask opening does not fit conventional assumptions. The conventional assumption is called physical optics approximation and is only useful when the distance from the mask is larger than W^2/λ, where W is the width of the object, and λ is the wavelength. More details are given in Chapter 2 on proximity printing. As part of the exact boundary conditions, there may be singularities at the edge of the mask opening. For illustration of the edge singularities, consult Ref. 15.

Figure 4.14 shows the fine structure captured in photoresist using exposure of a mask that is coated with resist at the absorber side. The resist is thus exposed while in perfect contact with the mask absorber, unlike resist exposure in conventional proximity printing with the resist in the proximity of the mask. Here, the resist development starts at the plane farthest from the mask and continues in the direction of the contact plane. The distance from the contact plane is preserved for measurement after development, and the diffracted image can be recorded in extremely close proximity to the mask absorber. In this figure, a 6.6λ-wide slit is used, and the distance from the

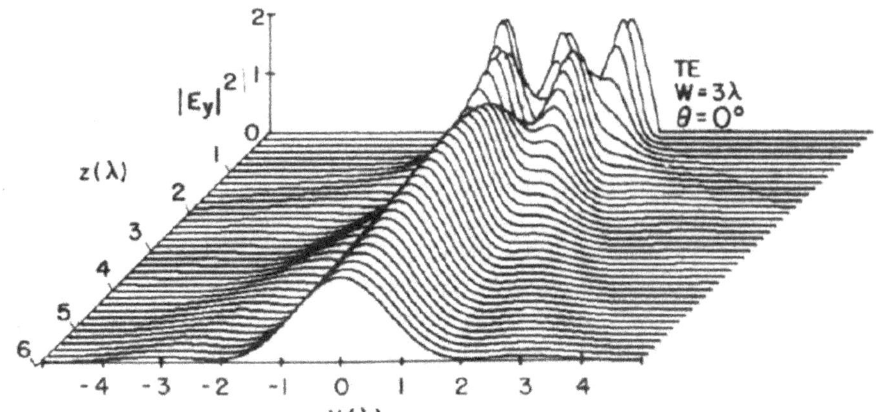

(a)

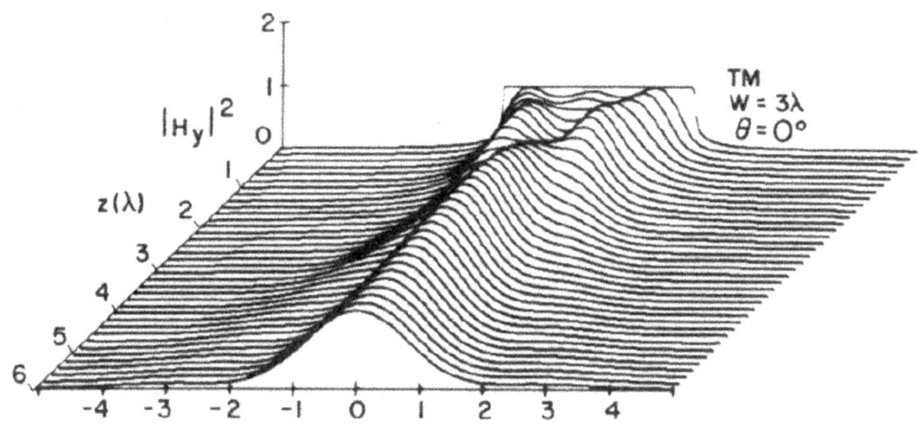

(b)

Figure 4.13 Electric field distribution at the mask plane of a 3λ-wide slit and its propagation into the image plane: (a) TE incidence and (b) TM incidence (reprinted from Ref. 15).

capture plane to the mask plane is 2.1λ. The incident light is unpolarized, containing equal TE and TM components.

Fortunately, such a highly accurate representation is needed in optical lithography only when the field in close proximity to the mask is of concern, such as in the case of proximity printing,[16] or when the mask pattern size is on the order of a wavelength or smaller, such as in the case of submicrometer 1X projection lithography or in 193-nm 4X reduction systems for 50-nm features or smaller. Attenuated-phase-shifting masks with absorbers that transmit light and shift their phase obviously do not follow Eq. (4.51). For most other purposes, the physical optics approximation for $E(x_o, y_o)$ is usually sufficient.

K, the transmission function of the lens [see Eq. (4.49)], is closely related to the kernel of Eq. (4.5) or Eq. (4.18). $G(\xi,\eta)$ is the pupil function, which is zero

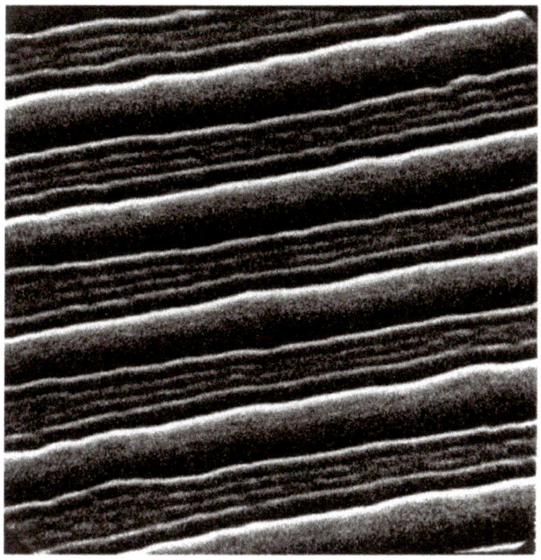

Figure 4.14 Experimental recording of an exposure intensity (electric field squared) distribution 2.1λ away from a 6.6λ-wide slit.

outside the lens aperture and is unity inside when there is no aberration. Otherwise, $G = 1 + \psi/r$, where ψ and r are defined in Eq. (4.27). The coordinates (x_o, y_o), (ξ,η), and (x_i, y_i) are depicted in Fig. 4.15. As mentioned earlier, photoresist responds to light intensity, not the electric field. Once the electric field distribution is determined, the intensity distribution simply follows from this equation:

$$I(x_i, y_i) = E^*(x_i, y_i)E(x_i, y_i). \tag{4.52}$$

4.1.5.2 Incoherent imaging from a mask pattern

With incoherent imaging, the phase relationship between different image points is completely random, as depicted by Fig. 4.16. Although the phase relationship within one infinitesimal object point and its conjugate image point remains deterministic, it is impossible to pinpoint the electric field deterministically. However, the intensity distribution is readily evaluated using

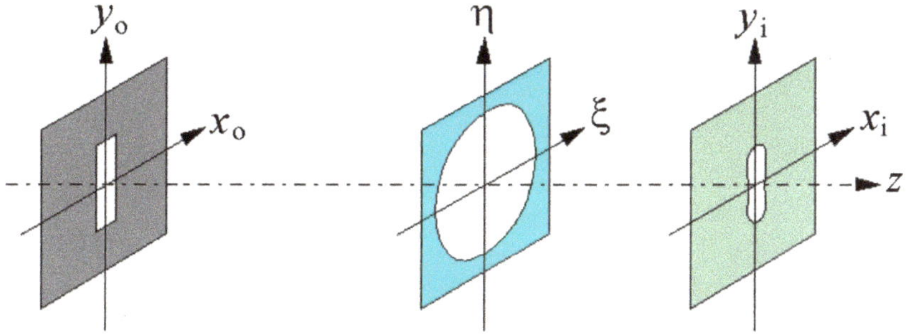

Figure 4.15 Object, pupil, and image coordinates.

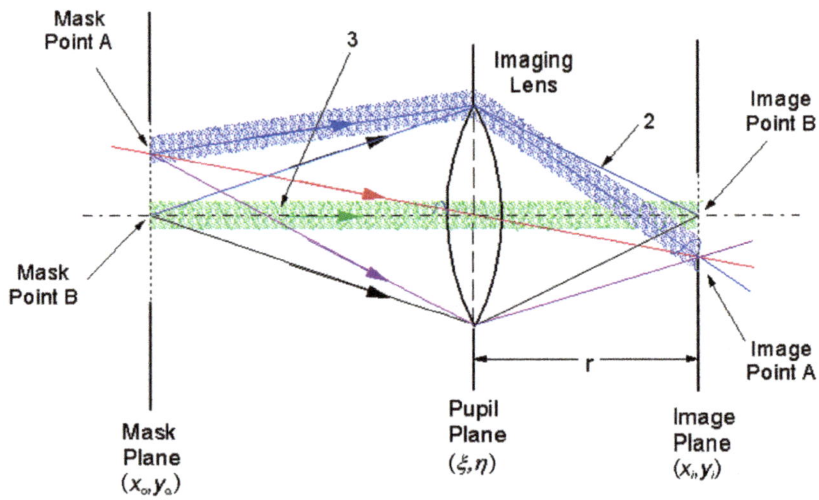

Figure 4.16 Incoherent imaging, where there is no deterministic phase relationship between any imaging beams.

$$I_i(x_i, y_i) = \int_{-\infty}^{\infty}\int_{-\infty}^{\infty} I_o(x_o, y_o)|K(x_i, x_o, y_i, y_o)|^2 dx_o dy_o. \qquad (4.53)$$

One can visualize the coherent imaging equation governing the imaging of an infinitesimal object. But with another object point, even if it is only a small, finite distance from the first object point, there is no deterministic phase relationship. The electric field between any two points does not interfere. The only way to account for the simultaneous presence of these points is to sum their contributions by intensity.

4.1.5.3 Partially coherent imaging from a mask pattern

In the actual practice of optical microlithography, the illumination of the mask is neither coherent nor incoherent but rather is partially coherent, meaning that the light waves between different image points hold a statistical relationship instead of being completely random. The situation is depicted in Fig. 4.17.

Evaluating the intensity distribution is not as straightforward as Eq. (4.49) or Eq. (4.53). The image intensity distribution from partially coherent illumination on a mask pattern based on Hopkins' analysis is[17,18]

$$I(x_i, y_i) = \int_{-\infty}^{\infty}\int_{-\infty}^{\infty}\int_{-\infty}^{\infty}\int_{-\infty}^{\infty} J_o(x_o, y_o; x'_o, y'_o) K^*(x_o, y_o; x_i, y_i) K(x'_o, y'_o; x_i, y_i) dx_o dy_o dx'_o dy'_o, \qquad (4.54)$$

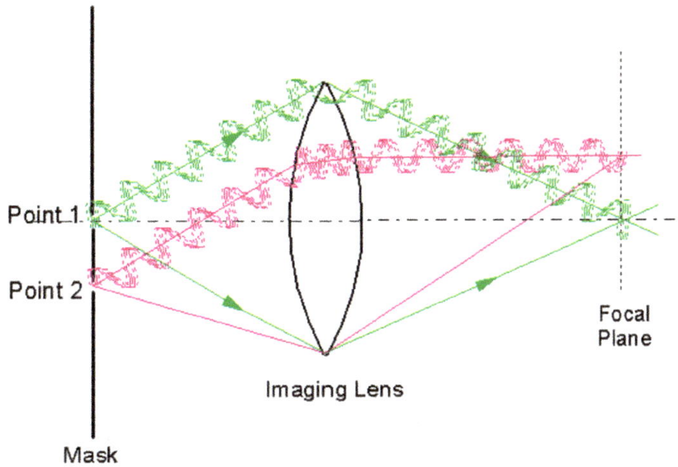

Figure 4.17 Partially coherent imaging, where there is a statistical phase relationship between any imaging beams.

where $J_o(x_o, y_o, x'_o, y'_o)$ is the mutual intensity at the mask plane, and K is the same transmission function as in Eqs. (4.49) and (4.53). The mutual intensity at the mask plane can be further decomposed into the two components B_o and M:

$$J_o(x_o, y_o; x'_o, y'_o) = B_o(x_o, y_o; x'_o, y'_o)M(x_o, y_o)M^*(x'_o, y'_o), \tag{4.55}$$

where B_o is the mutual intensity of the illumination, and M is the mask transmission function. B_o can be expressed as the Bessel function of the first kind of order 1, $J_1(u_c)$, where u_c is in units of NA_c, the numerical aperture of the condenser. K can also be expressed as $J_1(u_i)$, where u_i is in units of NA_i, the numerical aperture of the imaging lens. It turns out that when the actual calculation is performed, the aperture ratio $\sigma = NA_c / NA_i$ is often used. This ratio becomes a parameter to indicate the degree of coherence of the optical imaging system. With a coherent system, $\sigma = 0$, and with an incoherent system, $\sigma = \infty$. When $\sigma = 0$, B_o becomes unity, and Eq. (4.54) becomes Eq. (4.53). When $\sigma = \infty$, B_o is a delta function, and Eq. (4.54) becomes Eq. (4.49).

Though the physical meaning is clear, Eq. (4.54) is time consuming to evaluate numerically because it involves a four-dimensional integral. A good way to make the calculation faster and the mathematics easier to handle is to treat the electric-field propagation as if it were completely coherent with light emerging from an infinitesimally small light source, using Eq. (4.50). Partial coherence is introduced by making the source its actual finite size and integrating the image intensity from Eq. (4.52) over the area of the light source. Yeung[18] pioneered this concept for rigorous electromagnetic diffraction from high-NA imaging optics and through multilayered media, as in the resist and thin-film stack found in typical semiconductor substrates.

Using either Eq. (4.50) or (4.54) in the Fourier transform spatial frequency space can accelerate numerical computation. Not only do the convolution formulae become simpler, but the existing fast Fourier transformation algorithm also makes it extremely efficient to perform the Fourier and inverse-Fourier transformations. However, N, the total number of data points for integration, is often required to be an integer power of 2. In addition, a truely isolated feature must be approximated by a periodic multiplicity of itself, with a large distance separating the features.

4.1.6 Spatial frequencies

Spatial frequencies play an important role in microlithography imaging for two reasons:

1. They are key in understanding wavelength and subwavelength imaging. The working principles of the NA, illumination angle, phase-shifting masks, aperture engineering, etc., are most easily understood with arguments given in terms of spatial frequencies.
2. They enable faster computation of the diffraction formulae, as discussed in Section 4.1.5.3.

A given function in the 2D space can be decomposed into spatial frequencies using the Fourier integral:

$$\kappa(f, g) = \int_{-\infty}^{\infty}\int_{-\infty}^{\infty} K(x, y)e^{i2\pi(fx+gy)}dxdy. \tag{4.56}$$

The inverse-Fourier transformation restores $K(x, y)$:

$$K(x, y) = \int_{-\infty}^{\infty}\int_{-\infty}^{\infty} \kappa(f, g)e^{-i2\pi(fx+gy)}dfdg. \tag{4.57}$$

Applying Fourier transformation to both sides of Eq. (4.48) yields

$$E(f, g) = E_o(f, g)\kappa(f, g). \tag{4.58}$$

The physical meaning of Eq. (4.58) is clear. The spatial frequency distribution $\mathscr{E}(f, g)$ at the image plane is evaluated by multiplying the spatial frequency distribution at the mask plane $\mathscr{E}_o(f, g)$ with the pupil function $\kappa(f, g)$, which is simply unity inside the imaging lens aperture and zero elsewhere in an aberration-free system. The cutoff frequency $f_{cutoff} = g_{cutoff} = NA/\lambda$ for a lens with a circular pupil. The lens acts as a low-pass filter to relate the spatial frequency components at the mask plane to those at the image plane.

4.1.6.1 Spatial frequencies of an isolated line opening

We now illustrate, with a typical microlithography mask pattern, an isolated, infinitely long line opening represented by

$$E(x) = \begin{cases} 1, & -\frac{w}{2} \le x \le \frac{w}{2} \\ 0, & \text{elsewhere.} \end{cases} \quad (4.59)$$

Equation (4.56) produces its spatial frequency:

$$\mathscr{E}(f)/w = \frac{\sin(\pi w f)}{\pi w f}, \quad (4.60)$$

as shown in Fig. 4.18.

Hence, a simple line opening consists of a full spectrum of positive and negative spatial frequencies. The amplitude of these frequencies can be positive

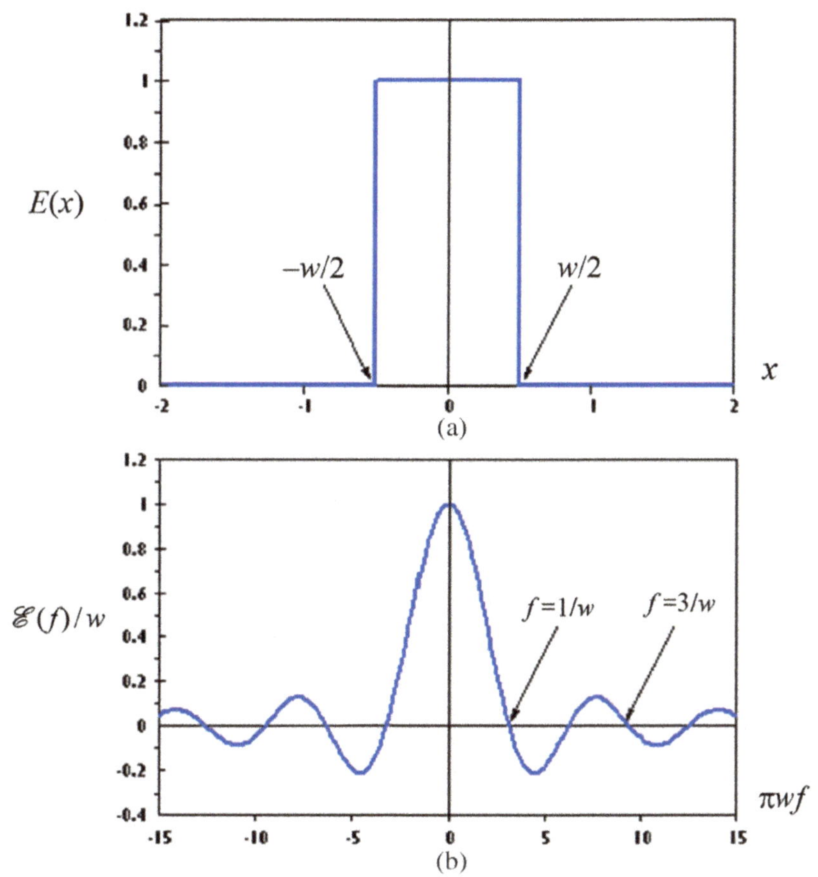

Figure 4.18 Electric field from an isolated line opening and its Fourier transform: (a) $E(x)$ and (b) $\mathscr{E}(f)$.

or negative. Because of the $1/f$ dependency, the significant frequencies in $\mathscr{E}(f)$ are distributed in the low-frequency region. The null points are $f = \pm n/w$, where n is a positive integer. In terms of the angular spectrum,[19]

$$\sin\alpha = \lambda f, \tag{4.61}$$

where α is the propagation angle of a plane wave corresponding to the spatial frequency f:

$$\sin\alpha = \frac{\lambda}{w}. \tag{4.62}$$

If the numerical aperture of the imaging lens sets the cutoff frequency at the first null positive point, then

$$NA = \frac{\lambda}{w}. \tag{4.63}$$

Modern microlithography aims at resolving features with w values smaller than λ, whereas the NA is seldom sufficient to reach a cutoff frequency higher than that at the first null point in the frequency spectrum. Let us consider $w = \lambda$; $NA = 0.3$, 0.6, and 0.9. The pupil functions are superimposed on the frequency distribution shown in Fig. 4.19(b). Equations (4.52), (4.57), and (4.58) are used to evaluate the resultant electric field distributions at the image plane in Fig. 4.19(a).

Note that, even at a very high NA, the resultant images look quite different from those on the mask because w is chosen to be greater than or equal to λ. Fortunately, microlithographers are not concerned with reproducing all gray levels in the object. The main purpose of microlithography is to control the image edge position, as stated in Section 1.3. The image is kept binary by taking advantage of the resist as a high-contrast recording material and the subsequent pattern transfer processes.

4.1.6.2 Spatial frequencies of line–space pairs

Another frequently used mask pattern in microlithography is the equal line–space pairs as represented by $E(x, y)$ in

$$E(x, y) = \begin{cases} 1, & np - \frac{w}{2} \le x \le np + \frac{w}{2} \\ 0, & np - \frac{w}{2} > x,\ x > np + \frac{w}{2} \\ n = 0, \pm 1, \pm 2, \ldots, & \end{cases} \tag{4.64}$$

where w is the width of the openings, p is the periodicity, and n is an integer from negative to positive infinity (in other words, an infinite number of equal line–space pairs). If the number of line–space pairs is truly infinite, then the spatial frequencies are discrete, with their amplitudes dictated by Eq. (4.60), as

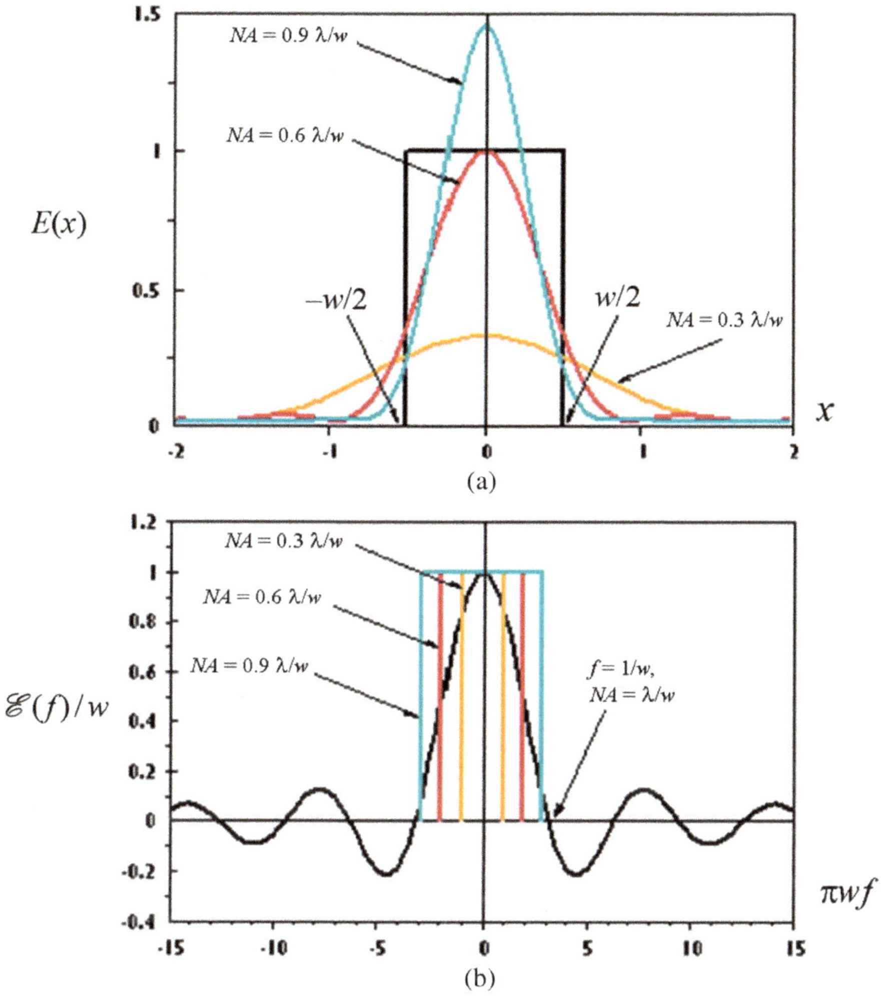

Figure 4.19 Pupil function applied to the line opening in Fig. 4.18: (a) $E(x)$ and (b) $\mathscr{E}(f)$.

shown in Figs. 4.20 and 4.21. Note that the distance between the frequency spikes is always $1/p$ regardless of w.

The spatial frequency, according to the definition above, is independent of the wavelength. It is purely a function of p with the amplitude related to w. When the wavelength is considered, the spatial frequency corresponds to a wave traveling in a direction as a function of p and λ, as shown in Fig. 4.22; that is,

$$\sin\alpha = \frac{\lambda}{p}. \tag{4.65}$$

When p is large with respect to λ, α is small and the beam propagates paraxial to the z-axis. The angle increases with smaller p until it equals λ.

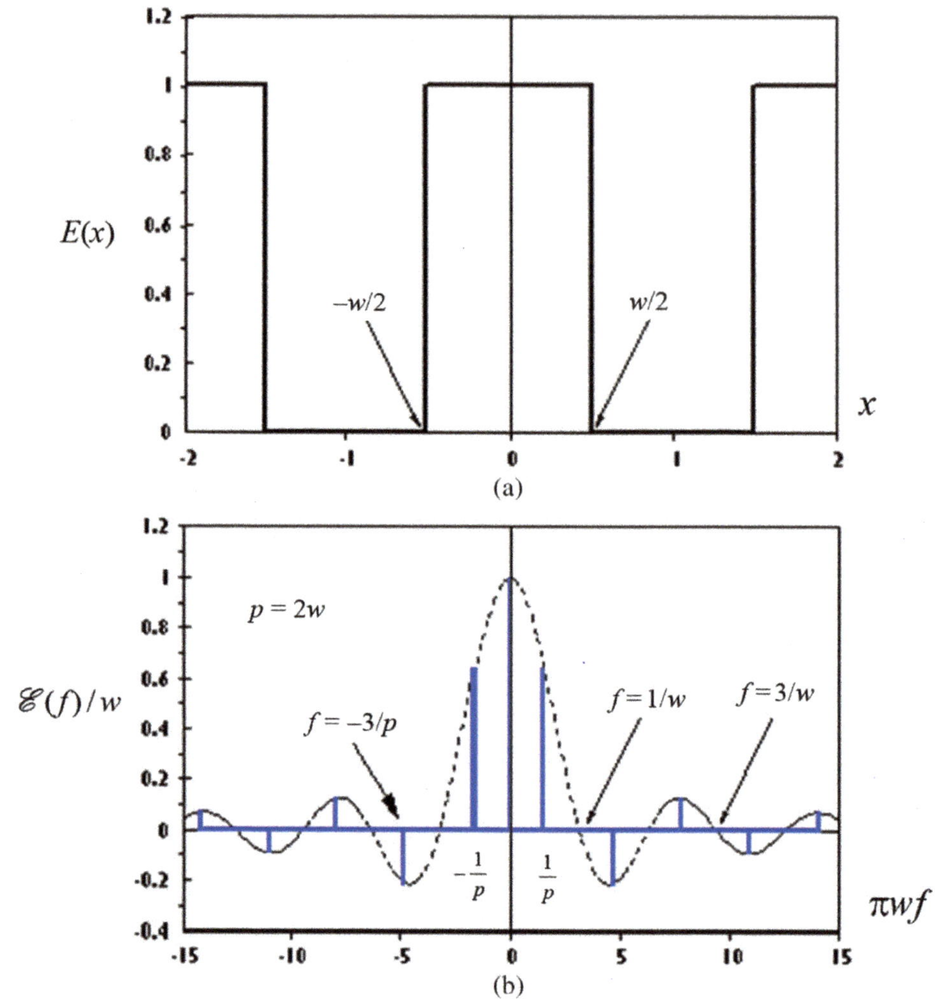

Figure 4.20 (a) Electric field of line–space pairs and (b) the corresponding spatial frequency distribution.

When p exceeds λ, the beam is no longer a regularly traveling wave, but an evanescent wave instead. The angle θ for the lens NA is independent of p and λ. This angle plays the role of cutting off the high spatial frequencies. When $\sin\theta$ is less than $\sin\alpha$, the lens aperture blocks the spatial frequency λ/p.

Putting all this together, when light is incident on a mask pattern, it is diffracted into a spatial frequency spectrum at node points $\pm n/p$. The frequencies lower than $1/\lambda$ travel toward the lens at angles $\alpha = \pm\sin^{-1}(\lambda/p)$. Those frequencies greater than or equaling $1/\lambda$ will not reach the lens. The lens numerical aperture $n\sin\theta$ further limits the spatial frequencies to $\pm NA/\lambda$. Note

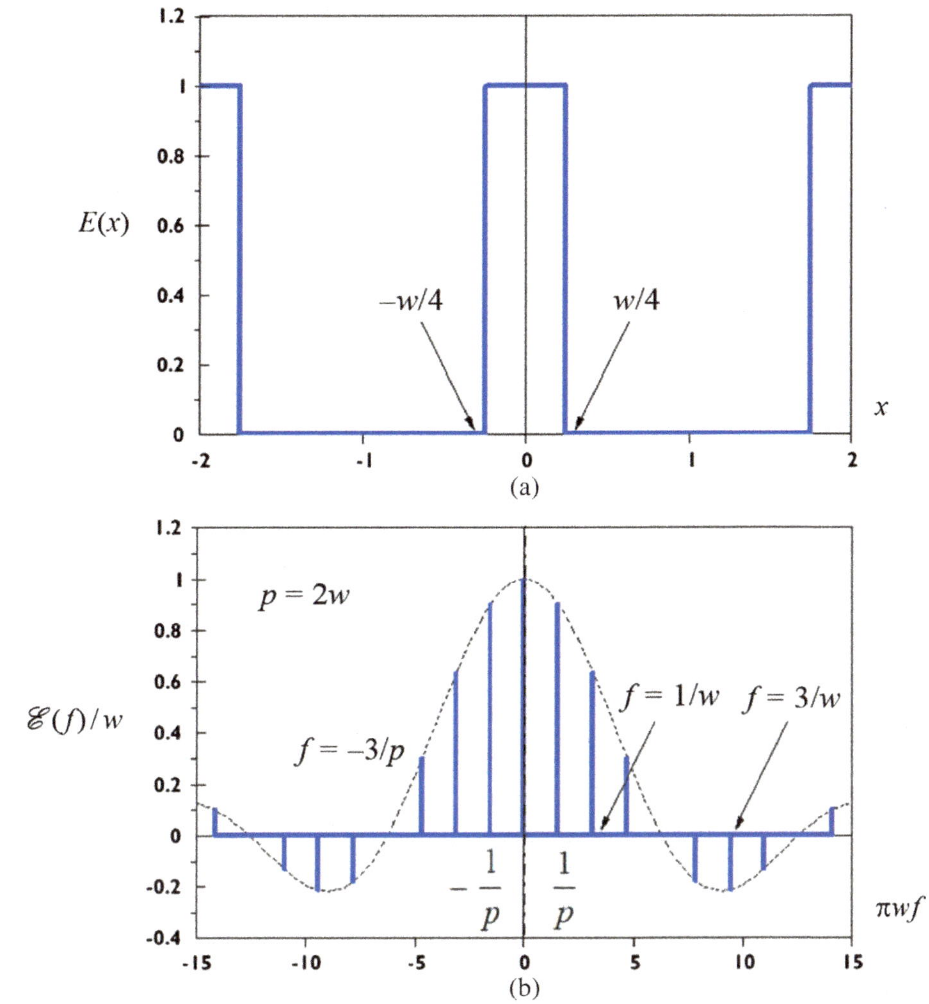

Figure 4.21 (a) Electric field of line–space pairs and (b) the corresponding spatial frequency distribution with line:space ratio = 1:3.

that the vacuum wavelength should be used; otherwise, the spatial frequencies are $\pm\sin\theta_{\text{medium}}/\lambda_{\text{medium}}$.

Analytically, the spatial distribution of the line–space pairs is

$$E(f) = \sum_{n=-\infty}^{\infty} \int_{np-w/2}^{np+w/2} e^{i2\pi fx}dx = \frac{\sin(\pi wf)}{\pi wf} \sum_{n=-\infty}^{\infty} e^{i2\pi npf}. \tag{4.66}$$

The first term is the envelope of the spikes. The summation part[20] is a series of delta functions centering at $f = \pm np$. Using the sum of a geometric series,

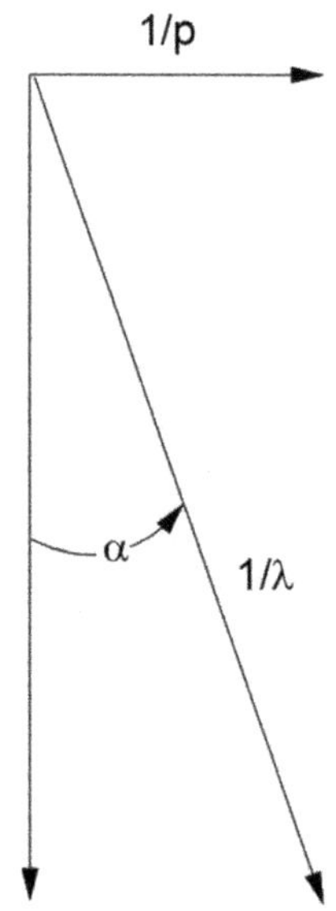

Figure 4.22 Relationship of the propagation angle α to λ and p.

$$\sum_{n=-\infty}^{\infty} e^{i2\pi npf} = \lim_{n\to\infty} \frac{e^{-i2\pi pf}\left[e^{i(4n+2)\pi pf} - 1\right]}{e^{i2\pi pf} - 1}$$
$$= \lim_{n\to\infty} \frac{\sin(2n+1)\pi pf}{\sin(\pi pf)} = \sum_{n=-\infty}^{\infty} \delta\left(f - \frac{n}{p}\right). \tag{4.67}$$

Therefore,

$$E(f) = \frac{\sin(\pi wf)}{\pi wf} \sum_{n=-\infty}^{\infty} \delta\left(f - \frac{n}{p}\right). \tag{4.68}$$

Just as in the case of an isolated line opening, one cannot afford for the high NA to transmit higher-order spatial frequencies because the lens of higher NA costs more. Most of the time, only the 0th and 1st orders pass through the lens, resulting in a pure sine-wave image plus a constant-intensity bias.

Figure 4.23(a) shows the image electric field distribution from the line–space pairs shown in Fig. 4.20. The NA values 0.3, 0.6, and 0.9 are used. With $NA = 0.3$, there is no structure in the image. It is simply a constant background of intensity 0.25. Referring to Fig. 4.23(b), the reason is obvious: the only spatial frequency accepted by the pupil according to $NA = 0.3$ is the 0th-order frequency. When the NA is increased to 0.6, the 1st-order spatial frequency is admitted, resulting in a sinusoidal image with an electric field that is a superimposed single sine wave with an average electric field amplitude of 0.5. Because the sum of the two 1st-order sine waves is larger than the

0th-order field, the total field at $x = 1$ is negative, resulting in the small positive peaks when Eq. (4.52) is applied to convert the electric field to intensity. Further increasing the NA to 0.9 does not change the intensity distribution because the cutoff frequency of the low-pass filter still excludes the 2nd-order frequencies.

When the transparent-to-opaque ratio is changed to 1:3 instead of 1:1, as shown in Fig. 4.24, the shape of the image intensity distribution in Fig. 4.24(a) is like that in Fig. 4.23(a), except that the magnitude is lower, resulting in an image of smaller openings than the 1:1 pattern at the same exposure level. This can be understood from the frequency spectrum shown in Fig. 4.24(b).

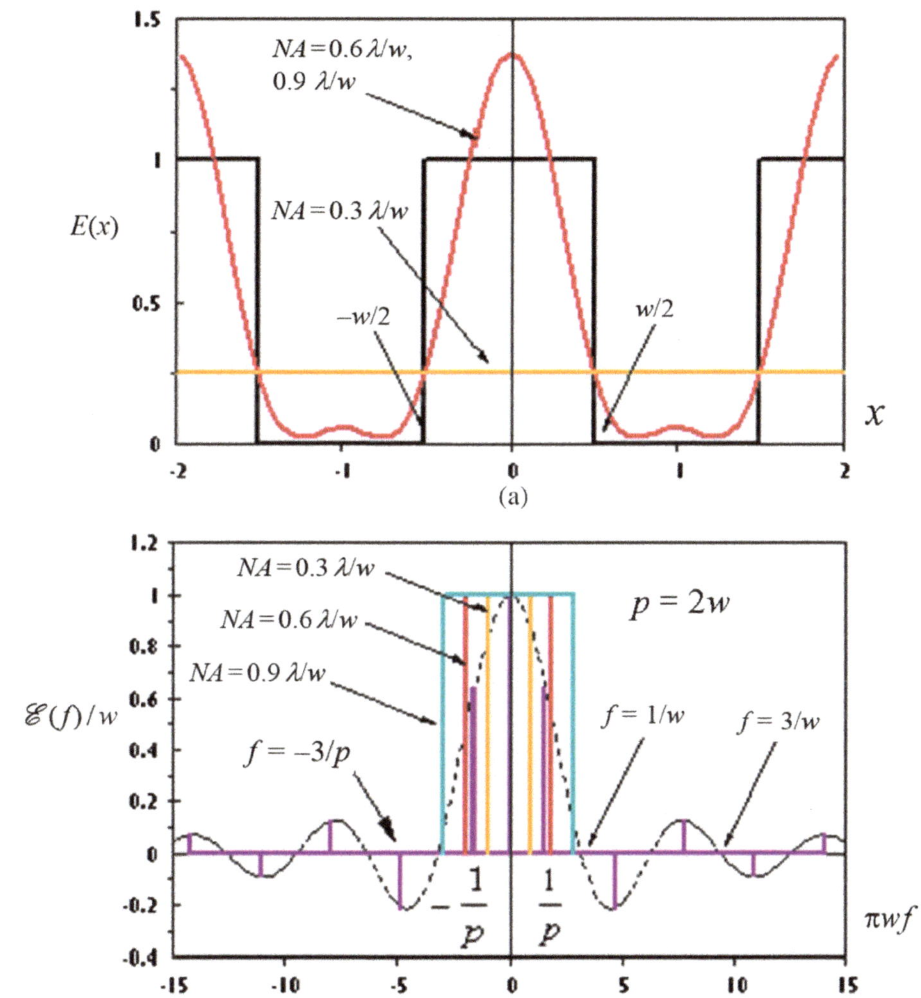

Figure 4.23 (a) Mask and image intensity distribution for 1:1 line–space pairs. (b) Corresponding spatial frequency distribution and the three different pupil functions.

Again, only the 0th- and 1st-order frequencies pass through the lens. The stronger 1st-order components bring down the combined intensity because they are negative.

With patterns much larger than the minimum feature size, higher-order spatial frequencies can be present. We now show 1:1 line–space pairs that are twice as large—equivalent to half the frequency of the line–space pairs shown earlier. As depicted in Fig. 4.25, NA = 0.3 suffices in collecting the 1st-order frequencies, and NA = 0.6 collects the 1st- and 2nd-order frequencies. However, there is no 2nd-order component in the original object. Hence, 0.3 and 0.6 NA produce the same image. With NA = 0.9, the 3rd-order frequencies are admitted, resulting in a broader but double-peaked image.

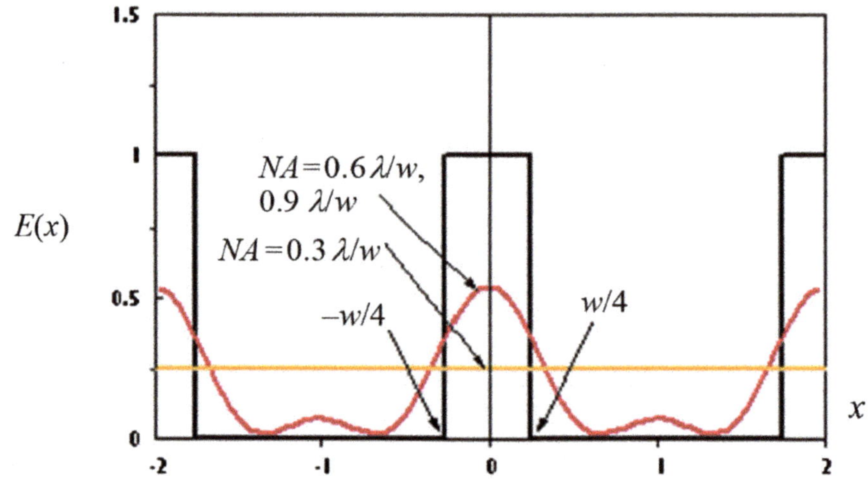

(a)

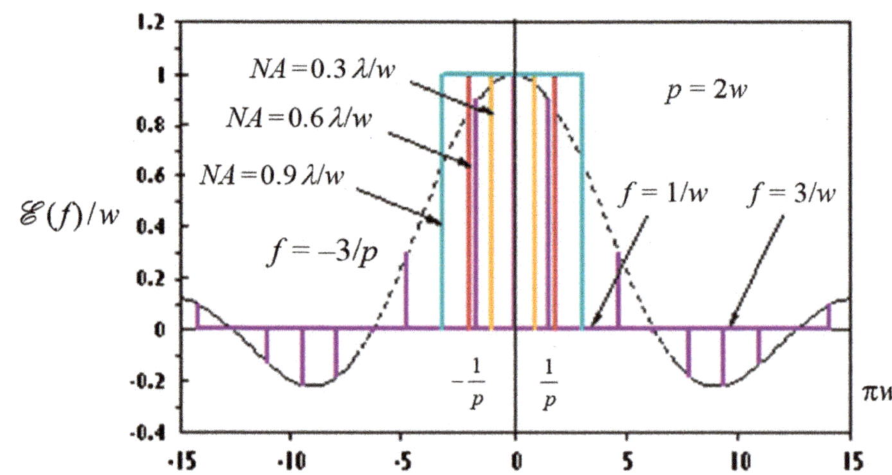

(b)

Figure 4.24 (a) Mask and image intensity distribution for 1:3 line–space pairs. (b) Corresponding spatial frequency distribution and the three different pupil functions.

4.1.6.3 Angular spectrum

Each spatial frequency f is associated with a pair of angular plane waves. The angles of these plane waves are

$$\theta = \pm\sin^{-1}(\lambda f), \tag{4.69}$$

where $\theta = 0$ is the direction of the optical axis. Using the conditions given in Sections 4.1.6.1 and 4.1.6.2, when $f = 1/w$ and $w = \lambda$, the angles of the plane waves are $\pm\theta = \pi/2$.

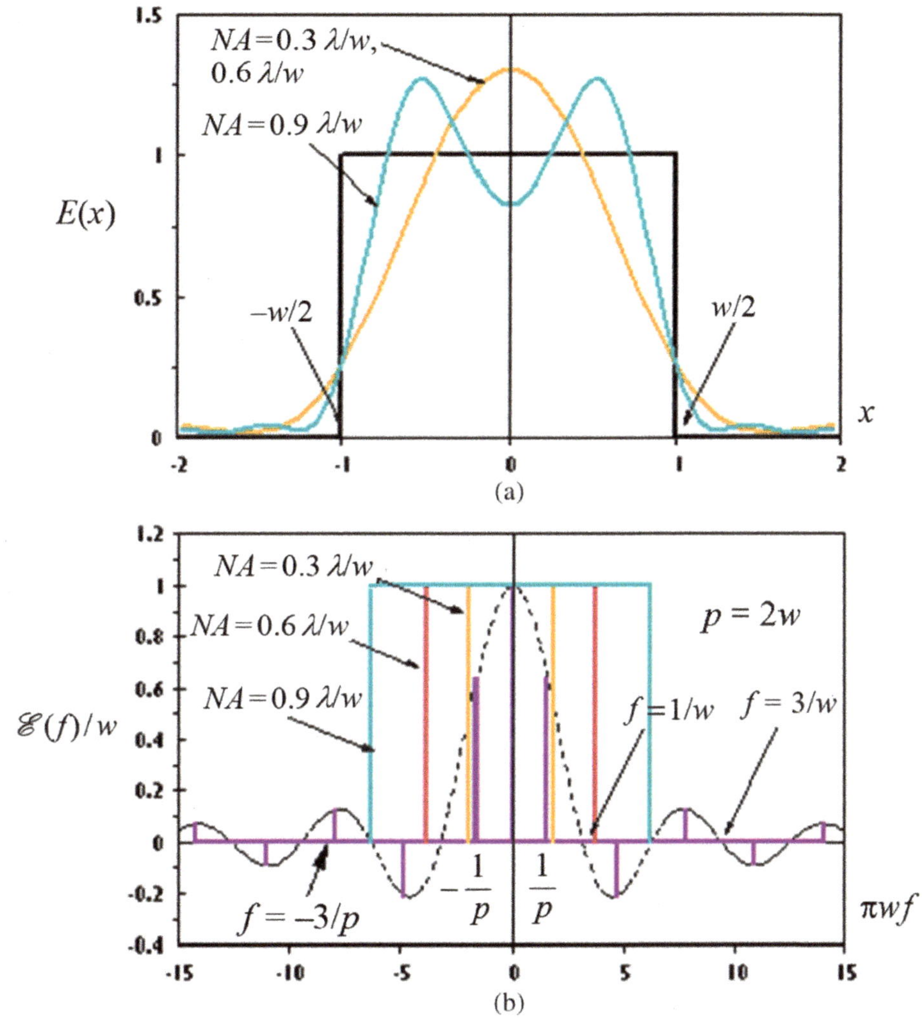

Figure 4.25 (a) Mask and image intensity distribution for 1:1 line–space pairs. (b) Corresponding spatial frequency distribution and the three different pupil functions with $w = 2$.

4.1.7 Imaging results

Two examples are given to compare the results from Eqs. (4.48), (4.53), and (4.54). The first example uses a test mask pattern called BigMaC,[21] which was designed for compact placement of a large number of feature combinations. There are minimally sized and large features with minimally sized and large spaces, long and short features, inside and outside corners, as well as vertical and horizontal features. The minimum feature size is chosen to be 0.4λ/NA for this demonstration, as shown in Fig. 4.26.

Figure 4.27(a) shows the diffracted image from the BigMaC mask at the focal plane. The solid line depicts the constant-intensity contour at a –0.6 log exposure. The other contour is evaluated at log intensity –0.5. The distance between the two contours corresponds to a 26% exposure latitude. The dotted line traces a ±10% linewidth control bracket of the original mask pattern. The log intensity level of the two image contours is chosen by centering them to the ±10% brackets. With such minimally sized features at 0.4 λ/NA, the diffracted image meets the linewidth-control criteria only at limited areas, for example, small parts of edges bb″, cc″, ef, hi, ij, jk, and kk′, with their symmetrical parts in the other three quadrants. Edge ee″ does not meet the criteria. The lengths of abb″a″ and cdd″c″ are severely shortened. The latter shortens less because it is farther from the other features. Due to uneven edge misplacements, abb″a″ appears to have shifted to the left, whereas cdd″c″ shifts to the right. The bend fghi is severely distorted. The centerline segment ee′e‴e″ suffers from oscillation in the edge position. When defocused to 0.4 λ/NA^2 as shown in Fig. 4.27(b), the image is even less useful. The image from a much lower illumination coherence corresponding to $\sigma = 0.8$ is shown in Fig. 4.28, where the linewidth oscillation is perceivably reduced. Optimizing the exposure to produce an acceptable section of ee″ and aa″ results in losing the other sections, especially the arm efghij. The situation at defocus = 0.4 λ/NA^2 is similar, although more severe.

Another example uses two adjacent openings of 3:1 height-to-width ratio as depicted in Fig. 4.29.[22,23] The features' size and separating distance is 0.987 λ/NA, which is large by modern standards. Even with such large features, corner rounding is easily observed in Figs. 4.30 to 4.32.[i] Reproducing sharp corners requires many high-order spatial frequencies. Incoherent diffraction from the two bars is shown in Fig. 4.30. The constant-intensity contours are separated by 1.5 dB. Log intensity is used so that the log slope of the optical image is independent of the intensity level. Level 0 contains all intensities higher than or equal to 1, which is the normalized incident intensity. Similarly, level 19 contains all intensity levels equal to or higher than that level. The diffracted image in Fig. 4.30(a) exhibits good straight edges, except for the rounding due

[i]This author is indebted to Shuo Yen Chou and Mengeng Chen, who modernized these figures using updated software.

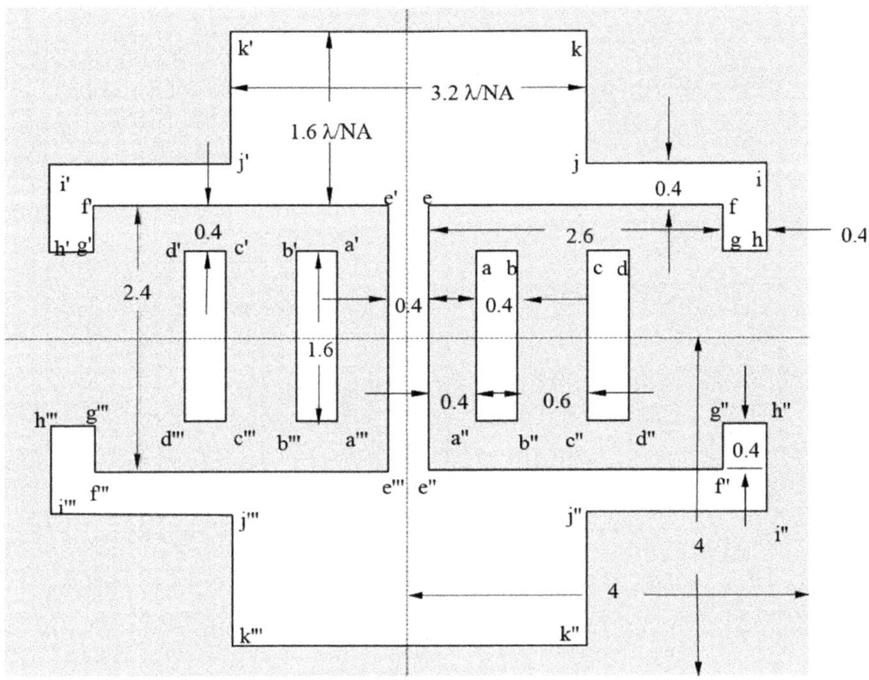

Figure 4.26 Big–small-feature and corner test (BigMaC) mask.

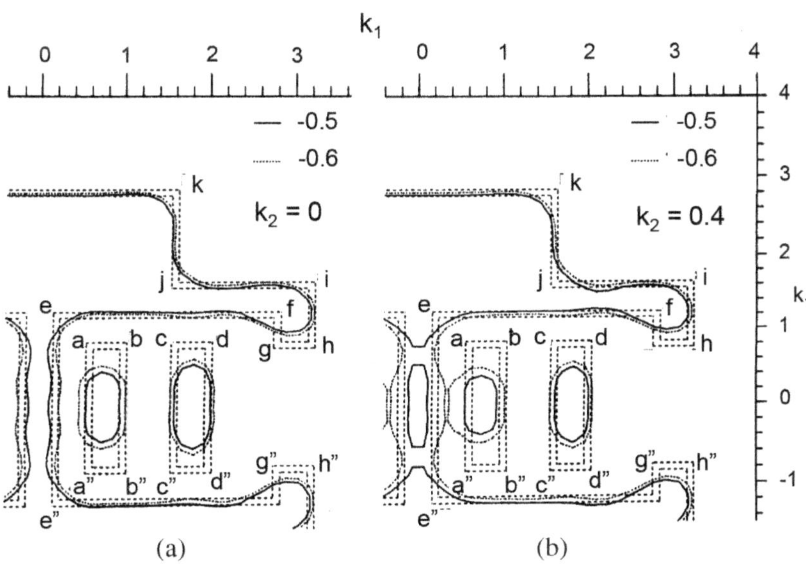

Figure 4.27 Diffracted image from the BigMaC mask at σ = 0.4: (a) defocus = 0 and (b) defocus = 0.4 λ/NA^2.

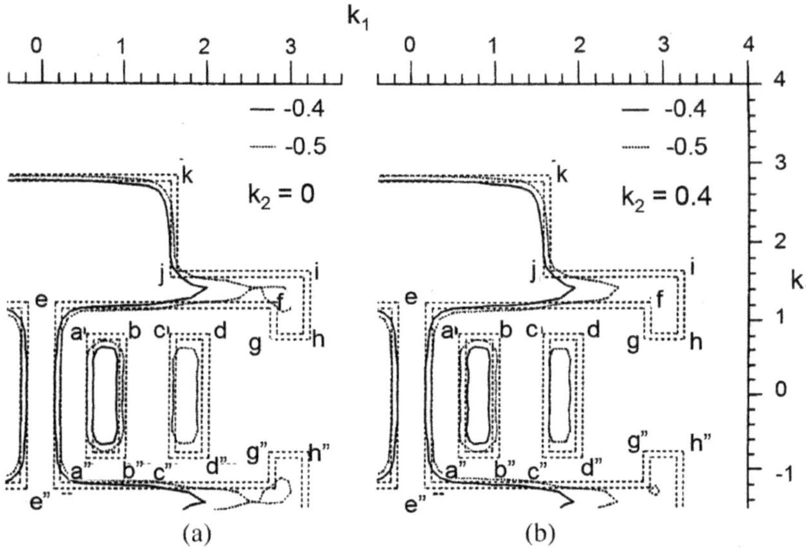

Figure 4.28 Diffracted image from the BigMaC mask at σ = 0.8: (a) defocus = 0 and (b) defocus = 0.4 λ/NA^2.

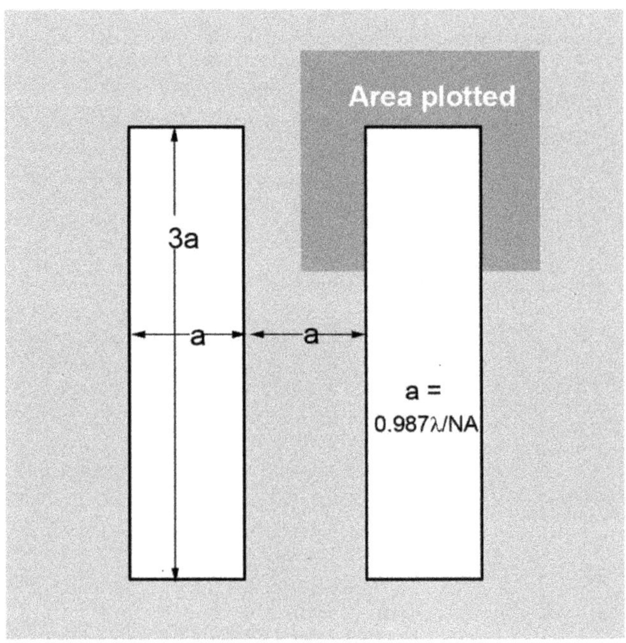

Figure 4.29 Mask pattern used for Figs. 4.30 to 4.32 (adapted from Ref. 22).

to loss of higher spatial frequencies. However, the separation of levels is wide, indicating poor linewidth control, because small exposure changes lead to a large linewidth variation. The situation is worse at the defocus plane.

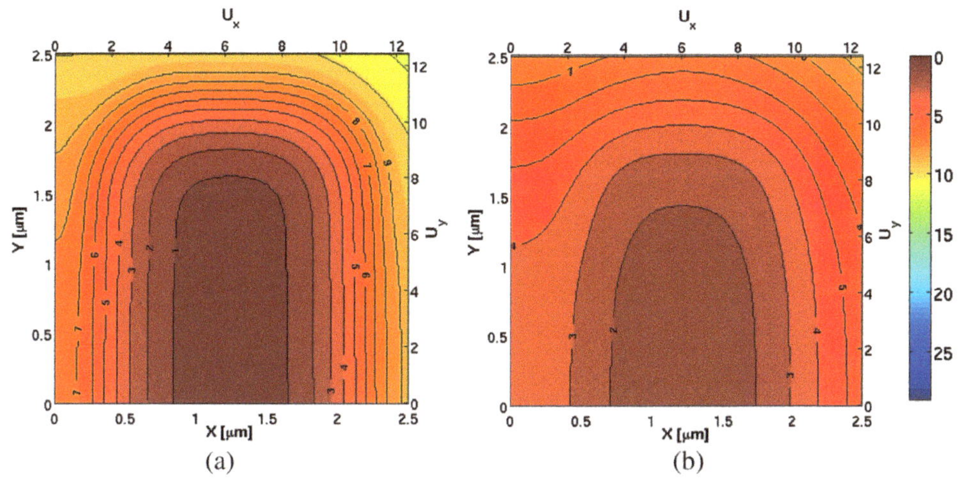

Figure 4.30 Diffracted image from the mask in Fig. 4.29. Illumination is completely incoherent, i.e., $\sigma = \infty$. (a) Defocus = 0 and (b) defocus = λ/NA^2.

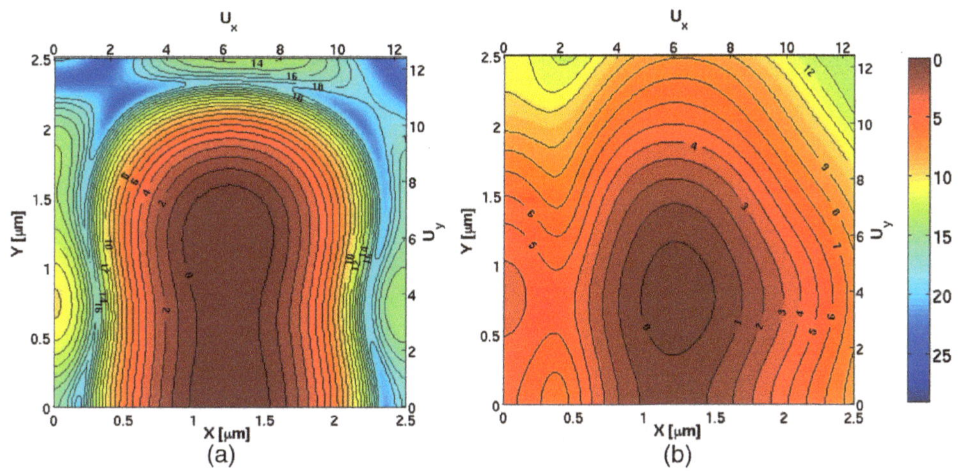

Figure 4.31 Same as Fig. 4.30, except that the illumination is completely coherent, i.e., $\sigma = 0$. (a) Defocus = 0 and (b) defocus = λ/NA^2.

Switching over to the coherent image in Fig. 4.31, the image contrast is improved, as indicated by the close spacing between exposure levels. However, the line edges wiggle, and there is a bright ghost line at the center. The waviness and influence of the ghost line are even worse with defocus.

The partially coherent image shown in Fig. 4.32 makes a good compromise by eliminating the ghost line while retaining edge straightness and image contrast. This example shows some of the qualitative behavior of 3D diffracted images. Actual optimization of σ needs to use the E-D methodology presented in Chapter 5.

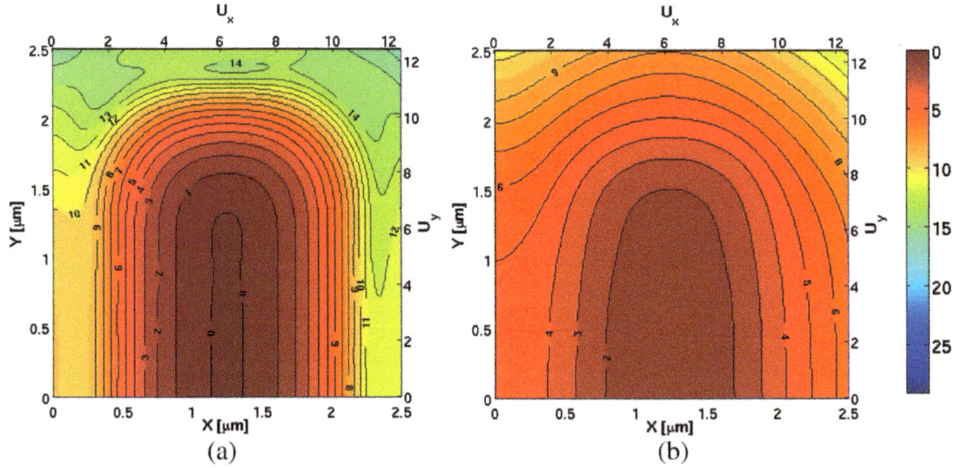

Figure 4.32 Same as Fig. 4.30, except that the illumination is partially coherent, i.e., $\sigma = 0.78$. (a) Defocus = 0 and (b) defocus = λ/NA^2.

4.2 Reflected and Refracted Images

As soon as the aerial image enters the resist film, it undergoes a series of changes. As depicted in Fig. 4.33, when the imaging wave reaches the air–resist interface, it refracts into the resist and reflects back to the air medium. To make the drawing easier to see, the image focus is set to the air–resist interface. In actual practice, the image should be focused on the photoresist. In any case, when the wave hits the resist–substrate interface, it refracts into

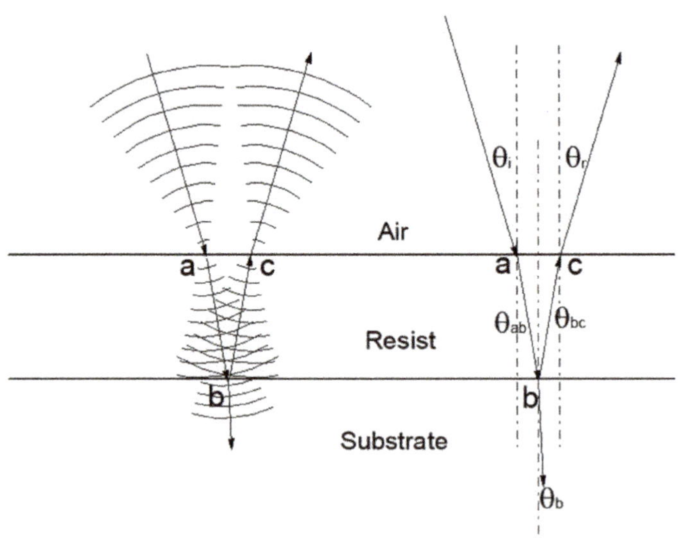

Figure 4.33 Refracted and reflected waves.

the substrate and reflects into the resist. The refraction angle obeys the law of refraction:

$$n_1 \sin\theta_1 = n_2 \sin\theta_2, \tag{4.70}$$

whereas the angle of reflection equals the angle of incidence:

$$\theta_{\text{incidence}} = \theta_{\text{reflection}}. \tag{4.71}$$

When the photoresist medium absorbs light, the refractive index is a complex number with an imaginary part that is the absorptive component. The angles of refraction and reflection still follow Eqs. (4.70) and (4.71). However, a phase shift at the interface is induced, accompanying the change of the amplitude in the refracted and reflected waves. As soon as the wave leaves the refracting surface, it is absorbed according to the attenuation coefficient α. Let the electric field distribution of the incident wave be

$$E_{\text{i}}(x, z) = e^{i\frac{2n_0\pi}{\lambda}(z\cos\theta_{\text{i}}+x\sin\theta_{\text{i}})}. \tag{4.72}$$

The reflected field distribution is then

$$E_{\text{r}}(x, z) = C_{\text{r}}e^{i\frac{2n_0\pi}{\lambda}(z\cos\theta_{\text{r}}+x\sin\theta_{\text{r}})}, \tag{4.73}$$

where the reflection coefficient C_{r} is a complex number dictating the amplitude and phase change at the air–resist interface, and n_0 is the refractive index of air. The electric field distribution from points a to b is

$$E_{ab}(x, z) = C_{ab}e^{i\frac{2n_0\pi}{\lambda}(z\cos\theta_{ab}+x\sin\theta_{ab})}, \tag{4.74}$$

where the refraction coefficient C_{ab} is also a complex number dictating the amplitude and phase change at the air–resist interface, and n_{resist} is the complex refractive index of the photoresist:

$$n_{\text{resist}} = n_{\text{resist, real}} + in_{\text{resist, imaginary}}. \tag{4.75}$$

Multiple reflections and refractions at the two interfaces take place until attenuation makes the remaining waves negligible, as illustrated in Fig. 4.34.

4.2.1 Methods to evaluate the reflected and refracted image from a mask

There are two methods to evaluate the reflected and refracted image from a given mask pattern. The first method traces the waves through all multiple reflecting and refracting surfaces shown in Fig. 4.34, following Eqs. (4.70) to (4.75). For a given mask pattern, the spatial frequency components are evaluated, then the higher spatial frequencies rejected by the lens's NA are

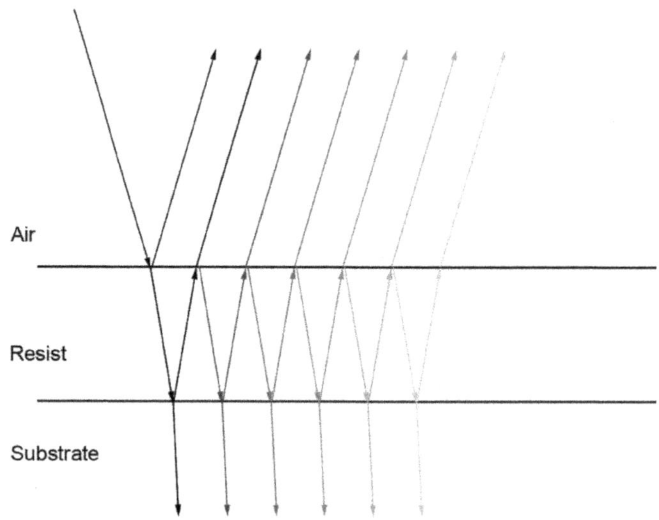

Figure 4.34 Multiple reflected waves in an attenuating medium.

removed. Each of the remaining frequency components is treated as a plane incident wave of angle θ_i to follow through reflections and refractions. Multiple reflections and refractions of plane waves through thin-film stacks are well treated in the literature. One simply applies these equations to each spatial frequency component, then recombines them with partial coherence, as was discussed in Section 4.1.5.3. When the resist bleaches or dyes (i.e., its complex refractive index is a function of exposure), the image is evaluated at the initial exposure, then the refractive indices are adjusted for the local exposure, and the reflected and refracted image is evaluated again until the change in refractive index stabilizes.

This approach is practiced by the UC Berkeley Electronic Research Laboratory,[24,25] Mack,[26,27] and many other researchers in this field. The difficulty with this approach is that when the thin-film stack consists of topography such as lines running over polysilicon gates, contact holes, metal lines, etc., the reflections and refractions must be traced at each interface, making it extremely tedious.

The most elegant approach to evaluating the reflected and refracted image in the presence of topography and the thin-film stack is to solve Maxwell's equations,[28] with boundary conditions given by the topography and the thin films. One can decompose the mask pattern into an angular spectrum of plane waves, then solve Maxwell's equations exactly for each plane-wave component before partial coherence recombination. Going one step further, boundary conditions can be set for the mask opening, thus removing the approximation in Eq. (4.51). Interested readers may consult the publications of Lin[15], Yeung,[29,30] and Barouch et al.[31,32]

4.2.2 Impact of multiple reflections on DOF

To understand the impact of multiple reflections on DOF, it is best to visualize the image changes caused by multiple reflections. Figure 4.35 shows the electric field distribution at the interfaces from a $k_1 = 0.35$ line opening illuminated with $\sigma = 0.5$ at a vertical incidence. Note that k_1 is used as a normalized lateral dimension, as is fully explained in Section 5.4. The beams are vertical but spread with a finite angle for the purpose of illustration. Hence, $\cos\theta$ is omitted by assuming a small angle of incidence. Otherwise, the optical distance na would become $na\cos\theta$.

At $z = a$, which is the air–resist interface where the image is focused, the electric field distribution is $E(x,0)$ and the reflected image is $R_{ads}E(x,a)$, where R_{ads} is the reflection coefficient of the first interface, a stands for the location of z, d is the downward direction of the incident light, and s is the state of polarization. Similarly, the transmitted electric field distribution is $T_{ads}E(x,a)$, where T_{ads} is the transmission coefficient at the air–resist interface. At the resist–wafer interface, the reflected electric field distribution is $R_{bds}T_{ads}E[x,a + (n + i\alpha)(b - a)]$, where R_{bds} is the reflection coefficient of the resist–substrate interface, α is the absorption coefficient of the multiply reflecting medium, and n is the medium's refractive index. The transmitted electric field distribution is $T_{aus}R_{bds}T_{ads}E[x, a+(n+i\alpha)(b-a)]$. The electric field distributions at the other points can be expressed similarly by multiplying the electric field distribution at a reference point by the appropriate reflection and transmission coefficients and adjusting the defocus distance in $E(x,z)$ according to the refractive index and absorption coefficient of the medium in which light has traveled.

In Fig. 4.35, 193-nm light is used. The thickness of the resist is 288 nm. The n and α values of the resist are taken to be 1.69 and 0.012, respectively, and the n and α values of the silicon wafer are 0.863 and 2.747, respectively. Hence, $r_{ads} = -0.257 - i0.003$, $T_{ads} = 0.743 - i0.003$, $R_{bds} = -0.385 - i0.656$, $T_{bds} = 0.615 - i0.656$, and $T_{aus} = 1.257 + i0.003$. Note that the outgoing

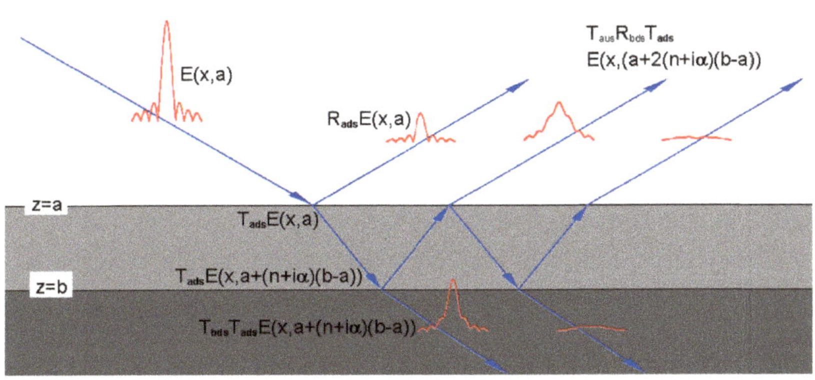

Figure 4.35 Electric field distribution at various interfaces (image courtesy of S.S. Yu).

electric field distribution at the second bounce is larger and wider than that at the first bounce. It is wider because the field distribution spreads out as the defocus distance increases, just as depicted in Figs. 4.39 and 4.40. It is larger because $T_{aus}R_{bds}T_{ads}$ is larger than R_{ads}.

Now, the resultant image is definitely of lower contrast at the best focus position when compared with the image without multiple reflections. Intuition suggests that a higher-contrast image may lead to a larger DOF. This is obviously not true when Eq. (4.21) is considered. When the enhancement of contrast is produced by a higher NA, the DOF decreases. In the case of multiple reflections, the DOF is also improved, despite the lower image contrast at the best focal point. The situation resembles the FLEX system,[33] in which the superposition of defocused images is achieved by multiple exposures with the wafer purposely offset at different stages of defocus. The averaging effect of these images takes the more-than-sufficient contrast in the focal plane to help the low contrast in the defocus planes.

Although both methods superimpose images at many defocus planes, multiple reflections are different from FLEX because the former superimposes coherently, whereas the latter superimposes incoherently. Multiple reflections should not be used to extend the DOF. In fact, they should be avoided at all costs because they are the strongest contributor of exposure nonuniformity, which leads to linewidth-control problems over topography, as is discussed in Section 6.6.2.5.

4.3 The Latent Image

The latent image is the radiation-induced change in the resist before it is developed into the resist image. A simple visualization is to directly relate the complex refractive index distribution $n(x,z)$ in the resist layer to the intensity distribution of the combined reflected and refracted images in the resist layer. The actual combination of these images is more complicated, as the refractive index change is a function of light intensity. One must combine them dynamically in time, as described above in Section 4.2.1.

The latent image is not readily visible since it mostly consists of refractive-index distribution instead of absorption distribution. For some resists with significant complex index change, the latent image can indeed be seen. For example, i- or g-line resists produce a visible latent image when drastically overexposed. Contrast enhancement resists, by definition, emphasize the production of a distinct latent image that is is readily visible.

However, it is desirable to view the latent image because this image offers an opportunity to see the aerial image captured in the resist media before it is locked into the developed resist image. The latent image is independent of many resist-processing characteristics and thus can settle doubts on whether the imaging tool or the resist is at fault when things go wrong. Being able to measure the latent image before development facilitates an *a priori* determination of the developing condition and the development end point to control the final developed image. Many attempts have been made to make the latent image visible. This is possible, but the image usually does not offer sufficient contrast to be useful.

4.4 The Resist Image

Resist exposure leads to modification of the resist's physical characteristics. Not only is its complex refractive index changed, but the dissolution rate of the resist in the developer is also modified according to the exposing image. The dissolution rate distribution $D(x, z)$ is directly related to $n(x, z)$ through an exposure–dissolution relationship, as shown in Fig. 4.36.

With a positive resist, higher exposure leads to a faster dissolution rate, and vice versa for a negative resist. Given $D(x, z)$, the resist image $R(x, z, z_{resist})$ can be evaluated. However, evaluating this image analytically is not trivial because removal of the resist is a function of the way the resist surface is exposed to the developer.

To further visualize the resist-development process, the sequence of developing an idealized dissolution-rate image is depicted in Fig. 4.37. The aerial image $I(x, y, z)$ through the depth of the resist is assumed to be completely zero in the opaque area of the mask and unity in the transparent area. $I(x, z) = 1$ when $-a < x < a$, and $= 0$ elsewhere.

Multiple reflections are disregarded by assuming a perfect antireflection coating at the air–resist and resist–wafer surfaces. Resist absorption and diffraction effects are also ignored. Therefore, the latent image is also a simple square function—zero in the opaque area and a constant value in the exposed area. The development rate $R(x, y, z)$ becomes $R(x, z) = 10$ nm/sec when $-a < x < a$, and $= 1$ nm/sec elsewhere. There is a finite unexposed development rate, and the resist is kept at low contrast for illustration purposes.

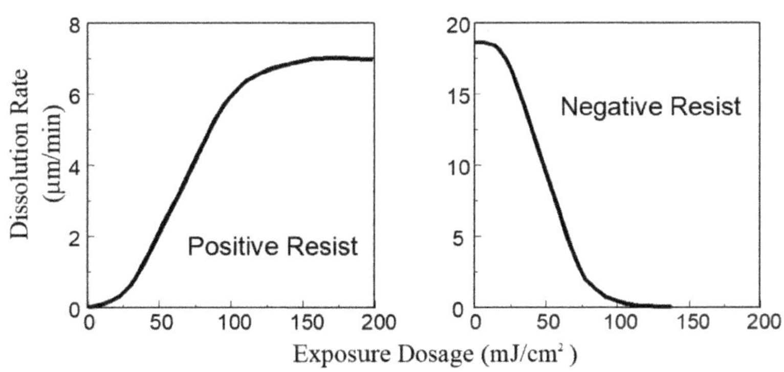

Figure 4.36 Resist dissolution as a function of exposure.

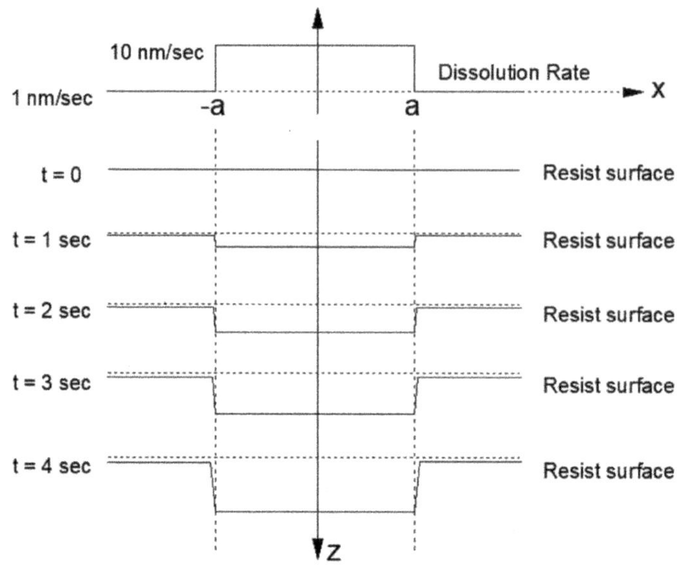

Figure 4.37 Development sequence of an idealized-dissolution-rate image.

Figure 4.37 shows that, because of the finite dissolution rate in the unexposed area, the resist not only loses thickness while being developed, but it is also removed sidewise as soon as the developer–resist interface in the vertical direction is created. Hence, the developed resist image has a natural overcut profile provided there is a finite dissolution rate in the unexposed area for positive resists or in the exposed area for negative resists.

When optical absorption is considered while ignoring the surface effects, the development rate decreases toward the bottom of the positive resist and vice versa for the negative one. The overcut profile is more exaggerated for positive resists, while there is a chance for compensation for negative resists, as shown in Fig. 4.38.

Instead of artificial aerial images, we now use the diffracted aerial image from a $k_1 = 0.64$ line opening illuminated with $\sigma = 0.5$ light to show the development process depicted in Fig. 4.39. The diffracted image is refracted in the resist media as shown. The diffracted image propagates in the air medium until it hits the air–resist interface at three positions. The top position shows the resist–air interface at $-1.8\ k_2$ from the focal point, the middle position is $-0.5\ k_2$, and the bottom position is $1\ k_2$. Note that k_2 is used as a normalized longitudinal dimension, as is fully explained in Section 5.4.

As soon as light reaches each air–resist interface, the propagation constant switches to the n and α of the resist; this is what is depicted in Fig. 4.39. The positive-resist dissolution rate curve in Fig. 4.36 is used for all three positions. Each family of the developed image curves represents changes in the resist image as a function of the development time. The normalized unit of

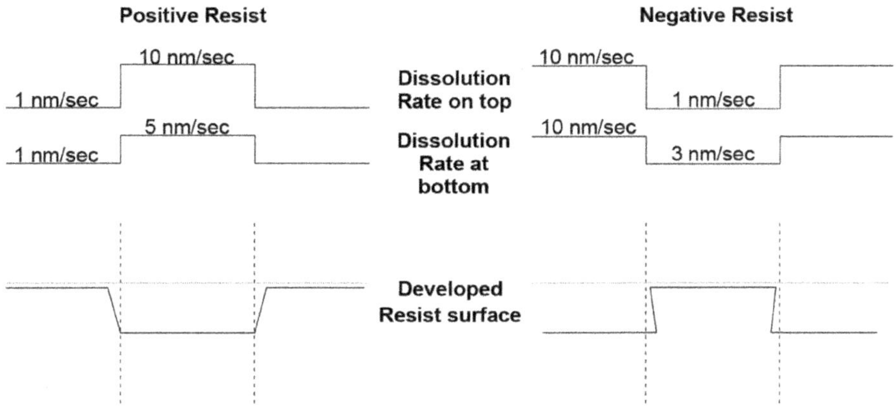

Figure 4.38 Absorptive positive- and negative-resist images.

dimension k_1 in the lateral direction and k_2 in the longitudinal dimension are later defined in Section 5.4.

Figure 4.39 shows that the developed resist image is a strong function of the refracted image. It is predominantly influenced by the refracted image on top, which is the first curve in each family of the refracted image. Also, the defocus resist images are not symmetrical, even though the refracted aerial images are symmetrical with respect to the focal plane marked in the figure.

The n and α values in Fig. 4.39(a) are typical of an i-line resist. The developed images look reasonable. The α value in Fig. 4.39(b) is 3 μm^{-1}. The

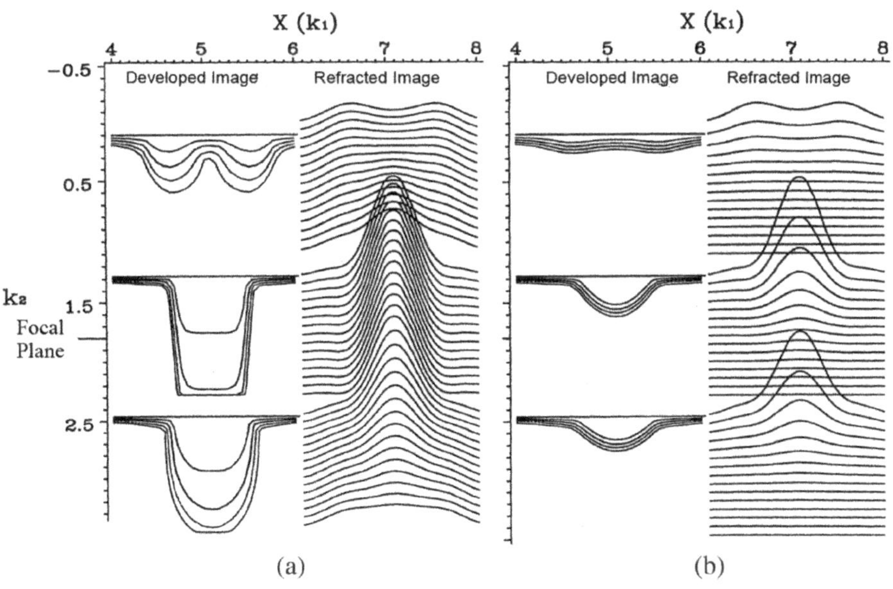

Figure 4.39 Development of resist (a) n_r = 1.5, α = 0.178 μm^{-1}, and (b) n_r = 1.5, α = 3 μm^{-1}.

corresponding developed image is extremely dependent on the refracted image at the resist surface. This is a desirable α value for top-surface imaging systems such as DESIRE.[34] On the other hand, the absorption is too high for a conventional resist system with resist thicknesses in the micrometer regime. Note that this is a typical level of absorption for 157-nm resist systems[35] and is one of the problems for 157-nm lithography. Nevertheless, when the resist thickness drops to 100 nm or below, this α value is about right to attain a 30% absorption in the resist layer.

When the resist absorption is reduced to 0, as shown in Fig. 4.40, the developed image in focus has a slightly more vertical profile than that in Fig. 4.39(a). However, the developed image at 1.0 k_2 is too wide because the undamped image is wider at the same threshold. Hence, having a completely absorption-free resist is not necessarily the most desirable situation, even if we could ignore the energy that the resist needs to absorb for chemical reactions.

Figure 4.40(b) shows the case with $n = 1$ and $\alpha = 0$. This resist consumes more DOF; therefore, at defocus = $-1.8\ k_2$ (in the refracted image), the refracted image at the resist bottom already becomes single peaked, whereas in Fig. 4.40(a) the refracted image is still double peaked. However, this situation is not bad. The developed resist image is closer to the focused image. The resist image only becomes worse at defocus = 1.0 k_2, where the refracted image at the bottom of the resist is farther away from focus. Because the developed image is more dependent on the refracted image on top, it is not much more severe than the corresponding image in Fig. 4.39(a). Naturally, when defocus is farther out, the difference will be more significant.

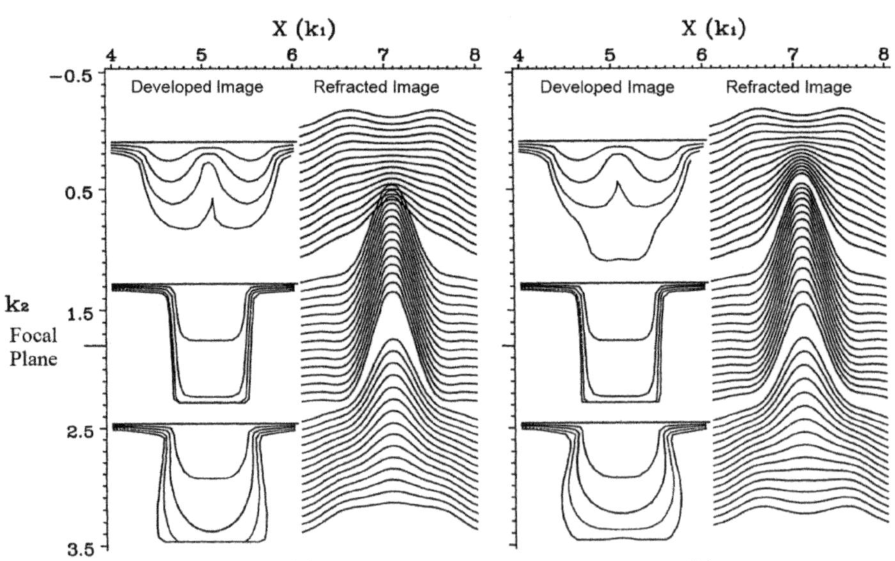

(a) (b)

Figure 4.40 Development of resist (a) $n_r = 1.5$, $\alpha = 0$, and (b) $n_r = 1$, $\alpha = 0$.

We just demonstrated how to simulate the developed image from the refracted image through the rate image. This simulation is possible, but very time consuming. The most accurate model method is to divide the resist space into cells and develop the resist cell by cell, according to the development rate of each cell. The development is not only dependent on the amount of exposure the cell received but is also a function of the number of facets in contact with the developer.

4.4.1 The *A*, *B*, *C* coefficients

The *A*, *B*, *C* coefficients were developed by Dill et al.[36] These coefficients characterize the resist exposure and development behaviors and are frequently used to quantify any given resist. As shown later in Eq. (4.123), the complex index of refraction of the resist n_{cr} is not merely a function of space, although this would be true if the resist stayed unchanged when receiving the diffracted light from the mask pattern through the lens. For a real photoresist, n_{cr} is a function of I_r, the light intensity that the resist receives. Therefore, the resist is bleached or dyed during exposure. On the other hand, I_r is always a function of n_{cr} regardless of bleaching or dyeing. Hence, I_r and n_{cr} are dynamically linked to each other in a loop. Let α be the absorption coefficient of the resist media. Then

$$\alpha = \frac{2\pi}{\lambda} n_i. \tag{4.76}$$

Lambert's law dictates that

$$\frac{dI(z)}{dz} = -\alpha I(z) \tag{4.77}$$

for light propagating in the z-direction. This takes care of the change of intensity as a function of α along the path of the light being absorbed. There must be another equation to govern the change of α as a function of intensity. Let us assume that α is composed of two parts, a constant term and a photoactive term, such that

$$\alpha = Am + B. \tag{4.78}$$

B is independent of the intensity of the incident light, whereas m is dependent on the incident intensity and is the concentration M of the photosensitive component of the resist, normalized to the initial concentration M_0 before exposure. A is the coefficient of m. Then the rate of change of m is

$$\frac{dm}{dt} = -Cm(t)I. \tag{4.79}$$

Now that I and α are dependent on each other, they must both be functions of space and time, as must be m as well. Eqs. (4.77) and (4.79) become

$$\nabla I(x, y, z, t) = -I(x, y, z, t)[Am(x, y, z, t) + B] \tag{4.80}$$

and

$$\frac{\partial m(x, y, z, t)}{\partial t} = -Cm(x, y, z, t)I(x, y, z, t), \tag{4.81}$$

respectively, where

$$\nabla I \equiv \frac{\partial I}{\partial x} + \frac{\partial I}{\partial y} + \frac{\partial I}{\partial z} \tag{4.82}$$

is the intensity gradient. Equations (4.80) and (4.81) are expressed in generalized coordinates. The original Dill equations treated I as if propagation was only in the z-direction. Once generalized, it is not trivial to derive an analytic solution. However, most simulation programs use these equations in either simplified or generalized forms to evaluate I and m numerically. To solve these two simultaneous equations, one needs to know the three coefficients (A, B, C) as well as the initial conditions [$I(x, y, z, t_0)$ and $m(x, y, z, t_0)$]. A and B have units of μm^{-1} and are known as the bleachable and nonbleachable absorption coefficients, respectively. C has units of cm^2/mJ and is known as the exposure rate constant. $I(x, y, z, t_0)$ is usually referred to as the aerial image, that is, the diffracted light from the mask openings through the imaging lens before the light passes through the photoresist. The normalized concentration m of the photosensitive component in the resist is generally homogeneous before patterned exposure.

For a given resist system, the A, B, C coefficients can be evaluated experimentally. The resist is coated on an optically matched substrate that has the same refractive index as the resist. Hence, light propagates without reflections and travels into the resist only once. The reflection at the air–resist interface can be calibrated by taking a reading with a blank index-matched substrate. The absorption coefficient is obtained by taking the ratio of the intensity of the incoming light to that of the outgoing light at the actinic wavelength for which a spectrophotometer is suitable:

$$\alpha = \frac{1}{d}\ln\left(\frac{I_{\text{in}}}{I_{\text{out}}}\right), \tag{4.83}$$

where d is the resist thickness. Then

$$A = \alpha_{\text{unexposed}} + \alpha_{\text{exposed}} \tag{4.84}$$

and

$$B = \alpha_{\text{exposed}}. \tag{4.85}$$

Evaluating C is more complicated. Dill et al. provide the following formula:[37]

$$C = \left\{ \frac{A + B}{AI_0 T(0)[1 - T(0)]} \right\} \frac{dT(0)}{dt}, \tag{4.86}$$

where

$$T(0) = \left(\frac{I_{\text{out}}}{I_{\text{in}}} \right)_{t=0}. \tag{4.87}$$

Dill et al. also proposed an iteration scheme to determine A, B, and C more accurately by performing a mean-square deviation fit to the experimental transmission curve $T(t)$.[37]

The A, B, C coefficients are good parameters with which to characterize the photoresist and simulate the resist image. They are readily provided with any commercially available resist system. The concern is in their accuracy. They are definitely a function of processing conditions. For example, a high-temperature, pre-exposure bake may remove some photoactive compound and make A and C smaller. Any baking may change the refractive index of the resist, affecting both A and B. Even if the resist and its processing are maintained very strictly, it is not trivial to perfectly match the indices of the resist and the substrate, especially when the index of the resist changes as a function of exposure. Some work must be done to accommodate the dynamic difference in the indices.

4.4.2 The lumped parameters

The A, B, C coefficients are useful for characterizing a given resist system and simulating the resist image. Once the coefficients are determined, one goes through the formulations suggested by Eqs. (4.127) through (4.130) to evaluate the final resist image. The process is physically sound, but tedious and time consuming. In most cases, once the aerial image is obtained, the resist image can be estimated with the lumped parameters, which consist of the resist thickness D, the absorption coefficient of the resist α, and its development contrast γ.

The lumped parameter model (LPM) was initialized by C. A. Mack[38] et al. in 1986. Later, Mack extended the model[39,41] to the enhanced lumped parameter model, which includes the effect of resist absorption by assuming a segmented development path in the vertical and horizontal directions. In 1996, Brunner and Ferguson,[40] assuming an exponential form of the aerial image near the threshold, obtained analytic expressions for the linewidth and sidewall angle of a resist feature for a nonabsorptive resist by analogizing to ray tracing in geometrical optics. However, if resist absorption is considered, the development path is also approximated by a segmented path, with an additional segment that is diagonal. Here we show that, in such a case, analytic expressions can also be obtained by straightforward derivation. Before doing this, we briefly summarize Mack's theory. The coordinate system is defined such that the z-axis points downward and is thus perpendicular to the resist top surface.

Mack's LPM makes the following assumptions:

- The aerial image is separable in its x- and z-dependence [i.e., $I(x,z) = I_x(x)I_z(i)$]. Then, the exposure energy at any point of the resist $E(x,z) = EI_x(x)I_z(z)$, where $I_x(\infty) = 1$ and $I_z(0) = 1$.

- Assuming a constant resist contrast γ, the resist local development rate is

$$r(x, z) = r_0 \left[\frac{E(x, z)}{E_0}\right]^{\gamma}, \tag{4.88}$$

where E_0 is the energy to clear in the large clear area.

- Assume a segmented development path that starts at $(x_0, 0)$ and proceeds vertically until $E(x_0, z + \Delta z) < E(x_0 + \Delta x, z)$, then proceeds horizontally to (x, z). The vertical development time from $(x_0, 0)$ to (x_0, z) is

$$T_z = \int_0^z \frac{dz'}{r(x_0, z')} = r_0^{-1} E_0^{\gamma} E^{-\gamma} [I_x(x_0)]^{-\gamma} \int_0^z [I_z(z')]^{-\gamma} dz'. \tag{4.89}$$

Similarly, the horizontal development time from (x_0, z) to (x, z) is

$$T_x = \int_{x_0}^x \frac{dx'}{r(x', z)} = r_0^{-1} E_0^{\gamma} E^{-\gamma} [I_z(z)]^{-\gamma} \int_{x_0}^x [I_x(x')]^{-\gamma} dx'. \tag{4.90}$$

Therefore, the development time from (x_0, z) to (x, z) is

$$T = T_z + T_x. \tag{4.91}$$

Since $x = 0$ is at the center of the mask feature, the x-coordinate of the final development point is in fact half of the developed resist linewidth (i.e., $x = w/2$). So, Eq. (4.91) can be regarded as an implicit function $E(x)$. After rearranging, we obtain

$$E^{\gamma} = r_0^{-1} t^{-1} E_0^{\gamma} [I_x(x_0)]^{-\gamma} \int_0^z [I_z(z')]^{-\gamma} dz' + r_0^{-1} t^{-1} E_0^{\gamma} \times [I_z(z)]^{-\gamma} \int_{x_0}^{x} [I_x(x')]^{-\gamma} dx'. \tag{4.92}$$

To derive the change of linewidth resulting from the change of exposure energy, we differentiate Eq. (4.92) with respect to E, resulting in

$$\frac{dE(x)}{dx} = \frac{E(x)}{\gamma D'} \left[\frac{E(0) I_x(0)}{E(x) I_x(x)} \right]^{\gamma}, \tag{4.93}$$

where we define $D' = r_0[I_z(D)]^{\gamma}T$ and relate energy to clear in the large clear area E_0 to the exposure energy, leading to zero linewidth $E(0)$ via $E_0 = E(0)I_x(0)$. Integrating on both sides, we finally obtain

$$\frac{E(x)}{E(0)} = \left\{ 1 + \frac{1}{D'} \int_0^x \left[\frac{I_x(x')}{I_x(0)} \right]^{-\gamma} dx' \right\}^{\frac{1}{\gamma}}. \tag{4.94}$$

- If the intensity of the aerial image in the neighborhood of the threshold is further approximated by

$$I(x, z) = I(x_0) e^{s(x - x_0)} e^{-\alpha z}, \tag{4.95}$$

where s is the image log-slope and α is the absorption coefficient, then the analytic expression for the resist linewidth can be derived, as shown below. One defines

$$n(x, z) \equiv \frac{1}{r(x, z)} = r_0^{-1} e^{-\gamma s(x - x_0)} e^{-\gamma \alpha z}, \tag{4.96}$$

where we have used $EI(x_0) = E_0$. Then, $\nabla n = n\gamma\,(-s\mathbf{x} + \alpha\mathbf{z})$. We now define a new coordinate system expanded by x_1 and z_1. The new orthonormal basis vectors are related to the old ones by

$$\begin{aligned} x - x_0 &= x_1 \cos\phi + z_1 \sin\phi \\ z &= -x_1 \sin\phi + z_1 \cos\phi. \end{aligned} \tag{4.97}$$

In the new coordinate system,

$$n(x_1, z_1) = r_0^{-1} e^{-\gamma\sqrt{s^2+\alpha^2}x_1}, \tag{4.98}$$

which only depends on x_1. It can be shown that $n\cos\theta_1$ is a constant, where θ_1 is the angle that the tangent vector of the development path makes with z_1. When $t = 0$, $x_1 = 0$, $z_1 = 0$, and

$$\cos\theta_1 = \cos\phi = \frac{s}{\sqrt{s^2+\alpha^2}},$$

we have

$$n\cos\theta_1 = r_0^{-1}\left(\frac{s}{\sqrt{s^2+\alpha^2}}\right).$$

That is,

$$\cos\theta_1 = \frac{s}{\sqrt{s^2+\alpha^2}} e^{\gamma\sqrt{s^2+\alpha^2}x_1}. \tag{4.99}$$

The differential of the development path in the x_1 direction is

$$\Delta x_1 = -r\sin\theta_1\Delta t = -r_0 e^{-\gamma\sqrt{s^2+\alpha^2}x_1}\sin\theta_1\Delta t. \tag{4.100}$$

Integrating both sides of the above equation, we obtain

$$x_1(t) = \frac{-1}{2\gamma\sqrt{s^2+\alpha^2}}\ln\left[\frac{s^2}{s^2+\alpha^2} + \left(\gamma\sqrt{s^2+\alpha^2}r_0 t\right)^2\right]. \tag{4.101}$$

The differential of the development path in the $\mathbf{z}_1$ direction is

$$\Delta z_1 = r\cos\theta_1\Delta t = r_0 e^{\gamma\sqrt{s^2+\alpha^2}x_1}\cos\theta_1\Delta t. \tag{4.102}$$

Integrating on both sides, we obtain

$$z_1(t) = \frac{1}{\gamma\sqrt{s^2+\alpha^2}}\tan^{-1}\left[\frac{\left(\gamma\sqrt{s^2+\alpha^2}\right)}{s} r_0 t\right]. \tag{4.103}$$

Transforming back to the old coordinate system, we finally obtain

$$x(t) = x_0 - \frac{s}{2\gamma\sqrt{s^2+\alpha^2}} \ln\left[\frac{s^2}{s^2+\alpha^2} + \left(\gamma\sqrt{s^2+\alpha^2} r_0 t\right)^2\right] + \frac{\alpha}{\gamma(s^2+\alpha^2)} \tan^{-1}\left[\frac{\gamma(s^2+\alpha^2)}{s} r_0 t\right] \tag{4.104}$$

and

$$z(t) = \frac{\alpha}{2\gamma(s^2+\alpha^2)} \ln\left[\frac{s^2}{s^2+\alpha^2} + \left(\gamma\sqrt{s^2+\alpha^2} r_0 t\right)^2\right] + \frac{s}{\gamma(s^2+\alpha^2)} \tan^{-1}\left[\frac{\gamma(s^2+\alpha^2)}{s} r_0 t\right]. \tag{4.105}$$

Also, we have

$$\cos\theta_1 = \frac{1}{\sqrt{1+\left[\frac{\gamma(s^2+\alpha^2)}{s} r_0 t\right]^2}}, \tag{4.106}$$

or

$$\tan\theta_1 = \frac{\gamma(s^2+\alpha^2)}{s} r_0 t. \tag{4.107}$$

So,

$$\theta(t) = \theta_1 - \phi = \tan^{-1}\left[\frac{\gamma(s^2+\alpha^2)}{s} r_0 t\right] - \sin^{-1}\left[\frac{\alpha}{\sqrt{s^2+\alpha^2}}\right]. \tag{4.108}$$

Choosing $t = T$ such that $r_0 T = D$, we obtain the final resist linewidth and resist sidewall angle.

In summary, the lumped parameter model[38,39,41] makes these assumptions:

- The resist development rate is related to the exposure by the γth power.
- The resist absorption is constant throughout its thickness and is independent of exposure time.
- The development process proceeds independently in the lateral and longitudinal directions.

With these assumptions, the exposure required to produce a resist feature can be related to the aerial image analytically such that

$$E(W/2) = E(0)\left[1 + \frac{1}{\gamma D_{\text{eff}}}\int_0^{W/2} I^{-\gamma}(x)dx\right]^{\frac{1}{\gamma}}, \tag{4.109}$$

where $E(W/2)$ is the exposure dosage required to develop to the resist image edge $x = W/2$, W is the width of the feature of interest, $E(0)$ is the exposure dosage corresponding to a feature width of zero, and $I(x)$ is the normalized intensity distribution of the aerial image. For a detailed derivation, consult Ref. 36.

The resist development contrast γ can be evaluated with

$$\gamma = \frac{dr/r}{dE/E}, \tag{4.110}$$

where r is the dissolution rate and E is the exposure. Hence, the physical meaning of γ is the slope of the dissolution rate versus the exposure curve normalized at the point of evaluation. That is, all values of r and E are taken at the intended exposure. When a log plot of r versus E is used, the slope is taken directly without having to normalize, as follows:

$$\gamma = \frac{d(\ln r)}{d(\ln E)}. \tag{4.111}$$

D_{eff}, the effective resist thickness named by Hershel and Mack,[41] is defined as

$$D_{\text{eff}} = r_0 t_{\text{dev}} e^{\gamma i(D)}, \tag{4.112}$$

where r_0 is the dissolution rate corresponding to an exposure E_0 that just clears the photoresist in a large, uniformly illuminated area. Furthermore, t_{dev} is the time it takes to develop the resist through the entire resist thickness D. Rearranging Eq. (4.112) reveals the physical meaning of D_{eff}:

$$D_{\text{eff}} = (t_{\text{dev}})\left[r_0 e^{\gamma i(D)}\right], \tag{4.113}$$

where the second term is the dissolution rate at a depth D if the resist is assumed to have a constant γ within the exposure range from the top to the bottom of the resist layer. Hence, D_{eff} physically means the equivalent resist thickness that the developer would open up if development were to take place at the development rate of $z = D$ for the duration required to clear the resist if the latter is illuminated with E_0.

So far, the second lumped parameter α has not been used explicitly. It is hidden in D_{eff}. Using

$$t_{\text{dev}} = \int_0^D \frac{dz}{r} = \frac{1}{r}\int_0^D e^{-\gamma i(z)} dz, \tag{4.114}$$

Hershel and Mack[41] derived

$$D_{\text{eff}} = \int_0^D \left[\frac{I(z)}{I(D)}\right]^{-\gamma} dz. \tag{4.115}$$

We perform the integration to expose the dependency of D_{eff} on α:

$$D_{\text{eff}} = \frac{1}{\alpha\gamma}\left(1 - e^{-\alpha\gamma D}\right). \tag{4.116}$$

Equations (4.109) and (4.116) are used together to predict the developed resist image from a given aerial image.

The accuracy of the lumped parameter model depends on the validity of the assumptions and the accuracy of the three parameters. In reality, the resist development rate can deviate from the γth power assumption. One should evaluate γ in the vicinity of the exposure dosage used to achieve the desired linewidth in order to minimize errors. The constant absorption assumption is generally acceptable. Unless the resist of interest is a contrast-enhancement resist that emphasizes its bleaching characteristics, typical photoresists have a small bleaching or dyeing component. The independent developments in the lateral and longitudinal directions do not seem to cause severe problems. In terms of the accuracy of the parameters, γ is the most important. D requires a resist thickness measurement that is not an issue. α can be taken from the A, B, C coefficients as

$$\alpha = \frac{1}{2}\left(\alpha_{\text{unexposed}} + \alpha_{\text{exposed}}\right) = B + \frac{A}{2}. \tag{4.117}$$

Figure 4.41 shows the impact of γ on the DOF of a 0.35-μm isolated line opening imaged at λ = 365 nm and σ = 0.3. The other two lumped parameters D and α are fixed at 0.9 μm and 0.2 μm^{-1}, respectively. The DOF is evaluated using the E-D tool methodology given in Section 5.3.2, setting the CD tolerance to ±10% and the exposure latitude (EL) to 10%. From Fig. 4.41, a relatively high γ of 5 reduces the DOF significantly from the aerial image prediction. The aerial image DOF potential is restored only when γ is increased to 18. Above 18, a higher γ cannot further improve the image. For i-line resists, γ rarely reaches 5. With chemically amplified resists, γ can be extremely high.

Figure 4.42 depicts the UVIII dissolution properties in CD-26, as reported by Conley et al.[41] With the additional supplemental lines between points 1 and 2,

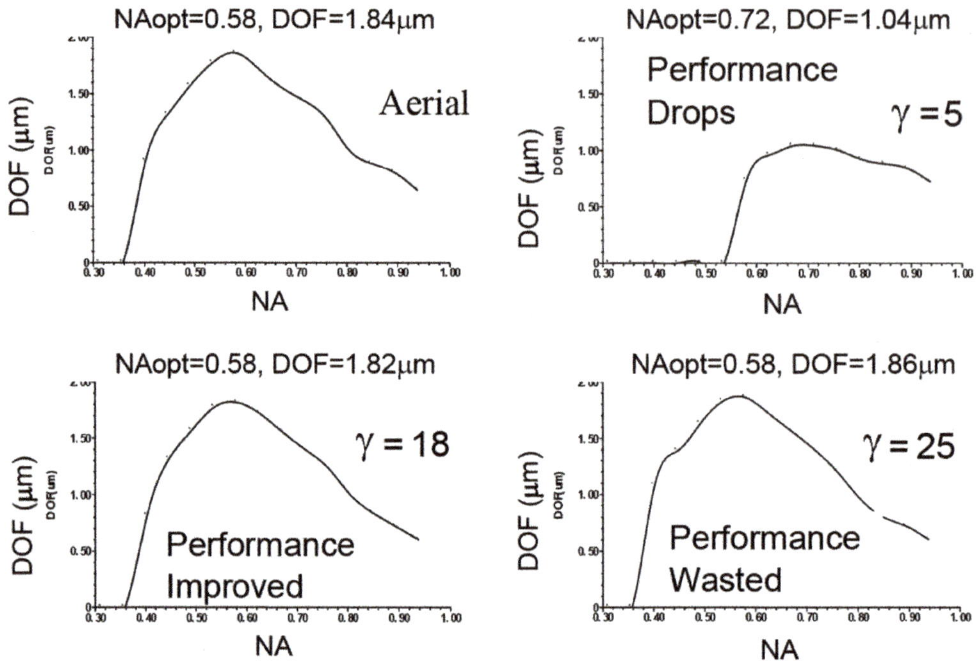

Figure 4.41 DOF versus NA at σ = 0.3, λ = 365 nm for a 0.35-μm isolated line opening, D = 0.9 μm, α = 0.2 μm^{-1}.

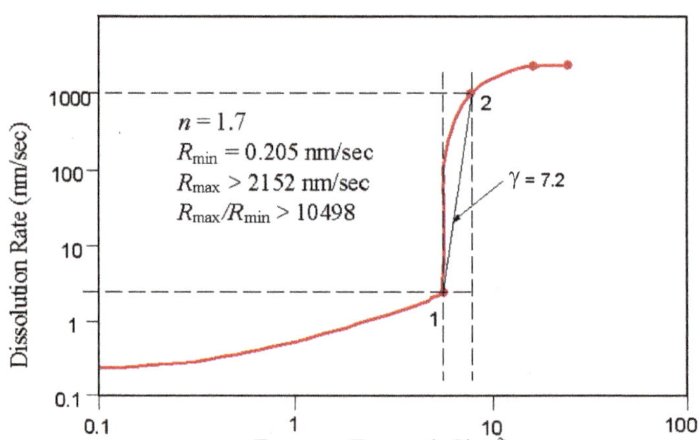

Figure 4.42 UVIII dissolution properties in CD-26 (reprinted from Ref. 42).

it is seen that the average γ between these two points is 7.2. If evaluated near point 2, the γ value is about 1, whereas if evaluated near point 1, γ is easily above 100. Katnani et al.[42] plotted γ explicitly for an e-beam chemically amplified resist. They demonstrated that γ is a function of the soft-bake temperature and time as well as the developer concentration and time, and that

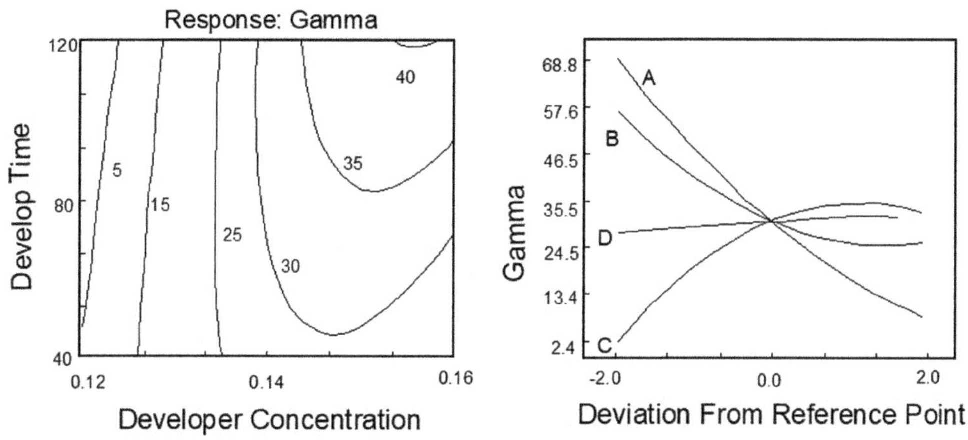

Figure 4.43 γ of an e-beam chemically amplified resist. In the right plot, curve A is the soft bake (90–130 °C), B is the bake time (2–6 min), C is the developer concentration (0.12–0.16 N), and D is the develop time: (40–120 sec) (adapted from Ref. 42).

the range of γ within their experimental parameters is from 2.39 to 68.6, as shown in Fig. 4.43.

The most useful way[43] to evaluate γ is by matching the simulated linewidth versus the exposure curve to the corresponding experimental curve. This type of γ is linewidth derived (LWD). The γ value that produces the smallest mean-square error between the two curves is chosen as the γ value of the resist at the given exposure and in the given development environment. Figure 4.44 shows the experimental curve and the simulated curves with a range of γ values. $\gamma = 4.45$ gives the best match. Figure 4.45 shows the E-D tree[ii] generated by simulation using LPM and the γ value from Fig. 4.44. The experimental data points are plotted in the E-D plane to check the accuracy of the simulation.

Figure 4.46 shows the effects of α on the DOF in the same situation as depicted in Fig. 4.41, but with γ fixed at 5. The impact of α on the DOF is less significant than in Fig. 4.41. However, the DOF seems to be improved by a larger α. This contradicts the knowledge that zero absorption produces a resist profile with less undercut, and that the resist image is closer to the aerial image.

Figure 4.47[44] depicts the difference between the aerial image and the developed resist image. Because the resist is simultaneously developed laterally and longitudinally, by the time the developer reaches the bottom, the developed edge at the bottom already clears the edge predicted by the aerial image. This additional bias penalizes the E-D window[iii] if the target CD is set to the mask dimension, which reduces DOF. When the resist absorption is high, light intensity attenuates more as it propagates toward the bottom.

[ii] E-D trees are discussed in Chapter 5.

[iii] E-D windows are discussed in Chapter 5.

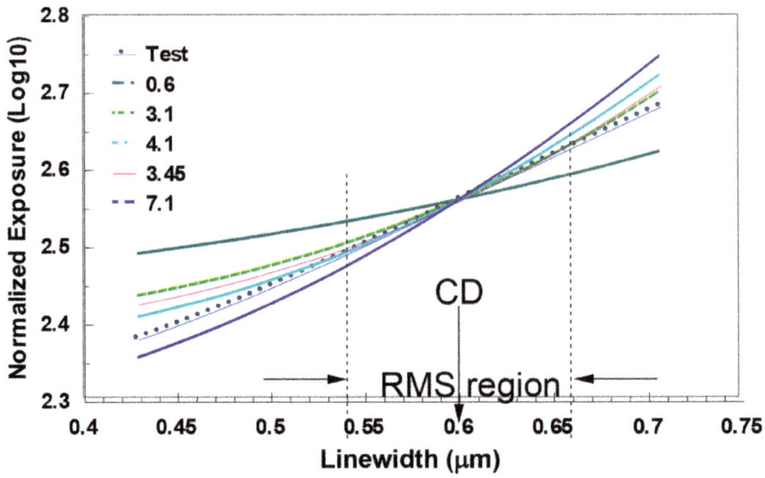

Figure 4.44 Evaluation of γ by matching experimental results.

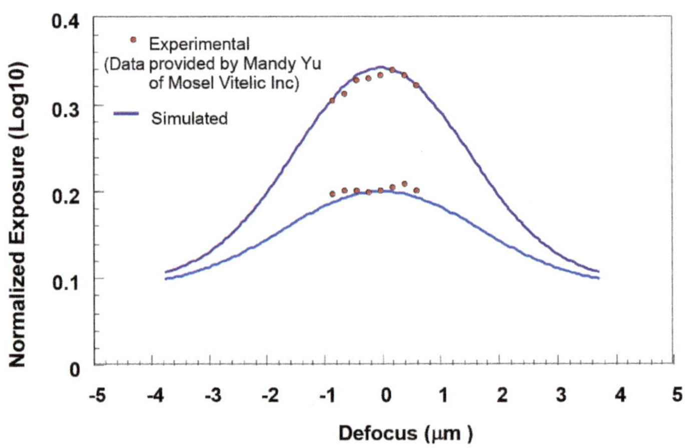

Figure 4.45 E-D tree simulated with γ derived from curve fitting compared with experimental data.

The light intensity at the bottom is now lower than before, making the resist profile more overcut, while the developed edge position at the top of the resist is unchanged because the exposure up there is unaffected by the increased resist absorption. Even though the profile is less appealing, if one strictly judges the resist image by the position of its edge at the bottom, the resist edge position is closer to that predicted by the aerial image than when the resist absorption is low and the DOF is larger. An additional advantage of high resist absorption is the reduction of multiple interference and standing waves. However, besides the undesirable overcut profile, the exposure time can increase manyfold, significantly lowering the wafer throughput.

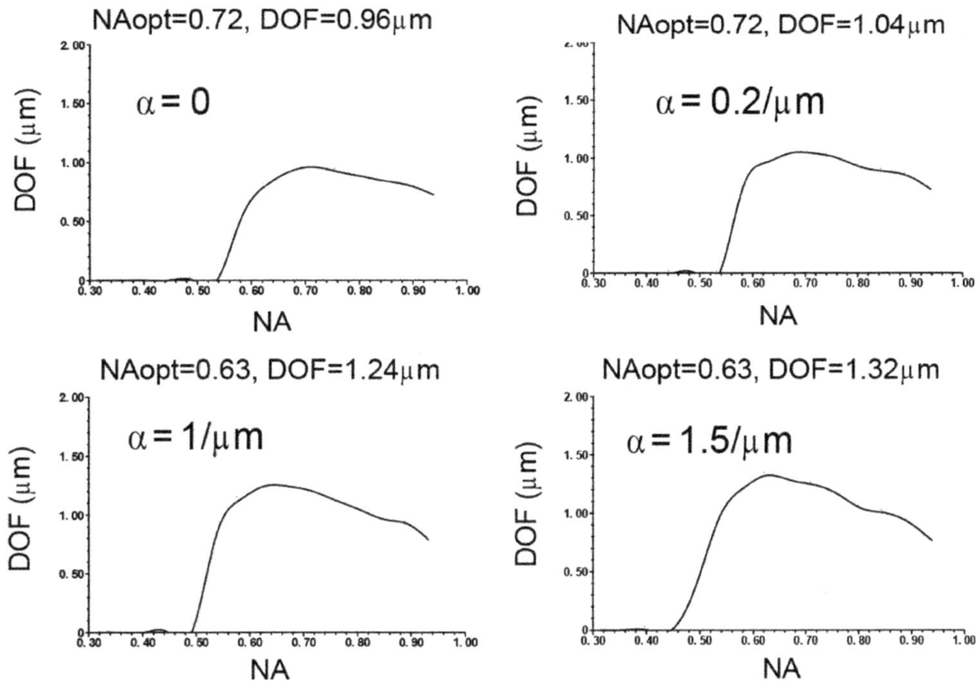

Figure 4.46 DOF versus NA at σ = 0.3, λ = 365 nm for a 0.35-μm isolated line opening, D = 0.9 μm, γ = 5.

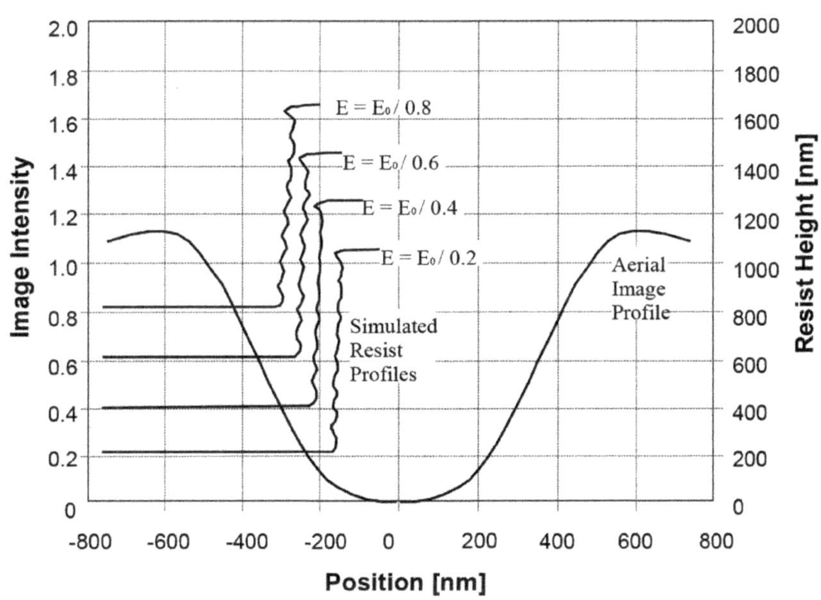

Figure 4.47 Comparison of aerial image profiles and resist profiles (reprinted from Ref. 44).

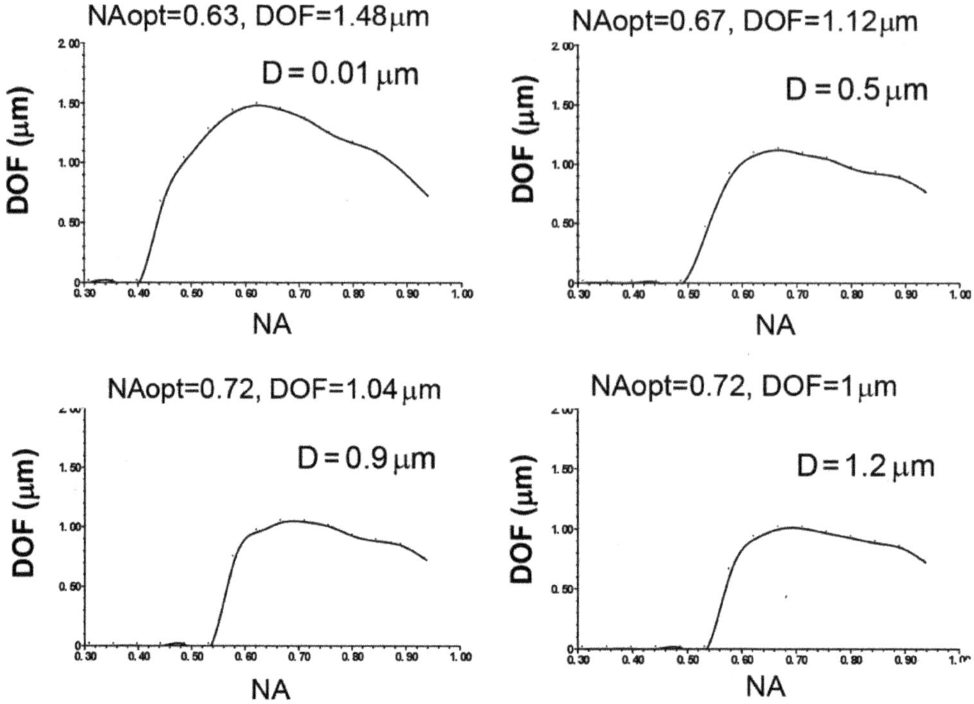

Figure 4.48 DOF versus NA at $\sigma = 0.3$, $\lambda = 365$ nm for a 0.35-μm isolated line opening, $\gamma = 5$, $\alpha = 0.2$ μm^{-1}.

Figure 4.48 shows the effects of D on the DOF in the same situation as depicted in Fig. 4.41, where α and γ are fixed at 0.2 μm^{-1} and 5, respectively. The resist thickness has a low significance in the context of LPM. The DOF is improved slightly only with unusable resist thicknesses in the vicinity of 0.01 μm. This does not mean that D is not important in imaging. The impact of D is to change the effective exposure in the resist induced by thickness changes. This translates to a change in the requirement of the exposure latitude. A larger exposure swing for two resist thicknesses to be delineated simultaneously requires a larger exposure latitude. Once the exposure swing is taken care of by specifying the correct amount of exposure latitude, the impact of D is similar to that of α.

When $D \to 0$, $\alpha \to 0$, and $\gamma > 25$, the resist image predicted with the LPM is practically identical to the aerial image. Figure 4.49 shows that the variation between two exposure tools is larger than the deviation of the results derived from the LPM.

We use a fully simulated resist image from PROLITH[iv] to check the effectiveness of LWD-γ.[45] Equal line–space pairs with $k_1 = 0.82$ and 0.9 are

[iv]PROLITH version 5.10. PROLITH is a registered trademark of Finle, Inc., a subdivision of KLA-Tencor.

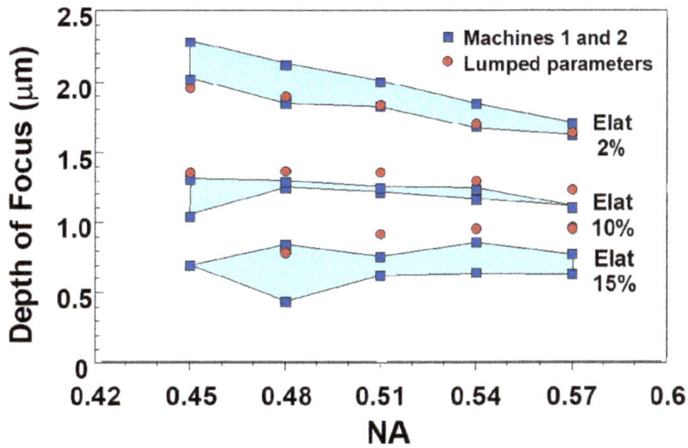

Figure 4.49 Lumped-parameter model and experimental results (Elat is the exposure latitude percentage).

exposed with 365-nm light at σ = 0.5 to an 800-nm-thick resist, which are typical resist and processing parameters on a planar silicon wafer. The required exposure dosage is plotted as a function of the developed resist-space width. This full simulation result is compared with an LPM-derived result using LWD-γ, as shown in Fig. 4.50. Very good matching is obtained.

4.4.3 β and η

The LPM works well for k_1 = 0.82 or greater. When matching is attempted for smaller k_1 values, the LPM fails. Figure 4.51 shows a comparison similar to that in Fig. 4.50, except that the feature size now ranges from k_1 = 0.362–0.411,

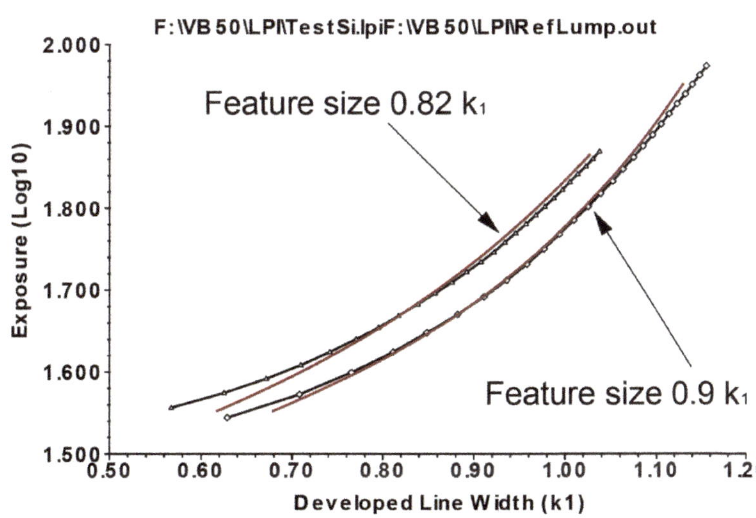

Figure 4.50 Linewidth-derived γ fits a lumped equation for full simulation.

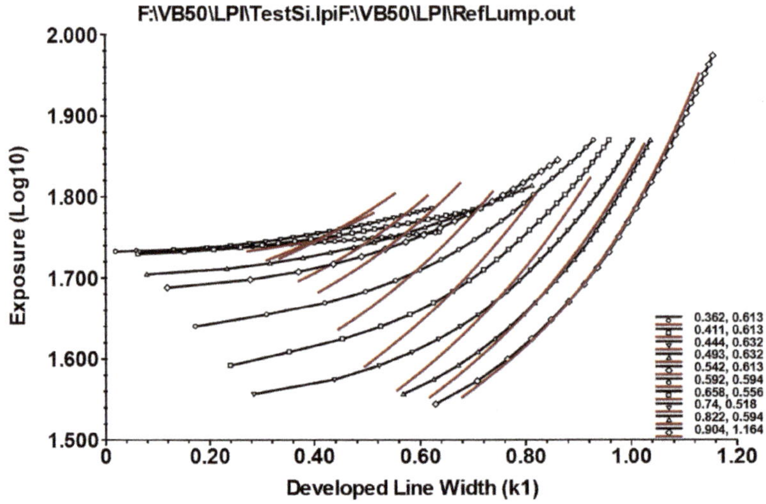

Figure 4.51 Lumped-parameter approximations applied to k_1 = 0.362 to 0.904.

0.444, 0.493, 0.542, 0.592, 0.658, 0.74, 0.822, and 0.904, with corresponding best-fitted LWD-γ values being 0.613, 0.613, 0.632, 0.632, 0.613, 0.594, 0.556, 0.518, 0.594, and 1.164. The poor-fitting LWD-γ at k_1 less than 0.8 is attributed to the Hershel–Mack lumped parameters ceasing to faithfully derive correct γ values.[41]

Instead of fitting the exposure versus the developed linewidth curve to that derived from the lumped parameters, one simply fits these curves to

$$E(w) = e^{\eta w}, \tag{4.118}$$

where E is the exposure dosage, w is the developed linewidth, and η is the new lithography performance indicator (LPI) called LWD-η.[46] From Eq. (4.118),

$$\eta = \frac{1}{w}\log_e E(w). \tag{4.119}$$

Hence, η is the log slope of exposure versus the developed linewidth; this log slope is different from the familiar log slope of intensity versus displacement in x. The accuracy of the η fitting is seen in Fig. 4.52. It is literally the log slope of each $E(w)$ curve at the nominal k_1 point. For example, the third curve from the right has a nominal k_1 value of 0.74. Even though the entire $E(w)$ curve covers a range of k_1 = 0.28 to 1.0, the log slope is the tangent at the position k_1 = 0.74. The nominal k_1 is the size of the feature to be printed, whereas the k_1 range in the $E(w)$ curve is the range of the developed linewidth.

Motivated by the need for a lumped parameter for all feature sizes, and encouraged by the excellent fitting result of Fig. 4.52, one can identify a new lumped parameter by performing a log–log fit of the $E(w)$ curve:

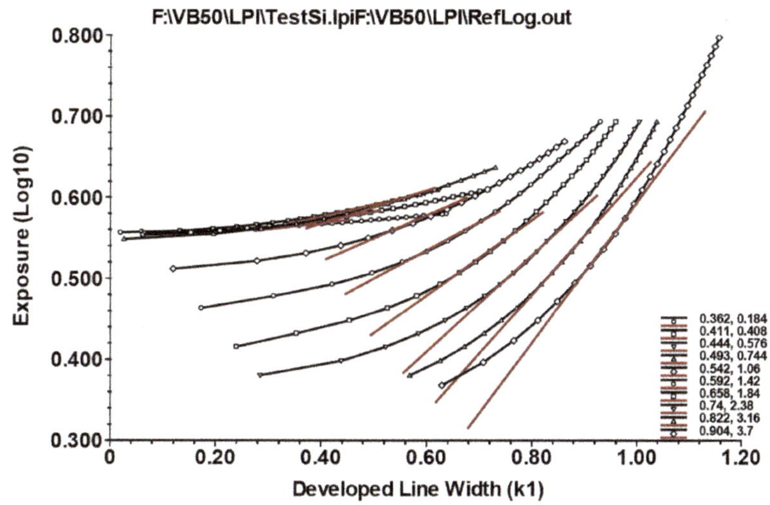

Figure 4.52 Fitting LWD-η from k_1 = 0.362 to 0.904.

$$\beta \equiv \frac{1}{w}\log_e[10\log_e E(w)], \tag{4.120}$$

which is taken from the analytic form of

$$E(w) = e^{0.1\exp(\beta w)}. \tag{4.121}$$

The fitting shown in Fig. 4.53 is much better than that of LWD-γ at small k_1 values. Hence, LWD-β can be used to derive a developed resist image from an aerial image in low-k_1 situations.

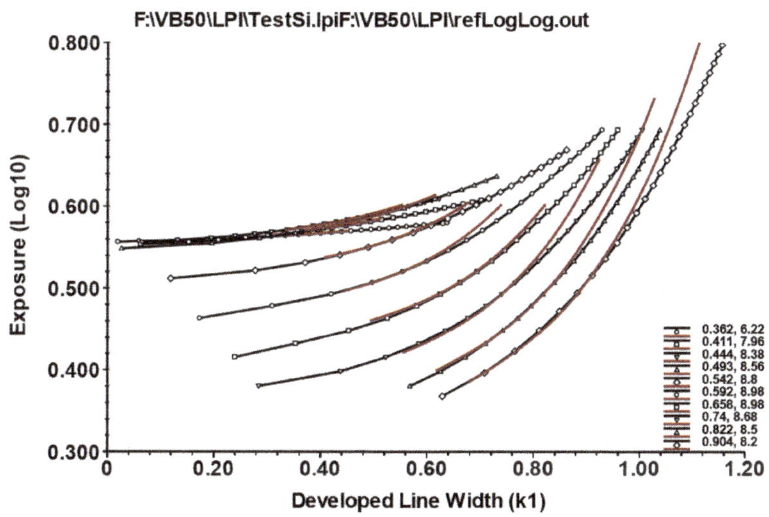

Figure 4.53 Fitting LWD-β from k_1 = 0.362 to 0.904.

While β is a better parameter with which to replace LPM in low-k_1 cases, η is useful as a lithography performance indicator, as shown in Ref. 45.

4.5 From Aerial Image to Resist Image

We now condense the discussions from Sections 4.1 to 4.4 into functional forms. We account for light radiating from the light source, going through the illumination system with properly optimized temporal and spatial coherence, then traveling through the mask, propagating through the many refracting and reflecting surfaces of a lens system, and inducing chemical interactions in the resist. Finally, we account for the resist latent image developing into a resist image. What follows are the parameters leading to an aerial image, refracted/reflected image, latent image, rate image, and finally a developed resist image.

- *Aerial image*—The aerial image is the image that is diffracted from the mask through the lens before the image has any interaction with the resist. In Eq. (4.122), the intensity of the aerial image I_a is a function of the feature size F_{size}, feature shape F_{shape}, and feature combinations F_{combi}; λ is wavelength; NA_i is the numerical aperture of the imaging lens; σ is the aperture ratio NA_c/NA_i between the condenser and the imaging lens; and the lens aberration coefficients are $a_1, \ldots a_n$:

$$I_a(x, y, z) = f_1(F_{shape}, F_{size}, F_{combi}, \lambda, NA_i, \sigma, a_1, \ldots a_n). \tag{4.122}$$

  This aerial image is assumed to be diffracted by an aberrated lens without consideration of the multiple reflections at the surfaces of the lens elements, or the refraction and reflections at the resist–wafer-substrate interfaces.

- *Refracted/reflected image*—The refracted/reflected image I_r in Eq. (4.123) includes these effects:

$$\begin{aligned} I_r(x, y, z, t) = f_2[&I_a, n_{cr}(x, y, z, I_r), n_{c1}, \ldots n_{cn}, \\ &S_1(x, y, z), \ldots, S_n(x, y, z)]. \end{aligned} \tag{4.123}$$

  where $n_c \equiv n_r + in_i$ is the complex index of refraction. The real part of the refractive index is n_r, which accounts for the changed wavelength in the media and phase shifting. The complex part of the refractive index is n_i, which accounts for light absorption in the film. The complex index of refraction of the resist layer is n_{cr}. The films above and below the resist layer are represented by n_{c1} through n_{cn}. S_1 through S_n are the lens surfaces. Notice that n_{cr} is not only a function of the space coordinates, but is also a function of I_r. This is the resist bleaching or dyeing phenomenon. The refractive index changes as a function of the incident light. This change proceeds to perturb the incident light in the resist, creating a feedback loop until bleaching or dyeing stops.

- *Latent image*—The exposure usually terminates before bleaching or dyeing saturates. This results in the expression of the latent image n_{cr} at the end of the exposure:

$$n_{cr}[x, y, z, I_r(t_{\text{end exposure}})]. \tag{4.124}$$

- *Dissolution rate image*—The resist dissolution rate can be directly linked to the localized exposure in the resist. This relationship is often plotted as the dissolution rate versus the exposure curve shown in Fig. 4.36. The latent image is also a direct indication of the localized exposure in the resist. Hence, the dissolution rate image is

$$\begin{aligned}\text{Rate}(x, y, z) = f_3(&I_r, d_t, \text{ resist variable, thermal history,}\\ &\text{resist development variables}).\end{aligned} \tag{4.125}$$

- *Resist image*—Given the dissolution rate distribution, Rate(x, y, z), the resist developer starts at the surface of the resist, carving out new surface shapes as the developer removes the resist differentially according to the rate distribution. The developer continues to adjust its advancement according to the new surface that is available for dissolution and the rate at the very point of dissolving. The resultant image after the developer is removed is the resist image, where THK is the thickness of the resist layer:

$$Z(x, y) = f_4[THK, \text{Rate}(x, y, z), t_{\text{dev}}]. \tag{4.126}$$

4.6 The Transferred Image

The resist image can be transferred to the wafer in four ways: etching, lift off, implantation, and electroplating. Etching can be further classified as isotropic or anisotropic. These transfer processes are discussed below.

4.6.1 Isotropic etching

With isotropic etching, the etchant removes the substrate under the resist etch mask at the same rate in all directions. As soon as the etchant penetrates the etch mask, the removal process goes laterally as well as vertically, as shown in Fig. 4.54. After the thin film is completely etched to the bottom, sideways etching continues, further enlarging the etched opening under the etch mask. Note that the profile of the etch image is always 45 deg because the simulation of the profile is strictly based on the static etch rate. During etch, the removal rate is a function of the static etch rate and the area exposed to the etchant. An

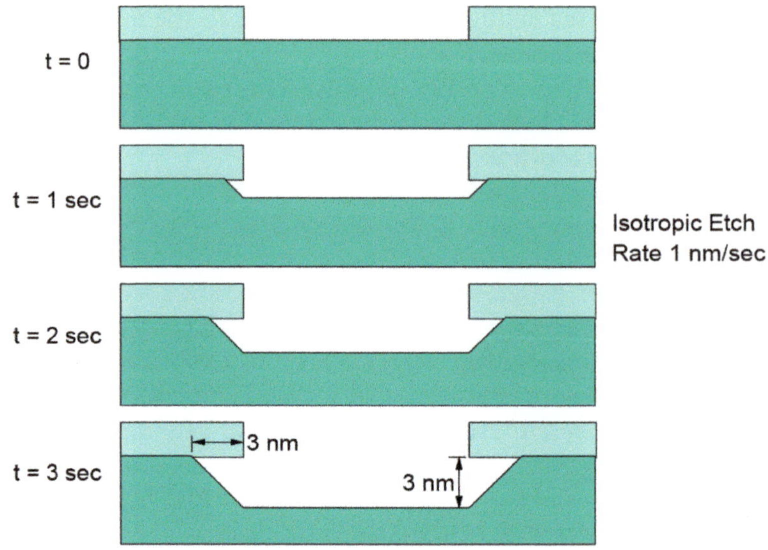

Figure 4.54 Isotropic etch sequence.

actual isotropic etch profile can have a spherical surface. The equation governing isotropic etching is

$$Z(x, y) = f_5(THK, \text{Rate}, \text{etch area}, t_{\text{etch}}), \tag{4.127}$$

where THK is the thickness of the substrate material to be etched. Instead of being a function of position, the dissolution rate is constant. Because of the removal of the sideways material, a small resist spot may be completely etched under and disappear. This is not always undesirable. However, small defects can be removed by such overetching. Wet etching is generally associated with isotropic etching, except for special etchants having preferential rates along the crystal lattices.[46] Certain types of dry etching such as plasma etching and ashing are isotropic. Isotropic etching is not suitable for high-density or high-aspect-ratio etched images. Anisotropic etching favoring the direction perpendicular to the surface of the substrate must be used.

4.6.2 Anisotropic etching

With anisotropic etching, material is removed faster in one direction than in the other direction. Generally, faster removal in the direction perpendicular to the substrate is preferred. Figure 4.55 shows the etch sequence of 5:1 anisotropic etching through a resist etch mask. The etched image is similar but not identical to that in Fig. 4.37. In the latter case, the resist protects the thin film underneath, while in the former case, the entire developing resist surface is gradually dissolved by the developer.

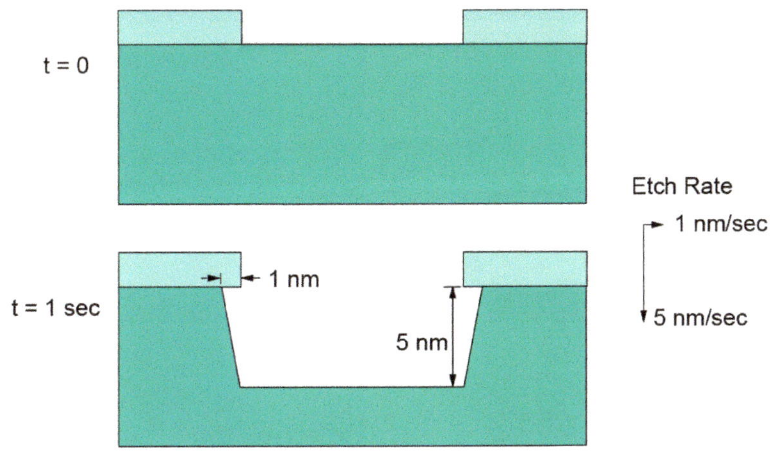

Figure 4.55 Anisotropic etching sequence.

In a way, surface removal in etching can be treated similarly to resist developing. In the case of isotropic etching, there is an abrupt layer of zero dissolution rate on top, and the dissolution rate in the bulk is uniform. With anisotropic etching, the dissolution rate distribution in the resist is still the same, but there is a directional preference in cell removal, where more cells are removed in one direction compared to the other direction. Similarly, when the etching surface is treated as a propagating wavefront, the wavefront moves faster in one direction than the other. The anisotropic etch equation is

$$Z(x,\ y) = f_5(THK,\ \mathrm{Rate}_{\mathrm{hor}},\ \mathrm{Rate}_{\mathrm{ver}},\ t_{\mathrm{etch}}), \tag{4.128}$$

where $\mathrm{Rate}_{\mathrm{hor}}$ is the rate of material deposition on the resist surface in the horizontal direction, and $\mathrm{Rate}_{\mathrm{ver}}$ is the rate in the vertical direction.

Anisotropic etching is used to achieve a high density and high aspect ratio. It is often necessary to produce a completely vertical profile. This cannot be achieved by just increasing the anisotropic etch-rate ratio. Passivation of the etched area helps to prevent sideways etching to produce the anisotropic effect.

4.6.3 Lift off

Lift off is an additive process that adds material to the resist openings, whereas etching is a subtractive process that removes material through resist openings. As shown in Fig. 4.56, the material to be added to the substrate surface is deposited indiscriminately on the resist image. The unwanted part over the resist-covered areas is removed with the resist by placing the substrate in a resist solvent. The resist image needs to have a vertical or undercut profile. Otherwise, the deposited material would cover the entire resist/substrate surface, preventing any solvent penetration. In addition, the

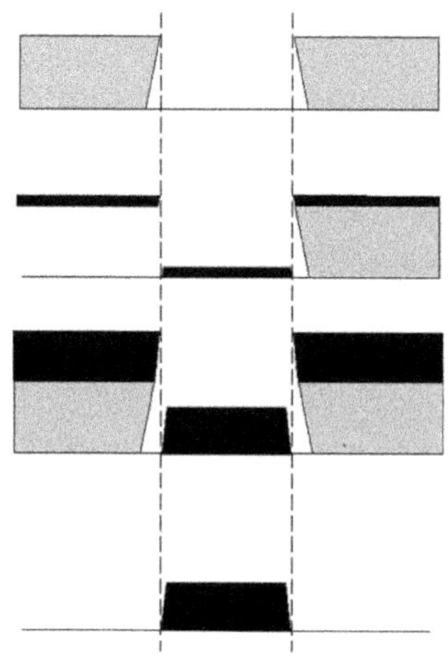

Figure 4.56 Lift-off pattern transfer.

deposition process must be directional (such as in evaporation) to promote the break between the deposited material on the top and bottom. Conformal deposition is not suitable for lift off.

Figure 4.57 shows an SEM micrograph of aluminum deposited over the resist image before lift off. Note that, as the deposited material builds up, it expands slightly in the lateral direction. The buildup affects the vertical profile because the opening that allows the deposited material to reach the substrate becomes smaller and smaller such that a vertical profile cannot be maintained. Figure 4.56 was drawn with this effect in mind. Figure 4.57 verifies the drawing.

The equation for lift off is like that for anisotropic etching, except that the profile of the lifted-off image is not dependent on its thickness. Naturally, the thickness of the resist must be larger than the thickness of the lifted-off

Figure 4.57 SEM micrograph of aluminum deposited over the resist image.

material, lest the opening close, thus preventing solvent penetration at the pattern edges. The lift-off equation is

$$Z(x,\ y) = f_5(\text{Rate}_{\text{hor}},\ \text{Rate}_{\text{ver}},\ t_{\text{deposition}}). \tag{4.129}$$

4.6.4 Ion implantation

In an ion implantation image transfer, the spatial distribution function $f(x,y,z)$ of implanted ions is given by[47]

$$f(x,\ y,\ z) = \frac{1}{(2\pi)^{3/2}\Delta R_p \Delta X \Delta Y} \exp\left\{-0.5\left[\frac{x^2}{\Delta X^2} + \frac{y^2}{\Delta Y^2} + \frac{(z - R_p)^2}{\Delta R_p^2}\right]\right\}, \tag{4.130}$$

where x and y are perpendicular to the incident z-direction of the ions. ΔX and ΔY are the standard deviations of the ion spreads in the x- and y-directions, respectively, and R_p is the so-called straggle in the incident direction, which is simply the standard deviation of ion penetration into the substrate. Equation (4.130) shows that, in addition to traveling in the incident direction, the ions have a transverse spread because of scattering in the resist and in the substrate. Therefore, at a resist edge, even though the ions are stopped in the incident direction by the resist at the resist side, the ions still penetrate under the resist. This happens because the ions that hit the substrate at the uncovered side of the edge are scattered sideways in both the covered and uncovered directions. Note that ion penetration is not perfectly isotropic; the ratio of transverse penetration to incident penetration is usually not unity. For example, Fig. 4.58 shows that the transverse-to-incident penetration ratio of phosphorus at 200 keV is about 1:2.

In a way, ion implantation is just ion-beam proximity printing. The resist image is a perfectly contacting mask that stops the ion beam. The beam

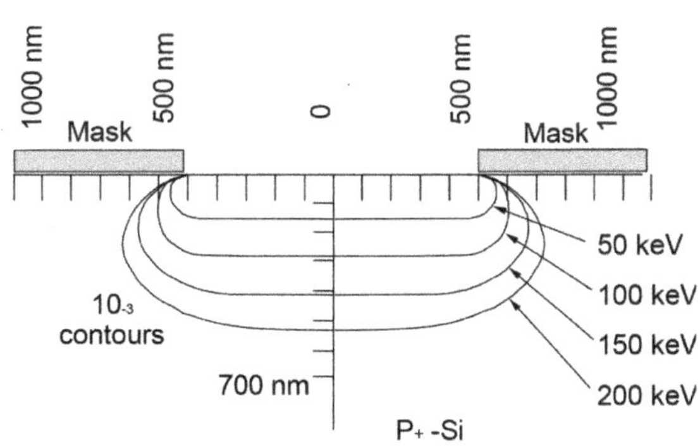

Figure 4.58 Equi-ion concentration curves at 0.1% intervals for phosphorus implantation into silicon at different implant energies.[48] A 1-μm mask cut is used. [Reprinted from Ref. 48 with permission; Copyright (1972) The Physical Society of Japan and The Japan Society of Applied Physics.]

that enters the unblocked substrate is scattered by the substrate material. This can be viewed as an ion-beam proximity effect, similar to the e-beam proximity effect, where the electrons are scattered by the material they penetrate. From this perspective, ion-beam lithography has become a high-volume production technology that has escaped our awareness. Of course, it can also be viewed as a pattern-transfer process, transferring the resist image to an implanted image.

4.6.5 Electroplating

With a plating base under an insulator pattern, metal may be plated on the plating base at the areas not covered by the insulator pattern. The insulator can be the resist image as well as the other insulator image transferred from the resist image. A schematic drawing of the electroplating process is shown in Fig. 4.59.

Like lift off, electroplating is an additive process. The metal image faithfully conforms to the insulator image as long as its thickness does not exceed the insulator thickness. Beyond that, because of the absence of constraints, the plated material grows sideways as well as vertically; it may be treated as an isotropic growth under most circumstances. The metal patterns may eventually be connected above the insulator if plating continues. Hence, the plated image is identical to the insulator image below the insulator surface. Above the insulator surface, a spherical wavelet can be integrated from the patterned areas to describe the 3D metallic surface.

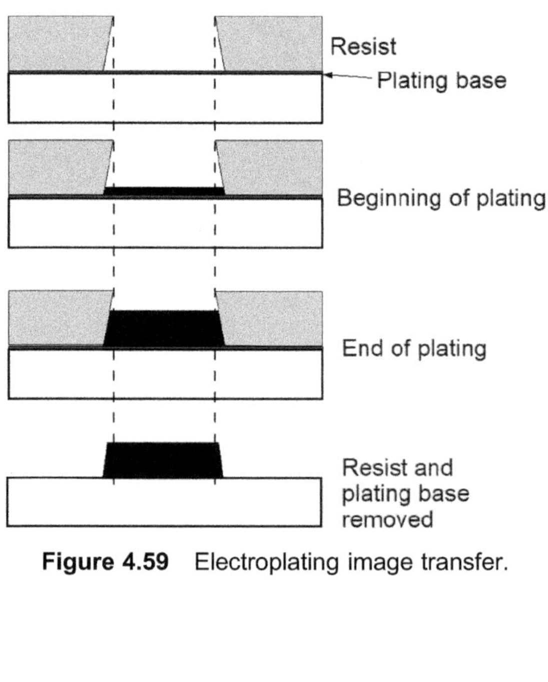

Figure 4.59 Electroplating image transfer.

References

1. C. J. R. Sheppard and H. J. Matthews, "Imaging in high-aperture optical systems," *J. Opt. Soc. Am. A* **4**(8), pp. 1354–1360 (1987).
2. M. Born and E. Wolf, *Principles of Optics, 6th Edition*, Cambridge Univ. Press, p. 435 (1980).
3. Eq. 17 of Section 5.2 in Ref. 2.
4. E. L. O'Neil, *Introduction to Statistical Optics*, Addison-Wesley, p. 50 (1963).
5. F. Zernike, *Physica I*, 689–704 (1934); English translation: "Diffraction theory of the knife-edge test and its improved form: the phase-contrast method," *J. Micro/Nanolith., MEMS, and MOEMS* **1**(2), pp. 87–94 (2002) [doi: 10.1117/1.1488608].
6. Section 9.2.2 of Ref. 2.
7. V. N. Mahajan, *Optical Imaging and Aberrations, Part II. Wave Diffraction Optics*, SPIE Press, p. 163 (2001) [doi: 10.1117/3.415727].
8. K. B. Doyle, V. L. Genberg, and G. J. Michels, *Integrated Optomechanical Analysis, 2nd Edition*, SPIE Press, Chapter 3 (2012) [doi: 10.1117/3.974624].
9. A. K. K. Wong, *Optical Imaging in Projection Microlithography*, SPIE Press (2005) [doi: 10.1117/3.612961].
10. V. Lakshminarayanan and L. S. Varadharajan *Special Functions for Optical Science and Engineering*, SPIE Press, Chapter 14 (2015) [doi: 10.1117/ 3.2207310].
11. Section 9.5 of Ref. 2.
12. J. W. Goodman, *Introduction to Fourier Optics*, McGraw-Hill, p. 108 (1968).
13. Sections 1.1.1 and 1.1.3 of Ref. 2.
14. M. P. Bachynski and G. Bekefi, "Study of optical diffraction images at microwave frequencies," *J. Opt. Soc. Am.* **47**(5), pp. 428–438 (1957).
15. B. J. Lin, "Electromagnetic near-field diffraction of a medium slit," *J. Opt. Soc. Am.* **62**(8), pp. 977–981 (1972).
16. B. J. Lin, "Optical Methods for Fine Line Lithography," Section 2.1 in *Fine Line Lithography*, R. Newman, Ed., North-Holland Publishing Co. (1980).
17. H. H. Hopkins, "On the diffraction theory of optical images," *Proc. Roy. Soc. A* **217**, p. 408 (1953).
18. M. S. Yeung, "Modeling high numerical aperture optical lithography," *Proc. SPIE* **922**, p. 149 (1988) [doi: 10.1117/12.968409].
19. J. W. Goodman, *Introduction to Fourier Optics*, McGraw-Hill, Section 3-7, p. 49 (1968).
20. A. Papoulis, *The Fourier Integral and Its Applications*, McGraw-Hill, Section 3-2, p. 45 (1962).

21. B. J. Lin, "Off-axis illumination—working principles and comparison with alternating phase-shifting masks," *Proc. SPIE* **1927**, p. 89 (1993) [doi: 10.1117/12.150417].
22. B. J. Lin, "Partially coherent imaging in two dimensions and the theoretical limits of projection printing in microfabrication," *IEEE Trans. Electron Devices* **ED-27**, p. 931 (1980). [Figures 4.30 to 4.32 were modernized by ShuoYen Chou and Minfeng Chen of TSMC.]
23. B. J. Lin, "Optical Methods for Fine Line Lithography," Section 2.3.2 in *Fine Line Lithography*, R. Newman, Ed., North-Holland Publishing Co. (1980).
24. W. G. Oldham, S. N. Nandgaonkar, A. R. Neureuther, and M. M. O'Toole, "A general simulator for VLSI lithography and etching processes: Part I - Application to projection lithography," *IEEE Trans. Electron. Dev.* **ED-26**, pp. 717–722 (1979).
25. K. K. H. Toh and A. R. Neureuther, "Three-dimensional simulation of optical lithography," *Proc. SPIE* **1463**, p. 356 (1991) [doi: 10.1117/12.44795].
26. C. A. Mack, "PROLITH: A comprehensive optical lithography model," *Proc. SPIE* **538**, p. 207 (1985) [doi: 10.1117/12.947767].
27. C. A. Mack and C. B. Juang, "Comparison of scalar and vector modeling of image formation in photoresist," *Proc. SPIE* **2440**, p. 381 (1995) [doi: 10.1117/12.209270].
28. Section 1.1.1 in Ref. 2.
29. M. S. Yeung and E. Barouch, "Three-dimensional nonplanar lithography simulation using a periodic fast multipole method," *Proc. SPIE* **2051**, p. 509 (1997) [doi: 10.1117/12.276030].
30. M. S. Yeung and E. Barouch, "Three-dimensional mask transmission simulation using a single integral equation method," *Proc. SPIE* **3334**, p. 704 (1998) [doi: 10.1117/12.310803].
31. E. Barouch, J. W. Cahn, U. Hollerbach, and S. A. Orszag, "Numerical simulations of submicron photolithographic processing," *J. Sci. Comput.* **6**(3), p. 229–250 (1991).
32. E. Barouch, B. Bradie, H. Fowler, and S. Babu, "Three-dimensional modeling of optical lithography for positive photoresists," *KTI Microelectronics Seminar Interface '89*, p. 123 (1989).
33. T. Hayashida, H. Fukuda, T. Tanaka, and N. Hasegawa, "A novel method for improving the defocus tolerance in step and repeat photolithography," *Proc. SPIE* **772**, p. 66 (1987) [doi: 10.1117/12.967035].
34. F. Coopmans and B. Roland, "DESIRE: a novel dry developed resist system," *Proc. SPIE* **631**, p. 34 (1986) [doi: 10.1117/12.963623].

35. R. R. Kunz, T. M. Bloomstein, D. E. Hardy, R. B. Goodman, D. K. Downs, and J. E. Curtin, "Outlook for 157 nm resist design," *J. Vac. Technol. B* **17**, pp. 3267–3272 (1999).
36. F. H. Dill, A. R. Neureuther, J. A. Tuttle, and E. J. Walker, "Modeling projection printing of positive photoresists," *IEEE Trans. Electron Dev.* **ED-22**, p. 456 (1975).
37. F. H. Dill, W. P. Hornberger, P. S. Hauge, and J. M. Shaw, "Characterization of positive photoresist," *IEEE Trans. Electron. Dev.* **ED-22**, p. 445 (1975).
38. C. A. Mack, A. Stephanakis, and R. Hershel, "Lumped parameter model of the photolithographic process," *Proc. Kodak Microelectronics Seminar 86*, pp. 228–238 (1986).
39. C. A. Mack, "Enhanced lumped parameter model for photolithography," *Proc. SPIE* **2197**, p. 501 (1994) [doi: 10.1117/12.175444].
40. T. A. Brunner and R. A. Ferguson, "Approximate models for resist processing effects," *Proc. SPIE* **2726**, p. 198–207 (1996) [doi: 10.1117/12.24906].
41. W. Conley, G. Breyta, B. Brunsvold, R. A. DePietro, D. C. Hofer, S. J. Holmes, H. Ito, R. Nunes, G. Fichtl, P. Hagerty, and J. W. Thackeray, "Lithographic performance of an environmentally stable chemically amplified photoreist (ESCAP)," *Proc. SPIE* **2724**, p. 34 (1996) [doi: 10.1117/12.24180].
42. A. D. Katnani, D. Schepis, R. W. Kwong, W. S. Huang, Z. C. H. Tan, and C. A. Sauer, "Process optimization of a positive-tone chemically amplified resist for 0.25-μm lithography using a vector scan electron-beam tool," *Proc. SPIE* **2438**, p. 99 (1995) [doi: 10.1117/12.210374].
43. B. J. Lin, "Signamization of resist images," *Proc. SPIE* **3051**, p. 620 (1997) [doi: 10.1117/12.276041].
44. T. A. Brunner and R. A. Ferguson, "Approximate models of resist development effects," *Proc. SPIE* **2726**, p. 198 (1996) [doi: 10.1117/12.240906].
45. B. J. Lin, "Lithography performance indicator (LPI) and a new lumped parameter to derive resist images from aerial images," *Proc. SPIE* **3677**, p. 408 (1999) [doi: 10.1117/12.350828].
46. M. P. Lepselter, "Beam lead technology," *Bell Sys. Tech. J.* **45**, p. 233–253 (1966).
47. S. K. Gandhi, *VLSI Fabrication Principles—Silicon and Gallium Arsenide*, John Wiley & Sons, p. 317 (1983).
48. S. Furukawa, H. Matsumura, and H. Ishiwara, "Theoretical considerations on lateral spread of implanted ions," *Jpn. J. Appl. Phys.* **11**(2), p. 134 (1972).

Chapter 5
The Metrics of Lithography: Exposure-Defocus (E-D) Tools

In microscopy, one is concerned with the closest distance between two objects that can be resolved in the image. The points of concern are quite different in lithography. A low-contrast image with a shallow intensity slope can be turned into a sharp edge with the photoresist. As mentioned in Section 1.3, the location of this edge is of prime interest and must be quantified against the operating parameters of the exposure tool during manufacturing because it determines the feature size that may be related to circuit switching speed, leakage, resistance, etc. In addition, the edge position also determines whether the image in a given layer can be effectively overlaid with previous or subsequent layers. The five primary parameters that can be adjusted in the field are exposure dosage, focal position, pattern alignment between layers, magnification, and rotation. The last three parameters are mostly related to overlay. Even though magnification can affect feature size in principle, its effect on overlay is much greater than on feature size. The first two parameters, exposure dosage and focal position, determine the process window of the lithographic system in a mutually dependent way. This mutual dependence and the ability to superimpose simultaneous requirements on different features are captured in exposure-defocus (E-D) tools, which are the backbone of lithography metrics.

5.1 The Resolution and DOF Scaling Equations

Section 3.2 introduced resolution and DOF scaling in projection printing using the proportional relationship to wavelength and inversely proportional relationship to the lens numerical aperture without introducing the proportional constant. Equations (4.12) and (4.16) used an arbitrary constant 0.5 to provide the physical meaning of these relationships. Here, we finally introduce the resolution and DOF scaling equation with rigorous proportional constants k_1 and k_3. The reason for the delay in presenting this equation is that the E-D tools discussed in this chapter can unambiguously determine these proportional constants.

The resolution of an optical projection imaging system can be described by the resolution scaling equation,[1]

$$W = k_1 \frac{\lambda_0}{NA}, \tag{5.1}$$

where W represents the minimum feature size or half pitch in the image to be delineated, λ_0 is the wavelength in vacuum, NA is the numerical aperture of the imaging lens, and k_1 is a proportional constant that varies from 0.25 to 1 (depending on many factors) and is discussed in the next section. Figure 5.1 depicts the relationship between the NA and the collection angle 2θ of the imaging lens: $NA = n \cdot \sin\theta$, where n is the refractive index of the medium between the object and the imaging lens.

Similarly, the scaling equation[1] for the DOF is

$$DOF = k_2 \frac{n\lambda_0}{NA^2}, \tag{5.2}$$

where k_2 is a proportional constant for the DOF. k_2 also depends on many factors,[2] including k_1.

As shown in Chapter 4, the relationships $W \propto \lambda_0/NA$ and $DOF \propto n\lambda_0/NA^2$ can be derived from diffraction theories[3,4] and are paraxial approximations. The large-angle resolution equation and DOF scaling equation[5] are

$$W = k_1 \frac{\lambda}{\sin\theta} \tag{5.3}$$

and

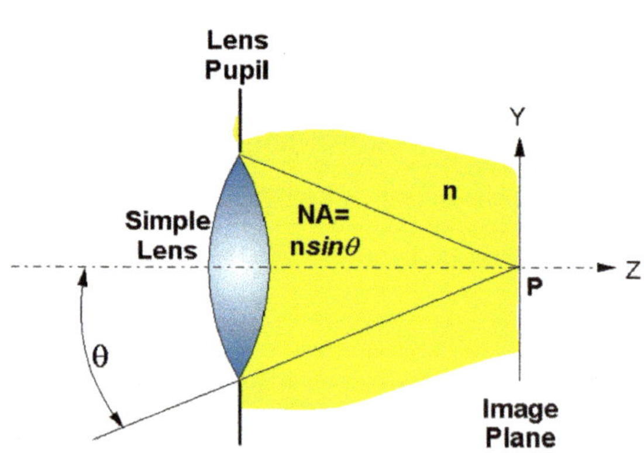

Figure 5.1 Definition of NA at the wafer side of the imaging lens, where *n* is the refractive index of the media between the mask and the imaging lens.

$$DOF = k_3 \frac{\lambda}{\sin^2(\theta/2)}, \tag{5.4}$$

respectively, where λ is the actinic wavelength. The relationships $W \propto \lambda/\sin\theta$ and $DOF \propto \lambda/\sin^2(\theta/2)$ hold for all illumination conditions. Obviously, k_1 must be adjusted for the degree of coherence in the illumination, and k_3 is a function of k_1 and many other parameters, just as in the paraxial case. Hence, k_1 is the coefficient for resolution; k_2 is the paraxial coefficient for DOF, and k_3 is the coefficient for DOF.

Several observations can be made from these equations:

1. Resolution is improved by increasing the $\sin\theta$ of the imaging lens. A higher $\sin\theta$ captures a larger range of spatial frequencies from the light diffracted from the features on the mask and thus can reproduce the original object with a higher resolution. Note that $\sin\theta$ and NA can be used interchangeably when $n = 1$, i.e., in non-immersion lithography.
2. The tradeoff for the higher resolution achieved with a higher $\sin\theta$ is a rapidly decreasing DOF. The different spatial frequency components quickly become out of phase with respect to each other as the image is defocused.
3. Resolution is improved by reducing the wavelength. A shorter wavelength reduces the spatial frequency in the diffracted light, making it easier for the imaging lens to capture more of it.
4. With wavelength reduction, the DOF is reduced at the same rate as the decrease in feature size. The loss of DOF is slower than that from the increase in $\sin\theta$ because the DOF is inversely proportional to the square of the latter.

Figure 5.2 depicts the light diffracted from a large object and a small object. The small object contains higher spatial frequencies that must be captured with an imaging lens of a larger NA. When the wavelength is reduced, the spatial bandwidth of both objects becomes narrower, making it possible for an imaging lens of smaller NA to resolve the objects. Figure 5.3 shows DOF versus resolution with λ and NA kept consecutively constant to illustrate the tradeoff in DOF for higher resolutions as well as the severity of the tradeoff with a wavelength reduction and NA increase.

5.2 Determination of k_1 and k_3 Based on Microscopy

It is desirable to establish the resolution limit and the DOF of a given imaging system. Hence, k_1 and k_3 must be evaluated. In the early stage of optical lithography, the definition of resolution was borrowed from microscopy.

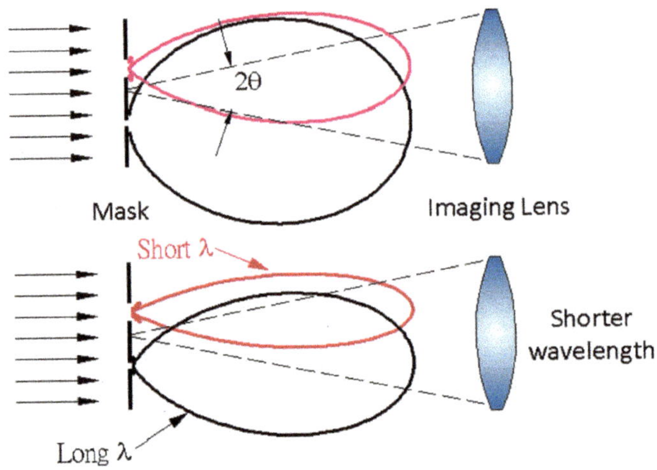

Figure 5.2 Diffracted light from two objects at two wavelengths, depicting the capture of spatial frequencies by the imaging lens.

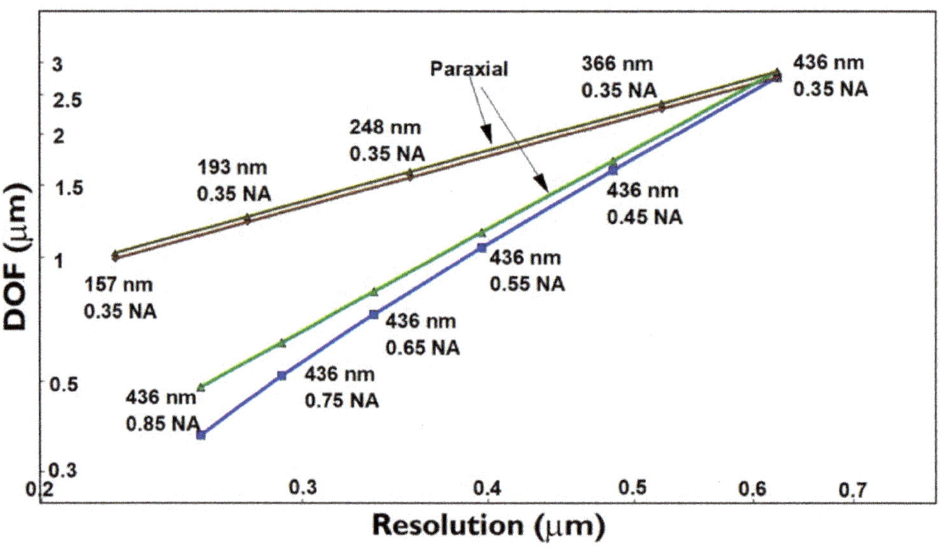

Figure 5.3 DOF versus resolution at a constant λ and a constant NA.

After all, a reduction projection imaging system can be viewed as an inverted microscope. Even with 1X projection or proximity printing systems, diffraction is the dominating phenomenon. In microscopy, two points are considered resolved if their images are separable.

With incoherent imaging, it is agreed that when the center of the Airy disk of a point source falls in the first zero of a second point source, the two points are resolved. Using a line source and a circular aperture, Goodman,[6] among many other authors, shows that the distance between the center and the first zero is 0.61λ/NA. This is the classical limit of resolution for incoherently illuminated objects.

For coherent illumination and partially coherent illumination, two points are considered resolved if the peak and valley intensities differ by an agreed-upon percentage[7] such as 26.5%, which is the same amount as when they are separated by 0.61λ/NA with incoherent illumination. Applying this criterion, the classical limit of resolution ranges from 0.5 to 0.8 λ/NA.

One can set the same criterion based on intensity difference to determine whether two points are resolved when defocus takes place. Again, besides the certainty of its dependence on $\lambda/\sin^2(\theta/2)$, the coefficient for DOF (k_3) is a strong function of the illumination, the normalized feature size k_1, the feature shape, the feature proximity, and the mask type, as well as the sensitivity and dynamic range of the detector. The difficulties of this classical derivation are in the arbitrariness and lack of generalization. In terms of arbitrariness, it may be coincidental that an average human eye resolves the incoherence distance because its sensitivity and dynamic range are suitable for detecting the 26.5% difference. The center of one source falling in the first zero with a second source is mathematically clean and physically illuminating, but there is no absolute reason that the resolution needs to stop there. With modern high-precision instrumentation, those classical limits can easily be exceeded.

In terms of lack of generalization, it is clear that k_1 and k_3 are dependent on many variables. In a circuit mask, there are many feature shapes, sizes, and proximity conditions that inevitably lead to different k_1 values. The best that one can do to determine the fabrication limit is to identify the case with the smallest k_1. However, as will be clear from Section 5.5.6.3, being able to resolve the most difficult part of the circuit pattern does not guarantee successful delineation of the easier part of the circuit. Each part of the circuit very often calls for different optimized exposure levels. When the exposure levels do not overlap, just resolving the most difficult parts still leaves some easier parts over- or underexposed.

5.3 Determination of k_1, k_2, and k_3 Based on Lithography

The lithographer's concern with resolution is not so much to resolve two adjacent lines, but to control the line edges to the specification of the product, as was described in Chapter 1. Namely, the edge placement must be within a window of tolerance, typically taken as ±5% of the critical dimension (CD) of the circuit.

5.3.1 E-D branches, trees, and regions

The edge placement criterion necessitates the knowledge of the ±10% CD boundary in the E-D space. Figure 5.4 shows these boundaries for one edge of a 0.26-μm isolated line opening using 248-nm light at $NA = 0.59$ and $\sigma = 0.5$.

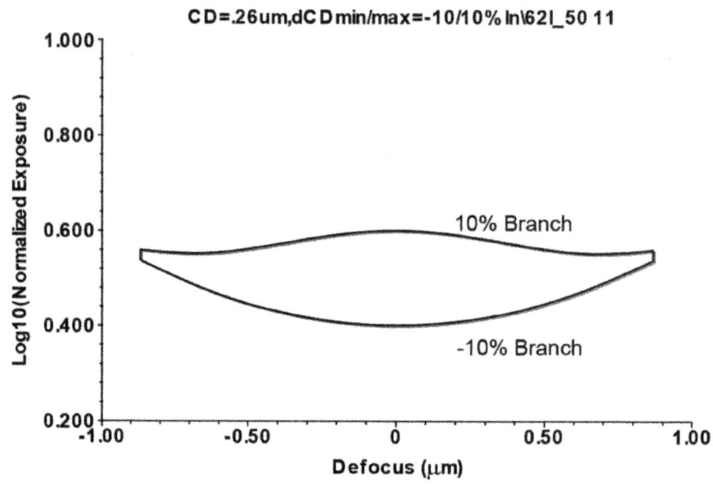

Figure 5.4 Constant-edge-position contours of a 260-nm line opening, 248-nm exposure, *NA* = 0.59, σ = 0.5.

These constant edge position contours are E-D branches. The +10% CD contours and the –10% CD contours form an E-D tree.[8] Any operating point in the E-D region bound by these two E-D branches will produce an image within the edge control budget. However, the branches for different object shapes or sizes are different. Hence, if there are three critical edges in the circuit, three E-D trees must be constructed. Only the common region enclosed by these branches is acceptable for the combination of features. Figure 5.5 shows the superposition of the E-D tree of one edge of the line opening above—that of an opaque space of the same size, and that of equal line–space pairs at a pitch of 0.52 μm. The region overlapped by all three trees is the common E-D region of this feature-shape combination. An operating point in the common region produces the image that meets the edge position requirement of all three edges of concern. The common E-D region is of

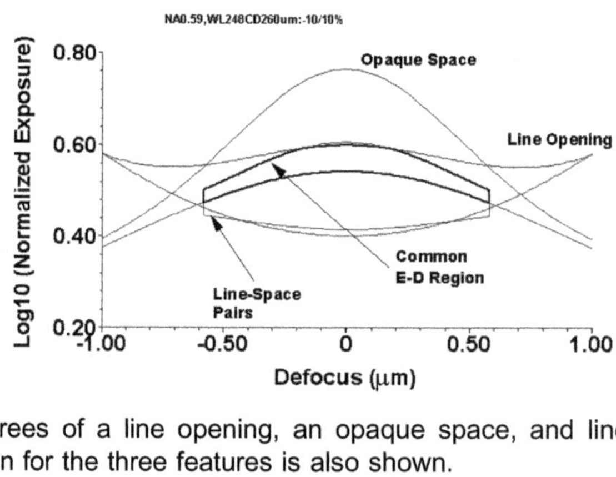

Figure 5.5 E-D trees of a line opening, an opaque space, and line–space pairs. The common E-D region for the three features is also shown.

concern not only for feature-shape combinations. Size combinations, mixtures of CD tolerances, non-uniformity of resist baking temperatures, bake time variations, development conditions, object at different lens field positions, and different fields, wafers, and wafer batches should also be subjected to the common E-D region characterization.

Strictly speaking, there are two E-D trees per feature because there are two edges in each feature. The common E-D region should be constructed from the superposition of six E-D trees instead of just three, as shown in Fig. 5.5. However, the lens used to image the above features is assumed to be aberration free. Therefore, the two edges of each feature move symmetrically in or out as a function of exposure and defocus, making the two E-D trees of each feature identical. The E-D tree of either edge can be used to represent the tree of the other edge and to represent the E-D tree plotted for a constant-line-width contour, in addition to a constant-edge-position contour. However, when there is asymmetry in the movement of the two edges, both edge trees must be used, and their common tree must be evaluated to assess the processing window.

Figure 5.6 depicts the same object as in Figure 5.4, but its edges are displaced by 0.01 μm to simulate lens distortion. The E-D trees of the edges are no longer identical to each other in terms of their relative position with respect to the ideal edge positions, resulting in two separate trees whose common region is much smaller than before. This situation is depicted in Fig. 5.7, where edges 1 and 2 are no longer at the center of the edge placement window. Edge 1 has more room for overexposure, and edge 2 has less room.

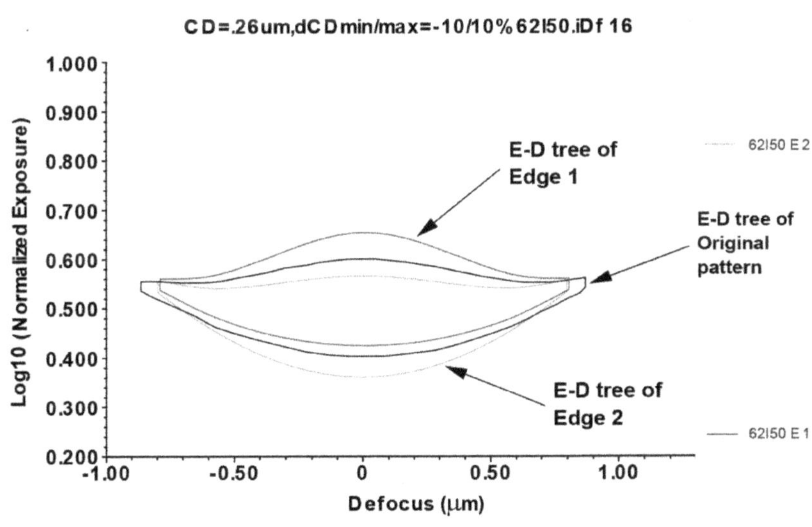

Figure 5.6 E-D trees of edge 1, showing the same line opening as in Fig. 5.4, and of edge 2, where the edges are displaced by 0.01 μm, showing that two trees must be used when the edges of a feature move asymmetrically.

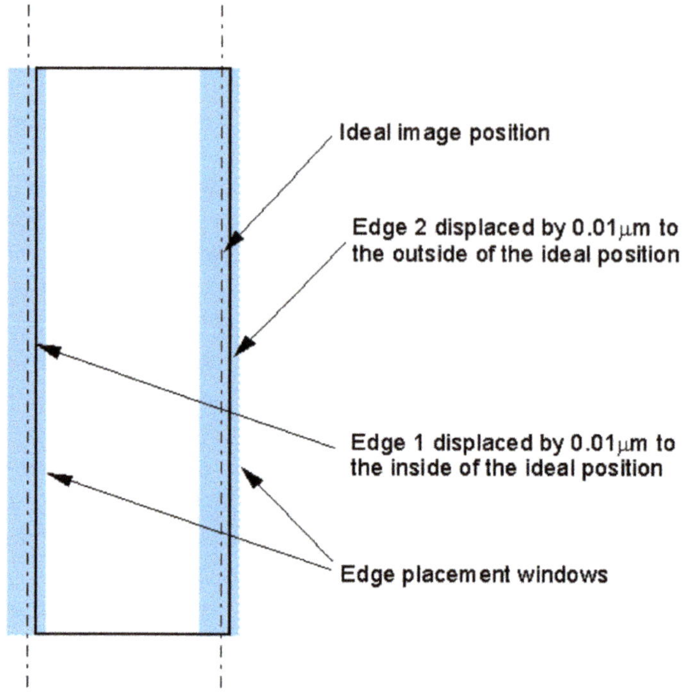

Figure 5.7 Image used in Fig. 5.6 with the asymmetric movement.

The consequence is reversed for underexposure. Hence, the E-D tree for edge 1 is higher than the line-width-based tree, and that for edge 2 is lower than the line-width-based tree.

5.3.2 E-D window, DOF, and exposure latitude

A rectangle can be fitted in the E-D region of Fig. 5.4, as shown in Fig. 5.8. The height of this rectangle is the exposure latitude (EL) of the given E-D tree, and the width is its DOF. For this rectangle, the EL is 36.7% and the DOF is 0.85 μm. If these parameters are maintained, the resultant edge placement meets the desired specification. The parts of the E-D region outside the rectangle are not as useful and cannot be counted towards the total EL or DOF. For example, point A (–0.6, 0.5) at log10 exposure = 0.5 and defocus = –0.6 μm cannot support an EL of 36.7%. Point B (–0.6, 0.46) is within the 36.7% EL but is outside the usable E-D region; therefore, it violates the edge placement specification.

One can fit another rectangle that has a width (i.e., DOF) of 1.2 μm. However, the EL is reduced to 20.2%. Consequently, there is a tradeoff between EL and DOF. Figure 5.9 depicts such a relationship. The EL is not the only parameter that changes with DOF; the E-D area, center of focus, and center of exposure change according to the size, shape, and location of the new rectangle.

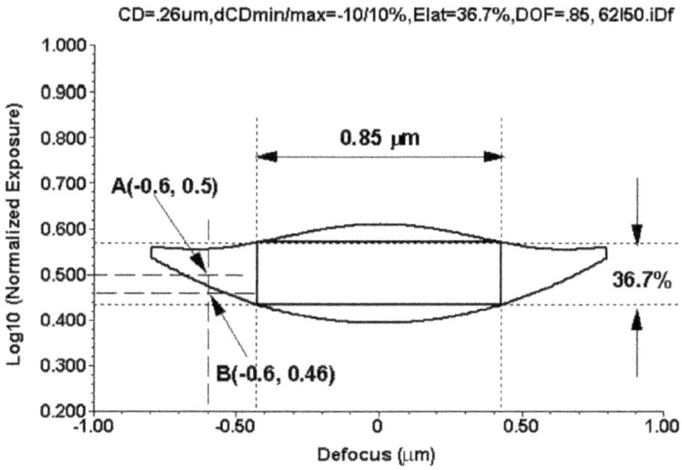

Figure 5.8 An E-D window constructed in the E-D tree of Fig. 5.4.

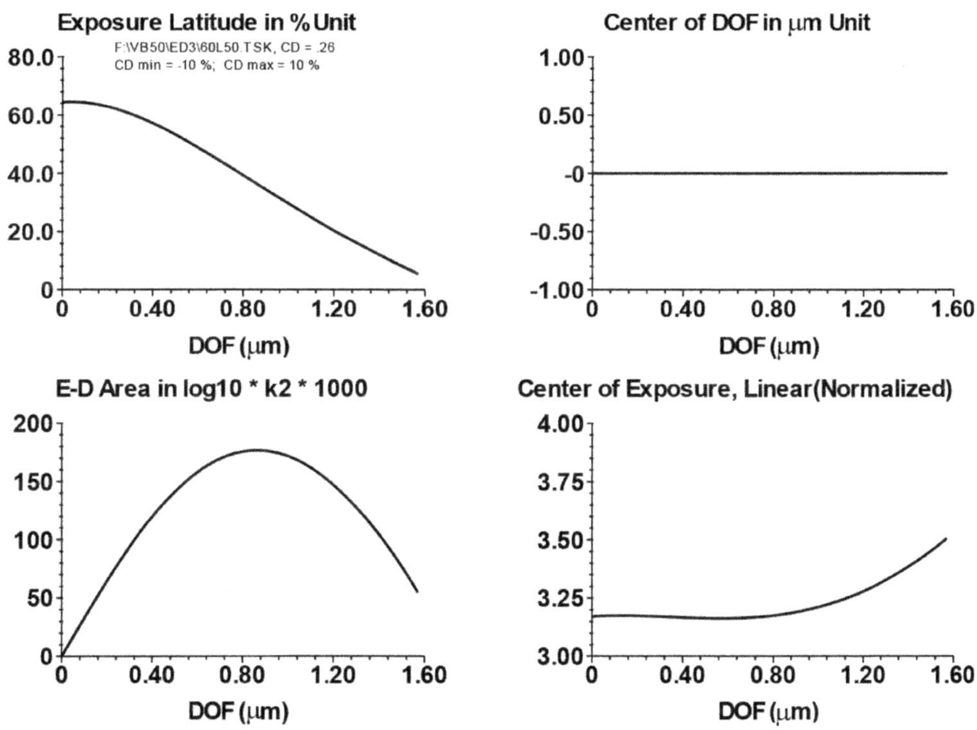

Figure 5.9 EL versus DOF, E-D area versus DOF, center of focus versus DOF, and center of exposure versus DOF with the parameters of Fig. 5.4.

5.3.3 Determination of k_1, k_2, and k_3 using E-D windows

Evaluation of k_1 and k_2 using the E-D methodology starts with constructing the E-D tree of each component, namely, each feature size, feature shape, feature orientation, proximity environment, field position, and processing condition. The E-D trees are superimposed to evaluate the common E-D region. A rectangle is fitted to the E-D region. Depending on the lithography environment, the rectangle may be of a fixed height to restrict the exposure tolerance or of a fixed width to restrict the DOF. When neither needs to be restricted, one may choose to use the rectangle of the largest E-D area for the largest possible processing window.

The situation leading to Fig. 5.5 is used as an example here. Rectangles are fitted to each individual tree as well as to the common tree, as shown in Fig. 5.10. Here, 10% EL is imposed; therefore, all rectangles are of the same height. Table 5.1 shows the dimension of these rectangles, with the line opening producing the largest DOF, followed by the line–space pairs, then opaque space. The exposure level required to achieve the largest DOF is different for each feature. The exposure level of the common tree and of the opaque space are similar because the opaque space is the limiting feature in this combination.

When the E-D windows are optimized for the E-D area, Fig. 5.11 results. The E-D values of the new E-D windows are shown in Table 5.2. Now, the ELs of the three features are no longer identical to each other and are much larger than in the previous case, at the expense of DOF. It is only coincidental that the EL of the common E-D window is 10%.

Using the first case, the imaging system can resolve $k_1 = 0.26/0.248 \times 0.59 = 0.62$ with $k_2 = 0.45/0.248 \times 0.59^2 = 0.63$ and $k_3 = 0.45/0.248 \times 0.31^2 = 0.17$. However, one must qualify the set of coefficients by identifying the feature combination, illumination, and EL. Because of the coincidence in the common E-D window, k_1, k_2, and k_3 for the optimized E-D area are also 0.62, 0.63, and 0.17, respectively. In general, the set can be different.

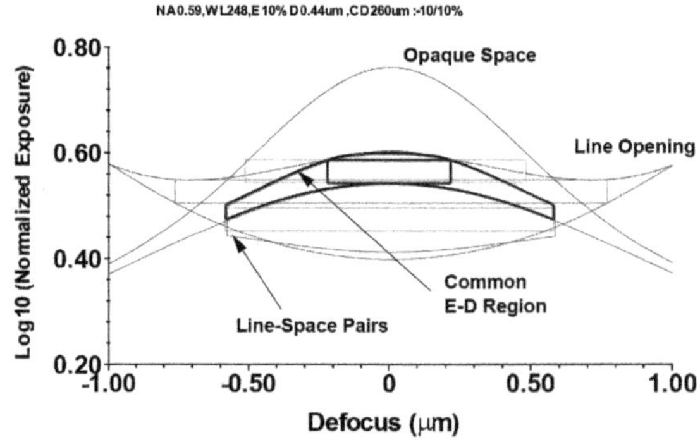

Figure 5.10 Individual and common windows fitted to the E-D tree in Fig. 5.5. All trees have 10% EL.

Table 5.1 Dimensions of the common and individual E-D windows in Fig. 5.10.

Feature	EL (%)	DOF (μm)	E-D area [EL(log10)·k_2·1000]
Common window	10	0.44	26.8
Line opening	10	1.53	93.8
Opaque space	10	1.00	61.2
Line–space pairs	10	1.16	71.4

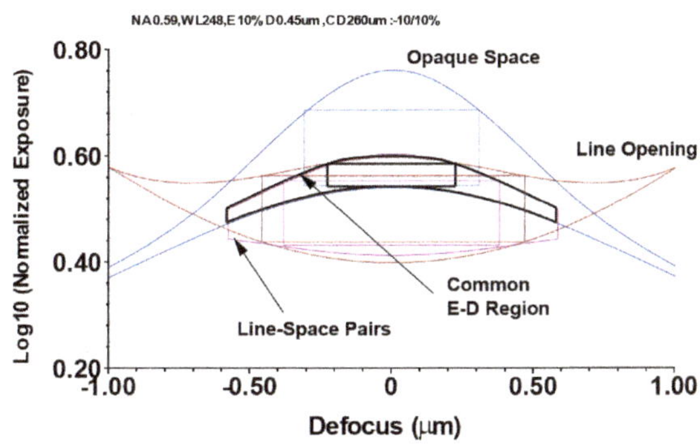

Figure 5.11 Same as Fig. 5.10, except that the optimized E-D area is the criterion for plotting the E-D windows.

Table 5.2 Dimensions of the common and individual E-D windows in Fig. 5.11.

Feature	EL (%)	DOF (μm)	E-D area [EL(log10)·k_2·1000]
Common window	9.73	0.45	26.8
Line opening	28.6	0.93	162.5
Opaque space	32.5	0.62	123.6
Line–space pairs	27.9	0.76	130.0

In these two cases, the resolution limit is not exceeded at $k_1 = 0.62$. With the same set of features, the resolution limit may readily be exceeded if any of the qualifications are changed. For example, the result for 15% EL is shown in Fig. 5.12. There is no common E-D window because the height of an E-D window with 15% EL is larger than the vertical distance between the two E-D branches of the common E-D tree. The E-D values are shown in Table 5.3.

5.4 k_1, k_2, and k_3 as Normalized Lateral and Longitudinal Units of Dimension

We have been using k_1, k_2, and k_3 as the coefficients for resolution and DOF, as in Eqs. (5.1) to (5.4). They can also be defined as normalized lateral and longitudinal dimensions:[2]

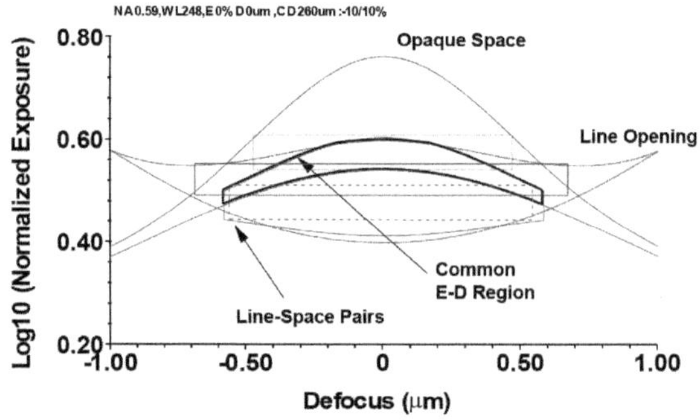

Figure 5.12 Same as Fig. 5.10, except that the EL is increased to 15%. There is no longer a common window for *EL* = 15%.

Table 5.3 Dimensions of the common and individual E-D windows in Fig. 5.12.

Feature	EL (%)	DOF (μm)	E-D area [EL(log10)·k_2·1000]
Common window	15	0	0
Line opening	15	1.37	125.3
Opaque space	15	0.94	85.8
Line–space pairs	15	1.11	101.7

$$k_1 = W\left(\frac{\sin\theta}{\lambda}\right) = W\left(\frac{NA}{\lambda_0}\right), \tag{5.5}$$

$$k_2 = Z\frac{\sin^2\theta}{\lambda}, \tag{5.6}$$

and

$$k_3 = Z\left(\frac{\sin^2(\theta/2)}{\lambda}\right). \tag{5.7}$$

Examples of the lateral dimension include resolution, linewidth, line length, edge position, distance between features, and any distance or location on the mask or wafer plane. Longitudinal dimensions include DOF, defocus, and film thickness. It is preferred to compare systems with the normalized dimensions. For example, a resolution of 0.26 μm at 248 nm and 0.59 NA is denoted by a resolution of 0.62 λ_0/NA, or simply 0.62 k_1 units. This is the same level of resolution as 0.38 μm at 365 nm and 0.59 NA or 0.34 μm at 248 nm and 0.45 NA. All are at $k_1 = 0.62$. A lower k_1 implies that a given set of imaging equipment can be made useful for future generations. Procurement of new equipment can be stretched out for cost savings.

5.5 The E-D Tools

The E-D tools consist of E-D branches, trees, forests, and windows. All of these tools, except E-D forests, have been covered in Sections 5.3.1 and 5.3.2. The E-D forest is simply a collection of E-D trees that can be too numerous for the human mind to handle. In this section, the method for constructing E-D trees is explained. Examples of E-D tool applications are also given.

5.5.1 Construction of E-D trees

E-D trees can be constructed from three types of image data, namely, linewidth as a function of exposure and defocus, edge position as a function of exposure and defocus, and image intensity distribution at a series of defocus planes.

5.5.1.1 E-D tree construction from E-D matrix linewidth data

An E-D tree is constructed from the linewidth as a function of exposure and defocus by evaluating the constant linewidth contours in the E-D space. Figure 5.13 shows the linewidth in E-D matrix data with the data points marked at the E-D locations. The constant linewidth contours are evaluated using a least-mean-square polynomial fitting. Fitting the data is preferrable to interpolating it directly unless the data is from simulation and there is absolutely no possibility of noise in the data. Most experimental data can benefit from least-mean-square fitting. Figure 5.14 shows the same data but

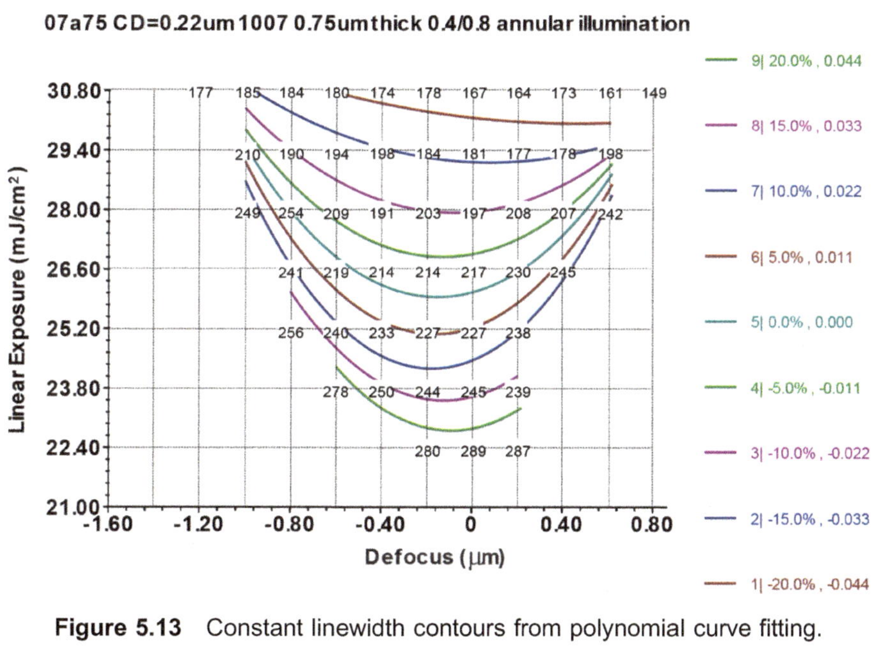

Figure 5.13 Constant linewidth contours from polynomial curve fitting.

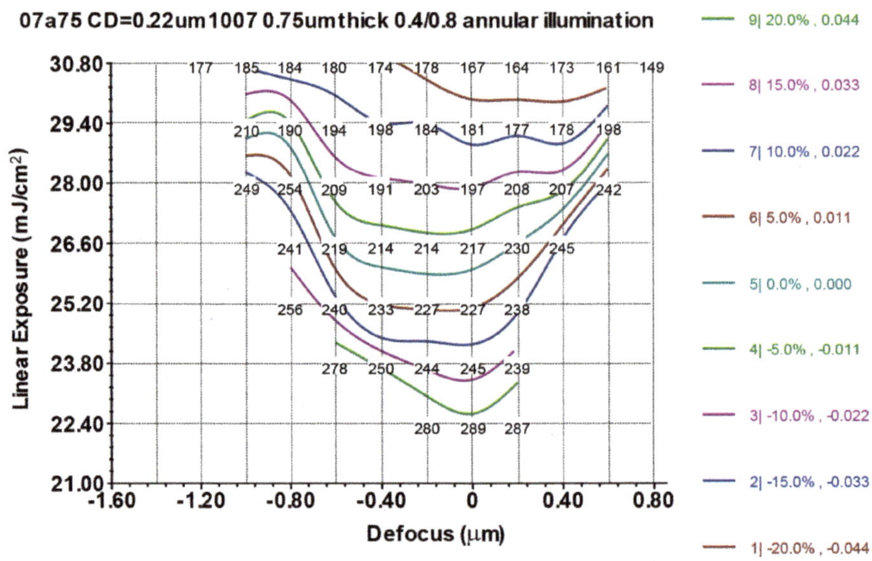

Figure 5.14 Same as Fig. 5.13, except that spline curve fitting is used.

with spline-fitted contours. Spline fitting forces the contours to pass exactly through all data points. Experimentally induced noise is preserved.

Now we discuss the numerical details involved in constructing Fig. 5.13. Take the –5% contour whose linewidth is the nominal CD (0.22 μm – 0.011 μm = 0.209 μm), marked as –5%, –0.011 on the chart. The data point is first linearly interpolated for exposure at each defocus point. For example, at z = –0.4 μm, 26.6 mJ/cm^2 produces a 214-nm feature, and 28 mJ/cm^2 produces a 191-nm feature. The CD interval can be made small in data collection so that linear interpolation can be used to evaluate the exposure dosage at the defocus points. In this example, the exposure dosage to produce 0.209 μm is 26.9 mJ/cm^2. After the linearly interpolated dosages at all defocus points are evaluated for the –5% contour, a polynomial fit is performed. The 26.9-mJ/cm^2 point becomes 27.16 mJ/cm^2 after polynomial fitting, as shown in the contour.

Following the argument of Section 5.5.2, the interpolation and polynomial fitting should be done using log exposure.

5.5.1.2 E-D tree construction from E-D matrix edge data

If edge position data as a function of the E-D coordinates is available, it is more desirable to construct edge-based E-D tree from the data. The procedure is like using linewidth E-D matrix data, except that, when curve fitting, one fits constant-edge-position contours to the data, as shown in Fig. 5.15. Fitting can be by least-mean-square polynomial fit or spline. In Fig. 5.15 only half of the data points are shown because the data is symmetrical in defocus.

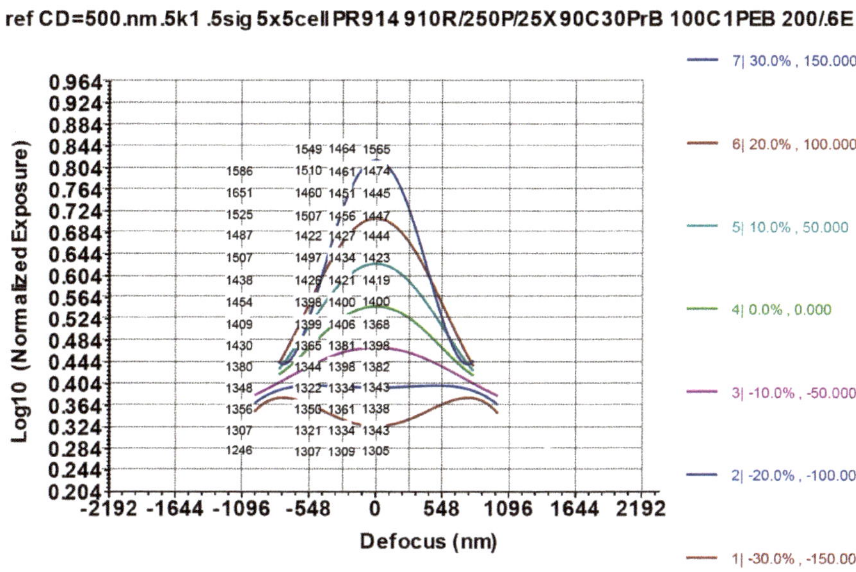

Figure 5.15 Constant-edge-position contours from polynomial curve fitting.

5.5.1.3 E-D tree construction from the intensity distribution

To construct an E-D branch from the intensity distribution of an image, we first identify the edge location for the branch of interest. For example, branch 0 is from the location of zero edge-position tolerance on the intensity distribution plot shown in Fig. 5.16. Branch 1 is for the edge at the higher limit, and branch –1 is for the edge at the lower limit. We find the intensity at

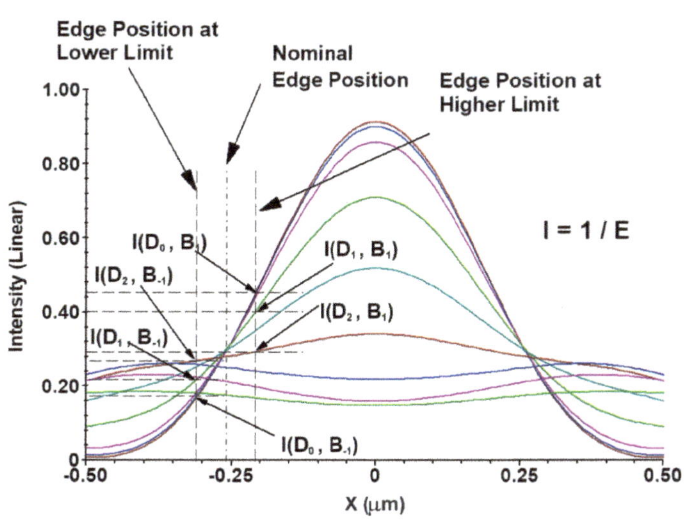

Figure 5.16 The intensity distribution curves at various defocus positions and identification of the $I(D,B)$ points.

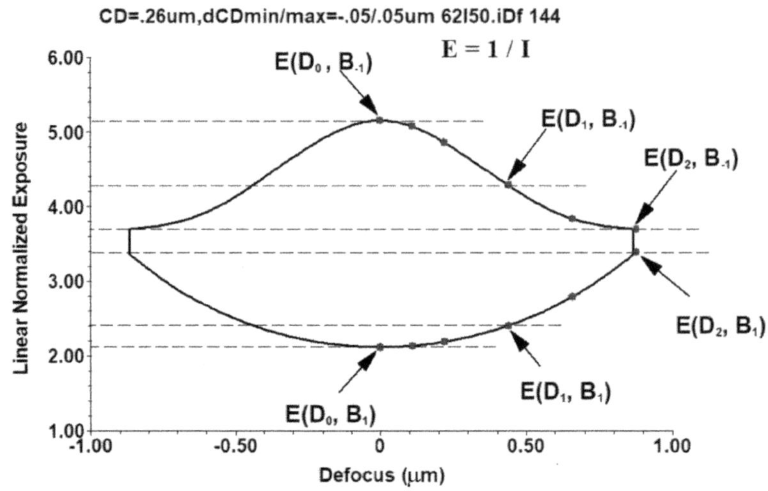

Figure 5.17 The E-D tree and branches composed with the *E*(*D*,*B*) points from the *I*(*D*,*B*) points in Fig. 5.16. *E* = 1/*I* was used to convert *I* to *E*, which is the apparent exposure explained in the text.

this edge position at each defocus position. Figure 5.16 uses three defocus positions to produce the points $I(D_0, B_1)$, $I(D_1, B_1)$, and $I(D_2, B_1)$ for branch 1 and three other points for branch –1. Here I stands for intensity, D for defocus, and B for branch. The intensity is converted to exposure by taking the reciprocal of intensity, producing the points $E(D_0, B_1)$, $E(D_1, B_1)$, and $E(D_2, B_1)$ in the E-D plane shown in Fig. 5.17. E-D branch 1 is obtained by joining all E-D points corresponding to branch 1. E-D branch –1 is obtained similarly.

The reciprocal relationship between E and I is based on a given resist sensitivity. It is different from the relationship $E = I \cdot t$, where I is the image intensity in mW/cm^2 and t is the time in seconds. The exposure unit is mJ/cm^2. Here, E is the apparent exposure seen by the operator of the exposure tool. Figure 5.18 shows an intensity distribution that has a peak of 100 mW/cm^2 to expose a resist of sensitivity 100 mJ/cm^2. Let the illumination on the mask be also 100 mW/cm^2. When an exposure time of 1 sec is used without a mask, the resist is completely exposed to be removed by the subsequent development. Thus, the blanket exposure intensity is 100 mW/cm^2, and the blanket exposure dosage is 100 mJ/cm^2. Now, with the intensity distribution shown in the figure, it takes 1.7 sec to delineate a 0.4-μm line for a negative resist, and 3.9 sec for a 0.6-μm line.

On the production floor, the exposure tool is typically set with an exposure dosage referring to the blanket exposure. To delineate a 0.4-μm line, the exposure dosage apparent to the process engineer would be 1.7 sec·100 mW/cm^2; to delineate a 0.6-μm line, the apparent exposure dosage would be 3.9 sec·100 mW/cm^2. With the resist sensitivity and peak

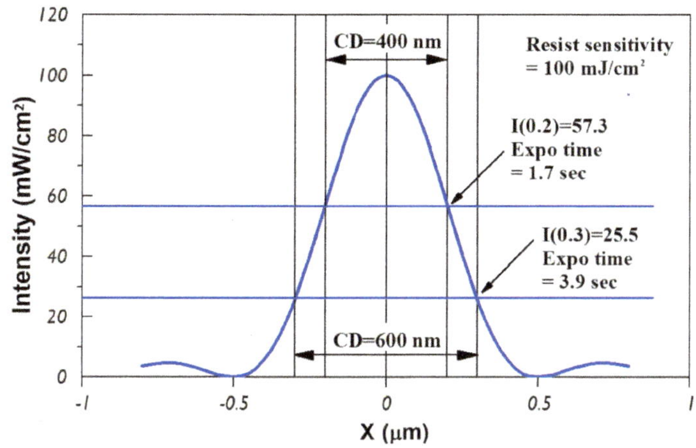

Figure 5.18 With the resist exposure threshold at 100 mJ/cm^2, to produce CD = 0.4 μm in a negative resist, the exposure time is 1.7 sec, while for CD = 0.6 μm, the exposure time is 3.9 sec. (1.7 = 1/0.573 and 3.9 = 1/0.255 are the normalized apparent exposure dosages for plotting Fig. 5.17.)

intensity normalized to 1 mW/cm^2 and 1 mJ/cm^2, respectively, the normalized apparent exposure would be 1.7 and 3.9, respectively, with the sec unit omitted. This normalized apparent exposure dosage is used in the relationship $E = 1/I$ in Figs 5.16 and 5.17.

5.5.2 The importance of using log scale in the exposure axis

Intuitively, one tends to plot the E-D trees in the linear scale for both exposure and defocus. However, because exposures should be represented by ratios, the log scale is the right way to plot E-D trees. This way, the horizontal and vertical distances in the E-D region can be compared directly. Let there be three exposures: E_1, E_2, and E_3. If plotted in log, the distance between any two points is $\log E_2 - \log E_1$, $\log E_3 - \log E_2$, and $\log E_3 - \log E_1$, indicating the log of ratios E_2/E_1, E_3/E_2, and E_3/E_1. If $E_2 = 2E_1$ and $E_3 = 2E_2$, then $\log E_2 - \log E_1 = \log E_3 - \log E_2 = \log 2$. However, $E_2 - E_1 = E_1$ and $E_3 - E_2 = 2E_1$. Therefore, the distance between E_3 and E_2 is identical to that between E_2 and E_1 on the log scale but is doubled on the linear scale. The EL determined from the distance between E values on the log scale is also identical:

$$EL = 2\frac{E_3 - E_2}{E_3 + E_2} = 2\frac{E_2 - E_1}{E_2 + E_1} = 2\frac{e^{(\log E_2 - \log E_1)} - 1}{e^{(\log E_2 - \log E_1)} + 1}. \tag{5.8}$$

Figure 5.19 shows that two E-D trees with an identical EL can appear to be different in size when plotted in linear scale in exposure. Conversely, the trees plotted in log scale correctly reflect the relative size of their EL and thus aid in visualizing a comparison of the process windows. The ellipses in these figures are explained in the next section.

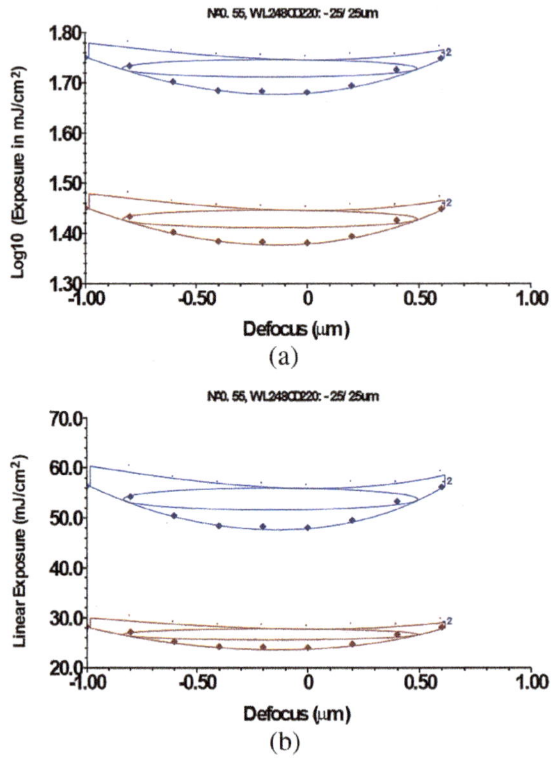

Figure 5.19 Two E-D trees with the same EL but different exposure dosages. (a) In log scale, the trees are identical, except for their vertical locations. (b) In linear scale, the tree with more exposure appears to be larger, thus giving a misleading visual impression.

5.5.3 Elliptical E-D window

The E-D window defined with rectangles indicates the absolute range of exposure and defocus allowed by the imaging process. It is too strict for manufacturing. The exposure and defocus variations in a heavy-wafer-flow situation often exhibit a Gaussian distribution. Violation of the rectangular boundary at the corners is tolerable. This leads to elliptical E-D windows as shown in Fig. 5.20. The EL and DOF defined by the two diameters of the ellipse are no longer absolute limits but are statistical in nature. They are related to a certain value of the standard deviation, such as 6σ. The actual value of the standard deviation is yet to be defined with a large number of CD-versus-exposure and defocus data. Figure 5.20 shows the E-D tree of a 220-nm line controlled to ±25 nm using $NA = 0.55$ and $\lambda = 248$ nm. E-D windows with 8% EL are plotted in elliptical and rectangular forms. The DOF defined by the elliptical window is 1330 nm, whereas that from the rectangular window is 1090 nm. The difference is 22%. When the exposure tolerance is changed to 15% as depicted by Fig. 5.21, the DOFs are 960 and 350 nm,

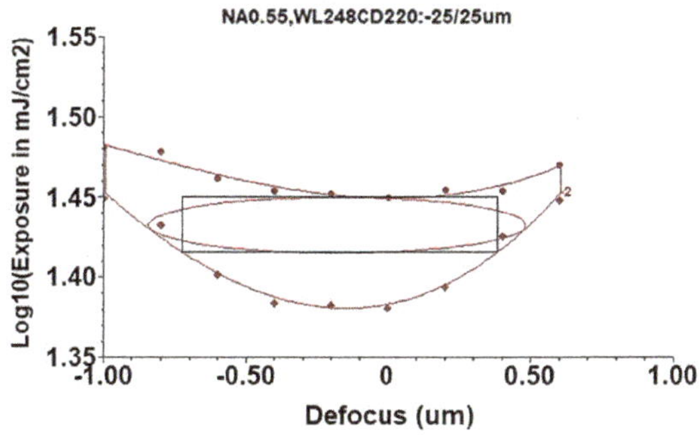

Figure 5.20 E-D tree of 220-nm line controlled to ±25 nm with *NA* = 0.55, λ = 248 nm. Elliptical and rectangular E-D windows with 8% EL are shown.

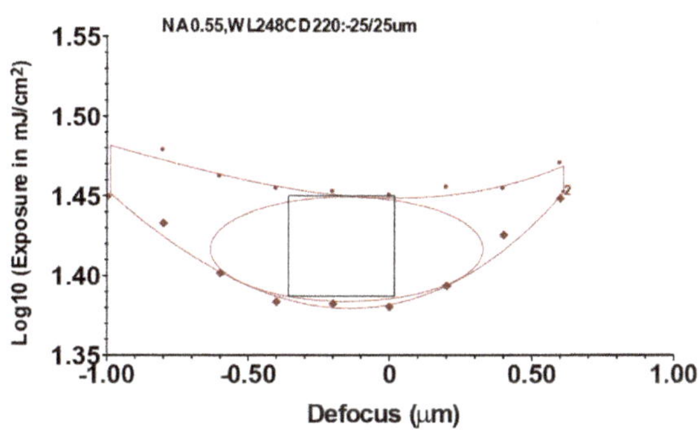

Figure 5.21 E-D tree of the feature used in Fig. 5.20, except that the elliptical and rectangular E-D windows are drawn with 15% EL.

resulting in a change of 267%. At an extremely small EL or DOF, the difference between the two types of window is insignificant, but near the maximum E-D area, the difference can be substantial.

A series of ELs is used to evaluate the corresponding DOF by constructing the respective E-D window. These EL-versus-DOF curves show the tradeoff between EL and DOF, i.e., their mutual dependence. Such a tradeoff curve for the feature used in Fig. 5.20 is shown in Fig. 5.22. The curve based on elliptical E-D windows has a sharper bend than that based on rectangular windows. This means that the common practice of defining EL at $DOF = 0$ and DOF at $EL = 0$ more closely resembles the case of the elliptical E-D windows.

Many authors have been erroneously plotting a perfect ellipse in the linear E-D space. A perfect ellipse only makes sense when plotted in the log space for exposure. The correct way is to work in the log space, construct the ellipse, then convert to the linear space. Fortunately, the range of exposures is usually small, making the error tolerable.

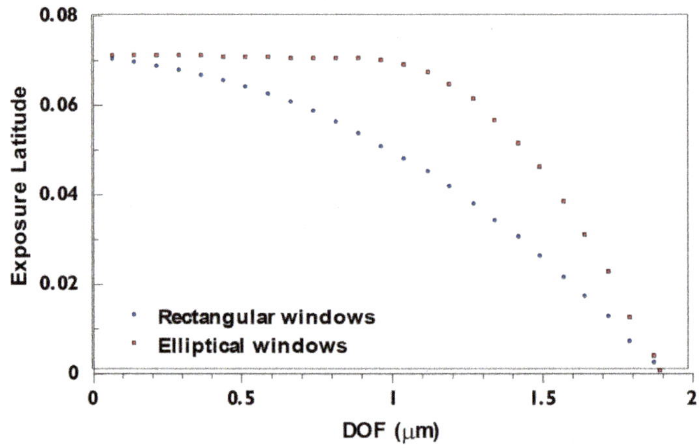

Figure 5.22 EL versus DOF for the 220-nm feature in Fig. 5.20. The curve taken from elliptical windows exhibits a squarer shape.

5.5.4 CD-centered E-D windows versus full-CD-range E-D windows

Adopting the concept of a statistical distribution of exposure and defocus, one assumes that the most popular occurrence of exposure and defocus is at the set point. Hence, setting the center of the ellipse to the $\Delta CD = 0$ contour ensures that the most frequently occurring linewidth is the nominal CD. This ellipse is the CD-centered (CDC) E-D window. What we previously defined was the CD-range (CDR) E-D window. Both are depicted in Fig. 5.23 with E-D windows of the maximum E-D area. The tradeoff of using the CDC window is a smaller EL and a smaller DOF. For example, the EL and DOF of the CDC window are 8% and 1.03 μm, respectively, compared with 8% and 1.33 μm of the CDR window.

Having a smaller E-D window is not the most severe problem with CDC windows. The common E-D window becomes meaningless because, regardless of which individual window is centered, there can only be one feature that

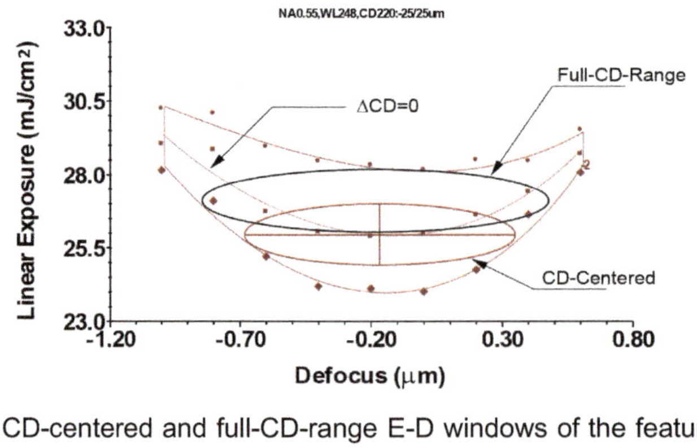

Figure 5.23 CD-centered and full-CD-range E-D windows of the feature in Fig. 5.20.

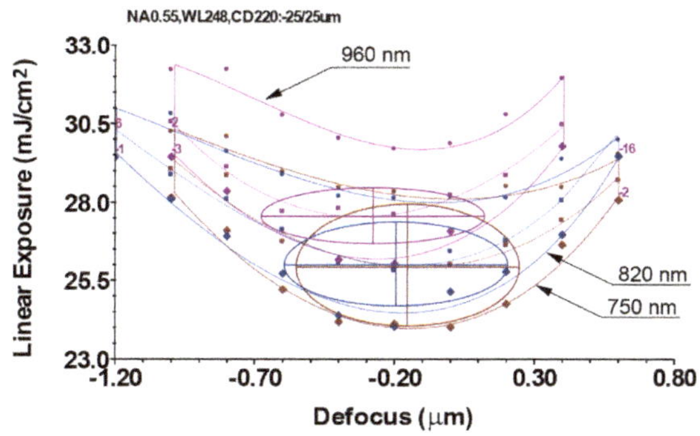

Figure 5.24 CDC E-D windows for three resist thicknesses. There is no common window in this case.

centers on the nominal exposure. A shown in Fig. 5.24, each E-D window of constant DOF centers at a unique (*E*, *D*) coordinate, even though those windows of the 750- and 820-nm resist thickness are much closer than that of the 960-nm resist thickness. Obviously, there is no common window in this case. In comparison, Fig. 5.25 shows the individual and common CDR E-D windows that share an identical DOF of 800 nm. The E-D parameters are shown in Table 5.4.

One way to center all windows to the nominal exposure is by mask biasing. Each feature is given a specific bias on the mask to result in the nominal CD at a given identical exposure. This is possible for 1D features.

5.5.5 E-D windows and CD control

The E-D window defines the EL and DOF of a given feature or a combination of them. When a point in the E-D space is outside of the window, the window-defined EL or DOF is violated. This E-D point cannot be considered for the EL and DOF. However, given two E-D trees enclosing an identical E-D window, the one with more area outside the window is statistically better for

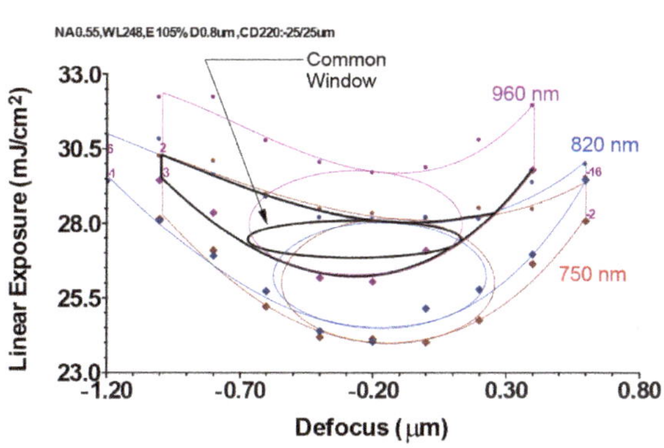

Figure 5.25 CDR E-D windows (individual and common) for three resist thicknesses.

Table 5.4 CDR and CDC E-D window parameters.

Resist thickness (nm)	E_{min} (mJ/cm^2)	EL %	D_{min} (nm)	DOF (nm)	E-D area [EL(log10)·k_2·1000]	E_{center} (mJ/cm^2)	D_{center} (nm)
Common	26.84	4.55	–671	800	19.29	27.46	–271
750	24.03	15.95	–548	800	67.75	25.95	–148
820	24.55	13.36	–578	800	56.70	26.19	–178
960	26.32	12.44	–671	800	52.78	27.95	–271
750 $CD_{centered}$	24.05	14.98	–554	800	63.60	25.85	–154
820 $CD_{centered}$	24.68	10.27	–595	800	43.54	25.95	–195
960 $CD_{centered}$	26.66	6.50%	–677	800	27.56	27.53	–277

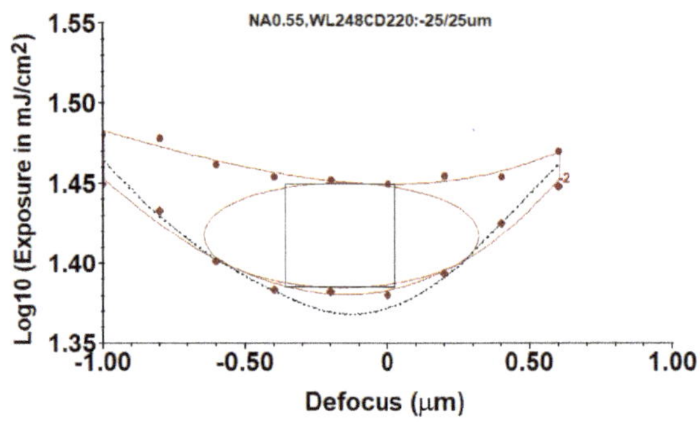

Figure 5.26 Modified E-D tree from Fig. 5.21. An extra E-D branch (shown in the dotted line) is artificially added to illustrate the difference in CD control with different E-D trees sharing an identical E-D window.

meeting the exposure or focus requirement. Thus, this tree indicates better CD control. For example, in the two E-D trees shown in Fig. 5.26, the artificially created E-D tree—defined by the lower branch in the dotted line and the upper branch that is shared with the real tree—has a larger area near focus. This artificially created tree provides better CD control near focus. Conversely, CD control is better for the real tree when the focus is not so accurate.

5.5.6 Application of E-D tools

The power of the E-D tools has many aspects. First, it provides a rigorous definition of the EL and DOF of a given optical imaging system as well as their mutual dependence. Second, the components can be superimposed for their common E-D window. Third, systems can be compared with identical criteria. These advantages lead to many important and interesting applications.

5.5.6.1 Replacement of Bossung curves

Bossung curves were first plotted in 1977 by John Bossung.[9] Because of their unique shape, these curves are also known as SMILE curves. The CD of a

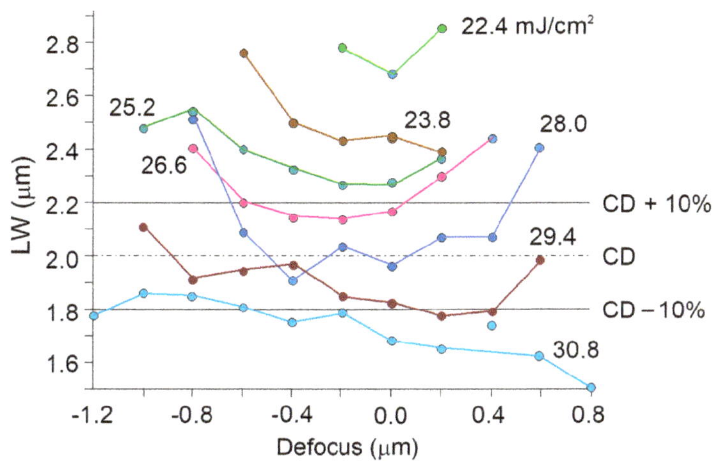

Figure 5.27 A Bossung plot in which the CD is plotted as a function of focus at different exposures.

given feature is plotted as a function of focus at different exposures in Fig. 5.27. The flatness of the curves indicates tolerance to defocus. The closeness of the curve shows a high dependence of CD on exposure. The CD boundary lines help to determine the acceptable range of exposure and defocus. The lines are useful for characterizing the behavior of a single feature as a function of the exposure and defocus settings. The irregularity of the curves indicates the presence of aberration, vibration, measurement noise, or other imperfect imaging conditions.

The same set of data can be plotted in the E-D space as an E-D tree, as shown in Fig. 5.28. The boundary of CD control is explicitly defined by the upper and lower E-D branches. A rectangle or ellipse can be draw within the boundary to quantitatively define the EL and DOF of the given single feature. Like Bossung curves, the closeness of the E-D branches indicates a smaller EL. The straightness of the branches indicates isofocal behavior. The imaging behavior can be better quantified and grasped using an E-D tree and E-D window.

Bossung curves cannot characterize the common behavior of two different situations, such as different features, as shown in Figs. 5.29 and 5.30. In Fig. 5.29, two sets of Bossung curves are used for two different features. The common imaging behavior of these two features cannot be shown with these plots. On the other hand, when the same sets of data are plotted in the E-D space as shown in Fig. 5.30, their E-D trees, windows, common trees, and common windows can readily show the EL and DOF of these situations.

The advantage of the E-D methodology becomes obvious for the case where the total imaging behavior of many situations must be quantified. Figure 5.31 shows the Bossung curves of six different situations. These plots cannot be superposed into a single chart. On the other hand, six E-D trees are

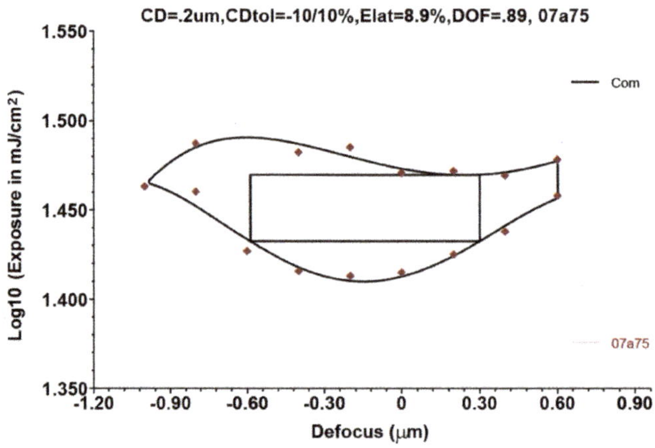

Figure 5.28 The data in Fig. 5.27 plotted in the E-D space as an E-D tree.

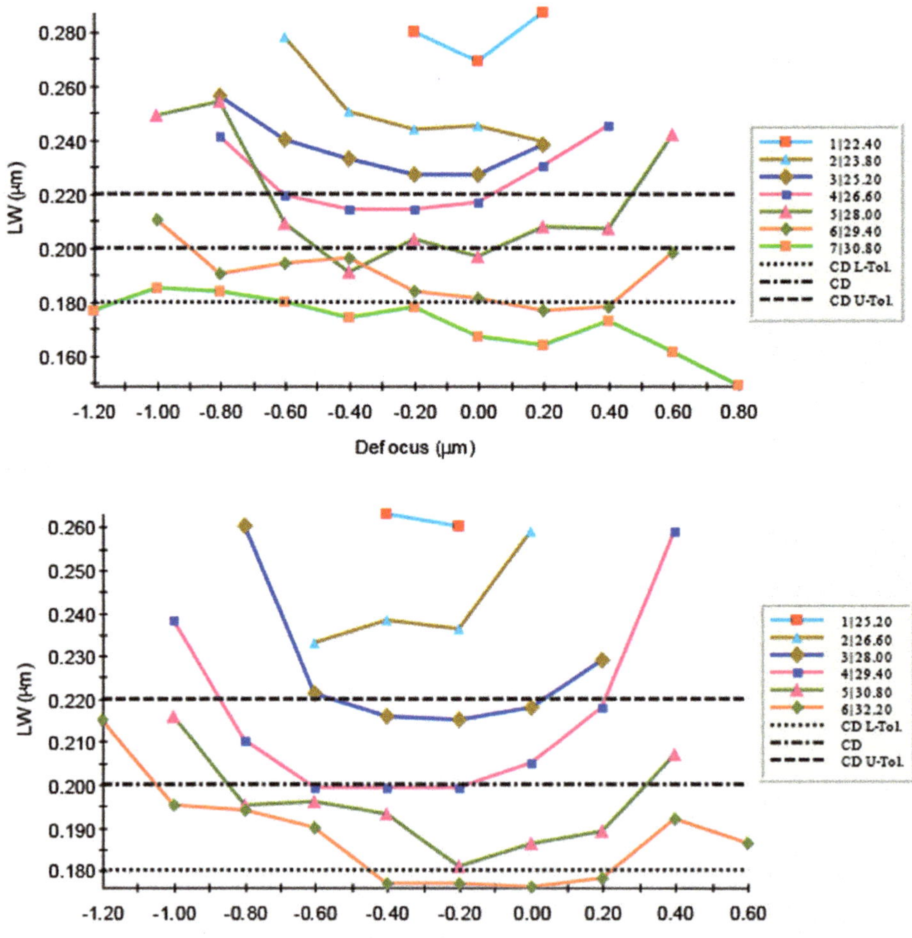

Figure 5.29 Two Bossung plots from two different datasets.

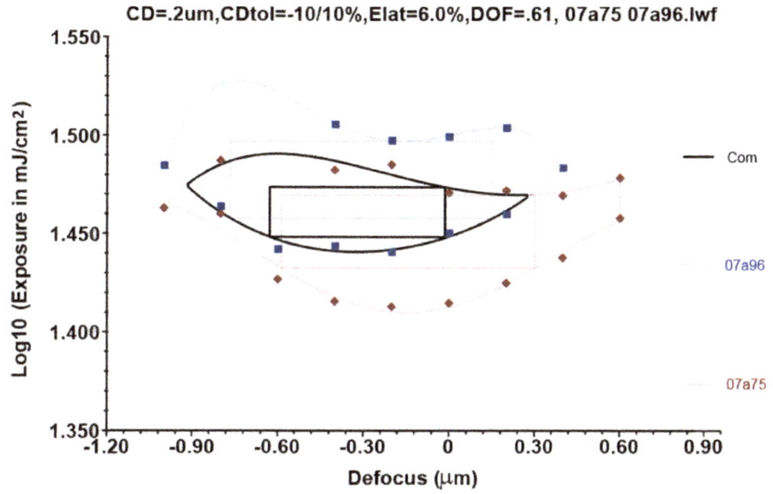

Figure 5.30 The data in Fig. 5.29 plotted in the E-D space as E-D trees.

plotted in the same E-D space in Fig. 5.32, together with their separate and common behaviors. In principle, there is no limit to the number of situations that can be considered together for common imaging characteristics. Figure 5.45 shows 50 E-D trees superimposed for their common process window.

5.5.6.2 Combination of feature types

Figures 5.5 and 5.10 are examples of the combination of line opening, opaque space, and line–space pairs. From these figures, one readily observes the distinct shapes associated with different features and their various locations in the E-D space. The individual trees suggest ample processing windows for each feature. However, because they have different exposure centers, the common E-D window is quite small, making it difficult to delineate all three features to the same linewidth criterion at an identical exposure. The E-D trees also show that the line-opening and line–space pairs can be biased as a group to produce a larger common window with the opaque space, or vice versa.

5.5.6.3 Combination of feature sizes

A common practice in circuit fabrication is to use the same gate length but to allow the separation between gates to differ. This is similar to mixing line–space pairs of different line:space (L:S) ratios. With a positive resist, the mask consists of opaque gate spaces of identical width. Here, the E-D trees and the common E-D window of a mixture of 1:1, 1:1.5, and 1:2 opaque:transparent ratios are shown in Fig. 5.33. The gate width is kept at 0.13 μm using $\sigma = 0.5$, $NA = 0.6$, and $\lambda = 193$ nm; i.e., the gate length is at $k_1 = 0.4$. Even though the

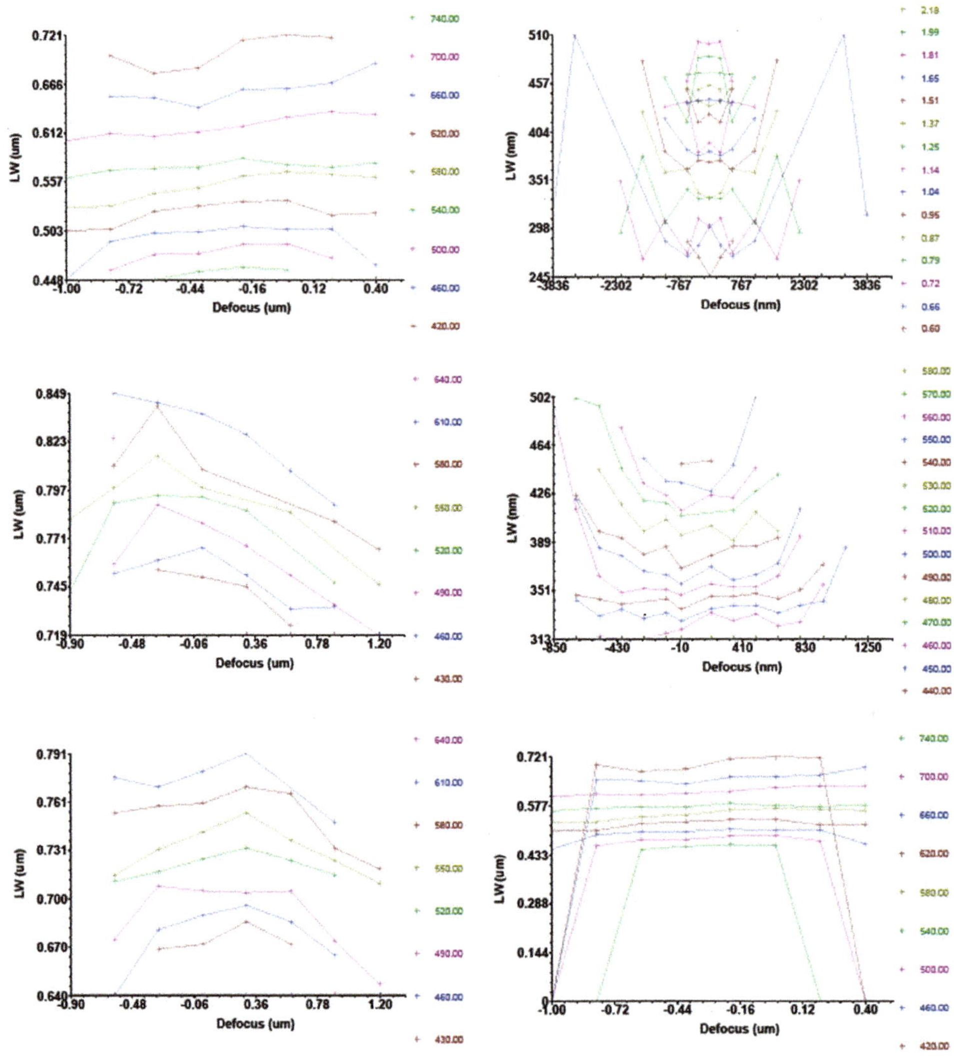

Figure 5.31 Six Bossung plots from different datasets.

gate lengths are identical, there is no common window because the separations between gates differ.

To find the NA/σ combinations for the common window of the three features in Fig. 5.33, all E-D trees from $NA = 0.3$ to $NA = 0.9$ at $\sigma = 0.86$ are plotted in Fig. 5.34. At $NA \geq 0.57$, the common E-D window appears. The DOF set at 8% EL is plotted as a function of NA in Fig. 5.35. The DOF of the individual trees is obviously much larger than that of the common tree. Furthermore, the 1:1 feature has a higher peak DOF than that of the 1:1.5 feature, which, in turn, has a higher peak DOF than that of the 1:2 feature. Intuitively, dense features are considered more difficult to print than features

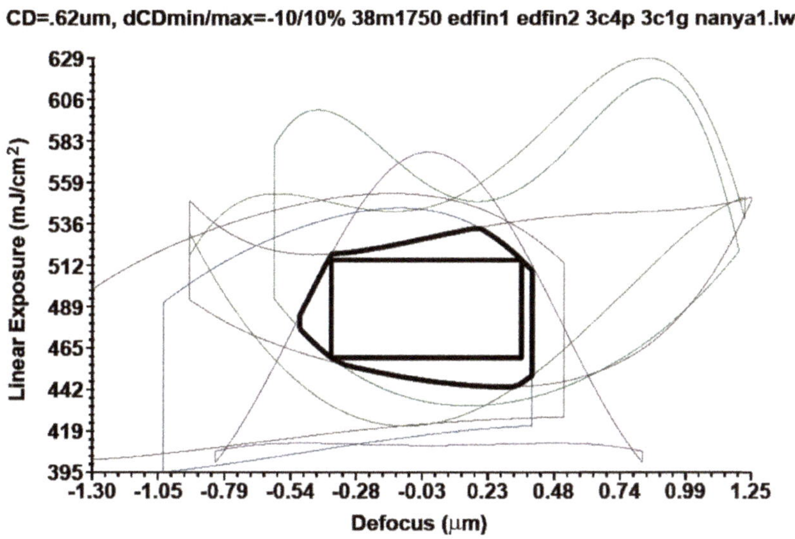

Figure 5.32 The data in Fig. 5.31 plotted in the E-D space as E-D trees.

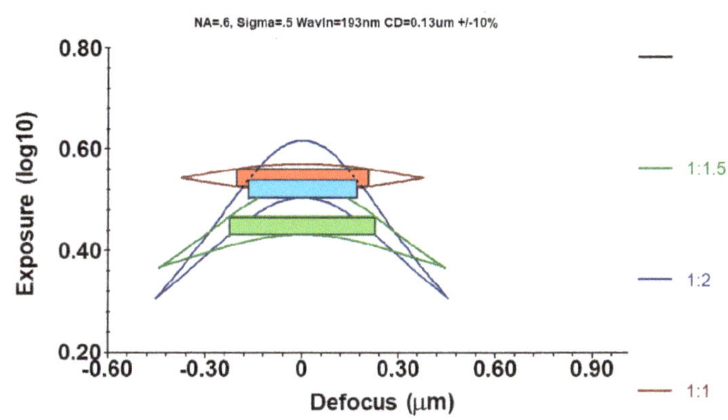

Figure 5.33 Combination of line–space pairs with L:S ratios 1:1, 1:1.5, and 1:2.

with larger separations. E-D tree analysis reveals that, even though the features with larger separations have a larger EL and hence more image contrast, at the focal plane, the DOF is smaller. The trees have more curvature and can support only smaller DOFs. Figure 5.36 plots DOF as a function of NA and σ. The DOF is optimized at $NA = 0.67$ and $\sigma = 0.7$.

Another common practice is to mix contact holes of different pitches. Figure 5.37 shows DOF versus σ from the E-D trees of 0.13-μm contact holes at $\lambda = 193$ nm and $NA = 0.64$, i.e., $k_1 = 0.43$, with pitches of 0.26, 0.325, and 0.39 μm combined on the same mask. This set of contact holes has hole: separation ratios of 1:1, 1:1.5, and 1:2, respectively. An attenuated phase-shifting mask with 6% transmission is used. The illumination is annular with a

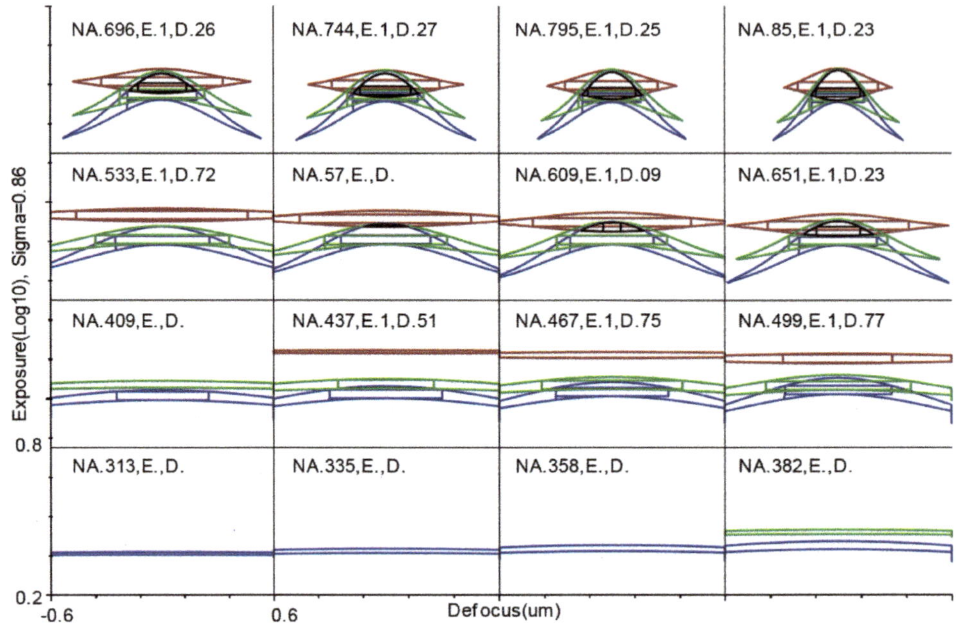

Figure 5.34 Same as Fig. 5.33 but plotted through *NA* = 0.3 to 0.9, σ = 0.86.

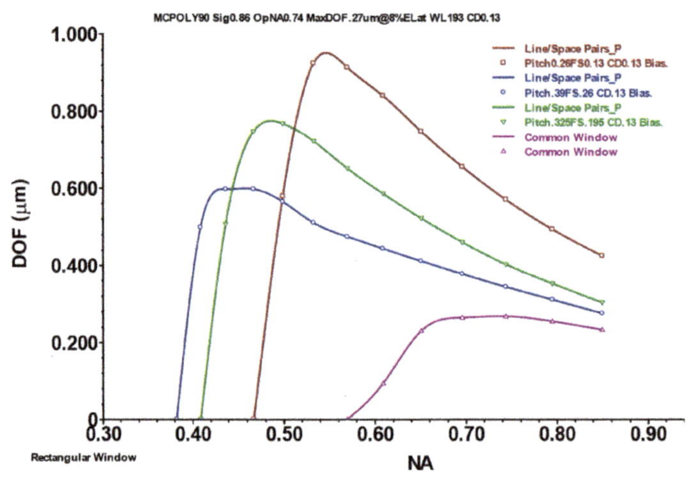

Figure 5.35 DOF versus NA with data from Fig. 5.34.

50% inside diameter. More discussions on annular illumination are found in Section 7.3.2. The E-D trees are constructed at $\sigma = 0.44$ to $\sigma = 1$ in 0.04 steps, and the E-D windows have an 8% fixed EL.

The target dimension of the contact holes is 0.13 μm controlled to ±10%. However, their actual size is 0.1 μm; i.e., they are biased by 30 nm. Without the bias, the DOF of these holes would be much smaller. These holes must be

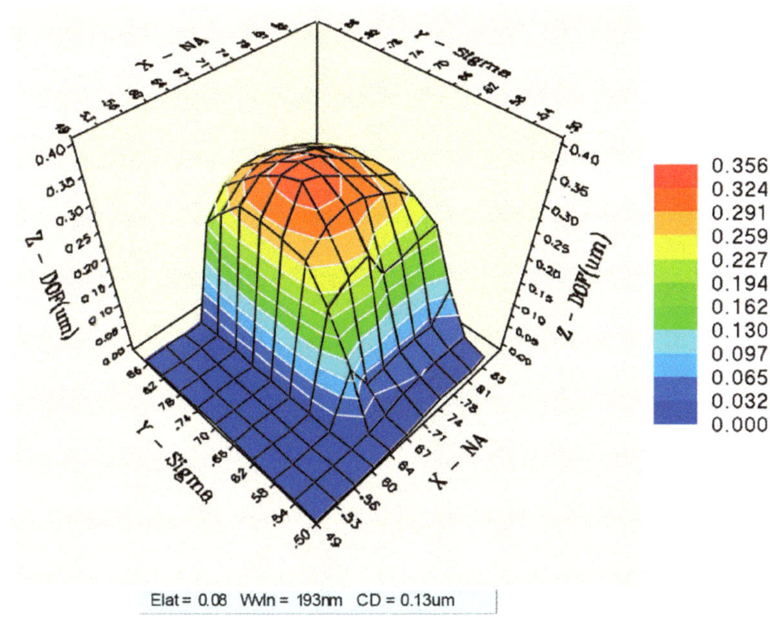

Figure 5.36 DOF versus NA and σ for the features in Fig. 5.34.

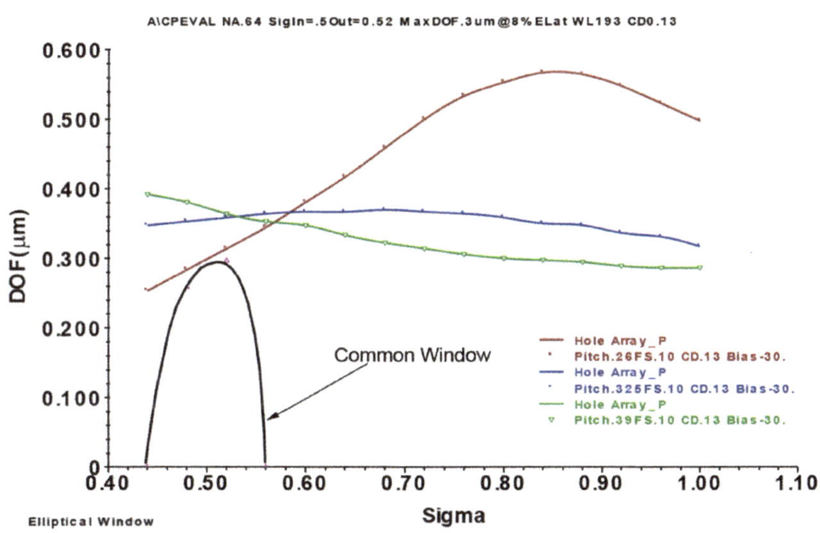

Figure 5.37 DOF versus σ from the E-D trees of 0.13-μm contact holes at λ = 193 nm and *NA* = 0.64. This set of contact holes has hole:separation ratios of 1:1, 1:1.5, and 1:2. 6% AttPSM is used. The illumination is annular with a 50% inside diameter.

overexposed to reach the target image size. To further improve the DOF, the bias can be individually adjusted. Readers are referred to Section 7.3.4 for detailed coverage of the topic.

5.5.6.4 Combination of cuts for 2D features

With 1D features, just one cut at the edge of the feature is sufficient to evaluate its E-D characteristics. Combining feature sizes and types still needs one cut per type and per size. If edge E-D trees are needed due to asymmetry of the feature layout or aberrations in the imaging system, the number of cuts may be increased to at most two per feature in the layout. With 2D features, the edge of the feature cannot be represented by just one or two cuts. Different locations or edges on one feature often must be maintained at the same edge control specification. For example, the width and length of a resist line must be kept within the CD tolerance specification.

Figure 5.38 shows the E-D trees of a 0.35-μm line that is 1.4-μm long illuminated at $\sigma = 0.7$ and 365-nm wavelength. Both the width and the length are controlled to ±35 nm of their nominal edge position. At $NA \leq 0.63$, even though each image edge can be controlled within specification at its own exposure dosage as depicted by the existence of its individual tree, there is no unique exposure dosage to satisfy both edges. In practice, because line ends are more susceptible to linewidth variations, a less stringent specification of CD tolerance is given, making the common E-D window larger. This compensation must be incorporated into the design rules.

In addition to line ends, the variation of edge position along a line in other 2D situations is also of concern. For example, the CD variation near the junction of a T, an L, or a disconnected T must be represented by many cuts at

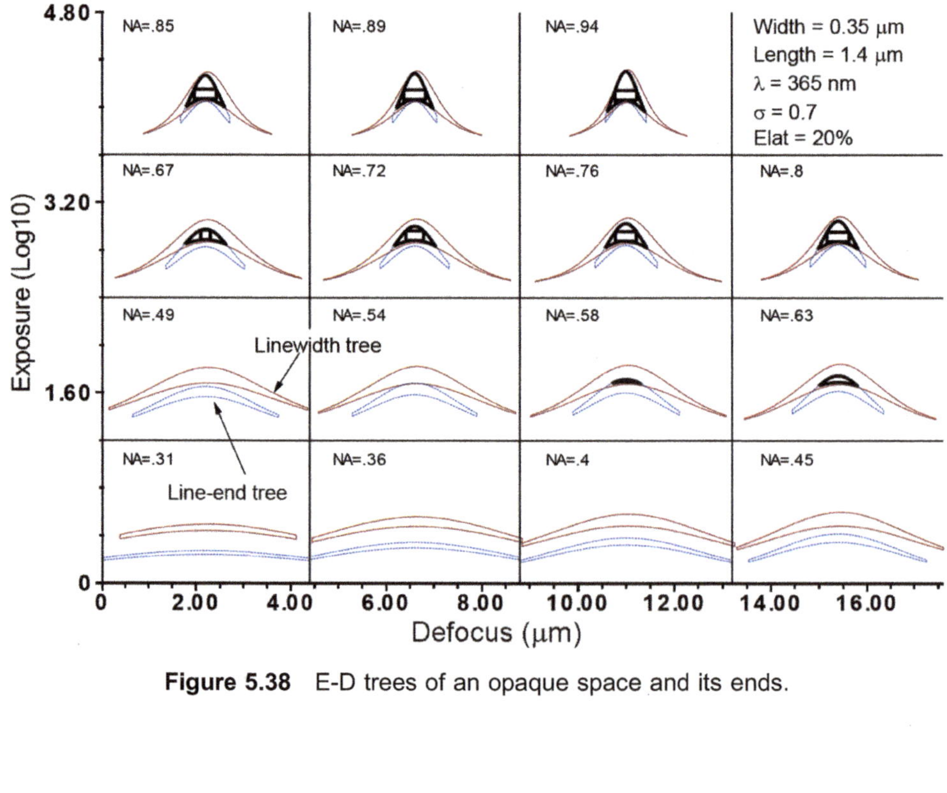

Figure 5.38 E-D trees of an opaque space and its ends.

the appropriate places. A typical application will be discussed in Section 7.7.2 on model-based optical proximity correction and is depicted in Figs. 7.101 and 7.102.

5.5.6.5 Combination of CD tolerances

E-D trees plotted at different CD tolerances can be combined to study the effect of different linewidth control requirements. Figure 5.39 shows the DOF evaluated with an elliptical E-D window of 10% EL for 0.1-μm positive resist lines in 0.3-μm pitch and 0.2-μm lines in 0.6-μm pitch. The CD tolerance set for the E-D branches is ±10 nm for either line group. It is seen that at low NA, the 0.1-μm line group gates the DOF, whereas when $NA \geq 0.7$, the DOF is gated by the wider line group. When only ±20 nm is required of the 0.2-μm line group, there is a DOF gain at a higher NA because the narrower line group is now gating the entire NA range, as shown in Fig. 5.40.

5.5.6.6 Combination of resist-processing tolerances

When the resist-processing conditions change, the E-D trees can take different shapes. From Section 6.6.2.5, the resist thickness plays an important role in linewidth control because of multiple reflections. Figure 5.41 shows the E-D trees and windows for a 220-nm line imaged with $NA = 0.55$, $\lambda = 248$ nm, and CD tolerance ±25 nm. The only parameter changed was the resist thickness. Each thickness supports an E-D tree with more than sufficient EL and DOF. However, the center of exposure and DOF change substantially, especially for the 960-nm thickness. If there were no reflection, the adjustment of the focus center would be the difference between 960 and 750 nm divided by the refractive index of the resist, resulting in a focus shift of about 112 nm, assuming a resist refractive index of 1.7. From Table 5.5, the shift of the center

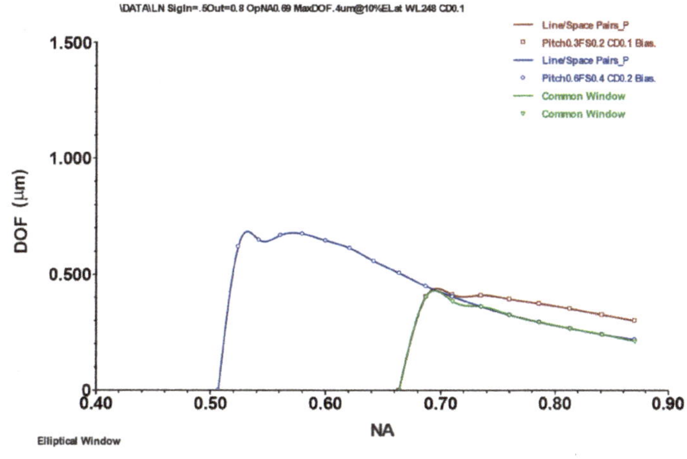

Figure 5.39 DOF versus NA for a 0.1-μm positive resist line in a 0.3-μm pitch and a 0.2-μm line in a 0.6-μm pitch. CD tolerance is set to ±10 nm.

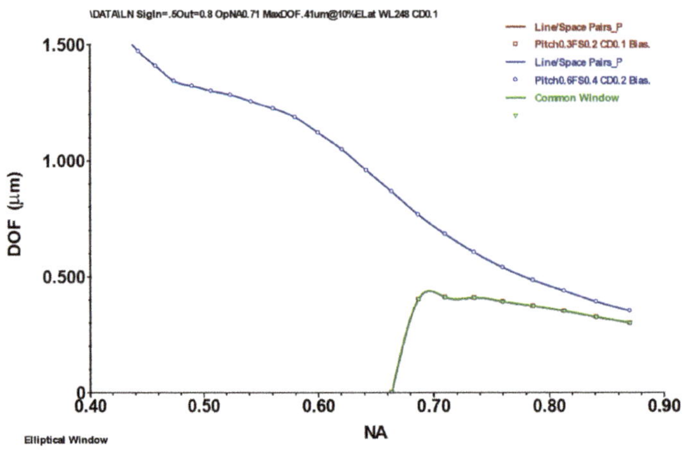

Figure 5.40 Same as Fig. 5.39, except that the CD tolerance is ±20 nm for the wider resist line.

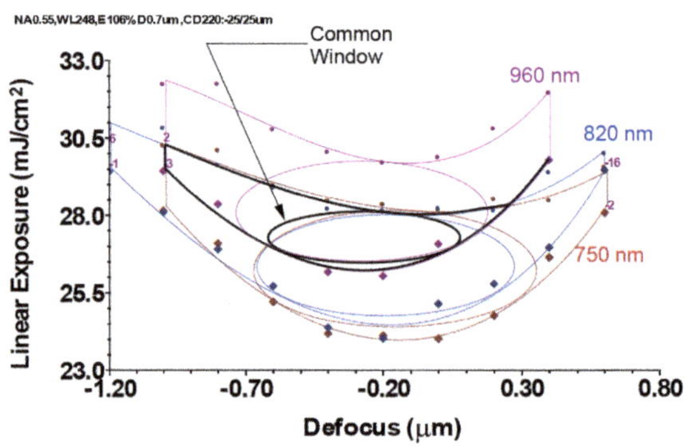

Figure 5.41 E-D trees and windows for the resist thicknesses 750, 820, and 960 nm. The feature is a 220-nm line, imaged with NA = 0.55, λ = 248 nm, and a CD tolerance of ±25 nm.

Table 5.5 E-D parameters for the windows in Fig. 5.41.

Resist thickness (nm)	E_{min} (mJ/cm^2)	EL (%)	D_{min} (nm)	DOF (nm)	E-D area [EL(log10)·k_2·1000]	E_{center} (mJ/cm^2)	D_{center} (nm)
Common	26.47	6.04	–619	697	22.31	27.27	–271
750	24.44	14.25	–676	1025	77.49	26.19	–164
820	24.80	12.33	–664	933	60.98	26.33	–198
960	26.56	11.47	–740	912	55.49	28.08	–284

of focus is 119.9 nm. Multiple reflections did not significantly shift the focus in this case. However, the exposure center and range are quite different between the 750-nm and 960-nm thicknesses. As a result, the EL and DOF of the common window are extremely reduced. Multiple reflections do affect exposure dosage, as predicted in Section 6.6.2.5.

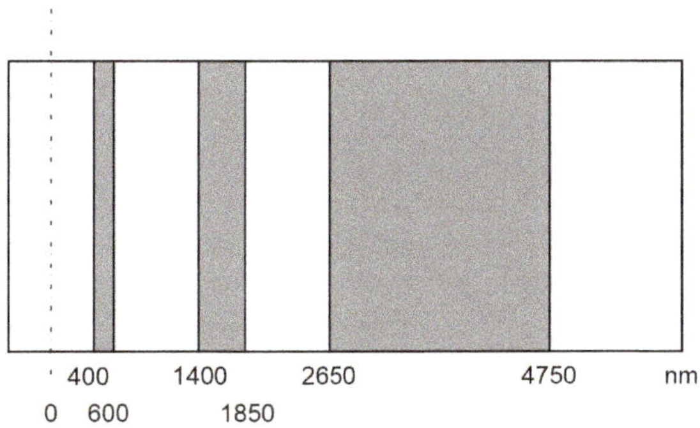

Figure 5.42 A six-edge pattern. The first edge was used for Figs. 5.43 and 5.44. The first five edges were used for Fig. 5.45.

Figure 5.42 shows a 400-nm edge in a six-edge mask pattern, using 250-nm wavelength, 0.5 NA, and $\sigma = 0.5$. Figure 5.43 is based on this first edge and shows the two E-D trees from 100 and 102 °C post-exposure bakes simulated with a typical resist. The CD tolerance is ±30 nm and the EL is 10%. Even though the shapes of the E-D trees are similar, the common window is much smaller because of the large disparity in the exposure centers.

Some processes are more critical than others. The post-exposure bake process shown in Fig. 5.43 induces more DOF reduction than the impact of the developing time, as shown in Fig. 5.44. Not only are the shapes of the E-D trees at the two developing times similar, but they are also very close in position. Hence, a developing time variation of ±1 sec is seen to be much less significant than a ±1 °C post-exposure bake temperature for this resist.

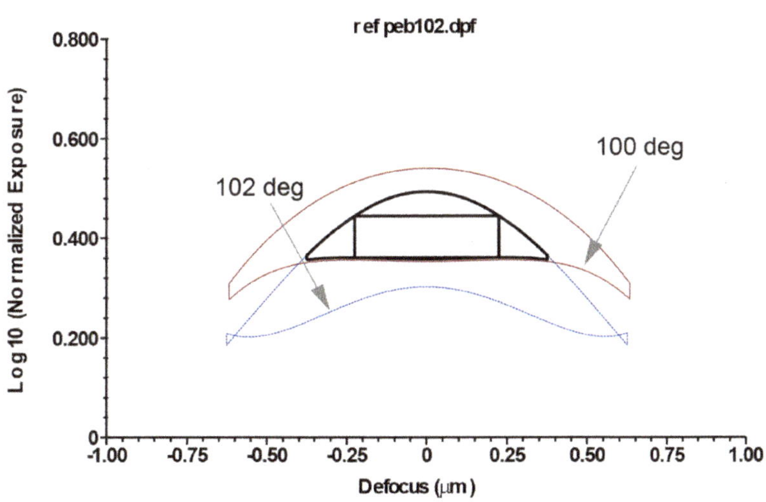

Figure 5.43 E-D trees from 101±1 °C post-exposure bake. Although the shapes of the E-D trees are similar, the common window is much smaller because of large disparity in the exposure centers.

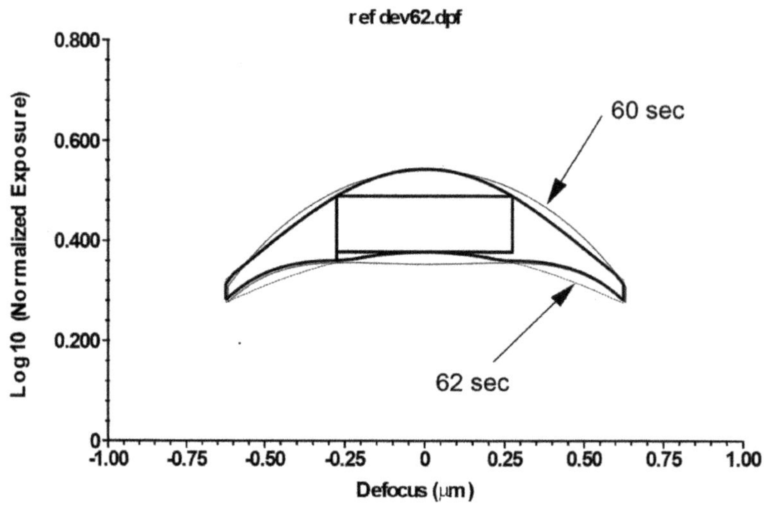

Figure 5.44 E-D trees from 61±1 sec developing time. The impact on the common window is less than that of the post-exposure bake in Fig. 5.43.

To evaluate the common E-D window realistically, all process tolerances must be considered. In Fig. 5.45, the E-D trees corresponding to the process tolerance in post-exposure bake temperature, time, prebake temperature, developing time, and resist thickness are plotted together for the first five

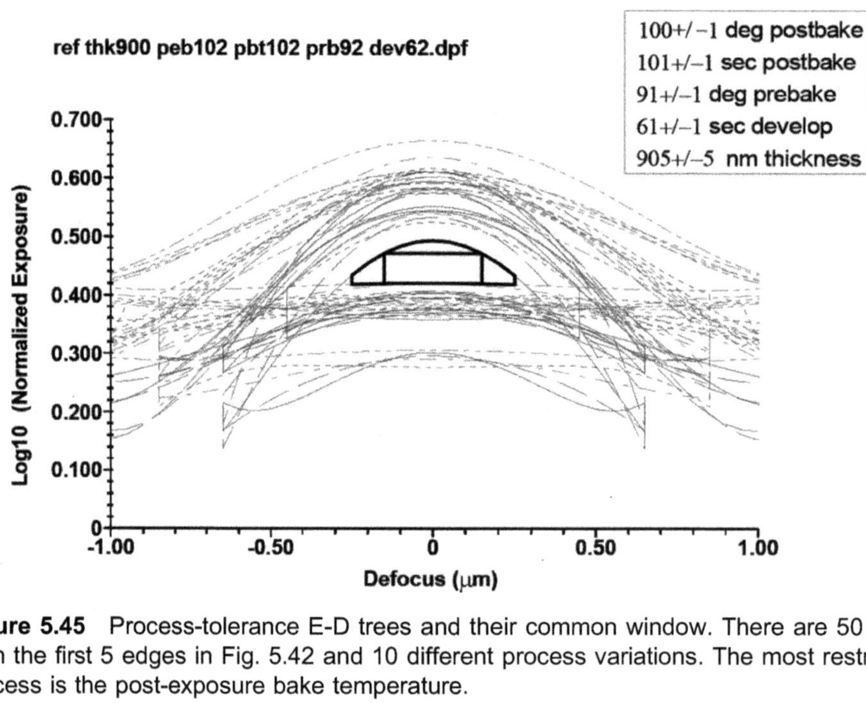

Figure 5.45 Process-tolerance E-D trees and their common window. There are 50 trees from the first 5 edges in Fig. 5.42 and 10 different process variations. The most restrictive process is the post-exposure bake temperature.

edges shown in Fig. 5.42. There are ten trees per edge, including the tree for the nominal processing condition. Hence, there are 50 trees in the figure. The common E-D window is much reduced. The most restricted tree is identified as the post-exposure bake tree. To improve on this, one must tighten the tolerance of the post-exposure bake temperature. Alternatively, the nominal baking temperature can be shifted for a larger common E-D window. There is a tradeoff between the imaging contrast reflected in the EL and the vertical position of the E-D window. Optimization of this tradeoff can be unambiguously achieved with E-D trees. One can work on the next restrictive processing condition until the largest common E-D window is obtained. Alternatively, one can use optical proximity correction to shift the position of the E-D trees. This topic will be further discussed in Chapter 7.

5.5.6.7 Combination of image field positions

Even the images of an identical feature can vary within a lens field in a number of ways: (1) The imaging field is tilted. (2) The field curvature of the imaging lens is not perfectly flat. (3) Lens aberrations such as spherical aberration, coma, and astigmatism all contribute to the field curvature. (4) The illumination is not completely uniform. (5) The mask and proximity correction may have their own CD signature and CD control residue. (6) The wafer might not be totally flat. (7) The topography on the wafer is often a problem. (8) If the exposure tool were a scanner, scanning errors could also contribute to field-dependent CD variation. In addition, EUV lithography suffers more from variations within the exposure field, as will be discussed in Chapter 9. Figure 5.46 and Table 5.6 show the E-D trees at nine exposure field positions of a 455-nm *y*-oriented feature using an $NA = 0.55$, $\lambda = 365$-nm stepper. Comparing the parts of the E-D area, one sees better imaging

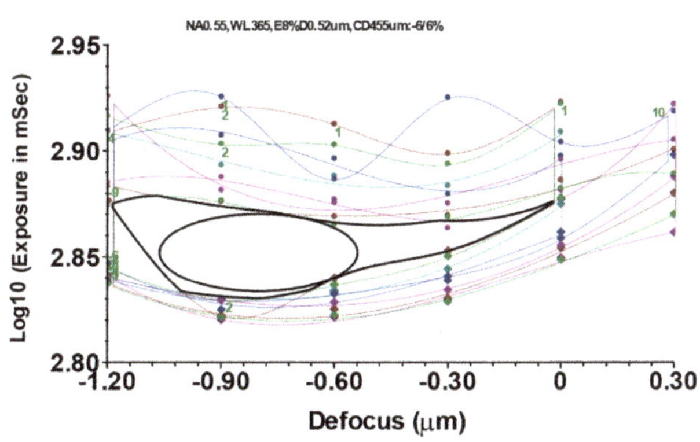

Figure 5.46 E-D trees of a 455-nm feature at nine field locations, with NA = 0.55, λ = 365 nm. Simply changing the location in the exposure field makes the common window much smaller.

Table 5.6 E-D window and common window at nine exposure-field positions of a 455-nm feature delineated at *NA* = 0.55, λ = 365 nm. The nine trees and the common window are shown in Fig. 5.46.

Field location	E_{min} (mJ/cm^2)	EL (%)	D_{min} (nm)	DOF (nm)	E-D area [EL(log10)·k_2·1000]	E_{center} (mJ/cm^2)	D_{center} (nm)
Common	22.70	8.39	–1062	523	15.81	23.65	–800
Left-Up	23.30	14.02	–1170	985	49.80	24.93	–678
Mid-Up	22.40	14.75	–1187	905	48.13	24.05	–734
Right-Up	22.10	12.19	–1182	925	40.64	23.45	–720
Left-Mid	23.00	14.00	–1191	985	49.74	24.61	–698
Mid-Mid	22.60	13.09	–1187	804	37.96	24.08	–784
Right-Mid	22.20	10.90	–1178	945	37.12	23.41	–706
Left-Down	22.90	11.40	–1182	1126	46.25	24.20	–619
Mid-Down	22.40	10.22	–1191	985	36.27	23.54	–698
Right-Down	22.10	10.38	–1178	945	35.36	23.25	–706

performance at the left and upper side of the field. The focus centers indicate that the best imaging field is lower in the middle column.

Without delving too much into what caused such a severe field dependence, one simplistic way to fix the problem is to modify the exposure according to the distribution of the exposure center. The resultant improvement can be seen in Fig. 5.47 and Table 5.7. The E-D trees are aligned in exposure, and the resultant E-D window has 9.97% EL and 865-nm DOF, compared with 8.39% EL and 523-nm DOF before the optimization. Many schemes can be used to vary the exposure within an exposure field. One straightforward method is to use a mask with a distribution of gray levels to compensate for the distribution of exposure centers for each E-D tree.

5.5.6.8 Setting the mask-making tolerance

To set the tolerance for mask makers,[10] three E-D trees are constructed corresponding to the maximum and minimum possible CD values on the

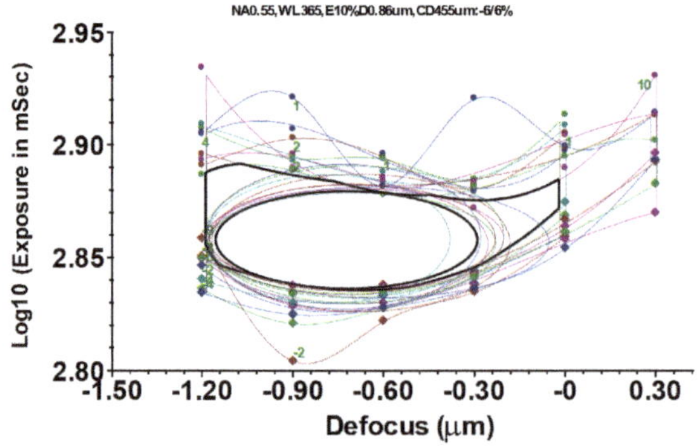

Figure 5.47 E-D tree at the nine positions, as in Fig. 5.48, except that the exposure at each field is modified according to the position of the exposure centers shown in Table 5.7.

Table 5.7 E-D window of the E-D trees from Fig. 5.47. After modifying the exposure at the nine field positions, the common window is much larger than that of Fig. 5.46.

Field location	E_{min} (mJ/cm^2)	EL (%)	D_{min} (nm)	DOF (nm)	E-D Area [EL(log10)·k_2·1000]	E_{center} (mJ/cm^2)	D_{center} (nm)
Common	22.86	9.97	–1154	865	31.05	24.00	–722
Left-Up	22.35	14.02	–1170	985	49.80	23.92	–678
Mid-Up	22.41	14.75	–1187	905	48.13	24.06	–734
Right-Up	22.53	12.19	–1182	925	40.64	23.90	–720
Left-Mid	22.54	14.00	–1191	985	49.74	24.12	–698
Mid-Mid	22.55	13.09	–1187	804	37.96	24.03	–784
Right-Mid	22.82	10.90	–1178	945	37.12	24.06	–706
Left-Down	22.66	11.40	–1182	1126	46.25	23.95	–619
Mid-Down	22.88	10.22	–1191	985	36.27	24.05	–698
Right-Down	22.75	10.38	–1178	945	35.36	23.94	–706

mask as well as the nominal CD value. We have chosen 0.18-μm positive resist lines at 0.45-μm pitch as an example. The exposure wavelength is 193 nm using annular illumination with $\sigma_{out} = 0.76$ and $\sigma_{in} = 0.5\sigma_{out}$. DOF versus NA is evaluated with 10% EL. Figure 5.48 shows the case with a 10-nm CD control range on the mask. The individual DOF-versus-NA curves are very close to each other. However, because of spreading of the exposure centers, the common DOF is reduced approximately by half and is completely lost at $NA < 0.48$.

When the mask CD range is relaxed by just 2 nm, the common DOF is almost completely gone, except for negligible values at $NA > 0.75$, as seen in

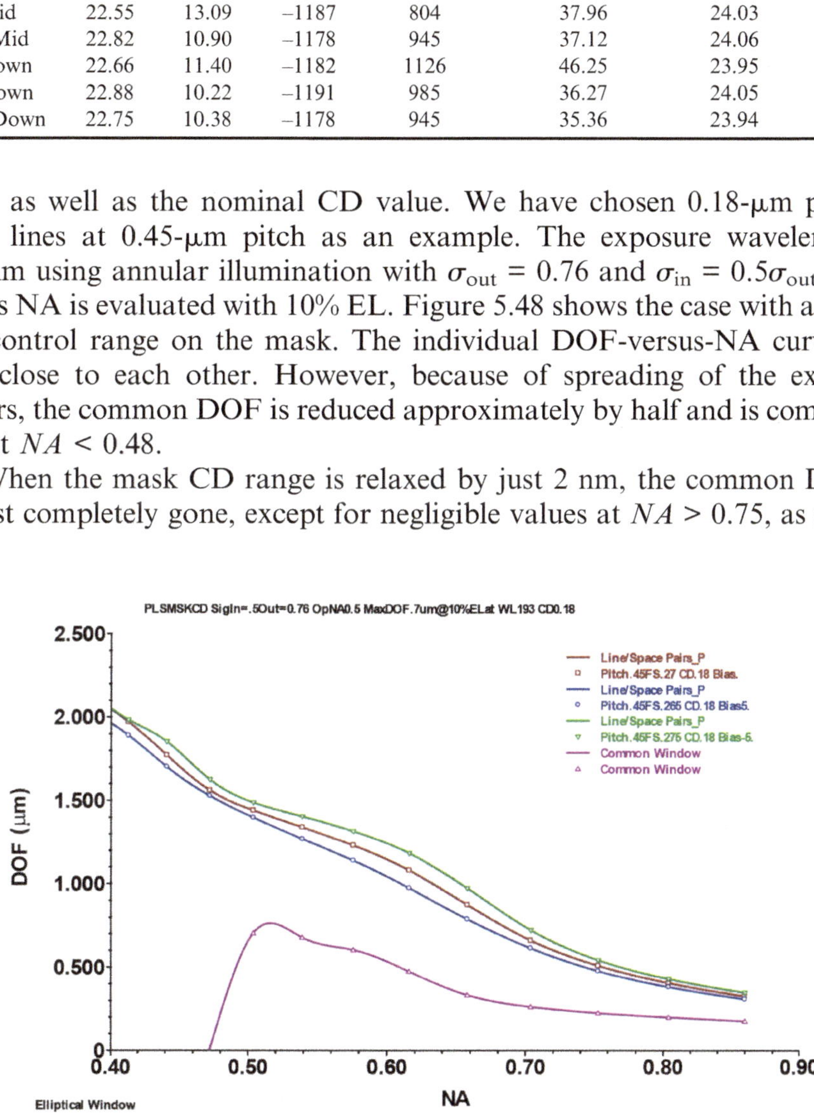

Figure 5.48 DOF versus NA from three E-D trees with 0.5-nm, and –5-nm mask biases. Although these curves look similar, the curve for the common window has a much smaller DOF due to the different positions of the centers of the individual E-D windows.

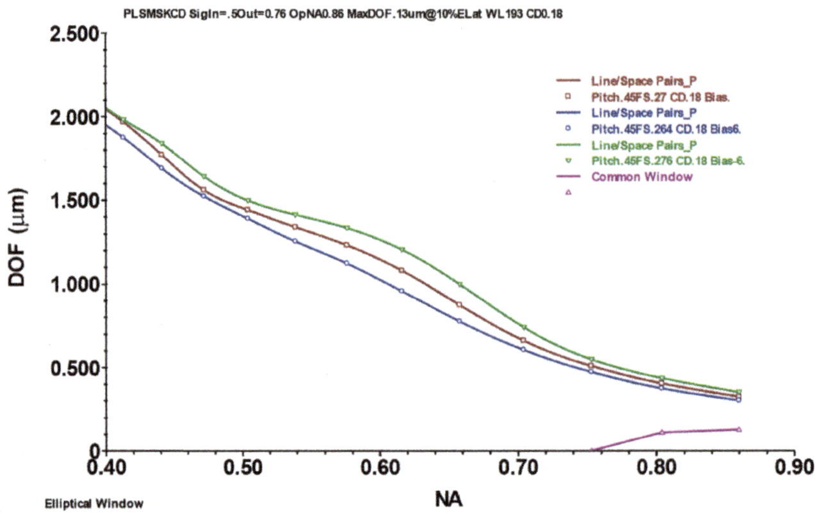

Figure 5.49 DOF versus NA from E-D trees with 0, 6-nm, and –6-nm mask biases. This slight change in the mask CD tolerance causes the common window to be lost.

Fig. 5.49. Tightening the mask CD range to 6 nm improves the common DOF significantly, as shown in Fig. 5.50.

The mask CD range is not the only mask parameter that needs to be controlled. Placement errors also contribute to reduction of the common E-D window. The same case as in Fig. 5.50 is now considered, except that the CD error is replaced by a placement error. Figure 5.51 shows the DOF-versus-NA curves of the three placement errors on mask, 0 and ±5 nm. The common

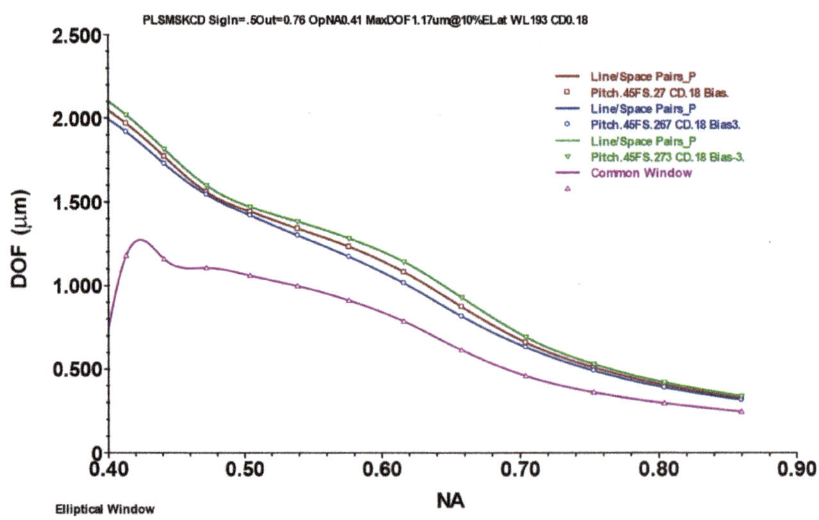

Figure 5.50 DOF versus NA from E-D trees with 0, 3-nm, and –3-nm mask biases, showing much improvement in the common window.

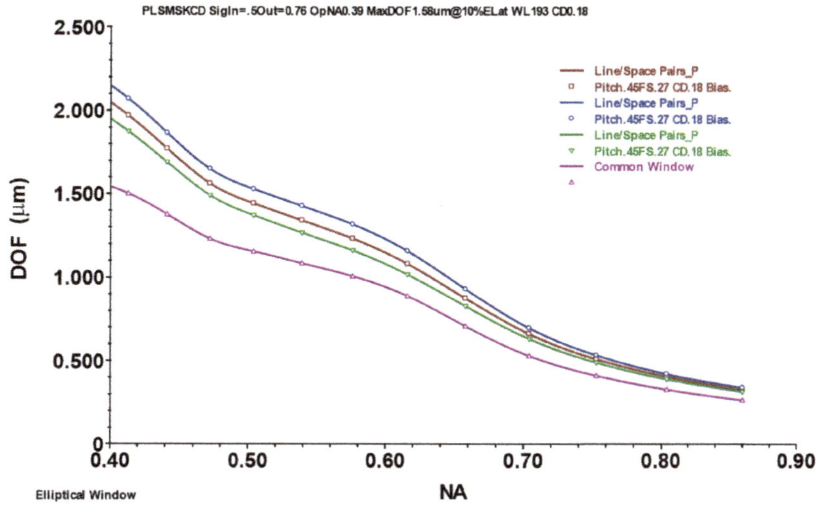

Figure 5.51 DOF versus NA from E-D trees with 0, 5-nm, and –5-nm mask placement errors. The reduction of the common window is less than that in Fig. 5.48.

DOF is again seen to be less than any of the individual DOFs. Compared with Fig. 5.48, the placement errors in this case are more tolerable than the CD errors.

Relaxing the placement error to ±8 nm results in the DOF-versus-NA curves shown in Fig. 5.52. There is clearly more loss of DOF with larger placement errors. Realistically, mask CD errors and placement errors co-exist. Figure 5.53 shows the combined effect of ±5-nm CD and mask placement

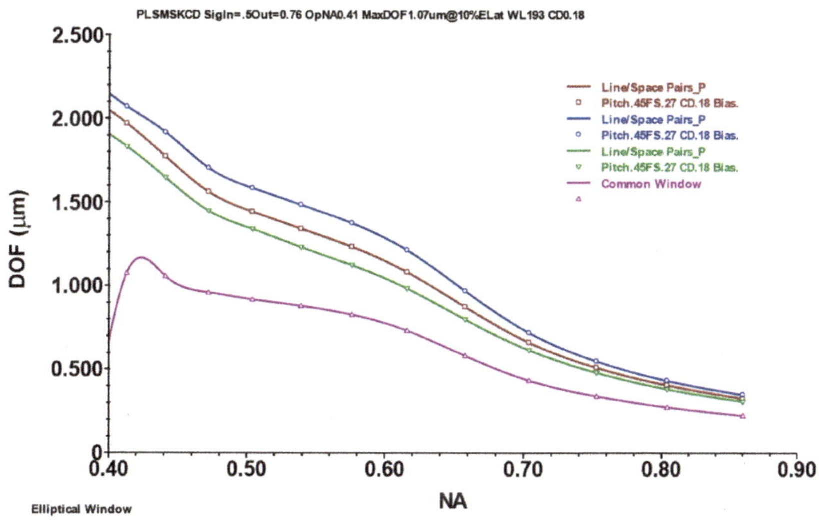

Figure 5.52 DOF versus NA from E-D trees with 0, 8-nm, and –8-nm mask placement errors. There is clearly more loss in DOF with larger placement errors.

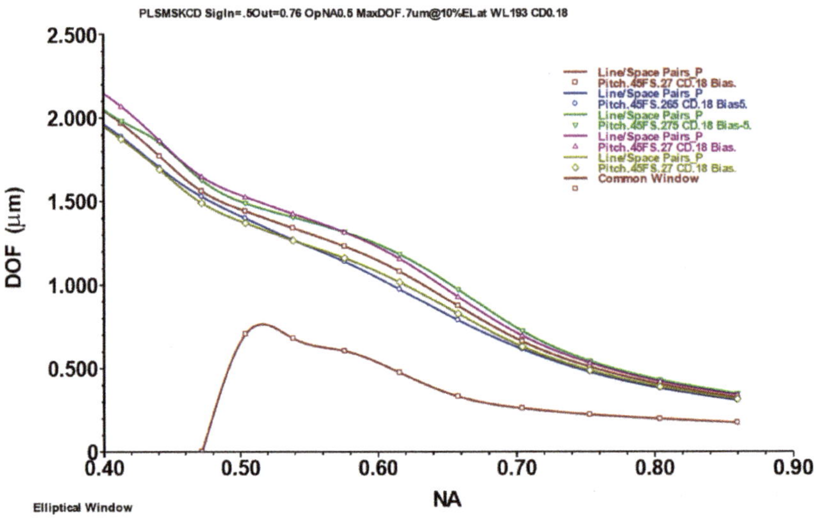

Figure 5.53 DOF versus NA from 5 E-D trees with 0, ±5-nm CD, and ±5-nm mask placement errors. The loss of E-D window is dominated by mask CD errors.

errors. In this case, the loss of the E-D window is dominated by mask CD errors.

5.5.6.9 Effects of phase-shifting mask errors

In addition to CD and placement errors, phase-shifting masks can further suffer from phase and transmission errors due to fabrication tolerances. Figure 5.54 consists of the E-D tree of five cases, namely, 100% transmission and 180-deg phase shift, 170-deg phase shift, 190-deg phase shift, 90%

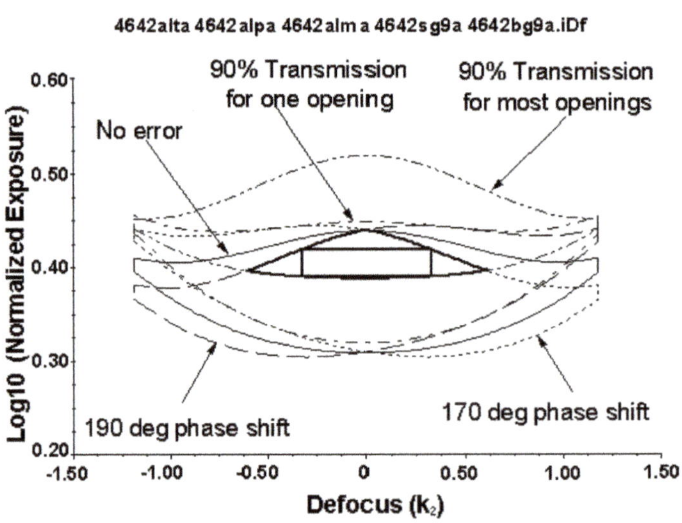

Figure 5.54 AltPSM with phase and transmission errors (k_1 = 0.46, σ = 0.42).

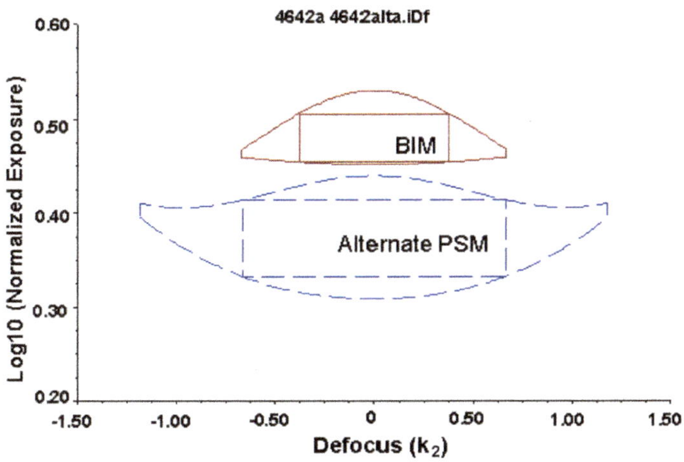

Figure 5.55 Perfect AltPSM taken from Fig. 5.54 and a perfect BIM. If one does not properly specify the AltPSM, the imaging performance of AltPSM can be worse than that of a simple BIM.

transmission at one of the openings, and 90% transmission at most of the openings. The common E-D window with these mask-making errors has EL = 7% and $DOF = 0.62k_2$. This is reduced from 18.4% and $1.33k_2$ of a perfect Alt PSM shown in Fig. 5.55. The perfect binary intensity mask (BIM) shown in the same figure has EL = 11.5% and $DOF = 0.75k_2$. Hence, if one does not properly specify the AltPSM, the imaging performance of this mask can be worse than that of a simple BIM. The CD tolerances of the BIM and the AltPSM are not included here.

References

1. B. J. Lin, "Where is the lost resolution?" *Proc. SPIE* **633**, p. 44 (1986) [doi: 10.1117/12.963701]. Note: B. J. Lin credited these two equations to Lord Rayleigh because of Rayleigh's criterion for resolution. The equation for DOF was derived by this author after consulting Eq. (8.8.27) in Ref. 3 and the equation for resolution.
2. B. J. Lin, "Optimum numerical aperture for optical projection microlithography," *Proc. SPIE* **1463**, p. 42 (1991) [doi: 10.1117/12.44773].
3. M. Born and E. Wolf, *Principles of Optics, 6th Edition*, Cambridge University Press, Section 8.5 (1980).
4. J. W. Goodman, *Introduction to Fourier Optics*, McGraw-Hill Publishing Co., Sections 6.2 and 6.3 (1968).
5. B. J. Lin, "New λ/NA scaling equations for resolution and depth-of-focus," *Proc. SPIE* **4000**, 759–764 (2000) [doi: 10.1117/12.389068].
6. Page 130 in Ref. 4.
7. Section 8.6, p. 424 in Ref. 3.

8. B. J. Lin, "Exposure-defocus forest," *Jpn. J. Appl. Phys.* **33**(12S) p. 6756 (1994).
9. J. W. Bossung, "Projection printing characterization," *Proc. SPIE* **100**, pp. 80–84 (1977) [doi: 10.1117/12.955357].
10. B. J. Lin, "The relative importance of the building blocks for 193 nm optical lithography," Lecture notes, First Seminar on 193 nm Optical Lithography in Taiwan, p. 403 (1998).

Chapter 6
Hardware Components in Optical Lithography

An optical lithography imaging system includes the following key components: a light source that provides the exposure photons with the desired energy spectrum; an illuminator that collects light from the source, adjusts its coherence and incident angles, and delivers light to the mask uniformly; a mask containing the circuit image to be replicated to the photoresist; an imaging lens of the desired NA and field size to reproduce the mask image in the resist by exposure; the resist layer coated on the wafer; the thin-film stack to be delineated by the pattern-transfer process or implantation; the wafer held by a chuck on a wafer stage that can be moved for alignment and field stepping; and alignment-viewing devices to guide the aligning movement of the stage. These components are described consecutively in the following sections.

6.1 Light Sources

There are two types of light sources for use in optical lithography exposure tools: the mercury arc lamp and the excimer laser. These light sources are bright and efficient in their respective wavelength spectra.

6.1.1 Mercury arc lamps

The mercury arc lamp has been the light source of choice for optical lithography because of its many usable emission lines in the near-UV(350–460 nm), mid-UV (280–350 nm), and deep-UV (200–280 nm) regions. There is even an emission line at 184 nm. This lamp's high brightness compared with other nonlaser sources is another reason for its popularity. Figure 6.1 shows the radiation spectrum of a typical mercury arc lamp. Figure 6.2 shows a schematic drawing of a mercury arc lamp. Pointed and rounded electrodes are enclosed in a quartz envelope that contains a noble gas and a drop of mercury. During ignition, electric discharge takes place between the two electrodes. As

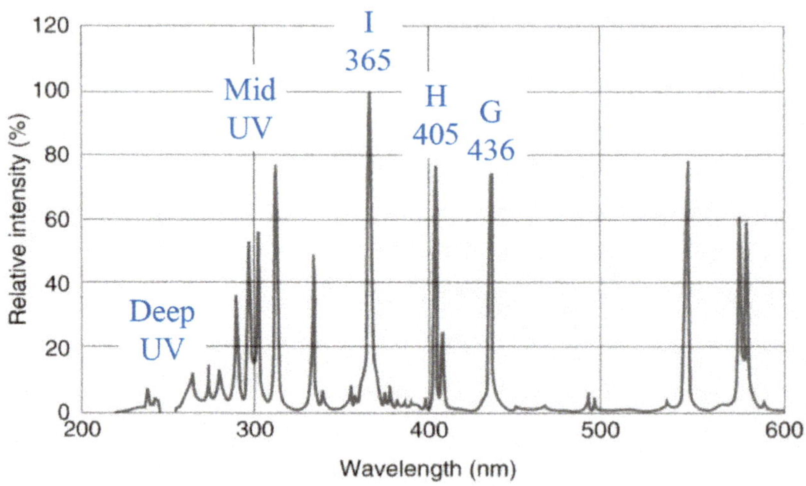

Figure 6.1 Radiation spectrum of a mercury arc lamp (reprinted with permission from Ushio).

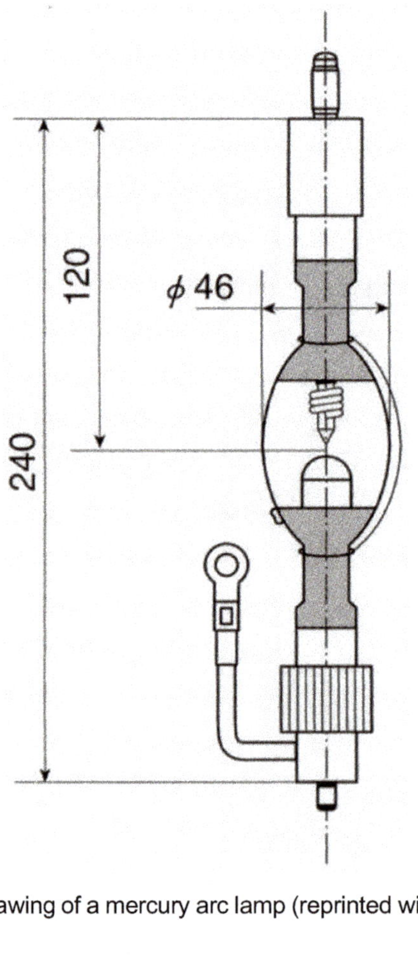

Figure 6.2 Schematic drawing of a mercury arc lamp (reprinted with permission from Ushio).

the heat builds up, the mercy is vaporized, and the radiation spectrum becomes dominated by that of the mercury. The pointed electrode becomes a bright radiating spot. The rounded electrode produces a spot of less brightness that is undesirable for diffraction-limited imaging because the illumination system is often designed for only a single point source.

The main problem with mercury arc lamps for monochromatic lithographic imaging is their low optical efficiency. The optical output power is typically 5% of the electrical input power; furthermore, no more than 1 or 2% of the optical energy is used after spectral filtering because the refractive projection optics requires a narrow bandwidth on the order of 5 nm. For deep-UV lithography, the mercury arc lamp is even more inefficient because of the low output of the 254-nm line. As the lamp ages, the electrode changes shape due to corrosion. As a result, the shape and size of the bright radiating spot changes, and the brightness is reduced. The inside wall of the lamp envelope may be contaminated by a light-absorbing deposit that further reduces the light output. Regardless, mercury arc lamps have been the workhorse of optical lithography for 436-nm g-line, 405-nm h-line, and 365-nm i-line systems. They have also been used for deep-UV broadband systems in 4X-reduction, deep-UV scanners,[i] 1X full-wafer deep-UV projection mask aligners,[ii] and 1X full-wafer deep-UV proximity mask aligners.[iii] By optimizing the gas mixture in the lamp and other parameters, the deep-UV spectrum can be enhanced. Figure 6.3 shows the radiation spectrum of an

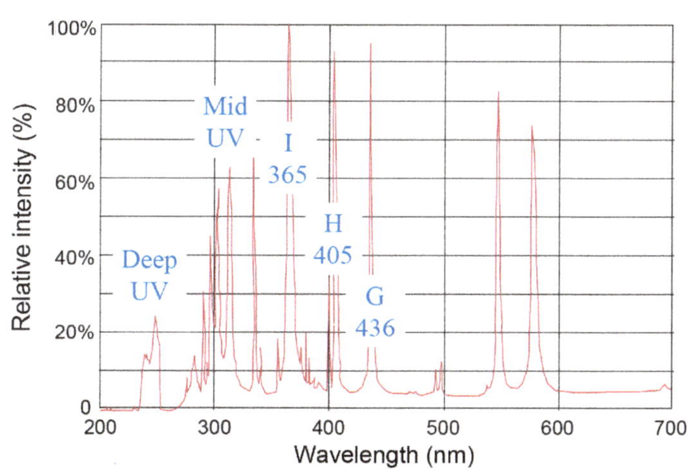

Figure 6.3 Radiation spectrum of an Ushio UXM-4000P Hg arc lamp optimized for the deep-UV, operating at 100 amp and 4 kW (reprinted with permission from Ushio).

[i]PerkinElmer Microscan 1 system.
[ii]PerkinElmer 500 and 600.
[iii]Canon deep-UV proximity mask aligner.

Ushio UXM-4000P Hg arc lamp that is optimized for the SVGL Microscan 1 deep-UV scanner. With such applications, the quartz envelope of the lamp needs to have high transmission in the deep UV.

6.1.2 Excimer lasers

When the exposure wavelength is shortened to the deep-UV regime, the low optical efficiency of mercury arc lamps makes it difficult to provide the required exposure power. Furthermore, in the vicinity of 250 nm there are very few optical materials with sufficient transmissivity and mechanical working properties to incorporate into achromatic imaging lenses. For deep-UV dioptric systems, fused silica is the only optical material used for the imaging lens; thus, no correction of chromatic aberration is possible, necessitating an illumination with a bandwidth in the picometer regime. Further reducing the wavelength makes the bandwidth requirement even tighter. A laser source is inevitable. However, most laser sources are highly coherent temporally and spatially. The speckles produced by such a source are undesirable for microlithography applications. Fortunately, the excimer laser happens to be extremely efficient. Because of its high gain, light needs to travel in the resonator only a few times before it is sufficiently amplified, making it inherently much less spatially coherent than other lasers. The selection of wavelengths such as 248 nm with KrF, 193 nm with ArF, and 157 nm with F_2 makes excimer laser a clear choice for deep-UV and shorter wavelengths.

The earliest application of excimer lasers to microlithography was in proximity printing.[1,2] Subsequently, these lasers have been demonstrated on a full-field all-reflective system[3] and a dioptric step-and repeat system.[4] A comprehensive treatment of the excimer laser, including its history, operation, design considerations, system, and maintenance, can be found in the work by Das and Sengupta.[5] Here, we examine the aspects of excimer lasers that are most relevant to understanding and practicing microlithography for manufacturing.

6.1.2.1 Operation principle of the excimer laser

Microlithography-related excimer lasers are pumped by electric discharge. During the discharge, KrF* excimers are formed. Once the excimers are present, spontaneous emission, stimulated emission, and quenching take place. Spontaneous and stimulated emissions are shown in the following reactions,

$$\mathrm{KrF}^* \rightarrow \mathrm{Kr} + \mathrm{F} + h\nu \tag{6.1}$$

and

$$\mathrm{KrF}^* + h\nu \rightarrow \mathrm{Kr} + \mathrm{F} + 2h\nu, \tag{6.2}$$

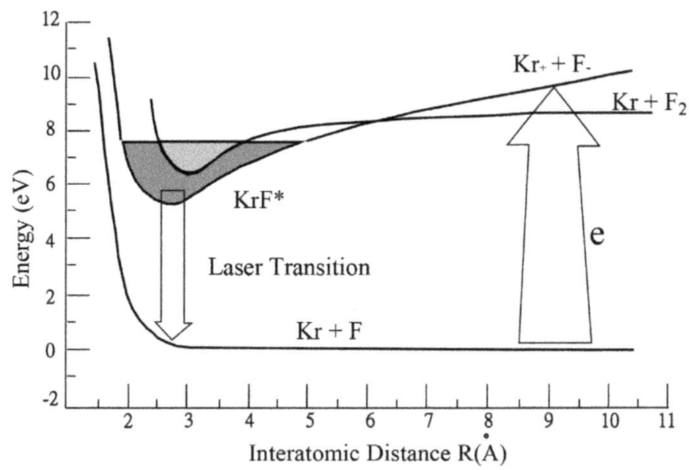

Figure 6.4 Energy diagram of a KrF excimer laser (adapted from Ref. 68).

respectively. The stimulated emission provides the desired optical output. Figure 6.4 shows the energy diagram of a KrF excimer laser. The $Kr^+ + F^-$ band supports stimulated emission, and the $Kr + F_2$ band supports spontaneous emission.

The discharge voltage is more than 10 kV with the gas in the resonator at several atmospheres in pressure. A pulsed power supply capable of such voltage at a repetition rate between 1 and 2 MHz is used. It consists of a series of π networks of saturable inductors and high-voltage capacitors controlled by a high-speed, high-voltage switch. A typical discharge duration is between 300 and 500 nsec. The optical pulse duration is about 10 to 20 nsec.

6.1.2.2 Bandwidth narrowing

The 248.35-nm pulse from an unregulated excimer laser has a bandwidth of a few hundred picometers. However, a much narrower bandwidth is required to reduce chromatic aberration for lithography. The bandwidth requirement is a function of the dispersion property of the optical material as well as the NA of the imaging lens. Table 6.1 shows the E95 and full-width-at-half-maximum (FWHM) spectral bandwidth requirements of 248-, 193-, and 157-nm lenses at different NAs. A higher NA requires a narrower bandwidth.

The E95 bandwidth is defined as the band where 95% of the entire spectral energy is located. The FWHM bandwidth is the full width of the spectrum at the half-maximum point. This bandwidth has been used to characterize laser bandwidth until E95 is shown to be a better quantity to keep track of. Figure 6.5 shows the FWHM and E95 bandwidths from the apparent laser output that is the raw data from the bandwidth-measuring equipment. This raw data is a convolution of the response spectrum of the measuring instrument. The actual laser output spectrum can be reconstructed by

Table 6.1 Bandwidth requirements of an imaging lens.

248 nm	<3 pm E95 at 0.35 NA	<1.4 pm E95 at 0.75 NA	<1 pm E95 at 0.85 NA
193 nm	1.2 pm E95 at 0.75 NA	0.75 pm E95 at 0.85 NA	0.5 pm E95 at 0.93 NA
157 nm	0.5 pm FWHM at 0.8 NA dioptric lens	1 pm FWHM at 0.8 NA catadioptric lens	

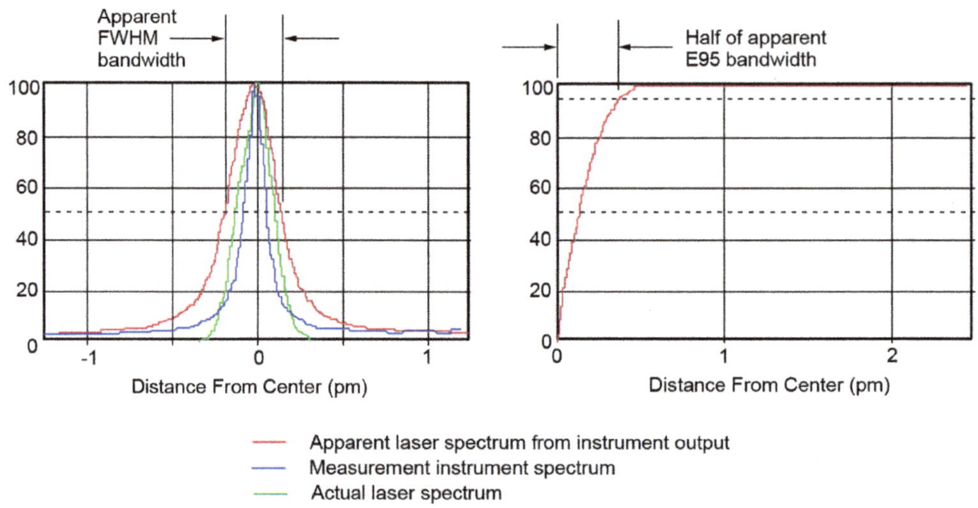

Figure 6.5 Laser output spectra: FWHM and E95 bandwidths.

deconvolution. The curves on the left-hand side of the figure indicate, consecutively, from the wide to the narrow widths at the 50% level: the apparent laser spectrum from the instrument output, the measurement instrument spectrum, and the deconvoluted actual laser spectrum. The right-hand side curve indicates the integrated power within the band defined by the distance from the center.

Table 6.2 shows the dispersion properties of fused silica and CaF_2 at 248 and 193 nm.[6] A higher dispersion calls for a narrower bandwidth in illumination. To maintain high efficiency, the wavelength-narrowing feature is incorporated into the laser resonator. Three wavelength-selecting elements

Table 6.2 Dispersion of fused silica and CaF_2.

	Fused silica (nm^{-1})	CaF_2 (nm^{-1})
248.4 nm	−0.00056[7]	−0.00037[8]
193.4 nm	−0.00157[9]	−0.00098[8,9]
157.6 nm	—	−0.0026[8,10]

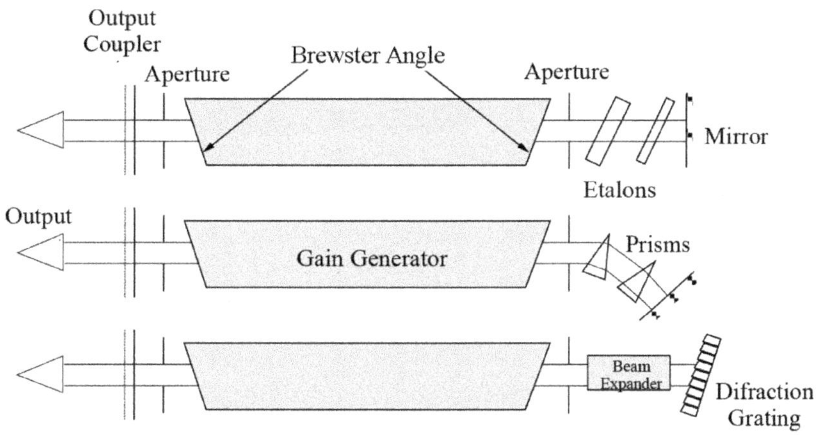

Figure 6.6 Bandwidth-narrowing schemes using basic wavelength-selecting elements.

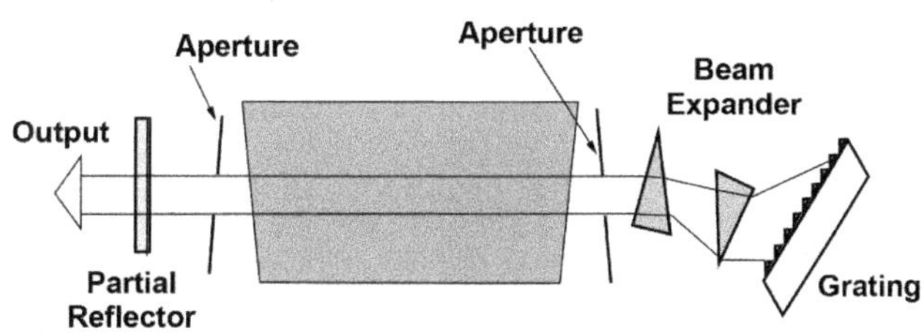

Figure 6.7 Bandwidth-narrowing system using prisms and a grating (adapted from Ref. 68).

have been attempted, namely, etalons, prisms, and gratings, as shown in Fig. 6.6. Etalons have high wavelength resolution but are highly sensitive to heat. Prisms do not provide sufficient resolution. The combination of a beam expander and a grating (as shown in Fig. 6.7) has been successfully employed.[11] Another combination uses a grating for coarse wavelength selection and an etalon for fine selection[12] to overcome the problem of etalon heating.

6.1.2.3 Spatial coherence

Although great efforts have been spent narrowing the bandwidth of the excimer output, its temporal coherence inevitably increases with a narrower bandwidth. More laser speckles are generated. The spatial coherence must be reduced to suppress speckle generation. Inherently, the excimer laser is already of low spatial coherence because only three or four light passes through the amplifying media are required, making it operate in numerous modes. The typical coherent length of excimer lasers is on the order of a few hundred micrometers (compared to meters for some highly tuned lasers). Nevertheless, further reduction of spatial coherence by scrambling is necessary. Early systems used a rotating mirror, fly's eye lens, or optical fiber to spread out the point source. Modern illuminators incorporate a light

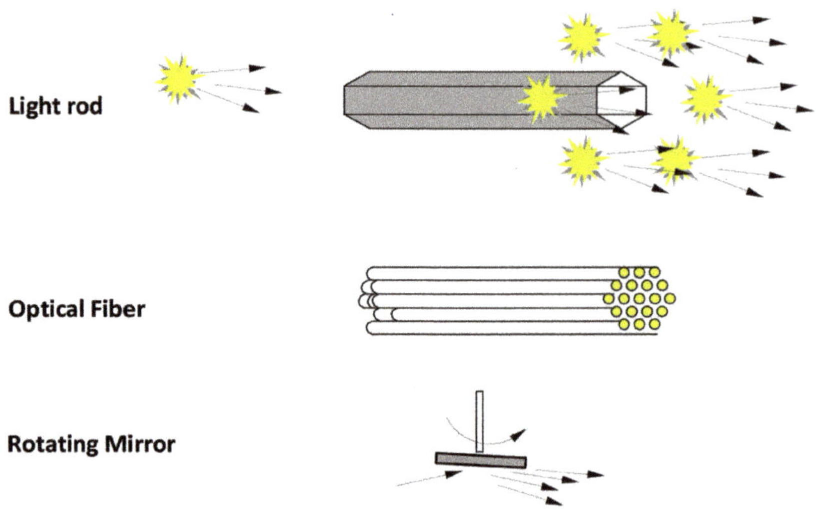

Figure 6.8 Methods to scramble the light source.

rod to internally reflect the light source many times to scramble the light and fill the entrance pupil. Figure 6.8 shows the early schemes that used a rotating mirror and an optical fiber, and a modern scheme that uses a light rod. The light rod is so effective that even conventional point sources such as mercury arc lamps can also benefit from it.

6.1.2.4 Maintenance, safety, and lifetime of excimer lasers

An excimer laser is more expensive to purchase and operate than a mercury arc lamp. At the very least, it requires maintenance of the laser chamber, the wavelength-stabilizing module, and the bandwidth-narrowing module. The laser also takes up more factory space.

One should be extremely careful with radiation-induced damage to the eyes. Unlike the mercury arc lamp, which has a visible output to alert bystanders, the output of KrF and ArF excimer lasers is completely invisible. Damage is realized only after the fact. Therefore, many safety interlocks are incorporated to prevent exposing the laser beam to humans in any circumstance.

Operating at many kilohertz, an excimer laser can deliver tens of billions of pulses per year. Its lifetime and maintenance are measured in number of pulses. A modern excimer laser for lithography has shown lifetimes on the order of 15 billion pulses for the chamber and 20 billion pulses for the wavelength-stabilizing and bandwidth-narrowing modules. Routine maintenance includes cleaning the F_2 trap and the output coupler. The former is required after several hundred new fills and the latter after 7 billion pulses. An excimer laser is usually fine-tuned monthly to remove wavelength drift and restore bandwidth and energy stability.

6.2 Illuminator

The illuminator of a typical scanner is shown in Fig. 6.9. The laser beam goes through the beam-pointing-and-steering unit and the variable attenuator to adjust its direction, position, and intensity before entering the illumination shaping optics. There, the beam is shaped for the type of illumination, such as axial, off-axis, Quasar, etc. To reduce the spatial coherence, the beam passes through the quartz rod, where its illumination energy is monitored. The coherence of the illumination is determined at the reticle masking assembly. Together with the imaging lens, these components constitute a Köhler illumination system. At the reticle masking assembly, the area outside of the printed area of the reticle can be masked off.

6.2.1 Köhler illumination system

A typical Köhler illumination system[13] is shown in Fig. 6.10. The key to Köhler illumination is that the source plane is conjugate to the pupil planes, which are, of course, conjugate to each other. Hence, the rays originating from the source point Q_s converge at point Q_{P1} at pupil 1. They converge again at Q_{P2}. The rays from other source points, represented by the other ray at the reticle opening $A_{M,}$ go through the imaging lens system, including pupil 2, to converge at the wafer at A_W. There are now two conjugate relationships, one linking the source plane and pupil planes, and the other linking the reticle plane, wafer plane, and reticle masking plane in the reticle masking assembly.

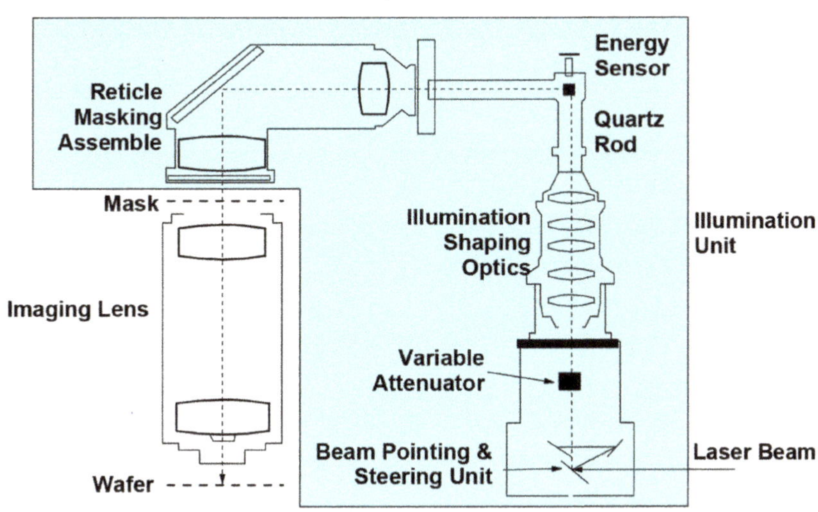

Figure 6.9 Schematic drawing of the illumination unit of an ASML scanner (reprinted with permission from ASML).

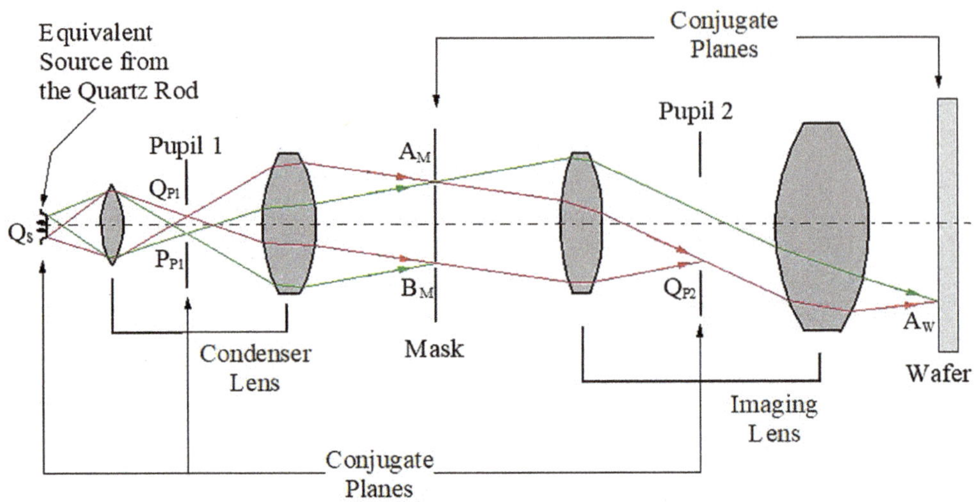

Figure 6.10 Schematic of a Köhler illumination system.

6.2.2 Off-axis illumination

The simplest way to produce off-axis illumination is to block the unwanted central illuminated area with a light-blocking stop. In fact, this method is very flexible because any arbitrary shape can be made. However, the illumination energy in the blocked area is wasted, which increases the exposure time and thus reduces the wafer throughput, increasing the per-wafer manufacturing cost.

Mating cones are used for producing off-axis light without energy loss, such as in the case of annular illumination. As depicted in Fig. 6.11(a), the illumination is on axis when the two cones are in contact with each other. When there is a separation between them, as in Fig. 6.11(b), the illumination

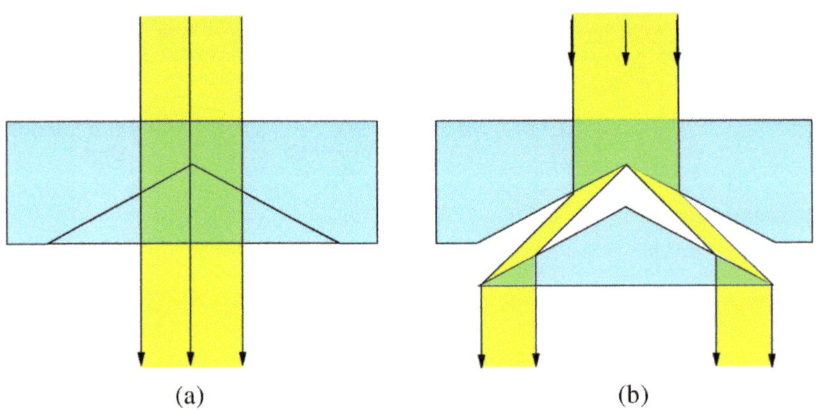

Figure 6.11 Schematic of mating cones that produce off-axis light without energy loss: (a) on-axis and (b) split illumination (reprinted with permission from ASML).

is split. The distance between the two beams is a function of the separation between the two wedges.

6.2.3 Arbitrary illumination

Modern illuminators must provide freeform[14] illumination settings instead of just simple off-axis illumination. The diffractive optical element (DOE) uses localized diffraction gratings to shape a uniform beam into any arbitrary distribution of orientations, as is required for optimum imaging. The different orientations become intensity distributions at the pupil plane of the imaging lens. The FlexRay® illuminator (from ASML) works similarly, except for replacing the cluster of gratings with a programmable micro-mirror array, as shown in Fig. 6.12. There can be thousands of micro-mirrors in a FlexRay® illuminator.

6.3 Masks

A photomask consists of patterned absorbers and/or phase shifters delineated on a transparent substrate. According to the presence or absence of the absorbers and phase shifters, as well as the number of levels of these patterning materials, the photomask can be classified generically into the six types shown in Fig. 6.13.

Binary intensity mask (BIM)—This is the most commonly used mask type. Absorbers are placed in various areas of the mask, according to the circuit pattern. The absorber completely blocks light as it hits the mask. Alternatively, light passes through the mask where no absorber exists. Note that many people refer to a BIM as a "binary mask." This is a misnomer because binary masks can be further subdivided into *binary intensity masks* and *binary phase masks*. Also note that BIM is sometimes confused with chrome on glass (COG); the former denotes the type of mask, and the latter denotes the type of mask blank. COG cannot be distinguished from a phase-shifting mask (PSM) because many PSMs consist of chrome and quartz, such

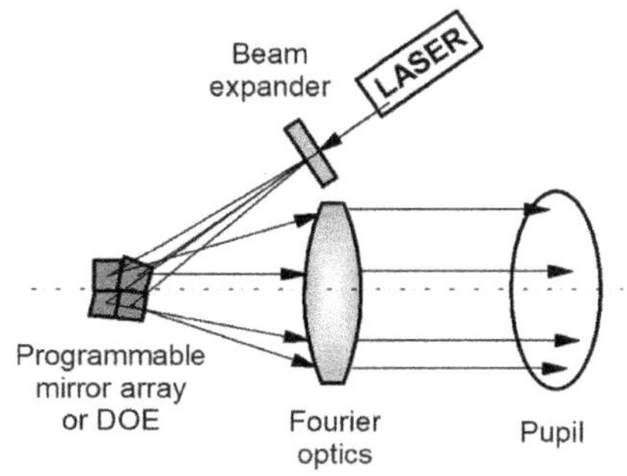

Figure 6.12 Working principle of a diffractive optical element (DOE) and a FlexRay® illuminator (reprinted with permission from ASML).

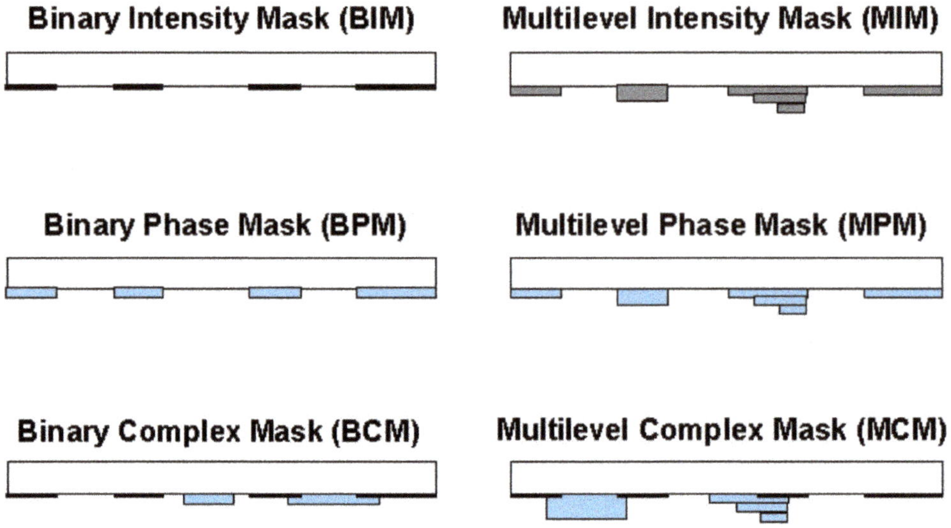

Figure 6.13 Types of photomasks.

as the AltPSM. The word "glass" is also a misnomer. Soon after projection printing became popular, the mask substrate became quartz instead of glass to take advantage of quartz's low thermal expansion coefficient. Deep-UV lithography definitely calls for a quartz substrate due to its transparency in the shorter wavelengths.

Binary phase mask (BPM)—There is no absorber on this mask. The only pattern-dependent variation possible for a BPM is a π phase shift. This is typified by the unattenuated (chromeless) PSM.

Binary complex mask (BCM)—The absorber controls the passing or blocking of light with its absence or presence. The phase shifter shifts the phase of light by 180 deg. There are phase-shifted and non-phase-shifted areas on the mask. BCM constitutes the bulk of PSMs, including the alternating, the attenuated, the subresolution-assist, and the rim types. The attenuated type is included because the distribution of the absorber/phase shifter is still binary, even though the absorber carries out the function of a phase shifter and does not completely absorb. However, when regular absorber patterns are included in these masks, they become a multilevel complex mask, as described below.

Multilevel intensity mask (MIM)—An intensity mask may contain many gray levels for specific applications, such as an exposure calibration mask that contains many calibrated gray levels to facilitate rapid and economical determination of the exposure dosage. MIMs may also be used for threshold leveling,[14] which modifies the background transmission on the mask according to the pattern density to produce large exposure-defocus (E-D) windows.

Multilevel phase mask (MPM)—The phase shifters produce many levels of phase on the mask—presumably, 0 deg, 90 deg, and 180 deg, or 0 deg, 60 deg, 120 deg, and 180 deg. This type of mask is very rarely used.

Multilevel complex mask (MCM)—Either the absorber or the phase shifter is multilevel. The latter is exemplified by the AltPSM in combination with multilevel phase-shifting edges.[16] In order to remove phase-shifting conflicts, extra phase-shifting edges are introduced and multilevel phase shifts of a small increment (such as 60 deg) are implemented at these edges to reduce their image contrast so that they are not printed. A multilevel absorber in an MCM is exemplified by AltPSMs having completely opaque areas in addition to the transmissive absorber/phase shifter. The need for MCM AttPSMs is given in Section 6.3.4.3.

Note that "PSM" is not a generic term. A PSM may be a BPM, BCM, MPM, or MCM. For a detailed description of the various types of PSMs, consult Section 6.3.4.3.

6.3.1 Mask substrate and absorber

Fused silica has been used as the substrate material for two reasons:

1. Its thermal expansion coefficient at 0.5 ppm/°C is an order of magnitude smaller than that of other candidates. Except for proximity printing, which requires matched thermal expansion coefficients between the mask and the wafer, it is very desirable to have thermal stability in the mask.
2. Fused silica is transparent from the visible region to below 200 nm. It is usable for g-line (435.83 nm), i-line (365.02 nm), KrF (248.35 nm) and ArF (193.39 nm) wavelengths. Even at the F_2 (157.63 nm) wavelength, where only a few fluoride compounds (such as CaF_2) transmit, fluorinated quartz is found to be acceptable as a mask substrate. If CaF_2 were needed, the cost of mask blanks in 157-nm lithography would be prohibitive. EUV masks for the 13.5-nm wavelength are extremely expensive for reasons other than the material for blanks. These masks will be covered in Chapter 9 on EUV lithography.

Like the mask substrate material, the mask absorber has also survived many generations of optical lithography. In the proximity printing days, the emulsion mask absorber quickly gave way to chromium because of the latter's hardness and smaller thickness. Chromium remains as the mask absorber of choice. A photoresist material[17] has been proposed for the mask absorber. By using the resist directly, the chromium deposition and delineation steps can be eliminated. Silicides such as $MoSi_2$ have also been proposed for mask absorbers because of their desirable processing characteristics and favorable absorption spectra. However, the inertia of low-defect chrome-on-quartz processing technology is simply too overwhelming. $MoSi_2$ is primarily used as the phase-shifting absorber for attenuated PSMs. It can be fine-tuned to exhibit 6% transmission and a 180-deg phase shift at the same thickness, making the AttPSM fabrication process like that of BIMs.

6.3.2 Pellicles

The pellicle is a transparent membrane[18] mounted on a frame that is attached to the mask substrate, as shown in Fig. 6.14. It is usually an organic polymer of a few micrometers in thickness. The 6-mm frame thickness is carefully chosen so that, if a particle that is not too large is attached to the pellicle, the particle is completely out of focus and can at most induce a tolerable exposure change in its locality. It is inconceivable that any mask would not be protected by a pellicle because great care has been taken to ensure that the mask induces no defect. If it is not protected, there will be repeating defects on every wafer at every exposure field. It is not easy to keep the mask defect free, even if it has been fabricated in this way. If only one foreign particle falls on the mask during transportation, loading/unloading to the exposure tool, the exposure process, storage, or any other type of handling, the damage is done. Hence, pellicles are indispensable.

When mounting the pellicle to the mask, care should be taken to prevent trapping or producing particles within the cavity. Pellicle mounting should not induce mask distortion. Remounting should be kept to a minimum. To reduce the number of remountings, mask-inspection tools are often designed for use without removing the pellicle. However, pellicle removal, and thus remounting, is inevitable when mask cleaning or repair is needed.

Glue is used to attach the pellicle frame to the mask. Great care must be taken in selecting the glue to prevent stress production, contamination, or generation of particles during actinic wafer exposures. In addition, the pellicle material must withstand prolonged exposures without reducing its transmissivity or smoothness. Sometimes tiny holes are drilled through the pellicle frame to prevent trapping of harmful vapor generated during exposure. However, this leaves the possibility of contamination from the outside.

6.3.2.1 EUV pellicles

It has been inconceivable to use pellicles in EUV lithography because transparent materials at the 13.5-nm wavelength are unavailable. However, lithographers have grown accustomed to having a pellicle to prevent fall-on defects on the mask. There have been intense efforts to develop the EUV pellicle technology. Details are given in Chapter 9 on EUV lithography.

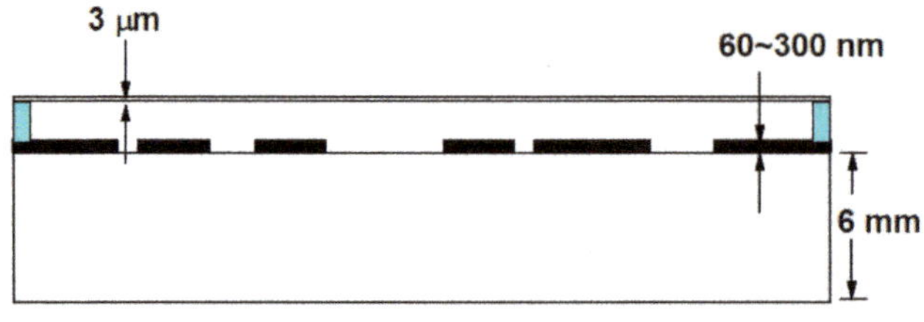

Figure 6.14 A mask and its pellicles.

6.3.3 Critical parameters for masks

6.3.3.1 CD control

The mean and standard deviation of the critical dimension (CD) designated by the circuit designer are important parameters for the mask. The CD mean must be as close to the desired value as possible. The desired value must be the target value on the wafer multiplied by the reduction factor of the imaging system. A slightly different value may produce a larger E-D window, depending on the feature shape, feature size, feature environment, optical train, resist characteristics, and processing conditions. This value is referred to as the mask bias. The standard deviation consumes the total wafer CD control budget.

6.3.3.2 Placement accuracy

Each feature on the mask must be placed at the desired position with respect to other features on the same mask, as well as on other masks whose images must be overlaid together. Hence, placement error contributes significantly to the overlay accuracy budget. Placement errors are often induced by the mask-making machine, mostly due to difficulties in controlling the position of the writing beam. Additionally, the interferometer-controlled table that moves the mask to each writing field can introduce residual errors, even though the interferometer control is supposed to have nanometer-scale precision.

6.3.3.3 Mask transmission and thermal expansion

The transmission of the mask blank should be above 90%. From the point of view of mask heating, a higher transmission is desirable. After reaching the material transmission limit, the only way to increase transmission is to reduce the mask thickness. In the early days of microlithography, mask thickness was on the order of 90 mils (2.3 mm). It was increased to 1/4 inch (6.35 mm) to reduce mask sag in a step-and-repeat exposure tool. The thickness has remained at 1/4 inch for decades.

Mask heating causes thermal expansion. With a thermal expansion coefficient of 0.5 ppm/°C, raising the temperature of the mask substrate by 1 °C expands a 150-mm square mask by 75 nm on each side. This can easily be corrected by fine tuning the magnification of the exposure tool, especially in a projection tool. In lensless systems, such as those using proximity printing, mask thermal expansion can be detrimental to the overlay because wafers at different stages of processing can exhibit subtle magnification changes. In x-ray proximity exposure tools, the mask must be thermally controlled to manage random magnification errors. Systematic errors are managed by preparing the mask according to the expected magnification change. The solution in e-beam proximity printing[19] is most unique. The tilting angle of the beam is adjusted to compensate for any predictable overlay errors, including magnification changes, with the advantage of covering only a small area in the field.

6.3.3.4 Mask reflectivity

One of the main concerns with chromium as a mask absorber is its high reflectivity at the side facing the imaging lens. This reflection can reduce image contrast, as seen in Fig. 6.15. Points a and b are the edges of an opening in the mask. The absorber can be found in the areas outside of a and b. We shall assume that these points image perfectly at their conjugate points a' and b' on the resist surface. In the best case, assuming no other thin-film layer on the wafer, the wafer itself serves as the reflection plane for points a' and b'. Point b' is now mirror imaged to point b'', which has a conjugate at b'''. The light ray that passes b'' and b''' hits the chromium part of the mask. This illumination is reflected back onto the photoresist in the areas that are supposed to be dark. The stray reflection is induced in the vicinity of the image edge that is most susceptible to stray-light-induced image deterioration. Hence, the image contrast is reduced. Figure 6.15(a) depicts ray tracing through a simple lens, and Fig. 6.15(b) depicts ray tracing through a double telecentric lens system. In both cases, the image contrast is affected by the mask reflectivity. More discussion on the telecentric lens system can be found in Section 6.4.1.5.

Figure 6.16 shows linewidth-versus-defocus plots at a range of exposure dosages for mask absorber reflectivities of 66%, 33%, and 10%. The 10% result clearly confirms better linewidth control and exposure tolerance. It is desirable to reduce absorber reflection to 10% or lower.

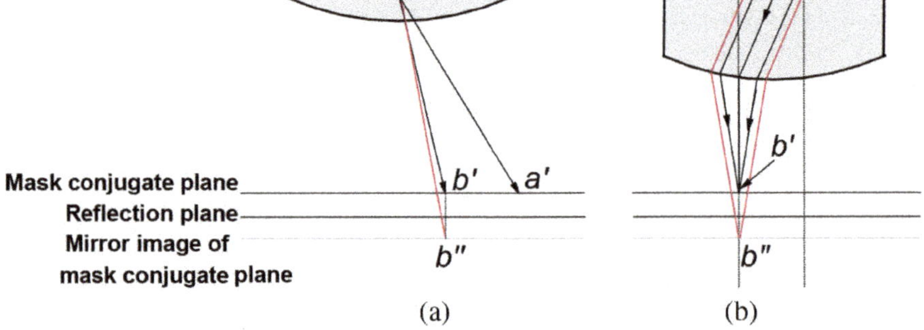

Figure 6.15 Stray light induced by a reflecting mask absorber, depicted by ray tracing through (a) a simple lens and (b) a double telecentric system.

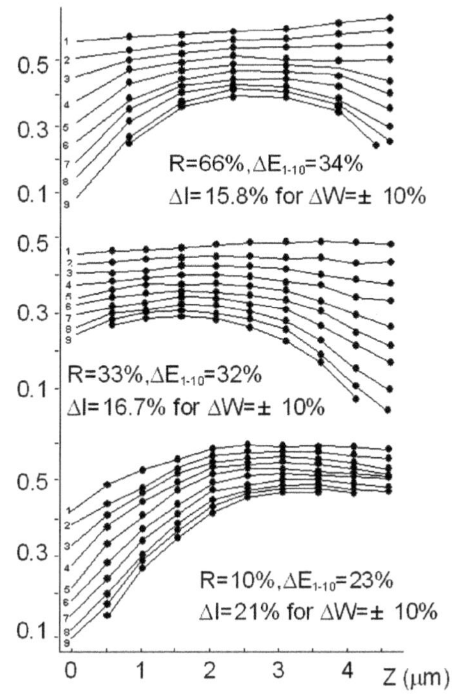

Figure 6.16 Linewidth versus defocus at different exposures for three levels of mask reflectivity.

6.3.3.5 Mask flatness

The flatness of the mask must be a small part of the DOF budget of the entire imaging system. With a reduction system, the flatness requirement is much relaxed because the DOF tolerance on the mask side is M^2, M being the reduction ratio. For example, with a mask flatness of 500 nm, the focusing error at the wafer side of a 4X reduction system is 32 nm. In addition to ensuring that the two surfaces of the mask blank are well polished and parallel to each other, the mask blank must have sufficient strength to support itself without sagging, which can induce DOF error.

6.3.3.6 Physical size

A standard photomask is $6 \times 6 \times 0.25$-inch in size. The thickness is chosen to minimize the sag that consumes the DOF budget, as mentioned in Section 6.3.3.5. The lateral dimensions are chosen to accommodate the field size requirement of the IC industry plus four sides outside the printable area. These sides are preferably opaque to frame the printable areas accurately, even though dynamically adjustable blades at four sides are provided on the exposure tool for coarse framing purposes. With the advent of step-and-scan systems and increasing pressures to increase the field size, 9×9-inch substrates have been considered. At one point, a 9-inch scanning range was reserved on ASML scanners in anticipation of future needs. However, moving to 9 × 9-inch substrates is not a simple matter. All mask-making equipment, including the writer, the track, the etcher, and the inspection and repair tools, must be changed. Also, it is not easy to agree on whether to use a 9 × 9-inch or 6 × 9-inch format. Mask makers prefer a square substrate for easier processing, but scanner makers prefer the rectangular substrate to minimize the size and weight of the reticle scanning stage.

6.3.3.7 Defect level

No printable defect is allowed on the mask for obvious reasons, as discussed in Section 6.3.2. The defect level is usually specified by the maximum allowable defect size as a function of its proximity to patterned features on the mask. A programmed defect mask (PDM), consisting of artificial defects varying in size and distance to patterned features, is often used to empirically determine defect printability for specification. Simulation is helpful to gain insight for designing the PDM. Because real defects are usually irregular in shape, their printability must be further qualified with a so-called aerial image monitoring system (AIMS®), which consists of a small-field imaging lens and an illuminator at an actinic wavelength to mimic the same $\lambda/NA/\sigma$ condition. Such an optical system is much easier to build compared to a scanner or stepper. The NA needs only to be $1/M$ of the exposure system. Therefore, 0.18 NA works for 0.72 NA of a 4X reduction system. The ultimate qualification involves exposing the mask in the production exposure system and inspecting the printed wafer with wafer-inspection tools.

6.3.4 Phase-shifting masks

A mask that contains phase-shifting areas with or without energy-absorbing areas is a PSM. It can be a BPM, BCM, or MCM, as defined in Section 6.3. Here, the types of PSMs and their configurations are discussed. The imaging performance and their comparisons are briefly given in Section 5.5.6.9. PSM applications and additional imaging performance details are given in Chapter 7, where their fabrication, inspection, and repair aspects are also discussed. An early overview article was published in 1990, when many different types of PSMs first started to emerge.[20]

6.3.4.1 Operating principle

Phase-shifting masks take advantage of the interference effect in a coherent or partially coherent imaging system to reduce the spatial frequency of a given object, to enhance its edge contrast, or to achieve both, resulting in a combination of higher resolution, greater exposure latitude, and greater DOF. The shifting of the phase is accomplished by adding an extra patterned layer of transmissive material on the mask, as shown in Fig. 6.17. When light propagates through the substrate and the extra layer, its wavelength is reduced

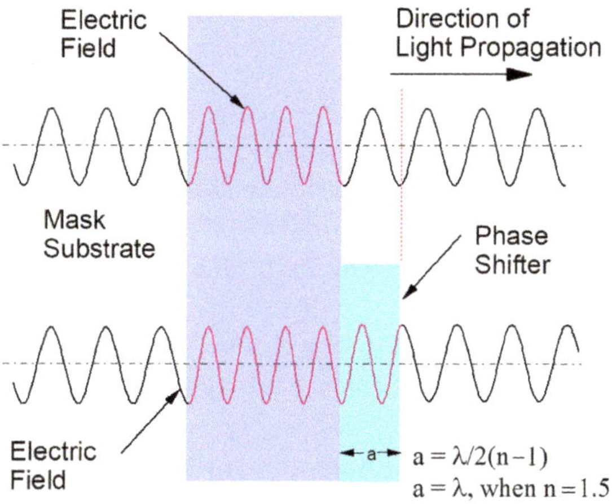

Figure 6.17 Principle of phase shifting.

from that in air by the refractive index of the substrate and then by the refractive index of the extra layer. Comparing the optical paths through the extra material and without the extra material, there is a difference of $(n - 1)a$, where n is the refractive index of the extra layer and a is its thickness. The phase difference θ is

$$\theta = a(n-1)\frac{2\pi}{\lambda}. \tag{6.3}$$

Usually, a phase shift of π is desirable, not only because the maximum difference in amplitude (namely, from 1 to -1) is achieved, but also because this is the part of the sine wave where the amplitude difference is least affected by small phase variations. In other words, $\partial\cos\theta/\partial\theta$ has its minimum at $\theta = \pi$. Setting $\theta = \pi$ in Eq. (6.3) yields

$$a = \frac{\lambda}{2(n-1)}, \tag{6.4}$$

which is the thickness required to induce a π shift. Because a popular phase shifter, SiO_2, has a refractive index of approximately 1.5, a now approximately equals the actinic wavelength. Note that (1) the phase shift can be any odd number of π, i.e., $(2m + 1)\pi$, where $m = 0, 1, 2, \ldots$, and that (2) phase shifting is relative. Phase shifting being relative means that either the high-refractive-index material in a phase-shifting mask or the low-refractive-index air path can be viewed as the phase shifter. To simplify the discussion, the extra layer is often called the phase shifter unless specifically defined otherwise.

6.3.4.2 An unflat BIM is not a PSM

Newcomers to the PSM technology often ask why a BIM with flatness variations larger than or close to $\lambda/2$ is not a PSM. Phase shifting is effective only when the illumination is substantially coherent. It starts to be ineffective for $\sigma > 0.5$. The region within which phase shifting is effective is on the order of the coherence length of the illumination. It is often a few micrometers at most, whereas the flatness variation on a BIM is gradual. The 0.5- to 2-μm thickness variation spreads over the entire area of the mask whose size is on the order of 200 millimeters. How a PSM improves the optical image is explained in the following section for each type of PSM.

6.3.4.3 PSM types and their mechanisms for imaging improvement

Instead of the generic names used in the beginning of Section 6.3, PSMs are generally named by the location or type of their phase shifters. The major types of PSMs are as follows.

Alternating phase-shifting mask (AltPSM)—This BCM system is characterized by the phase shifting of every other transparent element in a closely packed array.[21] The AltPSM approach is shown in Fig. 6.18(b) in comparison to the BIM in Fig. 6.18(a). The former electric field amplitude is now −1 at the shifted areas on the mask. This negative amplitude effectively reduces the spatial frequency of the electric field so that it is less prohibited by the lens transfer function and forms a higher-contrast amplitude image at the wafer plane. When this electric field is recorded by the photoresist, only the intensity, which is proportional to the square of the electric field amplitude, can be recorded. Hence, the reduced spatial frequency is doubled

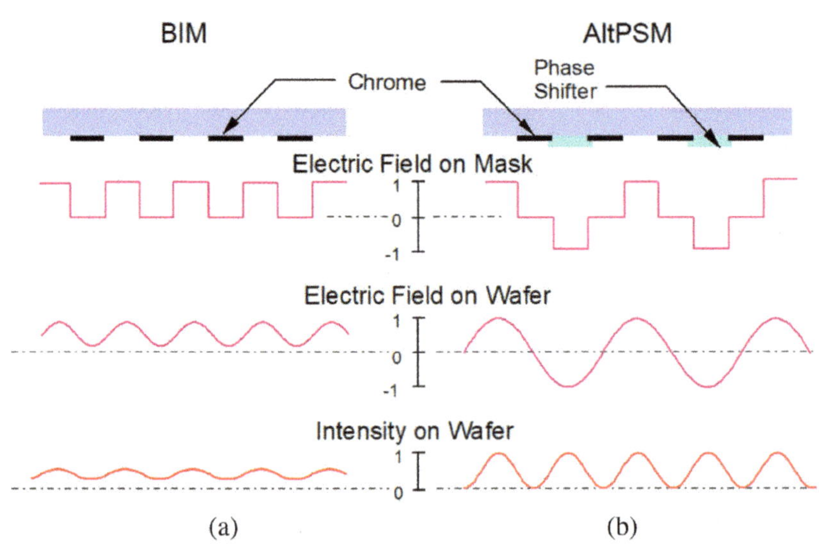

(a) (b)

Figure 6.18 Comparison of (a) BIM and (b) AltPSM mechanisms.

back to the original frequency, but the image results in a much higher contrast. In addition to the reduction of spatial frequencies, the fact that the electric field amplitude must pass through zero to –1 assures zero intensity at the wafer; thus, it helps to improve edge contrast. Hence, the AltPSM system benefits from reduction of spatial frequencies as well as enhancement of edge contrast. Figure 6.17 treats the illumination as completely coherent. However, in a real exposure system, the σ of the illumination is not 0, but rather is between 0.3 and 0.4.

If coherence is lost, an AltPSM becomes just a BIM. Figure 6.19 shows that with incoherent illumination the electric field fluctuates randomly in phase, and the spatial frequency-doubling effect shown in part (b) is not possible. Summing these into an intensity simply produces an intensity image identical to that in part (a).

The change in DOF of 0.35-μm line–space pairs at $\lambda = 365$ nm is shown in Fig. 6.20 as a function of σ. At $\sigma = 0.5$, DOF drops by 60% from the value at $\sigma = 0.3$. At $\sigma = 0.55$ and beyond, the image does not support any DOF at the given CD tolerance of ±10% and exposure latitude of 20%.

Subresolution-assisted phase-shifting mask (SA PSM)—The AltPSM opens the door for phase-shifting technology by demonstrating the potential of phase shifting. However, to be effective, it must have closely packed patterns. In actual circuit layouts there are many situations where critical dimensions are sufficiently far from any adjacent patterns to provide phase shifting. To provide phase shifting for isolated openings such as contact holes and line openings, another form of BCM uses subresolution phase shifters near isolated openings,[22] as shown in Fig. 6.21(a). The dimension of these

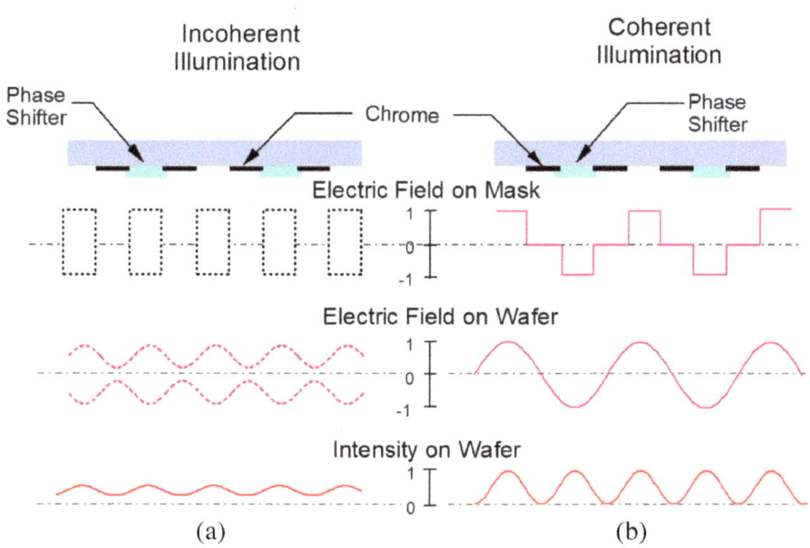

Figure 6.19 Comparison of (a) incoherent and (b) coherent illumination on an AltPSM.

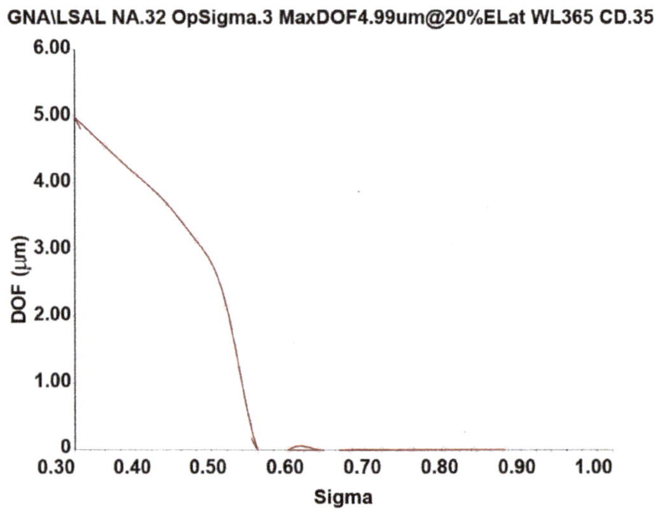

Figure 6.20 DOF of 0.35-μm line–space pairs at 365 nm for an AltPSM as function of σ.

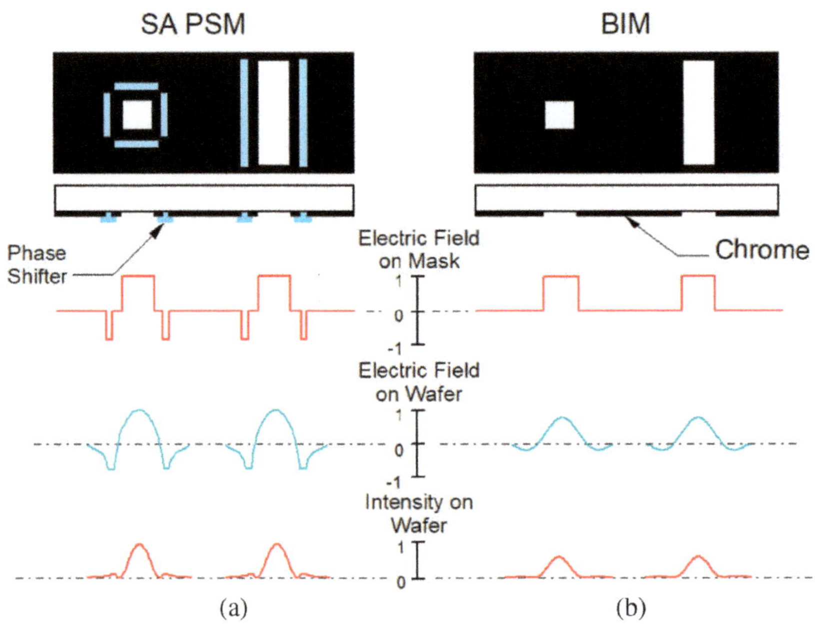

Figure 6.21 Comparison of (a) SA PSM and (b) BIM mechanisms.

phase shifters is below the resolution limit of the optical imaging system, so the subresolution phase shifters cannot be printed. Their sole function is to enhance the edge contrast of the pattern of interest.

Rim phase-shifting mask (rim PSM)—SA PSMs and AltPSMs are still limited by the inability to provide a phase shift to opaque patterns. Rim PSM,[23] yet another form of BCM [shown in Fig. 6.22(a)], overcomes such a

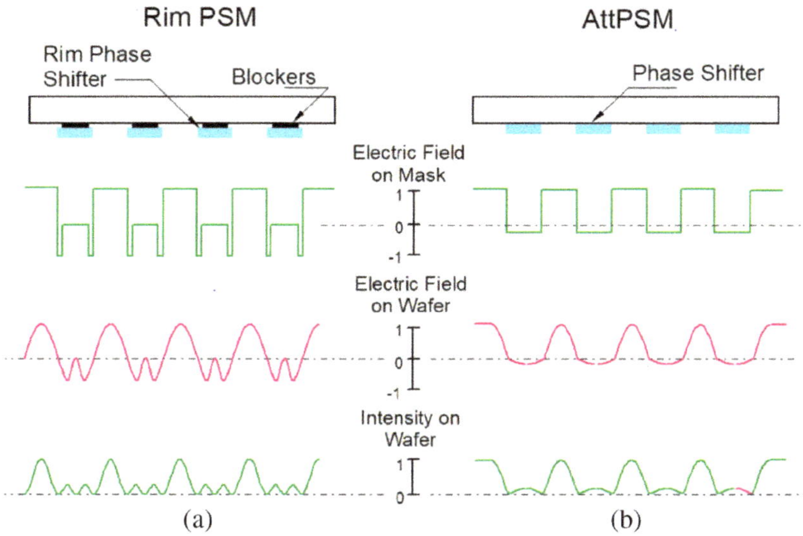

Figure 6.22 Comparison of (a) rim PSM and (b) AttPSM mechanisms.

problem and can be applied to an arbitrary mask layout. Here, phase shifting only takes place at the rim of the mask patterns. The center of the patterns is blocked by the absorber to prevent large areas of negative amplitude from producing bright areas where they are supposed to be dark. Again, bright areas result from negative or positive field amplitudes because the photoresist can only detect intensity that is proportional to the square of the electric field amplitude. Note that edge contrast enhancement is now the sole image-improving function of these phase shifters.

Attenuated phase-shifting mask (AttPSM)—The AttPSM[24,25,26] is a BCM that applies to arbitrary mask layouts, just as the rim PSM. It can be implemented on either a transmissive or a reflective mask. The dark areas of the mask can be phase shifted to π but with an attenuated amplitude to prevent producing too much light in these areas, as shown in Fig. 6.22(b). The negative amplitude provides the desired improvement in image edge contrast, and the attenuation prevents the negative amplitude from becoming too large and subsequently exposing the resist. Regular absorbers often need to be included at the borders of the mask plate to block light at the boundary of the exposure field and to provide regular reticle alignment marks. Therefore, an AttPSM often takes the form of a multilevel complex mask (MCM).

Phase-shifting edge (PS edge)—A phase-shifting edge[27] is simply the boundary between the phase-shifted and the unshifted regions. It is generically a BPM. A high-contrast dark line image following the boundary is produced, as shown in Fig. 6.23(a).

Covered phase-shifting edge (CPS edge)—This is like the phase-shifting edge, except that the edge is no longer just the boundary.[28] It is covered with

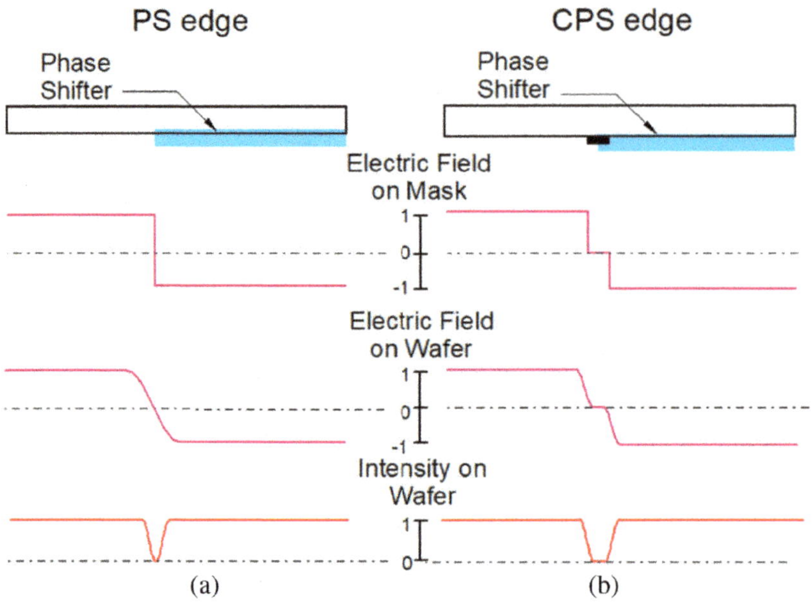

Figure 6.23 Comparison of (a) PS edge and (b) CPS edge mechanisms.

absorber as well, as shown in Fig. 6.23(b). The covered phase-shifting edge can be viewed as one-half of each opening in the AltPSM.

6.3.4.4 PSM Configurations

Starting with the basic PSM types, many different configurations can be built, as presented next.

Configurations from the phase-shifting edge—Two phase-shifting edges of opposite polarities can be combined to form an unattenuated (Utt) phase-shifted (PS) line. When the edges are close to each other, opaque images are formed, as shown in Fig. 6.24(b). The Utt PS lines with large edge separations can be combined periodically to double the spatial frequency, as shown in Fig. 6.24(a). These lines can be double exposed orthogonally to produce isolated opaque images,[27] as shown in Fig. 6.25. With a positive resist, the resultant image is a pedestal, and with a negative resist, it's a hole. Note that the image in the resist is rotated 45 deg. This is due to the iso-intensity lines in the 45-deg and 135-deg directions, which result from the orthogonal superposition of the intensity distribution of the PS edge as shown in Fig. 6.25.

Combination of AltPSM and rim PSM—The AltPSM is a strong phase-shifting scheme that works only on closely packed patterns, whereas the rim PSM is a weak phase-shifting scheme that works better on patterns with larger separations. It is possible to combine them on the same layout to take advantage of their complementary characteristics.[29,30] Figure 6.26 shows the combined layout and its fabrication steps. A three-level exposure is used to

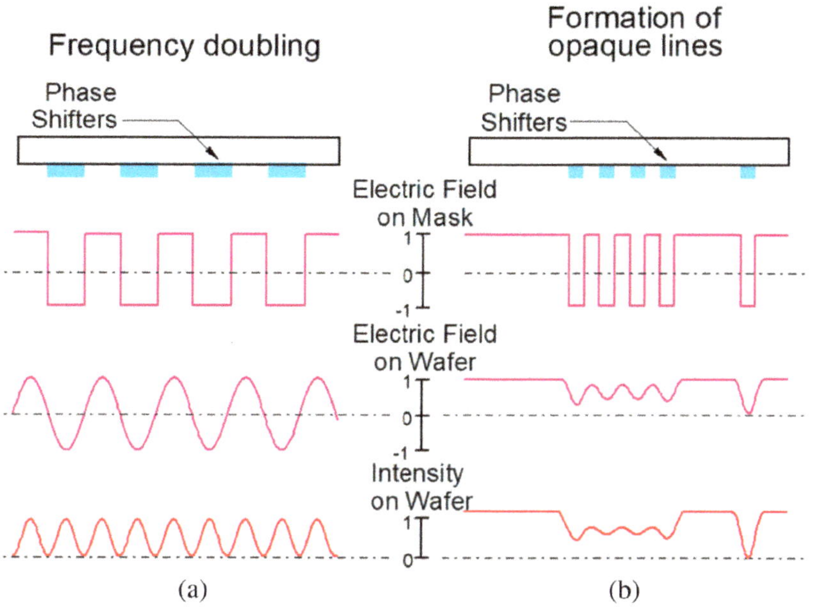

Figure 6.24 Configurations of PS edges using (a) frequency doubling and (b) opaque line formation.

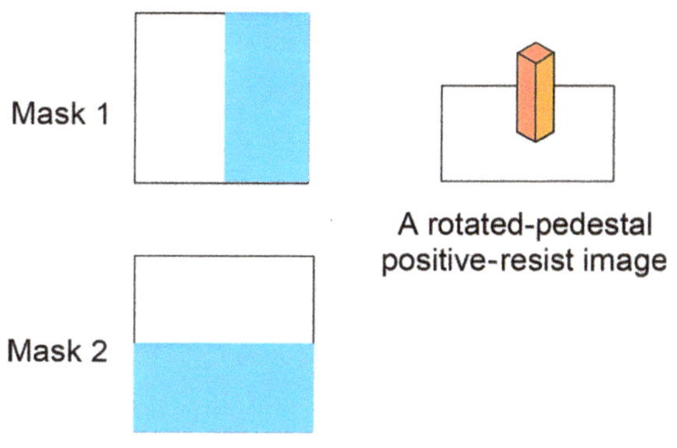

Figure 6.25 Double exposure of orthogonally displaced Utt PS lines.

write the mask. The partially exposed areas define the phase-shifter areas. The unexposed areas define the areas to be covered with chrome. The exposed areas become the transparent, unshifted areas.

Combination of CPS edge and BIM—The gate length of metal-oxide-semiconductor (MOS) devices is very critical, whereas the dimension elsewhere at the gate level is more forgiving. Therefore, the CPS edge can be combined with a BIM by double exposure to precisely control the gate length.[31] As shown in Fig. 6.27, mask 1 is a BIM consisting of the complete layout of the gate level, except that the gate length is intentionally made larger

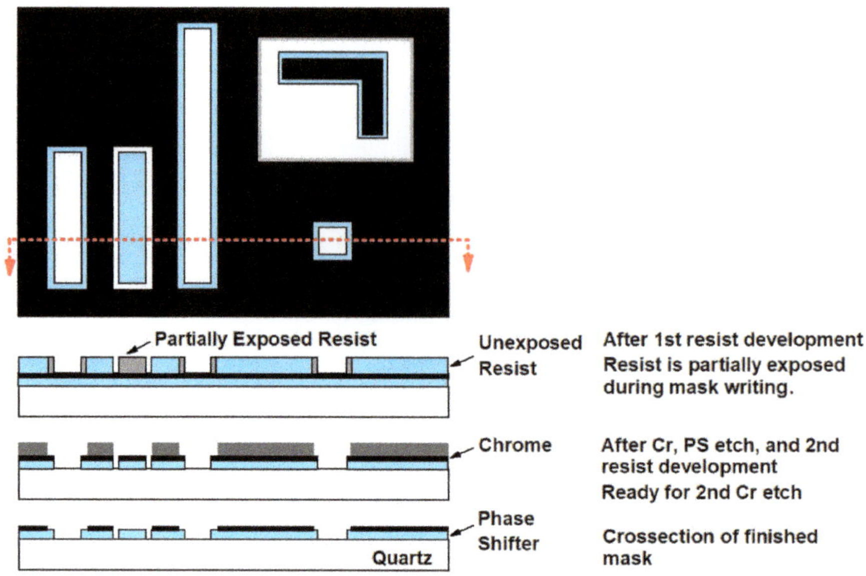

Figure 6.26 Combination of AltPSM and rim PSM and the mask fabrication sequence.

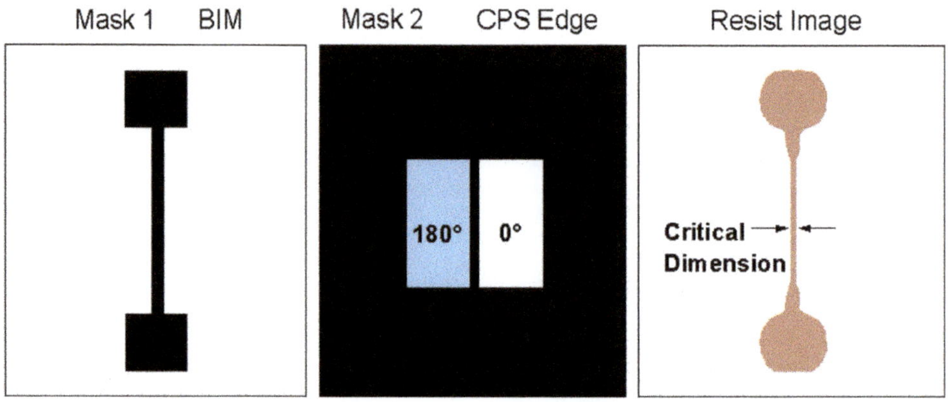

Figure 6.27 Combination of a CPS edge and a BIM.

so that the BIM can support the entire layout. After exposing mask 1, mask 2 (which contains the CPS edge at critical gates) is exposed over the enlarged gate areas defined by mask 1. This narrower pattern with higher contrast sets the gate length in the developed resist image and improves the linewidth control. This combination has propelled the company owning the patent to fame. Besides the obvious handicap of lost productivity with double exposure, a minimum distance between the CPS edges must be set, which limits the reduction of cell size. The higher cost and longer cycle time of the CPS mask also limit the manufacturing application of this technique.

The possible configurations are not exhausted with the above examples. Some combinations improve imaging performance. Some do just the opposite. It is important to evaluate imaging performance quantitatively, as is explained in Section 5.5.6.

6.4 Imaging Lens

The function of the imaging lens is to reproduce the mask pattern on the wafer with the given reduction ratio. This lens is a crucial part of the entire imaging system. Its power and quality directly impact the resolution of the imaging system.

6.4.1 Typical lens parameters

6.4.1.1 Numerical aperture

The power of the lens is indicated by its numerical aperture (NA), which is defined as the sine of the half angle that the lens aperture sustains from the mask multiplied by n, the refractive index of the medium between the imaging lens and the wafer. This relationship is depicted in Figs. 4.2 and 4.3 and is defined by Eq. (4.2). In a 1X system, the NA at the mask side (NA_M) and at the wafer side (NA_W) are the same. With a reduction system, the NA is smaller at the mask side by the reduction factor, as shown in Fig. 6.28. With 4X reduction, the NA at the object side is 25% of the NA at the image side. Even though the mask pattern is four times larger than its image, the solid angle sustained by the imaging lens at the mask side is only 25% of that at the wafer side. Note that the NA in optical microlithography usually refers to that at the wafer side because it is used to predict the resolution at the wafer.

The NA determines the angular extent of the imaging light. A larger NA captures a wider angular spectrum for imaging; therefore, it resolves a smaller

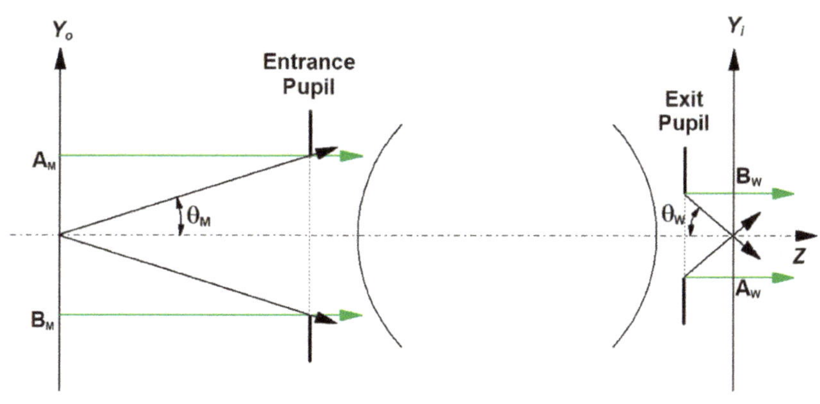

Figure 6.28 NA at the mask and the wafer sides. For mX reduction, $NA_W = M \cdot NA_M$.

feature. The resolution of a given imaging lens is related to the NA by the resolution scaling equation introduced in Chapter 5. According to this equation [Eq. (5.1)], the NA cannot be increased at will. First, it is progressively more difficult to keep the lens aberration free as the NA increases. A larger number of lens elements that require higher precision in the elements and in assembly makes the imaging lens extremely expensive. It has been estimated that the difficulty in designing and building a lens is proportional to the 5th power of the NA.[32] Second, a larger NA may lead to an unacceptable DOF, as predicted by Eq. (5.4), the scaling equation for DOF. The lack of DOF at large NAs and the loss of resolution at low NAs point to the existence of an optimum NA, which is discussed in Chapter 7.

6.4.1.2 Field size

The field size of a lens has a direct impact on how many chips can be exposed in one exposure step; this is discussed in Chapter 3. From an optical point of view, although the image field of a lens is circular, a microlithographic lens is usually specified in terms of certain rectangles or the largest square that can be used. For economic reasons, it is more important to populate the wafer with rectangular fields than to fully utilize the circular field. The unused areas are blocked off with movable blades at the mask plane. More precise framing can be fabricated in the mask.

Like the difficulty in designing and building lenses with large a NA, it is also very difficult to make diffraction-limited lenses with a large field size. The difficulty has been estimated to be proportional to the 3rd power of the field size.[32] Increasing the field size significantly worsens the lens distortion. Other aberrations such as astigmatism and coma are also field-size dependent.

6.4.1.3 Reduction ratio

The reduction ratio m of the lens has impacts on CD control and overlay accuracy as seen in Tables 3.1 to Table 3.3. It increases the DOF by m^2 and reduces the NA by m on the mask side. Making a reduction lens is more difficult than making a 1X lens because there is no longitudinal symmetry to take advantage of. The reduction ratio has been kept at five for a long time because five is a good compromise between difficulty of mask making, field size, and mask size. With step-and-scan mask aligners, m is taken to be four to accommodate the increased field size. Further reducing m can have a severe impact on CD control and overlay accuracy. Therefore, the tendency is to use larger mask blanks instead of further dropping m. Naturally, there are large-field low-resolution mask aligners that use m smaller than four.

6.4.1.4 Working distance

The working distance is the closest distance between the 1st lens surface and the wafer. It is usually between 5 and 10 mm. The distance must be small to keep the NA large, but a reasonable distance is required to facilitate damage-free and defect-free fast stepping. With immersion imaging, the working distance usually drops to below 3 mm so that the thermal control and inhomogeneity of the immersion fluid does not contribute significantly to imaging errors. More details on the working distance in immersion lithography are given in Section 8.5.2.

6.4.1.5 Telecentricity

When the principal ray of the imaging beam at any point in the image field is perpendicular to the image plane, i.e., the wafer plane, the system is telecentric at the image side. Being telecentric makes the lateral location of the image point insensitive to defocus, thus reducing overlay errors. Modern steppers are telecentric at both the mask side and the wafer side to ensure the stability of the reduction ratio.[33] Both types of telecentric system are depicted in Fig. 6.29.

6.4.2 Lens configurations

There are basically three types of imaging lens based on the type of lens elements. Dioptric systems consist of entirely refractive elements. All-reflective systems contain only reflective elements. Catadioptric systems mix refractive and reflective elements.

6.4.2.1 Dioptric systems

Most microlithographic lenses are dioptric. An example is given in Fig. 6.30. Here a 248-nm 0.8-NA 4X-reduction lens consisting of 20 fused-silica elements

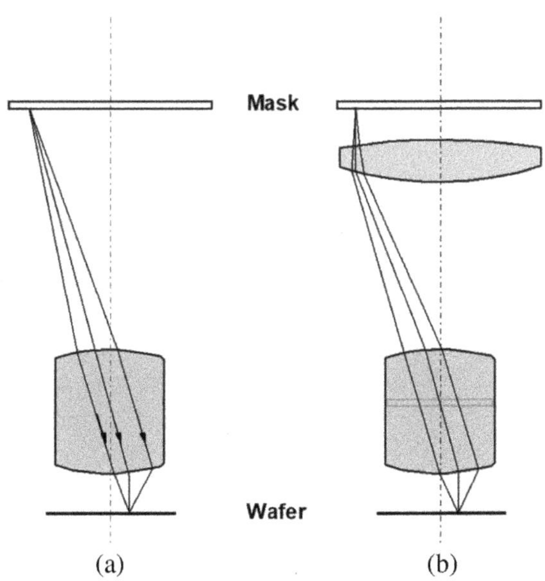

Figure 6.29 (a) Telecentric and (b) double telecentric imaging systems.

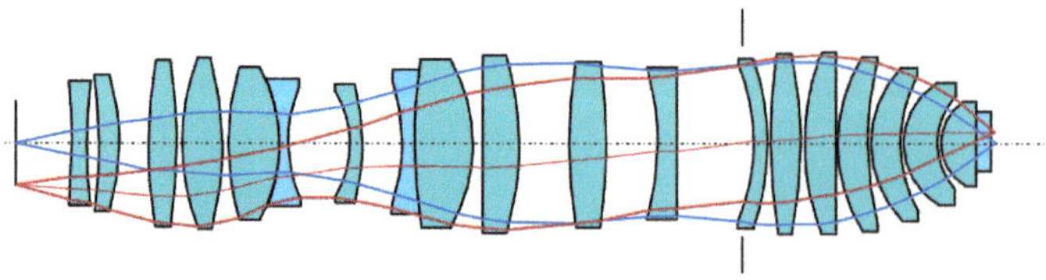

Figure 6.30 A 248-nm 0.8-NA 4X-reduction dioptric lens system (reprinted from Ref. 34).

is shown.[34] The advantage of dioptric systems is that they maintain a single optical axis from the mask to the wafer. This makes for easier assembly and alignment of the elements. However, because refractive elements are inherently dispersive, i.e., the refracting angles are wavelength dependent, many more elements and a mixture of lens material with mutually compensating dispersion characteristics are required to suppress chromatic aberrations. As the difficulties of chromatic aberration increase, the usable bandwidth of the exposure light is narrowed. For i-line steppers, the bandwidth is 6 nm, compared to 10 nm for g-line steppers. At 248-nm or shorter wavelengths, the practice is to strictly use a single type of material while placing the burden entirely on a very narrow bandwidth. Fortunately, with excimer lasers the bandwidth can be reduced significantly without inducing a substantial loss of energy. The bandwidth requirement in this case is between 0.1 and 10 pm, depending on whether a dioptric or catadioptric system is used, the number of transmissive materials, and the center wavelength.

6.4.2.2 Reflective systems

A well-quoted example of a reflective imaging system is shown in Figs. 3.4 to 3.6. The two concentric mirrors reproduce the mask pattern appearing in one region of the ring on another part of the same ring. Hence, if the mask is blocked by a plate with a circular slit that is a section of the ring and the same is done with the wafer, the mask and wafer can be scanned at the same speed to reproduce the image of the entire mask. Three plane mirrors mounted on a single block of glass are used to bend the light so that the mask and the wafer now face each other and can be moved in the same direction instead of being on the same plane and needing to be scanned in opposite directions. The configuration of two concentric mirrors is the only all-reflective imaging scheme that has been successfully used in manufacturing. Being a totally reflective system, its bandwidth is unlimited, except for the efficiency of the mirror coatings. This system has been used at wavelengths from the near UV to the extreme UV to demonstrate imaging and to develop exploratory photoresist systems. Hence, the advantages of a reflective system are freedom from chromatic aberration and a small number of optical elements.

The disadvantage of a reflective system (except one with of two concentric mirrors) is the difficulty in keeping all the optical elements on one straight optical axis. Correction of aberrations is also difficult because it almost always requires the introduction of optical elements, which results in an increase in the number of optical axes. The surface and assembly tolerances must be at least twice as good as those of a refractive system because any irregularity affects both the incident and the reflected beam on each element.

6.4.2.3 Catadioptric systems

There are both refractive and reflective elements in a catadioptric system. Usually, a spherical or aspherical mirror performs most of the imaging. Refractive elements are inserted for fine corrections. There may be planar reflective elements that bend the rays for better arrangement of the mask, the wafer, or other optical elements. This is exemplified by Fig. 3.12, which shows the 1X catadioptric lens used in Ultratech mask aligners. There is a main mirror and a refractive achromat. One folding prism redirects light from the mask to the achromat and the main mirror. The other folding prism sends the reflected light from the main mirror through the achromat to the wafer. Because the incoming and outgoing beams cannot share the same optical axis, only a semicircular field can be used, as shown in Fig. 6.31.

Figure 3.17 shows a catadioptric lens system with a beam splitter. It is like the one used in the SVGL step-and-scan mask aligner. Again, there is a mirror that is responsible for reduction imaging. The beam splitter keeps the incoming and outgoing beams to and from the mirror out of each other's way so that the entire circular field is available. There are also refractive elements in this system to correct for aberrations.

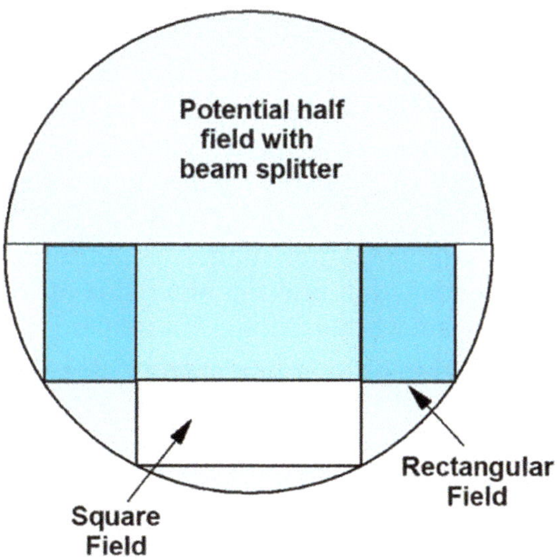

Figure 6.31 Semicircular lens field and the exposure fields from the lens system in Fig. 3.12. The entire circular field can be made available with a beam splitter.

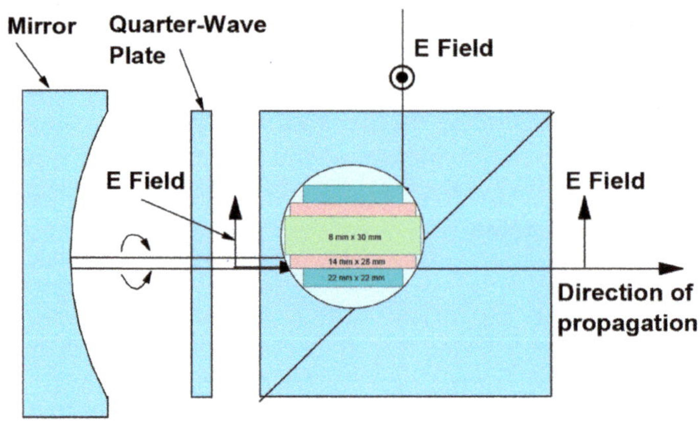

Figure 6.32 Working principle of a polarizing beam splitter in the catadioptric lens system in Fig. 3.17.

Several issues need to be considered when introducing a beam splitter into the lens system. First, a conventional beam splitter splits light into two paths. Hence, it reduces the intensity of each beam by half. One of the beams is not only wasted, but if not treated properly, it can contribute to stray light. The useful beam is split into two again when it travels back to the beam splitter before exposing the wafer. In total only 25% of the illumination is used, and the system can be prone to stray light. To overcome this problem, a polarizing beam splitter[35] is used. Its working principle is illustrated in Fig. 6.32. Light in the s-polarization mode is incident on the beam-splitting plane of the cube. The beam-splitting plane is composed of a multilayer stack to form a reflecting and refracting interface. This stack is designed to satisfy the Brewster condition for beams at a 45-deg incident angle such that the reflected beam is in the s-polarization mode and the refracted beam is in the p-polarization mode. Since the incident beam is s-polarized, it is completely reflected toward the quarter-wave plate. The beam becomes circularly polarized after leaving the quarter-wave plate. It is reflected by the mirror and enters the quarter-wave plate again. The beam that re-enters the beam splitter is now in the p-polarization mode; its reflection at the Brewster interface is zero and its refraction into the other side of the cube is 100%.

The second consideration when using a beam splitter is that it must be a very thick and very large optical element to have a large field size. The homogeneity and durability requirements of such an element are not easy to meet.

6.4.3 Lens aberrations

Lens aberration arises from the deviation of a spherical wavefront that converges to an image point, as discussed in Section 4.1.4 and shown in Fig. 4.8. The deviation of a spherical wavefront is caused by the following:

1. lens design residue,
2. imperfect surface configuration of the optical elements in the imaging lens, and
3. placement error of the optical elements during or after their assembly.

The deviation of a spherical wavefront can be quantified with Zernike polynomials or Seidel coefficients. Because of cylindrical symmetry, lens aberrations in optical microlithography are almost always specified in terms of Zernike polynomials, as covered in Section 4.1.4.2. Note that the $\sin(m\phi)$-dependent terms should be included because reasons 2 and 3 above induce asymmetry. Reason 3 can take place in the field due to lens heating or a change in the temperature or humidity of the lens environment. Mechanical impact to the lens also affects lens aberrations.

In general, spherical aberrations enlarge image points as if they are defocused. Coma induces displacement in two adjacent images. Astigmatism causes CD variation between features of different orientations. Spherical aberration is uniform across the lens field. Coma and astigmatism are field dependent. Distortion does not affect resolution, but it changes the image placement and is sensitive to field location. These characteristics help to sort out lens-induced and mask-induced distortions.

6.4.4 Lens fabrication

With modern electronic computation capability and the availability of lens-design software, today's lens design has become less formidable than fabricating a highly sophisticated lens. In fabrication, each refractive surface must be polished to 10~20 $m\lambda$ error. For mirror surfaces, the precision required is doubled because light passes through the imperfection twice.

When the elements are assembled, they must be centered to better than a fraction of a micrometer. Surface errors of individual elements are inspected with a wavefront interferometer and corrected before assembling. Subsystems assembled with several elements are also checked with the wavefront interferometer. Assembling errors will be evaluated using the interferometer result and a computer. The subassembly will be reworked if necessary. Because of the uncertainty of this trial-and-error approach, many lens manufacturers have developed proprietary high-precision assembly techniques that produce instant feedback to eliminate rework. Even so, manufacturing a high-NA imaging lens is a laborious process that requires a high skill level.

When demand exceeds supply for high-density semiconductor chips, the semiconductor factory capacity is usually limited by the production of mask aligners. Mask aligners are, in turn, limited by the production capacity of lens suppliers. The lens suppliers may depend on the availability of high-quality lens materials. At one time, SiO_2 was the gating material. Later, lens fabrication was gated by CaF_2, which is required to reduce absorption loss and to achromatize 193-nm imaging lenses. CaF_2 is also the sole refractive material for 157-nm lenses. The limitation of producing this material in quantity and quality to meet the more stringent requirement of 157-nm imaging is one of the key factors preventing 157-nm lithography from being further developed, despite the enormous development effort made by equipment and material suppliers.

6.4.5 Lens maintenance

The care taken in assembling the lens does not guarantee that the lens is free from damage or deterioration. There are many issues that users should keep in mind, as described below.

Prevention of mechanical shocks—Any mechanical shock or vibration that can change the assembly precision of the lens must be avoided. The greatest possibility of these disturbances occurring comes from transporting the exposure tool with the lens from its factory to the semiconductor manufacturing facility. Once there, locating the exposure tool, relocating it, or replacing the lens can also be problematic. Impact sensors are placed abundantly in the exposure tool to record any impact induced during transportation.

Prevention of environmental changes—It is well known that slight temperature and barometer changes in the lens environment can change the refractive index of the lens elements.[36] These changes can also affect the assembly precision if the lens mounting materials are not carefully engineered. Hence, the imaging lens in modern steppers or scanners are enclosed in environmental chambers that regulate the temperature to better than 0.1 °C. The lens itself is maintained at a designated pressure and humidity. It is crucial that the mask aligner's power stay on continuously so that these controls work without interruption. Otherwise, it takes a long time to re-stabilize or restore the system.

Surface contamination—A dust particle that falls on any of the lens surfaces becomes a scattering center that produces stray light and flare in the image, thus reducing the image contrast and, hence, the linewidth control. Therefore, the lens is assembled in a dust-free environment. In the field, the lens should not be subjected to any possibility of particulate contamination. When the photoresist outgasses, a layer of polymer material can accumulate on the front surface of the lens. This also must be avoided.

Radiation damage—The glues employed for lens elements, the antireflection coatings on the lens surfaces, and the lens materials themselves can deteriorate as exposure dosages accumulate. Extremely high-peak exposing pulses can accelerate the deterioration. Lenses have been known to require refurbishing at the lens factory or even replacement after a year or two. However, there is not much the mask aligner user can do to prevent this kind of damage, except to avoid unnecessary exposures and accidental introduction of high-peak pulses into the optical train.

6.5 Photoresists

The photoresist is a light-sensitive polymer that undergoes light-induced chemical reactions to produce a dissolution rate as a function of the exposure level. The difference in dissolution rate in the exposed and unexposed areas causes the resist to be removed selectively while being developed. The areas still covered with resist are protected when the wafer is subjected to etching, whereas the areas free of resist are etched.

The resist can be used in semiconductor processing in many ways, as shown in Fig. 6.33. It can serve as a protection mask for etching, in which case the areas exposed to the etchant are removed either isotopically or anisotopically. It can also serve as a lift-off mask, in which case an undercut profile in the resist is produced so that a material anisotopically deposited on the wafer is separated at the edges of the resist, resulting in the removal of the material with the resist when it is stripped. The resist can also be used as an ion-stopping mask to selectively implant ions into the semiconductor.

6.5.1 Classifications

The photoresist can be classified in many ways, namely, by polarity, by working principle, and according to the imaging configurations shown in Fig. 6.34.

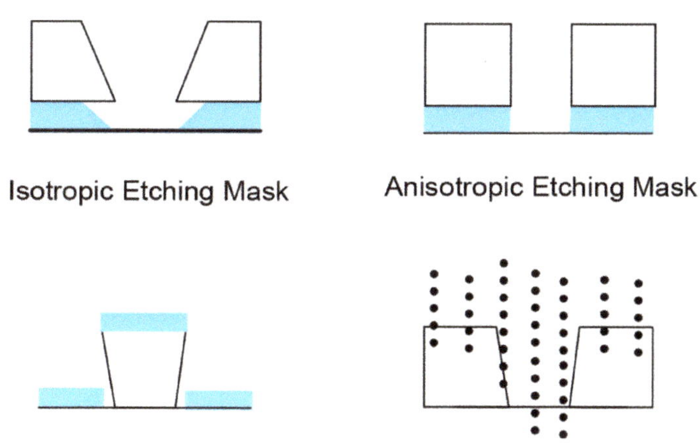

Figure 6.33 Various applications of resists.

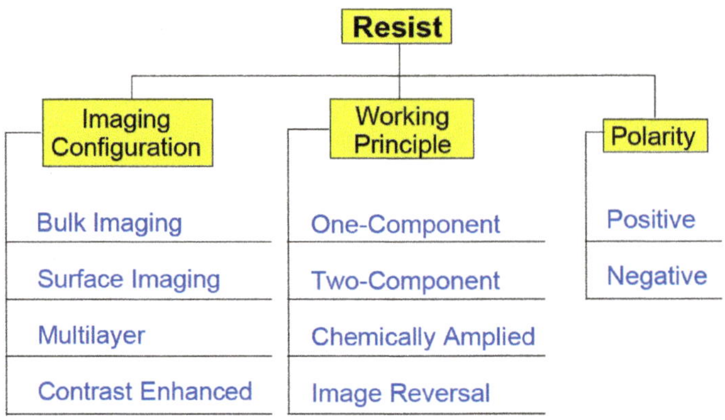

Figure 6.34 Classification of resist systems.

6.5.1.1 Polarity

There are positive and negative resists. The positive resist becomes more soluble in the exposure areas than the unexposed areas and vice versa for negative resist, as shown in Fig. 6.35. The difference between the resist and the latent image profiles in this figure is explained in Section 6.6.2. A positive resist is desirable in many respects, as explained below.

Optical imaging—The diffracted light from isolated openings, either lines or holes, always shows better defocus characteristics than that from isolated opaque features, such as isolated opaque lines or islands. The difference in aerial image between a hole and an opaque island of the same size is seen in Fig. 6.36. Note that these images were evaluated using axial illumination. The conclusion may be different under other imaging conditions.

Protection from ghost lines—As seen in Fig. 6.37, at a sufficient distance from focus, the aerial image degenerates into ghost regions, producing extra lines that are not on the mask. When the top surface of the resist is above the ghost regions, the unexposed positive resist prevents the exposed ghost images from developing. Note that this protection is lost if the top surface of the resist intersects the ghost image, as in the case of incorrect focus.

Freedom from shrinkage—Earlier negative resists exhibit significant shrinking after developing. Not only is the resist thickness reduced and its image wrinkled, but the stress can also bend the resist line away from its

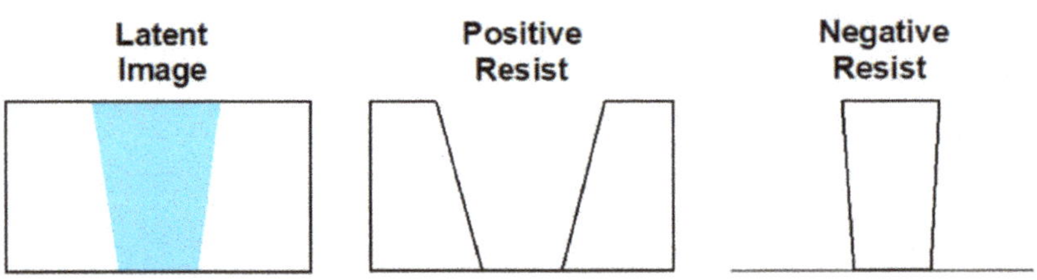

Figure 6.35 Positive and negative resists.

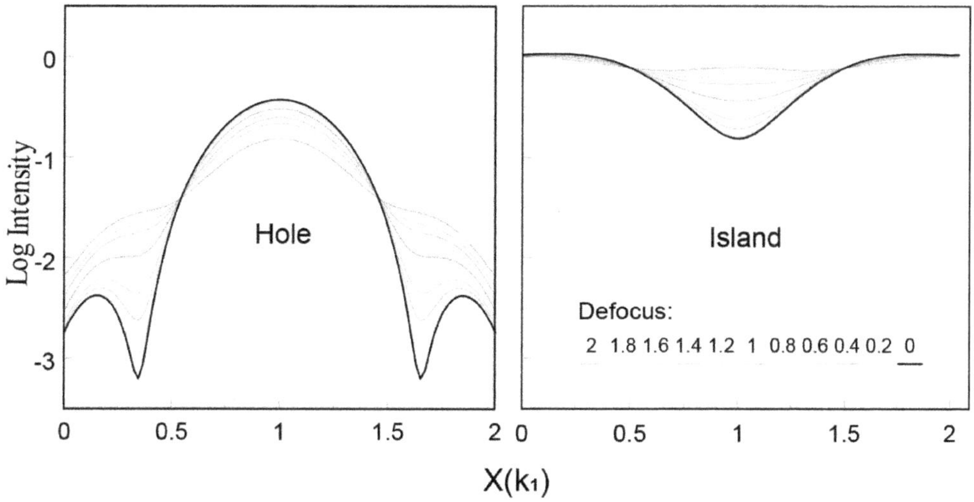

Figure 6.36 Aerial images of (left) a hole and (right) an opaque island.

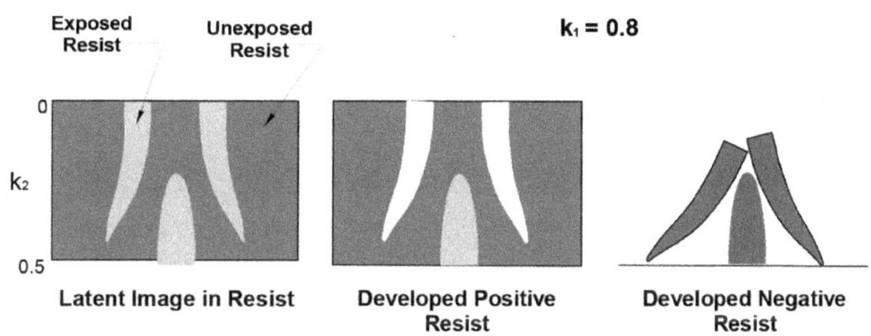

Figure 6.37 A positive resist prevents ghost regions from developing.

nominal position. Modern negative resists may not suffer from the shrinkage problem. However, the contrast of negative resists is usually lower than that of positive resists. In addition, negative resists tend to require solvent developers rather than aqueous developers. The latter is much more acceptable in a factory.

Negative resists are more desirable than positive resists in certain circumstances. For example, when high-resolution isolated resist lines need to be printed, negative resists support a larger DOF because the aerial image of an isolated line opening is much better than that of an isolated opaque space, as discussed above.

6.5.1.2 Working principle

One-component positive-resist systems—The one-component positive-resist system works based on the difference in dissolution rate induced by changes in

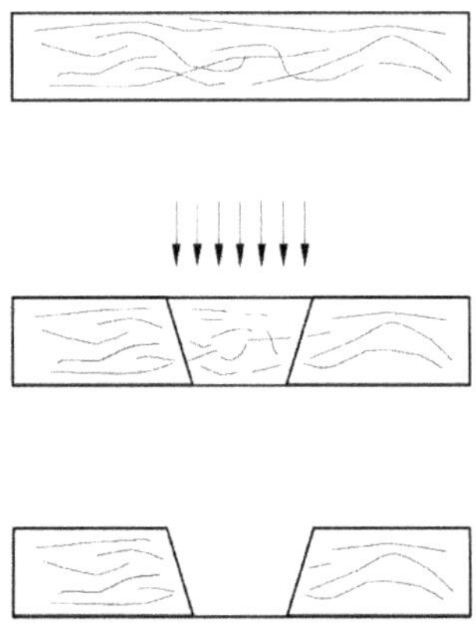

Figure 6.38 Chain scissioning of a one-component positive resist.

molecular weight. High-energy radiation chain scissions a larger polymer molecule into smaller ones, as shown in Fig. 6.38. The change can be several orders of magnitude, as reported by Moreau,[37] who also provides the following relationship between dissolution rate and molecular weight changes:

$$\frac{R}{R_0} = \left(\frac{M_0}{M}\right)^{-a}, \tag{6.5}$$

where R is the dissolution rate of the exposed resist, R_0 is the dissolution rate of the unexposed resist, M_0 is the initial molecular weight, M is the molecular weight that leads to R, and a is the developer's kinetic solubility parameter.

Polymethyl methacrylate (PMMA) is a good example of a one-component positive resist that is sensitivite to deep-UV, EUV, e-beam, and ion-beam radiations. Unfortunately, its sensitivity and dry-etch resistance are low.

One-component negative-resist systems—The one-component negative-resist system works based on the crosslinking of polymer chains upon radiation, as seen in Fig. 6.39. The crosslinked chains become harder to dissolve, thus making the resist behave like a negative resist. Moreau[38] shows the relationships between the negative-resist contrast γ and its polydispersity in single- or multi-component systems to be

$$\gamma = \left(\frac{2}{E_g}\right) \exp(-B^2) \tag{6.6}$$

and

$$B \equiv \frac{M_W}{M_n}, \tag{6.7}$$

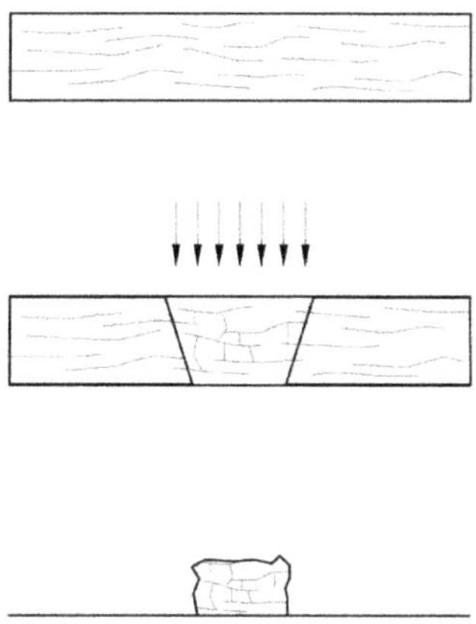

Figure 6.39 Crosslinking of a one-component negative resist.

where E_g is the gel point dose,[39] M_w is the weight-average molecular weight, and M_n is the number-average molecular weight.

Two-component positive-resist systems—A two-component positive-resist system consists of a base polymer and a sensitizer. The popular diazoquinone-sensitized novolak (DQN)[40] resist has a base polymer that is highly soluble in an alkaline developer. The sensitizer inhibits dissolution by as much as two orders of magnitude,[41] as shown in Fig. 6.40. After exposure, the sensitizer turns into a dissolution enhancer that can further increase the dissolution rate by another two orders of magnitude, as is also shown in the figure. The function[42] of each component in the DQN resist is shown in Fig. 6.41. The diazoquinone sensitizer, in the presence of actinic light and

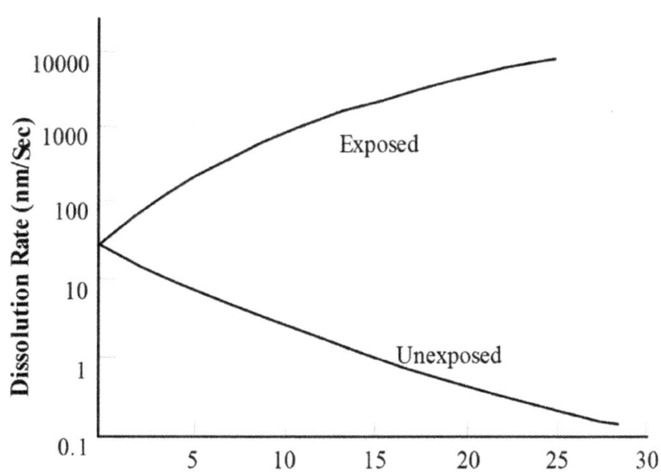

Figure 6.40 Effect of sensitizer concentration on a DQN resist [reprinted from Ref. 41 with permission; © (1980) IEEE].

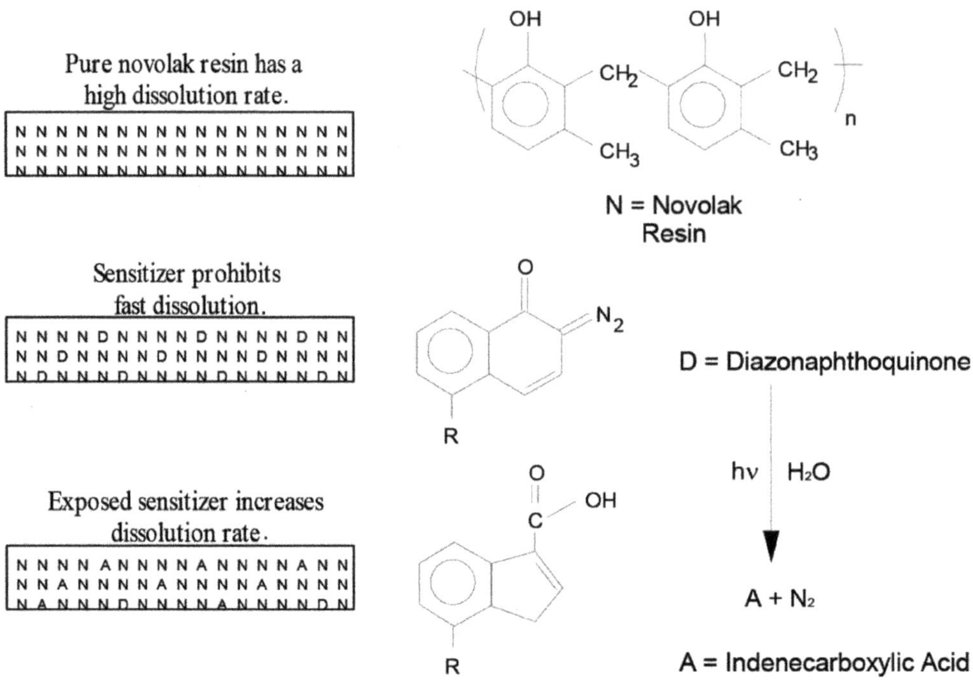

Figure 6.41 Function of each component in a DQN resist.

water found in the resist layer, produces a base-soluble indene carboxylic acid and nitrogen gas. The indene carboxylic acid is responsible for increasing the dissolution rate in the exposed areas. The imaging of the DQN resist is shown in Fig. 6.42.

Two-component negative-resist systems—Two-component negative resists are typified by those developed by Kodak, such as KTFR, KMER, and KPR. The working principle is shown in Fig. 6.43. The matrix resin of KTFR is cyclized poly(cis-isoprene), which is a cyclized rubber. The sensitizer is an azide

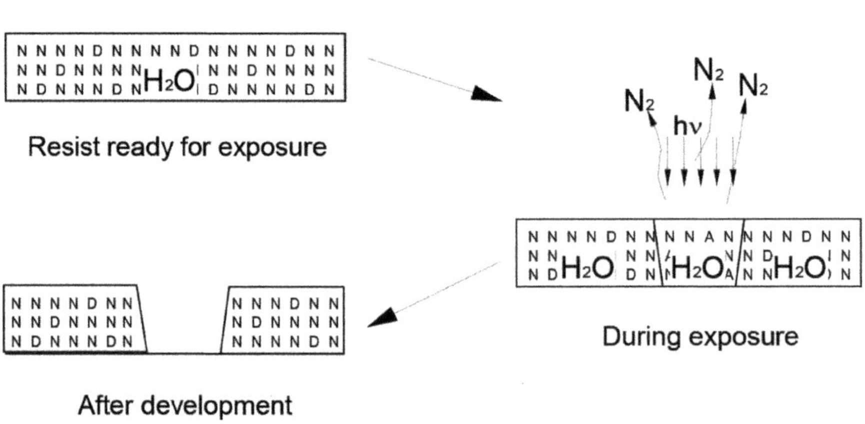

Figure 6.42 Imaging of a DQN positive resist.

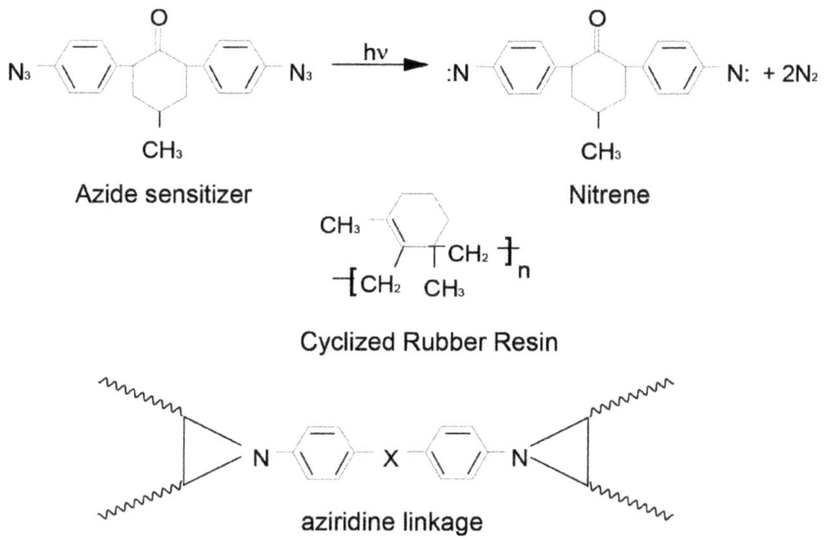

Figure 6.43 Crosslinking of a two-component negative resist (adapted from Ref. 42).

such as bis(arylazides). Upon actinic radiation, the azide releases nitrogen to form an extremely reactive intermediate, possibly a nitrene, which eventually crosslinks the resin. (Readers interested in pursuing this further are referred to Thompson et al. [43] and Moreau.[44] Figure 6.43 was composed based on the chemical formulae found in Ref. 43.) This type of resist is at least an order of magnitude more sensitive than the DQN resist. The adhesion is strong, and the crosslinked product is very durable. However, it does not work very well for high-resolution imaging because the resist swells while it is being developed and shrinks afterwards. The resist image is not only reduced in height but is also wrinkled. In the worst case, the resist line can take a zig-zag shape due to the tremendous stress it suffers. This type of resist is very popular for low-resolution applications that call for high sensitivity and good adhesion.

Chemical amplification systems—The chemically amplified resist (CAR) system is the ultimate scheme for higher sensitivity and thus higher wafer throughput. In principle, it needs to be photo-initiated to produce as few as a single proton, which can be used again and again to generate products consisting of new protons. The proton acts as a catalyst that facilitates chemical reactions, but it itself does not get consumed. Therefore, the sensitivity can be orders of magnitude higher than that of non-amplified systems. It has been estimated that one molecule of acid cleaves approximately 1000 carbonate units.[45,46] Fortunately, the reaction is somehow quenched to prevent the unexposed area from being thoroughly converted.

The t-boc deep-UV resist system[47] is given here as an example of chemical amplification. This resist system was such an impressive success that it was quickly adopted for production of 1-Mbit DRAM[48]—the most advanced application of 1X full-wafer-field scanners (reported in Section 3.2). The NA of these scanners stayed at about 0.18, which is not sufficient for 1-μm resolution using near-UV light. The extremely broadband capability of all-reflective systems is seamlessly adapted for deep-UV exposures. The CAR system provides extremely high deep-UV sensitivity to realize the manufacturing throughput potential of full-field optics.

As shown in Fig. 6.44, the actinic radiation gives rise to protons from the onium salt sensitizer. These protons react with the base resin to turn it from tertiary-butoxycarbonyl (t-BOC), a poly(p-t-butyloxycarbonyle-oxystyrene) (PBOCST), to p-hydroxystyrene (PHOST) and two volatile compounds, namely, carbon dioxide and isobutene. The conversion is most appreciable during the post-exposure-bake at temperatures above 180 °C. The proton remains after the reaction. Hence, it can catalyze other unreacted t-BOC molecules and, in principle, can be extensively used. PHOST is polar, whereas t-BOC is nonpolar. A polar solvent removes the PHOST (the exposed part), making a positive-resist system. A nonpolar solvent makes a negative resist by removing the opposite. Figure 6.45 shows the imaging of t-BOC from exposure to developing.

An important aspect in imaging CAR systems is the diffusion of the acid (proton) as it multiplies during the post-exposure bake (PEB). The diffusion and the quenching of the proton affect the sensitivity of the CAR. More diffusion increases the amplification ratio, thus raising the sensitivity of the CAR. CD can be affected by the resist blur caused by diffusion of the proton.

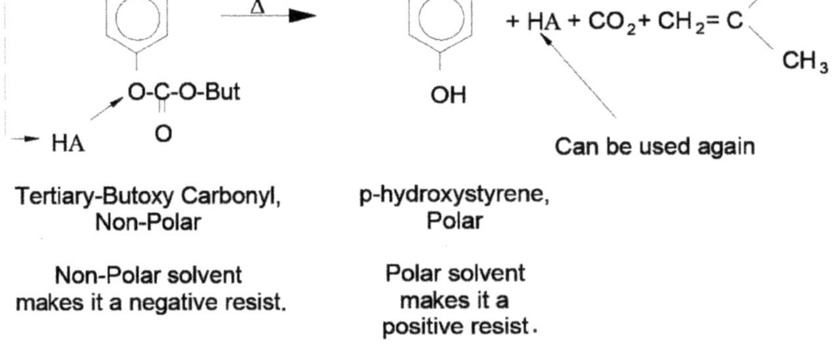

Figure 6.44 Photochemistry of a t-BOC chemically amplified resist (courtesy of C. G. Willson).

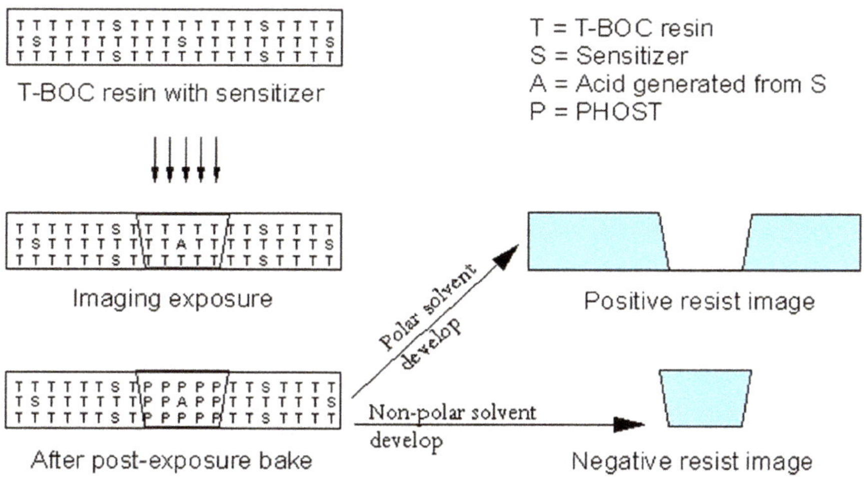

Figure 6.45 Imaging of a t-BOC chemically amplified resist.

The line-edge roughness (LER) in the resist image is also affected by diffusion. More diffusion tends to increase the LER.

In principle, the chemical amplification process continues until the resist is fully populated with protons. Therefore, the PEB must stop before the saturation of proton multiplication reduces the resist image contrast. Quenchers are incorporated to control the amplification process, which usually takes place during the PEB such that diffusion and quenching are working competitively.

A significant characteristic of CAR resists is their sensitivity to airborne molecular bases, such as ammonia or N-methylpyrrolidone (NMP), which are usually present in parts-per-billions (ppb) quantities in semiconductor factories. As an example of the concentration of airborne molecular bases, the concentration of NH_3 is 0.64 ppb, and that of NMP is 0.08 ppb.[49] It is expected that such low concentration levels are in a controlled environment.

Several phenomena involving airborne poisoning by molecular bases have been reported: (1) Due to neutralization of the acid catalyst, the resist sensitivity is lowered, making the exposure time difficult to control. (2) The resist–air interface is the most susceptible to sensitivity reduction, resulting in a T-top profile[50,51] for a positive CAR, as shown in Fig. 6.46. Figure 6.46(a) is the control resist image of image (c), where the resist was subjected to 15 minutes of exposure to urethane enamel vapor at an unreported concentration.[50] Figure 6.46(b) is the control resist image of image (d), where the resist was subjected to 15 minutes of 10 ppb of NMP.[51] The T-top is more severe in (d) than that in (b), indicating that the neutralization takes place closer to the resist–air interface.

These effects are functions of the type of airborne base, its concentration, and the time between post-applied bake and post-exposure bake. Thus, the

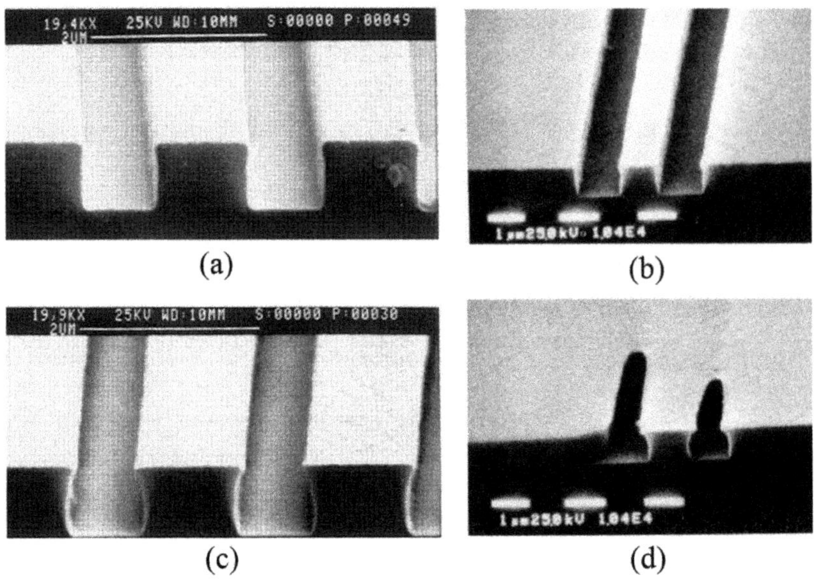

Figure 6.46 T-top profile from a contaminated CAR. (a) Control of (c) after 15 minutes in urethane enamel. (b) Control of (d) after 15 minutes in 10 ppb NMP before exposure (figures in left column reprinted from Ref. 50 and in right column from Ref. 51).

change in resist sensitivity is unpredictable unless the resist environment is under strict control. Reference 50 describes how the resist environment was controlled in manufacturing 1-Mbit DRAM by using charcoal-filtered air in the critical manufacturing areas where the coated and baked wafers were waiting for exposure. More recently, for smaller geometries, the environment in which the processes from exposure to post-exposure bake occur also must be filtered air. The resist latent image is stable only after PEB.

Image-reversal systems—The DQN[52] resist can be reversed into a negative-tone resist without the problems normally associated with conventional two-component resists, such as swelling, shrinking, low dry-etch resistance, and solvent developer. An image-reversal system is desirable for high-resolution imaging whenever a negative resist is called for, such as in the situations discussed in Section 6.6.1.1.

The working principle of image reversal is shown in Fig. 6.47. The novolak resin is turned to indene carboxylic acid just as in the positive imaging of DQN resists. However, due to a premixed base in the resist, the indene carboxylic acid is turned into indene during the PEB. Indene is insoluble in alkaline developers for DQN resists. One only needs to blanket expose the entire wafer to convert the unexposed resist into indene carboxylic acid to make it soluble in the alkaline developer. The resist dissolution characteristics are now reversed to a negative tone. The imaging of a reversed DQN resist is depicted in Fig. 6.48.

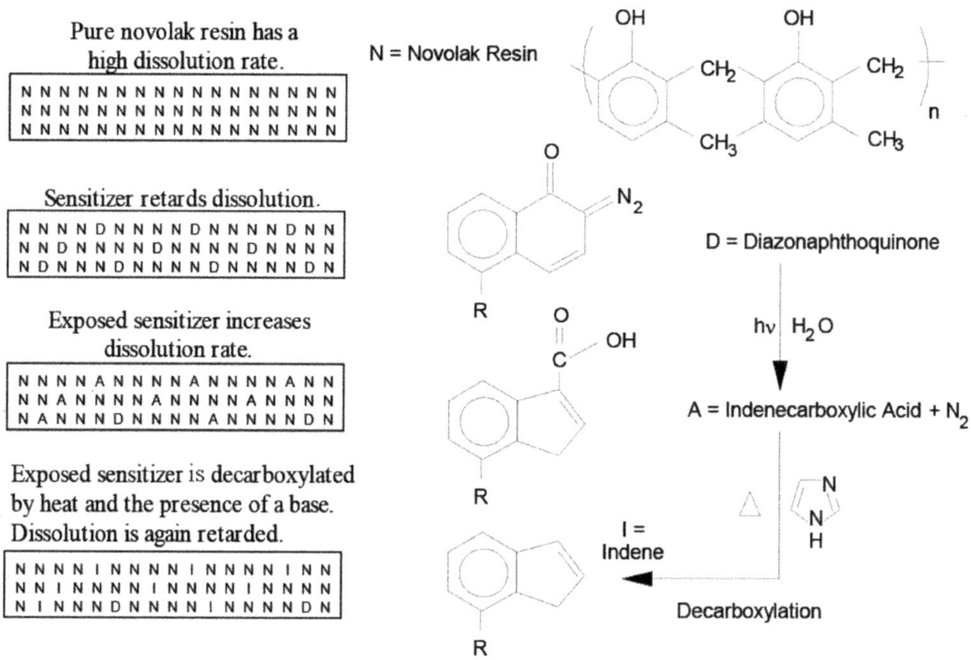

Figure 6.47 Function of each component in image reversal of DQN resists.

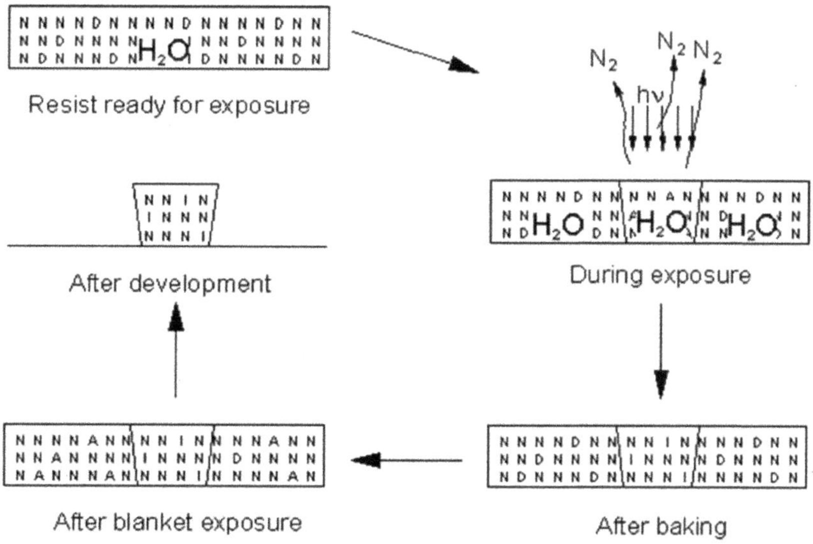

Figure 6.48 Imaging of a reversed DQN resist.

6.5.1.3 Imaging configurations

The resist system can be classified by the way image is formed in the resist according to the following imaging methods:

Bulk imaging—This is the most common type of imaging. The resist is exposed and developed through its entire thickness. The exposure process contributes to the developed resist profile, and the DOF of the imaging systems must support the thickness of the resist in addition to other components in the DOF budget, such as lens field curvature, wafer flatness, leveling, wafer topography, and focusing errors. The impact of the resist on the DOF will be discussed in detail in Chapter 7.

Top-surface imaging (TSI)—Only the top surface of the resist is exposed and developed, which makes for better planarization and takes up less of the DOF budget. The image is transferred to the bulk of the resist layer by another pattern transfer technique to produce a thick resist image with a vertical profile so that the resultant resist image is suitable for the subsequent semiconductor delineation processes exemplified in Fig. 6.33, wherever a thick and high-aspect-ratio resist image is needed. One important motivation for using TSI systems is the ability to isolate the topography on the wafer to prevent it from producing large exposure variations in the resist.

An example of a TSI system[53,54] is depicted in Fig. 6.49. Starting with a single layer of highly absorptive resist, light penetrates only the top part of the resist layer. This way, no reflected light is produced, thus isolating the latent image from the wafer topography. The resist is then subjected to a silylation process that selectively incorporates silicon into the exposed part of the resist. The subsequent oxygen reactive ion etching (RIE) process removes only the resist in the unsilylated part. Because the final resist image is void of resist in the unexposed areas, this is a negative TSI system. A positive TSI system is also possible. A more straightforward way to produce a positive TSI system would be to crosslink the latent image just as in the image-reversal process, to be followed by the silylation process, as shown in Fig. 6.50.

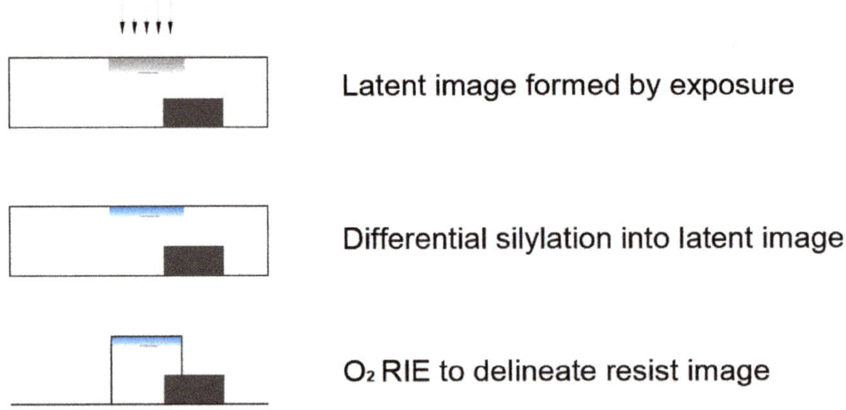

Figure 6.49 A DESIRE negative TSI resist system.

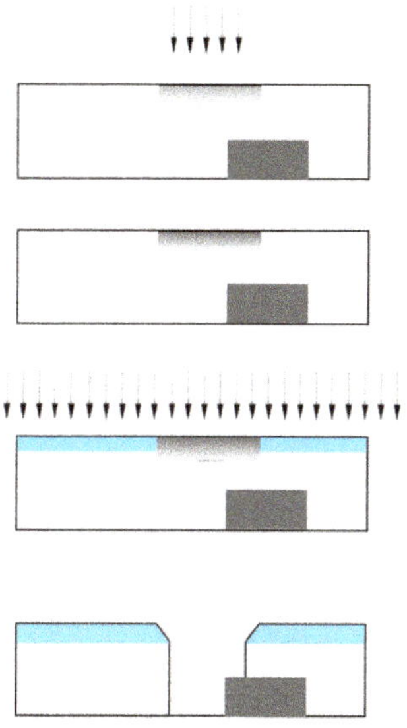

Figure 6.50 A positive TSI resist system using image reversal.

Multilayer—Similar to TSI systems, an important motivation for using multilayer systems is that they isolate the topography on the wafer to prevent it from producing large exposure variations in the resist.

There are two or three layers of material. The top layer is a photosensitive layer. In a three-layer system, after delineation of the image in the top layer, the image is transferred to the middle layer, which serves as a mask to delineate the bottom layer. A two-layer system uses the top layer as the photosensitive layer as well as the masking layer for the bottom layer. In multilayer systems, the top imaging layer is thin so that it is easy to image and does not take up a significant part of the DOF budget. The bottom layer is sufficiently thick for planarization over the topography on the wafer as well as for the subsequent semiconductor delineation processes exemplified in Fig. 6.33. Consult Ref. 55 for an extended coverage of multilayer resist systems.

Figure 6.51 shows a three-layer RIE-transfer resist system.[56] The bottom layer is a hard-baked resist that is desensitized and planarized. The middle layer is a thin inorganic layer that is resistant to O_2 RIE. The top layer is a photoresist suitable for the exposure wavelength. After normal exposure and when the development process to delineate the top imaging layer is finished, the resist image is transferred to the middle layer by etching. After that, an O_2 RIE step completes the delineation of the resist system.

A two-layer RIE-transfer resist system[57,58] is shown in Fig. 6.52. The top imaging layer contains an inorganic material such as Si or Ge_xSe_{1-x} and is a good O_2 etch mask. After photodelineation of the top layer, the image is transferred to the bottom layer by O_2 RIE.

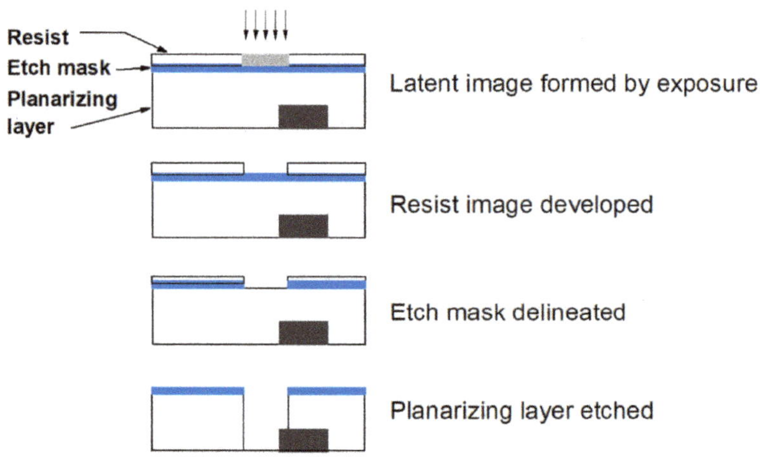

Figure 6.51 A three-layer RIE-transfer resist system.

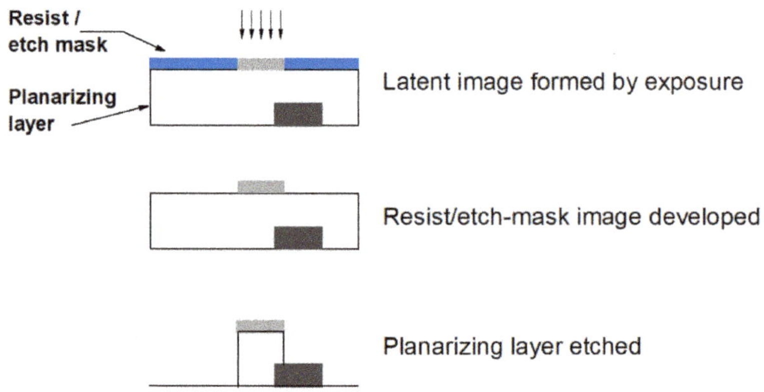

Figure 6.52 A two-layer RIE-transfer resist system.

Note that O_2 RIE is used as a generic term signifying high-selectivity polymer etching. A trace of other gases is often added to help improve the process. For example, a very small amount of fluorine-containing gas is included to clean up possible inorganic residue. This can reduce the etch selectivity between the polymer and the inorganic surfaces.

The transfer process need not be RIE. When the feature size to be transferred is larger than 0.5 μm, a deep-UV blanket exposure masked by a delineated top or middle layer produces a well-controlled proximity-printed image in the bottom planarizing layer. Of course, the bottom layer must be a deep-UV resist. Wavelengths shorter than those in the deep UV can be used to obtain a higher resolution or higher aspect ratio. The minimum feature size and aspect ratio can be predicted using Eqs. (5.3) and (5.4) to define the resolution limit of this system.

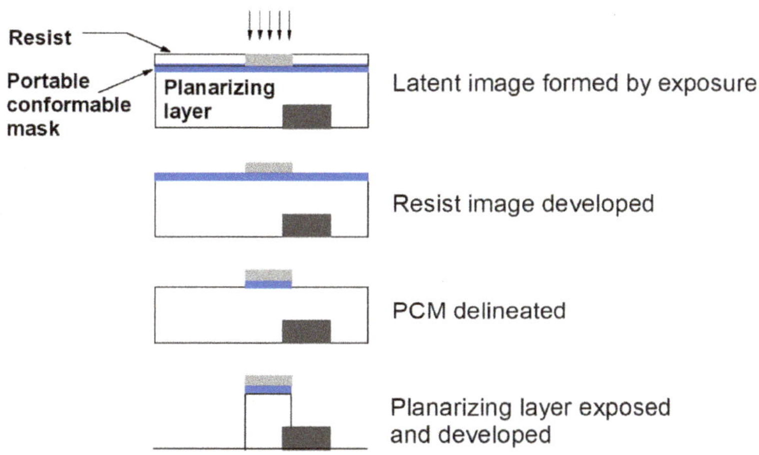

Figure 6.53 Three-layer phototransfer resist system.

Figure 6.53 shows the schematic of a three-layer phototransfer resist system. The three-layer phototransfer system is not popular due to the availability of a simpler two-layer phototransfer system,[59,60] shown in Fig. 6.54. The AZ1350 resist is used as the top imaging layer for near- or mid-UV exposure. This resist is highly absorptive in the deep UV, making it a good deep-UV mask.[17] PMMA, a well-known deep-UV[61,62] and e-beam[63] resist, is used for the bottom layer. The top layer acts as a conformable contact mask that is attached to the wafer, explaining why it is called a portable conformable mask (PCM) in Refs. 59 and 60.

As long as the feature size and aspect ratio are supported by the blanket-exposing wavelength, the phototransfer system is preferred because of the process simplicity and the low defect level. For features far below 0.5 μm,

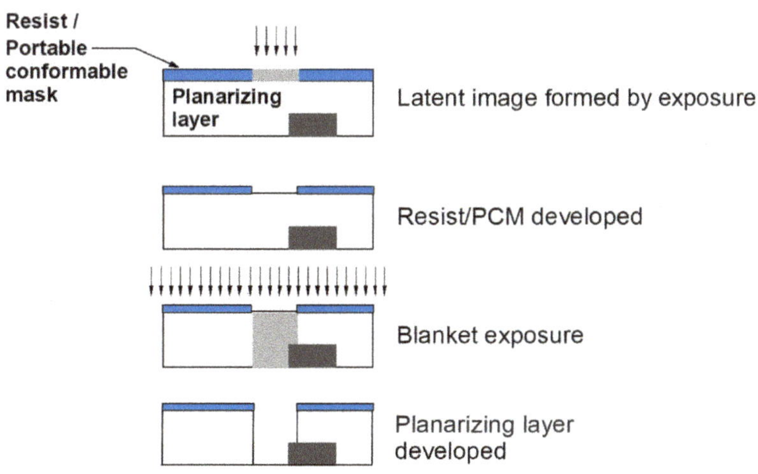

Figure 6.54 Two-layer phototransfer resist system.

the RIE-transfer system must be used. The choice between TSI and multilayer RIE-transfer systems depends on process control and cost. Advances in antireflection coatings for bulk imaging resist systems continue to be made. There is no urgent need to use TSI or multilayer systems until the feature sizes reach values on the order of single-digit or low-double-digit nanometers.

Contrast enhancement—Contrast enhancement is a two-layer phototransfer system that uses near-UV light to expose both the top and the bottom layers.[64] The top layer is highly absorptive and bleaches as it is exposed, so it also serves as an *in situ* portable conformable mask. Therefore, contrast enhancement is a one-exposure system. It also develops in a single step because the top layer readily dissolves in the aqueous developer of the bottom layer, which is a regular novolak resist. As soon as any part of the pattern is bleached, contrast enhancement allows light to penetrate for even more bleaching. This nonlinear process enhances the effective contrast of the resist system. Unfortunately, it also tends to favor larger features that bleach earlier, enhancing proximity effects. Figure 6.55 shows the working principle of the contrast-enhancement system and the enhancement of proximity effects.

6.5.2 Light interactions with a photoresist

6.5.2.1 Wavelength compression

As seen in Fig. 6.17, when light enters a higher-refractive-index medium, the wavelength is compressed by the index of refraction, as seen in

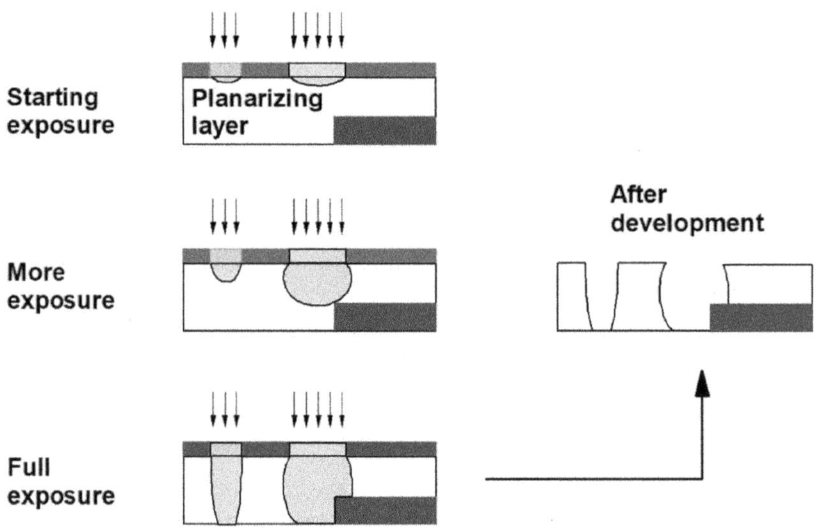

Figure 6.55 Working principle of the contrast-enhancement resist and the enhancement of proximity effects.

$$\lambda_2 = \left(\frac{n_1}{n_2}\right)\lambda_1 \tag{6.8}$$

and

$$D_2 = \left(\frac{n_1}{n_2}\right)D_1. \tag{6.9}$$

This is a desirable effect because the optical thickness D_2 of the resist film is now smaller, as given in Eq. (6.9), allowing the resist thickness to take up a smaller portion of the DOF budget.

6.5.2.2 Light absorption

A certain amount of light must be absorbed by the resist for it to acquire the energy to induce chemical changes. If there is too much absorption, the exposure light cannot penetrate the resist film and it becomes difficult to clear positive resists of the unwanted parts with the resist developer. It would also be difficult to crosslink the bottom of negative resists, causing feature dropping to ruin the resist image and to become a defect source. Even before feature dropping takes place, the resist profile would suffer severely. A quantitative optimization is given in Chapter 7. Here the effects of resist absorption are shown graphically.

6.5.2.3 Resist bleaching or dyeing

The resist may become more absorptive or less absorptive during exposure. The former is the dyeing phenomenon, and the latter is bleaching. Bleaching is preferred because it helps to increase the contrast of the resist image. However, the amount of dyeing or bleaching in workable resist systems is small unless something is designed specifically for it, such as the contrast-enhancement resist system.

6.5.2.4 Resist outgassing

Certain resists produce gaseous substances during exposure. For example, the diazoquinone-sensitized novolak resists release nitrogen gas during exposure, as shown in Fig. 6.41. In principle, outgassing can disturb the homogeneity of the refractive index in the vicinity of the resist surface that is critical for imaging. However, no ill effect has been reported, implying that the amount of outgassing is negligible for imaging. In the long run, if the released material accumulates on the lens surface, that surface must be cleaned. As to the effect on imaging, usually the specification for this accumulated material is more stringent than for bubbles in the resist. If the medium between the front element of the lens and the resist surface is a liquid, as in an immersion-exposure system, when the DOF is enhanced, the perturbation to the homogeneity of the media will be more severe and must be taken into consideration. This is discussed in Chapter 8.

6.5.2.5 Multiple reflections

When light is incident on an interface, it is refracted and reflected unless the refractive indexes of the two media are perfectly matched. Therefore, it undergoes multiple reflections and refractions at the resist–air and resist–wafer interfaces, as depicted in Fig. 6.56. In practice, there is a stack of thin films between the wafer and the resist. Each interface induces its own reflection and refraction, making the system very complicated to illustrate graphically. With only two interfaces shown in this figure, light bounces back and forth until the intensity is too low to be of concern.

Standing waves—Because of the interference of these light beams, standing waves are created, as illustrated in Fig. 6.57. Let's assume that the electric field of the incident light $\mathbf{E}_i$ is a plane wave perpendicular to the resist–wafer interface. In this case,

$$\mathbf{E}_i = e^{inkz}, \tag{6.10}$$

where n is the refractive index of the photoresist, $k \equiv 2\pi/\lambda$, where λ is the wavelength in vacuum, and z is the longitudinal distance the incident light is traveling. $z = 0$ refers to the resist–wafer interface. The electric field of the reflected beam E_r is represented by

$$\mathbf{E}_r = Re^{-inkz}, \tag{6.11}$$

where R is the complex reflection coefficient of the resist–wafer interface.

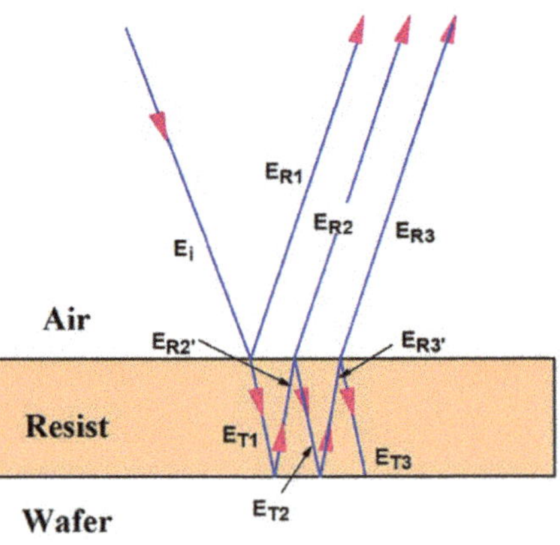

Figure 6.56 Multiple reflections (R) and refractions (T) of light in a photoresist. $\mathbf{E}_i$ is the incident electric field; $\mathbf{E}_r$ is the reflected field; $\mathbf{E}_T$ is the transmitted field at the air–resist interface; $\mathbf{E}_{R1}$ is the first reflected field from the air–resist interface; $\mathbf{E}_{R2'}$ is the reflected field from the wafer–resist interface; and $\mathbf{E}_{R2}$ is the continuation of $\mathbf{E}_{R2'}$ into air, and so forth.

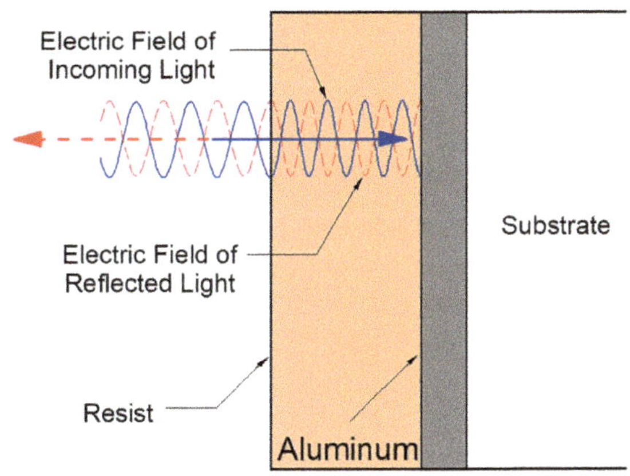

Figure 6.57 Standing waves on a metal layer in a photoresist produce no exposure at the metal surface.

When both beams are present, the total electric field **E** in the photoresist is

$$\mathbf{E} = \mathbf{E}_\mathrm{i} + \mathbf{E}_\mathrm{r} = (1 - R)e^{inkz} + 2R\cos nkz. \tag{6.12}$$

There are two components in **E**: the first component, represented by the 1st term, is a propagating wave, just like $\mathbf{E}_\mathrm{i}$ or $\mathbf{E}_\mathrm{r}$. The second component, represented by the 2nd term, is a standing wave. When we restore the usually omitted time function e^{-iwt} to the exponential terms in the wave equations [Eqs. (6.10) to (6.12)], the distinction between the propagating waves and the standing waves is even clearer. The standing wave does not have an e^{-iwt} component; therefore, it does not propagate in either direction but stands still. The resist exposure dosage is proportional to the intensity I:

$$I = |\mathbf{E}|^2 = (1 + R^2) + 2R\cos 2nkz. \tag{6.13}$$

Therefore, the periodicity p of the standing wave in the developed resist image is

$$p = \frac{\lambda}{2n}. \tag{6.14}$$

When the bottom surface of the resist does not reflect light, as in the case of a perfectly matched substrate, $R = 0$, and **E** and I are uniformly distributed. The resist image is at its best-possible profile. When R approaches -1, as in the case of a resist on aluminum, which is a good conductor and highly reflective, I varies from the two extremes, 4 and 0. At the resist–wafer interface, $z = 0$, and **E** and I both become zero. Hence, no matter how intense the incoming beam is, as long as the reflecting beam is from a highly conducting and reflective surface, the incident and reflected fields cancel each other, resulting in insufficient exposure at the interface.

With positive resists, the nodes of the standing wave block further development in the resist. At best, there is resist residue that must be ashed away at the resist–wafer interface. Before the advent of ashing, higher-concentration developers were used to increase the dissolution rate of the unexposed resist. The tradeoff is a reduction in the resist development contrast.

Negative resists exposed by strong standing waves lose adhesion at the interface because the resist there is not exposed and remains highly soluble. No adhesion promotion scheme can fix this problem unless the resist is exposed and thus crosslinked at the interface.

When the standing waves are not sufficiently severe to produce the above problems, they seem to be only cosmetically undesirable. This applies to isotropic and anisotropic etching, lift-off, and ion implantation, as seen in Fig. 6.58, where standing waves are drawn on one side of the resist profile for comparison. In isotropic etching, the most critical location for linewidth control is at the interface of the resist and the layer to be etched. The resist edge position here is strictly dependent on the refractive indexes of these two media and is independent of whether there is a tooth-shaped profile. For anisotropic etching with high etch selectivity, the arguments are similar to those for isotropic etching. When the selectivity is not so high and the resist is consumed during etching, the effect of standing waves is like that of a straight but more oblique profile drawn from one extreme of the standing wave at the bottom and from the other extreme of the standing wave on top, as shown in the figure. For ion-implant applications, the argument is similar. For lift-off

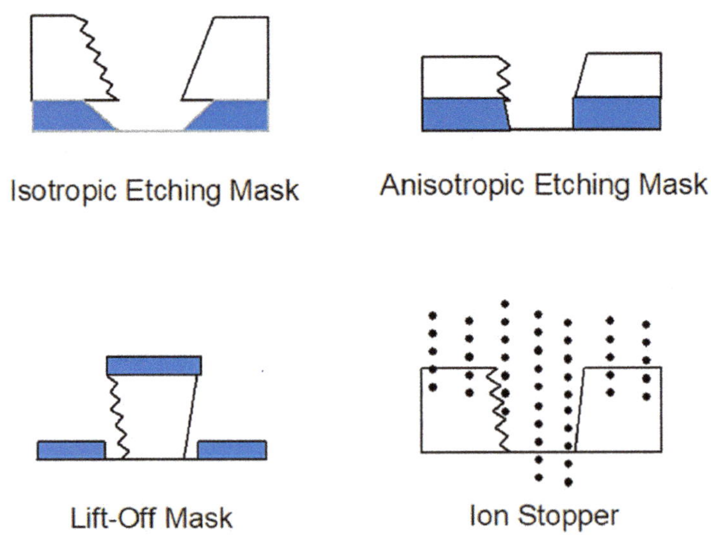

Figure 6.58 The effect of standing waves in resist applications. Standing waves are shown only on the left side of the resist profile.

applications, it is obvious that the tooth shape of the standing waves does not play any role in linewidth control.

Methods to control standing waves include the use of a PEB, a low-contrast developer, antireflection coatings, surface imaging, and multilayer resist systems. Eliminating standing waves with a PEB was first reported by Walker.[65] The standing waves in a diazoquinone novolak resist are eliminated with a 10-minute PEB at 100 °C, as shown in Fig. 6.59. The working principle is that the large longitudinal refractive-index gradient in the latent image is averaged out by diffusion at the PEB temperature. However, diffusion also takes place laterally. Removing the large gradient in the longitudinal direction also results in a reduction of the gradient in the latent image in the lateral direction. Hence, the resist profile is slightly more overcut, as can be seen in the figure. A smaller image gradient also leads to a smaller exposure latitude. Fortunately, the longitudinal-standing-wave-induced gradient is much larger than that of the lateral image edge. Otherwise, the PEB might carry too much tradeoff. Diffusion is less effective in reducing the longitudinal gradient at the resist–wafer interface because the direction of diffusion is now restricted to the resist side.

Other than for cosmetic reasons, using a low-contrast resist developer can reduce the skin effect (which slows down or blocks development into the bulk of the resist) and reduce the loss of adhesion from the substrate reflection. A high-contrast developer does not dissolve the unexposed positive resist; therefore, it makes the skin effect pronounced. When a developer can remove unexposed positive resist, the development contrast is dropped, but skins can more easily be removed. This is yet another tradeoff, as is the case of PEB.

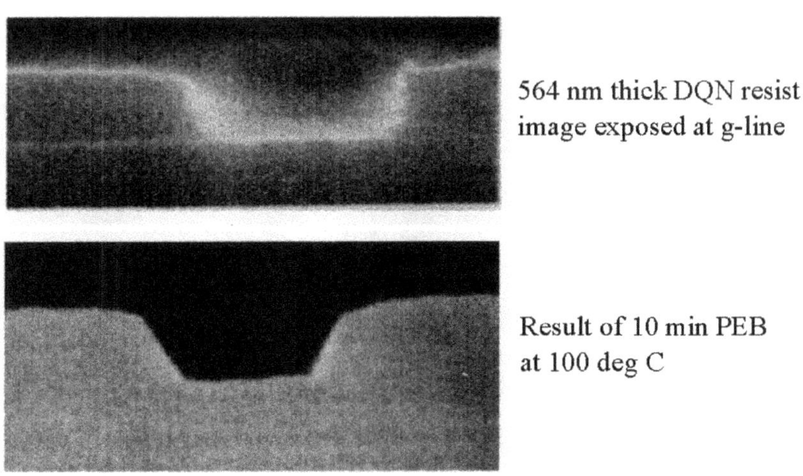

Figure 6.59 Standing-wave elimination by PEB [reprinted from Ref. 65 with permission; © (1975) IEEE].

The best method to eliminate standing waves is to remove substrate reflection, which will be discussed in Section 6.6.4. Other methods include surface imaging and use of the multilayer resist systems discussed in Section 6.6.1.3.

Stray light—Multiple reflections from the resist and the underlying films contribute to stray light in the imaging system, just like the reflection from the chromium area of the mask discussed in Section 6.3.3.4. Referring back to Fig. 6.15, one readily sees that film interfaces that are not in the conjugate plane of the mask reflect light to the chromium area to be reflected back, thus reducing the image contrast. This type of stray light is best treated with a reduction of mask reflectivity. Optimization of the resist and film thicknesses to minimize the total reflection is best when it is achieved by controlling the exposure tolerance, as given in Section 6.6.4.

Exposure nonuniformity from topography—In addition to standing waves, multiple reflections in the resist contribute to exposure variation as a function of resist thickness. Figure 6.60 shows reflectivity from the resist–air interface taken from a poly-Si/SiO_2 film stack. The reflectivity can change by 26% within a thickness of about 50 nm. Therefore, the resist exposure also changes by that amount because the resist-exposing light is the remainder of the incident light (after subtracting light that reflects from the air–resist interface) and the light that leaves the resist to enter the thin-film stack and substrate.

Generally, the resist coating thickness can be controlled to within a few nanometers and is not an issue. The primary cause of exposure nonuniformity is the variation in resist thickness over the topography, as shown in Fig. 6.61.

An exposure variation of 26% requires an imaging system that can support an exposure latitude better than 29%. The extra margin is used to allow for nonuniform exposure from illumination as well as inaccuracy in

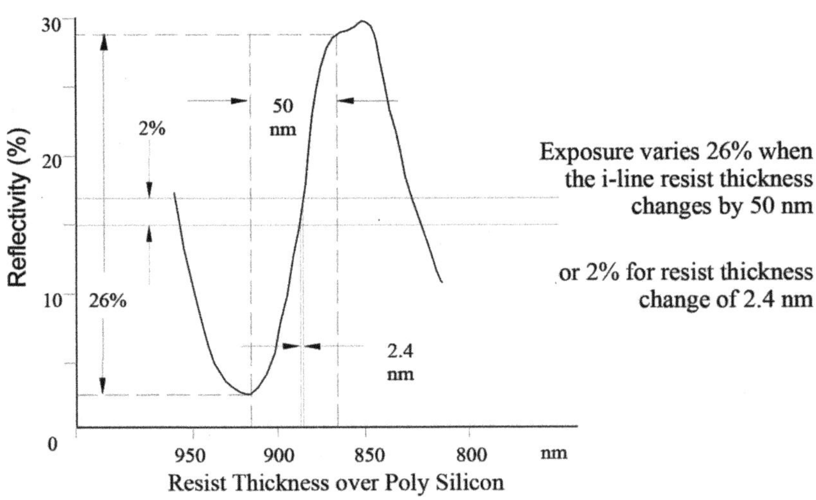

Figure 6.60 Reflectivity versus resist thickness can be very sensitive to resist thickness variations induced by the topography on the wafer.

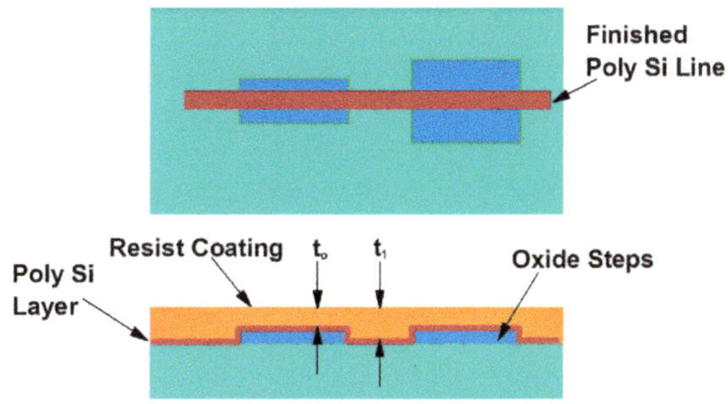

Figure 6.61 Resist thickness variation over the topography.

controlling the exposure shutter. As seen in Section 4.3.2 and subsequent examples, the exposure latitude needs to be contained within 15%, but preferably less than 10%. This requires close attention to the reflection variation produced by the resist and the thin-film stack. This topic is extremely important and is treated in Section 6.6.4.

6.5.3 Developed resist images

Figure 6.62 shows an aerial image that is narrower towards the bottom of the resist because of resist absorption of light. This promotes an overcut profile in the developed resist image. With a positive resist, the overcut is even more pronounced because, when the developer starts to penetrate the resist, the side of the resist image in the developed area is exposed to the developer. The upper side is subjected to the developer dissolution longer than the lower side. With a negative resist, the development effect compensates for the absorption effect in the profile of the resist image.

6.5.3.1 Development of simple aerial images

The detailed development stages of a 100-nm-wide and 200-nm-thick exposed resist image in a negative resist of 200-nm thickness is shown in Fig. 6.63. To

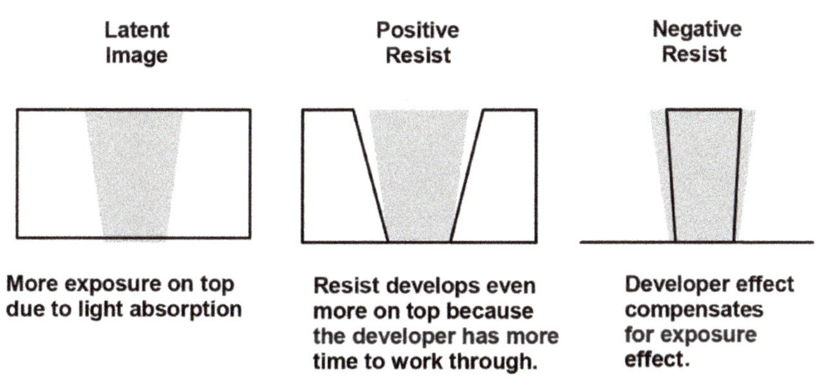

Figure 6.62 Profile of positive-resist and negative-resist images.

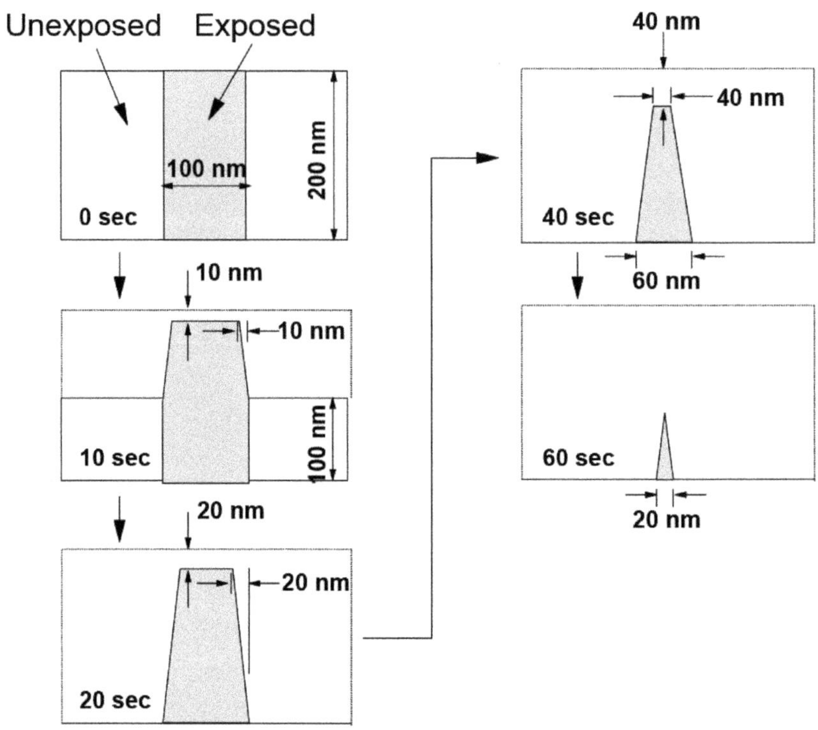

Figure 6.63 Negative development of a 100-nm-wide, 200-nm-deep exposed image.

simplify the illustration, the aerial image is not absorbed by the resist. Therefore, its width is constant along the 200-nm resist thickness. The dissolution rate is 10 nm/sec in the unexposed areas and 1 nm/sec in the exposed area. In the figure, the situations at $t = 0$, 10, 20, 40, and 60 sec are shown. Initially, the exposed resist image is 100-nm wide and 200-nm thick, as given.

After 10 sec, the unexposed background resist is uniformly removed by the developer and is thinned down to 100 nm. The top of the exposed area is lowered by 10 nm, while the sides of the exposed area are narrowed by 10 nm on each side at the top but are not yet removed at the development front of the unexposed resist.

At 20 sec, the unexposed area is completely cleared. The width of the developed image is 60 nm on top and 100 nm at the bottom. The thickness of the exposed area is now 80 nm.

At 40 sec, the exposed area is dissolved on top and at the sides, resulting in a 20-nm width on top and a 60-nm width at the bottom. The resist thickness is now 60 nm.

At 60 sec, the resist remaining is a wedge that is 20-nm wide at the bottom. Its height is only less than 80 nm because resist is removed simultaneously at the top and on two sides.

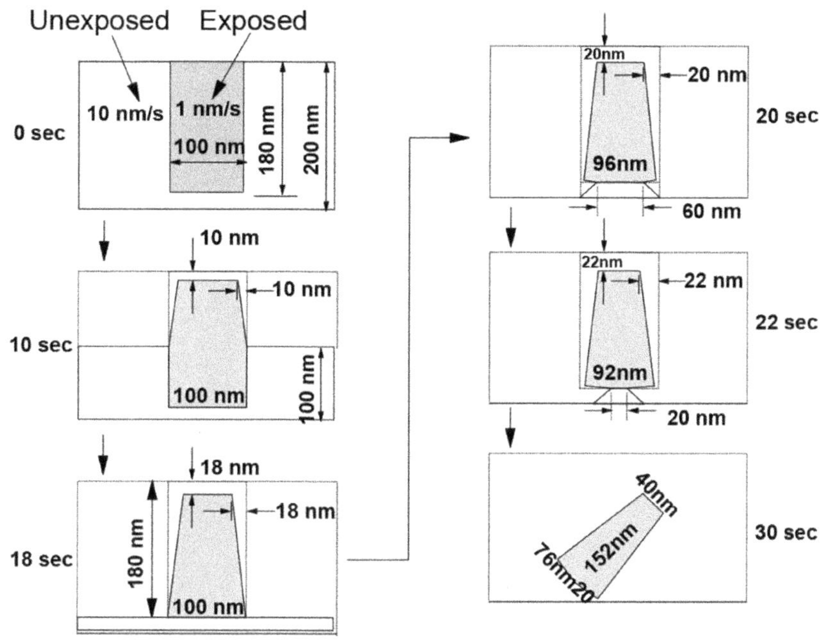

Figure 6.64 Development of a 100-nm-wide, 180-nm-deep exposed image in a negative resist of 200-nm thickness.

The situation in Fig. 6.64 is like that in Fig. 6.63, except that the resist was not exposed at the last 20 nm of its 200-nm thickness:

- At $t = 0$, 10, and 18 sec, the development proceeds as in the previous case.
- At $t = 20$ sec, the unexposed resist supporting the crosslinked structure is removed by the developer at the unexposed development rate, resulting in a thinning of this support.
- At $t = 22$ sec, the support is substantially corroded.
- At $t = 30$ sec, the crosslinked structure starts to lose its support and is separated from the substrate. It can be carried away by any flowing liquid or by air to rest on unintended areas, becoming a defect.

6.5.3.2 Development of diffracted aerial images

In reality, the aerial image is not that simple. Figure 6.65 shows the refracted and developed images of a $k_1 = 0.64$ line opening illuminated with $\sigma = 0.5$ at three defocus positions, as if a wafer with a coated resist is located at each of these positions. The focal plane is marked and is in the vicinity of $k_2 = 1.8$. The resist thickness is 1 in k_2 units. The diffracted image is initially simulated in air. When this image propagates into the resist, it continues to propagate in a new medium with n_r and α, where n_r is

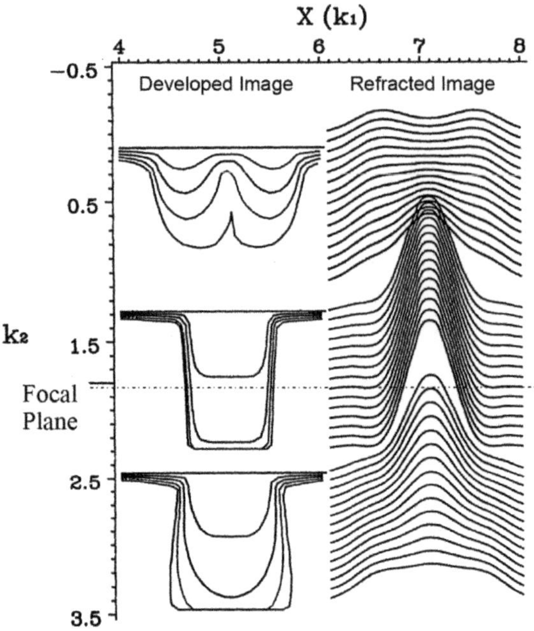

Figure 6.65 Refracted and developed images: $n = 1.5$, $\alpha = 0$.

the real part of the complex refractive index n_c of the resist, and the imaginary part n_i is given as

$$n_i \equiv \frac{\lambda}{2\pi}\alpha. \tag{6.15}$$

This image is strictly refracted. Reflections are ignored here. This refracted image is translated into a rate image and developed by simulation. The development contours are the developed surface at progressive development times. In addition to using Fig. 6.65 as a reference condition, an important observation can be made here. Although the wafer is placed symmetrically off focus from the best focus, the developed images at the two defocus locations are quite different because the refracted images at the starting surface for the developer are different. In the negative defocus region, the refracted image starts as a double-peak deteriorated image propagating into a single-peak focused image, whereas in the positive defocus region, the refracted image starts as single-peaked and degenerates into two peaks. This accounts for a focus shift when a resist is introduced. The loss in DOF in the negative region of defocus is made up by the gain in the positive region. Therefore, the net result in DOF is unaffected. The developed images flatten out in the last two wafer positions because the developer has reached the wafer surface. In the first position, the resist is not completely developed, even though the same development time is used for all three cases.

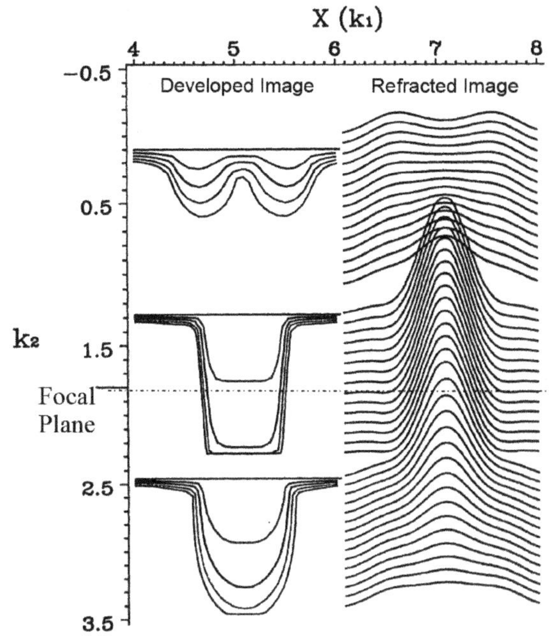

Figure 6.66 Refracted and developed images: $n = 1.5$, $\alpha = 0.178\ \mu m^{-1}$.

A small absorption is introduced, and its result is depicted in Fig. 6.66. In the negative defocus region, the resist is developed even less because less light reaches the bottom. In the focused region, the resist profile is slightly overcut, as predicted by Fig. 6.62. In the positive defocus region, the resist image has reversed its positive bias. This means that, with a slightly smaller absorption coefficient, the resist image can be made identical to the focused image. Therefore, there is an optimum amount of absorption.

When the absorption coefficient is increased to 3 μm^{-1}, the developed images do not clear the resist bottom in all three defocus positions, as shown in Fig. 6.67. However, this also means that reflection from the wafer substrate is completely annihilated. Although the resist is not suitable for bulk imaging, it is ideal for surface imaging.

6.5.4 Antireflection coating (ARC) (by B.J. Lin and S.S. Yu)

As pointed out in Section 6.6.2, controlling reflection from the resist–wafer interface is extremely important. There are many ways to accomplish this. The TSI or multilayer resist systems in Section 6.6.1 are optically simple, but more processing steps are required, and they are not yet ready for high-volume manufacturing. This leads to the need to extend the use of single-layer resist systems with thin-film-thickness engineering to the cases where the semiconductor device requirement is not violated. Extra polymer coatings that reduce reflection are also introduced. Hence, there are top antireflection

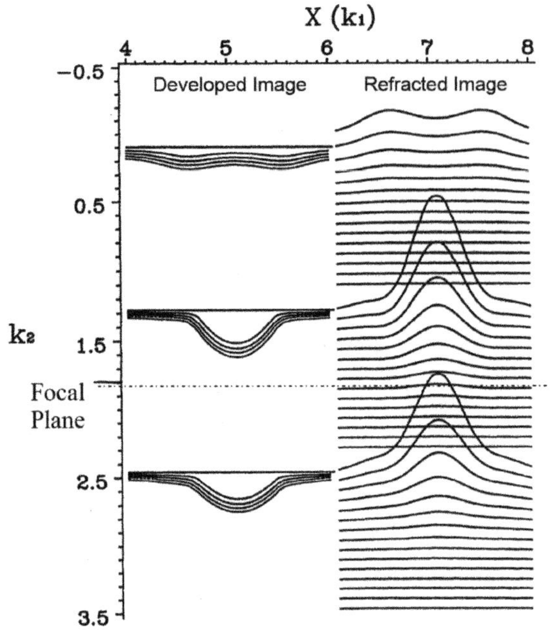

Figure 6.67 Refracted and developed images: $n = 1.5$, $\alpha = 3\ \mu m^{-1}$.

coatings (TARCs) and bottom antireflection coatings (BARCs). The former is applied to the top of the resist layer, while the latter is applied to the resist–wafer interface. BARCs can be a polymer film or an inorganic layer.

We use a simple case to understand ARC analytically. Take the situation shown in Fig. 6.68. The resist is coated directly on the wafer. The incident beam, whose electric field $\mathbf{E}_i$ is reflected and transmitted at the air–resist interface, produces $\mathbf{E}_{rAR1}$ and $\mathbf{E}_{tAR1}$. $\mathbf{E}_{tAR1}$ is reflected at the resist–wafer

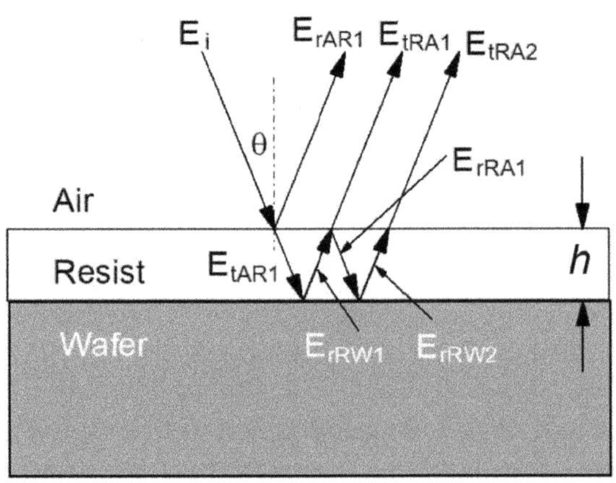

Figure 6.68 Multiple reflections in a single-resist layer coated directly on a plain wafer substrate.

interface, becoming $\mathbf{E}_{\mathrm{rRW1}}$, and then is reflected and transmitted at the resist–air interface, producing $\mathbf{E}_{\mathrm{rRA1}}$ and $\mathbf{E}_{\mathrm{tRA1}}$. Hence, the total reflected electric field $\mathbf{E}_{\mathrm{r}}$ is

$$\mathbf{E}_{\mathrm{r}} = E_{\mathrm{rAR1}} + E_{\mathrm{tRA1}} + E_{\mathrm{tRA2}} + \ldots. \tag{6.16}$$

We now denote the reflection and transmission coefficients at each surface as r_{AR}, t_{AR}, r_{RW}, r_{RA}, and t_{RA}, where r_{AR} is the reflection coefficient at the air–resist interface entering from air, t_{AR} is the transmission coefficient from air to resist at the air–resist interface, r_{RW} is the reflection coefficient at the resist–wafer interface entering from the resist, etc. Hence,

$$\frac{\mathbf{E}_{\mathrm{r}}}{\mathbf{E}_{\mathrm{i}}} = r_{\mathrm{AR}} + t_{\mathrm{AR}} r_{\mathrm{RW}} t_{\mathrm{RA}} \eta e^{i\delta} + t_{\mathrm{AR}} r_{\mathrm{RW}} r_{\mathrm{RA}} r_{\mathrm{RW}} t_{\mathrm{RA}} \eta^2 e^{i2\delta} + \ldots, \tag{6.17}$$

where

$$\delta = \frac{4\pi}{\lambda} nh \cos\theta_{\mathrm{R}}$$

is the phase introduced by each cycle of travel from the air–resist interface to the resist–wafer interface and back, n is the real part of the refractive index of the resist, h is the resist thickness, λ is the wavelength, and θ_{R} is the angle that the refracted beam makes with the normal of the reflecting surface. $\eta = e^{-2h/\cos\theta_{\mathrm{R}}}$ is the attenuation induced by each cycle of travel, and α is the absorption coefficient of the resist. Summing up the infinite series in Eq. (6.17), and recognizing that $r_{\mathrm{AR}} = -r_{\mathrm{BA}}$, $t_{\mathrm{AR}} = t_{\mathrm{RA}}$ $(n_{\mathrm{A}}\cos\theta_{\mathrm{i}}/n_{\mathrm{R}}\cos\theta_{\mathrm{R}})$, we obtain

$$\frac{\mathbf{E}_{\mathrm{r}}}{\mathbf{E}_{\mathrm{i}}} = \frac{r_{\mathrm{AR}} + \left(\frac{n_{\mathrm{R}}\cos\theta_{\mathrm{R}}}{n_{\mathrm{A}}\cos\theta_i} t_{\mathrm{AR}}{}^2 + r_{\mathrm{AR}}{}^2\right) r_{\mathrm{RW}} \eta e^{i\delta}}{1 + r_{\mathrm{AR}} r_{\mathrm{RW}} \eta e^{i\delta}} \tag{6.18}$$

and

$$\frac{\mathbf{E}_{\mathrm{r}}}{\mathbf{E}_{\mathrm{i}}} = r_{\mathrm{AR}} + \frac{t_{\mathrm{AR}} r_{\mathrm{RW}} t_{\mathrm{RA}} \eta e^{i\delta}}{1 - r_{\mathrm{RA}} r_{\mathrm{RW}} \eta e^{i\delta}}. \tag{6.19}$$

If there is a BARC or TARC, regardless of whether it is single layered or multilayered, the above formulae are still valid in their present form. We only need to interpret r_{AR}, t_{AR}, r_{RW}, t_{RA}, and r_{RA} properly. That is, for example, t_{AR} is the ratio of the electric field at the resist top to the electric field at the air bottom, even if a TARC is sandwiched in between. However, to find their numerical values in such a case, we need to trace reflection and transmission at all related interfaces, which becomes very tedious as the number of layers is increased.

In the following, we will adopt a different approach.[69,70] By employing the matrix formulation, which is derived by considering total fields inside each layer, the effect of multiple reflections at each interface is automatically taken into account, thus greatly simplifying the problem. In a matrix formulation, there are two matrices that govern the propagation of waves inside the thin-film stack. One is the boundary matrix $\mathbf{D}_{1,2}$, which relates electric field amplitudes at the two sides of an interface between, say layers 1 and 2:

$$\begin{bmatrix} E_1^+(d_1) \\ E_1^-(d_1) \end{bmatrix} = \mathbf{D}_{1,2} \begin{bmatrix} E_2^+(0) \\ E_2^-(0) \end{bmatrix} \tag{6.20}$$

with

$$\mathbf{D}_{1,2} = \frac{1}{t_{1,2}} \begin{bmatrix} 1 & r_{1,2} \\ r_{1,2} & 1 \end{bmatrix}, \tag{6.21}$$

where E_j^+ is the field amplitude of an electromagnetic wave propagating in the $+z$ direction, and E_j^- is the field amplitude of the one propagating in the $-z$ direction. Note that we have used the coordinate system in which the $+z$ direction is normal to and points towards the thin-film stack. The z-coordinate at layer j is z_j and ranges from 0 at the top of the layer to d_j at the bottom of the layer. Figure 6.69 shows the nomenclature of the layers and the direction of the z axis. Layer 0 denotes the resist layer. The negative indices indicate layers above the resist, and the positive indices indicate the layers below it. The first layer is designated by the index $-m$, and the last layer

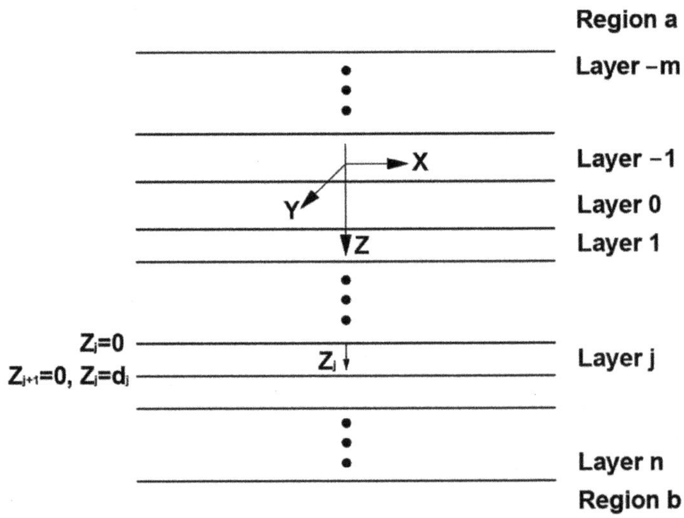

Figure 6.69 Coordinate system of the thin-film stack.

is designated by the index n. The region above layer $-m$ is denoted by a, and the region below layer n is denoted by b.

Free propagation inside a layer, say layer 2, is governed by another matrix $\mathbf{P}_2$, called the propagation matrix:

$$\begin{bmatrix} E_2^+(0) \\ E_2^-(0) \end{bmatrix} = \mathbf{P}_2 \begin{bmatrix} E_2^+(d_2) \\ E_2^-(d_2) \end{bmatrix}, \tag{6.22}$$

where

$$\mathbf{P}_2 = \begin{bmatrix} e^{-i\delta_2} & 0 \\ 0 & e^{i\delta_2} \end{bmatrix}. \tag{6.23}$$

As will be shown in Sections 6.6.4.1 and 6.6.4.2, the parameters of the boundary matrix and the propagation matrix, i.e., $r_{1,2}$ and $t_{1,2}$, are functions of the wave vector of the related layers:

$$\delta_2 \equiv \frac{2\pi}{\lambda_2} d_2 = k_2 d_2, \tag{6.24}$$

where λ_2 is the wavelength in layer 2, and k_2 is the corresponding wave number.

With the two matrices provided in Eqs. (6.20) and (6.22), we are now ready to derive reflectivity from the thin-film stack. First, considering wave propagation from layer a to layer b, we have

$$\begin{bmatrix} E_a^+ \\ E_a^- \end{bmatrix} = M \begin{bmatrix} E_b^+ \\ 0 \end{bmatrix}, \tag{6.25}$$

where $E_a^{\pm}$ is the electric field at the boundary of region a and layer $-m$, E_b^+ is the electric field at the boundary of layer n and region b, and M is the matrix product of all of the dynamic and propagation matrices that govern the wave propagation through all of the layers between a and b:

$$M = D_{a,-m} P_{-m} D_{-m,-(m-1)} \cdots D_{-1,0} P_0 D_{0,1} \cdots D_{n-1,n} P_n D_{n,b} = \begin{bmatrix} M_{11} & M_{12} \\ M_{21} & M_{22} \end{bmatrix}. \tag{6.26}$$

Note that in region b, there is no wave propagating in the $-z$ direction if the wave is incident from region a in the $+z$ direction.

For notational convenience, we rewrite $[D_{a,-m} \cdots D_{-1,0}]$ as $\mathbf{M}_{\text{res top}}^{-1}$ with

$$\mathbf{M}_{\text{res top}} = \begin{bmatrix} A_{11} & A_{12} \\ A_{21} & A_{22} \end{bmatrix} \tag{6.27}$$

and $[D_{0,1} \ldots D_{n,b}]$ as $\mathbf{M}_{\text{res bot}}$ with

$$\mathbf{M}_{\text{res bot}} = \begin{bmatrix} B_{11} & B_{12} \\ B_{21} & B_{22} \end{bmatrix}, \tag{6.28}$$

where "res top" stands for resist top, and "res bot" stands for resist bottom.

Consider the wave propagating in the $-z$ direction from the resist top to region a:

$$\begin{bmatrix} E_0^+(0) \\ E_0^-(0) \end{bmatrix} = \begin{bmatrix} A_{11} & A_{12} \\ A_{21} & A_{22} \end{bmatrix} \begin{bmatrix} 0 \\ E_a^- \end{bmatrix}, \tag{6.29}$$

leading to

$$r_{\text{res top}} \equiv \frac{E_0^+(0)}{E_0^-(0)} = \frac{A_{12}}{A_{22}}. \tag{6.30}$$

The field at the resist bottom is related to that in region b for a wave traveling in the $+z$ direction as follows:

$$\begin{bmatrix} E_0^+(d_0) \\ E_0^-(d_0) \end{bmatrix} = \begin{bmatrix} B_{11} & B_{12} \\ B_{21} & B_{22} \end{bmatrix} \begin{bmatrix} E_b^+ \\ 0 \end{bmatrix}. \tag{6.31}$$

Hence,

$$r_{\text{res bot}} \equiv \frac{E_0^-(d_0)}{E_0^+(d_0)} = \frac{B_{21}}{B_{11}}. \tag{6.32}$$

The reflectivity at the resist top $r_{\text{res top}}$ and resist bottom $r_{\text{res bot}}$ from inside the resist are two important quantities relating directly to the swing effect and thus play an important role in the CD control of optical lithography.

The reflectivity at the top surface of the entire thin-film stack with the incident wave from air is

$$r \equiv E_a^-/E_a^+ = M_{21}/M_{11}. \tag{6.33}$$

From Eq. (6.25),

$$\mathbf{M}_{\text{res top}} \begin{bmatrix} E_a^+ \\ E_a^- \end{bmatrix} = \mathbf{P}_0 \mathbf{M}_{\text{res bot}} \begin{bmatrix} E_b^+ \\ 0 \end{bmatrix} \tag{6.34}$$

and

$$\begin{bmatrix} A_{11} & A_{12} \\ A_{21} & A_{22} \end{bmatrix} \begin{bmatrix} E_a^+ \\ E_a^- \end{bmatrix} = \begin{bmatrix} e^{-i\delta_0} & 0 \\ 0 & e^{i\delta_0} \end{bmatrix} \begin{bmatrix} B_{11} & B_{12} \\ B_{21} & B_{22} \end{bmatrix} \begin{bmatrix} E_b^+ \\ 0 \end{bmatrix}. \tag{6.35}$$

Using Eqs. (6.30), (6.32), (6.33), and (6.35), we have

$$r = \frac{-A_{21} + A_{11}e^{i2\delta_0}r_{\text{res bot}}}{A_{22} - A_{12}e^{i2\delta_0}r_{\text{res bot}}}. \tag{6.36}$$

Thus, the reflectivity of the thin-film stack is expressed in terms of the matrix elements of $M_{\text{res top}}$ and the inside reflectivity at the resist bottom.

To calculate the electric field $E_j^{\pm}(z_j)$ of the TE wave inside each layer of the thin-film stack, we start from

$$\begin{bmatrix} E_a^+ \\ E_a^- \end{bmatrix} = \begin{bmatrix} 1 \\ r \end{bmatrix},$$

and then propagate it to any position of the thin-film stack by using the boundary and propagation matrices. First, we will derive the formula for $E_0^{\pm}(z_0)$, which is the electric field of the TE wave inside the resist. This formula is to be used for obtaining the analytical expression of the Poynting vector. To derive $E_0^{\pm}(z_0)$, we start by deriving $E_0^{\pm}(0)$.

Relate the electric field on top of the resist to that in region a:

$$\begin{bmatrix} A_{11} & A_{12} \\ A_{21} & A_{22} \end{bmatrix} \begin{bmatrix} E_a^+ \\ E_a^- \end{bmatrix} = \begin{bmatrix} E_0^+(0) \\ E_0^-(0) \end{bmatrix}. \tag{6.37}$$

From Eqs. (6.30), (6.36), and (6.37),

$$E_0^+(0) = \frac{\det \mathbf{M}_{\text{res top}}/A_{22}}{1 - r_{\text{res top}}e^{i2\delta_0}r_{\text{res bot}}} E_a^+. \tag{6.38}$$

For $E_0^-(0)$, start with the relationship between the electric field at the top of the resist and the electric field in region b:

$$\begin{bmatrix} E_0^+(0) \\ E_0^-(0) \end{bmatrix} = \begin{bmatrix} e^{-i\delta_0} & 0 \\ 0 & e^{i\delta_0} \end{bmatrix} \begin{bmatrix} B_{11} & B_{12} \\ B_{21} & B_{22} \end{bmatrix} \begin{bmatrix} E_b^+ \\ 0 \end{bmatrix}, \tag{6.39}$$

from which $E_0^+(0) = e^{-i\delta_0}B_{11}E_b^+$ and $E_0^-(0) = e^{i\delta_0}B_{21}E_b^+$, leading to

$$E_0^-(0) = r_{\text{res bot}}e^{i2\delta_0}E_0^+(0). \tag{6.40}$$

Finally,

$$E_0^+(z_0) = e^{i\zeta_0}E_0^+(0) \tag{6.41}$$

and

$$E_0^-(z_0) = e^{-i\zeta_0}E_0^- = e^{-i\zeta_0}r_{\text{res bot}}e^{i2\delta_0}E_0^+(0), \tag{6.42}$$

where $\zeta_0 = k_0 z_0$. We now have the analytic expressions for $E_0^{\pm}(z_0)$.

To evaluate the Poynting vector, the total electric field inside the resist is

$$\mathbf{E}_0 = \mathbf{E}_0^+ + \mathbf{E}_0^-. \tag{6.43}$$

The corresponding total magnetic field is

$$\mathbf{H}_0 = \frac{\mathbf{k}_0^+ \times \mathbf{E}_0^+ + \mathbf{k}_0^- \times \mathbf{E}_0^-}{\omega\mu_0}, \tag{6.44}$$

where $\omega \equiv 2\pi f$, with f being the frequency of the propagating wave and μ_0 being the permeability of the resist. The Poynting vector inside the resist is $\mathbf{I}_0 = \mathbf{E}_0 \times \mathbf{H}_0$. Since the electric field and the magnetic field involved are time harmonic, one considers only the time-averaged Poynting vector in the z direction:

$$\begin{aligned}\langle I_{0z}(z_0)\rangle &= \langle \mathbf{I}_0(z_0)\cdot\hat{\mathbf{Z}}\rangle \\ &= \frac{1}{2\omega\mu_\mathrm{v}}[[|E_0^+(z_0)|^2 - |E_0^-(z_0)|^2]k_0^R - 2\mathrm{Im}[E_0^+(z_0)E_0^{-*}(z_0)]k_0^I],\end{aligned} \tag{6.45}$$

where we have replaced μ_0 by μ_v, the permeability in vacuum. Note that k_0 is complex.

Now, we are going to derive the analytic expression of the swing curve. By substituting $E_0^\pm(z_0)$, obtained previously into Eq. (6.44), we obtain

$$\begin{aligned}\langle I_{0z}(z_0)\rangle &= \begin{bmatrix}[e^{-2k_0^I z_0} + e^{2k_0^I z_0}|r_\text{res bot}|^2 e^{-4k_0^I d_0}]k_0^R \\ -2|r_\text{res bot}|e^{-2k_0^I d_0}\sin[2k_0^R z_0 - 2k_0^R d_0 - \gamma_\text{res bot}]k_0^I\end{bmatrix} \\ &\quad\times\left|\frac{\det \mathbf{M}_\text{res bot}/A_{22}}{1 - r_\text{res top}e^{i2\delta_0}r_\text{res bot}}\right|^2\langle I_{az}^+\rangle,\end{aligned} \tag{6.46}$$

where we have used the Poynting vector of the incident wave,

$$\langle I_{az}^+\rangle = \frac{1}{2}\mathrm{Re}[\mathbf{E}_a^+ \times \mathbf{H}_a^{+*}]\cdot\hat{\mathbf{z}} = \frac{1}{2\omega\mu_\mathrm{v}}|E_a^+|^2 k_{az}^+. \tag{6.47}$$

To calculate energy absorbed inside resist, we use Poynting's theorem, which states that the energy ρ absorbed per unit volume per unit time equals the negative divergence of the Poynting vector at that point:

$$\begin{aligned}\rho \equiv -\nabla\cdot\langle \mathbf{I}_0\rangle &= \frac{1}{k_a}\begin{bmatrix}[2k_0^I e^{-2k_0^I z_0} - 2k_0^I e^{2k_0^I z_0}|r_\text{res bot}|^2 e^{-4k_0^I d_0}]k_0^R \\ +4k_0^R|r_\text{res bot}|e^{-2k_0^I d_0}\cos[2k_0^R z_0 - 2k_0^R d_0 - \gamma_\text{res bot}]k_0^I\end{bmatrix} \\ &\quad\times\left|\frac{\det \mathbf{M}_\text{res top}/A_{22}}{1 - r_\text{res top}e^{i2\delta_0}r_\text{res bot}}\right|^2\langle I_{az}^+\rangle,\end{aligned} \tag{6.48}$$

where we only care about the variation of the Poynting vector along the z direction. With its z_0 dependence, the term $4k_0^R|r_{\text{res bot}}|e^{-2k_0^I d_0}\cos[2k_0^R z_0 - 2k_0^R d_0 - \gamma_{\text{res bot}}]k_0^I$ in the above equation represents the standing wave inside the resist. The periodocity p of the standing wave is found to be $p = \pi/k_0^R = \lambda_0/2$. Assuming that the standing-wave term can be dropped after PEB due to diffusion of the photoacid, and that $|r_{\text{res bot}}|^2 e^{-4k_0^I d_0}$ as well as $|r_{\text{res top}}||r_{\text{res bot}}|$ are very small compared to 1, we obtain

$$\rho \approx \frac{1}{k_a} 2k_0^I k_0^R e^{-2k_0^I z_0}\left|\frac{\det \mathbf{M}_{\text{res top}}}{A_{22}}\right|^2 \times [1 + 2|r_{\text{res top}}|e^{-2k_0^I d_0}|r_{\text{res bot}}|\cos(\gamma_{\text{res top}} + 2k_0^R d_0 + \gamma_{\text{res bot}})]\langle I_{az}^+\rangle. \quad (6.49)$$

Considering ρ at the resist bottom, i.e., at $z_0 = d_0$, we have, from Eq. (6.49),

$$\rho = Ce^{-\alpha d_0}[1 + 2|r_r|e^{-\alpha d_0}\cos(\zeta d_0 + \gamma_r)]\langle I_{az}^{+I}\rangle, \quad (6.50)$$

where $C \equiv \frac{1}{k_a} \varsigma k_0^I |\frac{\det \mathbf{M}_{\text{res top}}}{A_{22}}|^2$, and

$$\alpha \equiv 2k_0^I,\ |r_r| \equiv |r_{\text{res top}}||r_{\text{res bot}}|, \text{ and } \gamma_r \equiv \gamma_{\text{res top}} + \gamma_{\text{res bot}}.$$

If the resist is operating in its linear response region, then the change of the resist CD is $\Delta w \propto \rho$. Upon integration, the generalized swing curve—i.e., the resist CD as an explicit function of the resist thickness [optical parameters of the thin-film stack are implicit in r_{res} (or $r_{\text{res top}}$ and $r_{\text{res bot}}$)]—can be obtained.

In the following, we derive elements of the boundary and the propagation matrices. We will solve the wave vector inside each layer of the thin-film stack first. Referring to Fig. 6.66, the z-axis is normal to the thin-film stack, and the xy-plane coincides with the surface of the thin-film stack. Since only the plane-wave solutions of electric and magnetic fields are considered, we assume for each layer j that $\mathbf{E}_j^{\pm} = \mathbf{E}_{j0}^{\pm} e^{i\mathbf{k}_j^{\pm}\bullet\mathbf{r} - i\omega t}$ and $\mathbf{H}_j^{\pm} = \mathbf{H}_{j0}^{\pm} e^{i\mathbf{k}_j^{\pm}\bullet\mathbf{r} - i\omega t}$, where $\mathbf{E}_j^{+}(\mathbf{H}_j^{+})$ propagates into the thin-film stack, and $\mathbf{E}_j^{-}(\mathbf{H}_j^{-})$ propagates away from the thin-film stack. To satisfy the boundary conditions on the interface between two layers, say region a and layer 1, the functional dependence of all waves on the interface should be the same, i.e., $\mathbf{k}_a^+ \cdot \mathbf{r} = \mathbf{k}_a^- \cdot \mathbf{r} = \mathbf{k}_1^+ \cdot \mathbf{r} = \mathbf{k}_1^- \cdot \mathbf{r}$ if $\mathbf{r}$ lies on the interface. Choosing $\mathbf{r} = \hat{\mathbf{x}}$, we have

$$\mathbf{k}_a^+ \cdot \hat{\mathbf{x}} = \mathbf{k}_a^- \cdot \hat{\mathbf{x}} = \mathbf{k}_1^+ \cdot \hat{\mathbf{x}} = \mathbf{k}_1^- \cdot \hat{\mathbf{x}}. \quad (6.51)$$

Similarly, choosing $\mathbf{r} = \hat{\mathbf{y}}$ results in

$$\mathbf{k}_a^+ \cdot \hat{\mathbf{y}} = \mathbf{k}_a^- \cdot \hat{\mathbf{y}} = \mathbf{k}_1^+ \cdot \hat{\mathbf{y}} = \mathbf{k}_1^- \cdot \hat{\mathbf{y}}. \quad (6.52)$$

Let $\mathbf{k}_a^+$ lie on the zx-plane, making it then the plane of incidence; $\mathbf{k}_a^+ \cdot \hat{\mathbf{y}} = 0$. Equations (6.51) and (6.52) can be applied repeatedly from between region a and layer 1 to between layer n and region b. We have $\mathbf{k}_j^\pm \cdot \hat{\mathbf{x}} = \mathbf{k}_a^\pm \cdot \hat{\mathbf{x}}$ and $\mathbf{k}_j^\pm \cdot \hat{\mathbf{y}} = 0$ for each layer j.

Inside each layer j, the electric field $\mathbf{E}_j^\pm$ (or the magnetic field $\mathbf{H}_j^\pm$) satisfies the homogeneous wave equation:

$$\nabla^2 \mathbf{E}_j^\pm - \mu_j \varepsilon_j \frac{\partial^2 \mathbf{E}_j^\pm}{\partial t^2} = 0. \tag{6.53}$$

Substituting plane-wave solutions into the above equation, we have

$$\mathbf{k}_j^\pm \cdot \mathbf{k}_j^\pm = \mu_j \varepsilon_j \omega^2. \tag{6.54}$$

The permittivity of layer j is $\varepsilon_j = \varepsilon_v N_j^2$, with $N_j \equiv n_j + i\kappa_j$ being the complex index of refraction of layer j. Assuming that $\mu_j = \mu_v$ (i.e., nonmagnetic), for each layer j, we have $\mu_j \varepsilon_j \omega^2 = (2\pi/\lambda_v)^2 N_j{}^2$, where ε_v is the permittivity in vacuum, μ_v is the permeability in vacuum, and λ_v is the wavelength in vacuum. From Eq. (6.54), $\mathbf{k}_j^\pm$ is also complex. Dividing $\mathbf{k}_j^\pm$ into its real and imaginary parts, we have $\mathbf{k}_j^\pm = \mathbf{k}_j^{\pm R} + i\mathbf{k}_j^{\pm I}$.

In the following, we will solve for $\mathbf{k}_j^\pm$ inside each layer j. This is not possible in general. However, here we only consider the special case where the material in layer a is nonabsorbing. In this case, $\mathbf{k}_j^+ \cdot \hat{\mathbf{x}} = (2\pi/\lambda_v) n_a \sin\theta_a$ if the angle of incidence at layer a is θ_a. Recalling that $\mathbf{k}_j^+ \cdot \hat{\mathbf{y}} = 0$, we can assume that $\mathbf{k}_j^+ = (2\pi/\lambda_v) n_a \sin\theta_a \hat{\mathbf{x}} + (k_{jz}^{+R} + ik_{jz}^{+I})\hat{\mathbf{z}}$. By employing Eq. (6.54), we have $\mathbf{k}_j^+ \cdot \mathbf{k}_j^+ = (2\pi/\lambda_v)^2 n_a^2 \sin^2\theta_a + (k_{jz}^{+R} + ik_{jz}^{+I})^2$, from which $k_{jz}^+ \equiv k_{jz}^{+R} + ik_{jz}^{+I}$ can be solved, i.e., $k_{jz}^+ = (2\pi/\lambda_v)\sqrt{N_j^2 - n_a^2 \sin^2\theta_a}$. Actually, there are two square roots for any complex number. For k_{jz}^+, we have chosen the one where $\mathrm{Re}[k_{jz}^+] > 0$, indicating that it propagates into the thin-film stack. The equation governing k_{jz}^- is exactly the same, so k_{jz}^- must correspond to another square root. In summary, we have

$$k_{jz}^\pm = \pm(2\pi/\lambda_v)\sqrt{N_j^2 - n_a^2 \sin^2\theta_a}. \tag{6.55}$$

Now, we are ready to derive the boundary matrix and the propagation matrix. For the system considered, it is well known that the TE wave and the TM wave are two linearly independent eigenfunctions that can be employed to expand the most general solution of the system.

6.5.4.1 TE wave

First, consider the TE wave where the electric field is perpendicular to the plane of incidence. Consider the interface between, say layers 1 and 2. Because the tangential component of the electric field is continuous, we have

$$E_1^+ + E_1^- = E_2^+ + E_2^-. \tag{6.56}$$

Because the tangential component of the magnetic field is continuous, we have $\hat{\mathbf{x}} \cdot \mathbf{H}_1^+ + \hat{\mathbf{x}} \cdot \mathbf{H}_1^- = \hat{\mathbf{x}} \cdot \mathbf{H}_2^+ + \hat{\mathbf{x}} \cdot \mathbf{H}_2^-$. From $\mathbf{H}_j^\pm = \mathbf{k}_j^\pm \times \mathbf{E}_j^\pm$, we have

$$k_{1z}^+ E_1^+ + k_{1z}^- E_1^- = k_{2z}^+ E_2^+ + k_{2z}^- E_2^-. \tag{6.57}$$

When $E_2^- = 0$, we may define $E_1^-/E_1^+ = r_{1,2}$ and $E_2^-/E_1^+ = t_{1,2}$. Thus, Eqs. (6.56) and (6.57) can be rewritten as

$$1 + r_{1,2} = t_{1,2} \tag{6.58}$$

and

$$k_{1z}^+ - k_{1z}^+ r_{1,2} = k_{2z}^+ t_{1,2}, \tag{6.59}$$

where we have used $k_{1z}^- = -k_{1z}^+$ and $k_{2z}^- = -k_{2z}^+$.

By solving Eqs. (6.57) and (6.58), we obtain

$$r_{1,2} = \frac{k_{1z}^+ - k_{2z}^+}{k_{1z}^+ + k_{2z}^+} \tag{6.60}$$

and

$$t_{1,2} = \frac{2k_{1z}^+}{k_{1z}^+ + k_{2z}^+}, \tag{6.61}$$

where k_{1z}^+ and k_{2z}^+ can be obtained from Eq. (6.55). In general, there exist both inward and outward propagating waves, i.e., $E_2^- \neq 0$. Nevertheless, we can also rewrite Eqs. (6.55) and (6.56) in terms of $r_{1,2}$ and $t_{1,2}$. In matrix form, we have

$$\begin{bmatrix} 1 & 1 \\ k_{1z}^+ & k_{1z}^- \end{bmatrix} \begin{bmatrix} E_1^+ \\ E_1^- \end{bmatrix} = \begin{bmatrix} 1 & 1 \\ k_{2z}^+ & k_{2z}^- \end{bmatrix} \begin{bmatrix} E_2^+ \\ E_2^- \end{bmatrix}. \tag{6.62}$$

Simplifying and substituting Eqs. (6.59) and (6.60), we have finally

$$\begin{bmatrix} E_1^+ \\ E_1^- \end{bmatrix} = \frac{1}{t_{1,2}} \begin{bmatrix} 1 & r_{1,2} \\ r_{1,2} & 1 \end{bmatrix} \begin{bmatrix} E_2^+ \\ E_2^- \end{bmatrix}. \tag{6.63}$$

6.5.4.2 TM wave

For the TM wave, by following exactly the same procedure, we obtain

$$\begin{bmatrix} H_1^+ \\ H_1^- \end{bmatrix} = \frac{1}{t_{1,2}} \begin{bmatrix} 1 & r_{1,2} \\ r_{1,2} & 1 \end{bmatrix} \begin{bmatrix} H_2^+ \\ H_2^- \end{bmatrix}, \tag{6.64}$$

where

$$r_{1,2} = \frac{(k_{1z}^+/\varepsilon_1) - (k_{2z}^+/\varepsilon_2)}{(k_{1z}^+/\varepsilon_1) + (k_{2z}^+/\varepsilon_2)} \tag{6.65}$$

and

$$t_{1,2} = \frac{2(k_{1z}^+/\varepsilon_1)}{(k_{1z}^+/\varepsilon_1) + (k_{2z}^+/\varepsilon_2)}. \tag{6.66}$$

The propagation matrix only accounts for phase changes of the plane wave during its propagation. Thus, it is of the same form for the TE and the TM waves. This confirms that in Eq. (6.23), $\delta_2 = k_{2z}^+ d_2$.

6.6 Wafer

From the lithographer's point of view, the wafer is a support for the light-sensitive medium, i.e., the photoresist. This support has several requirements. First, it must be sufficiently flat within the lens exposure field so that it does not take up a significant portion of the DOF budget. Second, the wafer surface should be smooth. Surface roughness not only affects the image quality by producing local image intensity variations and by increasing the level of stray light, but it also reduces the signal-to-noise ratio of the alignment mark. Third, the support should not contribute significantly to light reflection. Fortunately, with the advent of antireflection coatings, the impact of wafer reflectivity is manageable.

Wafer flatness is measured when the wafer is chucked down. Therefore, the wafer chuck makes an undeniable contribution to the wafer flatness. This is discussed in Section 6.8. As the wafer diameter increases, the wafer thickness and rigidity also increase, making it more difficult for the chuck to flatten the wafer. The wafer itself must be increasingly flat, even when it is free standing.

In principle, any deviation from parallelism between the top and bottom wafer surfaces can be compensated by tilting the wafer chuck. However, it is a good practice to make the surfaces as parallel as possible so that the tilting range of the wafer chuck is used for fine-tuning the flatness of the front surface of the wafer to within the lens field rather than being consumed in correcting for poor parallelism.

Figure 6.70 Orientation marks for wafers: with a flat, a notch, and a readable mark on the back.

Because roughness is not an optical issue for the back side of the wafer, most early wafers were not polished on the back side. However, in the sub-half-micrometer regime, wafers are polished on both sides to optimize the flatness of the front side.

To register the orientation of the wafer so that its crystal axis can be correctly aligned to build the devices properly and to coarse align the wafer in rotation, either a notch or a flat is made on the wafer, as shown in Fig. 6.70. A wafer notch takes up very little area on the wafer. However, its abruptness tends to induce accidental fracturing of wafers. A wafer flat is safer but consumes more wafer area and is slightly less accurate in orienting the wafer. To meet higher angular accuracy requirements for larger wafers, the length of the flat must be increased, consuming even more wafer area. As the wafer thickness increases (following the increase in wafer diameters), wafer breakage is less of a problem. Hence, notches are increasingly popular.

Modern technology makes it possible to mark the crystal orientation of the wafer without any cuts on the side of the wafer.[66] This development led to an opportunity to implement 450-mm wafers. Unfortunately, 450-mm wafers are not needed for mass production of semiconductors.

6.7 Wafer Stage

The function of a wafer stage is to hold the wafer firmly while positioning it in three dimensions and orienting it in three axes of rotation. The wafer stage must be capable of flattening and orienting the front surface of the wafer as each exposure area comes under the lens. The x and y movements step the wafer through each field position and align (in translation) the current masking level to the previously delineated masking level. The wafer stage tilts the wafer so that the largest flat area within the lens field is oriented perpendicular to the optical axis of the imaging lens. It also rotates laterally for rotational alignment of the wafer. The z movement is for focusing. Together, there are six degrees of freedom.

The wafer is held by vacuum to the chuck on the wafer stage. The contact surface of the chuck must be flat so that the bottom surface of the wafer is

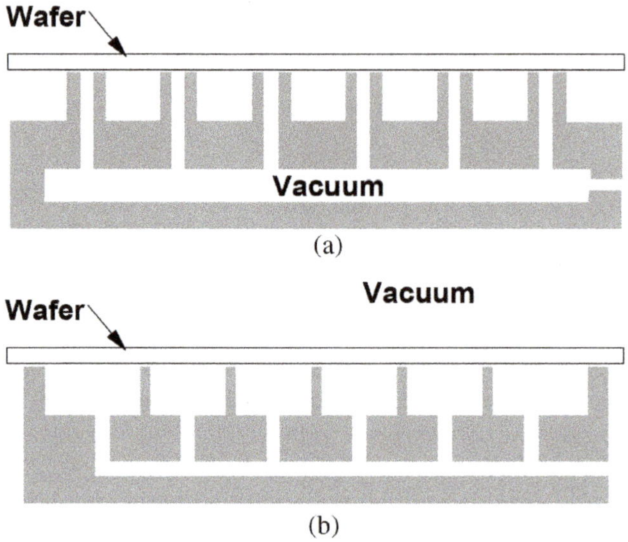

Figure 6.71 Low-contact-area wafer vacuum pin chucks.

flattened. This area also must be minimized so that particles accidentally attached to the bottom of the wafer do not induce flatness loss at the front surface of the wafer. Figure 6.71 shows two low-contact-area vacuum pin chucks for the wafer. Type (a) pulls the vacuum through the pins, whereas the pins in type (b) simply support the wafer. Although more difficult to create, the vacuum in type (a) does not deform the wafer. Figure 6.72 shows that the front-side wafer flatness is affected only when the wafer back-side contamination is in the vicinity of the pins. The front surface of the pins must be very flat to ensure planarization of the wafer bottom. However, to ensure that the front surface of the wafer is flat regardless of the flatness of its bottom, the pins may be made adjustable in height.[67] Various methods can be used to make the pins adjustable. Putting the pins on piezoelectric pedestals is

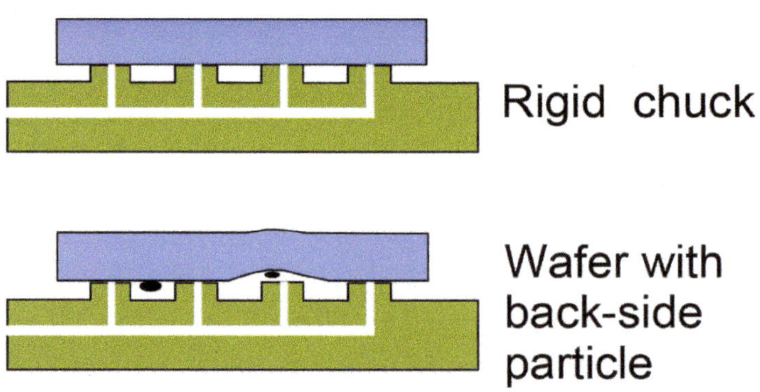

Figure 6.72 Mitigation of most of the wafer back-side contamination with a low-contact-area chuck.

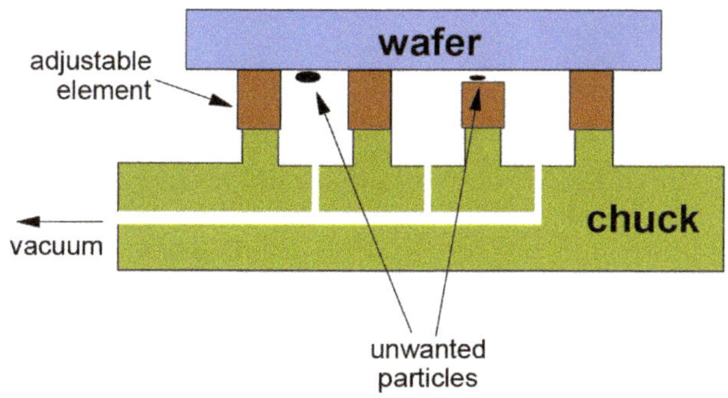

Figure 6.73 Height-adjustable pin chuck.

one example, as shown in Fig. 6.73. After chucking, the front surface of the wafer is flattened to a flat object such as an optical flat using interferometry while the chuck is at the metrology position. Then, the wafer can be stepped and scanned for imaging.

Mirrors are mounted on the sides of the wafer stage to monitor its translation and rotation for controlling the six degrees of freedom. The change in position of these sides is measured with an interferometer. Figure 6.74 shows the schematic of a typical interferometer. Highly coherent light from the laser is split into two beams by the beam splitter. One beam is reflected by a fixed reference mirror; the other beam is reflected by a mirror on the wafer stage. These two beams recombine at the beam splitter. Their interference is collected by the detector. As the wafer stage moves, the detected intensity changes from dark to bright, and vice versa, as a function of the relative position of the two mirrors. By precisely reading the intensity, the position of the moving mirror can be evaluated to better than $\lambda/64$.

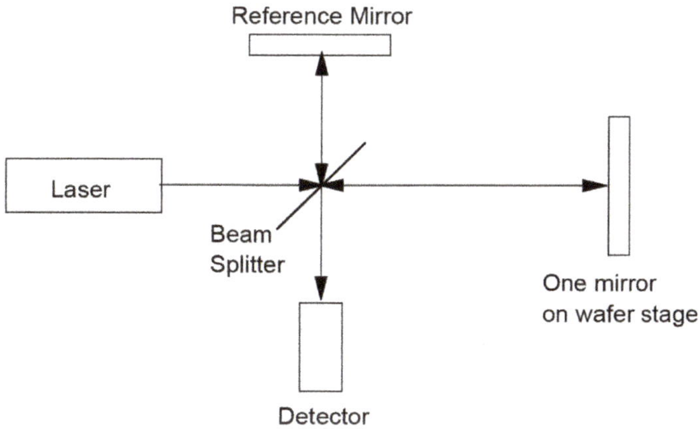

Figure 6.74 Laser interferometer used for monitoring the change in position of the sides of the wafer chuck.

6.8 Alignment System

The alignment system detects the location of the delineated masking layer, then directs the wafer stage to the desired position so that the exposed image correctly aligns with the delineated image. Instead of locating the delineated image, in general, the alignment marks imbedded in that level are located. Special alignment-viewing optics are provided for this purpose. Once the locations of the delineated alignment marks are determined, the lateral translation and rotation settings for the wafer stage are derived from them, referring to the position of the current masking level. The wafer stage is set accordingly, prior to exposure of the current masking level.

The accuracy required to locate the alignment mark is called, appropriately, *alignment accuracy*. A misleadingly similar but very different term is *overlay accuracy*, which includes the contribution from all overlay error sources, such as mask placement error, lens distortion, wafer distortion, and the tool-to-tool matching error of the exposure tools, in addition to the alignment error.

The overlay error is measured with a set of overlay monitor marks that print metrology patterns over the metrology patterns of the previous masking layer. The superimposed patterns from the two overlaying mask patterns are measured together. There are two types of overlay metrology patterns: box-in-box (BiB) and diffraction-based overlay (DBO) patterns, shown in Fig. 6.75. Box-in-box has been the workhorse for overlay monitoring for a long time. DBO has started to be adapted due to its potential for higher accuracy.

6.8.1 Off-axis alignment and through-the-lens alignment

The off-axis alignment method is an indirect method that locates the alignment-viewing optics outside the optical path of the imaging lens so that the alignment mark of the delineated masking level is viewed at a location outside of the optical path of the current masking level, as depicted in

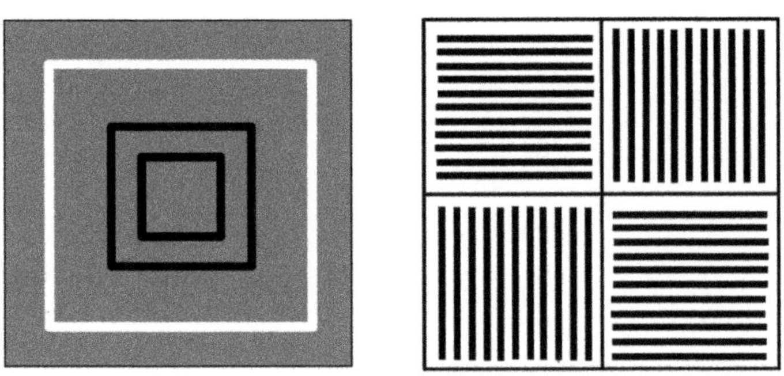

(a) (b)

Figure 6.75 Overlay-metrology patterns: (a) box-in-box (BiB) and (b) diffraction-based overlay (DBO).

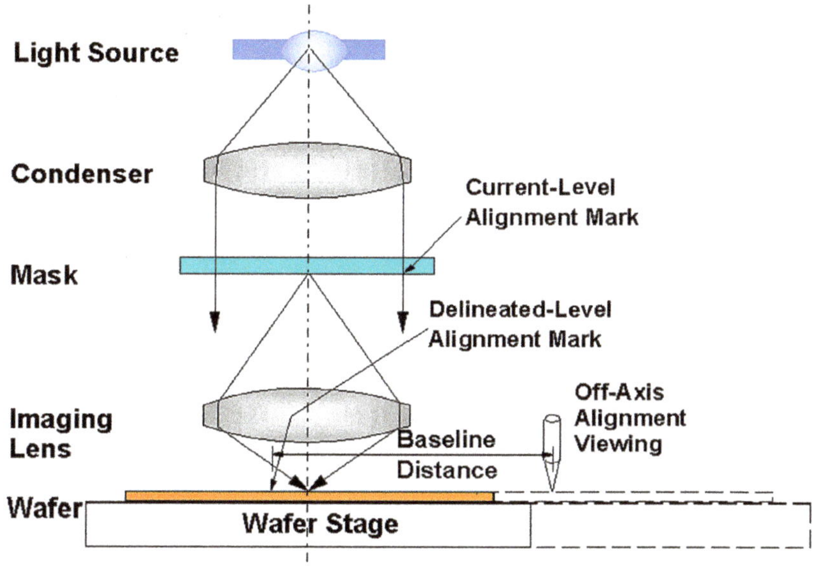

Figure 6.76 Off-axis alignment viewing.

Fig. 6.76. The distance from the alignment mark at the viewed position to its desired position for exposure is called the *baseline distance*. This distance must be included in the calculation of the desired lateral location and rotation of the wafer stage.

The through-the-lens alignment method is a direct method that incorporates the alignment-viewing optics in the optical path of the imaging lens to register the difference in the delineated masking level and the current masking level directly, as depicted in Fig. 6.77.

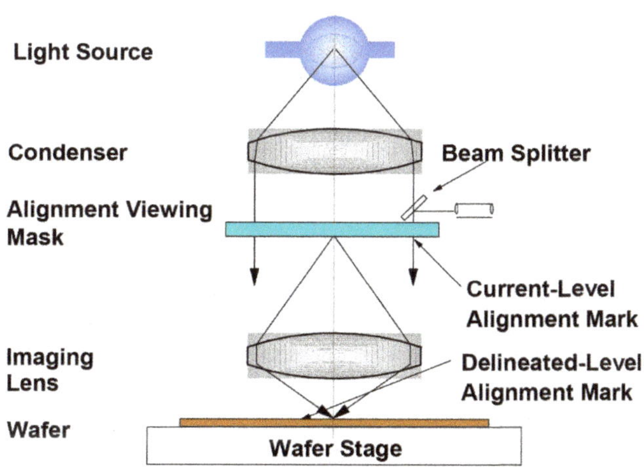

Figure 6.77 Through-the-lens alignment viewing.

By intuition, the through-the-lens method is better than the off-axis method because there is no baseline distance to consider. In reality, the following must be considered:

1. The viewing wavelength is difficult to determine. With actinic viewing, the alignment light can expose the photoresist, leading to unpredictable clearing of the resist in the vicinity of the alignment mark. In addition, the latent image of the exposed resist is a ghost image that can confuse the detection system, especially for sensitive detection schemes such as dark-field alignment detection. When a nonexposing wavelength is used, the chromatic aberration must be corrected, but this is not always easy to do.
2. The alignment-viewing NA cannot exceed the NA of the imaging lens, thus capping the resolution capability of the alignment optics.
3. Unless the imaging lens is designed for polychromatic illumination, as in all-reflective and catadioptric systems, it is not trivial to incorporate polychromatic alignment viewing. With monochromatic viewing, multiple reflections can create problems in the alignment viewing, just as in imaging the mask to the wafer.
4. Intrusion into the imaging optics makes the alignment-viewing optics more complicated than in off-axis viewing. In addition, the intrusion may introduce imaging aberrations.

With off-axis alignment, the only problem is a baseline error that is a strong function of how well the wafer stage can be monitored. With interferometers, the wafer stage can be tracked in x, y, z, and three angular axes to nanometer accuracy. Therefore, high-NA polychromatic non-actinic optics, bright-field imaging, dark-field imaging, or phase detection can be used for off-axis alignment—whichever is most suitable to detect the alignment mark on the wafer.

6.8.2 Field-by-field, global, and enhanced global alignment

Field-by-field alignment means that each field is aligned immediately prior to exposure. With global alignment, the entire wafer is aligned once for global translation and rotation, then the exposure fields are stepped without further checking of the alignment marks. Enhanced global alignment is like global alignment, except that a closer inspection of the alignment marks determines the positions for stepping the wafer.

Direct viewing facilitates field-by-field alignment but does not specifically call for field-by-field alignment. Global alignment using direct viewing has been done on some steppers and scanners. For field-by-field alignment, because each exposure field is tailored to each delineated field, one tends to think that this alignment method is the most accurate and that its lack of popularity is due to wafer throughput penalty. In actuality, the field-by-field alignment accuracy is limited to alignment mark detection, but the global alignment accuracy can be made much higher than the alignment mark detection limit by increasing the number of fields for global alignment. For example, let the 3σ alignment accuracy of alignment mark detection be 10 nm regardless of the alignment detection scheme, and the wafer-stage positioning accuracy be 2 nm. Also let 13 fields be aligned for global alignment. The field-by-field alignment accuracy in 3σ at the wafer is 10 nm, based on the ability to detect the alignment mark. For global alignment, the alignment accuracy is improved by the square root of the number of aligned fields. Hence, the accumulated alignment accuracy in 3σ is $10/\sqrt{13} = 2.77$ nm. After incorporating the 2-nm wafer positioning accuracy quadratically, the resultant alignment accuracy is 3.4 nm.

Maintaining the same alignment accuracy for field-by-field alignment, the only way to improve alignment is to repeatedly detect the same alignment marks n more times to produce a $\sqrt{n}$ improvement. The throughput penalty would be even higher than just one alignment mark reading per field. Fortunately, enhanced global alignment provides the advantage of aligning over a large number of alignment marks without severe penalties.

6.8.3 Bright-field and dark-field alignments

Bright-field and dark-field alignments refer to the range of observed spatial frequency from the alignment mark image. With bright-field illumination, all spatial frequencies—from the 0th order to the highest one within the NA of the alignment imaging optics—are used, whereas dark-field illumination sends the 0th-order beam away, and only the other orders are observed. Figures 6.78 and 6.79 illustrate the principles of bright-field and dark-field imaging, respectively.

Because of the removal of the 0th-order beam, dark-field imaging significantly increases the signal-to-noise ratio of alignment viewing. It is less susceptible to multiple interference, though not completely free from it.

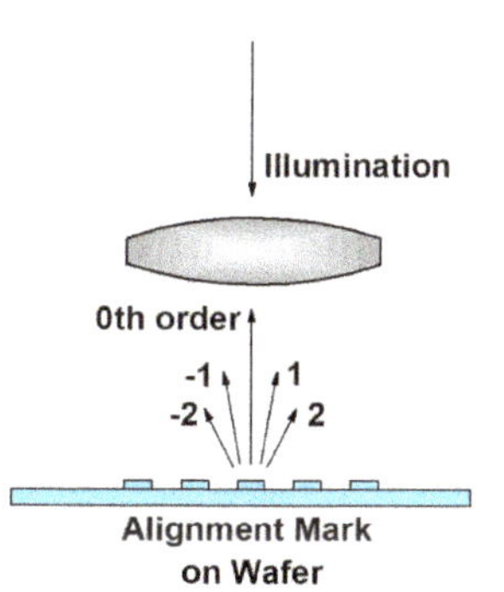

Figure 6.78 Bright-field imaging.

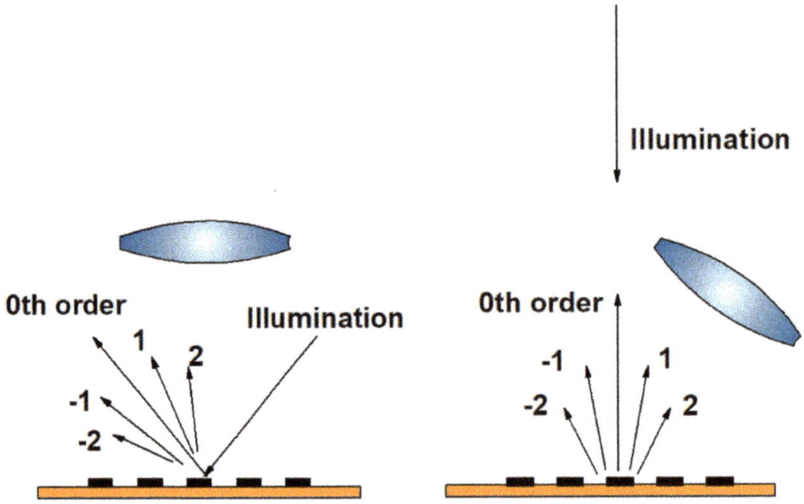

Figure 6.79 Dark-field imaging.

However, when the alignment is over grainy substrates, the grain boundary tends to show up with the alignment mark, confusing the alignment system. If the resist is exposed during the alignment process, the latent image can become a newly detected image that further confuses the alignment system.

6.9 Conclusion

The components in the optical imaging system are covered in this chapter. Equipped with this knowledge, we are ready to move on to the next chapter, where we cover ways to optimize these components and discuss the processes associated with them.

References

1. K. Jain, C. G. Willson, and B. J. Lin, "Ultrafast deep-UV lithography with excimer lasers," *IEEE Electron Dev. Lett.* **EDL-3**(3), pp. 53–55 (1982).
2. K. Jain, C. G. Willson, and B. J. Lin "Ultrafast high resolution contact lithography using excimer lasers," *Proc. SPIE* **334**, pp. 259–262 (1982) [doi: 10.1117/12.933585].
3. K. Jain and R. T. Kerth, "Excimer laser projection lithography," *Appl. Optic.* **23**(5), pp. 648–650 (1984).
4. V. Pol, J. H. Bennewitz, G. C. Escher, M. Feldman, V. A. Firtion, T. E. Jewell, B. E. Wilcomb, and J. T. Clemens, "Excimer laser-based lithography: a deep ultraviolet wafer stepper," *Proc. SPIE* **633**, p. 6 (1986) [doi: 10.1117/12.963697].

5. P. Das and U. Sengupta, "Krypton Fluoride Excimer Laser for Advanced Microlithography," Chapter 4 in *Microlithography—Science and Technology*, J. R. Sheats and B. W. Smith, Eds., Marcel Dekker, Inc. (1998).
6. Information on dispersion was provided by H. Feldmann at Carl Zeiss SMT AG.
7. I. H. Malitson, "Interspecimen comparison of the refractive index of fused silica," *JOSA* **55**(10), pp. 1205–1209 (1965).
8. R. Gupta, J. H. Burnett, U. Griesmann, and M. Walhout, "Absolute refractive indices and thermal coefficients of fused silica and calcium fluoride near 193 nm," *Appl. Optic.* **37**(25), pp. 5964–5968 (1998).
9. M. Daimon and A. Masumura, "High-accuracy measurements of the refractive index and its temperature coefficient of calcium fluoride in a wide wavelength range from 138 to 2326 nm," *Appl. Optic.* **41**(25), pp. 5275–5281 (2002).
10. J. H. Burnett, R. Gupta, and U. Griesmann, "Absolute refractive indices and thermal coefficients of CaF_2, SrF_2, BaF_2, and LiF near 157 nm," *Appl. Optic.* **41**(13), pp. 2508–2513 (2002).
11. Page 297 of Ref. 5.
12. R. Patzel, J. Kleinschmidt, U. Rebhan, J. Franklin, and H. Endert, "KrF excimer laser with repetition rates of 1 kHz for DUV lithography," *Proc. SPIE* **2440**, p. 101 (1995) [doi: 10.1117/12.209305].
13. M. Born and M. Wolf, *Principles of Optics, 6th Edition*, Cambridge University Press, p. 524 (1980).
14. M. Mulder, A. Engelen, O. Noordman, R. Kazinczi, G. Streukter, B. van Drieenhuizen, S. Hus, K. Gronlund, M. Degünther, D. Jürgens, J. Eisenmenger, M. Patra, and A. Major, "Performance of a programmable illuminator for generation of freeform sources on high NA immersion systems," *Proc. SPIE* **7520**, 75200Y (2009) [doi: 10.1117/12.837035].
15. B. J. Lin, "Optical projection printing threshold leveling arrangement," U.S. Patent 4,456,371 (1984).
16. C. A. Spence, "Method of optical lithography using phase shift masking," U.S. Patent 5,573,890 (1996).
17. B. J. Lin, "AZ1350J as a deep-UV mask material," *J. Electrochem. Soc.* **127**, pp. 202–205 (1980).
18. R. Hershel, "Pellicle protection of integrated circuit (IC) masks," *Proc. SPIE* **0275**, pp. 23–28 (1981) [doi: 10.1117/12.931869].
19. P. Nehmiz, W. Zapka, U. Behringer, M. Kallmeyer, and H. Bohlen, "Electron beam proximity printing: complementary-mask and level-to-level overlay with high accuracy," *J. Vac. Sci. Tech. B* **3**, pp. 136–143 (1985).
20. B. J. Lin, "Phase shifting and other challenges in optical mask technology," *Proc. SPIE* **1496**, pp. 54–79 (1990) [doi: 10.1117/12.29751].

21. M. D. Levenson, N. S. Viswanathan, and R. A. Simpson, "Improving resolution in photolithography with a phase-shifting mask," *IEEE Trans. Electron Devices* **ED-29**(12), pp. 1828–1836 (1982).
22. T. Terasawa, N. Hasegawa, T. Kurosaki, and T. Tanaka, "0.3-micron optical lithography using a phase-shifting mask," *Proc. SPIE* **1088**, p. 25 (1989) [doi: 10.1117/12.953131].
23. A. Nitayama, T. Sato, K. Hashimoto, F. Shigemitsu, and M. Nakase, "New phase shifting mask with self-aligned phase shifters for quarter micron photolithography," *IEDM Technical Digest*, p. 57 (1989).
24. H. I. Smith, E. H. Anderson, and M. L. Shattenburg, "Lithography mask with a π-phase shifting attenuator," U.S. Patent 4,890,309 (1989).
25. B. J. Lin, "The attenuated phase-shifting mask," *Solid State Technology*, p. 43, Jan (1992).
26. B. J. Lin, "The optimum numerical aperture for attenuated phase-shifting masks," *Microelectronic Engineering* **17**(1–4), p. 7985 (1992).
27. H. Jinbo and Y. Yamashita, "0.2 μm or less I-line lithography by phase-shifting-mask technology," *IEDM Technical Digest*, p. 825 (1990).
28. H. Jinbo and Y. Yamashita, "Improvement of phase-shifter edge line mask method," *Jpn. J. Appl. Phys.* **30**, p. 2998 (1991).
29. B. J. Lin, "Alternating rim phase shifting mask," U.S. Patent 5,403,682.
30. G. T. Dao, E. T. Gaw, N. N. Tam, and R. A. Rodriquez, "Method of fabrication of inverted phase-shifting reticle," U.S. Patent 5,300,379.
31. Y. T. Wang and Y. Pati, "Phase shifting circuit manufacture method and apparatus," U.S. Patents 5,858,580 (1999) and 6,228,539 B1 (2001).
32. J. S. Wilczynski, private communication.
33. A. Suzuki, "Double telecentric wafer stepper using laser scanning method," *Proc. SPIE* **538**, p. 2 (1985) [doi: 10.1117/12.947740].
34. W. Ulrich and R. Hudyma, "Development of dioptric projection lenses for deep-ultraviolet lithography at Carl Zeiss," *J. Micro/Nanolith., MEMS, and MOEMS* **3**(1), p. 87 (2004) [doi: 10.1117/1.1637592].
35. J. A. McClay, R. A. Wilklow, and M. Gregoire, "Polarization beamsplitter for photolithography," U.S. Patent 6,680,794 (2004).
36. F. Sporon-Fiedler and J. Williams, "Atmospheric pressure induced reduction errors in reduction stepper lenses," *Proc. SPIE* **538**, p. 86 (1985) [doi: 10.1117/12.947751].
37. W. M. Moreau, *Semiconductor Lithography: Principles, Practices, and Materials*, Plenum Press, Eq. 2-6-1 (1988).
38. Equation 4-1-9 of Ref. 37.
39. A. Reiser and E. Pitts, "Characteristic curve of crosslinking photoresists," *J. Photogr., Sci. and Eng.* **20**(5), p. 225–229 (1976).
40. O. Süss, Über die Natur der Belichtungsprodukte von Diazoverbindungen: Übergänge von aromatischen 6-Ringen in 5-Ringe," *Justis Liebigs Annalen der Chemie* **556**, p. 65, (1944).

41. D. Meyerhofer, "Photosolubility of diazoquinone resists," *IEEE Trans. Electron. Dev.* **27**(5), pp. 921–926 (1980).
42. L. F. Thompson, C. G. Willson, and M. J. Bowden, *Introduction to Microlithography, 2nd Edition*, American Chemical Society, Section 3.3.2 (1994).
43. Reference 42, pp. 160–164.
44. Reference 37, Chapter 4.
45. W. D. Hinsberg, S. A. MacDonald, N. J. Clecak, C. D. Snyder, and H. Ito, "Influence of polymer properties on airborne chemical contamination of chemically amplified resists," *Proc. SPIE* **1925**, (1993) [doi: 10.1117/12.154774].
46. D. R. McKean, U. Schaedeli, and S. A. MacDonald, "Acid photogeneration from sulfonium salts in solid polymer matrices," *J. Polymer Science: Part A: Polymer Chemistry* **27**, 3927–3935 (1989).
47. Reference 42, pp. 212–232.
48. J. G. Maltabes, S. J. Holmes, J. R. Morrow, R. L. Barr, M. C. Hakey, G. Reynolds, W. R. Brunsvold, C. G. Willson, N. J. Clecak, S. A. MacDonald, and H. Ito, "1X deep-UV lithography with chemical amplification for 1-micron DRAM production," *Proc. SPIE* **1262** (1990) [doi: 10.1117/12.20090].
49. U. Okoroanyanwu, *Chemistry and Lithography*, SPIE Press and J. Wiley & Sons, Chapter 7, p. 361 (2010) [doi: 10.1117/3.821384].
50. S. A. MacDonald, N. J. Clecak, H. R. Wendt, C. G. Willson, C. D. Snyder, C. J. Knors, N. B. Deyoe, J. G. Maltabes, J. R. Morrow, A. E. McGuire, and S. J. Holmes, "Airborne chemical contamination of a chemically amplified resist," *Proc. SPIE* **1466**, pp. 2–12 (1991) [doi: 10.1117/12.46354].
51. P. Rai-Choudhury, Ed., *Handbook of Microlithography, Micromachining, and Microfabrication I: Microlithography*, SPIE Press, Fig. 4.6 (1997) [doi: 10.1117/3.2265070].
52. H. Moritz, "Optical single layer lift-off process," *IEEE Trans. Elecron Devices* **ED-32**(3), pp. 672–676 (1985).
53. F. Coopmans and B. Roland, "DESIRE: a novel dry developed resists system, *Proc. SPIE* **631**, p. 34 (1986) [doi: 10.1117/12.963623].
54. F. Coopmans and B. Roland, "Enhanced performance of optical lithography using the DESIRE system," *Proc. SPIE* **633**, p. 262 (1986) [doi: 10.1117/12.963730].
55. L. F. Thompson, C. G. Willson, and M. J. Bowden, *Introduction to Microlithography*, ACS Professional Reference Book, Oxford Univ. Press, New York, Chapter 4 (1983).
56. J. M. Moran and D. Maydan, "High resolution, steep profile resist patterns," *J. Vac. Sci. Technol.* **16**, p. 1620 (1979).

57. M. Hatzakis, J. Paraszczak, and J. Shaw, "Double layer resist systems for high resolution lithography," *Proc. Int. Conf. Microlith. Microcircuit Eng.* **81**, p. 386 (1981).
58. K. L. Tai, W. R. Sinclair, R. G. Vadimsky, J. M. Moran, and M. J. Rand, "Bilevel high resolution photolithographyic technique for use with wafers with stepped and/or reflecting surfaces," *J. Vac. Sci. Technol.* **16**, p. 1977 (1979).
59. B. J. Lin, "Portable conformable mask: A hybrid near-ultraviolet and deep ultra-violet patterning technique," *Proc. SPIE* **174**, pp. 114–121 (1979) [doi: 10.1117/12.957186].
60. B. J. Lin, V. W. Chao, K. E. Petrillo, and B. J. L. Yang, "A molded deep-UV portable conformable masking system," *Polymer Eng. & Sci.* **26**(6), pp. 1112–1115 (1986).
61. W. Moreau and P. Schmidt, "Photoresist for high resolution proximity printing," *Electrochem. Soc. Ext. Abstr.* **138**, p. 459 (1970).
62. B. J. Lin, "Deep-UV lithography," *J. Vac. Sci. Technol.* **12**, pp. 1317–1320 (1975).
63. I. Haller, M. Hatzakis, and R. Srinivasan, "High-resolution positive resists for electron-beam exposure," *IBM J. Res. Develop.* **12**(3), pp. 251–256 (1968).
64. P. R. West and B. F. Griffing, "Contrast enhancement: a route to submicron optical lithography," *Proc. SPIE* **394**, p. 16 (1983) [doi: 10.1117/12.935119].
65. E. Walker, "Reduction of photoresist standing-wave effects by post-exposure bake," *IEEE Trans.* **ED-22**(7), pp. 464–466 (1975).
66. H. Oishi and K. Osakawa, "Production of notchless wafer," U.S. Patent 59,932,925 (1999).
67. B. J. Lin, T. S. Gau, and J. H. Chen, "Level adjustment systems and adjustable pin chuck thereof," U.S. Patent 7,659,964 (2010).
68. U. K. Sengupta, "Krypton fluoride excimer laser for advanced microlithography, *Opt. Eng.* **32**(10), 2410–2420 (1993) [doi: 10.1117/12.145969].
69. S.-S. Yu, B. J. Lin, A. Yen, C.-M. Ke, J. Huang, B.-C. Ho, C.-K. Chen, T.-S. Gau, H.-C. Hsieh, and Y.-C. Ku, "Thin-film optimization strategy in high numerical aperture optical lithography, part 1: principles," *J. Micro/Nanolith., MEMS, and MOEMS* **4**(4), 043003 (2005) [doi: 10.1117/1.2137967].
70. S. S. Yu, B. J. Lin, and A. Yen, "Thin-film optimization strategy in high numerical aperture optical lithography, part 2: applications to ArF," *J. Micro/Nanolith., MEMS, and MOEMS* **4**(4), 043004 (2005) [doi: 10.1117/1.2137987].

Chapter 7
Processing and Optimization

Now that we have provided sufficient background on imaging and tools, the processes that use them and the optimization of these processes are given in this chapter. We start with optimization of the exposure tools, then move on to resist processing, k_1 reduction schemes, and finally the control of critical-dimension (CD) uniformity and overlay accuracy.

7.1 Optimization of the Exposure Tool

The optimization of an exposure tool involves the optimization of its numerical aperture (NA), illumination setting, exposure/focus, depth-of-focus (DOF) budget, and monitoring of focus, as well as its throughput optimization in field size and exposure routing.

7.1.1 Optimization of the NA

Most modern exposure tools and steppers allow users to adjust the NA of the imaging lens because the NA cannot be preset in the factory when the resolution is extended to the low-k_1 regime. The optimum NA[1,2] is not the highest one in the range of available settings on the tool, but rather is a function of feature size and shape combinations, illumination, etc. It is important to set the NA to the best value to take advantage of the capability of the imaging system.

From Eq. (5.3), the resolution scaling equation, $\sin\theta$ must be sufficiently large to sustain the resolution of a given feature. However, Eq. (5.4) dictates that a large θ reduces the DOF. Therefore, in a dry system where the space between the last lens surface and the resist is not immersed in a liquid, the NA of its imaging lens has a medium value whose DOF is maximum. Figure 7.1 shows the DOF as a function of the NA at $\sigma = 0.82$ for 0.20-μm resist lines separated by 0.20-μm spaces. The DOF is evaluated with the exposure-defocus (E-D) forest methodology covered in Chapter 5, based on 8% EL and CD bounds at ±15 nm. The DOF quickly rises from 0 to 1.04 μm at $NA = 0.557$, then gradually decreases as the NA increases. Figure 7.2 depicts

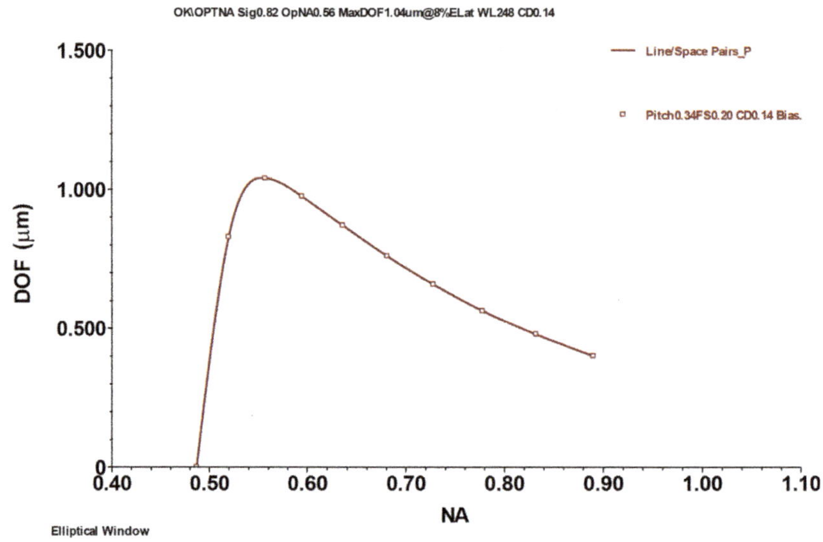

Figure 7.1 NA optimized for DOF with 0.14-μm resist lines : 0.20-μm spaces, λ = 248 nm, σ = 0.82, EL = 8%, CD_{tol} = ±15 nm.

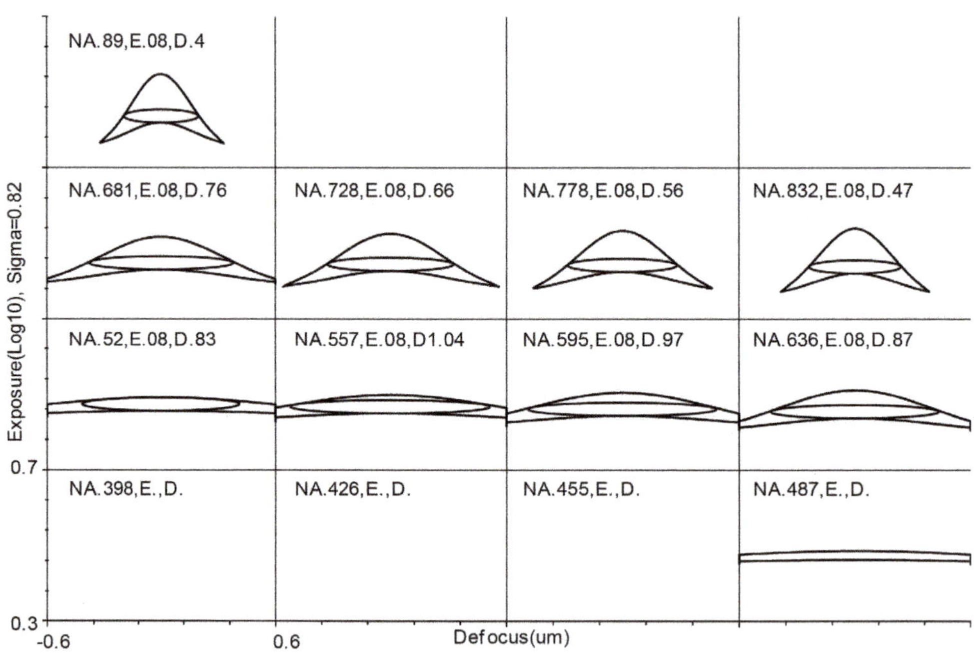

Figure 7.2 E-D windows of Fig. 7.1.

the E-D trees at the NA values evaluated. At NA = 0.455 and below, the resolution is too low to support any E-D tree. An E-D tree can be constructed at NA = 0.487, but there is insufficient branch separation to support the E-D

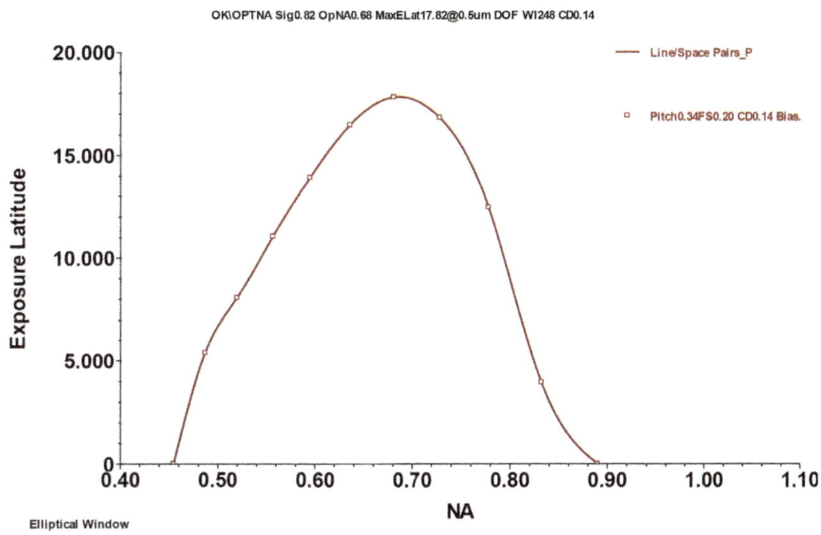

Figure 7.3 NA optimized for EL with 0.14-μm resist lines: 0.20-μm spaces, λ = 248 nm, σ = 0.82, DOF = 0.5 μm, CD_{tol} = ±15 nm.

window specified for 8% EL. In this case, the E-D trees become broader, trading off with increased curvature at the branches. The tradeoff favors the DOF in the beginning, resulting in the maximum DOF at 1.04 μm. Then, it starts to favor the EL at the expense of the DOF, resulting in the gradual reduction of DOF at higher NAs.

The NA can be optimized using other criteria than just the DOF. There can be a different optimum NA for the EL, as shown in Fig. 7.3, using the same feature and illumination condition as in Figs. 7.1 and 7.2, except that the DOF of the E-D windows is kept at 0.5 μm. The optimum NA for a maximum DOF is 0.557, whereas that for a maximum EL is 0.681. When the NA is low, the image has low contrast, which leads to a small EL. When the NA becomes larger, the EL in the focal plane increases, but the curvature of the E-D trees and their shrinking in the defocal direction make it difficult to sustain a large EL at a given DOF, as seen in Fig. 7.4. At NA = 0.89, so much shrinking has occurred that no E-D window with DOF = 0.5 μm can fit into the E-D tree.

The NA can be optimized for the E-D area, as shown in Figs. 7.5 and 7.6. The NA is now optimized at 0.636, which happens to be between the NA that optimizes the DOF and the NA that optimizes the EL. The argument is similar. At a low NA, the low image contrast leads to narrow E-D regions. At a high NA, the image contrast is high in focus but quickly vanishes.

Among the DOF, EL, and E-D area, which should be optimized? One should first optimize for the DOF at an EL that can be sustained by the resist processing, such as by using antireflection coatings or multilayer resist systems. Typically, one takes the percentage of exposure range derived from

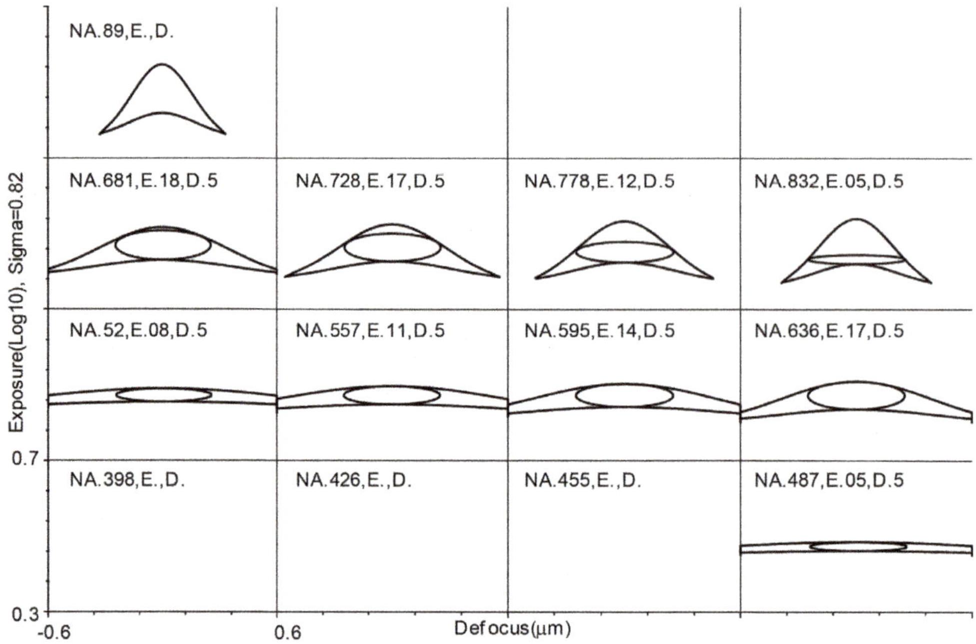

Figure 7.4 E-D windows of Fig. 7.3.

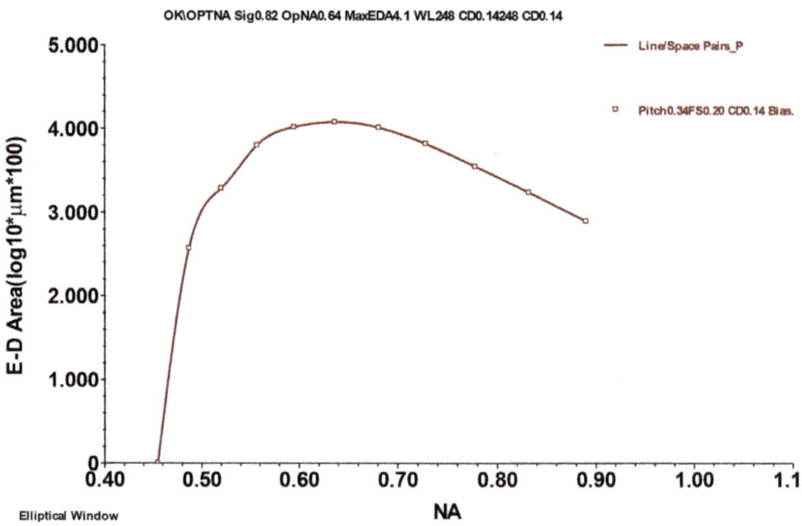

Figure 7.5 NA optimized for the E-D area, with 0.14-μm resist lines: 0.20-μm spaces, λ = 248 nm, σ = 0.82, DOF = 0.5 μm, CD_{tol} = ±15 nm.

the resist swing curve within the range of topography in the given exposure field and combines it with the percentage of illumination nonuniformity, as well as the percentage of dosage control inaccuracy, to evaluate the EL.

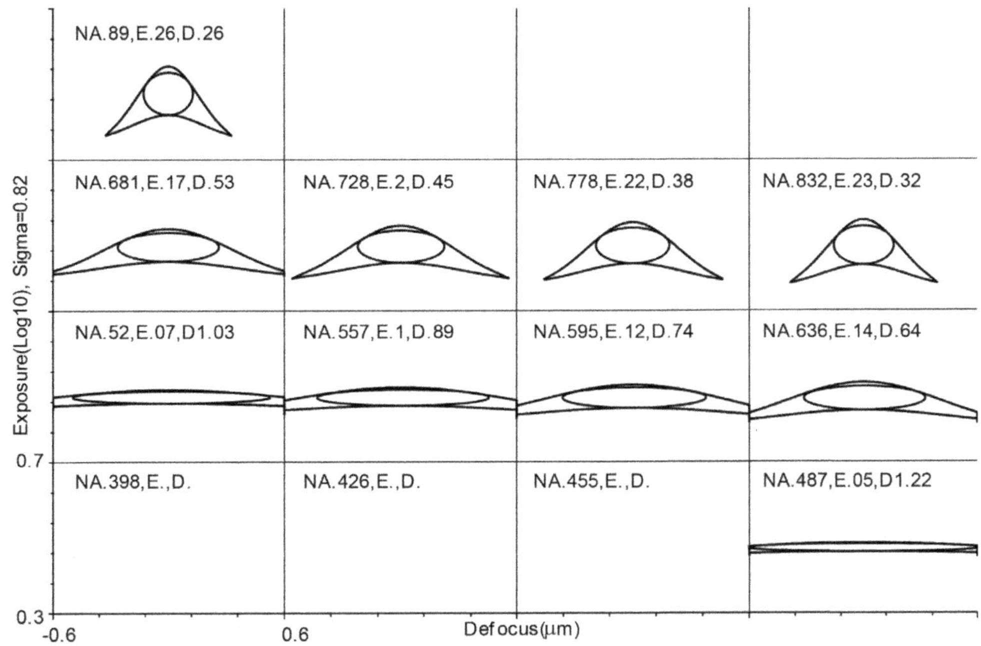

Figure 7.6 E-D windows of Fig. 7.5.

Usually, 8% is acceptable. If the DOF at the optimum NA does not meet this requirement, this imaging system cannot handle the given task. If the DOF at only one or two NA settings slightly exceeds the requirements, these NA settings should be used. If the DOF exceeds the requirement at many NA settings within the NA-adjustable range, then optimizing for the EL should be attempted because a larger EL helps to control the CD. If there are many NA settings exceeding the DOF and EL requirements, then the NA that produces the largest E-D area should be selected because this largest processing window is statistically the easiest to meet; thus, a higher lithography-imaging yield is attained.

7.1.2 Optimization of illumination

The illumination settings on the exposure tool play a role in image performance that is equally important as the NA setting on the imaging lens. First, the type of illumination must be decided, namely, disk, annular, or multipole illumination. Once decided, the settings must be optimized, such as the σ setting for disk illumination (DKI), σ_{in} and σ_{out} for annular illumination, and pole size, shape, and location for multipole illumination. New illuminations using diffractive optical elements (DOEs) or FlexRay® can achieve the full potential of illumination optimization beyond the above simple setups. Furthermore, the mask pattern can be optimized iteratively with illumination optimization to achieve source mask optimization (SMO), which will be discussed in Section 7.3.5 on optical proximity correction. For this section, we will confine the discussion to simple illumination setups such as disk, annular, or multipole illuminations.

With DKI, the DOF, EL, and E-D area obviously change as a function of σ. However, the existence of an optimum σ cannot be as easily visualized as when using the resolution and DOF scaling equations to visualize the existence of an optimum NA. With off-axis illuminations, such as annular and multipole, the existence of an optimum illumination angle is directly related to the working principle of these systems, as explained in Section 7.3.2.

Figure 7.7 shows the change in DOF as a function of the σ setting of the disk illumination. The DOF is indeed dependent on σ. In this case, all chosen features favor a large σ, with the denser features optimizing at σ in the regime of 0.8. The optimum σ for the common E-D window is even more pronounced, centering at around 0.82.

Figure 7.8 is similar to Fig. 7.5, except that annular illumination with $\sigma_{in} = 0.67\sigma_{out}$ is used. The DOF peaks for the 1:1.5, 1:2, and 1:3 lines are even more obvious than in the case of DKI, and they are larger. Unfortunately, the DOF of common windows is smaller than in the former case. It can be improved with optical proximity correction but is limited by the features with larger spaces in both cases.

Section 7.1.1 shows a way to further improve the DOF. Following optimization of the illumination, the NA can be optimized at this illumination setting, as shown in Fig. 7.9. In this case, the DOF is improved from 0.21 to

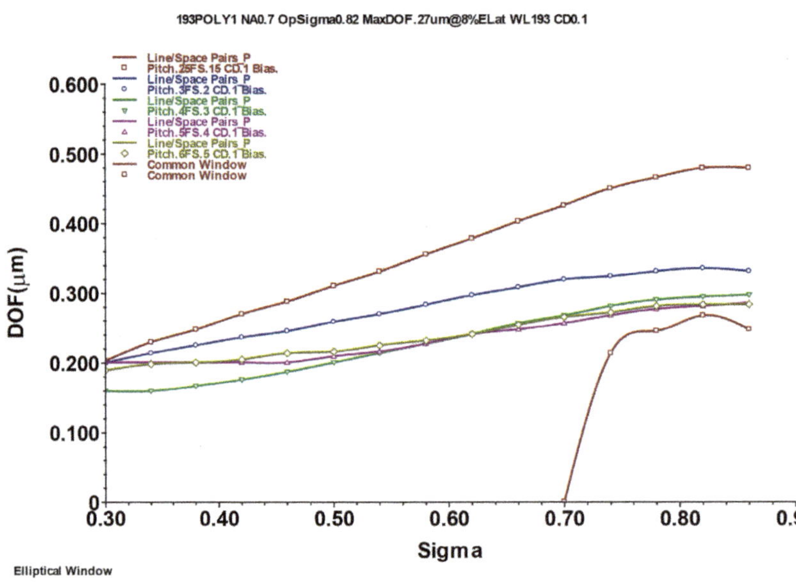

Figure 7.7 DOF as a function of σ under DKI for positive-resist line:space ratios of 1:1.5, 1:2, 1:3, 1:4, and 1:5, and their common window. The resist linewidth is 100 nm, λ = 193 nm, *NA* = 0.7, *EL* = 8%.

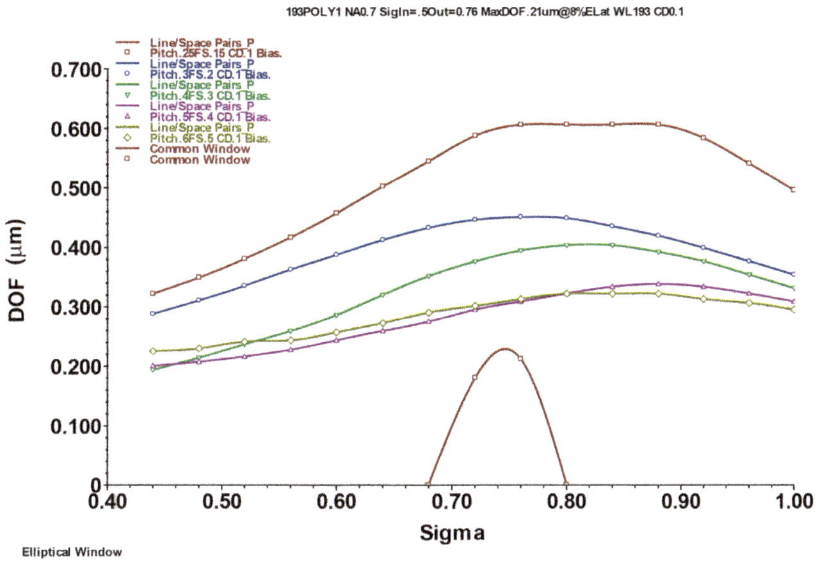

Figure 7.8 Same as Fig. 7.7, except using annular illumination with σ_{in} = 0.67σ_{out}.

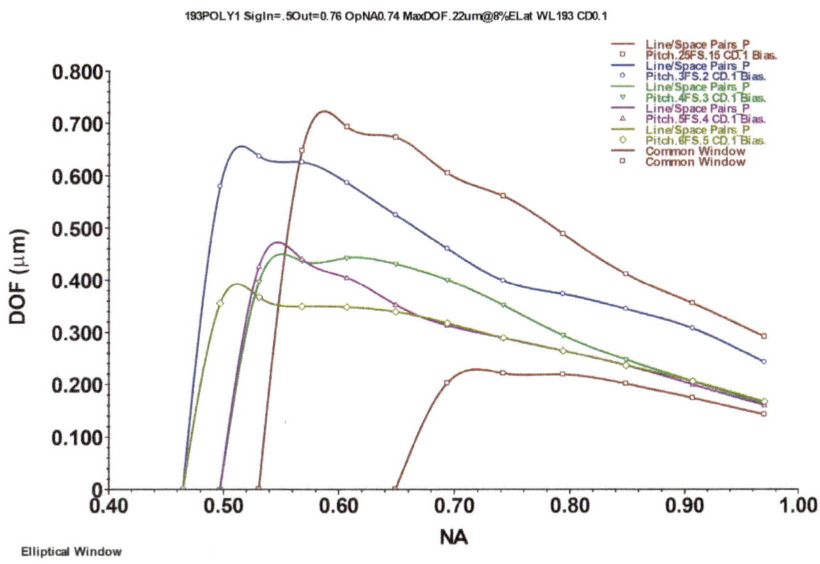

Figure 7.9 Same as Fig. 7.8, except that the NA is optimized at the σ settings evaluated from Fig. 7.8, with σ_{in} = 0.5 and σ_{out} = 0.76.

0.22 μm with the NA changing from 0.7 to 0.74, which is only slightly improved. In principle, one can repeat the NA- and σ-optimization processes alternately until the final optimum is reached. In the examples shown in Figs. 7.6 and 7.7, the convergence is fast. Figure 7.10 shows DOF as a function of NA and σ plotted together, using the situation studied in Fig. 7.1.

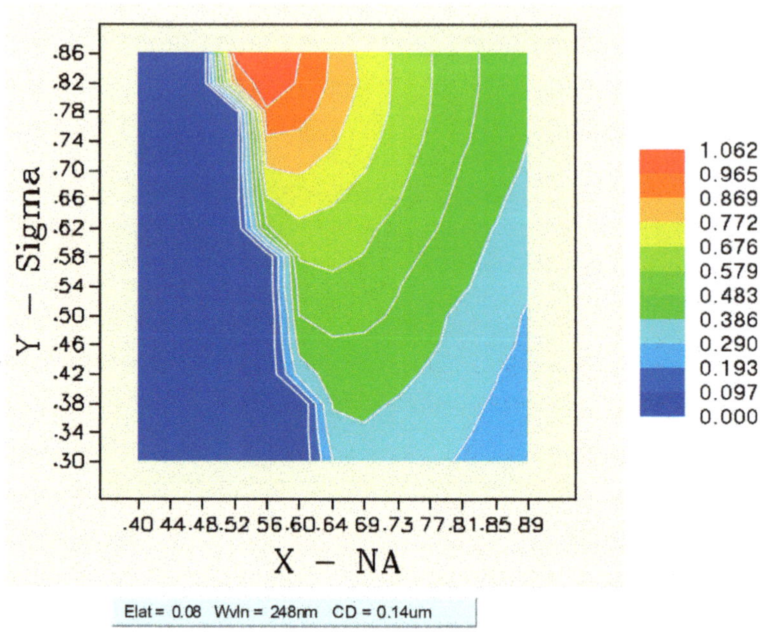

Figure 7.10 DOF as a function of NA and σ for the feature in Fig. 7.1.

7.1.3 Exposure and focus

Exposure dosage and focus are the two most important field-adjustable parameters. These parameters must be carefully and correctly set. Sections 5.3.2 and 5.3.3 cover the need to definine the EL and DOF together in an E-D window because these two quantities are mutually dependent. The center of exposure and focus is the center of the E-D window. However, which E-D window should one use in a processing environment: the E-D window of a predominant representative feature, the common E-D window of all or some pertinent features, or the E-D window of one or a set of test features placed in every mask set?

First, let us examine the difference between the usual practice and the E-D window approach. Intuitively, one finds the best focus from a typical feature whose exposure looks reasonably decent. On a product wafer, setting the center of exposure and focus is more complicated because the proximity environment needs to be taken into account. Figure 7.11 shows the E-D trees and windows of a 140-nm opaque line with 120-, 160-, 200-, 300-, and 400-nm transparent spaces, delineated with 193-nm light, using $NA = 0.6$ and annular illumination $\sigma_{in} = 0.3$ and $\sigma_{out} = 0.6$. The EL is set at 5% and CD tolerance is ±9 nm. The exposure and focal center of each feature combination and those of the common E-D tree are clearly different from each other. Their numerical values are listed in Table 7.1. In this case, the exposure center of the E-D window is within 0.01 mJ/cm^2 of that of the common window and the focal

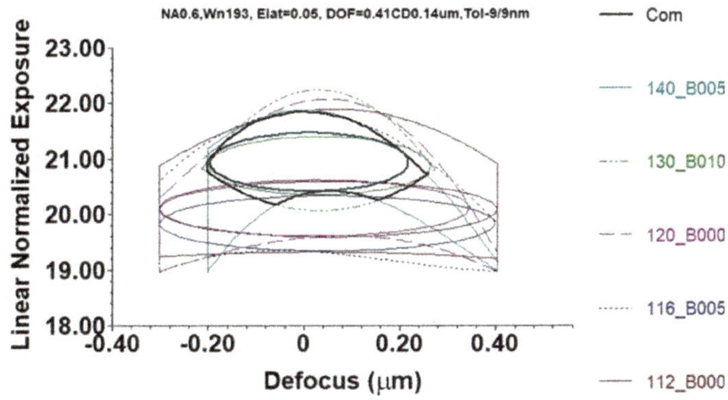

Figure 7.11 E-D windows of 140-nm opaque lines with 120-, 160-, 200-, 300-, and 400-nm transparent spaces when λ = 193 nm, NA = 0.6, σ = 0.6/0.3. EL = 5%, ΔCD = ±9 nm.

Table 7.1 Exposure center, focal center, and DOF of the five feature combinations in Fig. 7.11.

Feature	Exposure center (mJ/cm^2)	Focal center (nm)	DOF (nm)
Common window	20.94	11	413
140 nm	20.09	50	704
140 nm:160 nm	19.83	48	699
140 nm:200 nm	20.1	21	646
140 nm:300 nm	20.87	31	465
140 nm:400 nm	20.94	9	418

center, only 2 nm higher. Instead of monitoring all of the features, one can monitor the 140 nm : 400 nm line–space pairs and adjust the centers accordingly. This line:space ratio is the gating feature for a combination of the five features. Therefore, its E-D behavior is close to that of the common window.

7.1.4 DOF budget

The DOF budget can be considered in two ways: as the DOF budget provided and as the DOF budget consumed. The provided budget is determined by imaging parameters such as wavelength, bandwidth, NA, illumination, feature shape, feature size, feature combination, CD tolerance, phase and intensity on the mask, multiple reflections, and EL, as was elaborated in Chapter 5. The DOF budget is consumed by imperfection of the components in the imaging train, such as mask nonflatness, mask tilt, mask topography, mask chucking error, proximity effects, lens aberrations, lens heating, illumination nonuniformity, resist thickness, thickness nonuniformity (in the TARC, resist, BARC, and thin-film stack on the wafer), topography, wafer nonflatness, wafer tilt, wafer chucking error, and focusing error.

The provided budget is best allocated in at least two imaging media, i.e., the coupling medium from the lens to the resist and the resist medium. This is particularly important with immersion lithography in which the refractive index of the coupling medium exceeds 1. An extensive treatment is given in Chapter 8. $DOF_{diffrac}$ (defined in Section 8.4.3) provides the DOF budget, and $DOF_{required}$ (defined in Section 8.4.4) consumes the DOF budget. We now discuss the imperfections that constitute $DOF_{required}$.

7.1.4.1 Components of $DOF_{required}$

Mask contributions—Contributions from the mask include its flatness, roughness, topography of the absorber or phase shifter, mask tilt due to nonparallelism, chucking, and scanning imperfections. All vertical imperfections are reduced by $1/M^2$ for the DOF budget at the wafer side. Given $M = 4$ for scanners, the 1/16 reduction factor is not so effective in modern-day lithography. For example, the polysilicon layer of the 32-nm logic node with 45-nm half pitch, and with the available DOF, i.e., $DOF_{diffrac}$ consumed by the resist thickness, is only 100 nm. The best mask flatness presently commercially available is 500 nm. The chromium absorber thickness is 200 nm, while the phase shifter in quartz at the 193-nm wavelength is also on the order of 200 nm. The equivalent vertical variation on the wafer is now on the order of 57 nm—more than half of the available DOF budget.

Subsequent mask processing can affect the flatness of a mask blank. For example, the stress on the mask is changed after the absorber is delineated. The mask blank flatness is also affected by mounting of the mask pellicle. Fortunately, the static imaging field of a scanner is 26 mm × 8 mm instead of the rectangular area within the circular lens field, as in a stepper. It is possible to actively compensate for mask imperfections by adjusting the tilt of the mask dynamically during scanning. Improvement on the order of 25% is often seen.

Lens effects—The lens effects from consuming the DOF budget are generally categorized as image plane deviations (IPDs). Basically, an IPD contains the variation of the Zernike Z_4 term at different field positions. Field curvature, astigmatism, and coma all contribute to consumption of the DOF budget. Some of these effects can be artificially avoided. For example, astigmatism-induced DOF deficiency can be alleviated by assigning the critical dimensions in one direction only. It is known that some semiconductor manufacturers align their critical gates in one dimension. Of course, reducing the effect of astigmatism is only one of the reasons to align the gates. In addition, the aberrations of a heated lens when the exposure tool is running at its full capacity are different from those evaluated when the lens is in a testing environment. The effect on DOF can be as large as 30 nm.

Illumination nonuniformity has a direct impact on the DOF budget. This is obvious from E-D trees. A larger EL is required to accommodate the illumination nonuniformity, trading off DOF. When the illumination wavelength drifts, the change of wavelength causes a focus shift. The DOF can also be affected.

Thicknesses—The resist thickness consumes a major part of the DOF budget. Take the 32-nm node as an example again. The resist thickness required is on the order of 100 nm. It consumes 100/1.75 nm of the DOF budget in a dry exposure system, where 1.75 is the refractive index of the resist. In a water-immersion system, the resist consumes 100 × 1.44/1.75 nm of the DOF budget, where 1.44 is the refractive index of the immersion fluid at the 193-nm wavelength. [For the reasoning behind multiplying by the refractive index of the coupling medium, refer to Eq. (8.12)].

Film nonuniformity consumes the DOF budget physically and optically. Nonuniformities in the TARC, resist, BARC, and thin-film stack on the wafer dislocate the vertical position of the imaging layer. Optically, these thickness nonuniformities perturb multiple reflections on the interfaces, creating exposure nonuniformity, and lead to consumption of the DOF budget. Topography affects exposure and focus uniformity, as explained in Section 6.6.2.5. In principle, chemical mechanical polishing (CMP) should have flattened the topography. However, at the 90-nm node, the CMP residual topography can be on the order of 50–100 nm.

Wafer nonflatness, wafer tilt, and wafer chucking error obviously consume the DOF budget. For sub-100-nm lithography, double-side polished wafers are primarily used because polishing the back side improves its flatness so that the front surface is less distorted when the wafer is chucked down. A 30% reduction in focus variation with a double-sided polished wafer has been reported by Choi.[3]

Similar to mask flatness, the wafer can deteriorate from its original flatness after repeated patterning and film deposition, especially with film stress caused by thermal cycles. CMP helps but does not completely eliminate the wafer flatness deterioration.

Fortunately, active compensation of wafer flatness imperfections can be performed on the scanner just like compensating for mask flatness imperfections. Hence SEMI proposes a site-flatness quality requirement (SFQR) to characterize wafer flatness. SFQR is defined as being site-based and front-surface-referenced, using a least-squares reference plane within the site and reporting the range of flatness within the site. The size of the site should be chosen as the size of the exposure slit, i.e., 26 mm × 8 mm.

In fact, when performing a scanning exposure, the exposure slit/slot moves continuously across a field and levels dynamically. As pointed out by ASML,[4] for the exposure slit/slot to be more closely correlated to the residual focus error after leveling, it is more appropriate to use the moving average (MA), defined as the average of the residual defocus of all points exposing a point P in a field:

$$\langle \Delta z_s(x_s) \rangle = h_s^{-1} \int_0^{h_s} \Delta z_s(x_s, y_s) dy_s, \tag{7.1}$$

where Δz_s is the residual defocus after leveling when P is exposed by $P_s(x_s, y_s)$ of the exposure slit, and h_s is the slit/slot height. The leveling is optimized to take care of the local topography inside the region of the exposure slit.

Another important component of DOF consumption is focusing error. Focusing error is the error committed by the focusing sensor and the vertical positioning of the mask, the lens, and the wafer on the exposure tool. Once the best focus is determined, the focusing sensor and the positioning devices must consistently maintain each exposure at this best focus. Failure to do so results in focusing error.

Focus sensing can be done several ways, namely, using optical sensing, air gauge sensing, and capacitive gauge sensing, as shown in Figs. 7.12, 7.13, and 7.14, respectively. Each of these sensing methods has advantages and disadvantages. Optical sensing uses an obliquely incident beam on the wafer surface at grazing incidence. The beam is detected at the other end. Any change in the vertical position or tilting of the wafer changes the position of the detected beam. The air gauge is a small tube placed in proximity to the wafer surface. Compressed air exiting the tube changes pressure as a function of the closeness of the wafer surface. By monitoring the pressure change, one can detect the vertical position of the wafer. The capacitive gauge places a small capacitive probe near the wafer surface, and the capacitance changes as a function of the vertical position of the wafer. The latter is obviously affected by the electrical property of the films and features on the wafer. For the air gauge, rushing air has a thermal effect on the wafer and is also a contamination source. Optical sensing can be affected by the optical property of the films on the wafer. Grazing incidence tends to minimize this effect.

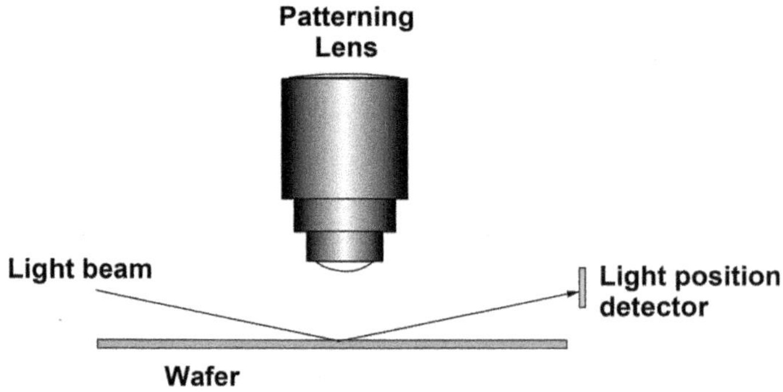

Figure 7.12 Optical focus sensing.

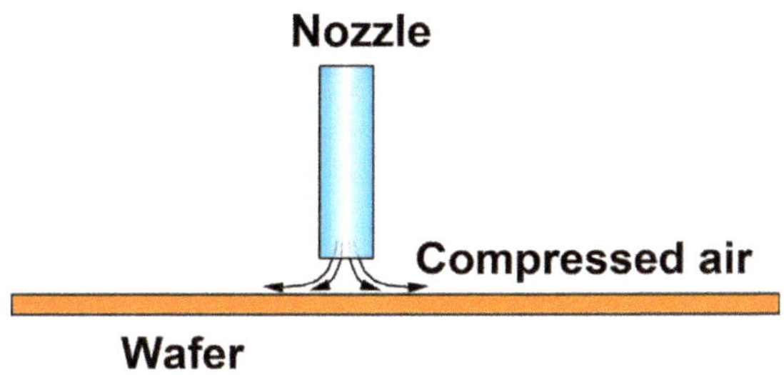

Figure 7.13 Focus sensing using an air gauge.

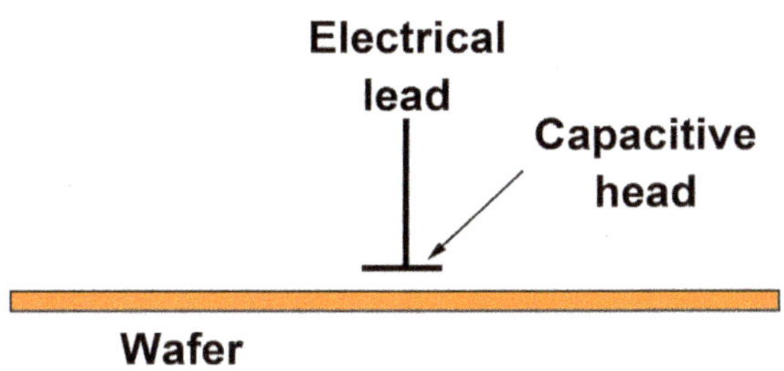

Figure 7.14 Focus sensing using a capacitive gauge.

Hence, all scanner suppliers have adopted optical sensing. Before the shakeup of optical exposure tool suppliers, these companies included GCA, PerkinElmer (which later became SVGA), Ultratech, Optimetrix, Censor, and Electromask. All three types of sensors have been adopted by one supplier or another. Currently, due to stringent focus control requirements, the air gauge is used to supplement optical focus sensing.

Common E-D window—One must make sure that all of the parameters that determine the provided DOF budget are used to set the best focus. These parameters are wavelength, bandwidth, NA, σ, feature shape, feature size, feature combination, CD tolerance, phase and intensity on the mask, proximity correction, multiple reflections, and EL, as listed in Section 7.1.4. If the Bossung[5] curve is used, the best focus can only be evaluated on one critical feature, which is often insufficient. Figure 7.15 shows the E-D trees of a 0.5-μm line–space and a 0.5-μm isolated line opening, imaged at λ = 365 nm, NA = 0.48, σ = 0.5. The center of DOF for the former feature is at –231 nm and that for the latter feature is at 308 nm. Taking the common E-D window of the two features, the center of DOF is at 154 nm. The two features are quite different. Frequently monitoring the critical features can be tedious and unproductive. For daily focus monitoring, one should identify certain test patterns that can precisely indicate whether the best focus of the critical features is met. For example, line-end shortening (LES) is very sensitive to defocus, so it can serve as a good test pattern for monitoring focus.

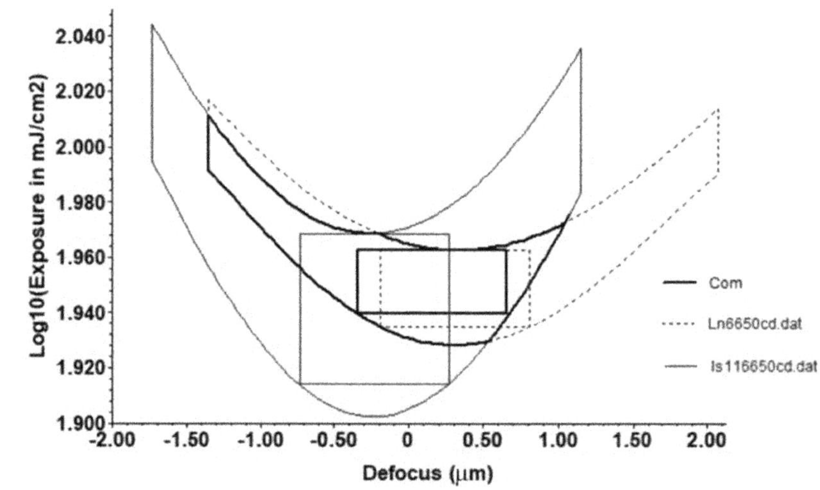

Figure 7.15 Determination of the best focus for two 0.66-k_1 features with L:S = 1:1 and an isolated line opening, imaged at λ = 365 nm, NA = 0.48. The center of DOF for the lines and spaces is –231 nm, the center of DOF for the isolated line opening is 308 nm, and the common E-D window is 154 nm.

7.1.4.2 Focus monitoring (by S.S. Yu)

Focus monitoring builds on focus sensing. The wafer image is monitored for the accuracy of the best-focus setting. First, one must define the best focus, then, one must determine the way to evaluate it.

Ausschnitt[6] made attempts to facilitate focus monitoring with a model-based approach by first calibrating the relationships of CD versus exposure dosage and CD versus focus, then finding the corresponding focus through the measured CD. By using line-end shortening (LES) as the change in CD (ΔW), the sensitivity to focus can be made even higher. With proper arrangement of the mask features, such a LES can be measured by overlay metrology tools. This method has been commercialized.[7]

Since CD depends on exposure dosage E and focus F, E and F should be solved simultaneously. Therefore, we need to measure the CD of at least two kinds of patterns. In Ausschnitt's original approach, W (a function of E and F) is assumed to be of the following form:

$$W[k] = a_0[k] + a_1[k]E + (a_2[k] + a_3[k]E)F^2, \tag{7.2}$$

where k is used to indicate the pattern type, and $a_0[k]$, $a_1[k]$, $a_2[k]$, $a_3[k]$ are to be characterized by a series of Bossung curves. Then, if the CD of two kinds of patterns is measured, E can readily be solved by eliminating F, as in the following:

$$\frac{W[1] - a_0[1] - a_1[1]E}{a_2[1] - a_3[1]E} = \frac{W[2] - a_0[2] - a_1 2E}{a_2[2] - a_3[2]E}. \tag{7.3}$$

F follows once E is obtained. However, since the response of CD to defocus is mainly quadratic, the sign of defocus cannot be determined unambiguously. Additional marks are added to resolve the defocus sign problem. In addition to the sign problem, the center of focus for different patterns may not be identical. Equation (7.2) should be modified as follows and Eq. (7.1) adjusted accordingly:

$$W[k] = a_0[k] + a_1[k]E + (a_2[k] + a_3[k]E)\,(F - F_0[k])^2. \tag{7.4}$$

If the center of focus for the two chosen patterns is the same, then E and F can be solved by Eqs. (7.2) and (7.3). If not, E and F can still be solved since we know the center of focus of the two chosen patterns. However, the expression of the analytical solution is much more complicated. This method is only for focus monitoring, not for accurate determination of the center of focus of a group of patterns.

The monitoring scheme using a PSM does not have the sign problem.[8] In its normal implementation such as in AltPSM, the phase difference of two phase shifters next to a dark line is π to maximize the DOF. However, if the phase difference is not exactly π, then the image of the dark line will be displaced laterally when there is defocus. A clever implementation of this idea realizes the phase-shift focus monitor (PSFM), as shown in Fig. 7.16. Setting the phase difference to $\pi/2$ maximizes the detection sensitivity $\Delta x/\Delta z$. By putting phase shifters besides the bars in the BiB overlay mark, one can convert Δz in defocus to Δx (or Δy) in overlay errors.

The sensitivity can be further enhanced by using a phase grating.[9] The corresponding methodology is called the phase-grating focus monitor

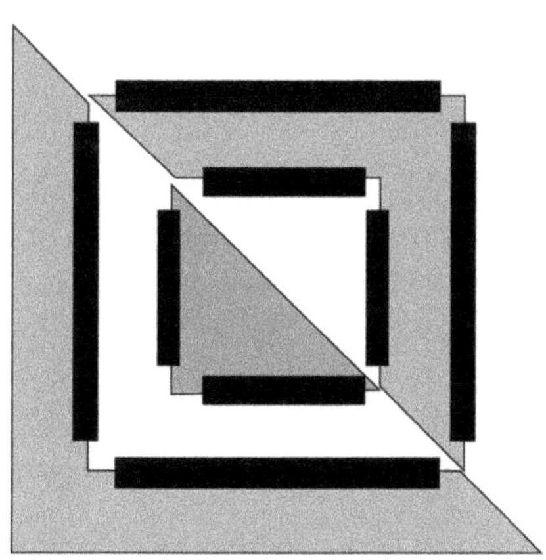

Figure 7.16 Phase-shift focus monitor (PSFM) pattern (reprinted from Ref. 8).

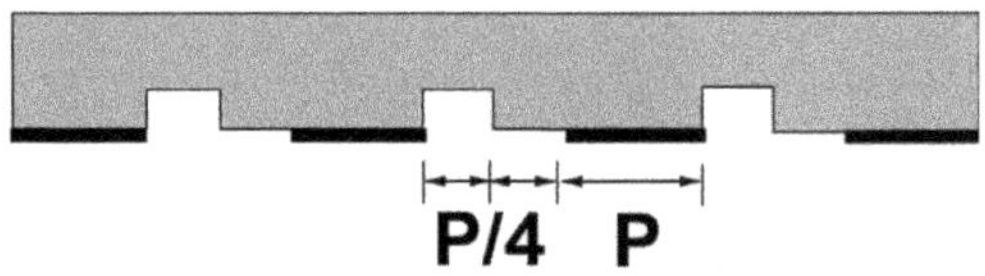

Figure 7.17 Phase-grating focus monitor (PGFM) pattern (reprinted from Ref. 9).

(PGFM), as illustrated in Fig. 7.17. The sensitivity of the PGFM is reported to be 3.03^{-1} for ArF scanners—more than two times higher than that of PSFM at 7.14^{-1} under the same processing conditions.[10] Because it is easier to handle the theoretical aspects of the PGFM, we use the PGFM to illustrate the underlying principle of focus monitoring with a PSFM.

Assuming that the incident light is a plane wave corresponding to the disk illumination at $\sigma = 0$, the amplitude $A(f)$ of various diffraction orders is calculated by the Fourier transform of the mask intensity and phase distributions [i.e., $M(x) = 1$ from $x = 0$ to $x = p/4$, then $M(x) = e^{i\xi}$ from $x = p/4$ to $x = p/2$, and $M(x) = 0$ for the remaining part of the period]:

$$\begin{aligned} A(f) &= \int_{-\infty}^{\infty} M(x)\, e^{-i2\pi f x} dx \\ &= \sum_{n=-\infty}^{\infty} \int_{np}^{np+\frac{p}{4}} e^{-i2\pi f x} dx + \sum_{n=-\infty}^{\infty} \int_{np+\frac{p}{4}}^{np+\frac{p}{2}} e^{i\xi} e^{-i2\pi f x} dx \\ &= (1 + e^{i\xi} e^{-i2\pi f \frac{p}{4}}) \int_{0}^{\frac{p}{4}} e^{-i2\pi f x} dx \sum_{n=-\infty}^{\infty} (e^{-i2\pi f p})^n \\ &= \sum_{m=-\infty}^{\infty} (1 + e^{i\xi} e^{-i\pi \frac{m}{2}}) \frac{p}{4} e^{-i\pi \frac{m}{4}} \text{sinc}\left(\frac{m}{4}\right) \frac{1}{p} \delta\left(f - \frac{m}{p}\right), \end{aligned} \tag{7.5}$$

where $\text{sinc}(z) = \frac{\sin(\pi z)}{\pi z}$, and $\sum_{n=-\infty}^{\infty} (e^{i2\pi f})^n = \sum_{m=-\infty}^{\infty} \delta(f - m)$.

$A(f)$ is nonzero only when $f = m/p$, where m is an integer. This indicates that there are discrete diffraction orders. The factor $(1 + e^{i\xi} e^{-i\pi m/2})$ in Eq. (7.5) can be used many ways. When $\xi = -\pi$, as in the case of PSFM, the factor is 0, $1 + i$, and $1 - i$ for $m = 0$, 1, and −1, respectively, assuming that only the three lowest diffraction orders are collected by the lens aperture. When PGFM is used, $\xi = -\pi/2$, and the factor becomes $1 - i$, 0, and 2 for the same respective values of m.

The wafer image is also a grating. Since the PGFM image is formed by the 0th and the −1st orders instead of the +1st and −1st orders, the image propagates obliquely along the optical axis. The intensity distribution of the grating image is given by

$$I = |A_0|^2 + |A_{-1}|^2 + 2\mathrm{Re}\left[A_0 A_{-1}^* e^{i(Gx+\Delta qz)}\right], \tag{7.6}$$

where

$$I = |\mathbf{E}_0 + \mathbf{E}_{-1}|^2,$$

$$\mathbf{E}_0 = A_0 e^{i\mathbf{k}_0 \bullet \mathbf{r}} \hat{\mathbf{y}},$$

$$\mathbf{E}_{-1} = A_{-1} e^{i\mathbf{k}_{-1} \bullet \mathbf{r}} \hat{\mathbf{y}},$$

$$\mathbf{k}_0 = k\hat{\mathbf{z}},\ k = 2\pi/\lambda,$$

$$\mathbf{k}_{-1} = \mathbf{k}_0 - \mathbf{G} - \Delta\mathbf{q},$$

$$\mathbf{G} = G\hat{\mathbf{x}},$$

$$G = 2\pi/p,$$

$$\Delta\mathbf{q} = \Delta q\hat{\mathbf{z}},$$

and

$$\Delta q = k - \sqrt{k^2 - G^2}.$$

A_0 and A_{-1} denote the amplitude and phase of the diffraction orders corresponding to $f = 0$ and $f = -1/p$, respectively. The vectors $\mathbf{E}_0$, $\mathbf{E}_{-1}$, $\mathbf{G}$, $\mathbf{q}$, and the angle θ are depicted in Fig. 7.18.

Let ϕ_0 be the phase of A_0, i.e., $A_0 = |A_0|e^{i\phi_0}$. By requiring $Gx + \Delta qz + \phi_0 - \phi_{-1}$ to be a constant, the sensitivity of the PGFM $\Delta x/\Delta z$ is $\Delta q/G$. Note that $G/k = \sin\theta$, where θ is the angle between the 0th- and -1st-order beams. Hence, the sensitivity of a PSGM is $\tan(\theta/2)$. The maximum possible sensitivity that can be achieved theoretically occurs when θ equals the aperture angle of the imaging lens.

In addition to using a PSFM or PGFM, single-beam off-axis illumination (SB-OAI) has been proposed to achieve the highest degree of image obliquity.[11] SB-OAI can be realized with an opening at the upper side of a double-sided chrome mask, along with the conventional overlay mark at the lower side of the same mask, located with its center directly beneath the edge of the opening. Therefore, the overlay mark sees only half of the employed illumination, as shown in Fig. 7.19. Note that a PSM is not needed. Note also that the mark is applicable to ArF, KrF, i-line, and any other wavelengths used in microlithography because the overlay mark is much larger than the wavelength. In addition, because the methodology is impervious to lens aberrations, the focus response is nearly linear. The effect of lens aberrations

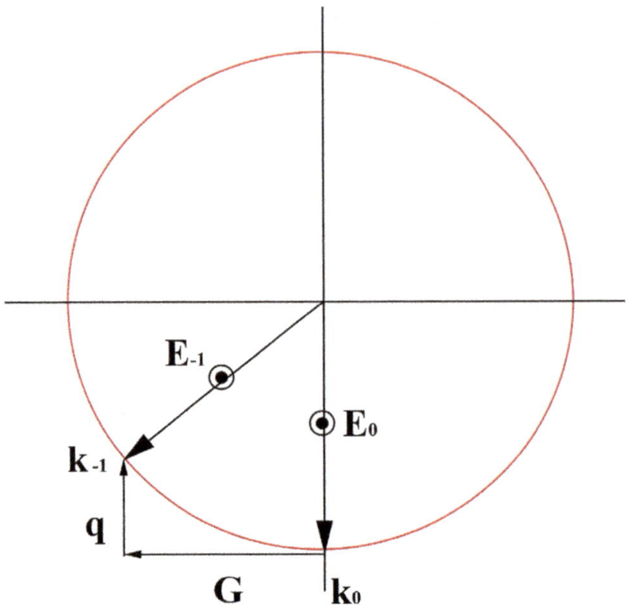

Figure 7.18 The vectors $\mathbf{E}_0$, $\mathbf{E}_{-1}$, **G**, **q**, and the angle θ.

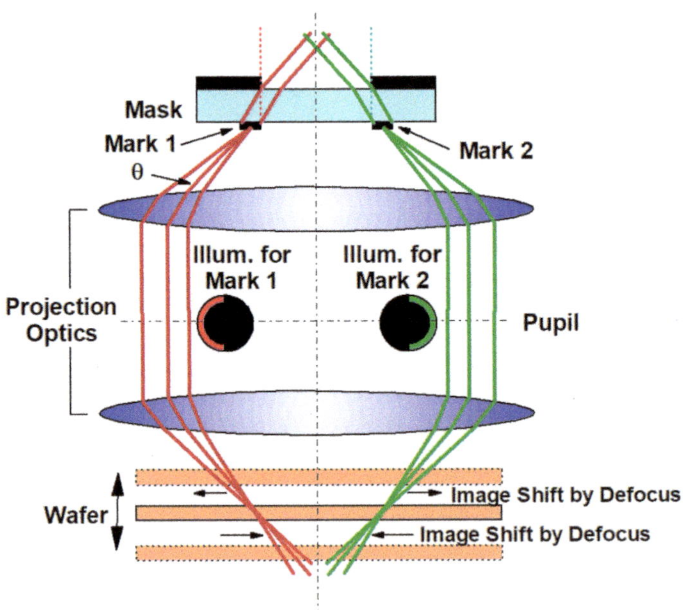

Figure 7.19 Marks on a double-sided chrome mask for one-beam imaging (reprinted from Ref. 11).

can be embedded in the calibration curve because only a small portion of the projection lens is utilized.

PFSM, PGFM, and SB-OAI call for specific illuminations, the former two requiring small-σ disk illumination, and the latter requiring OAI. These methods cannot be used with production masks having a wide variety of illumination conditions. Optical CD (OCD) metrology [often referred to as scatterometric CD (SCD) metrology] can measure not only the lateral dimension but also the profile of the grating image in terms of the side-wall angle (SWA), which changes monotonically with defocus near the best focus. Hence, the realization of an *in situ*, BIM, highly precise, polarity-unambiguous, one-shot-focus monitoring metrology is promising, provided there is sufficient room on the mask to accommodate the OCD test pattern.

Presently, the best focus-monitoring method uses part of a length-sensitive pattern to identify the value of F in Eq. (7.4). The monitoring scheme allows one to monitor the best focus for a combination of critical patterns in the real layout, λ, bandwidth, NA, σ, and OPC settings imaged with the production mask on production wafers.

7.1.5 Exposure tool throughput management

The throughput of an exposure tool has a direct impact on the economy of manufacturing. The per-exposure cost, or cost of ownership (CoO) is usually calculated from the cost of the tool divided by the available hours of the tool within the depreciation period and by the wafer-per-hour throughput:

$$CoO = \frac{P}{N \times U \times A \times W}, \tag{7.7}$$

where P is the capital cost of the tool, N is the number of hours within the depreciation period, U is the utilization, A is the availability, and W is the wafer-per-hour throughput.

A major component of wafer throughput is the exposure time. For a stepper, this is the time when the reticle[i] and the wafer are static, and the shutter is open. For a scanner, strictly speaking, the exposure time is the time during which the reticle and wafer are scanned in synchronization and the shutter is open. However, time is needed for the stages to accelerate to the full scanning speed before the shutter is opened, then decelerate in preparation for wafer-stage stepping. For convenience of calculation, the time during which both the reticle and the wafer stages are moving is classified as the exposure scanning time. The time during which the reticle remains stationary while the wafer is stepped into a new position for scanning is classified as the stepping time.

[i]The word *reticle* is often used for a mask used in an exposure tool. *Mask* is a more generic term. These words have been used interchangeably by many authors.

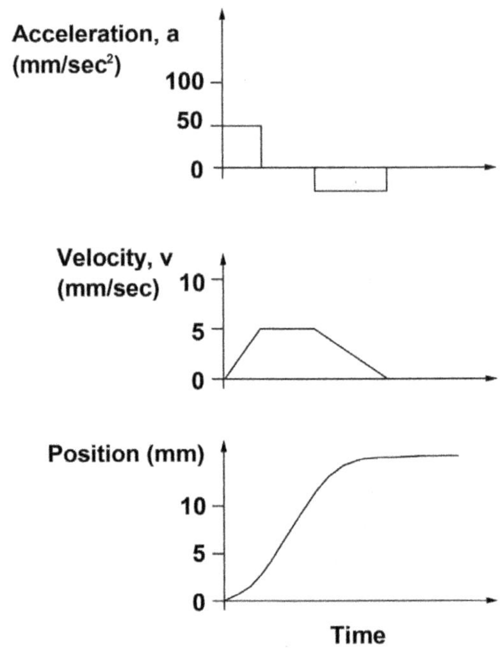

Figure 7.20 Acceleration, velocity, and position and of an early wafer stage.

Figure 7.20 shows the wafer stage position, velocity, and acceleration of an early stepper from the 1980s. Note that, for modern steppers, the stage velocity can be two orders of magnitude higher, and the acceleration is in the m/sec^2 regime. For the early stepper, the wafer stage is accelerated from a static position to reach a constant velocity, then decelerated to zero velocity. For modern steppers, the acceleration does not remain constant; the jerk is on the order of hundreds of m/sec^3. For scanner operation, both the reticle and the wafer follow similar motions, except for mX speed magnification at the reticle stage, where m is the reduction ratio of the imaging lens. To optimize the scanner exposure time, the velocity v must be high. Velocity is inversely proportional to the width of the scanning slot and the illumination intensity. Also, the acceleration a must be very high such that the time during acceleration and deceleration is a small portion of the time. For stepper operation, it is more important to take the wafer stage to its exposure position as quickly as possible. The total time of movement must be minimized with the best combination of stage acceleration and velocity.

Figure 7.21 shows the stage position, velocity, and acceleration of a deep-UV stepper. Its maximum stage velocity is $v = 0.32$ m/sec, maximum acceleration is a = 4 m/sec^2, and jerk h = 200 m/sec^3. The curves for the three stepping distances that the stage traveled (11.4 mm, 18.3 mm, and 31.2 mm, are shown. The equations used for these calculations are as follows:

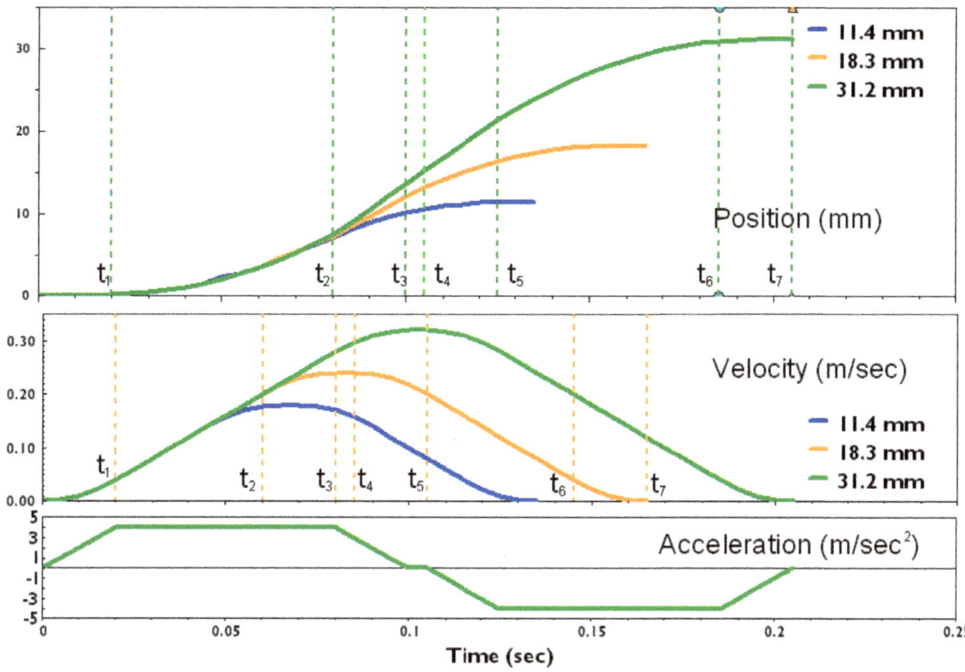

Figure 7.21 Position, velocity, and acceleration of a modern wafer stage.

When $0 \le t \le t_1$,

$$\begin{aligned} a &= ht,\ a_1 = ht_1, \\ v &= \int_0^t a\,dt = \frac{1}{2}ht^2,\ v_1 = \frac{1}{2}ht_1^2, \\ s &= \int_0^t v dt = \int_0^t \frac{1}{2}ht^2 dt = \frac{1}{6}ht^3,\ s_1 = \frac{1}{6}ht_1^3. \end{aligned} \tag{7.8}$$

When $t_1 < t \le t_2$,

$$\begin{aligned} a &= ht_1 = a_1 = a_2, \\ v &= v_1 + ht_1(t - t_1),\ v_2 = v_1 + ht_1(t_2 - t_1), \\ s &= s_1 + v_1(t - t_1) + \frac{1}{2}ht_1(t - t_1)^2,\ s_2 = s_1 + v_1(t_2 - t_1) + \frac{1}{2}ht_1(t_2 - t_1)^2. \end{aligned} \tag{7.9}$$

When $t_2 < t \le t_3$,

$$
\begin{aligned}
&a = ht_1 - h(t - t_2),\ a_3 = ht_1 - h(t_3 - t_2),\\
&v = v_2 + ht_1(t - t_2) - \frac{1}{2}h(t - t_2)^2,\ v_3 = v_2 + ht_1(t_3 - t_2) - \frac{1}{2}h(t_3 - t_2)^2,\\
&s = s_2 + v_2(t - t_2) + \frac{1}{2}ht_1(t - t_2)^2 - \frac{1}{6}h(t - t_2)^3,\\
&s_3 = s_2 + v_2(t_3 - t_2) + \frac{1}{2}ht_1(t_3 - t_2)^2 - \frac{1}{6}h(t_3 - t_2)^3.
\end{aligned}
\tag{7.10}
$$

When $t_3 < t \leq t_4$,

$$
\begin{aligned}
&a = 0,\ a_4 = 0,\\
&v = v_3,\ v_4 = v_3,\\
&s = s_3 + v_3(t - t_3),\ s_4 = s_3 + v_3(t_4 - t_3).
\end{aligned}
\tag{7.11}
$$

When $t_4 < t \leq t_5$,

$$
\begin{aligned}
&a = -h(t - t_4),\ a_5 = -h(t_5 - t_4),\\
&v = v_4 - \frac{1}{2}h(t - t_4)^2,\ v_5 = v_4 - \frac{1}{2}h(t_5 - t_4)^2,\\
&s = s_4 + v_4(t - t_4) - \frac{1}{6}h(t - t_4)^3,\ s_5 = s_4 + v_4(t_5 - t_4) - \frac{1}{6}h(t_5 - t_4)^3.
\end{aligned}
\tag{7.12}
$$

When $t_5 < t \leq t_6$,

$$
\begin{aligned}
&a = -h(t_5 - t_4) = a_5 = a_6,\\
&v = v_5 + a_5(t - t_5),\ v_6 = v_5 + a_5(t_6 - t_5),\\
&s = s_5 + v_5(t - t_5) + \frac{1}{2}a_5(t - t_5)^2,\ s_6 = s_5 + v_5(t_6 - t_5) + \frac{1}{2}a_5(t_6 - t_5)^2.
\end{aligned}
\tag{7.13}
$$

When $t_6 < t \leq t_7$,

$$
\begin{aligned}
&a = a_6 + h(t - t_6),\ a_7 = a_6 + h(t_7 - t_6),\\
&v = v_6 + a_6(t - t_6) + \frac{1}{2}h(t - t_6)^2,\ v_7 = v_6 + a_6(t_7 - t_6) + \frac{1}{2}h(t_7 - t_6)^2,\\
&s = s_6 + v_6(t - t_6) + \frac{1}{2}a_6(t - t_6)^2 + \frac{1}{6}h(t - t_6)^3,\\
&s_7 = s_6 + v_6(t_7 - t_6) + \frac{1}{2}a_6(t_7 - t_6)^2 + \frac{1}{6}h(t_7 - t_6)^3,
\end{aligned} \tag{7.14}
$$

where t_1, t_2, t_3, etc., mark the times that a changes. The stepping time at various step sizes is calculated by Eqs. (7.8)–(7.14) and is shown in Fig. 7.22.

The 31.2-mm stepping distance in Fig. 7.21 is the minimum distance for which the maximum velocity is reached. Time must be allowed to decelerate the stage to zero velocity for exposure. This stepping distance actually already exceeds the field size of the imaging lens at 22 mm^2. For the 18.3-mm stepping distance, only 0.24 m/sec passes before the stage must be decelerated. This time is even shorter for the 11.4-mm distance: 0.18 m/sec. This means that a larger step size makes the total wafer stepping time shorter. The average speed of the three stepping distances is 0.152, 0.111, and 0.0846 m/sec, respectively. Hence, for high throughput, the area covered in each exposure step should be as large as possible. Figure 7.23 shows a wafer exposed with a 11.4 × 18.3 mm^2 area and a 22.8 × 18.3 mm^2 area. The former area requires 122 steps of 11.4-mm distance and 9 steps of 18.3-mm distance. The total stepping time for this wafer is 17.955 sec. With the double-area field, the total stepping time is only 12.165 sec.

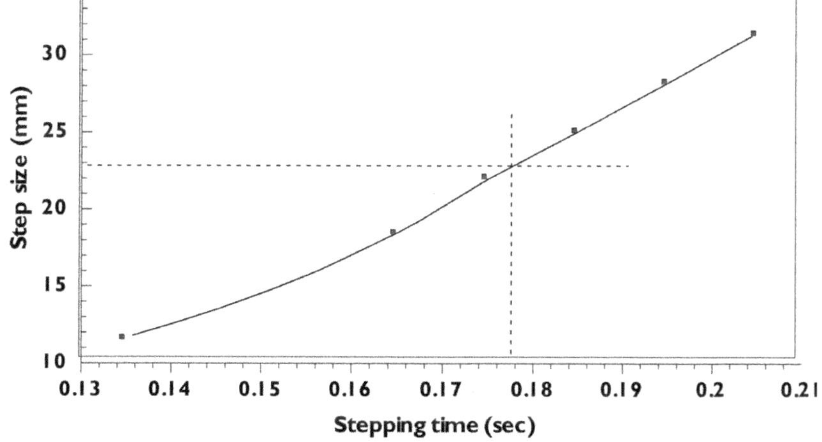

Figure 7.22 Calculated step size versus stepping time.

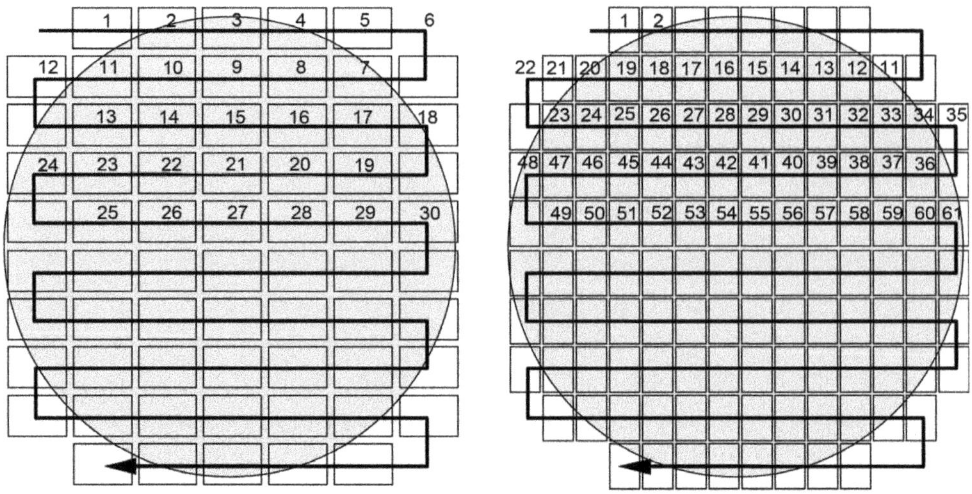

9 short steps 60 long steps
12.165 sec stepping time

122 short steps 9 long steps
17.955 sec stepping time

Figure 7.23 Exposure steps from different field sizes.

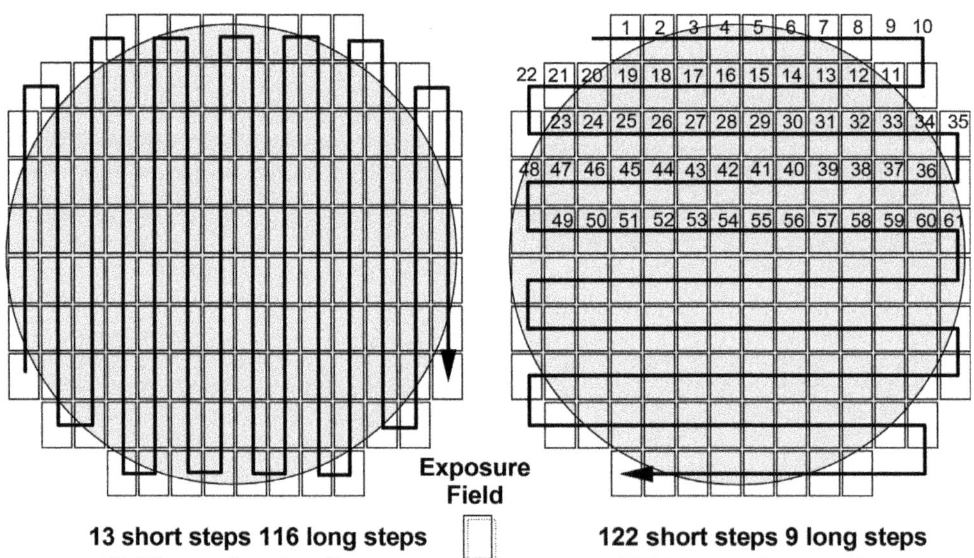

13 short steps 116 long steps
21.03 sec stepping time

(a)

122 short steps 9 long steps
17.955 sec stepping time

(b)

Figure 7.24 Two routing schemes.

Next, let us look at the proper orientation of a rectangular exposure field for throughput. Although the average speed of stepping through the shorter side is lower than through the longer side, the total stepping time depends on the distribution of long and short steps. Figure 7.24 shows the same wafer layout exposed in two different routing sequences. The routing in part (a) consists of 13 short steps and 116 long steps, whereas the routing in part (b) consists of 122 short steps and 9 long steps. The total stepping time for part (a) is 21.03 sec and for part (b) is 17.955 sec.

Mask field utilization (MFU) plays a very important role in wafer throughput. The exposure field of a modern-day scanner is 26 mm × 33 mm. If the dies on the mask cannot fill the entire exposure field, more steps must be taken to expose the entire wafer; otherwise, throughput suffers. When the die size is extremely unfavorable for filling the exposure field, the throughput can only be on the order of 40% of the most favorable situation.

Figure 7.25(a) shows[12] the velocity and acceleration of the wafer and the reticle stages specific to the scanner operation. Jerk is introduced to speed up acceleration in order to minimize the acceleration and deceleration times. Figure 7.25(b) shows the change in direction of the wafer stage after each exposure. It also shows the wafer stepping in x taking place while the wafer and reticle stages are still decelerating in y after the scanning exposure. This way, the wafer stepping time is absorbed in the scanning time.

Because the lens is a very expensive part of the exposure tool, letting it idle while the wafer stage moves away for alignment and wafer flatness measurement is wasteful. To get around this problem, tools with two wafer stages have been built. One wafer stage moves under the lens for exposure while the other stage is moving in the metrology area outside of the lens to measure the position of the alignment marks and the topography of the wafer surface. The key challenge of such a twin-stage system is to keep the two stages under laser interferometer surveillance without interruption.[13,14] Figure 7.26 shows how such a twin-stage system works. Wafer 1 is

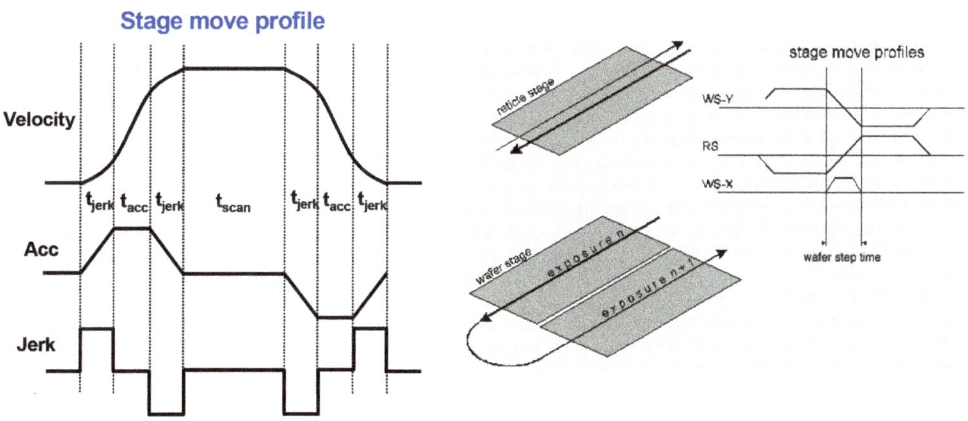

(a) (b)

Figure 7.25 (a) Velocity, acceleration, and jerk of a step-and-scan wafer and the reticle stages. (b) Change in direction of the wafer stage after each exposure and the wafer stepping in x (reprinted with permission from ASML).

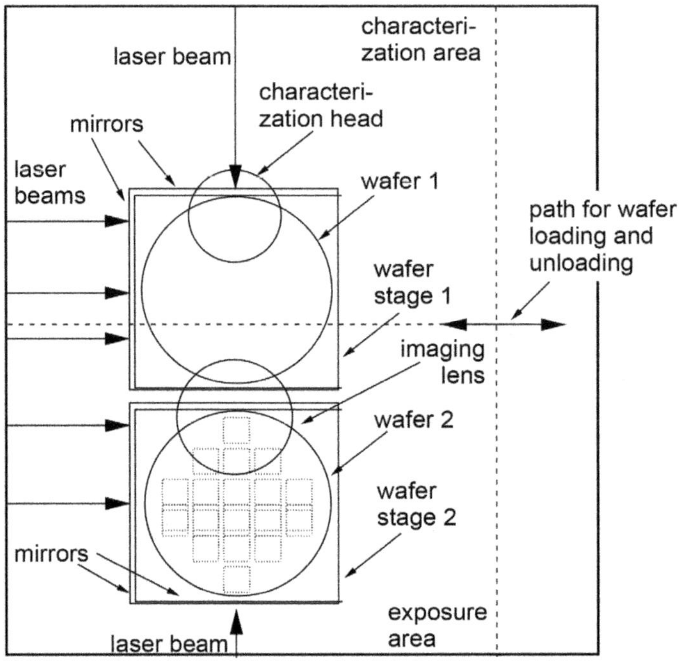

Figure 7.26 Twin-stage exposure station.

characterized while wafer 2 is exposed. After exposure, both wafer stages are moved all the way towards the top edge of the characterization area so that wafer stage 2 can move out along the loading/unloading line for wafer unloading. Stage 1 now moves to the exposure position. While it is stepping for exposure, stage 2 is loaded with a wafer and comes into the characterization area to start topography and alignment measurements. Once the wafer is characterized, the laser beam surveillance is unbroken. At least one vertical beam and two horizontal beams take turns monitoring the position of the wafer stage until it is moved out for unloading.

ASML implemented the twin-stage concept in their 12-inch scanner lineup. The commercial name is TWINSCAN. Many commercial publications and conference proceedings from ASML on TWINSCAN can readily be found.

7.2 Resist Processing

The optimization of resist processing includes resist coating, various types of resist baking, and resist developing.

7.2.1 Resist coating

The most popular way to coat the photoresist is by spin coating. The wafer, after being primed with an adhesion promoter, is held by vacuum on a spin chuck. Resist is applied to the wafer while it is stationary or slowly spinning. After a sufficiently large puddle is formed on the wafer, the chuck starts accelerating to a constant spin speed between several hundred and several thousand revolutions per minute (rpm). Spinning stops after the resist thickness and uniformity stabilizes. Then, the wafer is sent for prebaking. An edge bead removal step is usually performed while the wafer is still spinning. More discussion on the edge bead is in Section 7.2.1.4.

A good resist coating has the following properties: high thickness accuracy and uniformity, good surface planarity, strong adhesion to the substrate, and a low defect level.

7.2.1.1 Defects

The first step in achieving a low defect level is to have a thoroughly clean substrate surface. It is preferred to protect the surface from contamination instead of cleaning it prior to the coating process. If there is a nonwetting particle on the surface, it will induce a pinhole in the resist. A particle that is coated usually shows interference comet tails towards the circumstance of the wafer if a spin-coating process is used. This perturbs the local exposure because of multiple reflections from the nonuniform resist surfaces. The particle itself introduces a reflectivity that differs from that of the substrate. If the particle remains after resist development and stripping, it blocks the ensuing etching.

7.2.1.2 Resist adhesion

The photoresist polymer generally does not adhere well to Si-based substrates and many other types of solid films used in the semiconductor industry. Although the theory of adhesion is outside the scope of this book, we will shed some light by considering surface tension. The surface tension of the resist and that of the substrate cannot differ by 20 dyne/cm^2 (or more) so the resist can adequately wet the wafer.[15] Good wetting is necessary for good adhesion but does not absolutely dictate it. We can also reason that hydrophilic (water wetting) materials adhere well to other hydrophilic surfaces but not to hydrophobic (water repelling) ones, and vice versa. This is supported by the observation that the Langmuir–Blodget technique,[16,17] in which the coating is built up by pulling the substrate on its side from a monolayer film suspended in a tank. By repetitively pulling and dipping the substrate from and into the tank, the monolayer wraps itself back-to-back and front-to-front to build up a substantially thick coating. One side of the monolayer is hydrophobic while the other side is hydrophilic, suggesting that the same types of surfaces are more likely to adhere to each other.

The wafer surface must be treated with an adhesion promoter before resist coating. Hexamethyldisilazane (HMDS) is a popular adhesion promoter. It is most effectively applied in a heated vacuum chamber using an inert carrier gas. Applying HMDS in the liquid phase has also been practiced but is less effective than gas-phase application primarily because it is easier to elevate the reaction temperature in the gas phase.

7.2.1.3 Resist thickness

The spun resist thickness is related to the polymer concentration C, molecular weight M, and spin speed ω by the following equation:[18]

$$thk \propto \frac{C^{\beta} M^{\gamma}}{\omega^{\alpha}}, \tag{7.15}$$

where β and γ are usually close to 2, and α is close to 0.5. Therefore, the resist concentration is commonly used to change the resist thickness grossly, and the spin speed is used to fine tune it.

7.2.1.4 Resist uniformity

The resist coating process is generally considered as planarizing rather than conforming. This is true only for small features on the wafer. If there are large features, the resist over these features has an identical thickness to the resist over the wafer background, as shown in Fig. 7.27. Let the resist thickness over a planar wafer be a; then, thickness a is maintained on large planar regions of the wafer. With high feature density, the situation is similar. Thickness a is maintained from the top of the topography. Unfortunately, the resist thickness is now $a + b$ at the bottom of the topography, where b is the height of the topography. At medium feature density, the resist is partially planarizing while its thickness on the top of the topography is less than a. For a totally isolated small feature, the resist thickness at its top may be only slightly larger than $a - b$.

The presence of topography on the wafer further complicates film planarity from radial resist pileup against protrusions. The most significant impact of these pileups is alignment mark detection errors.

Resist accumulates significantly at the edge of the wafer. This piled-up edge is generally called the edge bead. Its thickness can be many times more than the thickness in the main area of the wafer. The width of the edge bead is usually a few millimeters. The edge bead is undesirable in many ways. (1) The extra resist tends to go under the wafer near the edge, making it difficult to be contacted uniformly to attain uniform wafer chucking or heat transfer. In

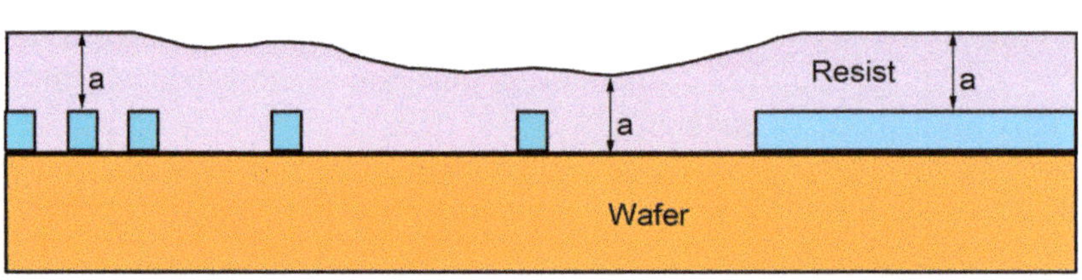

Figure 7.27 Resist planarization on the wafer topography.

addition, the extra thickness cannot support good imaging at the edge of the wafer. (2) The edge bead prolongs the resist drying and stripping times. (3) It tends to contact the wall of wafer carriers or robot wafer grips. Some resist may break loose and become a defect source. (4) For full-wafer contact printing, the edge bead prohibits uniform, full-wafer contact to the mask because it is much thicker than the resist. Edge beads can be removed in a modern resist coater. A small knife-edge jet of solvent is applied to the edge of the wafer while it is spinning, removing the resist at the edge of the wafer.

Striation in the resist film is exemplified by irregular radial lines that are easily seen by examining the resist surface in reflection. Striation is usually caused by a rapid loss of solvent due to either use of the wrong solvent or improper airflow management, both of which can affect film uniformity in many other ways.

7.2.1.5 Conserving resist material

Resist coating is a very wasteful process. The puddle formed at the wafer prior to spinning is usually about one-half of the diameter of the wafer, and its thickness is in the order of 1 mm. Most of the material is discarded by centrifugal force. Only a layer on the order of 100 nm to 1 μm is left for imaging. For imaging sub-0.1-μm resolution patterns, the resist thickness can be 100 nm or less. In a 200-mm diameter wafer, for example, the material applied to the wafer is about 8 ml and the resultant material used is about 2 nl. Only 1/250 of the material is used. Many techniques have been employed to reduce the initially applied thickness to a few milliliters. However, this still involves two orders of magnitude of waste. It is difficult to cover the entire wafer completely if the area of application before spinning is further reduced. Reducing the applied thickness results in a nonuniform spun-material thickness because the resist tends to form a high-concentration layer or even a skin at the top before spinning takes place.

There are many reasons to save the resist material. In addition to the initial expense, there are costs associated with transportation and storage of the material and disposal of the excess material. The extra burden to the environment alone justifies a more efficient technique to coat the resist.

Many methods to deliver a distributed and thin coating to the wafer have been tried. Ausschnitt and Hutchital[19] used a vapor deposition technique to deposit a uniform layer of resist material on the wafer. However, it is not easy to simultaneously control the thickness and uniformity. Spraying can deposit a thin layer of polymer material to a surface, just as in spray painting. However, the uniformity is totally inadequate to meet semiconductor manufacturing specs. Spraying followed by spinning suffers from a solvent evaporation problem that is much worse than that in puddle-and-spin.

A spray-and-spin-in-vapor (SASIV) technique[20,21] overcomes the above problems. The wafer is kept in a solvent-rich environment when the resist is

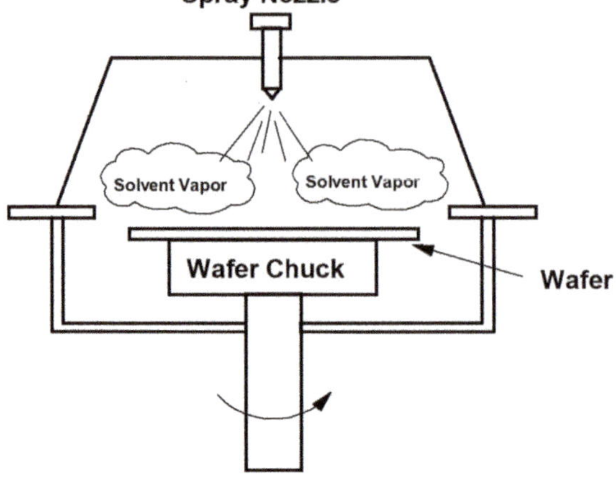

Figure 7.28 The SASIV technique.

sprayed on the wafer to reach a thickness slightly greater than the target thickness. Excess material is spun off while still in the vapor and is removed after the resist planarizes over the topography while the wafer is stationary. This can result in savings by a factor of 50 or more over conventional methods and can improve uniformity, especially for the resist that has piled up on the protruding topography. Figure 7.28 shows the SASIV concept, which has been experimentally verified, as shown in Fig. 7.29. Using simple laboratory equipment, a 20X resist savings has been demonstrated,[22] as shown in Fig. 7.30.

The essential sequence of SASIV is as follows:

1. Introduce solvent vapor to saturate the wafer ambient.
2. Spray resist onto the wafer in the vapor-rich ambient to a thickness slightly greater than the spun-material thickness.

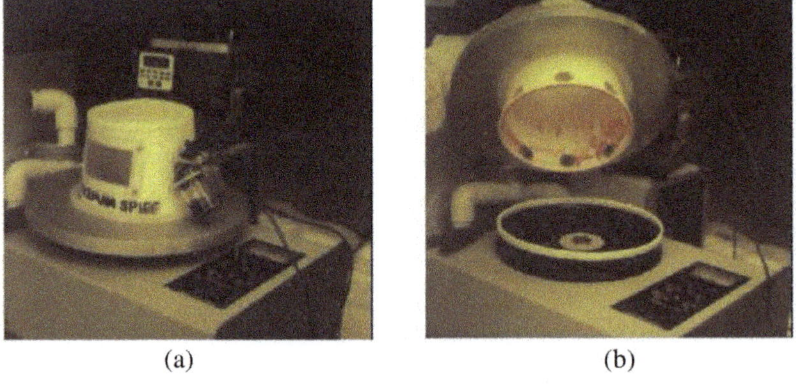

(a) (b)

Figure 7.29 The experimental apparatus for SASIV. (a) Vapor chamber is closed to introduce vapor and resist. (b) Vapor chamber is opened to load/unload the wafer.

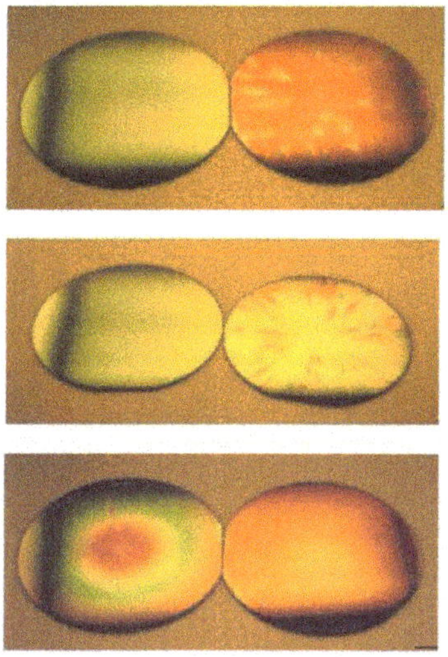

Left: Control wafer, 1150 nm resist, 113 nm variation at 5 points
Right: Low vapor pressure, no delay, 1020 nm thk, 113 nm variation

Left: Control wafer, 1150 nm resist, 113 nm variation at 5 points
Right: Low vapor pressure, with delay, 951 nm thk, 59 nm variation

Left: Low vapor pressure, 966 nm thk, insufficient delay, 320 nm variation
Right: High vapor pressure, 1054 nm thk, sufficient delay, 35 nm variation

Figure 7.30 Resist thickness uniformity on a wafer with/without vapor/delay. 1 ml of resist was needed to puddle a 3-inch wafer without SASIV; 0.05 ml was used with SASIV.

3. Wait 30 seconds until the stress in the resist (induced by the spray) is released.
4. Spin the resist layer (in the vapor) for the desired thickness.
5. Hold the wafer stationary in the vapor to allow the spun resist to level out the piled-up resist near the edges of the topography on the wafer.
6. Remove the vapor.
7. Bake the wafer as usual.

The stress-releasing delay can be eliminated to enhance wafer throughput by using a proximity-dispense-in-vapor (PROXIV) technique.[23] Instead of spraying the resist, it is gently spread on the wafer as it flows through small nozzles under slight hydraulic pressure while the wafer chuck is slowly rotating. A solvent vapor environment is maintained to prevent the thin resist layer from drying until it is spun off to reach the desired thickness.

7.2.2 Resist baking

Baking serves many purposes in resist processing. The pre-coating bake improves adhesion by driving out surface-absorbed water at the wafer. The pre-exposure bake [also known as post-apply bake (PAB)] removes the resist solvent after coating and fine-tunes the photosensitivity. Diffusion of the photoactive compound during the post-exposure bake (PEB) before development helps to eliminate standing waves in the resist, although doing so also dilutes the lateral image contrast. With chemically amplified resist systems, the PEB activates the catalyst reaction to convert the resist at the exposed

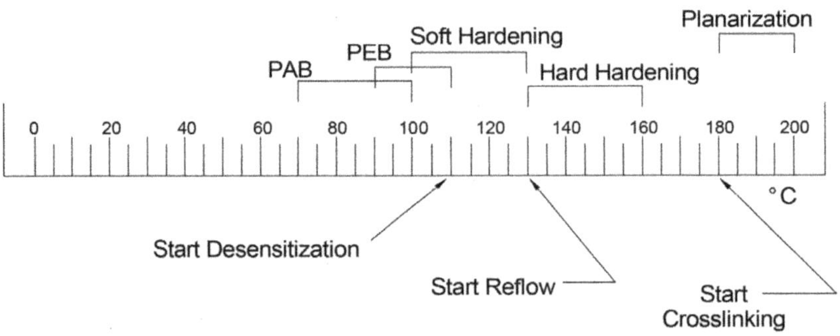

Figure 7.31 Effect of temperature on a typical DQN resist.

areas. The post-development bake hardens the resist to improve its adhesion in preparation for the substrate delineation processes to follow. Figure 7.31 shows the effect of temperature on a typical DQN resist.

7.2.2.1 Pre-coating bake

The pre-coating bake should not be performed separately. Right after a SiO_2 film is grown from an oven (e.g., a metal film from the evaporator, a sputtering machine, etc.), it is most desirable to prime the wafer with an adhesion promoter before the wafer is exposed to any water vapor. This not only ensures the freedom from water adsorption, but also eliminates a processing step.

7.2.2.2 Post-apply bake (pre-exposure bake)

The most important function of the post-apply bake (PAB) is to remove the photoresist solvent after spin coating. The baking temperature must be sufficiently high to remove the solvent in a reasonably short time. However, high temperatures may deactivate the photoactive compound and reduce the photoresist sensitivity. The development contrast of the resist is also affected. Figure 7.32 shows the remaining resist thickness as a function of the PAB temperature and time.[24] The resist development rate at the same exposure dosage is reduced when a higher PAB temperature or a longer PAB time is used. However, the development contrast is improved with the higher PAB temperature.

7.2.2.3 Post-exposure bake

Initially, the post-exposure bake (PEB) was used to remove the standing wave in the resist latent image, as shown in Fig. 6.59. When feature sizes are large compared with the wavelength, removing the standing waves is mostly cosmetic. The escalation of the exposure latitude requirement induced by multiple reflections cannot be eliminated with the standing wave. When feature sizes are reduced with respect to the wavelength, i.e., k_1 becomes

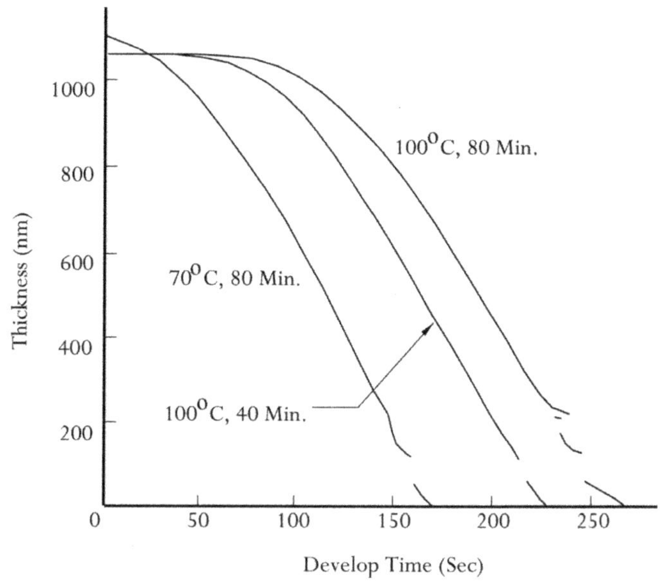

Figure 7.32 Effect of PAB on DQN-resist-development characteristics [reprinted from Ref. 24 with permission; © (1977) International Business Machines Corporation].

smaller, the PEB becomes critical. It reduces the lateral image contrast while the diffusion of the photoactive compound takes place in the longitudinal and lateral dimensions. With the advent of chemically amplified resists, the PEB has become a necessity, as was explained in Section 6.6.1.2, Fig. 6.45, and Fig. 6.46. Fortunately, most chemically amplified resists have a high γ value compared with DQN resists, as shown in Fig. 4.43, to make up for the loss in lateral contrast.

The PEB is usually the most critical processing parameter to control. Figure 7.33 shows 50 E-D trees (only 30 are labeled) constructed from five different types of edges under five different process tolerances. The five edges are ee″, aa″, bb″, cc″, and dd″, taken from a big-small-feature & corner (BigMaC) test mask,[25] as depicted in Fig. 4.26. A detailed description of the BigMaC mask was given in Section 4.1.7. The processing conditions of the E-D trees are centered at:

- resist thickness = 910 nm
- PAB = 90 °C
- PEB = 100 °C
- PEB time = 100 sec
- development time = 60 sec

Five E-D trees are constructed using these five central reference conditions. Another five E-D trees are constructed using a resist thickness

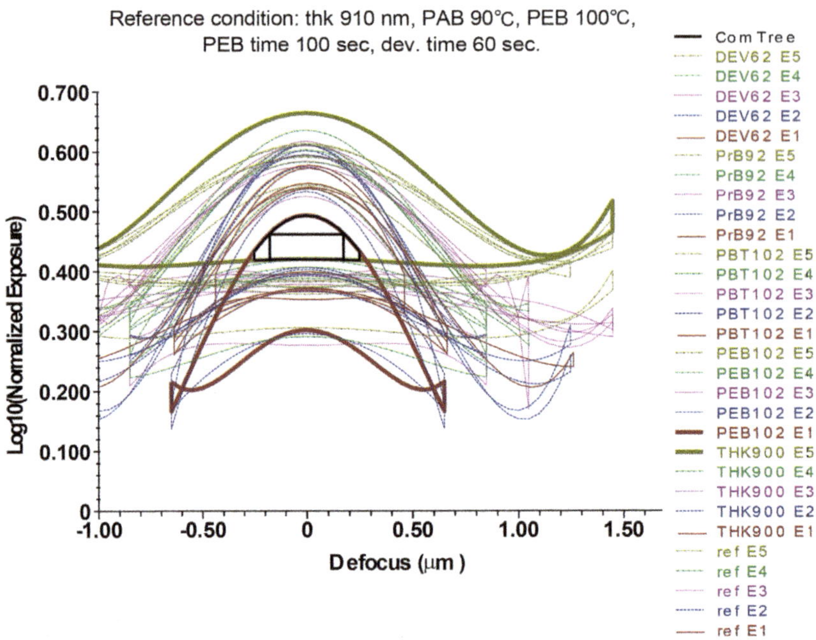

Figure 7.33 E-D tree of different processing conditions. PEB102 edge 1 is the upper limit; THK900 edge 5 is the lower limit. The common E-D window is 10% EL, 0.35-μm DOF.

of 910 nm, but the other processing conditions remain unchanged. Similarly, five more E-D trees are constructed for PAB 92 °C, PEB 102 °C, PEB time 102 sec, and develop time 62 sec, respectively, resulting in the 50 E-D trees. The common E-D window of these trees is the actual usable E-D region for the feature combination and process tolerances. The two critical conditions that bound the common E-D tree are: (1) PEB 102 °C for edge ee″ as the top branch and (2) 900-nm resist thickness for edge dd″ as the bottom branch. The PEB bounds are more pronounced than the thickness bounds.

To relax the PEB bounds, one can pose a more restrictive condition on the PEB tolerance. e.g., instead of a tolerance of ± 1 °C, reduce the tolerance to ±0.5 °C or smaller. However, it may be difficult to control the baking temperature and uniformity at will. If the PEB tolerance needs to remain at ± 1 °C, then an adjustment can be made to the position of edge ee″ to move the bounding E-D tree upwards for a larger window, as shown in Figs. 7.33 and 7.34. In other words, optical proximity correction is used to compensate for PEB-induced errors. Gau et al.[26] applied this concept for a different purpose—they used the PEB to reduce the optical proximity effect.

7.2.2.4 Hard bake

The hard bake is used after resist development to prepare the wafer for the subsequent image transfer processes. It serves to improve the adhesion of the

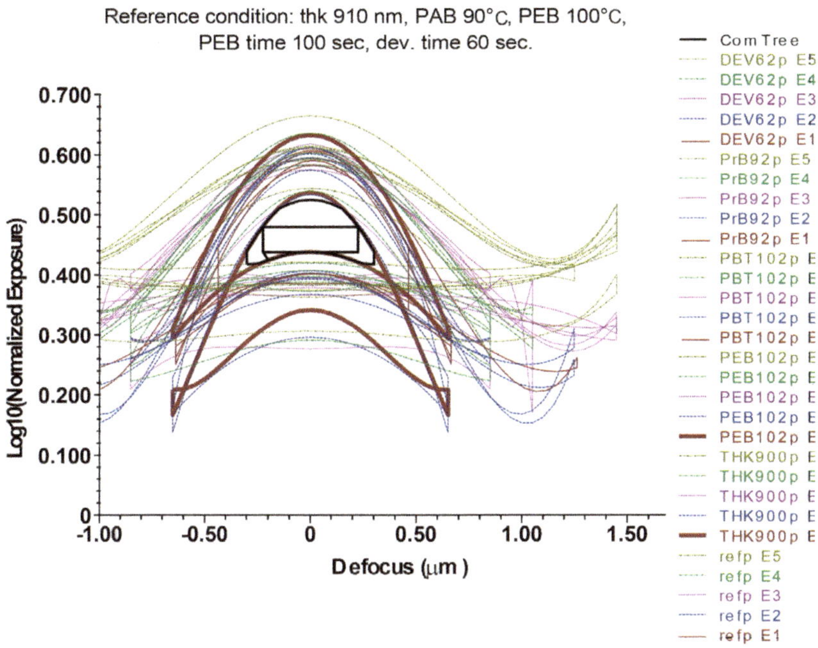

Figure 7.34 The PEB restriction in Fig. 7.33 eased with proximity correction. Edge 1 moved by 15 nm. The common E-D window is 10% EL, 0.45 μm DOF.

resist and to harden it. The baking temperature is important. When the baking temperature exceeds the PEB temperature, the resist starts to desensitize. A further increase in temperature makes the resist reflow. An even higher temperature crosslinks some resists, such as those containing the novolak resin, rendering it chemically inert.

If the resist profile is to be preserved, the hard bake is usually carried out below the reflow temperature. When a higher temperature is necessary, a hardening step must be included. Typical hardening steps consist of irradiation with deep-UV light[27] or treatment with a non-eroding plasma.[28] These steps are effective for small features because either treatment hardens the skin of the resist image. When a wide resist line or a large area of resist is to be hardened, the treatment must be prolonged to ensure complete penetration of the entire thickness of the resist. This may be impractical for resists thicker than 2 μm, especially for applications in fabricating magnetic read-write heads for disk memory.

The solution is a mold-hardening technique[29,30] that completely hardens the resist in the bulk. The hardened resist can be taken to complete carbonization at temperatures exceeding 500 °C in a nitrogen environment without distorting the resist image profile. The mold-hardening process is shown in Fig. 7.35. After the DQN resist image is delineated normally, polymethyl methacrylate (PMMA) is spin coated to completely cover the

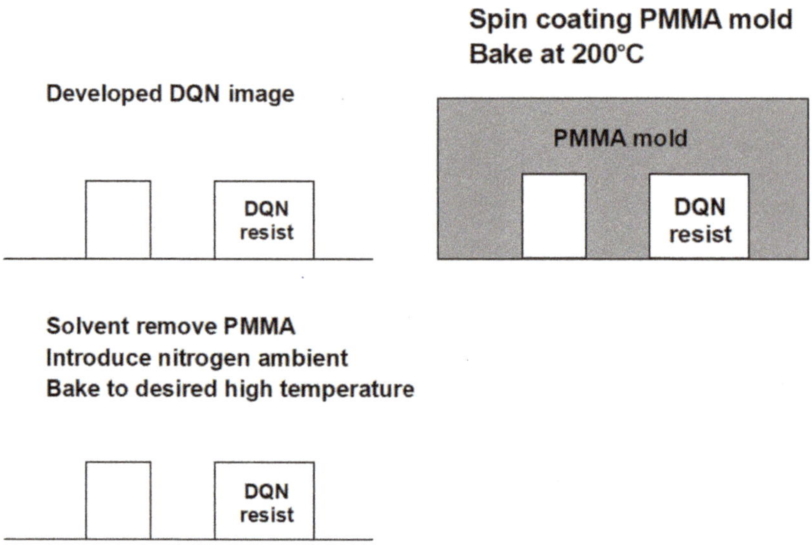

Figure 7.35 The mold-hardening process.

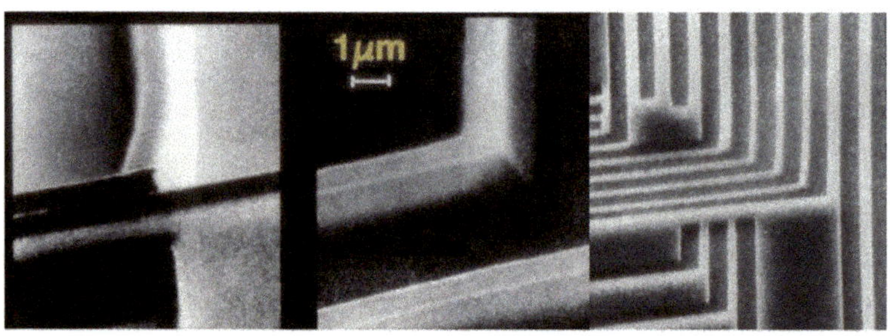

Figure 7.36 An AZ® resist image after mold hardening.

resist image, using a thickness greater than that of the resist image. This PMMA layer holds the DQN image to evenly distribute the stress induced by baking. After the DQN resist is completely hardened at 200 °C and above, the PMMA layer can be removed with solvent. The hardened DQN resist is robust for any subsequent process, including further baking in nitrogen ambient to temperatures exceeding 500 °C. The AZ® resist image after hardening is shown in Fig. 7.36.

7.2.3 Resist developing

The resist-developing step is one of the most crucial steps in the resist-imaging procedure. If this step is not carried out properly, the effort to obtain the aerial and the latent images are wasted. The development process is governed by the resist dissolution rate, which is a function of the resist sensitivity, absorption coefficient, exposure, and baking conditions. A typical rate-versus-exposure

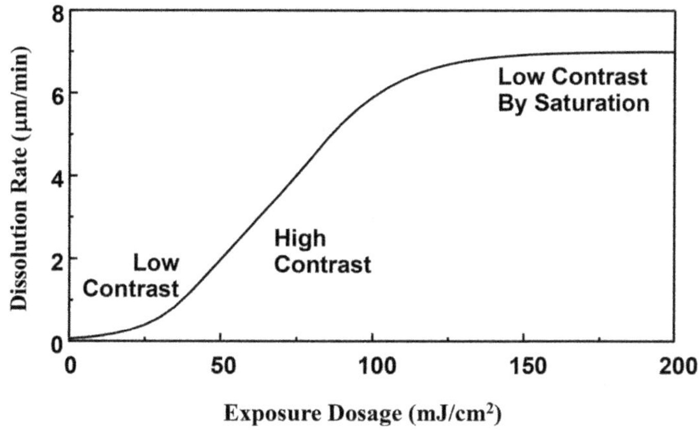

Figure 7.37 Bulk dissolution rate versus exposure of a typical DQN resist.

curve is seen in Fig. 7.37. This curve usually consists of a low-contrast region and a high-contrast region. Complete depletion of the photoactive compound flattens the dissolution rate at extremely high exposures.

The developer removes the resist at the developer–resist interface according to the local dissolution rate. The next instant the developer is facing a new interface and continues to locally remove the resist according to the dissolution rate. This takes place in three dimensions in a process very similar to the diffusion process, except that this diffusion-like process has a new surface to diffuse into at each infinitesimal time increment. The development process is also likened to the wet etching process, except that the latter is generally isotropic and is not guided by the 3D distribution of the dissolution rate. The development phenomenon is illustrated in Figs. 6.63 and 6.64 by strictly applying the dissolution-rate-versus-new-surface rule. To better simulate a realistic image, the following two phenomena need to be considered. (1) In addition to the dissolution rate consideration, the ability of the developer to wet the resist surface may decrease inside small openings, just as it does in partially developed contact holes or small line openings. (2) The developer may be depleted locally depending on whether the exposed area to be removed is large. This is a development proximity effect similar to the etching proximity effect.

- *Immersion developing*—The wafer is entirely submerged in the developer. This is usually done with an entire batch of wafers. The developer environment is static. In some cases, the developer circulates in the developer tank to refresh. Nevertheless, circulation cannot overcome the problem of developer nonwetting. Incorporating a surfactant in the developer to reduce surface tension is desirable.

  The immersion-developing process used to be the primary process in manufacturing. Due to the advent of track-processing systems, immersion development is giving way to the two other processes that are more adaptable to wafer-processing tracks.

- *Puddle developing*—A developer puddle is produced at the wafer in the developing station on a track. The developing station is like a resist-coating station that uses a wafer chuck that can spin at thousands of rpm. Instead of delivering resist to the wafer, the nozzle delivers the developer instead. After the development endpoint is reached, the developer is spun off and the wafer is spray rinsed with de-ionized water at a medium speed. The wafer is completely spun dry at a high speed to complete the process.

  Puddle developing is the track-adaptable form of immersion development. Thus, developer nonwetting and development proximity effects remain. The differences between immersion development and puddle development are: (1) In puddle development, each wafer must be processed separately, increasing the total processing time of a wafer batch. (2) Puddle development is wasteful of developer because the developing puddle is spun off and is not reused. (3) Development proximity effects cannot be overcome with circulation as in the case of immersion development.
- *Spray developing*—The wafer is developed in a developing station on the wafer track just as in puddle development. However, the developer is sprayed vigorously onto the wafer while the chuck is spinning at a high speed, thus refreshing itself. If the spray has sufficient energy, surface tension can be reduced, allowing better wetting of the developer to the resist. Hence, spray developing has the elements of high-quality developing. The tradeoff of spray development is its excess use of developer, which result in cost and chemical-disposal problems.

Because of the tradeoffs of these three development methods, there is no clear choice. Each method must be adopted by optimizing the balance of performance versus economy. The need for good surfactants is obvious. One important requirement of the surfactants, in addition to chemical compatibility with the developer and the resist, is that it can be completely removed, leaving no trace residue to affect the performance of the finished electronic device. Marriott[31] reports the resist contrast in terms of exposure versus developing time, as shown in Fig. 7.38. It is seen that spray developing produces the highest development contrast, followed by puddle developing.

7.2.4 Aspect ratio of the resist image

After the resist is developed, i.e., during the drying process, liquid is removed from the resist image. As the liquid leaves the resist image, the capillary force pulls adjacent features together.[32] If the adhesion of the resist to the substrate is insufficient, the resist peels. Alternatively, if the resist is not sufficiently

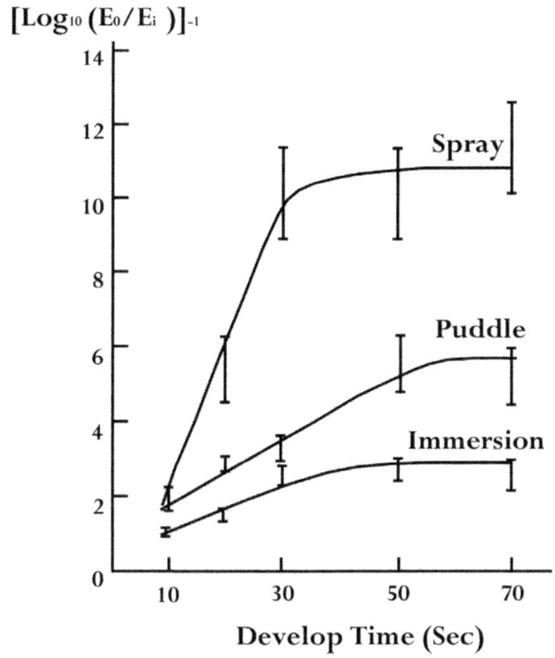

Figure 7.38 Effect of the development method on the resist contrast (reprinted from Ref. 31).

rigid, it bends or collapses due to elastic and inelastic deformations, respectively, as shown in Fig. 7.39.

Note that only isolated line pairs or the outside lines of a group collapse. The capillary's pull on the inside lines is balanced on both sides of the lines, so the forces there cancel out. The right side of Fig. 7.39 shows some exceptional resist deformations inside the array of resist lines. The pressure causing resist collapse is described by the following equation.[33] This pressure P affects the area $h \times l$, where h is the height of the resist and l is its length:

$$P = \frac{2\sigma \cos\theta}{d}, \tag{7.16}$$

where σ is the surface tension of the rinsing liquid, θ is the contact angle of the rinsing liquid, and d is the distance between two resist lines, as shown in Fig. 7.40. The pressure on the resist sidewall increases with the surface tension of the rinsing liquid and the sidewall angle of the rinsing liquid to the resist. The former is related to the liquid itself; the latter is related to the hydrophilicity of the liquid. When two resist lines are closer to each other, d is small, and P becomes larger.

The displacement δ of the resist is given by combining Eqs. (4) to (7) in Ref. 33:

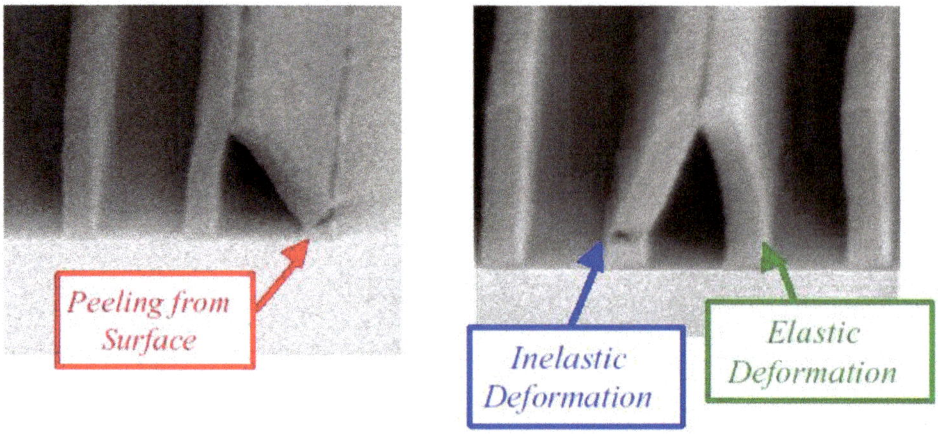

Figure 7.39 Resist collapse during liquid removal can cause peeling from the surface as well as inelastic and elastic deformation [reprinted from Ref. 32 with permission; © (1992) The Physical Society of Japan and The Japan Society of Applied Physics].

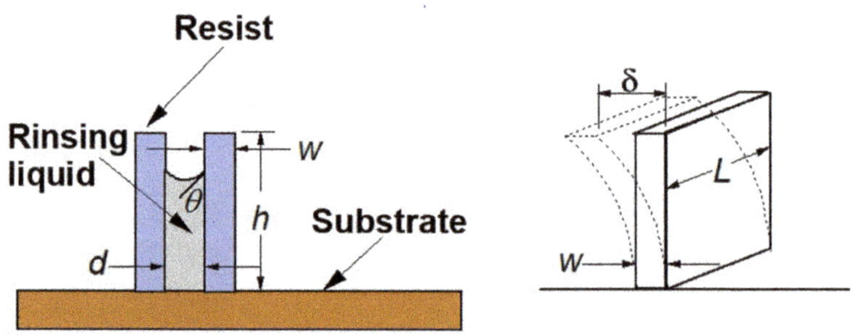

Figure 7.40 Contact angle and radius of curvature of the rinsing liquid [adapted from Ref. 33 with permission; © (1992) The Physical Society of Japan and The Japan Society of Applied Physics].

$$\delta = \frac{3\sigma \cos\theta}{E \times d} \left(\frac{h}{w} \right)^3 h, \tag{7.17}$$

where E is Young's modulus. When δ exceeds a limit, the resist peels, breaks, or bends sufficiently to touch a neighboring line. In any case, it collapses. From Eq. (7.17), the onset of collapse is proportional to the surface tension and the cosine of the contact angle, and is inversely proportional to Young's modulus E. The onset is also proportional to the third power of the resist height-to-width aspect ratio modified by an additional dependence on the absolute value the resist height.

It is very unusual for a resist image to exceed a 3:1 height:width aspect ratio without collapse. This ratio has been used as the rule of thumb to determine the resist thickness as a function of the minimum feature size.

Therefore, the resist thickness decreases as the feature size is reduced for each technology node. The etching technology must be developed to handle the smaller resist thickness.

7.2.5 Environmental contamination

Chemical amplification resists (CARs) are extremely sensitive to alkaline-containing vapor in the environment. This is because only a very few protons are needed to initiate chemical amplification during PEB. If the environment has trace amounts of alkaline-containing vapor, such as amine, after exposure and before PEB, any annihilation of protons is amplified. The amount that can affect the chemical amplification process is on the order of parts per billion. Therefore, in the CAR processing area, heavy filtration of amines by charcoal filters must be implemented, requiring significant space and cost. In the early days of CAR development, the resist exposure-development characteristic was erratic, and by no means were CARs worthy of manufacture until the root cause was identified.[34]

7.3 k_1 Reduction

From the scaling equations presented in Section 5.1, the approaches to improving resolution and DOF are clear. For resolution, the approaches are to increase the NA and reduce λ and k_1. Equations (8.3) and (8.7) further point to the increase in the refractive index of the coupling medium between the lens and the resist. Mathematically, this refractive index increase can be imbedded in changes to the NA and λ. For DOF, the scaling equations alert us to the tradeoff between resolution and DOF with changes in NA and λ. Reduction of k_1 is unique in the sense that it can improve resolution and DOF together. Because the imaging tool and imaging medium do not need to change, k_1 reduction is potentially an economical solution. The tradeoff in k_1 reduction is the increase of the mask error-enhancement factor (MEEF). Hence, k_1 is an indicator of the processing capability of a manufacturing site and its potential to cut costs. However, a true measure of manufacturing prowess is k_1/cost. Using k_1/cost, one need not blindly pursue the lowest k_1 among all manufacturing companies. MEEF and yield are also included in the considerations. k_1 reduction is often referred to as a resolution-enhancement technique (RET). However, the scope of k_1 reduction is much broader than resolution enhancement.

In the following sections, several k_1 reduction techniques are covered, namely, phase-shifting masks, off-axis illumination, scattering bars, and optical proximity corrections (OPCs). Several approaches can be taken before starting these k_1 reduction techniques. Optimization of NA and illumination as covered in Sections 7.1.1 and 7.1.2 must be used to realize the k_1 potential of a given imaging system before implementing k_1 reduction techniques. However, this is often done in conjunction with OPC, which is one of the k_1 reduction techniques.

Reducing reflections on the mask and the wafer restores lost contrast in the imaging system, thus facilitating k_1 reduction.[35] Reducing vibration in the imaging system also restores lost resolution.[36] These reduction methods are covered in subsequent sections.

7.3.1 Phase-shifting masks

The working principle of many types of PSM and their combinations is given in Section 6.3.4. Here, the performance and optimization of PSMs is covered. AltPSM, AttPSM, and chromeless PSM have potential for manufacturing and are included.

7.3.1.1 AltPSM

The AltPSM started the PSM movement. Being a strong PSM that has the capability to double the printable spatial frequency on the wafer and improve the image contrast significantly, the AltPSM draws much attention and is the subject of many studies. (Its configuration and working principle are given in Section 6.3.4.3.) However, it is the most difficult and expensive PSM to use in practice. Figure 7.41 shows the ideal patterns for AltPSMs, namely, equally spaced lines within the coherence range and, similarly, evenly spaced holes. These structures lend themselves to straightforward phase assignments with no possibility of creating phase-shifting conflicts.

The difficulties and conflicts associated with AltPSMs are depicted in Fig. 7.42. The brick layout in Fig. 7.42(a), once popular for DRAM, has a phase conflict between the ends of the bricks positioned along each row of bricks. If the phase of the bricks in a row is alternated, there are conflicts between rows. One way to solve these conflicts is to assign 120-deg shifts instead of 180-deg shifts; however, this reduces the image contrast. Mask

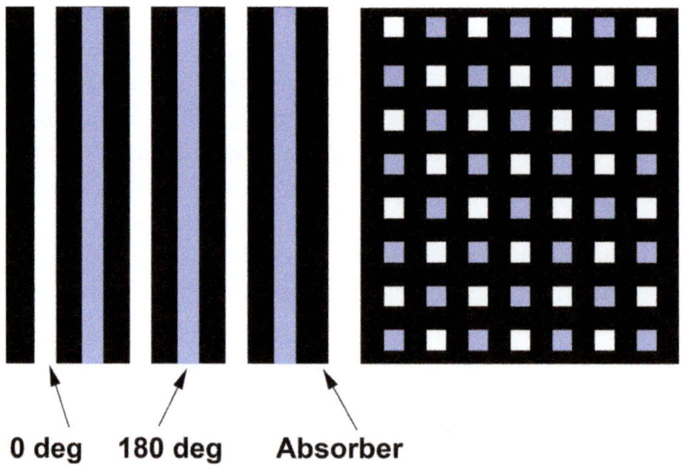

Figure 7.41 Ideal patterns for the AltPSM.

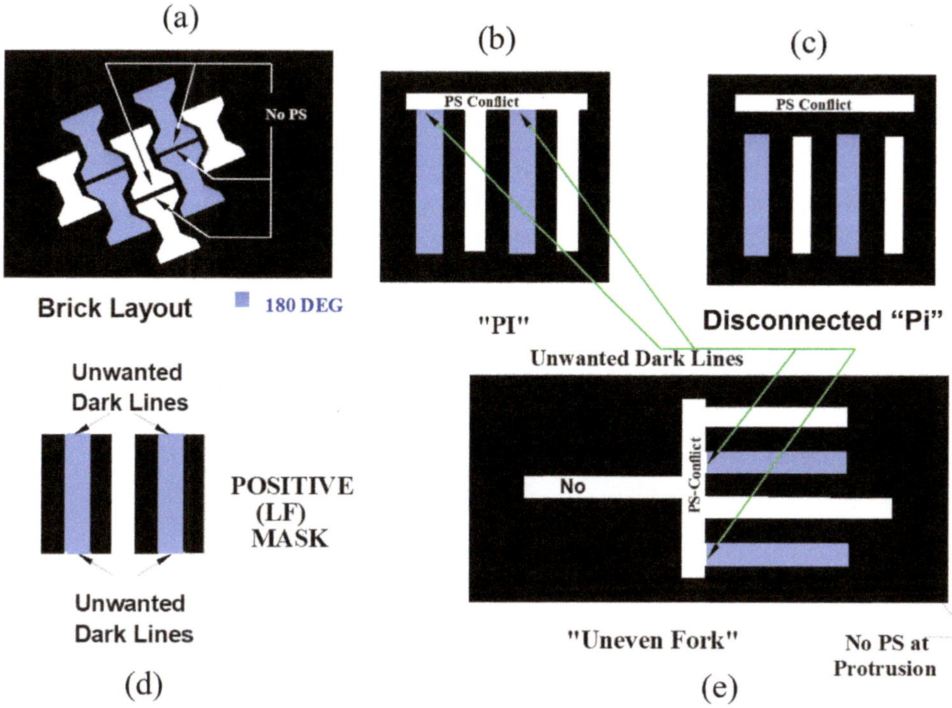

Figure 7.42 Difficulties and phase conflicts with AltPSM.

fabrication also becomes more difficult because the imaging sensitivity to a phase change around 120 deg is higher than that around 180 deg. The "π" ("pi") and disconnected "π" in Figs. 7.42(b) and (c) have conflicts at the horizontal line. Even in the well-behaved line–space pattern shown in Fig. 7.42(d), if a positive, i.e., a light-field (LF), mask is used, redundant dark lines will be delineated at the phase-shifting edges that were inadvertently created. The uneven-fork structure in Fig. 7.42(e) has an isolated opening that suffers from a lack of phase-shifting support, in addition to the conflicts in the "π" structure.

It is understandable that phase-shifting conflicts are unavoidable in arbitrary 2D patterns because there is only one parameter to handle two degrees of freedom. Figure 7.43 demonstrates the phase-shifting conflicts in a metal layer of an IC. On the leftmost side, the line openings are kept without conflict, but when they degenerate into more complicated 2D patterns, conflicts arise.

A dark-field mask, i.e., one in which the metal lines are transparent, consists of openings in a positive resist. Electroplating patterns the metal lines. A subtractive process such as etching can be used to delineate the metal lines so that a light-field mask can be used with a positive resist. This way, the

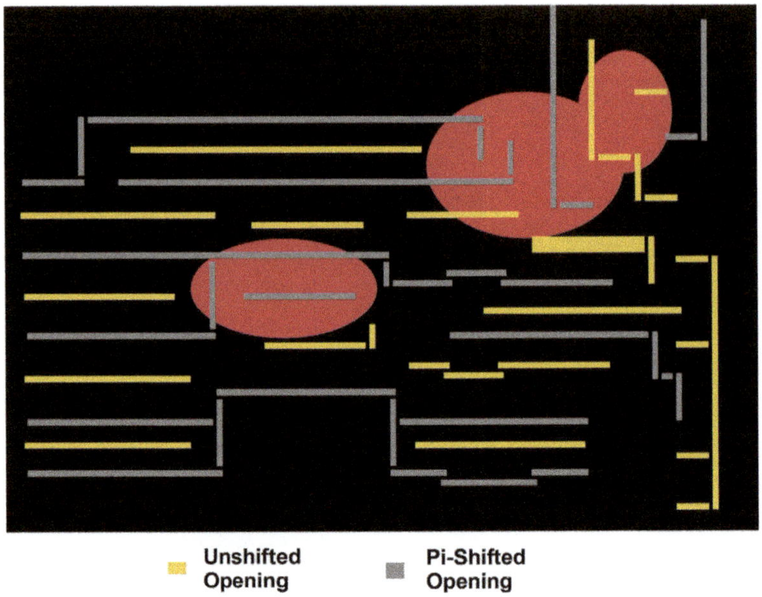

Figure 7.43 Dark-field alternating phase-shifting conflicts in a metal layer.

phase-shifting conflicts can be eliminated with two exposures, as shown in Fig. 7.44. One of the exposures uses the layout and phase-shifting scheme shown, with phase shifting taking place at each opaque line. The edges resulting from the boundary between phase shifts can be trimmed with another mask. The process is illustrated in Fig. 7.45. The layouts of the two masks are superimposed in Fig. 7.46 to show the relative position of the edges of the two layouts. The trim patterns are intentionally placed to trim off some line edges in some cases. In other cases, they are intentionally placed slightly

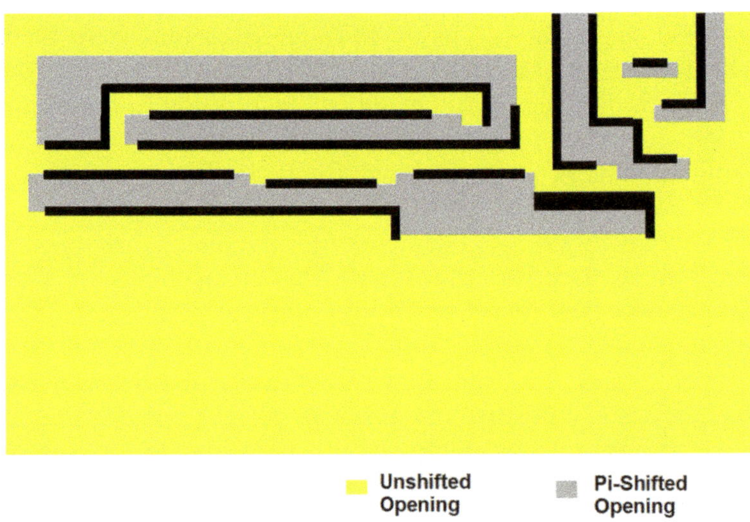

Figure 7.44 Bright-field AltPSM from the upper half of Fig. 7.43.

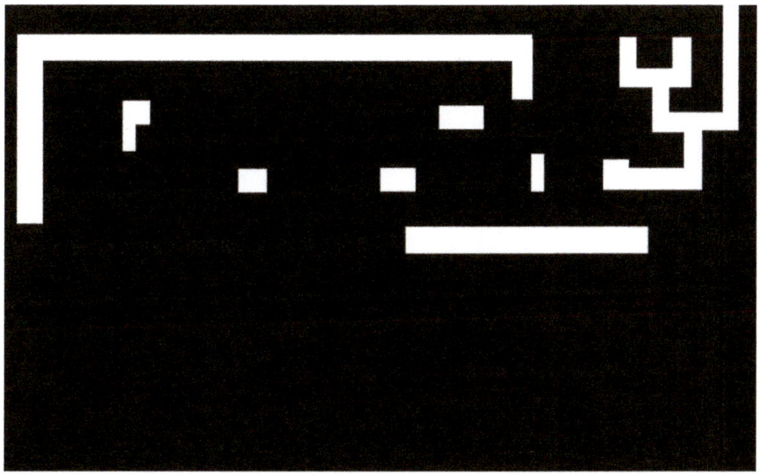

Figure 7.45 Trim mask complementing the mask in Fig. 7.44 to remove the unwanted phase-shifting edges.

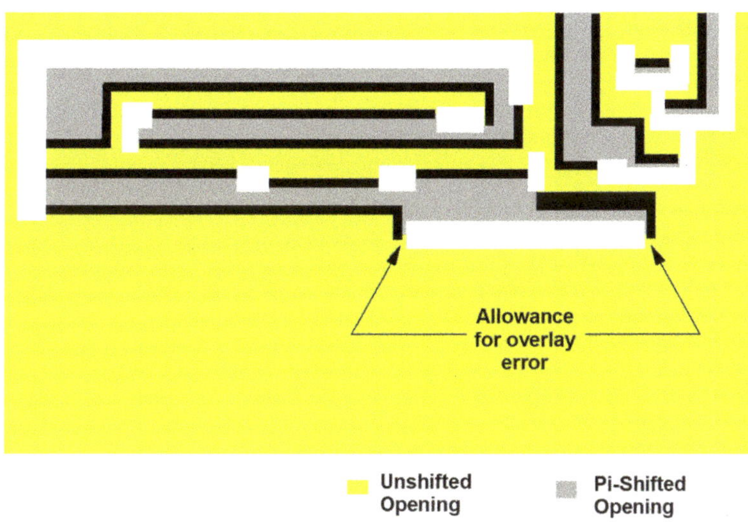

Figure 7.46 The two mask layouts in Figs. 7.44 and 7.45 superimposed.

away from the metal line pattern, leaving a whisker of a metal line from the phase-shifting edge image. These placements are done in anticipation of overlay errors between the two exposures.

A double-exposure scheme[37] has been adopted by several companies for manufacturing the logic gate. This scheme defines the poly-layer pattern in a BIM but with the gates enlarged, as exemplified by Mask 1 in Fig. 7.47. Mask 1 superimposes Mask 2, a covered-edge (CE) phase-shifting trim mask that looks like Fig. 7.44. The gate length is now defined by the trim mask and can be much shorter due to the high image contrast produced by the PSM,

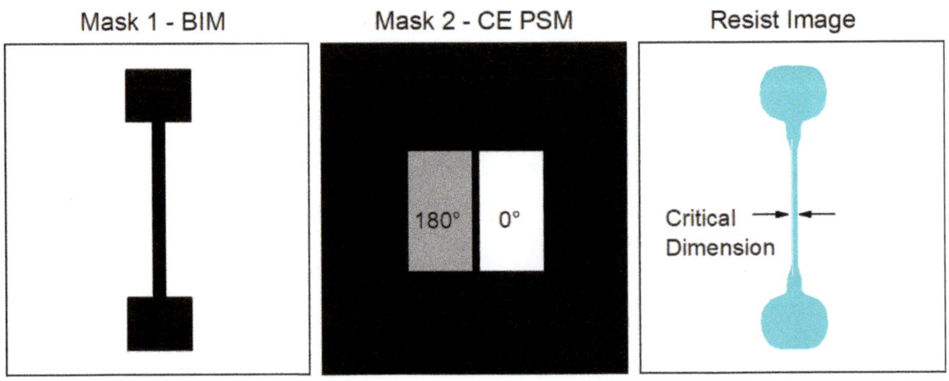

Figure 7.47 Covered phase-shifting edge with a double exposure for the gate level.

which results in narrower gates with better CD uniformity. The type of phase shifting is classified as covered-edge phase shifting according to Section 6.3.4.3. The effect of AltPSM is felt only when the phase-shifting edges get too close to each other.

The imaging performance of dark-field AltPSMs is represented by the DOF versus NA plot in Fig. 7.48. The parameters are as follows: alternately phase-shifted gratings with 1:1, 1:1.5, and 1:2 opening:space ratios; disk illumination at $\sigma = 0.3$; CD tolerance = ±10%; and EL = 8%.

In addition to phase-shifting conflicts, this type of dark-field layout presents another problem for the AltPSM. The shifted and unshifted openings are unevenly spaced, posing a proximity problem that is more severe than with a BIM. The problem is depicted in the following figures. Take 0.35-μm equal line–space pairs, using an AltPSM at λ = 365 nm; its DOF as a function of NA is shown in Fig. 7.49. This is based on 20% EL with ±10% CD

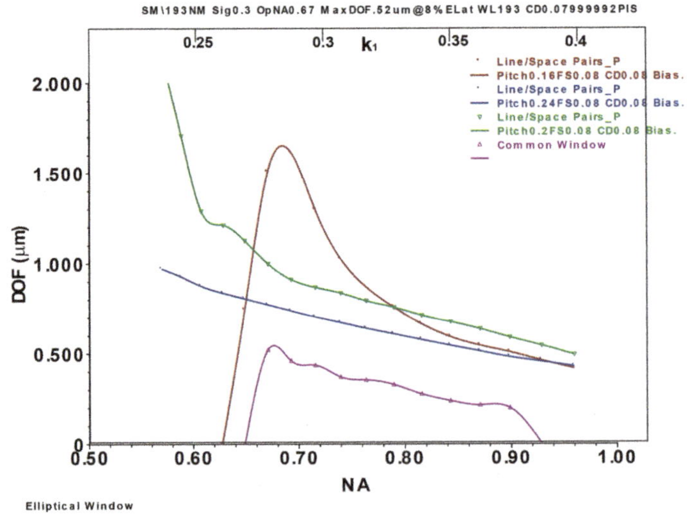

Figure 7.48 DOF versus NA for 80-nm 1:1, 1:1.5, and 1:2 line-opening:opaque-space ratios at 193 nm for an AltPSM.

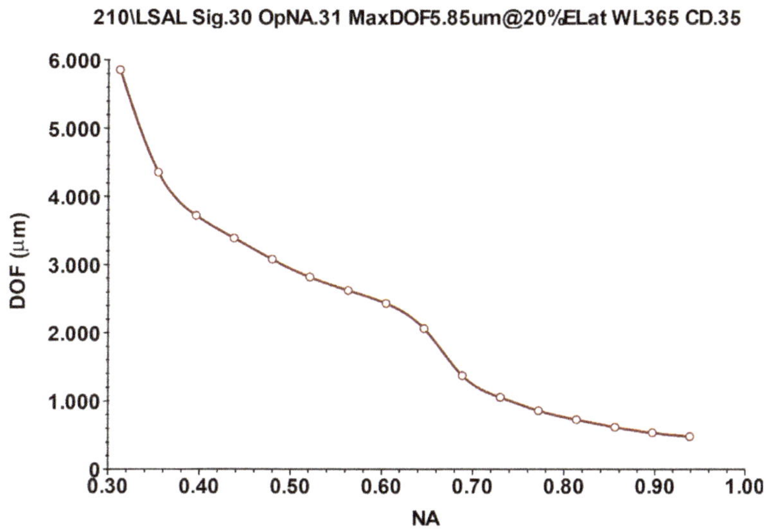

Figure 7.49 DOF versus NA for 0.35-μm line–space pairs at 365 nm for an AltPSM.

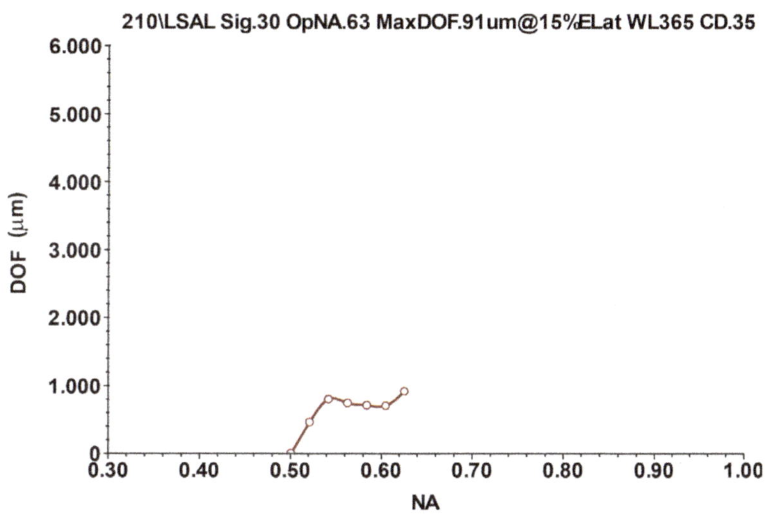

Figure 7.50 DOF versus NA for 0.35-μm 1:1 and 1:2 line–space pairs at 365 nm for an AltPSM.

tolerance. The DOF is very large, except when the NA is larger than 0.65. By simply introducing another set of line–space pairs (1:2 line opening to opaque space), the DOF drops to below 1 μm, as seen in Fig. 7.50. Either feature supports a large DOF, except when they must be imaged together without individual adjustments to the print bias, as seen in Fig. 7.51.

There is no reason that the mask features cannot be made larger or smaller than the target CD on the wafer by using biasing. Figure 7.52 shows the same set of features as Fig. 7.50, except that the 1:2 feature is now biased by 80 nm, i.e., making the line openings larger by 80 nm. Figure 7.53 shows that the E-D

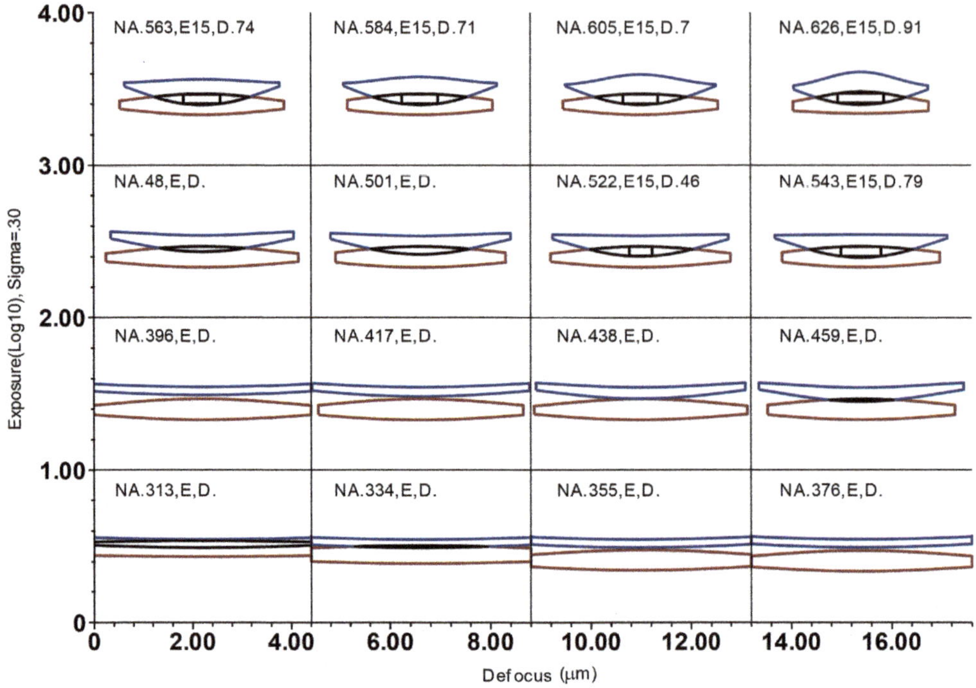

Figure 7.51 E-D windows of Fig. 7.50.

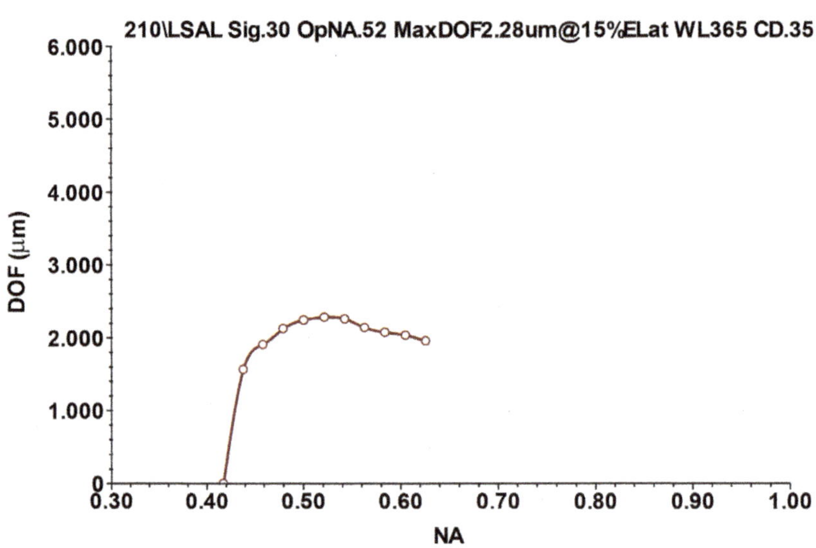

Figure 7.52 DOF versus NA for 0.35-mm 1:1 and 1:2 line–space pairs at 365 nm, AltPSM, 80-nm bias for 1:2.

windows of the 1:2 feature are shifted lower towards the E-D windows of the 1:1 feature, making larger common windows. This is a preliminary form of optical proximity correction.

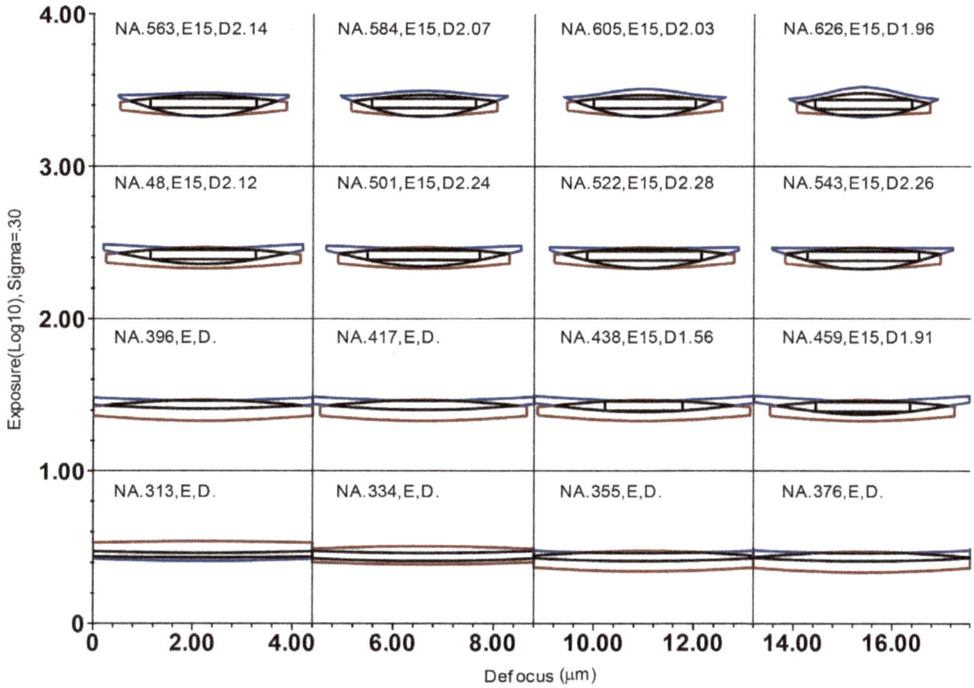

Figure 7.53 E-D windows of Fig. 7.52.

The optical proximity effect is not the only effect that can impact the imaging performance of an AltPSM. The fabrication tolerance of PSMs also has a strong influence. Figure 7.54 shows the E-D windows of $k_1 = 0.46$ line–space pairs from a BIM and an AltPSM. The larger E-D window belongs to the AltPSM. The latter indeed performs better than the former. However, when the fabrication tolerance of $\Delta_{\mathrm{phase}} = 180$ deg ± 10 deg and $\Delta_{\mathrm{transmission}} = 95 \pm 5\%$, the common E-D window of four imperfect E-D trees superimposed

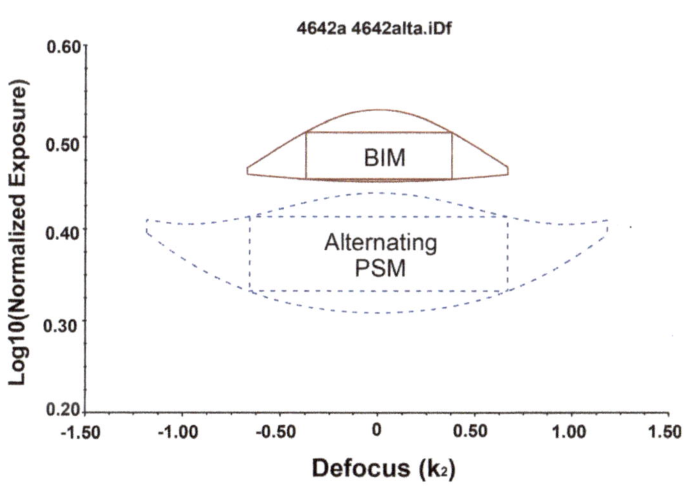

Figure 7.54 AltPSM improvement at $k_1 = 0.46$, $\sigma = 0.42$.

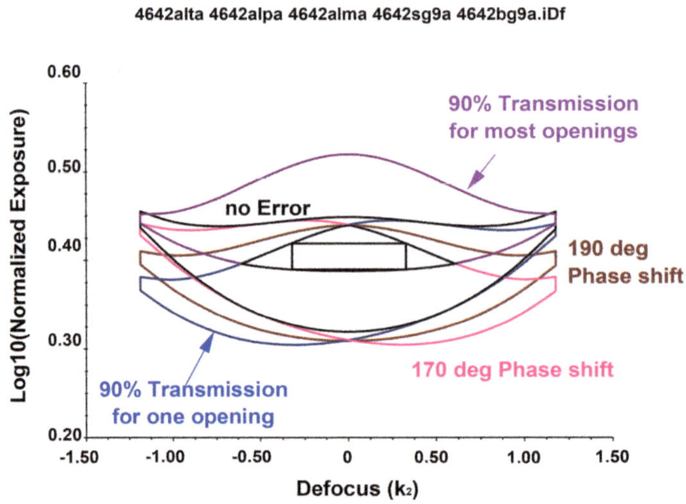

Figure 7.55 AltPSM with phase and transmission errors. The performance is worse than for a BIM.

on the perfect E-D tree is much smaller than the perfect E-D tree by itself, as shown in Fig. 7.55. The assumption for the BIM is perfect transmission and phase consistency. Lateral patterning errors using either mask have been assumed to be identical.

7.3.1.2 AttPSM

The attenuated PSM (AttPSM) is the most widely used PSM for manufacturing. Its configuration and working principle are given in Section 6.3.4.3. The popularity of AttPSMs is mainly due to their simple process and accommodation of arbitrary mask pattern. Commercial mask blanks employ the MoSi absorber that provides the required 6% transmission and π phase shift. So, the delineation of the phase-shifting pattern is like that of a BIM, namely, expose, develop, and etch. However, it is desirable to cover the area outside the patterned area with chromium so that the reticle information for the exposure tool is on chromium instead of on a phase-shifting material. Therefore, commercial mask blanks come with a chromium layer above the MoSi layer. Needless to say, the resist layer is pre-coated on the chromium layer in the mask blank factory just as in the case of BIMs. The processing of such a composite blank is still simple. The chromium image can be used as the hard mask to delineate the MoSi image, followed by a second patterned exposure to remove the chromium at the phase-shifting areas. Because the continuity of the chromium layer is broken by the mask patterns, electrons can be trapped in the isolated conducting areas, thus producing charging effects. Hence, the second exposure uses an optical mask writer instead of an e-beam.

The 6% transmission was defined in the early phase of AttPSM development.[38] Too little transmission makes the phase-shifting effect negligible; too much increases stray light. What's worse is that too much transmission also produces ghost lines at the side lobe of the optical image.

Unlike an AltPSM, which has the potential to double the resolution, an AttPSM is a weak resolution-enhancement technique (RET); it merely enhances the edge contrast. Fortunately, it can be applied to any arbitrary layout, in contrast to the restriction with AltPSMs due to phase conflicts. In the early 1990s, while people were still exploring ways to overcome the restriction of AltPSMs, the rim PSM was invented.[39] This invention immediately revived interest in PSMs. The lithography community has dedicated much effort to developing rim PSMs since then—until the rediscovery of the AttPSM[40,41] for x-ray lithography.

Figure 7.56 is an analysis of the imaging performance based on the E-D methodology for an AttPSM and a rim PSM using an isolated line opening whose width in k_1 typically ranges between 0.3 and 0.7. FDOF stands for the figure of merit for the depth of focus (DOF). From Eq. (5.2),

$$k_2 = DOF\frac{NA^2}{n\lambda_0}, \tag{7.18}$$

where n is the refractive index of the imaging medium, and λ_0 is the imaging wavelength in vacuum. Here k_2 is treated as a dimensionless normalized DOF. Also, from Eq. (5.1),

$$k_1 = MFS\frac{NA}{\lambda_0}, \tag{7.19}$$

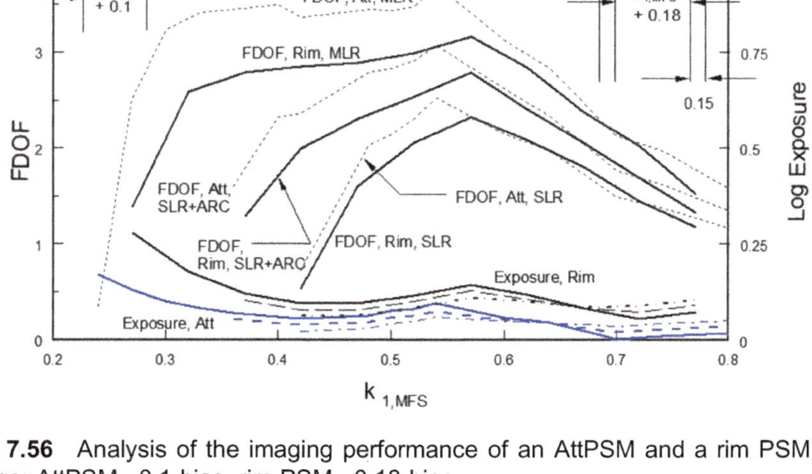

Figure 7.56 Analysis of the imaging performance of an AttPSM and a rim PSM for line openings: AttPSM –0.1 bias, rim PSM –0.18 bias.

where k_1 is treated as a dimensionless, normalized minimum feature size (MFS):

$$FDOF \equiv \frac{k_2}{k_1^2} = DOF \frac{\lambda_0}{nMFS^2}, \tag{7.20}$$

which is the normalized DOF with respect to the MFS with respect to the imaging wavelength in vacuum.

Figure 7.56 shows that the AttPSM attains a larger DOF with a smaller mask bias and requires less exposure within this wide k_1 range. The three groups of DOF curves—single-layer resist (SLR), SLR+ARC, and multilayer resist (MLR)—represent ELs of 30%, 20%, and 10% respectively. These exposure latitudes were used to construct the E-D trees and windows for the comparison. Not only does the rim PSM main feature have a 0.18 k_1 bias against 0.1 k_1 with AttPSMs, but it needs an additional mask area for a rim width of 0.15 k_1, thus severely impacting the packing density. AttPSMs also have the inherent advantage of process simplicity and process control.

Figure 7.57 shows the imaging performance of contact holes with these two types of PSM using just the 10% EL. The AttPSM excels again in DOF, image intensity, and packing density. It is no surprise that the use of AttPSMs for contact holes is widespread.

The improvement of the contact-hole DOF with 8% AttPSM over BIM is shown in Fig. 7.58. The holes are 0.35 μm in size and the wavelength is 365 nm. On-axis illumination at σ = 0.42 is used. The improvement is about 20% over the range of evaluated NA values. The –50 nm bias achieves a larger DOF in both the AttPSM and the BIM. The EL is set at 20%.

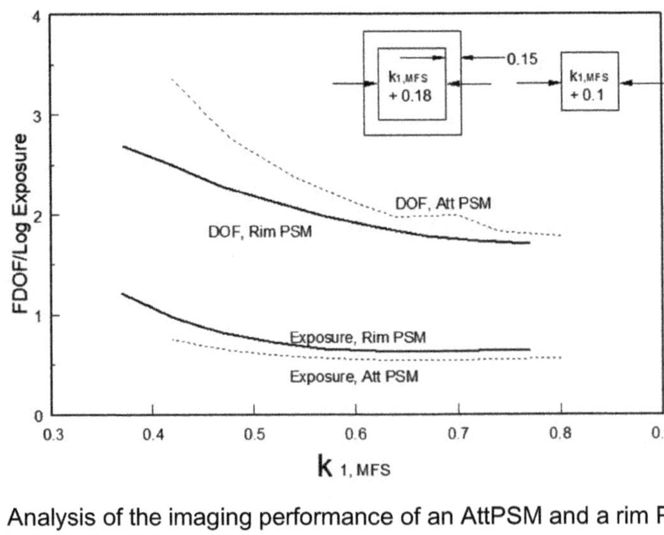

Figure 7.57 Analysis of the imaging performance of an AttPSM and a rim PSM for contact holes: AttPSM –0.1, Rim PSM –0.18 bias.

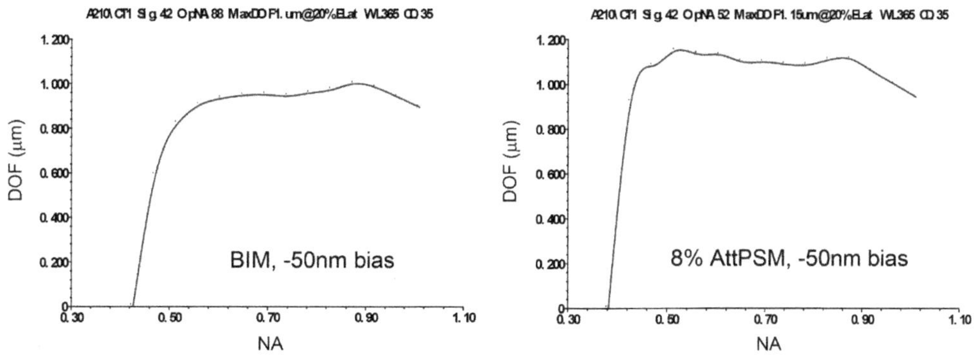

Figure 7.58 Comparison of 0.35-μm (left) BIM PSM and AttPSM (right) contact holes with NA optimization, *EL* = 20%.

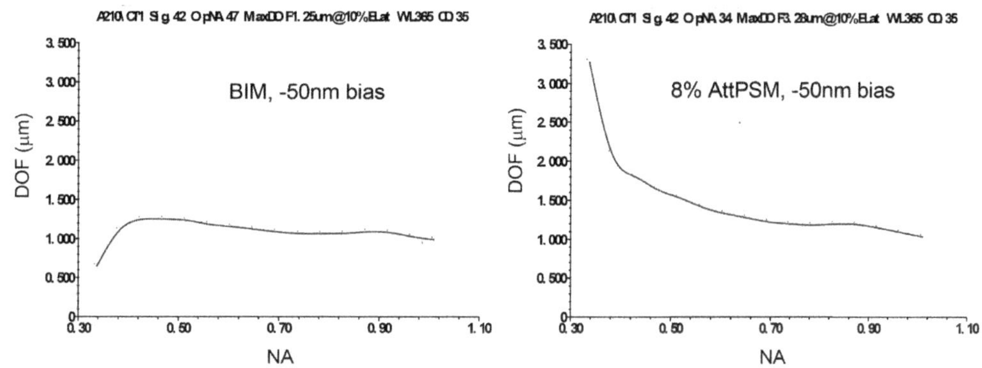

Figure 7.59 Same as Fig. 7.58, except with *EL* = 10%.

When the EL is changed to 10%, the DOF improves with BIMs and AttPSMs. At $NA = 0.34$, corresponding to $k_1 = 0.355$, the improvement in DOF with an AttPSM is dramatic, as shown in Fig. 7.59. However, this potential to improve the DOF is seldom taken advantage of for two reasons. (1) Using such a low k_1 significantly increases the MEEF, making CD control difficult. (2) The resist system does not have sufficient contrast to realize the aerial image.

The resist contrast phenomenon is illustrated in Fig. 7.60, which turns the aerial image result into that of the resist image by using lumped parameters. The parameters chosen for this resist are $\alpha = 1.12\ \mu m^{-1}$, $\gamma = 5$, and resist thickness = 770 nm, which are fairly typical values. The extraordinary gain in DOF disappears with the resist image.

Changing α to 2 does not seem to help, as seen in Fig. 7.61(a) However, when γ is set to 25, which is rarely achieved in commercial resist systems, the dramatic gain in DOF is restored, as seen in Fig. 7.61(b).

The AttPSM is often used with axial illumination at a low σ for contact holes. For line–space patterns, the AttPSM works best combined with off-axis

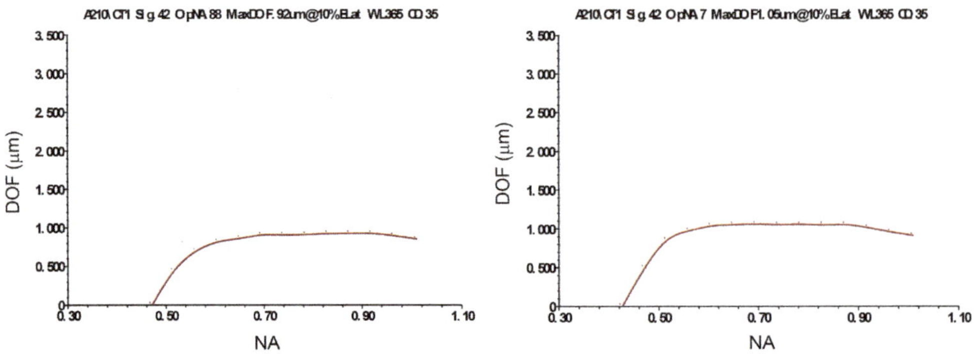

Figure 7.60 Same as Fig. 7.59, except that lumped parameters are used to simulate the resist image instead of using an aerial image; resist thickness = 0.77 µm, α = 1.12, γ = 5.

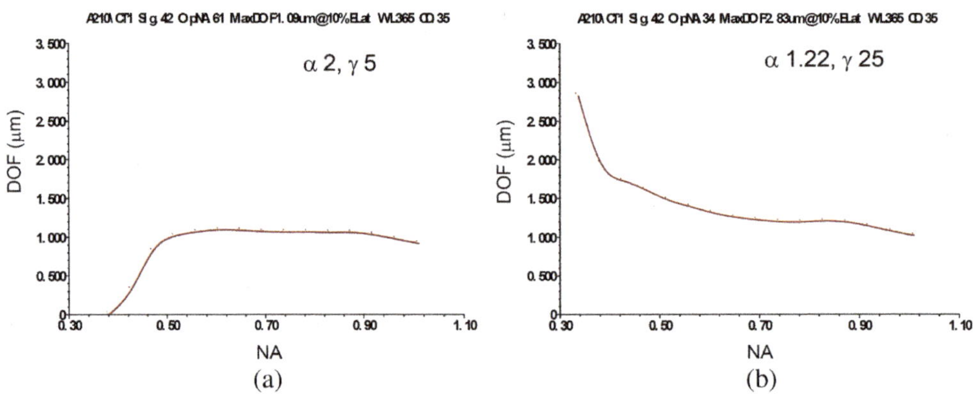

Figure 7.61 The AttPSM part of Fig. 7.59 with (a) α changed to 2 and (b) γ changed to 25.

illumination to enhance the DOF of closely packed lines and subresolution assist features to improve the isolated lines. This line–space application using off-axis illuminations and AttPSM appears innumerously in Chapters 7 to 9.

7.3.2 Off-axis illumination

Off-axis illumination (OAI)[42,43] was proposed in the early 1990s to supplement or replace the phase-shifting mask technology. OAI has the advantage of lower cost, but it has its own limitations. The following subsections explain the working principle of OAI, provide an analysis of the technique, and simulate its imaging performance.

7.3.2.1 Conceptual illustration

OAI employs shifting of the diffraction angle of a given object to improve the resolution and DOF. It can best be illustrated with a simple grating object. The grating, being periodic, contains discrete spatial frequency components, namely, the 0th, $\pm 1^{st}$, $\pm 2^{nd}$, ... orders. When the minimum resolvable

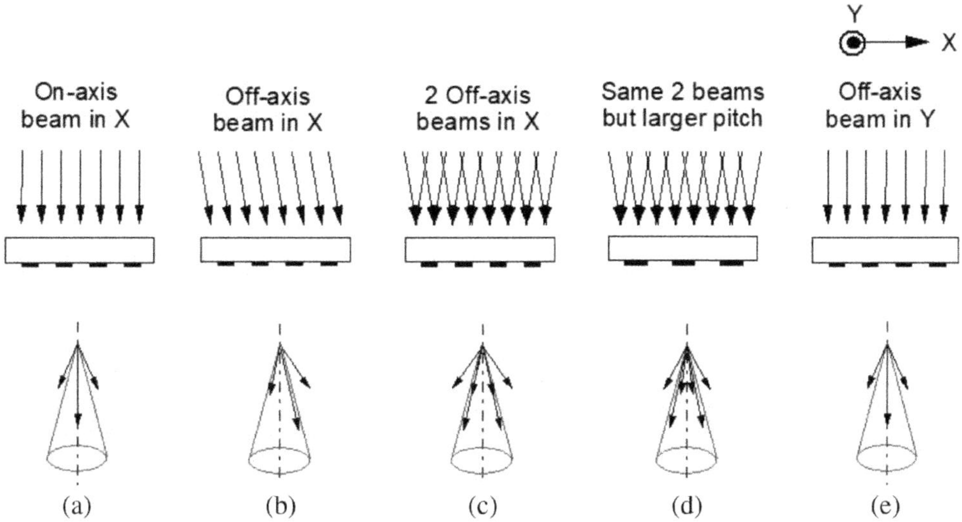

Figure 7.62 (a) On-axis and (b)–(e) off-axis illuminations affecting the 0th- and $\pm1^{st}$-order spatial frequency vectors.

periodicity is of interest, only the 1st-order frequencies are preserved so that the resolution potential of the imaging lens can be fully utilized. This situation is shown in Fig. 7.62(a), where DKI is illustrated. The spatial frequency spectrum consists of the vertically oriented 0th-order beam and the $\pm1^{st}$-order beams. An image of the fundamental frequency is formed with these three beams. Figure 7.62(a) depicts a normally incident, on-axis illumination, known as three-beam imaging. The angle of the $\pm1^{st}$-order beams is a function of the periodicity (pitch) of the grating. A smaller pitch produces a higher spatial frequency and thus larger diffraction angles in the $\pm1^{st}$-order beams. When the feature size is too small, i.e., the spatial frequency is too high, the angle of the diffracted beams becomes larger than the acceptance angle of the imaging lens, and the $\pm1^{st}$-order beams are rejected. Only the 0th-order beam passes. The image becomes a structureless uniform beam.

The electric field on a BIM is simply taken as 1 and 0 at the transparent and opaque parts of the mask, respectively. The actual field is obviously much more complicated when exact electromagnetic diffraction is considered.[44] Some of the findings in Ref. 44 are shown in Fig. 4.13 and are experimentally demonstrated in Fig. 4.14. Here, the simple assumption of a binary image from the mask suffices to explain the working principle of OAI. The intensity at the mask plane is proportional to the square of the simplified electric field and is also 1 and 0, in complete accordance with the electric field.

In Fig. 7.62(a), the intensity at the image plane is uniform, just as if no object were present, because the $\pm1^{st}$-order beams are not accepted by the lens. Only the 0th-order beam, which has no information content, passes through the lens. The intensity of the image is reduced because the energy of the rejected beams is not recovered.

Figure 7.62(b) shows the situation of a single collimated illumination beam that is obliquely incident on the mask and thus is off-axis. The three beams shown in Fig. 7.62(a) are now tilted by the incident angle of the illumination. When the angle is adjusted to make the 0th-order and one of the 1st-order beams symmetrical with respect to the optical axis, the largest angular spread between the two beams is possible with neither of these beams being cut off by the acceptance angle of the lens; thus, the highest resolution is achieved. This is known as two-beam imaging. Note that the other 1st-order beam is cut off, resulting in a lower exposing intensity.

In Fig. 7.62(c), two symmetrically tilted beams are used. When the angle of the illumination is optimized for a given periodic object, the 0th order of the left beam coincides with the 1st order of the right beam, and the 0th order of the right beam coincides with the -1st order of the left beam, as shown in the figure. The image consists of a single-frequency component and is well reproduced.

When objects of a lower spatial frequency, as shown in Fig. 7.62(d), are illuminated by the same set of beams as in Fig. 7.62(c), the spatial frequency vectors no longer coincide, producing additional spatial frequencies. These frequencies can induce worse imaging characteristics compared those in on-axis illumination. Figure 7.62(e) shows that tilting the beam in *X* does not help to resolve the spatial frequencies in *Y*.

7.3.2.2 Analysis of three-beam and two-beam images*

7.3.2.2.1 Analysis of three-beam images

Figure 7.63 is a detailed rendering of Fig. 7.62 without 7.62(e). The notation for the various angles is designated here to start the image analysis. The analytic forms reveal many special characteristics of OAI. The mth-order diffraction angle θ_m follows the grating diffraction equation:[62]

$$n_{tran} \sin\theta_m = n_{inc} \sin\theta_{inc} + m\lambda_0/P, \tag{7.21}$$

where n_{tran} is the refractive index of the transmitting medium, and n_{inc} is that of the incident medium.

First, let's examine the case of axial illumination, as shown in Fig. 7.63(a). The 0th-order and $\pm 1^{st}$-order beams are also shown in Fig. 7.64 and in Eqs. (7.22) to (7.24):

$$E_0(x, z) = A_0 e^{ikz} \tag{7.22}$$

$$E_1(x, z) = A_1 e^{ik(x \sin\theta_1 + z \cos\theta_1)} \tag{7.23}$$

*Extensive insights for this section have been contributed by S.Y. Chou.

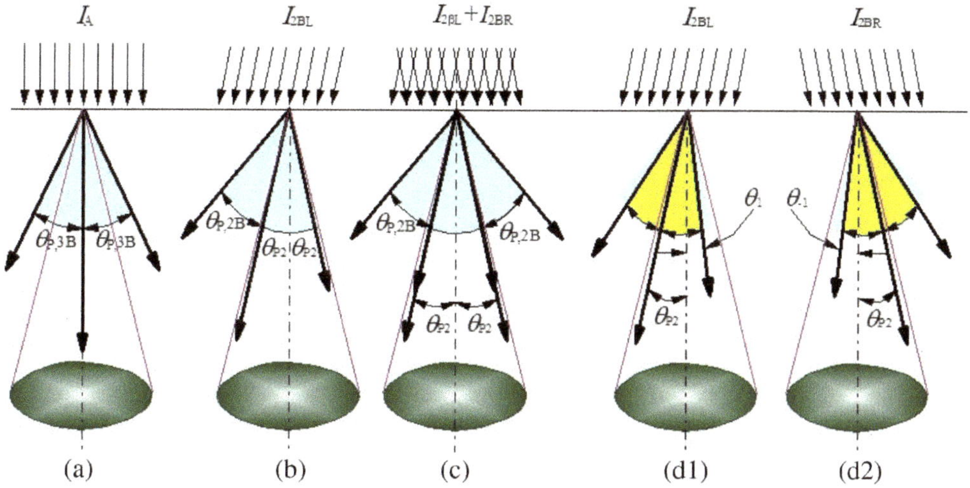

Figure 7.63 Detailed presentation of Fig. 7.62 without 7.62(e). The angles are named according to the discussion in Section 7.3.2.2. I_A is on-axis illumination; I_{2BL} is illumination pointing to the left, and I_{2BR} is illumination pointing to the right. $\theta_{P,3B}$ and $\theta_{P,2B}$ are determined by the grating diffraction equation and are not identical.

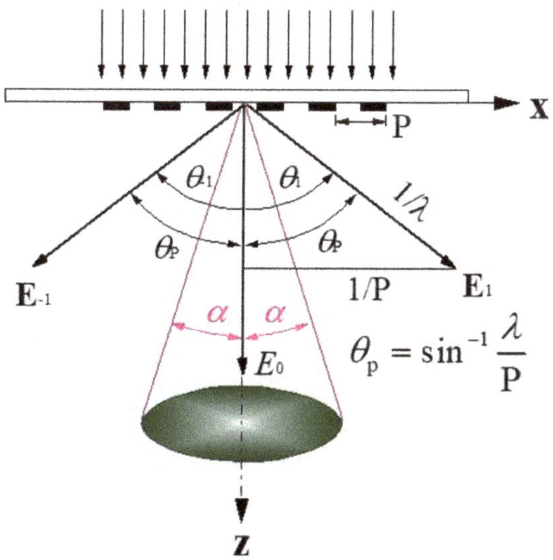

Figure 7.64 Three beams diffracted from a grating that makes $\theta_1 = \theta_{-1} = \theta_P$. α is the aperture angle of the imaging lens.

$$E_{-1}(x, z) = A_{-1}e^{ik(x\sin\theta_{-1}+z\cos\theta_{-1})}. \tag{7.24}$$

In this chapter, we only consider imaging in air, whose refractive index is close to 1: $n_{\text{tran}} = n_{\text{inc}} = 1$. For the 0th-order beam ($m = 0$), $\sin\theta_0 = \sin\theta_{\text{inc}}$. For the 1st-order beams ($m = \pm 1$), $\sin\theta_{\pm 1} = \sin\theta_{inc} \pm \lambda/P$. For three-beam imaging, $\theta_{inc} = 0$, $\sin\theta_1 = \lambda/P$, and $\sin\theta_{-1} = -\lambda/P$.

The diffraction angle of the 1st-order beam is θ_P, which is strictly a function of the incident wavelength λ and the periodicity P of the grating according to

$$\theta_1 = -\theta_{-1} = \theta_P \equiv \sin^{-1}\frac{\lambda}{P}. \tag{7.25}$$

After setting $A_1 = A_{-1}$, the diffracted field E_{3B} and the resultant intensity I_{3B} of this three-beam system are

$$\begin{aligned} E_{3B} &= E_0 + E_1 + E_{-1} \\ &= A_0 e^{ikz} + A_1 e^{ik(x\sin\theta_1 + z\cos\theta_1)} + A_1 e^{ik(x\sin\theta_{-1} + z\cos\theta_{-1})} \\ &= A_0 e^{ikz} + A_1 e^{ikz\cos\theta_P}\left(e^{i2\pi x/P} + e^{-i2\pi x/P}\right) \\ &= A_0 e^{ikz} + 2A_1 e^{ikz\cos\theta_P}\cos(2\pi f x), \end{aligned} \tag{7.26}$$

where $f = 1/P$ is the spatial frequency of the image; and

$$\begin{aligned} I_{3B} &= E_{3B}^* E_{3B} \\ &= [A_0 e^{-ikz} + 2A_1 e^{-ikz\cos\theta_P}\cos(2\pi f x)][A_0 e^{ikz} + 2A_1 e^{ikz\cos\theta_P}\cos(2\pi f x)] \\ &= A_0^2 + 4A_1^2\cos^2(2\pi f x) + 2A_0A_1\cos(2\pi f x)[e^{-ikz(1-\cos\theta_P)} + e^{ikz(1-\cos\theta_P)}] \\ &= A_0^2 + 4A_1^2\cos^2(2\pi f x) + 4A_0A_1\cos(2\pi f x)\cos[kz(1-\cos\theta_P)] \\ &= A_0^2 + 2A_1^2 + 2A_1^2\cos(4\pi f x) + 4A_0A_1\cos(2\pi f x)\cos[kz(1-\cos\theta_P)]. \end{aligned} \tag{7.27}$$

The fundamental spatial frequency in the image is

$$f_{1,3B} = \frac{1}{P}, \tag{7.28}$$

and that of the second harmonic is

$$f_{2,3B} = \frac{2}{P}, \tag{7.29}$$

corresponding to the pitch of the fundamental and the second harmonic frequencies,

$$P_{1,3B} = \lambda/\sin\theta_P \tag{7.30}$$

and

$$P_{2,3B} = \lambda/2\sin\theta_P, \tag{7.31}$$

respectively. To estimate the DOF, take $|kz(1-\cos\theta_P)| < \pi/2$ as the upper bound of the DOF; then,

$$\begin{aligned} & kz(1-\cos\theta_P) < \pi/2 \\ & DOF_{\text{upper bound}} < \frac{\pi}{2k(1-\cos\theta_P)} = \frac{\lambda}{4(1-\cos\theta_P)}. \end{aligned} \tag{7.32}$$

7.3.2.2.2 Two OAI beams

We now consider the two off-axis illuminated beams. When the incident beams are tilted by an angle θ, then

$$\theta_0 = \theta. \tag{7.33}$$

From the grating diffraction equation,

$$\sin\theta_1 = \sin\theta + \lambda/P, \tag{7.34}$$

$$\sin\theta_{-1} = \sin\theta - \lambda/P, \tag{7.35}$$

$$E_0(x, z) = A_0 e^{ik(x\sin\theta_0 + z\cos\theta_0)}, \tag{7.36}$$

$$E_1(x, z) = A_1 e^{ik(x\sin\theta_1 + z\cos\theta_1)}, \tag{7.37}$$

and

$$E_{-1}(x, z) = A_{-1} e^{ik(x\sin\theta_{-1} + z\cos\theta_{-1})}, \tag{7.38}$$

as shown in Fig. 7.65.

Let I_{2BR} be the intensity of the combination of E_0 and E_{-1} that passes through the imaging lens. θ_0 is now pointing to the right:

$$\begin{aligned} I_{2BR}(x,z) &= [E_0(x,z)+E_{-1}(x,z)]^*[E_0(x,z)+E_{-1}(x,z)] \\ &= A_0^2 + A_{-1}^2 \\ &\quad + A_0A_{-1}(e^{-ikx\sin\theta_0 - ikz\cos\theta_0}e^{ikx\sin\theta_{-1} + ikz\cos\theta_{-1}} + e^{ikx\sin\theta_0 + ikz\cos\theta_0}e^{-ikx\sin\theta_{-1} - ikz\cos\theta_{-1}}) \\ &= A_0^2 + A_{-1}^2 \\ &\quad + A_0A_{-1}[e^{-ikx(\sin\theta_0 - \sin\theta_{-1}) - ikz(\cos\theta_0 - \cos\theta_{-1})} + e^{ikx(\sin\theta_0 - \sin\theta_{-1}) + ikz(\cos\theta_0 - \cos\theta_{-1})}] \\ &= A_0^2 + A_{-1}^2 + 2A_0A_{-1}\cos[kx(\sin\theta_0 - \sin\theta_{-1}) + kz(\cos\theta_0 - \cos\theta_{-1})]. \end{aligned} \tag{7.39}$$

Similarly, let I_{2BL} be the intensity of the combination of E_0 and E_1 that passes through the imaging lens. θ_0 is now pointing to the left:

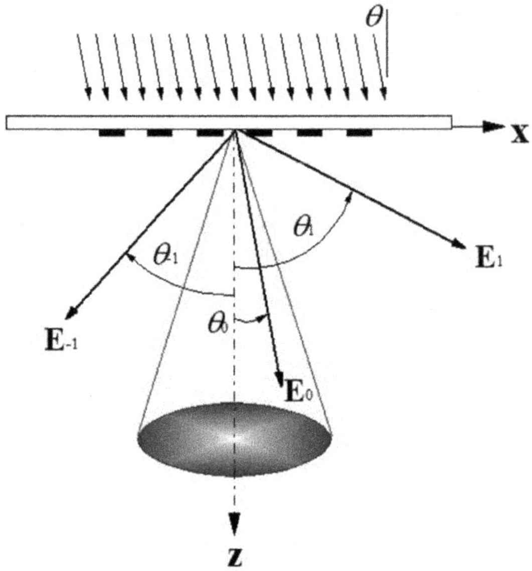

Figure 7.65 Tilting the beams by θ_0 for two-beam imaging.

$$\begin{aligned}I_{2BL}(x,z) &= [E_0(x,z)+E_1(x,z)]^*[E_0(x,z)+E_1(x,z)]\\ &= A_0^2+A_1^2\\ &+A_0A_1(e^{-ikx\sin\theta_0-ikz\cos\theta_0}e^{ikx\sin\theta_1+ikz\cos\theta_1}+e^{ikx\sin\theta_0+ikz\cos\theta_0}e^{-ikx\sin\theta_1-ikz\cos\theta_1})\\ &= A_0^2+A_1^2\\ &+A_0A_1(e^{-ikx\sin\theta_0+ikx\sin\theta_1-ikz\cos\theta_0+ikz\cos\theta_1}+e^{ikx\sin\theta_0-ikx\sin\theta_1+ikz\cos\theta_0-ikz\cos\theta_1})\\ &= A_0^2+A_1^2+2A_0A_1\cos[kx(\sin\theta_0-\sin\theta_1)+kz(\cos\theta_0-\cos\theta_1)].\end{aligned} \tag{7.40}$$

7.3.2.2.3 Two symmetrical beams

When the beams are rotated counterclockwise so that θ_0 and θ_1 are symmetrical with respect to the optical axis,

$$\theta_0 = \sin^{-1}\frac{\lambda}{2P} \text{ and } \theta_{-1} = -\sin^{-1}\frac{\lambda}{2P},$$

$$\cos\theta_0 - \cos\theta_{-1} = 0,$$

$$\sin\theta_0 - \sin\theta_{-1} = \frac{\lambda}{P},$$

and

$$I_{2BR}(x, z) = A_0^2 + A_{-1}{}^2 + 2A_0A_{-1}\cos\left(\frac{2\pi}{P}x\right). \tag{7.41}$$

This situation is depicted in Fig. 7.63(b). The spatial frequency of this two-beam image is

$$f_{2B} = \frac{1}{P}, \tag{7.42}$$

and its pitch is

$$P_{2B} = P. \tag{7.43}$$

The DOF is infinite and telecentric because I_{2BR} is independent of z, as shown in Fig. 7.66.

Similarly, when the beams are rotated clockwise to be symmetrical such that

$$\theta = -\sin^{-1}\lambda/P,$$

$$\theta_0 = -\sin^{-1}\frac{\lambda}{2P}\text{ and }\theta_1 = \sin^{-1}\frac{\lambda}{2P},$$

and

$$I_{2BL}(x, z) = A_0^2 + A_{-1}{}^2 + 2A_0A_{-1}\cos\left(\frac{2\pi}{P}x\right). \tag{7.44}$$

The spatial frequency is again $1/P$. I_{2BL} is independent of z, and the DOF is also infinite, as shown in Fig. 7.66.

In order for the two beams making angles $\theta_0 = \sin^{-1}(\lambda/2P)$ and $\theta_{-1} = -(\lambda/2P)$ to pass through the imaging lens, the lens needs an acceptance angle of $2\sin^{-1}(\lambda/2P)$, i.e., with a numerical aperture of $\lambda/2P$. Whereas, in the three-beam case, the acceptance angle of the imaging lens has

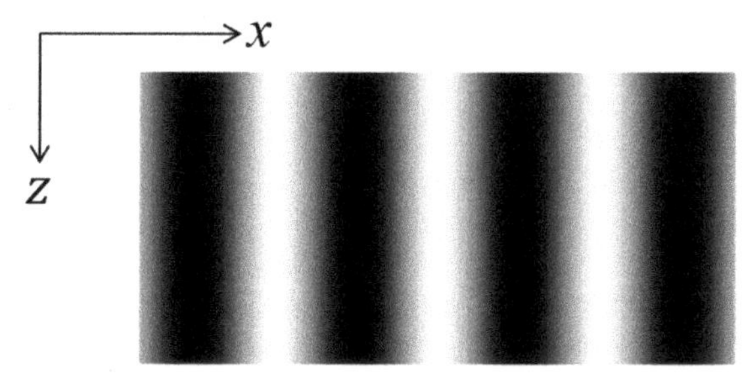

Figure 7.66 Telecentric image with an infinite DOF. The two beams are symmetric with respect to the optical axis (image courtesy of S.Y. Chou).

to be $2\sin^{-1}(\lambda/P)$, i.e., with a numerical aperture of λ/P, which is twice as large. Therefore, with two-beam imaging, a given imaging lens can resolve a spatial frequency that is twice as high as that possible with three-beam imaging. The resolution power of two-beam imaging is doubled. This is illustrated by the numerical example in Section 7.3.2.2.6.

7.3.2.2.4 Two asymmetrical beams due to non-optimized pitches

For a mask with a mixture of spatial frequencies, such as in many production IC designs, the off-axis angle is always optimized for the highest spatial frequency. Hence, the illumination is set at this angle. For pitches larger than the optimum pitch, the two beams are not symmetrical with respect to the optical axis of the imaging lens, as shown in Figs. 7.63(d1) and 7.63(d2), where the conditions of the illumination to the left and to the right are depicted separately. Let us analyze the two cases one by one.

In the case of I_{2BR}, corresponding to Fig. 7.63(d1), from Eq. (7.39),

$$I_{2BR}(x,z)=A_0^2+A_{-1}{}^2+2A_0A_{-1}\cos\left[kx(\sin\theta_0-\sin\theta_{-1})+kz(\cos\theta_0-\cos\theta_{-1})\right].$$

Let θ in Eq. (7.25) remain at $\theta = \sin^{-1}\lambda/2P$; now the grating has a larger pitch: $P_1 > P_2$:

$$\theta_0 = \sin^{-1}\lambda/2P. \tag{7.45}$$

θ_{-1} in Eq. (7.35) becomes

$$\sin\theta_{-1} = \sin\theta_0 - \lambda/P_1. \tag{7.46}$$

Substituting Eqs. (7.45) and (7.46) into Eq. (7.39),

$$\begin{aligned}I_{2BR}(x,z)&=A_0^2+A_{-1}{}^2+2A_0A_{-1}\cos\left[kx(\sin\theta_0-\sin_{-1})+kz(\cos_0-\cos_{-1})\right]\\&=A_0^2+A_{-1}{}^2+2A_0A_{-1}\cos\left[kx\lambda/P_1+kz(\cos\theta_0-\cos\theta_{-1})\right]\\&=A_0^2+A_{-1}{}^2+2A_0A_{-1}\cos2\pi\left[f_1x+\frac{z}{\lambda}(\cos\theta_0-\cos\theta_{-1})\right]\\&=A_0^2+A_{-1}{}^2\\&\quad+2A_0A_{-1}\cos2\pi\left[f_1x+\frac{z}{\lambda}\sqrt{1-(\lambda/2P)^2}-\sqrt{1-(\lambda/2P-\lambda/P_1)^2}\right].\end{aligned} \tag{7.47}$$

From Fig. 7.63(d1), $\cos\theta_0-\cos\theta_{-1}<0$; i.e., this image travels from the left to the right as z increases. The angle of travel is a function of P_1. When $P_1 = P$, the image is telecentric.

Next, to evaluate I_{2BL}, let $\theta=-\sin^{-1}\lambda/2P$. Then $\theta_0=-\sin^{-1}\lambda/2P$, $\sin\theta_1=\sin\theta_0+\lambda/P_1$, and

$$
\begin{aligned}
I_{2BL}(x,z) &= A_0^2 + A_{-1}{}^2 \\
&\quad + 2A_0A_1\cos[kx(\sin\theta_0 - \sin\theta_1) + kz(\cos\theta_0 - \cos\theta_1)], \\
&\quad A_0^2 + A_{-1}{}^2 + 2A_0A_1\cos[kx\lambda/P_1 + kz(\cos\theta_0 - \cos\theta_1)] \\
&= A_0^2 + A_{-1}{}^2 + 2A_0A_1\cos 2\pi\left[-f_1x + \frac{z}{\lambda}(\cos\theta_0 - \cos\theta_1)\right] \\
&= A_0^2 + A_{-1}{}^2 \\
&\quad + 2A_0A_1\cos 2\pi\left[-f_1x + \frac{z}{\lambda}\sqrt{1-(\lambda/2P)^2} - \sqrt{1-(-\lambda/2P+\lambda/P_1)^2}\right].
\end{aligned}
\tag{7.48}
$$

From Fig. 7.63(d1), $\cos\theta_0 - \cos\theta_1 < 0$; i.e., this image travels from the right to the left as z increases.

With only I_{2BR} or I_{2BL}, neither image is telecentric any longer. The two images travel in opposite directions because of the difference in the sign of kz. This case of I_{2BL} is depicted in Fig. 7.67.

7.3.2.2.5 Combining two asymmetrical beams

This image shift is not acceptable for overlay control. As a result, the illumination is made symmetrical by combining I_R and I_L incoherently:

$$
\begin{aligned}
I_{2BA} &= I_{2BR} + I_{2BL} \\
&= 2A_0^2 + 2A_1^2 \\
&\quad + 2A_0A_1\cos 2\pi\left[f_1x + (z/\lambda)(\sqrt{1-(\lambda/2P)^2} - \sqrt{1-(\lambda/2P-\lambda/P_1)^2})\right] \\
&\quad + 2A_0A_1\cos 2\pi\left[-f_1x + (z/\lambda)(\sqrt{1-(\lambda/2P)^2} - \sqrt{1-(-\lambda/2P+\lambda/P_1)^2})\right] \\
&= 2A_0^2 + 2A_1^2 + 2A_0A_1\cos 2\pi(f_1x + Cz) + 2A_0A_1\cos 2\pi(-f_1x + Cz) \\
&= 2A_0^2 + 2A_1^2 \\
&\quad + 2A_0A_1\begin{bmatrix}\cos(2\pi f_1x)\cos(2\pi Cz) - \sin(2\pi f_1x)\sin(2\pi Cz) \\ +\cos(-2\pi f_1x)\cos(2\pi Cz) - \sin(-2\pi f_1x)\sin(2\pi Cz)\end{bmatrix} \\
&= 2A_0^2 + 2A_1^2 + 4A_0A_1\cos(2\pi f_1x)\cos(2\pi Cz),
\end{aligned}
\tag{7.49}
$$

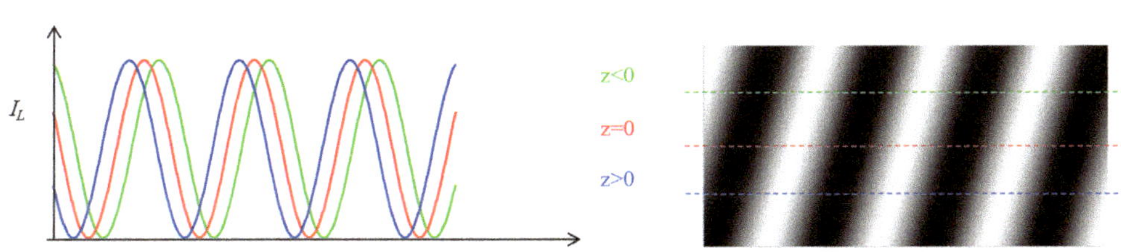

Figure 7.67 The image shifts to the left as z increases when the illumination is I_L (image courtesy of S.Y. Chou).

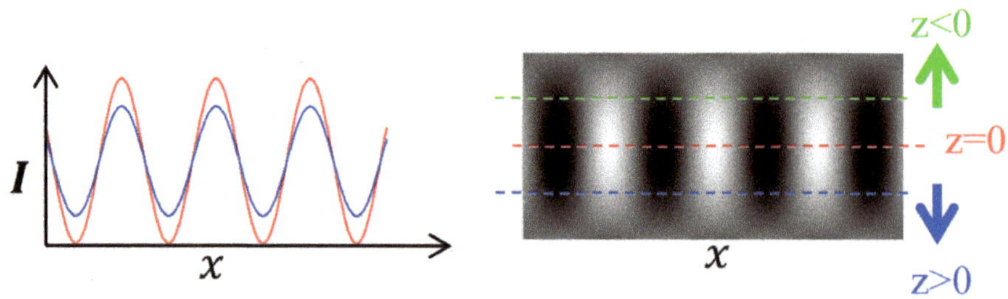

Figure 7.68 Combining I_{2BR} and I_{2BL} incoherently results in a telecentric image with a finite DOF (image courtesy of S.Y. Chou).

There is no longer a phase modification from z but an intensity modification instead. The intensity is telecentric, but the image intensity varies along the z axis. Therefore, the spatial frequency f_{2BA} is unchanged, but the DOF is finite, as shown in Fig. 7.68.

How does DOF_{2B} compare with DOF_{3B}? DOF_{2B} is bound by

$$|2\pi Cz| < \pi/2 \Rightarrow DOF_{2BA} < \left|\frac{1}{4C}\right|. \tag{7.50}$$

Using Eq. (7.32),

$$\frac{DOF_{2BA}}{DOF_{3B}} = \frac{1-\cos\theta_P}{|\cos 0.5\theta_P(1-\cos\Delta\theta_P) - \sin 0.5\theta_P \sin\Delta\theta_P|}. \tag{7.51}$$

Again, taking the worst case when $\Delta\theta_P = 0.5\theta_P$,

$$DOF_{ratio} = \frac{DOF_{2BA}}{DOF_{3B}} = \frac{1-\sqrt{1-(\lambda/P)^2}}{2\left|\sqrt{1-(\lambda/P)^2} - \sqrt{1-(\lambda/2P)-(\lambda/P_1)^2}\right|}. \tag{7.52}$$

When $P = \lambda$, i.e., the resolution of the grating is pushed to the limit,

$$DOF_{ratio} = \frac{1/2}{\left|\sqrt{\frac{3}{4}} - \sqrt{1-(1/2-\lambda/P_1)^2}\right|}. \tag{7.53}$$

When $P_1 = P$, $DOF_{ratio} = \propto$. This is expected because the two beams become symmetrical. When $P = \propto$, the DOF ratio also becomes $\propto$.

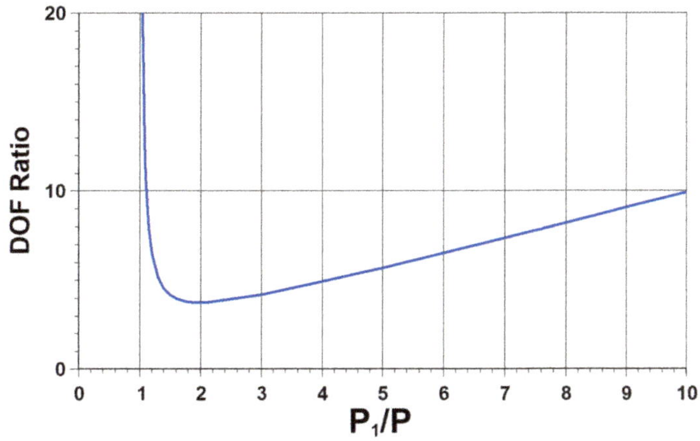

Figure 7.69 The ratio of two-beam DOF to three-beam DOF as a function of normalized periodicity.

The DOF ratio as a function of P_1 is plotted in Fig. 7.69. There is a minimum of 3.73 at $P_1/P = 2$. This indicates that the DOF of the two-beam system is at least 3.73 times larger than that of the three-beam system. When P_1 is within 1.1 of P, the DOF improvement is more than 10X.

7.3.2.2.6 Numerical example of OAI in 1X and reduction systems

The largest $\sin\theta$ of commercial imaging lens in lithography is 0.93, which is used in this section to demonstrate OAI numerically. Figure 7.70 shows a 1X system where the angular spread of the two beams is $\sin^{-1}(0.93)$; i.e., $\sin\theta_{P2} = \lambda/2P = 0.93$. This angle fits the acceptance angle α of the imaging lens. $P = 0.93(\lambda/2)$. The incident angle of the illumination is $\sin^{-1}\theta_{P2} = 68.43$ deg.

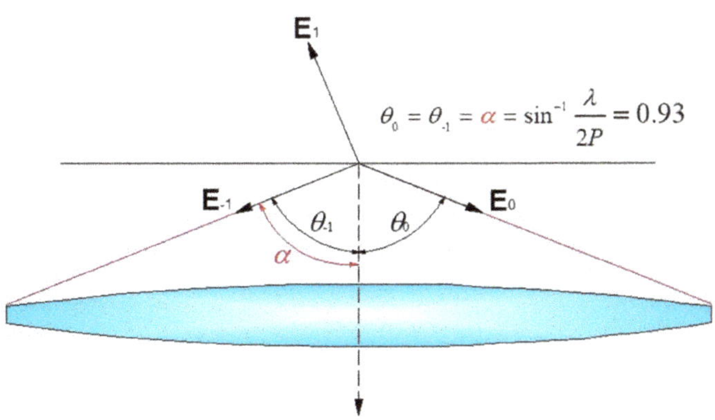

Figure 7.70 The diffraction angles θ_1 and θ_0 in an NA = 0.93, 1X two-beam imaging system.

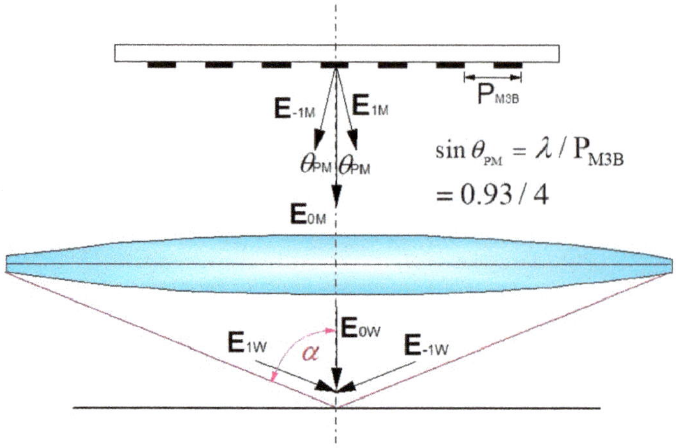

Figure 7.71 Three-beam imaging in a 4X system with $NA_W = 0.93$, $NA_M = 0.93/4$ resolves $P_W = \lambda/0.93 = 1.08\lambda$.

In a 4X reduction system, the NA at the mask side is 0.93/4 = 0.23. The sine of the +1st-order and –1 st-order diffracted angles of the $\mathbf{E}_{-1M}$ and $\mathbf{E}_{1M}$ beams are –0.23 and 0.23, respectively. The corresponding imaging beams are accepted by the 0.93-NA imaging lens in a three-beam system. The resolvable pitch is 1.08λ, as shown in Fig. 7.71. This resolution is the same as that of the 1X system in three-beam imaging.

Turning to a two-beam system, we can reduce P_M to make the sine of the $\mathbf{E}_{0M}$ and $\mathbf{E}_{1M}$ beams ±0.93/2. The resolution is now doubled to $P_W = 0.54\lambda$, as shown in Fig. 7.72.

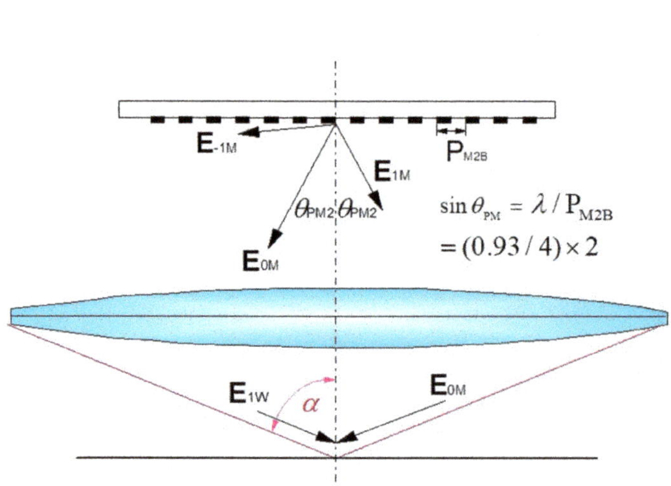

Figure 7.72 In the same system as in Fig. 7.71, two-beam imaging resolves $P_{M2B} = 2\lambda/0.93 = 2.15\lambda$. $P_W = 0.54\lambda$.

7.3.2.3 3D illumination on 2D geometry

When 2D geometry is present, the OAI beams are directed along a conical surface to illuminate a ring area in the aperture of the imaging lens, as shown in Fig. 7.73(a). When only vertical and horizontal lines are present, the best OAI takes the form of four small, illuminated areas at the intersection of the ±45-deg lines with the ring, as shown in Fig. 7.73(b). This way, each beam is off axis to either the *x*- or the *y*-oriented lines on the mask. With 45-deg lines on the mask, only one of the two diagonal beam pairs is off-axis to either aa' or bb', as shown in Fig. 7.73(c). The gain in imaging performance produced by the off-axis beams is diluted by the other on-axis beams. In addition, the gain in imaging performance for mask patterns with an identical pitch but different orientations is inconsistent because the diagonal distance between the poles is larger than the vertical or perpendicular distances. The case of ring OAI in Fig. 7.73(a) is a continuous combination of off-axis and on-axis beams, plus the effect of changing the optimizing pole distances. Therefore, the imaging performance in this configuration is a compromise between on-axis illumination and OAI with diagonal quadrupoles. So far, the discussion has been qualitative. The actual imaging performance with partially coherent light must be simulated and carefully compared.

Disk illumination (DKI), ring illumination, various quadrupole illuminations, and a disk–OAI combination are shown in Fig. 7.74. DKI is on axis, characterized by the aperture ratio $\sigma \equiv NA_{\text{condenser}}/NA_{\text{imaging lens}}$. At the lens pupil, σ = 1. For ring OAI, the ring is characterized by σ_{in} and σ_{out}. For

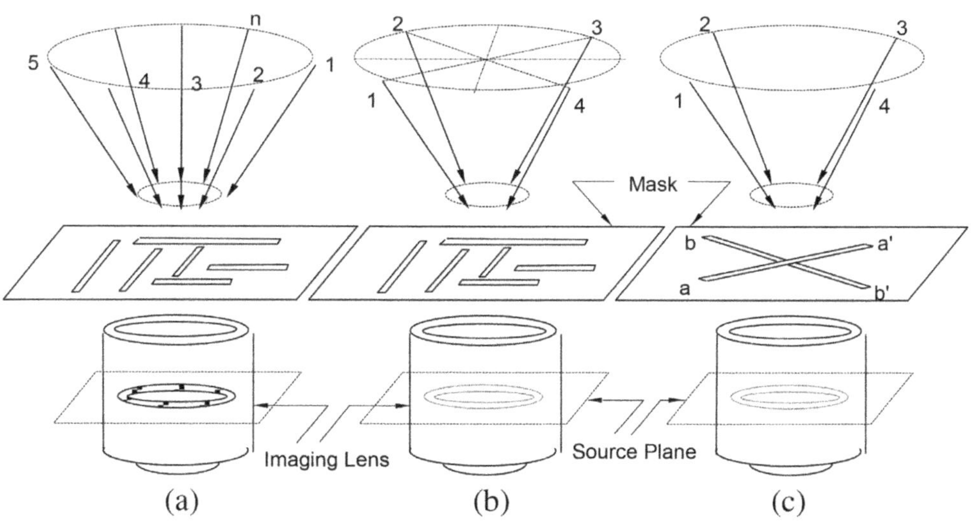

Figure 7.73 Illumination and pattern orientations. (a) Ring illumination, where each line orientation sees a mixture of off-axis and on-axis illuminations. (b) Quadrupole illumination, where all beams are off axis with respect to all vertical and horizontal lines. (c) Quadrupole illumination with beams 2 and 3 off axis with respect to aa′ but on axis with respect to bb′.

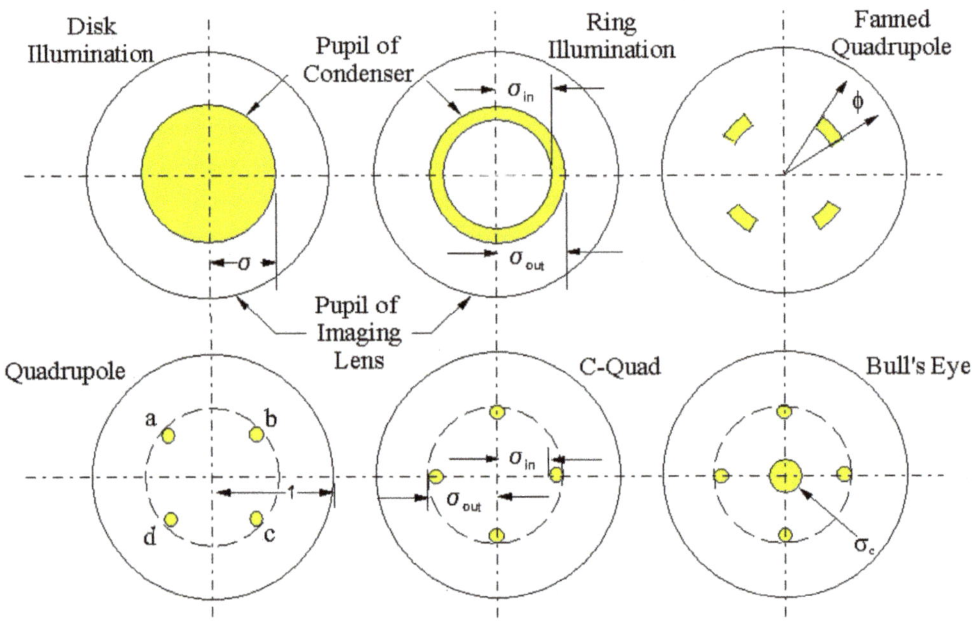

Figure 7.74 Various illumination schemes.

quadrupoles, the σ_{in} and σ_{out} terminology is still followed. The diameter of the poles can readily be derived from the two σ values. The angular spread ϕ is used to characterize the size of the fanned quadrupole. This type of quadrupole is often used for convenience because it can readily be formed from the basic ring illumination. The bull's-eye illumination combines a center disk with quadrupole. It compromises the imaging performance of isolated and dense features to strike a desired balance. C-Quad illumination simply lays out the quadrupoles according to the orientations of a cross. Dipole illumination is characterized by either the horizontally or vertically oriented poles in C-Quad; it is not shown in the figure.

What is the difference between the imaging performance of quadrupole and C-Quad illumination? Taking a vertical line object in a quadrupole illumination setup, poles *a* and *b* are off axis, as are poles *c* and *d*. Alternatively, with a horizontal line object, poles *a* and *d* as well as *b* and *c* are off axis. Therefore, there are no on-axis poles to dilute the image contrast. However, the horizontal and vertical distance from the center of the poles to the *x*- or *y*-axis is only $\sigma/\sqrt{2}$, where σ is the distance from the center of the poles to the center of the lens pupil. With C-Quad illumination, the horizontal and the vertical distance of the poles to the axes is the full length of σ, despite the dilution of the contrast from the perpendicular pair of poles. The tradeoff between these two types of quadrupole illumination is between resolution and contrast.

Dipole illumination produces high resolution and contrast without compromise. However, if a 2D pattern is used, it must be split into two

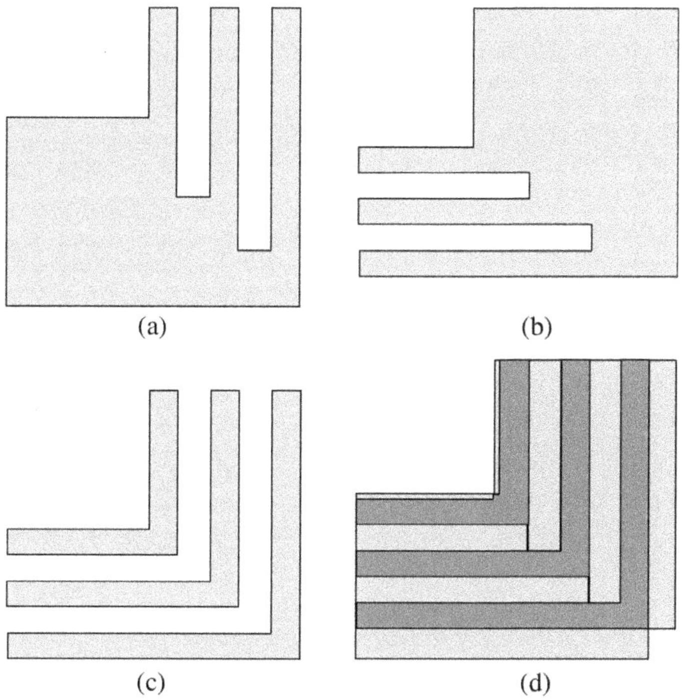

Figure 7.75 Splitting of L-shaped patterns for double-dipole illumination: (a) and (b) the split patterns, (c) the target pattern, and (d) the pattern made by superimposing the exposures from (a) and (b).

masks with the vertically oriented patterns exposed by the horizontally oriented dipole and the horizontally oriented patterns exposed by the vertically oriented dipole in a second exposure. This illumination setup is called double-dipole illumination.[45] Splitting of L-shaped patterns is illustrated in Fig. 7.75, where parts (a) and (b) are the two split patterns. Figure 7.75(c) is the target pattern, whereas Fig. 7.75(d) is the resultant pattern after combining the exposures from parts (a) and (b). If the pattern alignment is not perfect, the corners will have slight complications. Fortunately, the corner area has been known to be restricted for critical dimensions. One drawback of double-dipole illumination is that two exposures reduce the throughput, thus increasing the cost.

7.3.2.4 Simulation results

The imaging performance of OAI systems from simulation are given in this section. Figure 7.76 shows constant-intensity contours using line–space pairs with a 0.47 k_1 half-pitch as the object. Physically, the 248-nm wavelength at $NA = 0.54$ is used. The images from the two types of illumination are illustrated: DKI at $\sigma = 0.8$ and ring illumination with $\sigma_{min} = 0.4$ and $\sigma_{max} =$

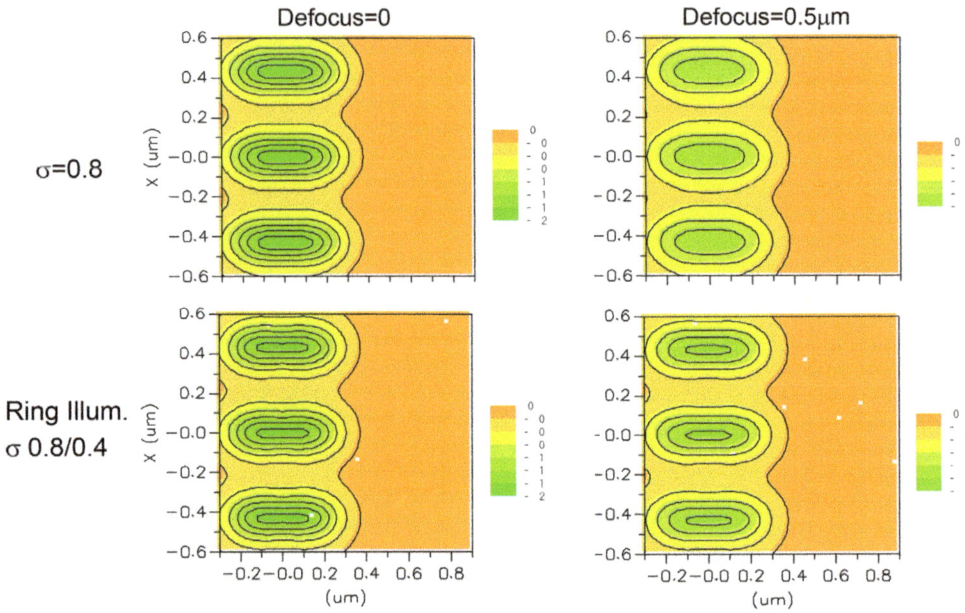

Figure 7.76 Constant-intensity contours of disk illumination (top row) and ring illumination (bottom row) for line–space pairs with 0.47-k_1 pitch. The physical parameters are λ = 248 nm, NA = 0.54.

0.8. At zero defocus, the image contrast exhibited by the contour density is higher. At 0.5-μm defocus, the improvement from OAI is even larger.

The improvement in DOF is shown quantitatively in Fig. 7.77, still comparing disk and ring illuminations. Three σ_{in}:σ_{out} ratios, namely, 1:3, 1:1, and 2:3, of the ring illumination are included. For disk illumination, σ approaches the optimum value around 0.8 slowly and insignificantly. Within the three ring illuminations, the narrowest ring with σ_{in}:σ_{out} = 2:3 produces the most significant improvement in absolute DOF at the peak. The change in

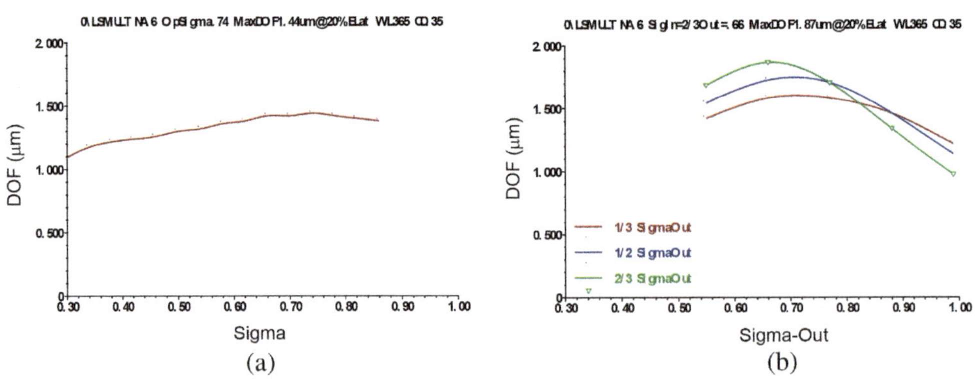

Figure 7.77 DOF versus σ and σ_{out} for (a) disk and (b) ring illuminations, respectively, on 0.35-μm 1:1 line:space, λ = 365 nm, E_{tol} = 20%, σ_{in}:σ_{out} = 1:3, 1:1, 2:3.

DOF as a function of σ_{out} is also more significant. In addition, the optimum σ_{out} shifts lower as the ring width narrows. At σ_{in}:σ_{out} = 1:3, the DOF versus σ_{out} behavior approaches that of DKI. The tradeoff for better imaging performance is higher coherence. Before OAI was implemented with a zoom lens, i.e., when OAI was achieved by a blocking aperture in the pupil plane, loss of exposure energy was also a tradeoff.

From Section 7.3.2.1, we know that only one pitch is optimized for a given illumination angle. In Fig. 7.78 additional non-optimized pitches are included to demonstrate the situation of simultaneously delineating multiple pitches. For the 1:1.5 line–space pairs, 950 nm DOF can be achieved at σ_{out} = 0.84. However, the DOF is limited to 470 nm by the features with larger openings. When all features are considered together, the common DOF is only 260 nm. Figure 7.79 shows the E-D windows that are used to produce Fig. 7.78. The windows for individual features are indeed very large. Only the common window suffers.

In terms of feature shapes, Fig. 7.80 shows the E-D tree of line–space pairs, a line opening, an opaque space, a hole, and a pedestal using disk illumination in part (a) and quadrupole illumination in part (b). The line–space pairs with disk illumination have an extremely narrow and shallow E-D tree. Changing to an AltPSM helps to enlarge the E-D tree but does not enlarge them as much as quadrupole illumination. With disk illumination, even though the E-D trees are very narrow, those for 1D features are close to each other. With quadrupole illumination, they are more widely separated; i.e., the proximity effect is stronger. All features tend to straighten up with quadrupole illumination, allowing for a larger DOF.

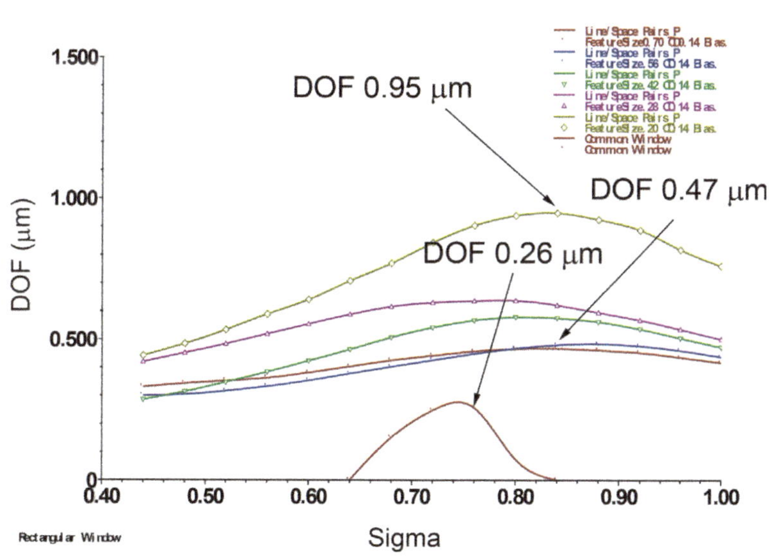

Figure 7.78 DOF versus σ_{out} for 1:1.5, 1:2, 1:3, 1:4, and 1:5 line–space pairs with ring illumination. CD is opaque space at 140 nm. λ = 248 nm, *NA* = 0.65, *EL* = 8%, 6% AttPSM, σ_{in} = 0.5, σ_{out} = 0.76.

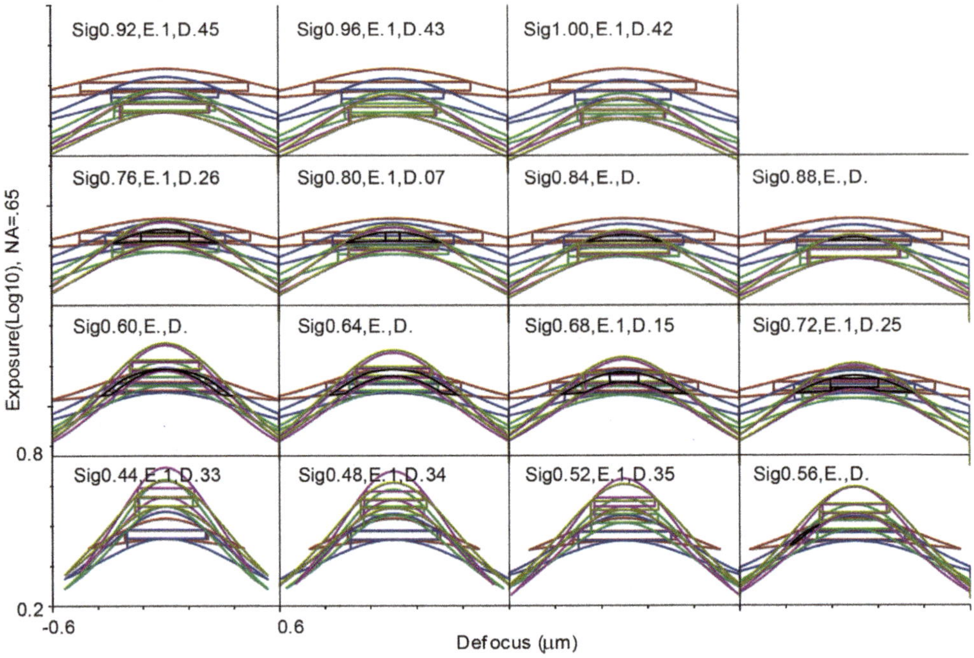

Figure 7.79 E-D windows of the 1:1.5 to 1:5 line–space pairs with ring illumination from Fig. 7.78.

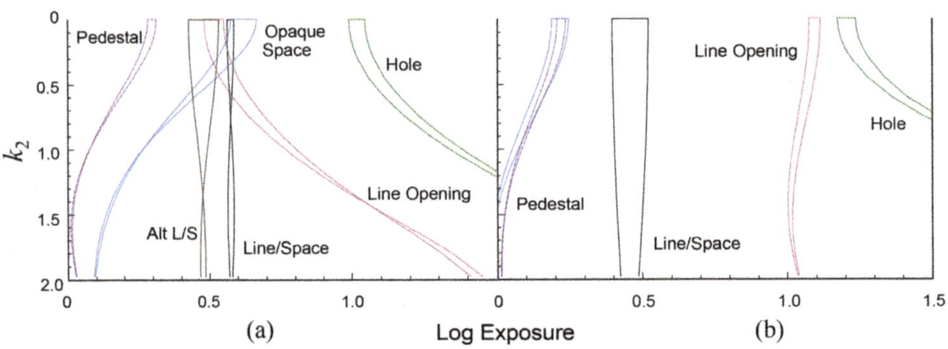

Figure 7.80 E-D trees for five 0.36-λ/NA representative features using: (a) disk illumination with σ = 0.5 and (b) quadrupole illumination with four σ = 0.04 squares at the cross sections of 450 lines and the σ = 0.94 ring.

Using the same five features but at 0.48 k_1, the effects of BIM, AttPSM, and two quadrupole illuminations are compared, as shown in Fig. 7.81. An AttPSM makes the E-D tree of the openings wider and straighter. It also makes the trees wider for the opaque space and the pedestal but is more tilted. Quadrupole illumination with σ_{in} = 0.44, σ_{out} = 0.52 does not improve the E-D trees over BIM or AttPSM with disk illumination, except for a slight improvement in the line–space pairs. The more widely spread quadrupoles

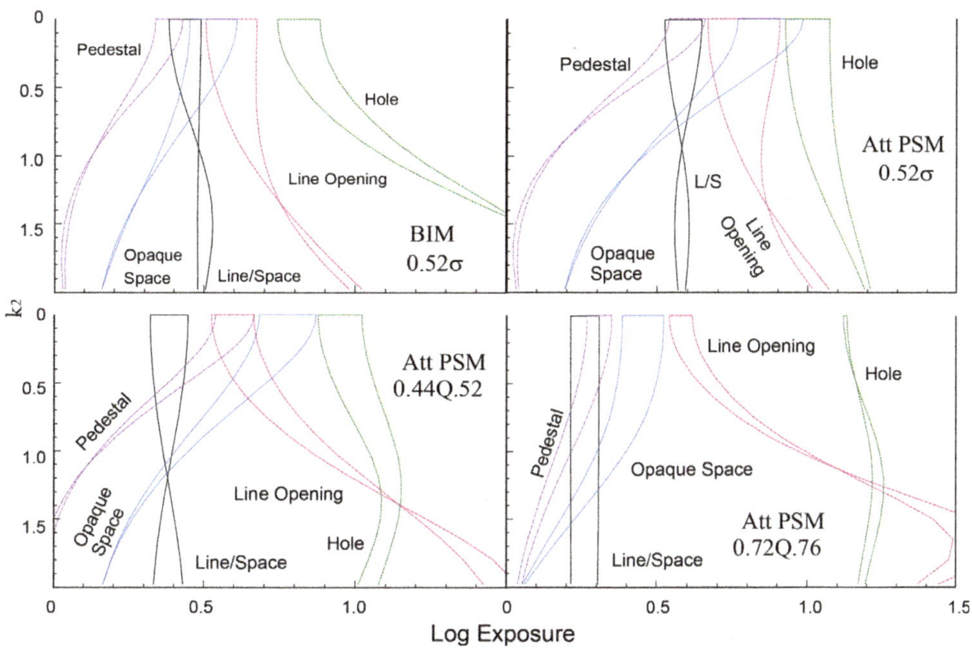

Figure 7.81 E-D trees of five features at 0.48 k_1 with disk and quadrupole illuminations from a BIM and a 6% AttPSM.

produce an extremely large DOF for the line–space pairs, at the expense of the image performance of the other features. In this setup, AttPSM with disk illumination is preferred.

Figure 7.82 shows the E-D trees of different quadrupoles at (a) k_1 = 0.48 and (b) k_1 = 0.36 for line–space pairs. In part (a) the quadrupoles are optimized

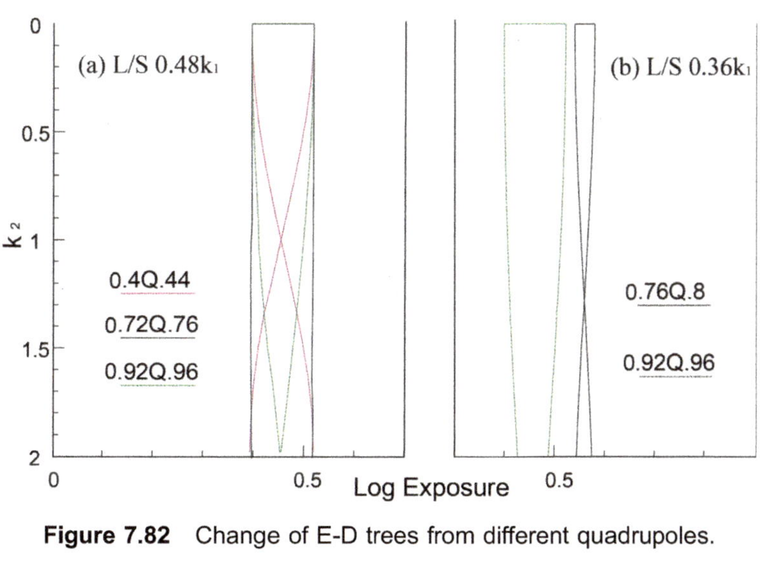

Figure 7.82 Change of E-D trees from different quadrupoles.

at 0.72Q0.76. Larger or smaller σ values cannot achieve iso-focus. For part (b), because of the low k_1, even 0.92Q0.96 cannot bring the image to iso-focus.

The BigMaC mask shown in Fig. 4.26 is used here to demonstrate the effects of OAI and PSM. For ring illumination as shown in Fig. 7.83, the DOF is better than for BIM at either value of σ. The pattern fidelity at the corners and line ends combines the favorable aspects of a BIM at either σ value.

A severe problem with quadrupole illuminations is exemplified in the ripples in the SEM image of Fig. 7.84. The ripples are typified by enlargement at the line ends and corners, followed by oscillation from wide to narrow toward the center of the lines. The rightmost SEM micrograph shows a superposition of the resist image and a simulated image to verify the fidelity of the simulation. The ripples there are most likely due to light interference of the two diffracted beams that start from opposite poles. Unlike typical

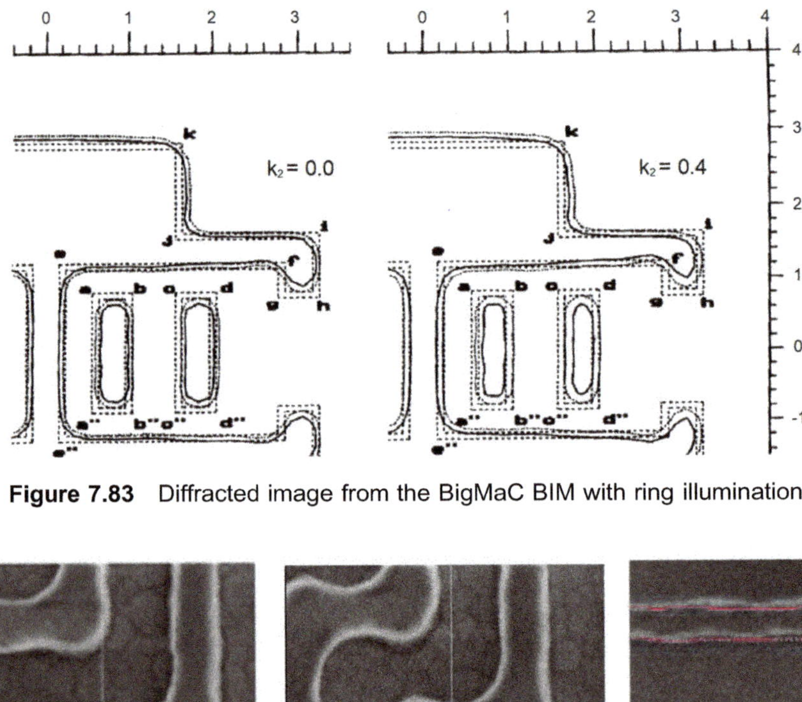

Figure 7.83 Diffracted image from the BigMaC BIM with ring illumination.

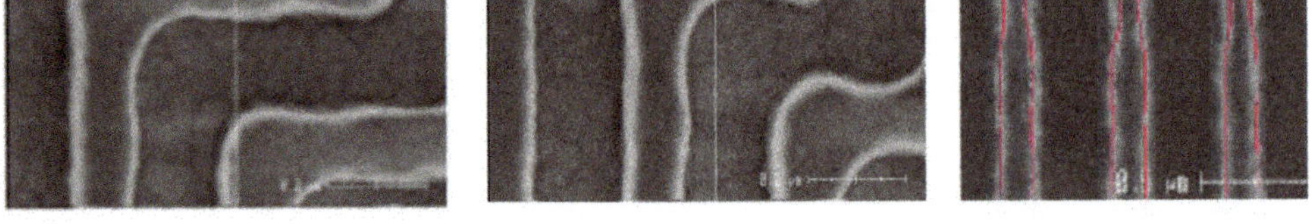

Figure 7.84 Ripples in the resist image under unoptimized OPC conditions (images courtesy of R.G. Liu).

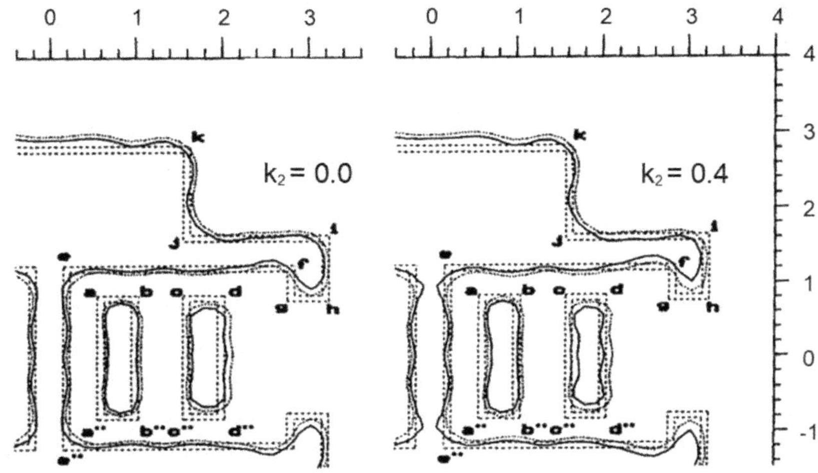

Figure 7.85 Diffracted image from the BigMaC BIM with quadrupole illumination.

line-edge variations, OPC on ripples is not very effective; even when optimized for ripples, OPC cannot fully remove them.

Ripples were predicted early on from simulations using the BigMaC mask, as shown in Fig. 7.85. The ripples are severe along the edges k′, k, j, i, h, g, f, e, and e″ and their symmetrical counterparts (e.g., k′ is the mirror image of k to the left of the vertical center line). The outer-line opening cdc″d″ also suffers from ripples. The ripples are different from the waviness of edge ee″ in Fig. 4.27(a). The latter is confined only to the segment between e and e″.

The image in Fig. 7.86 from an AltPSM does not have ripple defects. The only deficiency shown in Fig. 7.86 is a shortening of the outer opening cdc″d″ due to lack of a phase-shifting neighbor.

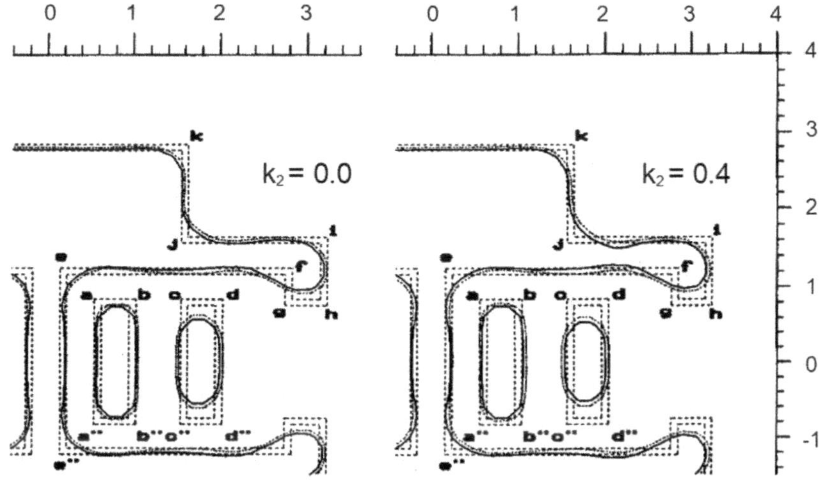

Figure 7.86 Diffracted image from the BigMaC AltPSM at σ = 0.4.

7.3.2.5 Comparison of OAI and AltPSM

OAI and AltPSM have many similar functional characteristics, namely, both are able to double the resolution of the imaging lens, both have good iso-focus capability, and neither is able to impact isolated features. But there are also some differences to explore.

A grating with $k_1 = 0.4$ line–space pairs is used to show the DOF improvement of OAI and AltPSM. Here, k_1 is used as a normalized lateral dimension related to a physical dimension W by $k_1 = W(NA)/\lambda$. Similarly, k_2 is used as a normalized longitudinal dimension related to a physical dimension Z by $k_2 = Z\,(NA)^2/\lambda$. Figure 7.87(a) shows the log intensity distributions and the E-D tree of 0.4σ on-axis illumination with a BIM, evaluated at $k_2 = 0$, 0.2, 0.4, 0.8, 1.2, 1.6, 2, 2.4, and 2.8. At such a low k_1, low image contrast and shallow log slope in the intensity distributions are seen. The E-D branches are too closely spaced to support even 10% EL. Alternately phase shifting the mask produces a DOF of 0.8 k_2 for 26% EL, as seen in Fig. 7.87(b). An even higher k_2 of 2.6 can be obtained by using optimized quadrupole OAI. An iso-focus situation is created as seen by the identical log slope and edge location in the intensity distribution and in the almost horizontal E-D branches in Fig. 7.87(c).

Other than the imaging performance, the most significant difficulty encountered with an alternating PSM is phase-shifting conflicts with arbitrary 2D patterns, as discussed in Section 7.3.3.1. Shifting to a light-field mask to avoid phase-shifting conflicts necessitates a second exposure with a cutting mask to remove these lines. Hence, there are issues of overlay accuracy, reticle changing, throughput, and cost.

OAI is less expensive to use than AltPSM when the one-time cost of OAI is amortized over the cost of all AltPSMs. However, an inherent problem with OAI is that some of the illumination energy for high-resolution features is blocked by the finite acceptance angle of the lens NA. This leads to not only a longer exposure time, but also an exaggerated proximity effect. The tradeoffs of the three RETs are summarized in Table 7.2.

7.3.2.6 Combination of OAI and AltPSM

The working principle of an AltPSM is restated here [see Fig. 7.88(a)] to show why AltPSM and OAI cannot be combined. For an AltPSM, the phase of every other opening on the mask is shifted by 180 deg. The electric field behaves as if its spatial frequency were reduced to half of the original spatial frequency with no 0th-order component. After passing through the lens, which would cut off the original frequency but not the reduced one, the frequency is doubled back in the resist image, which responds only to the intensity that is proportional to the square of the electric field amplitude. OAI cannot be combined with AltPSM to achieve an even higher resolution, as illustrated in Fig. 7.88(b). At the highest resolution achievable by either AltPSM or OAI, the off-axis beam turns one of the two phase-shifted beams out of the

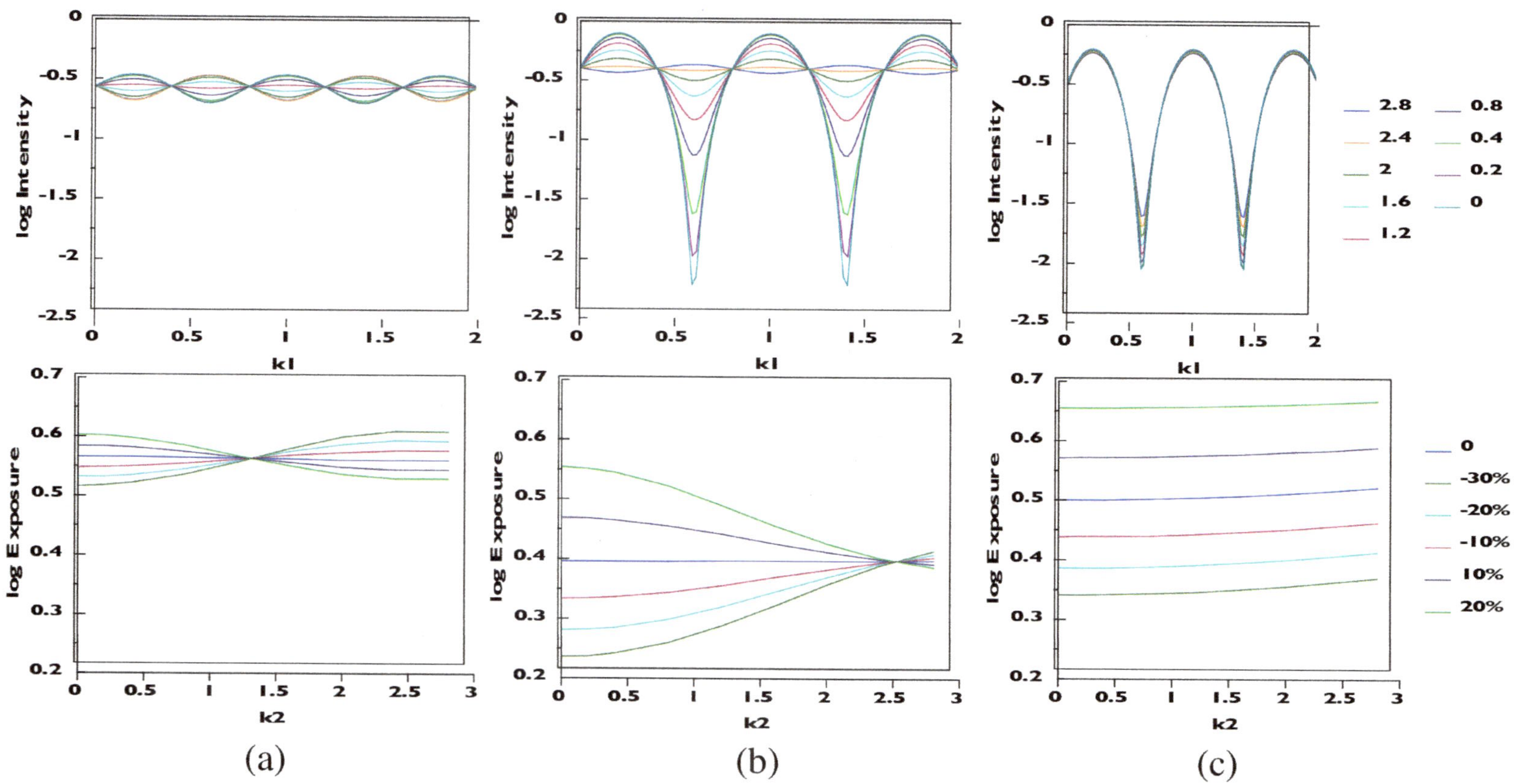

Figure 7.87 Comparison of (a) BIM (0.4σ) disk illumination, (b) AltPSM (0.4σ) disk illumination, and (c) BIM OAI for 0.4 k_1 lines and spaces, using 0.02 radius quadrupoles at the cross sections of 45-deg lines and a σ = 0.98 circle.

Table 7.2 Comparison of ring illumination, quadrupole illumination, and AltPSM.

	Ring illumination	Quadrupole illumination	AltPSM
Minimum feature size	0.4 k_1	0.3 k_1	0.3 k_1
OPC required	Yes	Yes	Yes
Applicable patterns	High-density L/S in any orientation	High-density L/S in X and Y orientations	High-density L/S in any orientation
Combination with low-density patterns	Use scattering bars	Use scattering bars	Combine with AttPSM
Cost	One-time cost of exposure tool	One-time cost of exposure tool	Extra cost for each mask

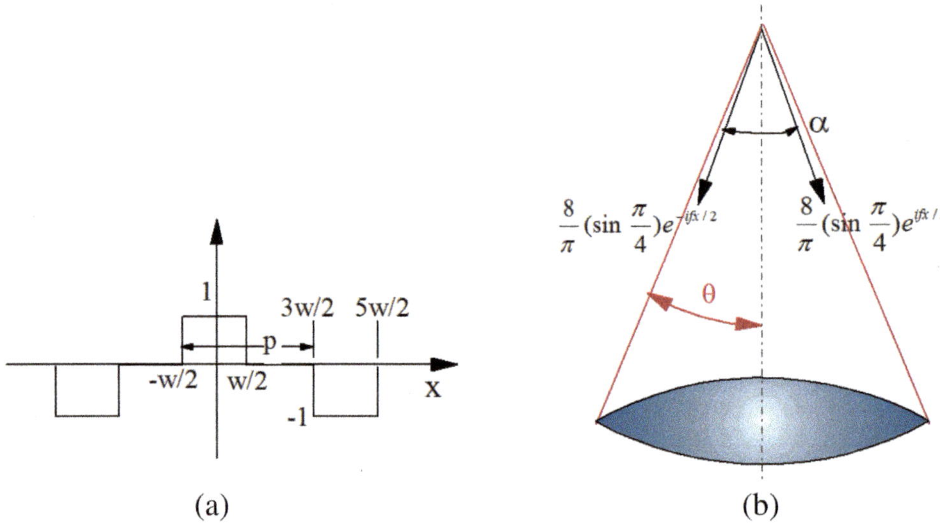

Figure 7.88 Electric field and spatial frequencies of axial illumination through an AltPSM. OAI cannot tilt the two imaging beams for any gain in resolution. (a) The electric field after passing through the AltPSM. (b) The two spatial frequency vectors from the electric fields through the AltPSM. If OAI is used, one of these vectors will be outside the acceptance angle of the imaging lens and the image will be just a uniform propagating wave that cannot reproduce the grating object.

acceptance angle, while the remaining beam has a low spatial frequency that cannot reconstruct the original frequency of the object.

Analytically, the electric field on the BIM is split into two parts, the unshifted part $E_u(x)$ and the phase-shifted part $E_s(x)$. The latter is shifted by π:

$$\begin{aligned} E_u(x) &= 1, \ 2np - \frac{w}{2} \leq x \leq 2np + \frac{w}{2}, \ n = 0, \ \pm 1, \ \pm 2, \ \ldots \\ &= 0, \ \text{elsewhere}, \end{aligned} \tag{7.54}$$

$$E_s(x) = -1,\ 2np - \frac{3}{2}w \le x \le 2np + \frac{3}{2}w,\ n = 0,\ \pm 1,\ \pm 2,\ \ldots \\ = 0,\ \text{elsewhere}, \tag{7.55}$$

and

$$E_0(x) = E_u(x) + E_s(x). \tag{7.56}$$

Introducing an x translation in $E_s(x)$, the Fourier transformation of $E_0(x)$ normalized to w is

$$E_0(\xi) = \frac{\sin \pi\xi w}{\pi\xi w}(1 - e^{i2\pi p\xi}) \sum_{n=-\infty}^{\infty} \delta\left(\xi - \frac{n}{2p}\right). \tag{7.57}$$

Note that the delta functions are now centered on $n/2p$ instead of n/p as in the case of BIM. That is, the spatial frequencies are now reduced by a factor of 2. Also, $E_0(\xi) = 0$. Hence, there is no 0th-order component. When the lens NA limits n to ± 1, the image electric field becomes

$$E_i(x) = \frac{8}{\pi}\left(\sin\frac{\pi}{4}\right)(e^{ifx/2} + e^{-ifx/2}). \tag{7.58}$$

There are only two beams symmetrically displayed with respect to the z-axis. Imposing OAI on AltPSM leads to

$$E_i(x) = [e^{ikx\sin\varphi} + e^{-ikx\sin\varphi}]\left[\frac{2}{\pi}(e^{ifx/2} + e^{-ifx/2})\right]. \tag{7.59}$$

There is no 0th-order beam to be tilted. OAI either keeps the two beams within the lens acceptance angle or moves one beam out of it. Hence, it either does not change the resolution or reduces it. The resolution cannot be further doubled. The situation is depicted by Fig. 7.88.

7.3.3 Scattering bars

Figure 7.89 shows that it is desirable to pack the layout with dummy patterns to reduce the spread of pitches. The packing material should not be printed. Subresolution assist features serve the purpose. These are generally referred to as scattering bars (SBs) or subresolution scattering bars (SSBs). There is room in the medium-size and large pitches to insert dummy features. Figure 7.90 shows SSBs inserted between features wherever space allows.

7.3.3.1 Imaging improvement from scattering bars

The width of a SSB and its distance from the main feature can be fine-tuned to optimize the imaging performance. A larger width provides stronger pitch

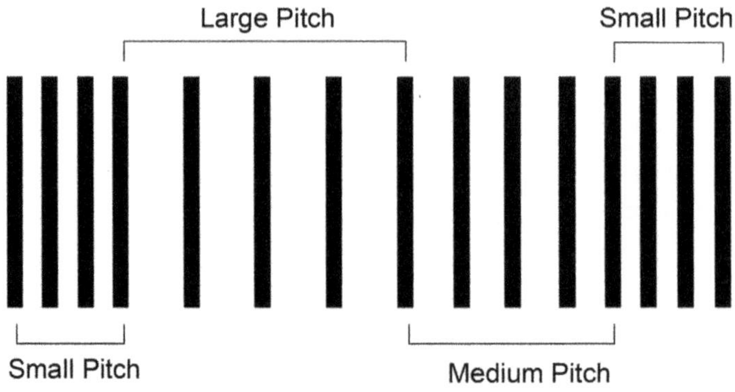

Figure 7.89 Pitch distribution.

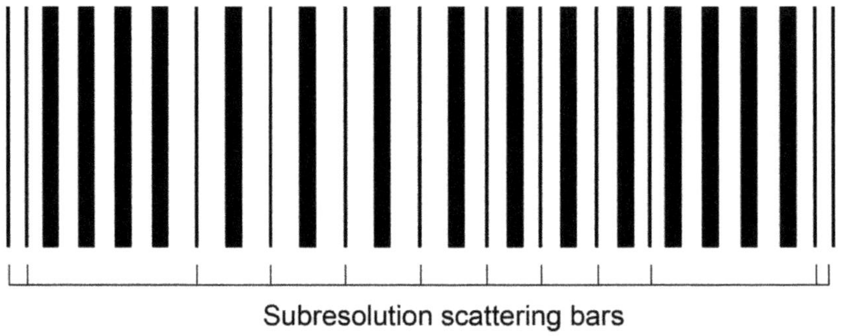

Figure 7.90 SSBs added to reduce the pitch dispersion.

filling, but it must be optimized against SSB printing and sufficient space to insert the bars. Figure 7.91 shows the intensity distribution of a main feature surrounded by two scattering bars under quadrupole illumination. The side lobes of the intensity distribution must be kept within an allowed limit to prevent its printing. The imaging improvement using SSBs can be depicted with E-D windows, as shown in the same figure. When there is no SB, the DOF at $EL = 7\%$ is 0.6 μm. With a SSB of typical width W and spacing S, the DOF is extended to 0.9 μm. Further customizing W and S (see Fig. 7.91, top right) brings the DOF to 1.26 μm.

7.3.3.2 Restricted pitch

The OAI angle is usually set near the minimum pitch to gain DOF there. The DOF gradually decreases as the pitch increases to an unacceptable level. As soon as space between features allows, SSBs are inserted. Normally, there is an ideal distance from the SSB to the edge of the feature that the SSB is intended to enhance. This is an edge SB (ESB). However, this distance is quite large, and the DOF already drops beyond acceptability with a much smaller pitch. A central SB (CSB) is placed at the center of two tight features,

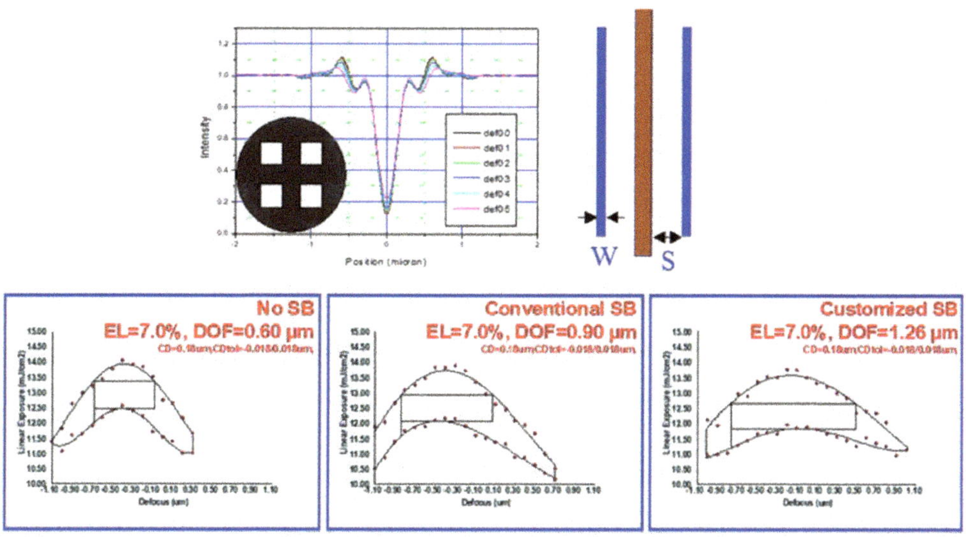

Figure 7.91 (top left) Intensity distribution of a main feature and the two SSBs shown on the top right. The quadrupoles are seen in the lower left corner of the top left image. (lower left) The E-D tree and window when there is no SSB. (lower middle) Same as lower left, except with conventional SSBs. (lower right) The same as lower left, except with customized SSBs (images courtesy of R.G. Liu).

regardless of whether the position is optimum, provided it does not create an unresolvable pitch or become part of the printed image. There may still be a small range of pitches for which the space between features is so small that even a CSB cannot be used. This is the range of restricted pitch that should be avoided in the circuit design. Figure 7.92 shows the DOF of a range of pitches

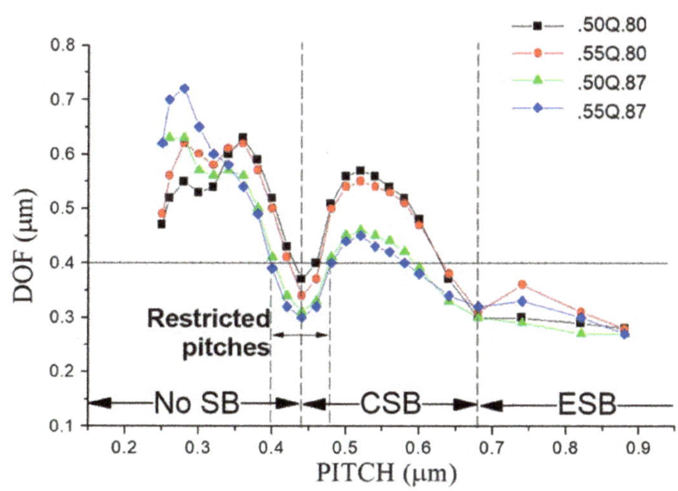

Figure 7.92 DOF versus pitch using four different quadrupole illumination settings and a SSB on an AttPSM; λ = 248 nm, *NA* = 0.7, *EL* = 8% (image courtesy of R.G. Liu).

from an AttPSM using four different quadrupole illumination settings. The zones for the CSB and ESB are marked. The desired 0.4-μm DOF line is drawn. The DOF falls below 0.4 μm in the pitch range between 0.4 and 0.48 μm. This is the range of the restricted pitch with AttPSM 0.55Q0.87, and SSB at 248-nm wavelength, 0.7 NA, and 8% EL. The other quadrupole settings are better. Note that the DOF is less sensitive to changes in σ_{min} and that the only benefit of a larger σ_{max} is a larger DOF at very small pitches.

7.3.3.3 2D features

Adding a SSB to 2D features is neither as straightforward nor as effective as adding one to 1D features. Figure 7.93 shows the design of two polygates with an enlarged portion for the contact holes. The original design is perturbed at the edges, which have been corrected for optical proximity effects (see Section 7.1.1). Scattering bars are placed near the pattern edges but cannot wrap around the features without breaking to prevent ghost images at the SSB corners.

7.3.3.4 Mask-making concerns

The broken SSBs pose a risk of resist falling off during mask making. Short SSBs are vulnerable to fall-off. When the SB is too narrow, resist fall-off also occurs. Therefore, the width of the SSB must be carefully optimized. A large width leads to printing all the way through the resist, while a small width promotes fall-off of the SSB resist during mask making.

The small size of the SSB also poses a problem for mask inspection. Scattering bars should not be mistaken for defects by the mask inspection tool. The size of a printable defect usually falls under a dimension like that of SSBs. Therefore, the short scattering bars shown in Fig. 7.93 can be mistaken for defects.

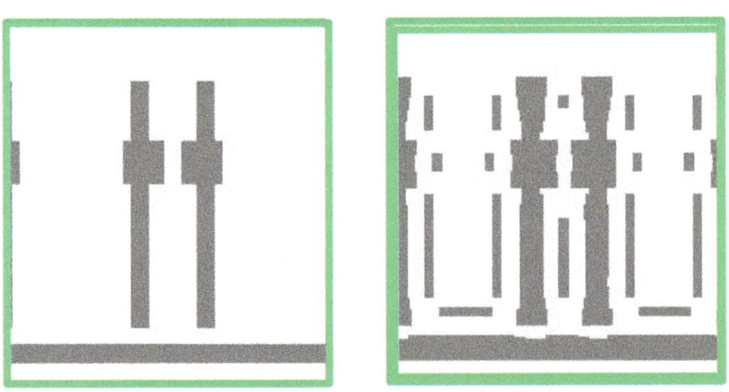

Figure 7.93 Scattering bars with optical proximity correction (images courtesy of R.G. Liu).

7.3.3.5 Full-size scattering bar

By definition, a full-size SB (FSB) has dimensions like those of the main features. Therefore, FSBs are printable and must be removed with an additional trim mask. Is it worth the extra cost? Figure 7.94(a) is identical to Fig. 7.92, which illustrates quadrupole illumination and SSB. Figure 7.94(b) displays a similar DOF-versus-pitch plot with identical imaging conditions, except that a FSB is used instead of a SSB. In the "No-SB zone" there is no change, as is expected. In the zone where a SB can be inserted, the FSB clearly outperforms the SSB. However, it does not help the first group of restricted pitches. Figures 7.94(c) and 7.94(d) show the E-D trees for pitches of 260, 450, and 700 nm, the common E-D tree, and the corresponding E-D windows for SSB and FSB, respectively. The feature size is 113 nm $CD_{\mathrm{tol}} = \pm 9$ nm, $\lambda = 248$ nm, $NA = 0.7$, $EL = 8\%$. The E-D trees and corresponding windows are identical at pitches of 260 and 450 nm because no SB is used. The E-D window for FSB at $P = 700$ nm is larger than that for SSB. The resultant common DOF is also larger for FSB—480 nm versus 390 nm. Whether it is worth the extra cost to use a FSB depends on whether the DOF required to support mask flatness, mask topography, lens field curvature, focusing tolerance, wafer leveling, wafer flatness, resist thickness, wafer topography, etc., can be met with the 390-nm DOF.

7.3.3.6 Hollow subresolution scattering bars and subresolution assist PSMs

The SSB looks very similar to a subresolution assist PSM when the SSB is hollow; i.e., a hollow SSB (HSSB) can assist in imaging transparent features such as holes and line openings. Figure 7.95 shows these two types of features. The only difference between them is phase shifting in the case of SA-PSM. The two features share the same space limitation; i.e., there must be sufficient space for the assist feature. SA-PSM helps the isolated features more than HSSB. For spaces suitable for a CSB, SA-PSM is more vulnerable to printing through. The combination of OAI with SSB has been well studied. Combining OAI with SA-PSM requires more investigation.

7.3.4 Optical proximity correction

When k_1 is large, all types of features can be imaged together with a large common E-D window. While k_1 is being reduced, the common window becomes smaller and smaller, even to the point of nonexistence. Note that optical proximity correction (OPC) is not a bona fide k_1 reduction technique because it does not increase the image slope or increase the spatial frequency like PSM or OAI. Like using a SSB or FSB, OPC is a low-k_1 enabler. After the k_1 reduction techniques take care of the image slope and/or the spatial frequency, there may still be no common E-D window. OPC facilitates the

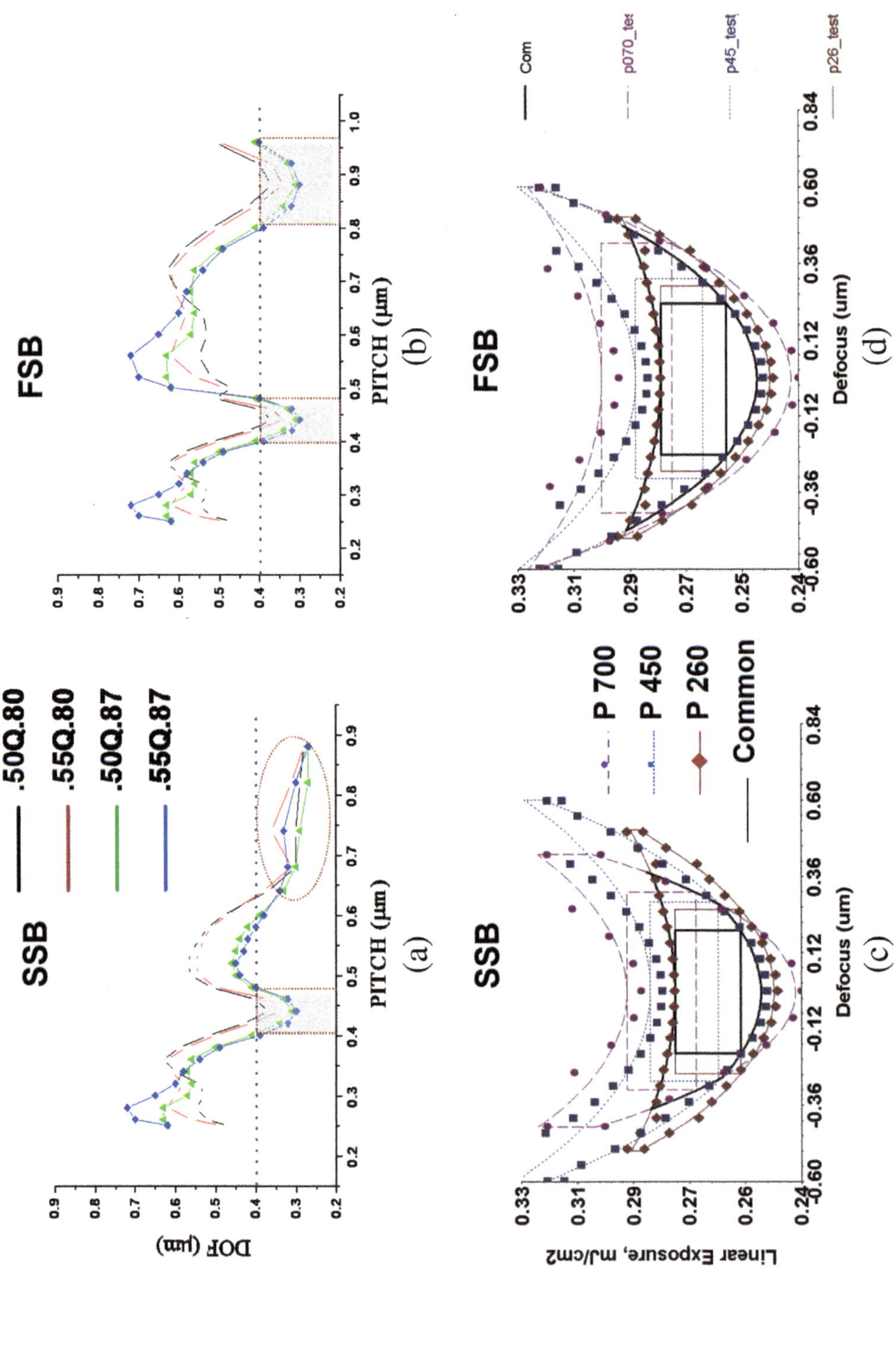

Figure 7.94 Imaging performance comparison using a SB and a FSB (images courtesy of R.G. Liu).

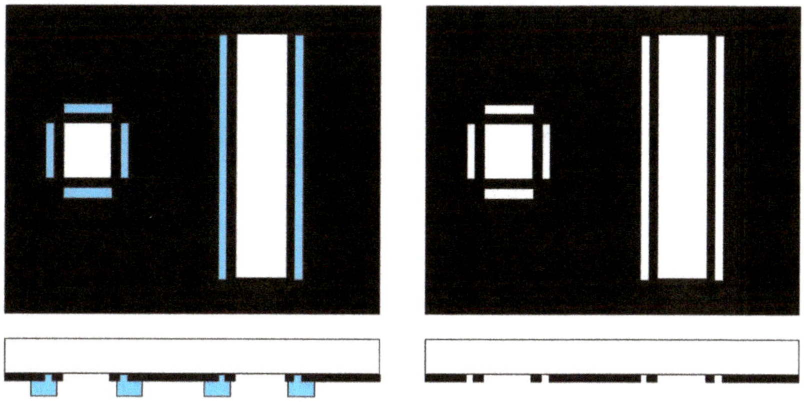

Figure 7.95 (left) Subresolution assist PSM and (right) hollow subresolution scattering bars.

production of the common window, even if it does not enlarge the individual E-D windows.

There are many ways to bring the individual E-D trees together to improve the common E-D window—by changing the exposure level of individual features, by changing the feature size or the pattern edge position, or by distorting the pattern edge. Initially, attempts were made to avoid changing the mask pattern. Writing the mask already took a long time, and making the pattern even more complicated was simply impractical. The pattern shapes started to be sorted. Early on, it was sufficient to use only holes at the hole contact layer; then the choice was further narrowed to single-size contact holes. Next, the range of contact hole densities was restricted. For the noncontact layers, only lines and spaces were allowed, then only critical line openings or critical opaque spaces were allowed, then only single-size CDs were allowed; finally, the range of pitches was restricted. However, despite the above discussion, line-end shortening is the most problematic issue. The solution was to extend the line ends on the mask, then enlarge the line ends with hammerheads (see Section 7.3.4.2). Pretty soon, complicated shape corrections were implemented. We will first introduce the cause of proximity effects, then present the methods used to correct them, spanning from the past to the future.

7.3.4.1 The proximity effect

When the feature size and pattern density approach $k_1 < 0.6$ for optical imaging or approach the submicrometer level for resist developing, etching, or chemical mechanical polishing, the proximity effect sets in. The optical proximity effect (OPE) is mostly due to diffraction, which induces a finite

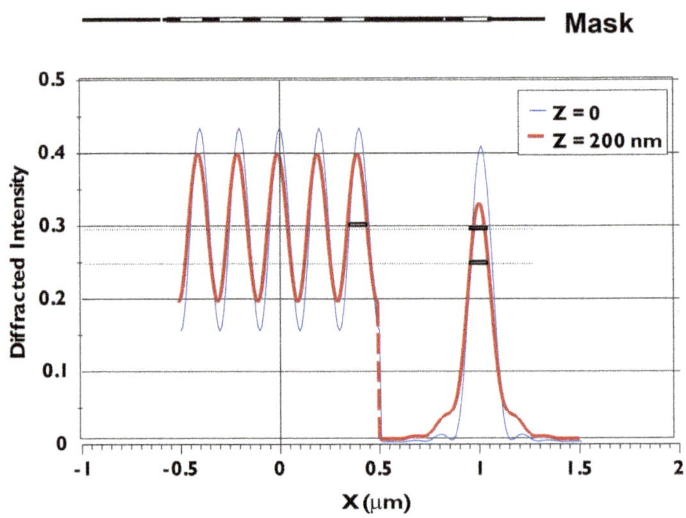

Figure 7.96 Proximity effects due to diffraction demonstrated with 100-nm openings, at λ = 193 nm, NA = 0.75, σ = 0.8.

spread in the intensity distribution that can substantially influence the intensity distribution of close-by neighbors, as shown in Fig. 7.96. The diffracted intensity distribution of five 100-nm line openings in 100-nm spaces and one 100-nm line opening in a 500-nm space is simulated for defocus = 0 and 200 nm, using λ = 193 nm, NA = 0.75, and disk illumination at σ = 0.8. This corresponds to k_1 = 0.39. Because of the close proximity of the equal line–space pairs, the intensity level is raised. When the exposure threshold is chosen to be 0.3, the image linewidth of the equal line–space pairs is within the design limit, but the image linewidth of the line with a large space is smaller, especially at defocus. Increasing the exposure by lowering the intensity threshold to 0.24 brings the defocus image of the line with a large space to the same width as the equal line–space pairs that are in focus, but the defocused line–space pairs become wider.

The proximity effect is not only an optical effect; e-beams produce a finite spread in the resist that can be larger than the resolution of the beam because of forward and backward scatterings, resulting in Gaussian distributions in both forward and backward directions. Close-by neighbors are thus influenced. In developing and dry or wet etching, closely spaced openings consume the resist developer or etchants more slowly than sparsely spaced openings. Figure 7.97 shows three levels of etchant consumption that lead to three levels of etch rate. Where the etchants are depleted sooner, the etch rate is lower.

Microloading also occurs during the chemical mechanical polish (CMP). Here, in addition to chemical depletion, the pressure difference due to the difference in pattern density is also a cause of the CMP proximity effect.

Aside from 1D OPE, there are 2D OPEs, such as corner rounding and line-end shortening, as shown in Fig. 7.98. These 2D OPEs also result from the finite

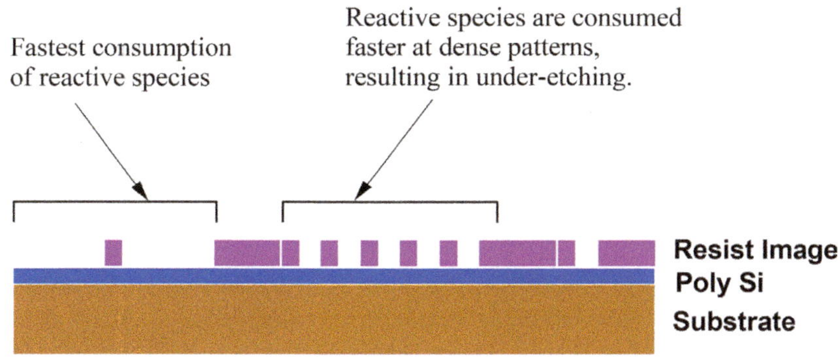

Figure 7.97 Proximity effects due to microloading during the etch.

pupil size of exposure tools. Although rounding of corners is irremediable with OPC if the NA of the imaging lens is not allowed to increase, OPC is done for lithographic reasons rather than for pattern fidelity (e.g., making sufficient poly-endcaps to avoid leakage or using image areas for landing of contacts or vias to reduce contact resistance). Figure 7.98 also shows other OPEs that are independent of being 1D or 2D. These are local CD changes, which when reaching the extreme become missing patterns. Patterns touching each other is also a derivative of CD changes. In fact, the idea of OPC is to counter the local

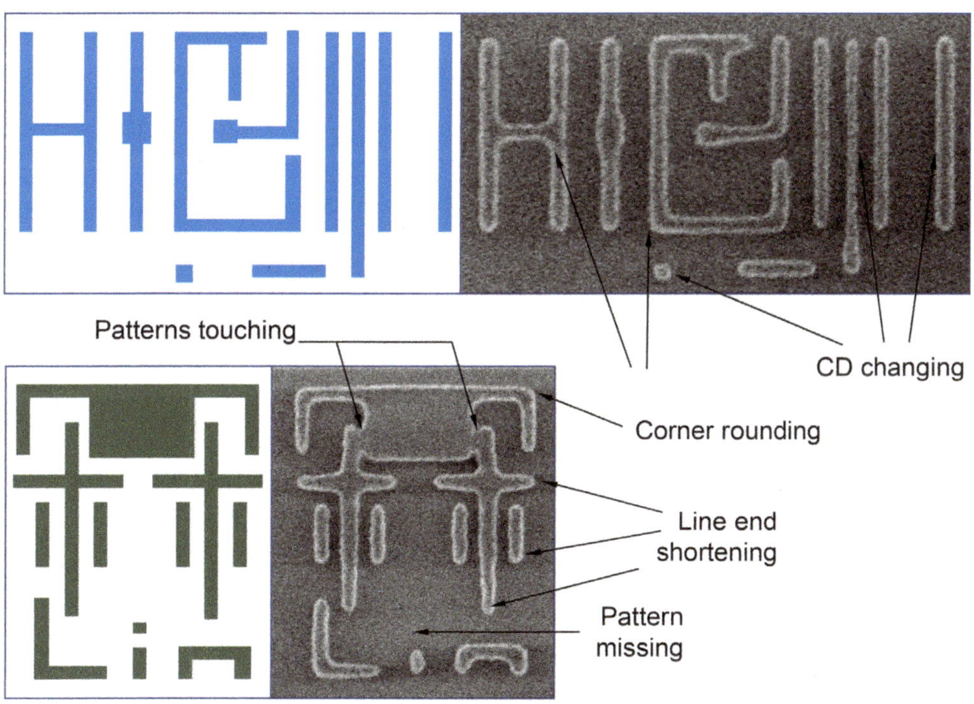

Figure 7.98 2D and other proximity effects (images courtesy of R.G. Liu).

CD changes from OPEs so that the pattern edges or areas meet the circuit requirements. Incidentally, the pattern shown in the bottom of Fig. 7.98 was chosen for study by the author's OPC colleagues because it includes patterns resembling the author's Chinese and English last names.

7.3.4.2 Edge corrections

The most straightforward way to adjust the position of the image edge is to modify the pattern edge at the mask. Figure 7.99 shows how the mask pattern edge is modified to bring the edge of the wafer image as close to the desired position as possible. The line ends are enlarged with a so-called hammerhead to maintain the linewidth and line length. The inside corner is carved in, while the outside corner is beefed up. The line edge facing a line end is moved slightly away from the line end. The islands are enlarged to compensate for the reduced image size. Even though the loss is averted, the size of the islands is still much smaller in the wafer image.

The basis of edge correction lies in relocating the edge of the mask pattern such that the image edge falls on the desired position. Figure 7.100 shows how to determine the direction in which to move the pattern edge.

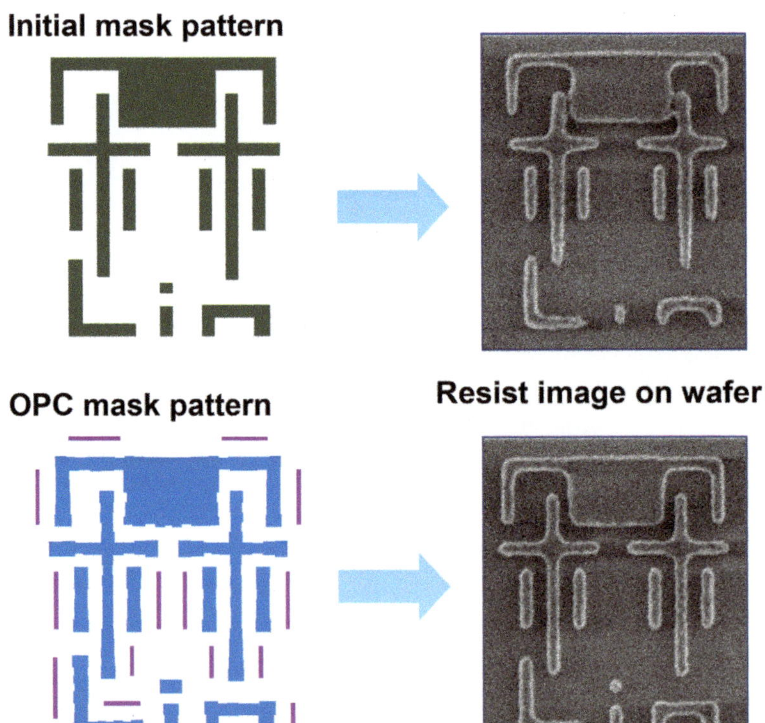

Figure 7.99 OPC by edge bias (images courtesy of R.G. Liu).

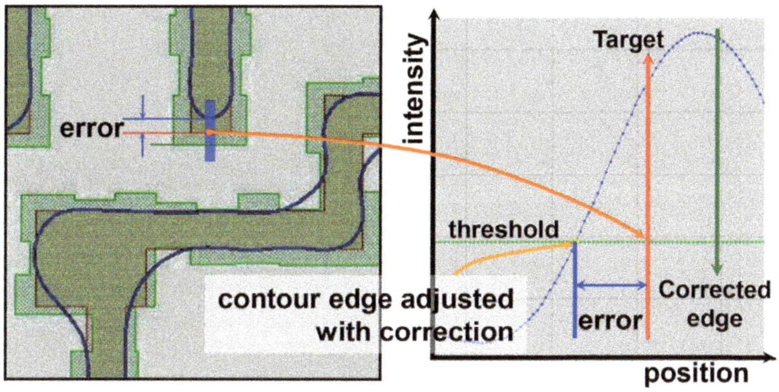

Figure 7.100 Correction by moving the edge to meet the threshold (image courtesy of R.G. Liu).

The exact amount of movement needed is related—but not equal—to the amount of edge error. There are three ways to assign the changes at the mask pattern edge: rule-based, model-based, and manual fine tuning. The latter is suitable only for highly repeating cells, such as those used in memory circuit fabrication.

7.3.4.3 Edge correction by rule-based OPC

Rule-based OPC follows experimentally determined OPC rules. Such rules include the bias table for 1D patterns, and serif or hammerhead rules for 2D patterns. With the help of design rule checking (DRC) in CAD tools, OPC can be performed automatically. For correction efficiency, OPC rules are established by considering only the pattern itself or its nearest-neighboring patterns. Thus, OPC is only applicable for correcting loosely distributed application-specific integrated circuits (ASICs) or patterns with a fixed environment, such as memory cells. Figure 7.101 shows an OPC bias table, which indictates the change in features according to their width and space, and their distance from a neighboring feature. The magenta color indicates an addition to the feature, and the light green color indicates a subtraction.

OPC rules can become quite complicated. Take the hammerhead rule, for example. Figure 7.102 illustrates the simple rule that, when the space between a line end and another feature is larger than 0.20 μm, a hammerhead with the dimension shown should be added. However, the added hammerhead is not allowed to protrude into the 0.20-μm zone. Hence, the inclusion of the partial hammerhead in the rightmost part of the figure.

7.3.4.4 Edge correction by model-based OPC

Model-based OPC uses a model to predict the wafer image from the mask pattern, using an equation similar to Eq. (4.53). The model is a transfer function connecting the intensity at the mask plane to the intensity at the wafer plane:

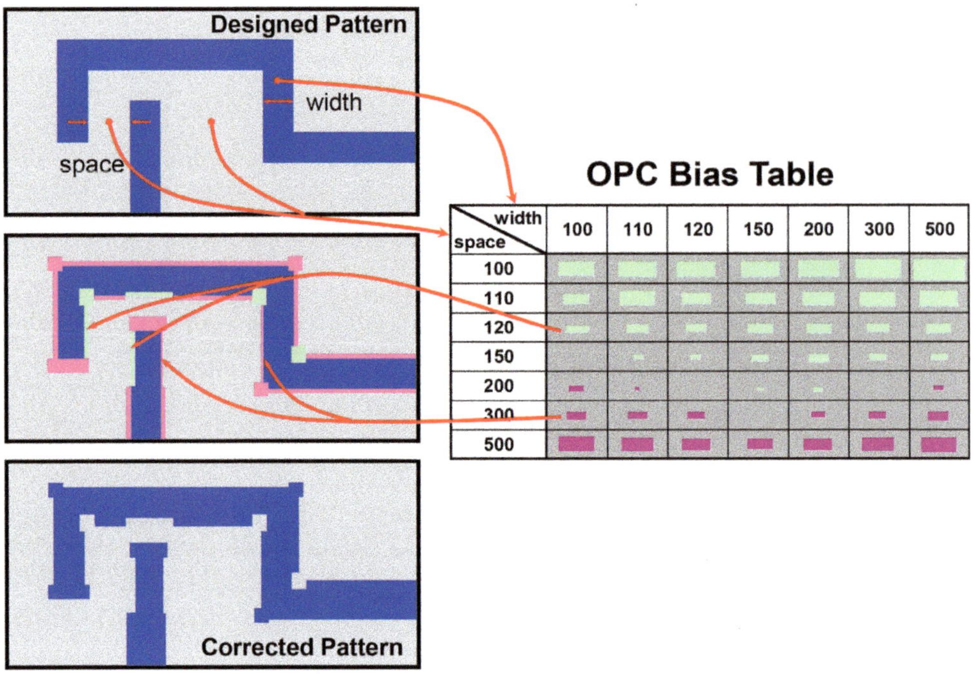

Figure 7.101 Rule-based OPC from a lookup table (image courtesy of R.G. Liu).

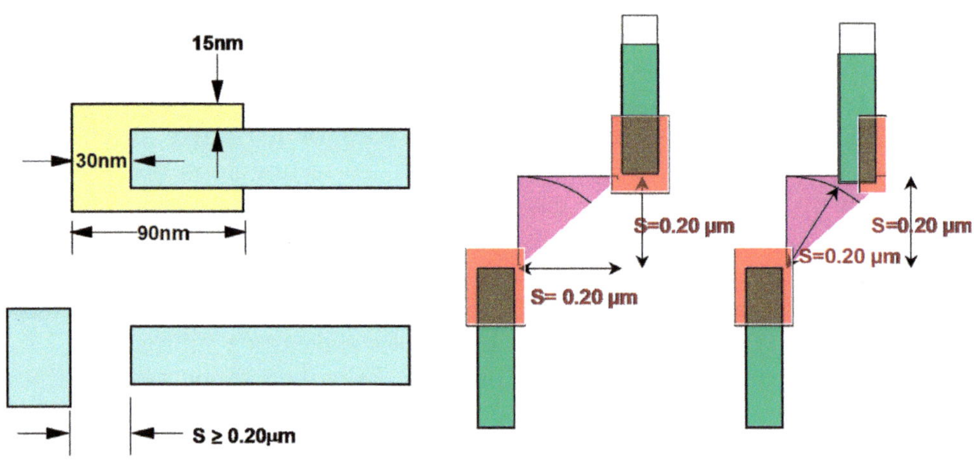

Figure 7.102 Illustration of the hammerhead rule as an example of rule-based OPC (images courtesy of R.G. Liu).

$$I_i(x_i, y_i) = \int_{-\infty}^{\infty}\int_{-\infty}^{\infty} I_o(x_o, y_o)H(x_i, x_o, y_i, y_o)dx_o dy_o. \tag{7.60}$$

Actually, I_i can also be governed by Eq. (4.48) or Eq. (4.54), depending on the degree of coherence of the illumination. However, the functional connection is more than just an optical connection. The transfer function includes the effect of anything that can affect the final image. After the formation of the optical aerial image, the resist exposure, baking, and development behaviors take place, followed by etching. The transfer function in Eq. (7.60) must encompass all of these effects. Theoretically, the governing equation of each phenomenon must be used in the following steps:

1. transfer I_o through the aerial image stage,
2. propagate through multiple reflections and absorptions in the resist,
3. create the acid in the resist from the exposure,
4. amplify the acid through PEB if the resist is a CAR,
5. diffuse the acid during PEB,
6. describe the time development behavior of the resist needed to produce the after-development image (ADI), and
7. account for the etching behavior needed to transfer the resist image to the after-etch image (AEI).

At the aerial image generation stage, the governing equations can be Eqs. (4.48), (4.53), or (4.54). All other governing equations are assumed to be in the form of Eq. (4.53). In reality, it is impractical to use so many stages of calculation due to the computation time and lack of exact analytical description of the physical phenomena. One would use experimental results to establish an empirical model to transfer the mask image into the wafer image through Eq. (7.60). This is the starting point of model-based OPC.

The model is generated by fitting kernels to the experimental data. As shown in Fig. 7.103, representative patterns with relevant points at the pattern edges are measured. The data are used to evaluate the kernels for OPE.

For lack of a theoretically derived closed-form expression, a set of eigenfunctions with adjustable coefficients is used to represent the transfer function H. These eigenfunctions are called kernels, as they are used as components of H in Eq. (7.60). Figure 7.104 shows a typical kernel and examples of four other kernels. The coefficients for the kernels are evaluated by fitting the kernels to experimental data. Hundreds to thousands of data points must be taken in the case of using many high-order kernels. Thus, model fitting is a time-consuming process requiring very careful and accurate metrology.

The accuracy of the model is judged by the fit between the resist image and the model-generated image. Figure 7.105 shows the image edge contour generated from the given model, superimposed on the resist image, confirming both the accuracy of the model and the fidelity of the SEM tool used to take the resist image. Contour plots are time consuming. Other ways to determine

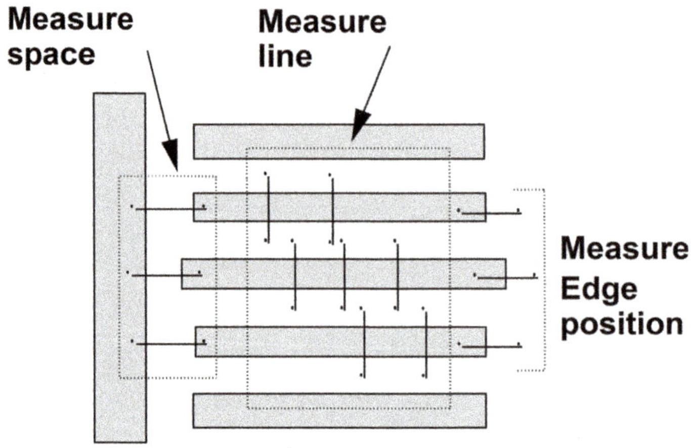

Figure 7.103 OPC test site. The values measured from the wafer image are used to evaluate the kernels (image courtesy of R.G. Liu).

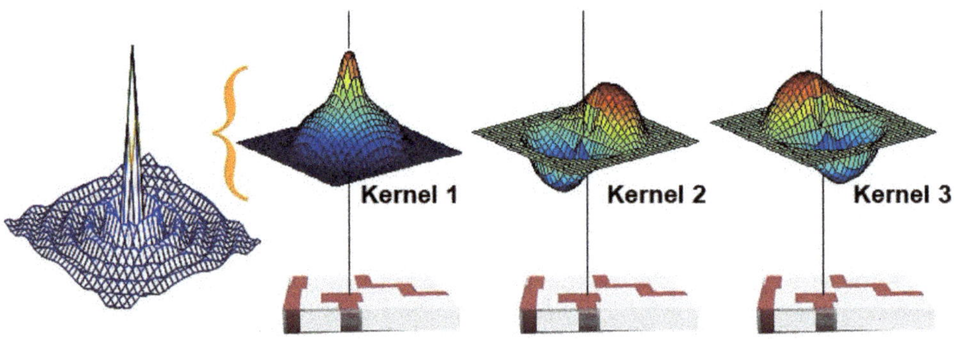

Figure 7.104 Model and kernels (images courtesy of R.G. Liu).

the fitting accuracy involve plotting the CD and edge intensity as a function of pitch for 1D patterns and plotting the line-end position as a function of separation.

Once the model is established, the first-pass wafer image from the mask can be calculated and compared with the target image, which is usually the image edge position at selected target points. The mask pattern is adjusted to reduce the difference, calculated, and compared again until the difference is acceptable, as depicted in Fig. 7.106. The whole process starts with the layout from the designer: $I_o^0(x_o)$. This layout is used in two operations. First, the target point distribution $I_T(x_i)$ is generated from the given layout. The target points are points along the pattern edges on the layout that will be compared with corresponding points in the wafer image $I_i^0(x_i)$, which is generated by applying Eq. (7.60) to $I_o^0(x_o)$ with the transfer function $H(x_o, x_i)$; this is the second operation on $I_o^0(x_o)$. The first corrected layout $I_o^1(x_o)$ is made based on

Figure 7.105 Model-generated edge contour superimposed on the resist image (image courtesy of R.G. Liu).

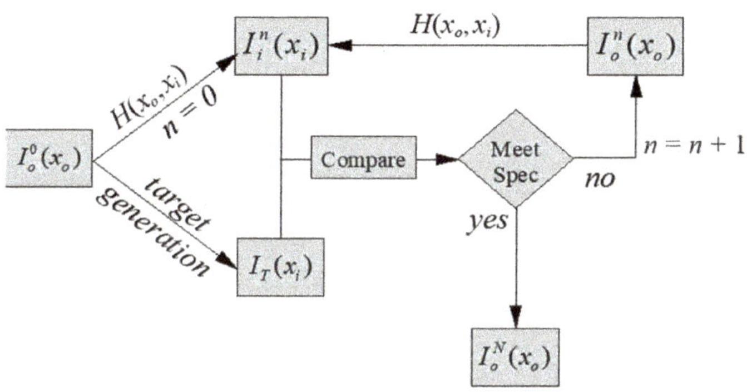

Figure 7.106 Block diagram showing the flow of model-based OPC (image courtesy of R.G. Liu).

the difference from the comparison and goes through the transfer function to generate $I_\mathrm{i}^1(x_\mathrm{i})$ for comparison with $I_\mathrm{T}(x_\mathrm{i})$ for the second round. If the comparison shows an acceptable difference, the new layout $I_\mathrm{o}^1(x_\mathrm{o})$ is accepted as the optical-proximity-corrected layout for mask making. If the difference is not acceptable until $n = N$, then $I_\mathrm{o}^N(x_\mathrm{o})$ is the accepted optical-proximity-corrected layout.

The target point is an important part of model-based OPC. Figure 7.107 shows how a pattern is dissected at its edges. There is one target point in each dissection. Note that the dissections and the target points need not be equally spaced. The edge section may move in or out depending on the difference between the location of the diffracted wafer image and the target image. In this figure, target point 1 in the first dissection is moved out. The amount of

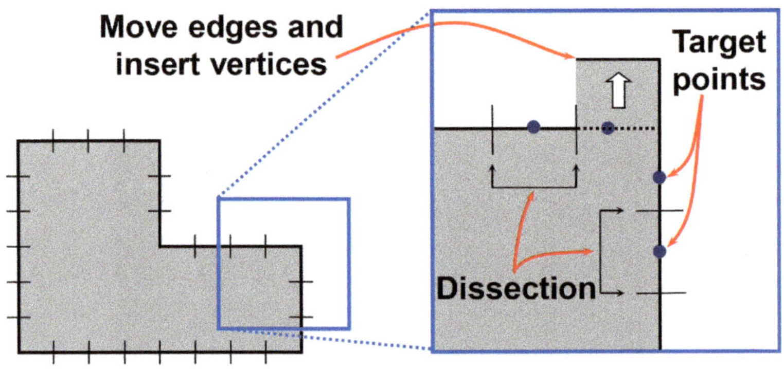

Figure 7.107 Dissections and target points (image courtesy of R.G. Liu).

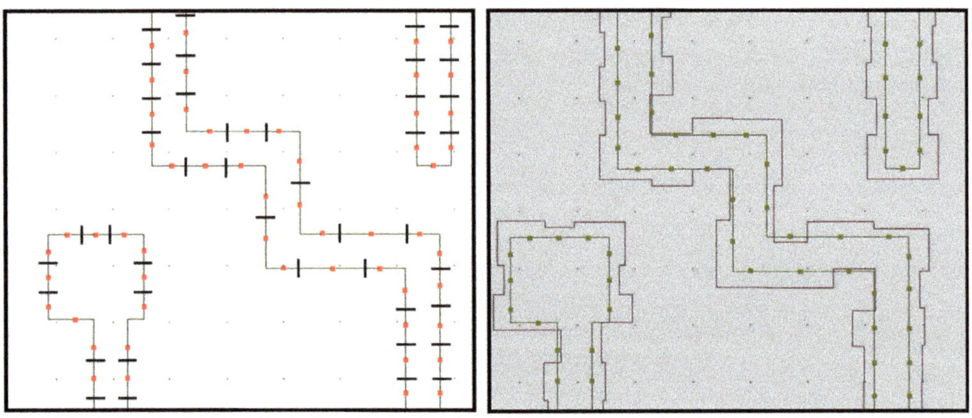

Figure 7.108 Correction vectors applied to an entire pattern (images courtesy of R.G. Liu).

movement indicated by the correction vector is a function of the difference in the edge positions according to the model.

Figure 7.108 shows the pattern from Fig. 7.107, the correction vectors applied to the entire pattern, and the resultant optical-proximity-corrected layout. The placement of the dissections and target points is very important. An incorrect placement leads to slowly converging iterations. If the iterations converge, the OPC improvement may meet the specs only at the target points, but not between them. Too many dissections prolong the cycle time. Days of running time at hundreds of central processing units (CPUs) are often required for complicated layouts, even with optimally placed target points.

Reducing the OPC time while maintaining correction accuracy is very demanding. In every node advancement, the pattern density increases by a factor of 2. If the computation efficiency is not improved, the OPC time will increase by a factor of 2. One way to reduce the OPC time is to take advantage of the hierarchy in the original design, as illustrated in Fig. 7.109.

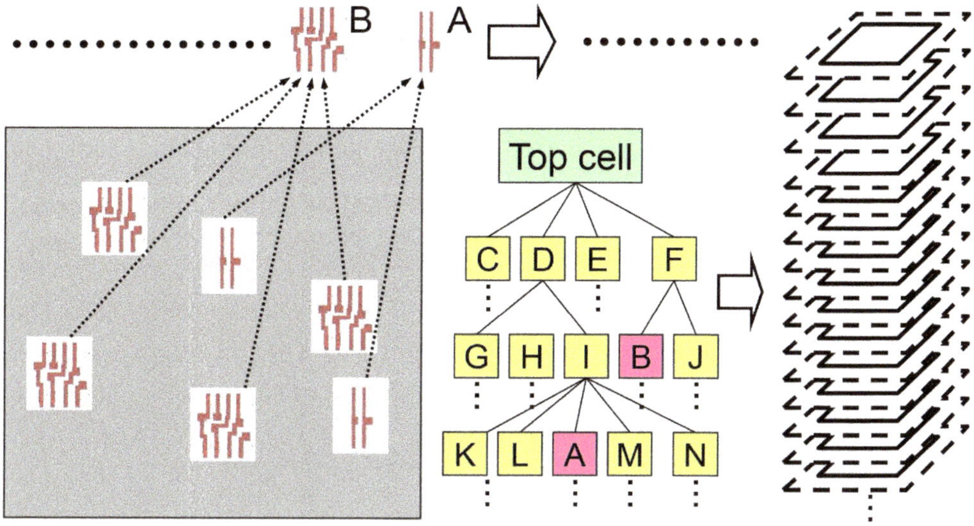

Figure 7.109 Use of a hierarchy to easily correct identical patterns in one step for all layouts (image courtesy of R.G. Liu).

One seeks the identical patterns in the layout and corrects one pattern for all other patterns. Unfortunately, the hierarchy in the design was established to improve the efficiency of the design, not for OPC. A usable hierarchy for OPC must include the neighborhood of the identical patterns within the OPE range. The hierarchy also must be identical for all representative corrections. Finally, the algorithm for handling hierarchy for OPC is not yet mature, and hierarchical OPC is still not very dependable.

7.3.4.5 Local-dosage OPC

So far, edge corrections have been covered. However, this is not the only correction method. One can correct for proximity effects by local-dosage OPC (LD-OPC) just as in the case of direct-write e-beam, where the beam current or dwell time can be set point-by-point. For optical projection, a local dosage variation is imbedded in the mask. Figure 7.110 shows the combination of dense and isolated openings taken from Fig. 7.96. A gray-level mask having 12.9% absorption in the dense area is used to lower the intensity distribution of the dense line openings by 0.056. The openings in the image are now much closer in dimension to each other at the new threshold of 0.244 from 0.3 in Fig. 7.96. This technique is called threshold leveling.[46]

A more sophisticated LD-OPC method is shown in Fig. 7.111. Subresolution half-tone features are properly distributed on the mask to strengthen the outside corners and to thin down the inside corners. Line ends and contact holes are beefed up. All corrections are within the confinement of the original openings. This way, features do not run into each other as in the case of edge-correction OPC. The distribution is automatically generated with

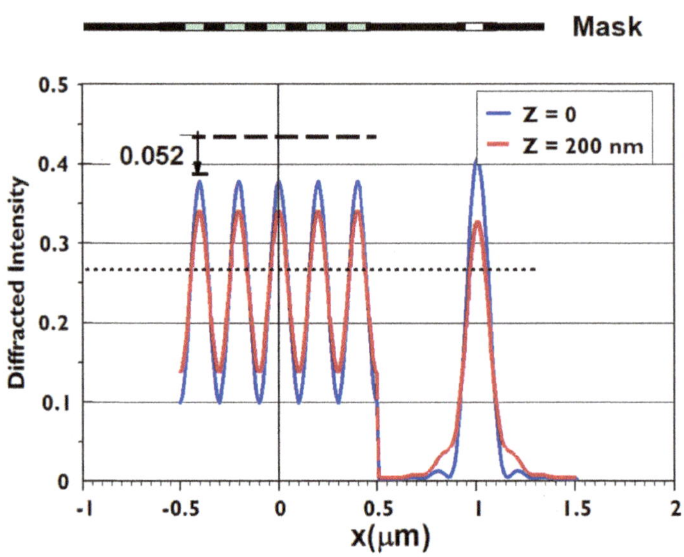

Figure 7.110 Threshold leveling.

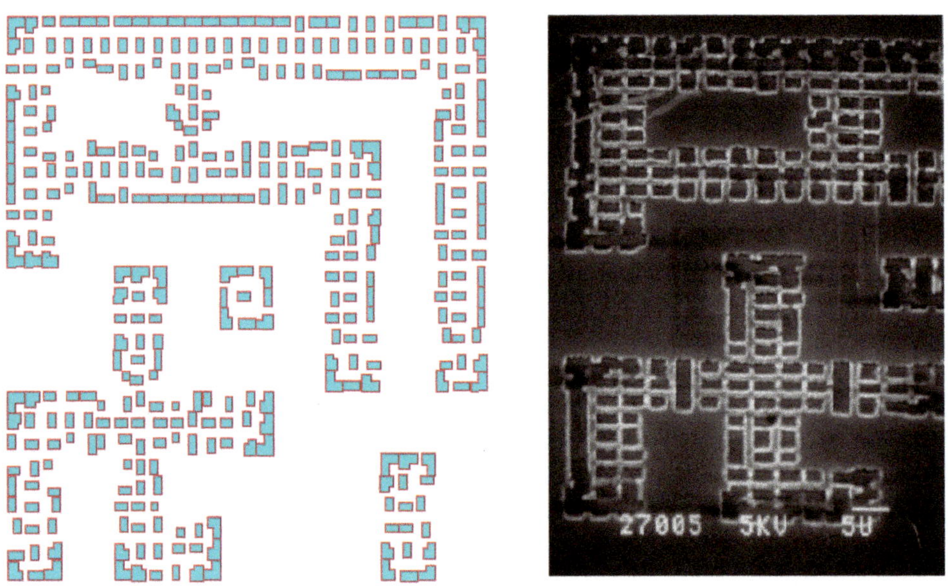

Half-tone mask design Fabricated half-tone mask

Figure 7.111 Half-tone mask used for local-dosage correction.

an algorithm given in Ref. 47. Figure 7.112 shows a printed result from the half-tone corrected mask. The 0.58-k_1 optical-proximity-corrected image is better than the 0.78-k_1 uncorrected image.

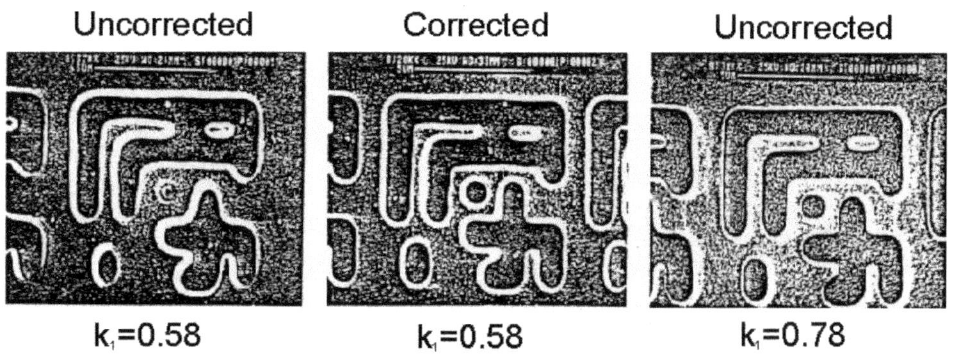

Figure 7.112 Half-tone LD-OPC results.

It is surprising to find the need for OPC even with the relatively high-k_1 imaging condition. In ideal conditions, $k_1 = 0.58$ does not need much correction. However, during the early days of optical lithography, lens aberration and stray light were comparatively high. They tend to reduce the contrast of the optical image, resulting in equivalence to lower-k_1 images. LD-OPC restores the lost resolution to produce a much better image. Similarly, a stringent surface accuracy and smoothness requirement at the 13.5-nm EUV wavelength leads to severe stray light and difficult aberration correction. The equivalent k_1 in EUV images is worsened and can use some restoration.

7.3.4.6 Full-depth OPC

Most OPC methods optimize the wafer image at the focal plane (FP-OPC) and take any improvements from the off-focus image. To get the most benefit from OPC, the entire range of defocus to be imaged should be considered. To make this point, Fig. 7.113 shows the E-D trees of three different resist thicknesses: 750, 820, and 960 nm. The feature is a 250-nm resist line delineated at $NA = 0.55$, $\lambda = 248$ nm, ring illumination σ_{in}:σ_{out} = 0.4:0.8. The E-D trees are evaluated at $CD_{tol} = \pm 10\%$ and $EL = 4\%$. The common E-D windows clearly do not line up with the center of the individual E-D trees. This means that the center of focus is different for each resist thickness. These resist thicknesses occur simultaneously due to the topography on the wafer. Full-depth OPC (FD-OPC) is performed by adjusting the bias of the features on the mask to center the E-D window while keeping the same 250-nm CD. The DOF is now 1.04 μm instead of 0.4 μm.

Figure 7.114 shows another example of the need for FD-OPC. Equal line–space pairs and an isolated line opening, both at $k_1 = 0.66$, are imaged together. Disk illumination at $\sigma = 0.5$ is used with $\lambda = 365$ nm, $NA = 0.48$, CD = 0.5 μm ± 10%, and $EL = 5.3\%$. The trees have a center of focus at 154, –231, and 308 nm, as indicated in the figure.

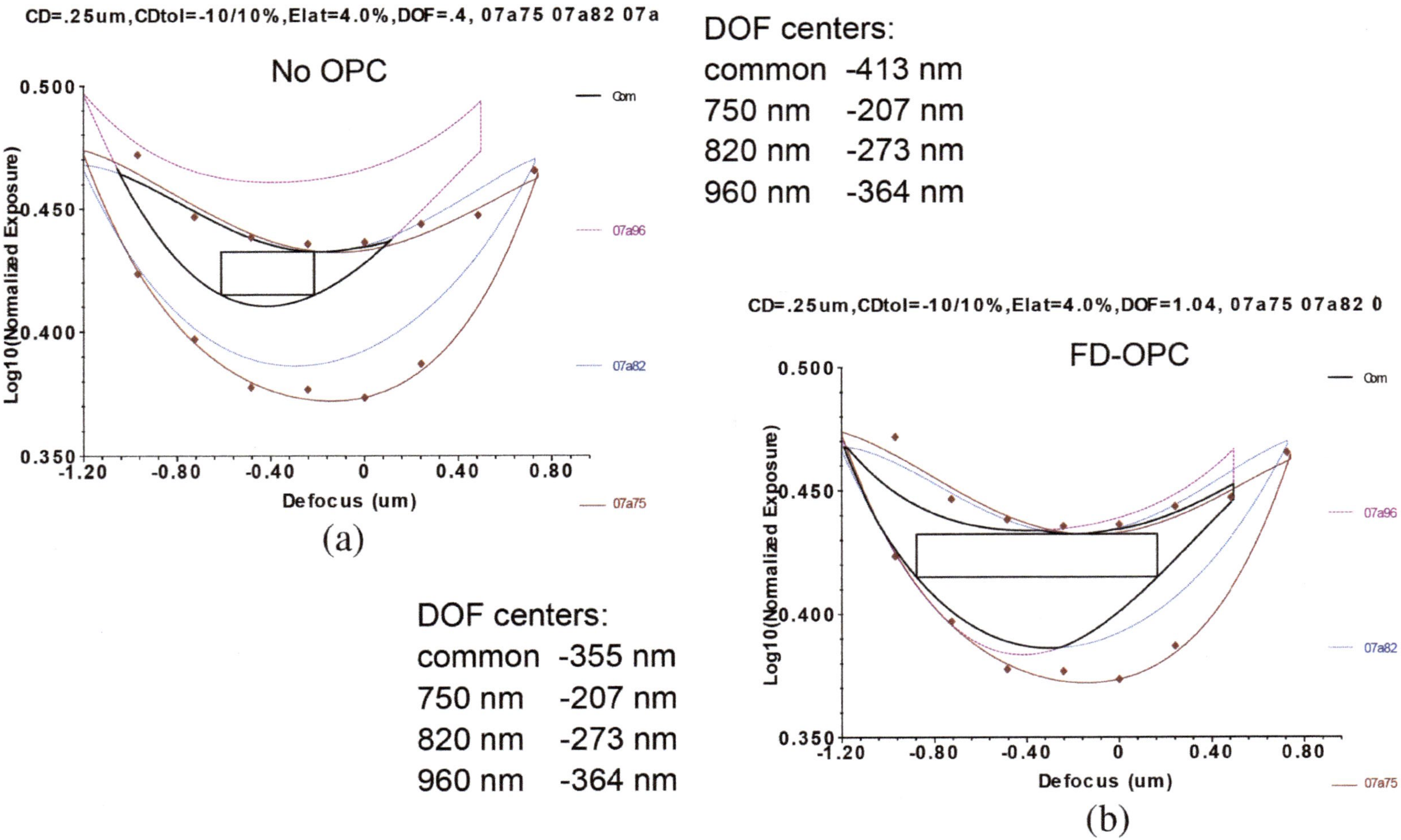

Figure 7.113 E-D windows of three resist thicknesses. The E-D windows do not center at best focus.

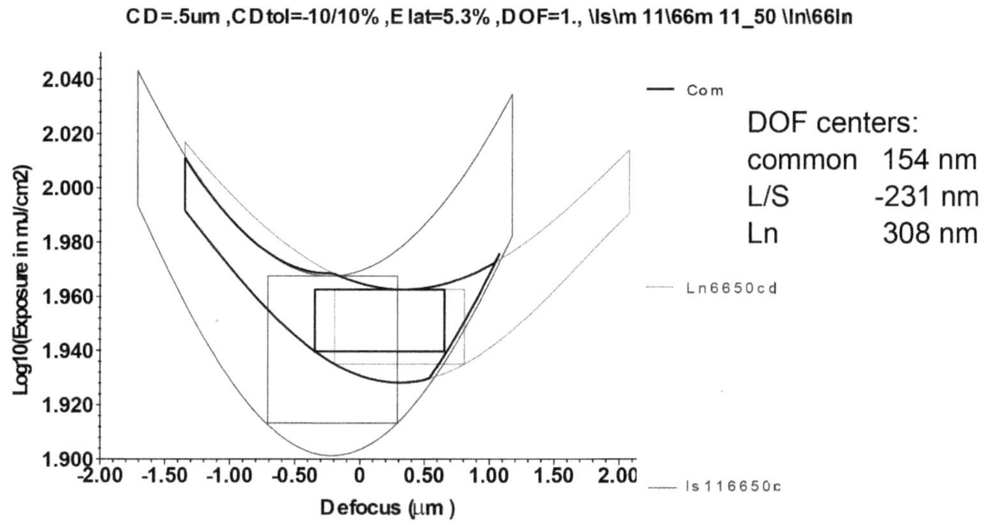

Figure 7.114 E-D window for two 0.66 k_1 features: line:space = 1 and an isolated line. The separation of their centers of focus is as large as 539 nm.

Performing FD-OPC is more time consuming than FP-OPC and FD-OPC. Instead of generating the wafer image at the focal plane, one must take as many image planes as necessary and evaluate the difference between each image and the target image. Adjustment of the mask pattern edge is now based on minimizing the differences among all of these image planes.

7.3.4.7 Correcting the after-etch image (AEI)

The electrical circuit performance is based on the image that is transferred from the resist image, which is usually the etched image. Therefore, OPC should be based on the after-etch image (AEI). However, it is not easy to extend model-based OPC to AEI for the following reasons. (1) The range of microloading types of proximity effects is much larger than that of optical kernels. For the latter, a few micrometers are sufficient to include in the 2D integration of Eq. (7.60), whereas the former case requires a range at least an order of magnitude larger, thus putting a tremendous burden on the OPC cycle time and cost. (2) During the development phase of a manufacturing technology, the lithography and etching processes usually go through independent iterations. If the OPE model includes both lithography and etching, frequent re-establishment of the model is required, placing a heavy burden on time and resources.

The most popular practice for extending OPC to the AEI is to establish a model for the ADI for model-based OPC and to build a correction table from the resist image to the etched image for rule-based correction from the calculated ADI.

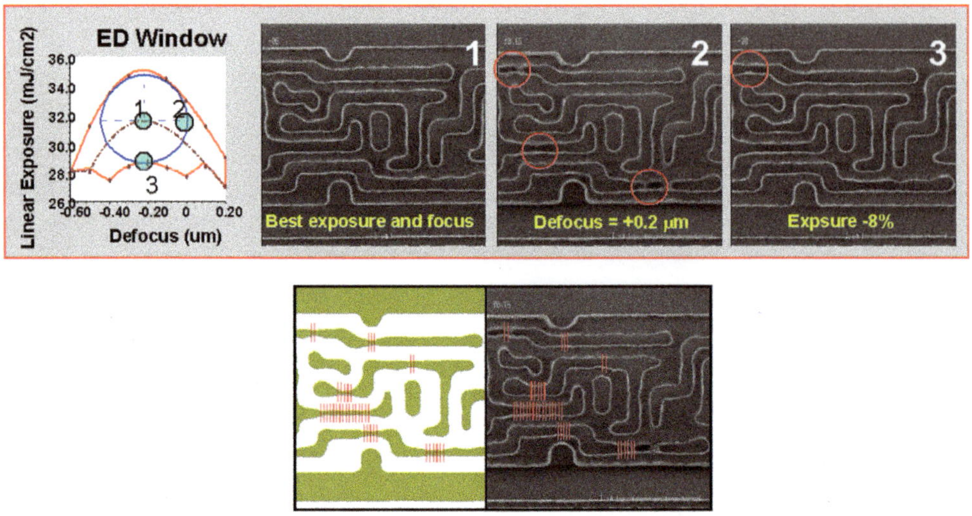

Figure 7.115 Series of resist images in the top row indicating hot spots near the boundary of defocus and exposure. The markings in the lower row show the detected hot spots on the left and the corresponding locations on the resist image on the right (images courtesy of R.G. Liu).

7.3.4.8 Hot-spot checking

After OPC, but before sending the data for mask making, it is preferable to perform a hot-spot check to fine tune the corrected mask pattern. The reason is that the OPC spec is based on the summation of the deviation from the target image at all target points. This addresses the fidelity of a single image edge but not the distance between nearby edges. There may be situations where two edges get too close. Even though the design rule before OPC addresses feature size and the distance between edges on the design, OPC changes them. The optical-proximity-corrected image must be examined. Figure 7.115 shows the formation of hot spots at the boundary of the defocus and exposure. In the lower row, the hot spots identified with a hot-spot check are marked on the optical-proximity-corrected image and the resist image.

7.4 Polarized Illumination

As the feature size of integrated circuits shrinks to the subwavelength regime, polarization of the illumination becomes very important. A lot of resolution can be lost with the wrong polarization. Therefore, polarization can be considered as another k_1-restoration technique. The logical placement of polarized illumination in this book should be here. However, this topic was extensively explored in the first edition in the chapter on immersion lithography. In order not to drastically disturb the original organization of this book, this discussion and its update remain in Chapter 8.

7.5 Multiple Patterning

With the RET discussed in Section 7.3 and exemplified in Section 7.3.2.2.6, the smallest pitch that can be resolved with optical lithography is $\lambda/2$. One can also realize the resolution capability by

$$\theta = \sin^{-1}\frac{\lambda}{2P}, \tag{7.61}$$

where P is the periodicity of the grating.

By the time the 28-nm node came into use, optical lithography was using $NA = 1.32$, $\lambda = 193$ nm, with elaborate source mask optimization, pushing k_1 to the regime of 0.28, which is very close to the theoretical limit of 0.25. The only way to shrink further is to use multiple exposures to split the pitch. Unfortunately, the photoresist remembers previous exposures and cannot prevent image interference even though the exposures are separated. One must resort to multiple patterning, i.e., patterning each exposure all the way to etching, stripping the resist, and coating a flash layer of resist to receive the next exposure. Before EUV lithography was ready for use, the multiple-patterning technique (MPT) carried semiconductor manufacturing through many nodes.

7.5.1 Principle of the multiple-patterning technique (MPT)

The idea of using MPT to split the pitch is illustrated in Fig. 7.116. Newer-generation masks are split into two or more levels—as many levels of patterning as required. Each of these masks would have a minimum pitch that is acceptable for the existing technology node. The steps are as follows: (1) expose the first mask, (2) develop the resist, (3) etch the pattern into the wafer, (4) strip the resist, and (5) coat a new layer of resist to delineate the next mask

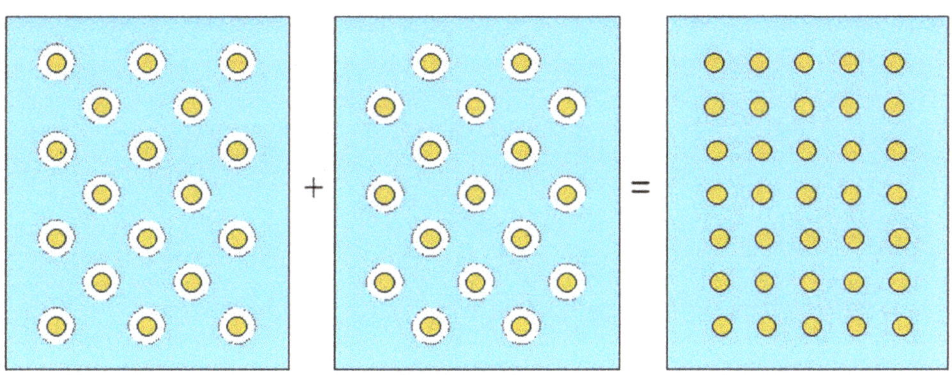

Figure 7.116 Pitch splitting to double the pitch by combining patterns with shrunken features.

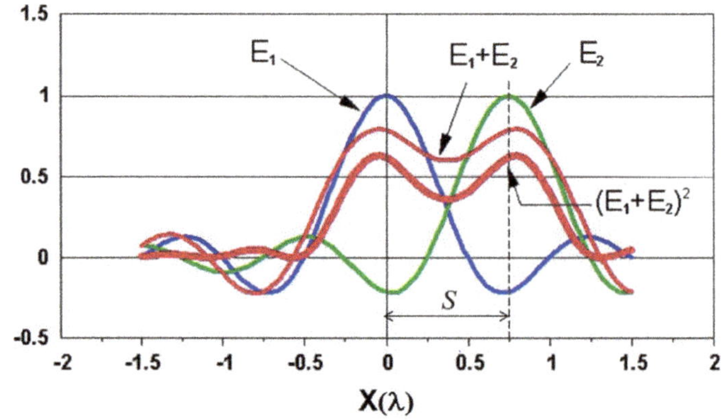

Figure 7.117 Coherent summation with S = 0.75λ.

pattern. During the first patterning, the feature size will be shrunk to the dimension suitable for the new generation.

Why not simply multiply expose the mask and etch after all the exposures are done? This would save a lot of etching and resist costs because the resist memory of residual exposures can accumulate to ruin the image, even though simply splitting the exposure can remove optical interference effects. Let us first examine the optical interference effect. Figure 7.117 shows the electric field distribution $E_1(x)$ as a sinc function of x centering at $x = 0$. This is a good representation of the diffraction from a line opening. Another electric field that centers at $x = 0.75$ is plotted beside the first one. The distance between the centers of these two fields is S. With coherent illumination, the two fields just superimpose and are plotted as $E_1 + E_2$. The combined electric field has only a slight dip in the center, indicating a tremendous loss in image contrast. The two lines are barely resolvable. In fact, when the two fields get closer to $S = 0.67\lambda$ as shown in Fig. 7.118, there is no longer any dip, and the two lines

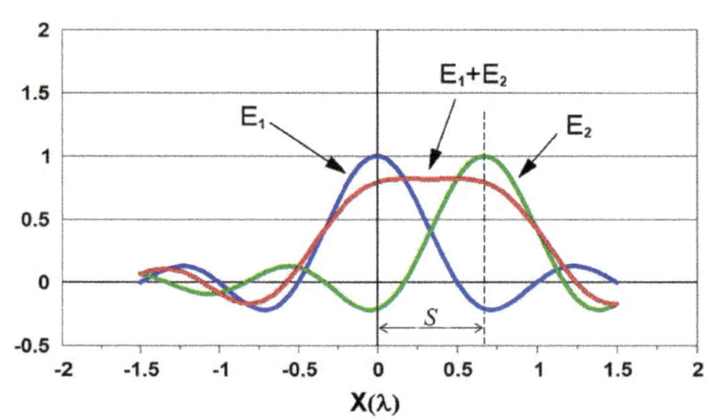

Figure 7.118 Coherent summation with S = 0.67λ.

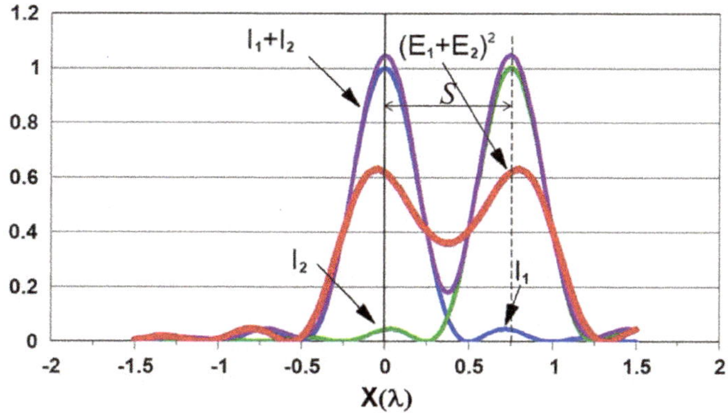

Figure 7.119 Incoherent summation with S = 0.75λ.

cannot be resolved. The intensity distribution $(E_1 + E_2)^2$ is also plotted (see Fig. 7.119) because the resist only responds to the light intensity, not the electric field, and the resist image is determined by the intensity distribution.

The resolution can be improved by splitting the exposure into two exposures. Figure 7.119 shows the image intensity of the two features superimposed incoherently. This is the situation of multiple exposure. The superimposed intensity of the two curves is much smaller than in the coherent case, which is shown for comparison. The resolution continues to improve until $S = 0.5\lambda$, as shown in Fig. 7.120.

The two lines become absolutely unresolvable at $S = 0.42\lambda$, as shown in Fig. 7.121. Note that the loss of resolution is still related to the stored exposure for pattern 1, which reduces the contrast when pattern 2 is exposed.

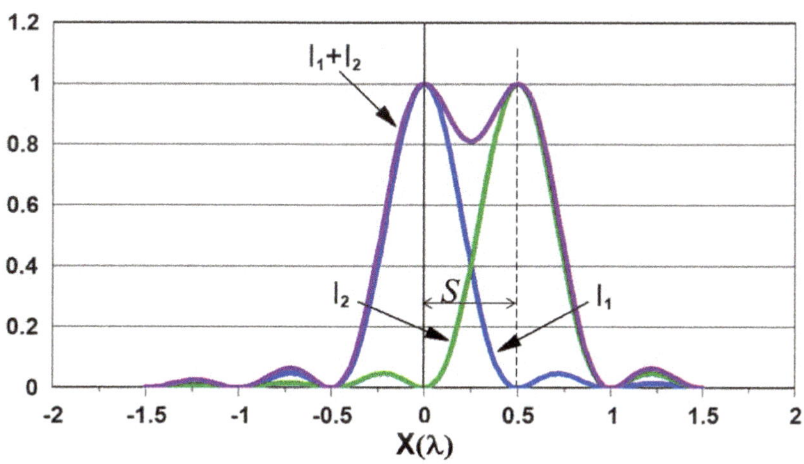

Figure 7.120 Incoherent summation with S = 0.5λ.

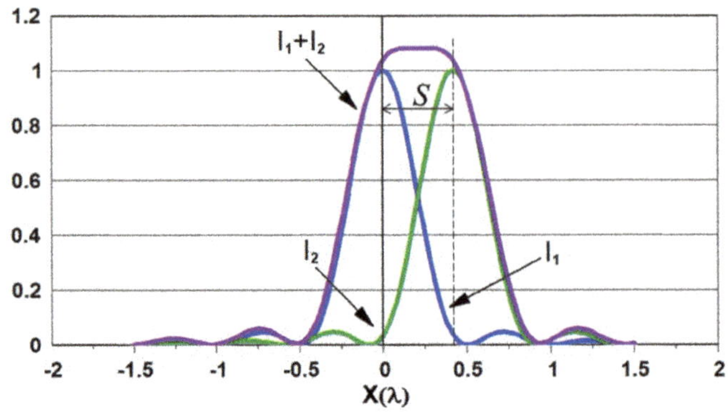

Figure 7.121 Incoherent summation with S = 0.42λ.

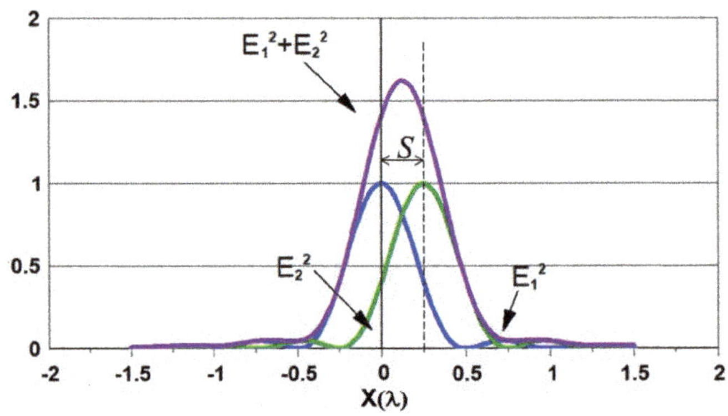

Figure 7.122 Incoherent summation with S = 0.25λ.

When $S = 0.25\lambda$, it is at the half-point between the origin and the first zero of the sinc^2 curve. The combined intensity has a sharp peak directly in the middle of the two individual sinc^2 curves, as shown in Fig. 7.122.

Now, to completely eliminate the resist memory, the 1st resist is stripped after serving for patterned exposure, development, and as an etch mask. There is no longer any resist to remember the 1st pattern. The 2nd resist is applied to the wafer and is delineated independent of the prior patterning. The sequence is shown in Figs. 7.123 through 7.125.

It looks as though the two lines will still merge when the center of the image at the right-hand side is moved closer to the center of the left-hand-side image by 0.5λ, which is not necessarily the resolution limit of double patterning. Figure 7.126 shows that with a proper etch bias, good position control of the wafer stage, and an accurate overlay of the two exposures, the two lines can get closer.

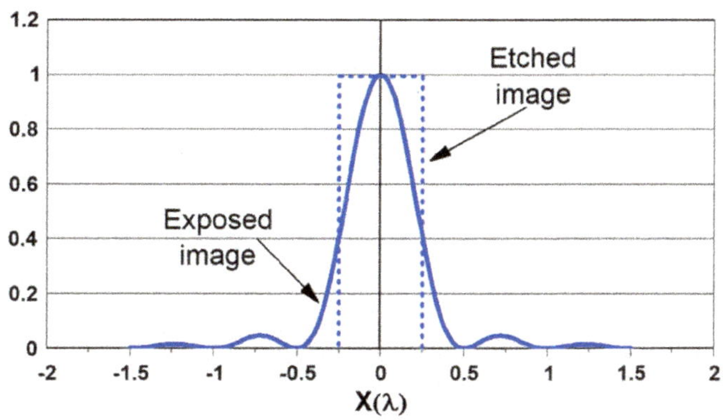

Figure 7.123 Double-patterning step 1. The resist is exposed with the 1st mask and is developed, and the wafer is etched.

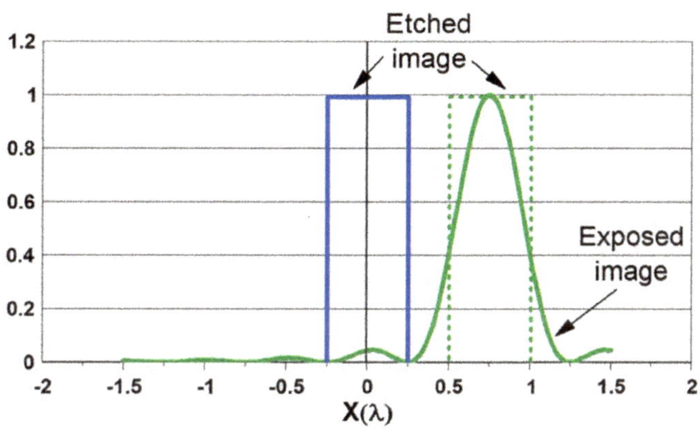

Figure 7.124 Double-patterning step 2. The 1st resist is stripped, then the 2nd resist is coated and delineated.

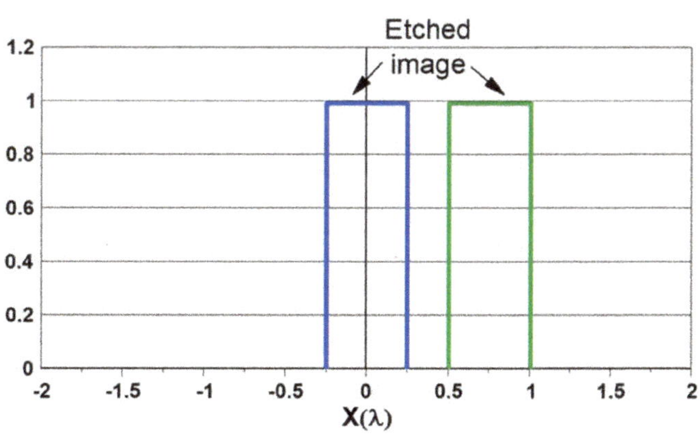

Figure 7.125 Double-patterning step 3. The 2nd resist is developed, the wafer is etched, and the resist is stripped.

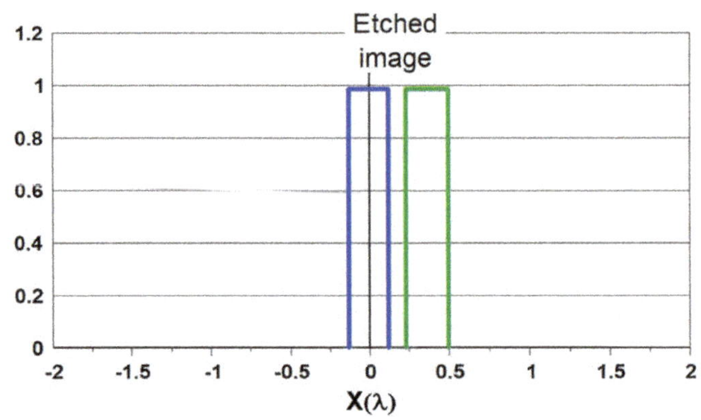

Figure 7.126 Double-patterning RET. The double-patterning technique (DPT) can further increase resolution by increasing the etch bias. Better overlay and wafer stage accuracy must be improved to support the better resolution.

7.5.2 MPT processes[48,49]

To fully describe a MPT process, one would name the lithography steps and etching steps by number. Thus, a MPT process can be 2L2E, 3L2E, 3L3E, pronounced two litho two etch, three litho two etch, three litho three etch, etc. 2P2E, 3P2E, 3P3L, etc., processes have also been used, where 2P means double pattern.

MPT can be used for splitting the pitch, but that is not its only role. It is also often used for pattern trimming. It is further used in the spacer technique, which is also capable of reducing the pitch. Figure 7.127 demonstrates a 3L3E process performing the pitch split and the line-end-cutting processes. This layout consists of three closely spaced lines. Their pitch is not resolvable with a single exposure, so two of the three lines are on Mask 1. The other two lines are on Mask 2. In the final pattern, the horizontal end of the T is too close to the vertical thick line of the L. Even though the T and the L are split into two masks, a third mask is still needed to trim them and produce an acceptable separation between them.

The effectiveness of line trimming can be seen in Fig. 7.128. With a single exposure (1L1E), several resist line ends are joined in the after-develop image (ADI), even though the after-etch images (AEIs) are separated. Using a second exposure to cut the lines (2L1E), the ADI images are well separated, but the shapes are not impressive. Resorting to cutting the etched lines (2L2E), the line ends are free from rounding and shrinking, and the lines are straight through out.

Figure 7.129 shows the triple-patterning process using the spacer technique to define subresolution pitches and features. This technique is usually called self-aligned double patterning (SADP). The same layout as that in Fig. 7.127 is used. For the three lines with unresolvable pitch, two bars with

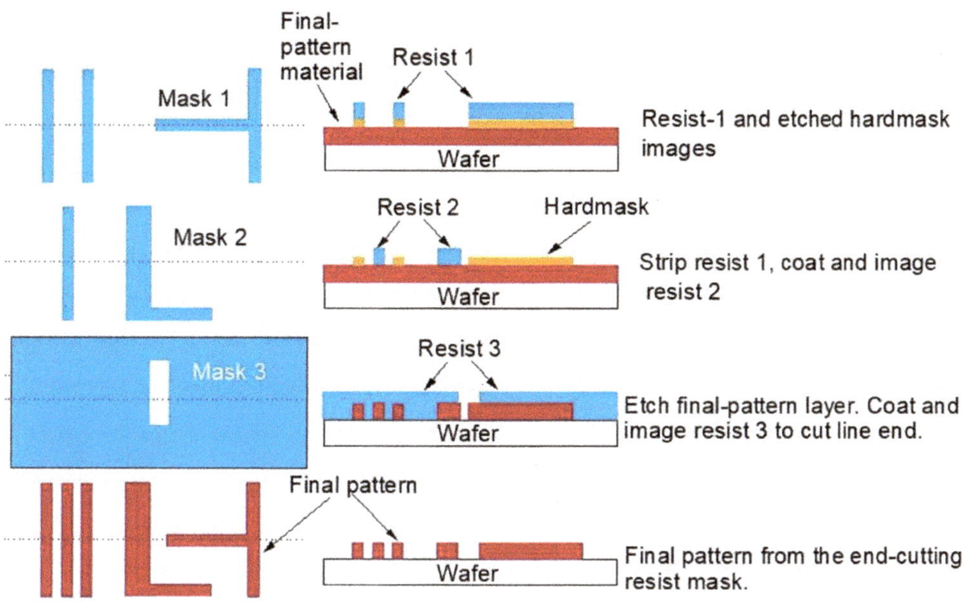

Figure 7.127 Triple patterning using split-pitch and line-end-cutting processes.

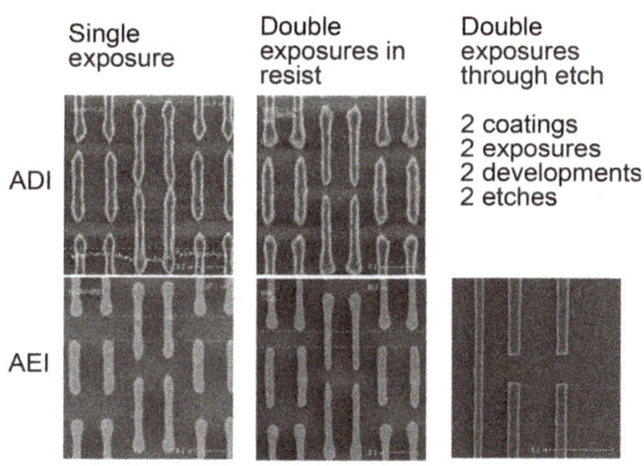

Figure 7.128 Comparison of line end images using single exposure, double exposure, and double patterning (images courtesy of K.S. Chen).

an acceptable separation are first produced. These bars serve as the spacer hosts. A conformable deposition of a spacer material is deposited on the hosts. A subsequent anisotropic etch removes the covering spacer material from the hosts but leaves the sidewall material to be used as the etch mask. Sidewalls inevitably are loops. When the circuit design does not call for loops, the unwanted sidewalls must be removed. Therefore, the second mask is a trim mask that removes the unwanted sidewalls. The third mask adds the thick

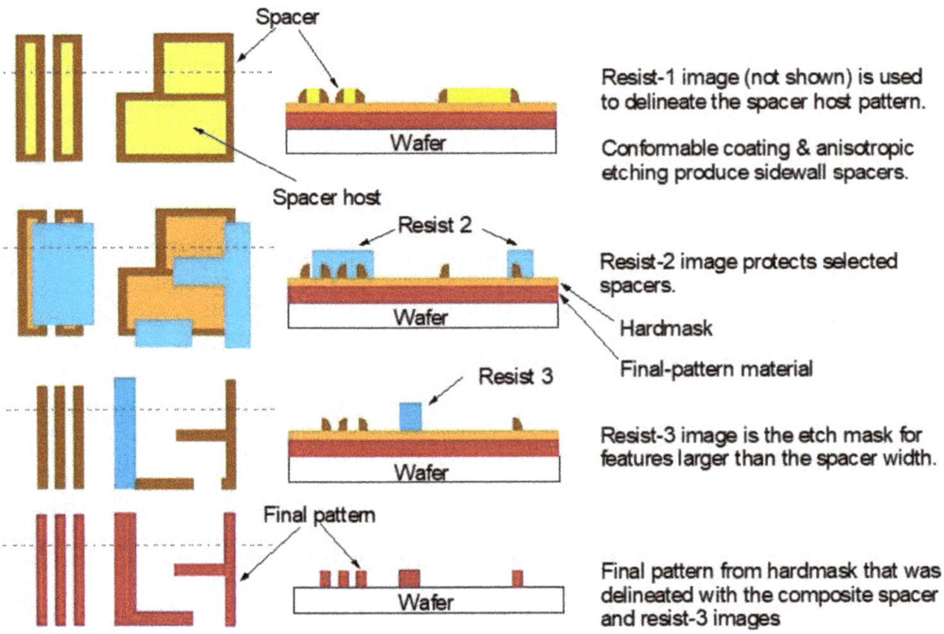

Figure 7.129 Triple patterning using spacers.

vertical line in the L because all sidewalls are of the same width. Any lines with a different width must be produced with an additional mask.

The spacer MPT can place the fine features in the same mask, eliminating overlay errors between masks. However, spacer trimming may leave some imperfect line ends. The sidewall deposition process is often more expensive. In terms of exposure productivity, spacer MPT and pitch-splitting MPT often use a similar number of masking layers.

Difficulties in coating the bottom antireflection layer (BARC) for the 2nd patterning in pitch-splitting MPT can lead to different heights of the etch mask. A placement error between the patterns from the two masks can lead to erroneous spaces between the features. Using a negative mask makes the handling of BARC 2 easier, but the CD can be affected by an incorrect overlay between the two masks, as shown in Fig. 7.130.

Etching is expensive, but there are creative ways to save an etch. For example, Fig. 7.131 shows that Resist 1 can be crosslinked to render it insensitive to further exposures. The example of a crosslinking process that uses mold hardening is given in Section 7.2.2.4.

7.5.3 MPT layouts[48,50]

Some examples of ways to lay out split-pitch masks are presented next. Figure 7.132 shows Masks A and B, which are split for the densely packed horizontal lines. Mask C is used to cut the dense horizontal lines.

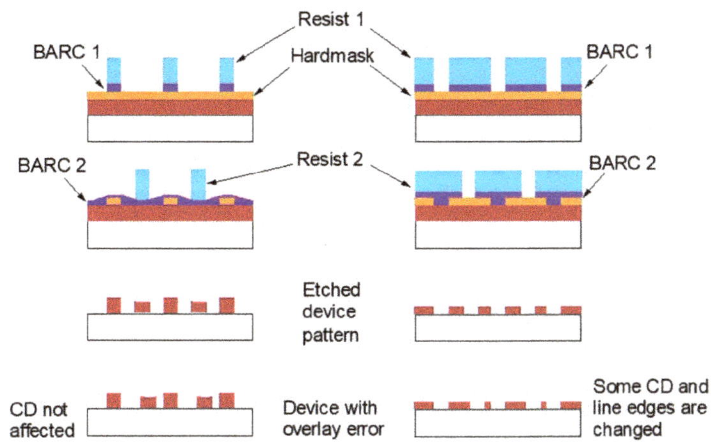

Figure 7.130 Realistic consideration of pitch-spitting DPT.

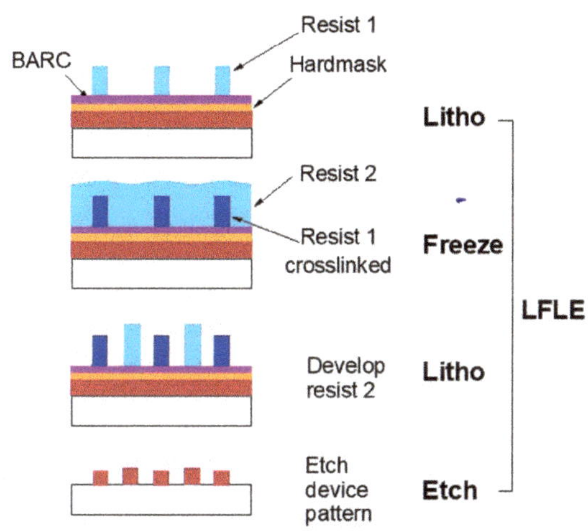

Figure 7.131 Eliminating one etch for lines by crosslinking Resist 1, then coating Resist 2 on top of it.

The split mask can be optimized with dummy assist features, as shown in Fig. 7.133, where OPC was applied with contours around the features and hollow scattering bars, and "hollow" indicates the openings in a dark-field mask. The contoured features are also openings.

7.5.4 G-rule for the double-patterning technique (DPT)

Not all layouts can be split into only two masks. For cost-related reasons, the least number of masks should be used. It is advantageous to alert the designer to this number during the circuit pattern layout, not when the design is completed. To this end, the G-rule ("G" means odd number in Chinese) has

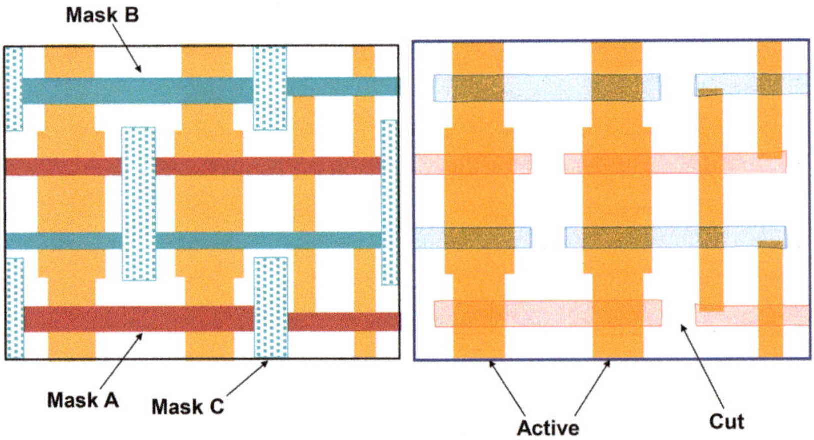

Figure 7.132 Split pitch + line-end cutting with three masks (images courtesy of C.K. Chen).

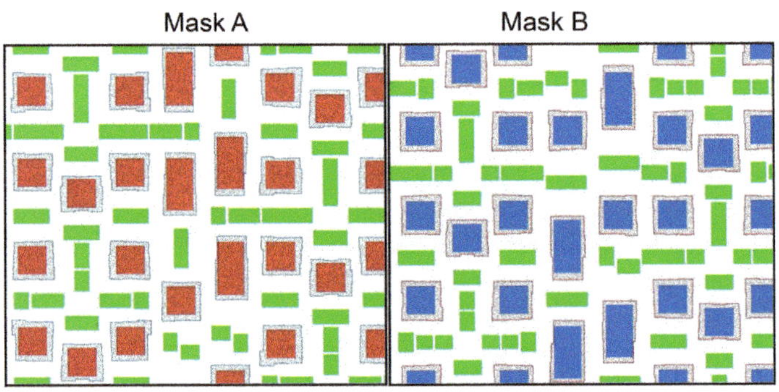

Figure 7.133 Split masks with hollow SBs and OPC (images courtesy of C.K. Chen).

been invented. This rule dictates that any loop linking a forbidden distance between circuit elements should contain an even number of elements to allow for splitting into two masks. Figure 7.134 shows a circuit layout linked into loops. P_3 is far from its adjacent elements and is not linked into the loops. P_7 links only to P_2. The layout can be split into one of the two masks. The P_2-P_4-P_5-P_6 loop has an even number of elements and can be split into two masks. Loop P_1-P_2-P_4 violates the G-rule, as there is no way to split this group into two masks.

7.5.5 Pack-unpack

The pack-unpack technique[51,52] can be a MPT or a multiple-exposure technique. It uses a second resist exposure to handle the cross-interference issue. This double-exposure technique removes the less favorable pitch. In the case of contact holes, either with AltPSM or OAI, the less favorable pitches

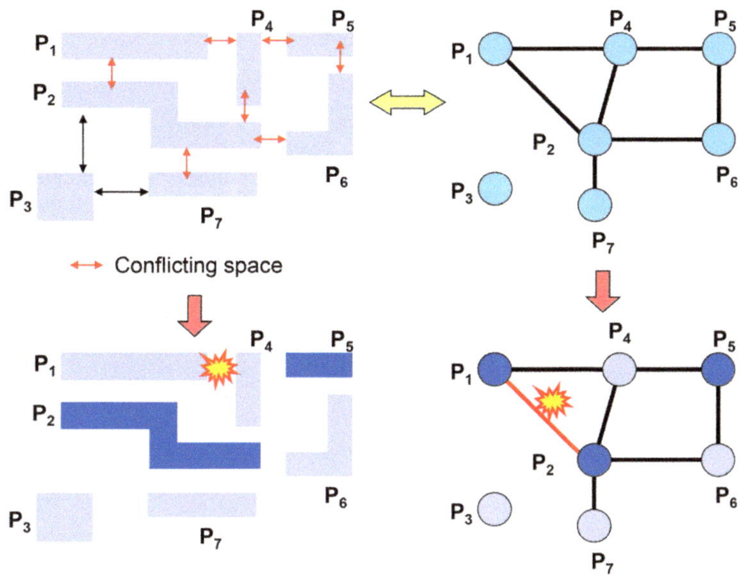

Figure 7.134 Demonstration of G-rule violations (images courtesy of C.K. Chen).

are the large ones all the way to the isolated holes. These pitches suffer from a small DOF. Hence, a mask with padded holes filling the gaps left by the desired holes is first exposed, followed with exposure of a second mask containing unpacking features to fill the unwanted holes or reopen the desired holes. The packed mask, the unpacking mask, and the final image are shown in Fig. 7.135(a). The resist image from the first exposure is covered with a second resist layer, as shown in Fig. 7.135(b). This double-exposure technique not only produces a larger DOF, but the alignment of the second exposure is less critical than in many other double-exposure techniques because its function is only to remove the unwanted holes. Figure 7.135(c) shows the final resist SEM image of a double-exposure system using a second resist that can be coated over the first resist image without damaging it. The alcohol-based resist system reported in Ref. 51 does not affect the first resist image. It images well, as seen in the two SEM images at two magnifications.

7.5.6 Resolution-doubling theory illustrated

There are many stages of resolution doubling. The first stage involves OAI. Figure 7.136(a) shows the on-axis diffraction that produces the 0th-order, $\pm 1^{st}$-order, and many higher-order beams. When the lens pupil is sufficiently large, these spatial frequencies are captured by the imaging lens. Figure 7.136(b) shows that when the lens pupil is only half of that in part (a), the $\pm 1^{st}$-order beams cannot be captured, so the object is not resolved. Figure 7.136(c) shows that the illumination can be tilted for two-beam

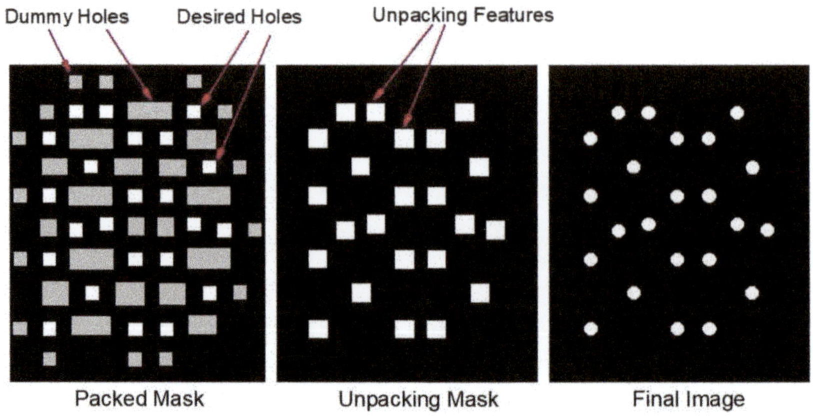

Figure 7.135(a) Packed mask, unpacking mask, and final image.

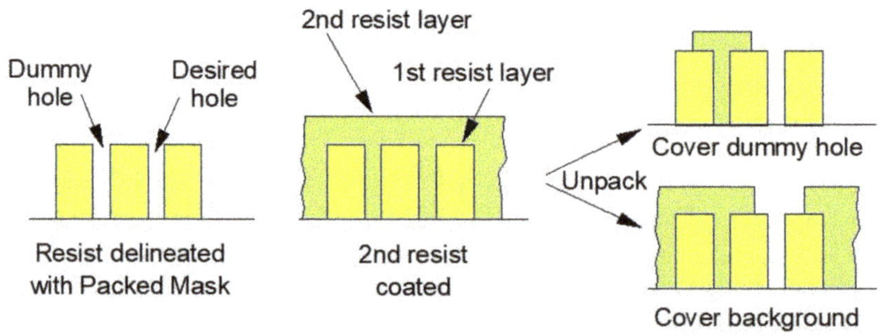

Figure 7.135(b) The pack-unpack process.

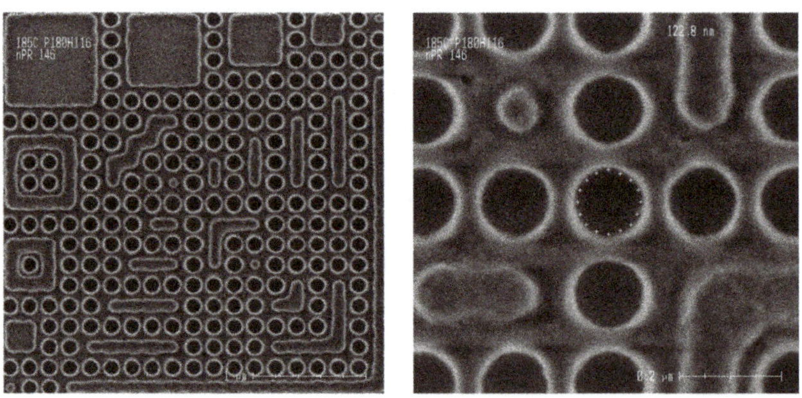

Figure 7.135(c) The desired and the unpacked holes using two complementary resist systems (reprinted from Ref. 51).

imaging, which is explained in detail in Section 7.3.2. As semiconductor scaling marches on, even resolution doubling cannot meet the need. The two beams required for two-beam imaging still cannot be captured by the imaging

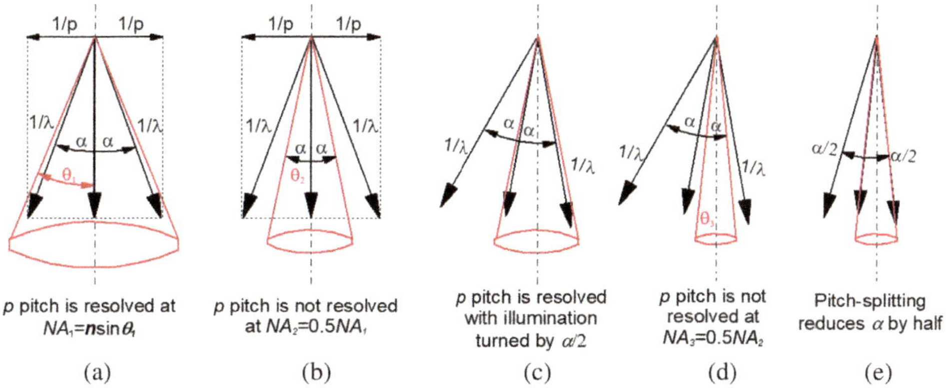

Figure 7.136 Resolution doubling with OAI and with pitch splitting.

lens, as shown in Fig. 7.136(d). Pitch splitting keeps the spatial frequency of the two beams within the capture range of the imaging lens, as shown in Fig. 7.136(e). Resolution doubling is accomplished by patterning the resolvable patterns twice with a displacement between them. The tradeoff is more processing steps and higher cost. Of course, MPT can further extend DPT when overlay accuracy is maintained, as discussed in Section 7.5.7.

7.5.7 Overlay considerations of MPT[49,50]

With the double-patterning technique, the lowest k_1 value that can be achieved can be reduced by a factor of 2, and with another pitch splitting, by another factor of 2. This is a total of four technology nodes, which is too good to be true. There are two discouraging aspects of MPT. First, the photo and etch costs are doubled for each 2X improvement. In addition, all of the split layers must align with each other. Overlay accuracy is always worsened with each addition of splitting. When several split layers are aligned to each other, secondary and higher orders of alignment inaccuracy occur. Figure 7.137 shows a 2P2E layer aligned to a 3P3E layer, which is aligned to the 1st 2P2E layer. The overlay accuracy is assumed to be 1.5% of the CD—a very tight overlay. A2, B3, B1, and B2 can all be directly aligned to A1, so their

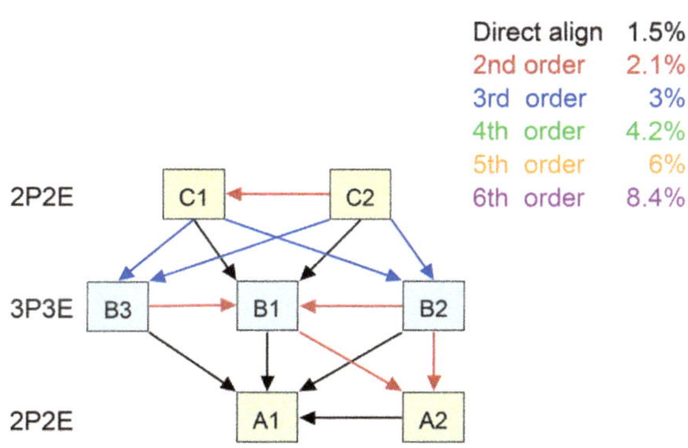

Figure 7.137 Overlay tree of 2P2E aligned to 3P3E on 2P2E.

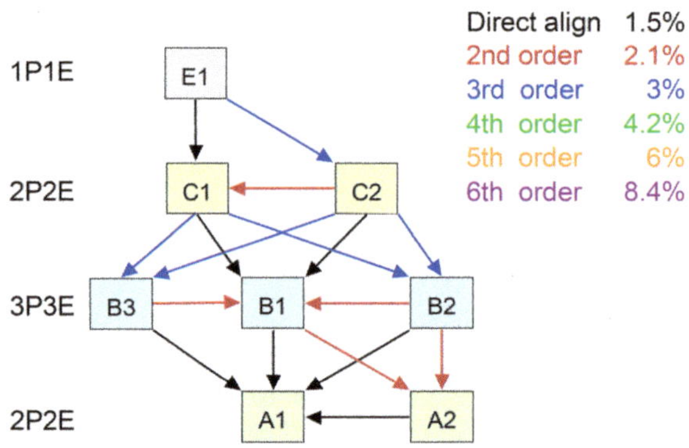

Figure 7.138 1P1E aligned to the configuration of Fig. 7.137.

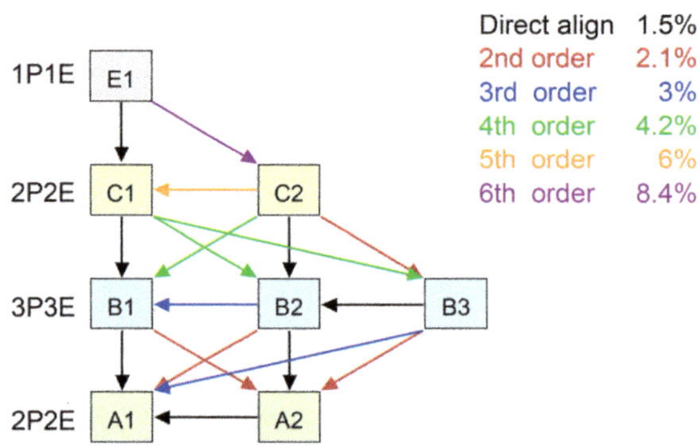

Figure 7.139 A 6th-order overlay error introduced by an unoptimized overlay tree.

alignment accuracy is 1.5%. In this case, B3, B1, and B2 can be aligned to A2 through A1. The alignment accuracy is 1.5% multiplied by $\sqrt{2} = 2.1\%$. C1 and C2 are directly aligned to B1, so their mutual alignment is a 2nd-order inaccuracy. Their alignment to B3 is a 3rd-order inaccuracy.

Figure 7.138 shows that when a 1P1E layer is aligned to C1 in the 2P2E layer, its alignment to C2 is a 3rd-order error because C2 is aligned to C1 by the 2nd-order error. Note that the alignment can be as bad as the 6th order if the alignment tree is not managed correctly, as shown in Fig. 7.139.

7.5.8 Overcoming throughput penalty with double imaging

All double-imaging techniques have a potential to lose at least 40% of the throughput from the exposure tool, and perhaps less than 40% if the tool is specifically optimized for double exposure. This will not happen unless another competing lithographic system can outperform in imaging and cost. Moreover, the overlay error between the two exposures consumes the overlay error budget. Proposals have been made to overcome these problems. To reduce the overlay error, the wafer can remain on the wafer stage while the mask is changed. Because of the 4X magnification at the mask, the alignment of the mask is more accurate than that of the wafer. However, a mask change is time consuming, so one mask change serves to expose two wafers without compromising the alignment accuracy. This method is called ABBA, meaning that mask A is replicated to wafer 1, mask B is replicated to wafers 2 and 3, and so forth. The highest exposure-tool productivity is achieved with the AAAA…BBBB exposure scheme; however, this scheme also has the largest overlay error.

To reduce the reticle alignment time for the ABBA scheme, one can implement a long reticle stage that is capable of holding two masks simultaneously and that allows a scanning distance slightly more than the length of two masks.[53] The two masks are pre-aligned to each other on the common stage. Therefore, the wafer needs to be aligned only once to just one of the masks. Between exposures A and B, the reticle stage is simply shifted from one predetermined position to another. The masks do not have to be unloaded and loaded. The load/unload position of the reticle stage should be at the middle of the stage so that each reticle can complete one scan cycle back and forth without incurring extra time for travel. This is illustrated in Fig. 7.140.

To completely remove the double-imaging throughput penalty and the overlay error component, two illuminators are used for the two masks on the long reticle stage. A long beam splitter having a small cross section slightly larger than the width of the scanning slot combines the two beams from the two masks into the imaging lens,[54] as shown in Fig. 7.141. This way, only the number of illuminators and the power of the light source need to be doubled. The increase in cost is much less than the cost of using two scanners to maintain the single-exposure throughput.

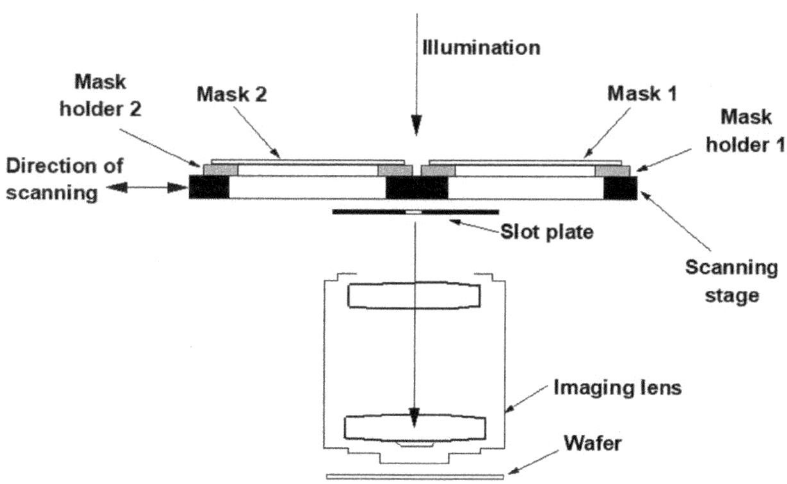

Figure 7.140 Scanner using a dual-mask-scanning stage to save the cost of double exposure.

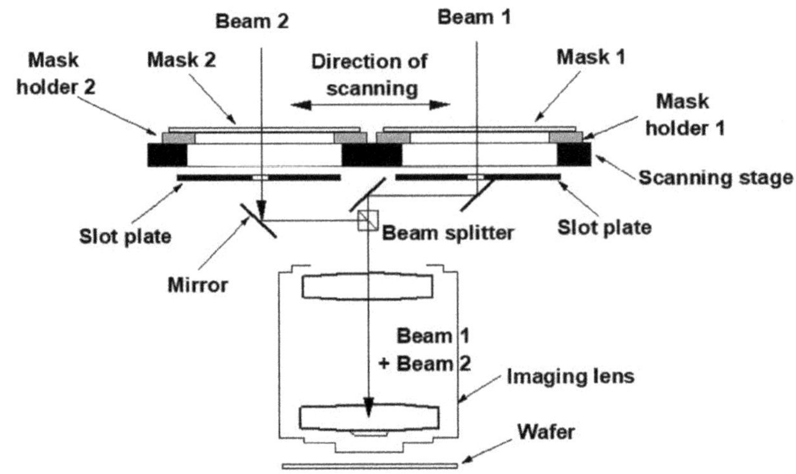

Figure 7.141 Scanner using two illuminators, two masks, one long beam splitter, and one imaging lens.

This technique applies to all double-imaging schemes, except for double patterning. As discussed earlier, because of the accumulation of residual resist exposure, the contrast of the resist must be higher than that used for a single exposure. If this high contrast is developed, the double-mask technique can make the cost of double patterning identical to that of single patterning. The impact will be tremendous.

7.6 CD uniformity (*by S.S. Yu*)

To accommodate more devices in an exposure field, the feature size and the spacing between the features should be as small as possible. However, since there are so many features in an exposure field and so many exposure fields in a wafer, CD uniformity (CDU) should be sufficiently high to ensure a reasonable device yield. In general, CDU becomes worse when the CD becomes smaller. The purpose of process tuning is to find an operating point that compromises between small feature size and CDU to ensure an economically optimum device density versus device yield using the current semiconductor manufacturing technology.

CDU is a good indicator of the processing capability of a semiconductor manufacturing fab. It is determined by two factors: the processing window and process fluctuations. The larger the processing window the less sensitive the CD is to process fluctuations. To enlarge the processing window, k_1-reduction techniques (resolution enhancement techniques) such as OAI or PSM, and low-k_1-enabling techniques such as OPC or MPT should be employed. To reduce process fluctuations, we need to address the imperfections from masks, exposure tools, and processes.

Next, we will perform an analysis to identify contributors to CD nonuniformity (CDNU) and their relative importance so that we can find efficient ways for CDU improvement.

7.6.1 CD nonuniformity (CDNU) analysis

All imperfections in masks, exposure tools, and processes contribute to CDNU. In addition to lithography-related imperfections, previous processing steps such as chemical vapor deposition (CVD) and chemical mechanical polishing (CMP), and subsequent processing steps such as etching all impact CDNU.

7.6.1.1 Linear model for CDU contributions

Before doing CDU analysis, we will introduce a linear model for CDU contribution that makes possible decomposition of CDU into its contributors, based on a general observation. From these components, possible contributors to CDNU might be found.

Assuming that there are only two contributors to CDNU, we have, for each CD denoted as W, in a specific CD distribution,

$$W = \overline{\overline{M_A}}\,\overline{\overline{M_B}}\,W^0, \tag{7.63}$$

where $\overline{\overline{M_A}}$ and $\overline{\overline{M_B}}$ are two operators that stand for two independent contributors to CDNU, and W^0 is the CD when $\overline{\overline{M_A}}$ and $\overline{\overline{M_B}}$ are not present.

For useful lithographic processes, $\overline{\overline{M_A}}$ and $\overline{\overline{M_B}}$ deviate only slightly from the perfect CD; i.e., $\overline{\overline{M_A}} \approx \overline{\overline{I}} + \overline{\overline{\Delta M_A}}$, $\overline{\overline{M_B}} \approx \overline{\overline{I}} + \overline{\overline{\Delta M_B}}$, etc., where $\overline{\overline{I}}$ is the identity operator. Substituting these imperfect CD relationships into Eq. (7.63), and keeping only the terms that are to the first order of $\overline{\overline{\Delta M_A}}$ or $\overline{\overline{\Delta M_B}}$, we have

$$\Delta W \approx \overline{\overline{\Delta M_A}}W^0 + \overline{\overline{\Delta M_B}}W^0, \tag{7.64}$$

where $\Delta W = W - W^0$. In semiconductor manufacturing, the total CD error tolerance is on the order of 10% of the target CD value. Therefore, we may say that each contribution to CDNU amounts to only a small portion (say 5%) of the total CD error tolerance. The error committed in Eq. (7.64) by neglecting higher-order terms is at most 0.25%, which is only about 0.25 nm for the 90-nm node.

7.6.1.2 Geometrical decomposition

Because circuit patterns are first delineated on a mask and then exposed field-by-field on a wafer in current semiconductor manufacturing, the CD at any location on a wafer should exhibit the spatial characteristics of a field as well as of a wafer—the so-called intrafield and interfield signatures, respectively. In the following, subscript B will be used to indicate interfield quantities, while subscript A indicates intrafield quantities.

Extraction of the intrafield signature of the CD error distribution, i.e., CDNU on a wafer, can be accomplished by choosing a set of coefficients c_i such that

$$\sum_k \left[\Delta W(\mathbf{r}_k) - \sum_i c_i P_{i,A}(\mathbf{r}_{k,A})\right]^2$$

is minimized. Here, $\Delta W(\mathbf{r}_k) = W(\mathbf{r}_k) - W^0(\mathbf{r}_k)$, with $W^0(\mathbf{r}_k)$ being the target CD value at $\mathbf{r}_k$. In the above expression, the position vector $\mathbf{r}_k$ is decomposed into its interfield and intrafield components, i.e., $\mathbf{r}_k = \mathbf{r}_{k,B} + \mathbf{r}_{k,A}$ with $\mathbf{r}_{k,B}$ being the vector from the wafer center to the center of the field where $\mathbf{r}_k$ is located. Summation over k is performed over each location on the wafer where the CD is measured. Figure 7.142 depicts the interfield, the intrafield, and the field vectors. Summation over i is performed over the intrafield basis functions $P_{i,A}$. Since the field has a rectangular shape, the Legendre polynomials $L_\alpha(x_{k,A})L_\beta(y_{k,A})$ are a good choice for $P_{i,A} = P_{(\alpha,\beta),A}$, where $\mathbf{r}_{k,A} = x_{k,A}\hat{\mathbf{x}} + y_{k,A}\hat{\mathbf{y}}$. The number of intrafield basis functions employed will depend on the number of sampling points in a field.

Once the values of c_i are found, the intrafield signature of the CD distribution is obtained as

$$\Delta W_A(\mathbf{r}_{k,A}) = \sum_i c_i P_{i,A}(\mathbf{r}_{k,A}). \tag{7.65}$$

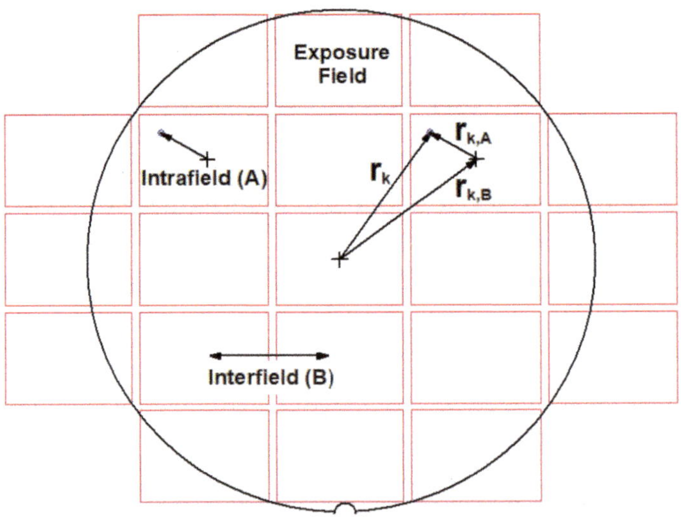

Figure 7.142 Intrafield, interfield, and field vectors.

The interfield signature follows readily such that for each $\mathbf{r}_k$,

$$\Delta W'_B(\mathbf{r}_k) = \Delta W(\mathbf{r}_k) - \sum_i c_i P_{i,A}(\mathbf{r}_{k,A}). \tag{7.66}$$

Practically, we are only interested in the slowly varying part of the CD error distribution. The rapidly varying part generally results from edge roughness. It is a random error for which it is impossible to compensate. Therefore, we also expand the interfield signature in terms of interfield basis functions, just as we did for the intrafield signature:

$$\Delta W'_B(\mathbf{r}_k) = \Delta W_B(\mathbf{r}_k) + \Delta W''_r(\mathbf{r}_k), \tag{7.67}$$

with

$$\Delta W_B(\mathbf{r}_k) = \sum_j c_j P_{j,B}(\mathbf{r}_k), \tag{7.68}$$

where summation over j is performed over the interfield basis functions $P_{j,B}$, and $\Delta W''_r$ is the residual after expansion. Since the wafer has a circular shape, the Zernike polynomials are a good choice for $P_{j,B}$. Note that the set of c_j values is the same for all possible values of $\mathbf{r}_k$.

The above argument assumes that the intrafield signature for each field is the same. In real situations, this is not the case because of lens heating, lens transmission hysteresis, transient effect of sensors, focus difference between scanning up and scanning down, etc. We extract such a field-to-field CD error variation ΔW_F by averaging $\Delta W''_r$ inside each field. Then we obtain $\Delta W''_r(\mathbf{r}_k) = \Delta W_F(\mathbf{r}_{k,B}) + \Delta W_r(\mathbf{r}_k)$, where ΔW_r is the residual after complete CDNU geometrical decomposition.

In fact, the interfield signature from the scanner and from the track cannot be decomposed unambiguously. The reason we identify ΔW_F with $\Delta W''_r$ rather than $\Delta W'_B$ is based on our experience that the interfield contribution from the scanner is much smaller than that from the track.

In summary, the decomposition of each individual CD error in a CD error distribution is achieved according to

$$\Delta W(\mathbf{r}_k) = \Delta W_A(\mathbf{r}_{k,A}) + \Delta W_B(\mathbf{r}_k) + \Delta W_F(\mathbf{R}_{k,B}) + \Delta W_R(\mathbf{r}_k). \tag{7.69}$$

The above decomposition is good for steppers. For scanners, one generally further extracts the following from the intrafield CD error signature: the CD error signature in the slit/slot (x) direction and in the scan (y) direction [denoted as $\Delta W_{A,\text{slit}}(x_{k,A})$ and $\Delta W_{A,\text{scan}}(y_{k,A})$, respectively]. To obtain the CD error signature in these directions, for each x and each y value, find the average CD error at all possible y and x positions. Figure 7.143 shows the decomposition of CD error distribution ΔW into ΔW_A, ΔW_B, ΔW_F, and

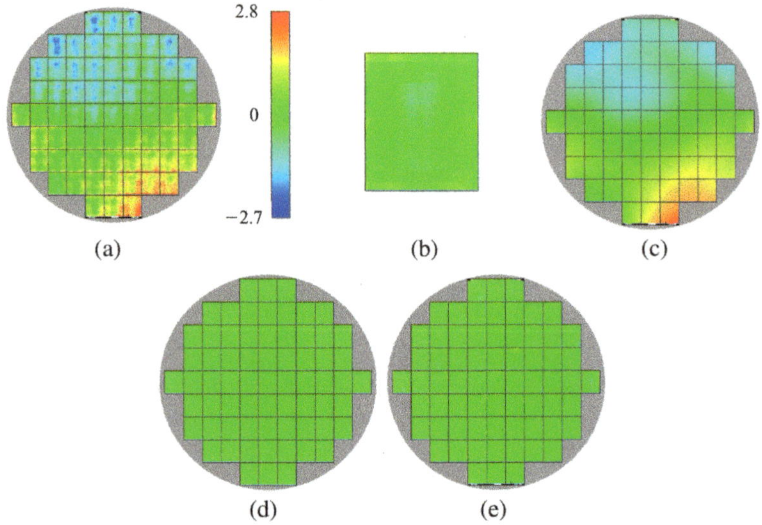

Figure 7.143 An example illustrating the geometrical decomposition of CD error distribution on a wafer. ΔW, ΔW_A, ΔW_B, ΔW_F, and ΔW_r, are shown in parts (a), (b), (c), (d), and (e), respectively (reprinted from Ref. 59).

ΔW_r, as shown in figure parts (a), (b), (c), (d), and (e), respectively. Here, CDs of a specific mask feature in an array of $49(x) \times 13(y)$ inside each field are measured for all 67 fields on a wafer. The number of intrafield basis functions employed is 49×13, i.e., the set of Legendre polynomials $L_\alpha(x_{k,A})L_\beta(y_{k,A})$ with (α, β) from (0,0), (0,1), ..., to (48,12). The number of interfield basis functions employed is 36, i.e., the set of Zernike polynomials $Z_j(\mathbf{r}_k)$ with j from 1 to 36. Increasing the number of interfield basis functions from 36 to 100 changes little of the interfield signature ΔW_B; e.g., 3σ of ΔW_B changes only by 0.01 nm. Note that all CD error signatures are plotted in terms of mosaic, with the color of each block determined by its local CD error value. No interpolation or smoothing in data or figures is made. The CD error signature in the slit/scan direction for the intrafield CD error signature in Fig. 7.143 is shown in Fig. 7.144.

One should not evaluate the intrafield signature ΔW_A or the interfield CD signature at $\mathbf{r} \neq \mathbf{r}_k$. This is because polynomials of very high degrees are employed to extract these signatures, so spurious oscillations may take place.

Although interfield basis functions are not orthogonal to intrafield basis functions, by following the above procedure to extract the intrafield component first, CDU geometrical decomposition can be performed without ambiguity. No iteration is needed.

Our method of finding the intrafield signature is through the projected image, and it differs from the method proposed by Wong et al.,[55,56] who average the CD error at each intrafield location over all interfield locations,

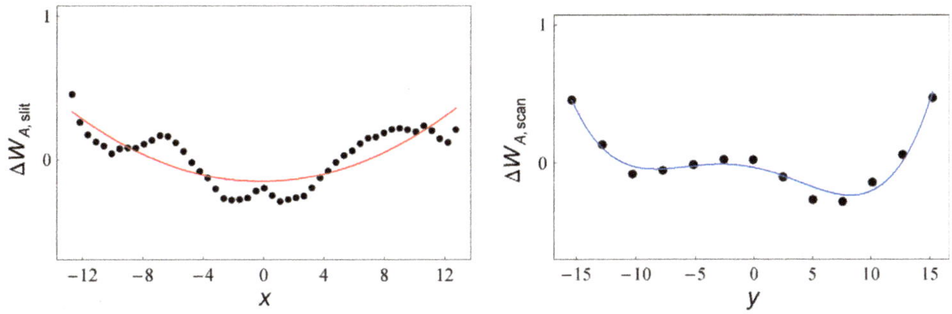

Figure 7.144 CD error signature in the (left) slit/slot (*x*) and (right) scan (*y*) directions for the intrafield CD error signature of Fig. 7.143 (reprinted from Ref. 61).

which is a good approximation if the average of the interfield signature at each intrafield point is negligible.

Vanoppen et al.[57] use a similar method to analyze the CD error distribution taken with electrical measurement. However, they employ polynomials for both the interfield and the intrafield basis functions.

Geometrical decomposition of the CD error distribution provides not only clues about the contributors to CDNU but also the information necessary for CDU improvement through active CD error compensation, as will be detailed in Section 7.6.2.

7.6.1.3 Physical decomposition

Geometrical decomposition of the CD error distribution can identify the individual CDNU contributors that have different spatial characteristics. However, there are many CDNU contributions, all of which lead to the intrafield or intrafield signature of the CD error distribution. To distinguish them, we need to perform independent tests. With the help of physical models, CDNU can be further decomposed. This is physical decomposition.

In general, imperfections of the mask contribute to the intrafield signature, while imperfections of the track contribute to the interfield signature. Imperfection of the scanner contributes to both signatures.

The contributors to the intrafield signature are CD errors on the mask, as well as illumination imperfection, lens aberrations, flare variation, and slit/slot nonuniformity on the scanner. The contributors to interfield signatures are temperature variation on hot plates, coating nonuniformity, exposure variation between fields, scanning-speed variation between fields, and field-to-field focusing variations.

To separate the contribution of the mask from that of the scanner, we need to (1) perform independent mask CD measurements and (2) find the mask error enhancement factor (MEEF) under the employed process conditions. Then, the mask contribution to the intrafield signature is

$\Delta W_B^{\mathrm{mask}} = MEEF \times \Delta W^{\mathrm{mask}}/M$, where ΔW^{mask} is the mask CD error, and M is the reduction ratio of the exposure tool.

For other contributors, e.g., lens aberrations, we need more deliberate models implemented in lithographic image simulators. This will be described in other sections.

The contributors to the interfield signature are thickness nonuniformities from the wafer substrate (including the TARC, resist, and BARC layers), temperature nonuniformity of the hot plates for post-apply and post-exposure bakes, development process nonuniformity, polishing, and etching process nonuniformities.

According to the linear model, all components of the intrafield and the interfield signatures are simply added together.

7.6.1.4 CDU summation

There are always arguments about how to sum different contributors to CDU, using an arithmetic sum or a root-mean-square sum. The general rule is that if the errors are systematic, use an arithmetic sum, and if the errors are random, a root-mean-square sum should be applied to evaluate the standard deviation.

However, there are many contributors. Each contributor has its own functional dependence. For example, 1D proximity errors depend on the pattern geometry or the pitch of the pattern. Mask-related contributors depend on the intrafield locations. Track-related contributors depend on the interfield locations. Aberration-related contributors depend on both the pattern and intrafield locations. The CDNU from all of these contributors certainly should be classified as systematic if their signature is fixed.

If we directly sum these contributors, the resulting CDU is too pessimistic. This CDU indicates that a pattern with the largest bias error and the highest aberration sensitivity—located at an intrafield location with the largest lens aberration and at an interfield location with the highest temperature for PEB—is equally probable to occur as a pattern with a medium bias error and average aberration sensitivity located at inter- and intrafield positions with average lens aberration and an average PEB temperature. It does not reflect the real situation in manufacturing.

For those CD errors that are to be directly summed, their contributors (at least) should be of the same functional dependence, i.e., they should have the same independent variable. Therefore, it is more reasonable to statistically sum the contributors with various functional dependences, i.e., by root-sum-square, even if these contributors are systematic in other senses.

Next, we discuss how to combine the CDNU of different kinds of patterns. Note that patterns with the same characteristics but at different intrafield/interfield locations are classified as the same kind of patterns. In the following, we will derive formulae that are very useful when calculating the 3σ of the total CD errors by combining CD errors from different contributors.

First, consider the simplest case of two groups of CD data. For group 1 and group 2, the individual CD is denoted as $W_{1,i}$ and $W_{2,i}$, the total number of CDs is denoted as as n_1 and n_2, and the mean CD is denoted as M_1 and M_2, respectively. When they are regarded as a single group, with the individual CD denoted as W_i, the total number of CDs denoted as n, and the mean CD denoted as M, the standard deviation σ of W_i can be found as follows:

$$n\sigma^2 = \sum_{i=1}^{n} [W_i - M]^2 \tag{7.70}$$

$$= \sum_{i=1}^{n} [(W_{1,i} - M_1) + (M_1 - M)]^2 + \sum_{i=1}^{n} [(W_{2,i} - M_2) + (M_2 - M)]^2$$
$$= n_1\sigma_1^2 + n_1(M_1 - M)^2 + n_2\sigma_2^2 + n_2(M_2 - M)^2,$$

i.e.,

$$\sigma^2 = \frac{n_1\sigma_1^2 + n_2\sigma_2^2}{n} + \frac{n_1(M_1 - M)^2 + n_2(M_2 - M)^2}{n}. \tag{7.71}$$

For K groups of data, the above formula can be extended as

$$\sigma^2 = \frac{1}{n}\sum_{k=1}^{n} n_k\sigma_k^2 + \frac{1}{n}\sum_{k=1}^{n} n_k(M_k - M)^2. \tag{7.72}$$

If the number of CD samples for each group is the same, we obtain

$$\sigma^2 = \frac{1}{K}\sum_{k=1}^{K} \sigma_k^2 + \frac{1}{K}\sum_{k=1}^{K} (M_k - M)^2, \tag{7.73}$$

i.e.,

$$\sigma^2 = \overline{\overline{M}}[\sigma_k^2] + \overline{\overline{\sigma^2}}[M_k], \tag{7.74}$$

where the operators $\overline{\overline{M}}$ and $\overline{\overline{\sigma^2}}$ indicate the mean and σ squared, respectively, of a collection of numbers, i.e.,

$$(3\sigma)^2 = \overline{\overline{M}}[(3\sigma_k)^2] + \overline{\overline{(3\sigma)^2}}[M_k]. \tag{7.75}$$

In other words, the total 3σ square equals the mean of 3σ square plus the 3σ square of the mean. Note that the above equation is exact since it is derived without any approximation.

In Eq. (7.75), the second term also contributes and cannot be neglected. A good example of this is seen when we find the 3σ of the CDNU of 1D through-all-pitch patterns. With Eq. (7.75), the impact of 1D OPC errors on the total CDNU can be easily calculated.

In a fab, when we talk about CDNU, we generally do not mean the CDNU of a single pattern, but rather the total CDNU of a product, which includes CD error variations from within the field, from within the wafer, from wafer to wafer, and from lot to lot of all patterns of that product. When Eq. (7.75) is applied recursively, the total CDNU in terms of 3σ of a product can readily be found.

7.6.2 CDU improvement

CDU improvement is very important because it is directly related to yield improvement, especially for low-k_1 optical lithography. There are several ways to improve CDU. One is to use RET to increase the process window. Another is to improve the imaging performance of process tools, such as the track and the exposure tool.

Here, we will introduce yet another method, i.e., the method of active compensation, which utilizes the high controllability of the exposure dosage on the exposure tool or the temperature on the hot plates of the track. Active compensation is cost-effective, quick to implement, and has little impact on the existing lithographic process.

7.6.2.1 Active compensation with exposure tools

Modern scanners provide many ways to manipulate the exposure at the interfield and intrafield levels. For interfield levels, the exposure at each field can be assigned to compensate for systematic interfield CDNU. This can be accomplished in many ways. One can adjust the scanning speed, the width of the scanning slot, the laser pulse rate, or the light attenuator in the illumination system on a field-by-field basis. For intrafield levels, the width of the scanning slot along its length and the scanning speed can be adjusted dynamically within each scanning field. In principle, the combined CD fingerprint from the mask, the track, the etcher, etc., can all be corrected.

7.6.2.2 Active compensation on tracks

The example of track active compensation is a CD optimizer that is used for correcting interfield CDNU caused by the post-exposure bake (PEB) of the resist.

For the purpose of the PEB, a multizone hot plate called a chilling precision hot plate (CPHP) is commercially available. With a CPHP, the temperature of each heating zone can be set independently, allowing some flexibility when adjusting the temperature distribution of the hot plate.

However, for commercially available chemically amplified resists, especially those for ArF exposure, the PEB sensitivity can be as high as 10 nm/°C. This means that the temperature uniformity should be better than 0.2 °C to achieve interfield CDU of less than 2 nm. Adjusting the temperature distribution to achieve such uniformity is a difficult task. The main concern is that the thermal environment must be kept identical to that of a real production wafer.

Even if we can measure the entire temperature history during the PEB in the correct thermal environment by using commercially available wireless sensor wafers, we still do not know how to optimize the temperature distribution because the link between the CD and the temperature history is still not accurately known. An example of the impossibility of optimizing the temperature distributions is seen when matching the temperature of different heating zones in the rising stage or in the steady state.

Hence, we propose a CD optimizer that directly uses resist as the temperature sensor, with its response to the temperature revealed in the final CD after the resist process. To prove that our method is feasible, we first expose and process wafers with the CPHP for the PEB set at different temperatures. It is shown that the average CD for each wafer varies linearly with the PEB temperature in a sufficiently large temperature range. Figure 7.145 demonstrates the linearity on a wafer. Next, we find the response of each heating zone by (1) measuring the baseline CD distribution W^0 of a wafer processed with the CPHP for PEB at the baseline setting. Then, (2) for each heating zone i, we measure the perturbed CD distribution W_i of a wafer processed with the temperature of only the heating zone i deviated from the baseline setting by ΔT_i.

After evaluating the response of each heating zone, we can subtract the baseline CD distribution from the perturbed CD distribution to obtain the basis function $P_i = W_i - W^0$. Repeating the procedure for each heating zone i produces the basis functions. Figure 7.146 shows these basis functions for a seven-zone CPHP designed for ACT12®.

The CDNU is obviously $\Delta W = W^0 - W^t$, where W^t is the target CD distribution. To determine how much temperature should be adjusted for each

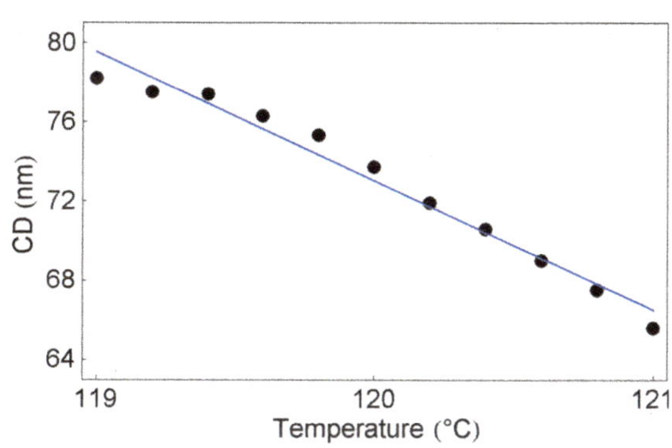

Figure 7.145 Average CD on a wafer versus temperature setting of the CPHP for the PEB process of that wafer.

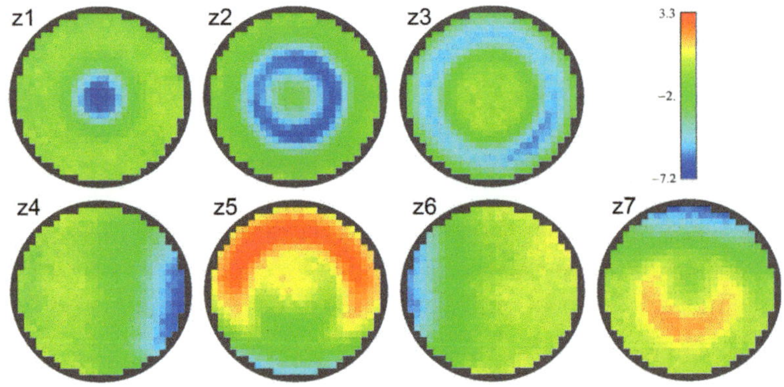

Figure 7.146 Basis functions for a seven-zone hot plate, obtained by choosing ΔT_i = 0.8 °C for each heating zone.

heating zone in order to correct the CDNU, we expand the CDNU in terms of the basis functions, i.e.,

$$\Delta W = \sum_i c_i P_i, \tag{7.76}$$

where the c_i values are expansion coefficients. If the temperature of each heating zone i is adjusted by $-c_i \Delta T_i$, the resulting CD distribution should be optimal under the given degrees of freedom, i.e., the number of heating zones in the CPHP. The predicted CD distribution after correction is

$$W^c = W^0 - \sum_i c_i P_i. \tag{7.77}$$

Note that W^0, W_i, P_i, W^t, and W^c are all scalar fields; i.e., each specified location on a wafer has a corresponding CD.

The CD-optimizer method is illustrated with experimental results shown in the two previous figures and the one that follows. Figure 7.145 shows the average CD on a wafer as a function of the temperature setting of the CPHP used for the PEB of that wafer. The experiment was done by increasing the temperature from 119 °C to 121 °C in steps of 0.2 °C. If the first two data points are ignored, the CPHP has the potential to achieve temperature controllability better than 0.1 °C. The abnormal behavior of these points is probably due to data taken before thermal equilibrium was reached. Figure 7.146 shows the basic functions for a seven-zone hot plate. These functions were obtained by setting ΔT_i = 0.8 °C for each heating zone. Figure 7.147 shows the CDU improvement achieved with a CD optimizer. The original CD map is shown in part (a), while the predicted CD map after correction is shown in part (b). The actual corrected CD map is shown in part (c). After correction, the average of the interfield CD distribution remains

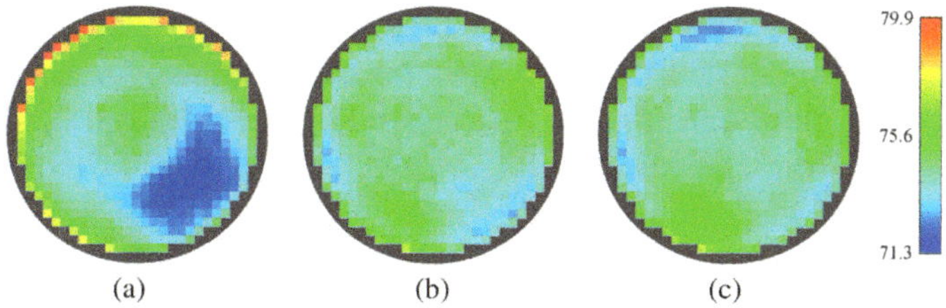

Figure 7.147 CDU improvement achieved by employing the proposed CD optimizer.

unchanged, while the 3σ is improved from 4.64 to 1.86 nm in part (b) and to 1.96 nm in part (c).

7.7 Alignment and Overlay

Semiconductor fabrication is a multilayer process in which the layers must be aligned to each other. Section 6.9 explained that there is alignment accuracy and there is overlay accuracy. Having high alignment accuracy is a necessary condition, while high overlay accuracy in the integrated circuit (IC) is the ultimate requirement for high yield and high performance of the IC.

7.7.1 Alignment and overlay marks

The original alignment mark is a cross, as shown in Fig. 7.148(a). With direct alignment, the mark on the alignment detector is usually a hollow cross so that the solid lines do not overlap for better viewing. Initially, visual alignment through a microscope objective was used. As soon as the interferometer wafer stage was implemented, global alignment referring to a fixed target on the stepper was preferred, eliminating the need for direct alignment. Matching the wafer- and mask-alignment marks is no longer necessary. Modern-day alignment marks take the form of four gratings in four quadrants, two vertical and two horizontal, as shown in Fig. 7.148(b). There is no reason that

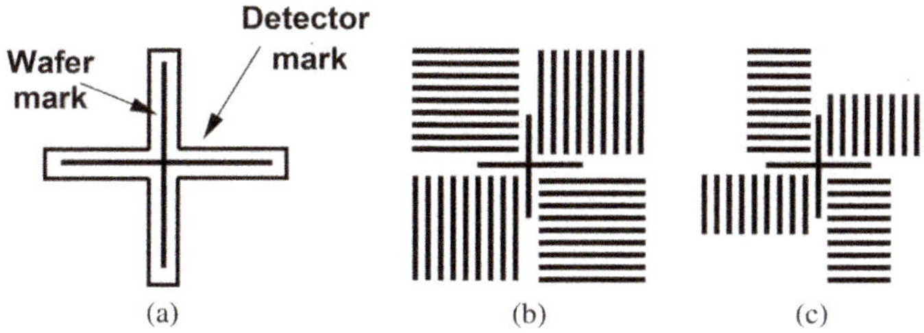

Figure 7.148 Grating alignment marks: (a) wafer and detector marks, (b) mark on wafer, and (c) possible mark on detector.

the gratings with the same orientation cannot have different pitches. On the detector side, a similar arrangement can be used, as shown in Fig. 7.148(c). Detection of alignment can now be automated by the strength of the combined diffraction signal from the wafer and the detector. To save valuable wafer area, the alignment marks are placed in the wafer scribe line where the dies are cut.

The alignment mark needs to be printed on only the current masking layer. Overlay marks must be printed on both the current layer and the previous layer so that the overlay of these two layers can be measured. The box-in-box[59] (BiB) mark and its variation, the bar-in-bar mark, both shown in Fig. 7.149, have been used for a long time. Overlay measurement with BiB is intuitive. The horizontal and the vertical distances between the boxes on the 1st and the 2nd masks indicate the overlay accuracy.

In the era of single-nanometer overlay accuracy, the BiB mark starts to approach its limit. Therefore, a diffraction-based overlay (DBO) mark has been introduced (Fig. 7.150).[60,61] The DBO mark on the wafer looks like the grating alignment mark in Fig. 7.148(b). For overlay measurement, unlike the case with alignment marks, another mark must be provided, as shown in Fig. 7.149(a). This additional mark is similar to the mark on the 1st mask, except that the gratings in each quadrant are moved slightly, each moving in one of

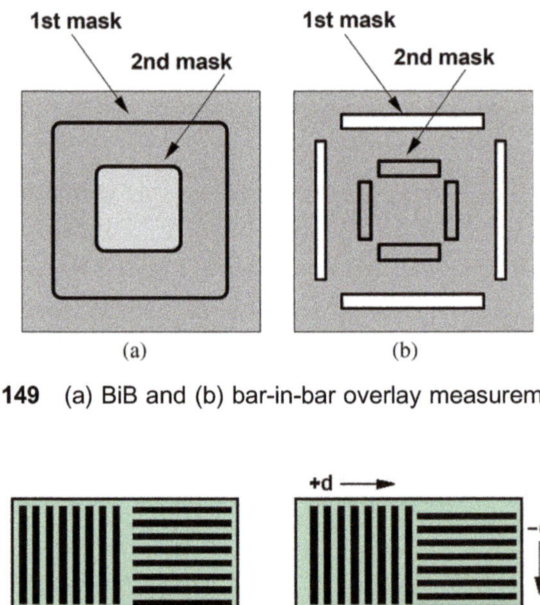

Figure 7.149 (a) BiB and (b) bar-in-bar overlay measurement marks.

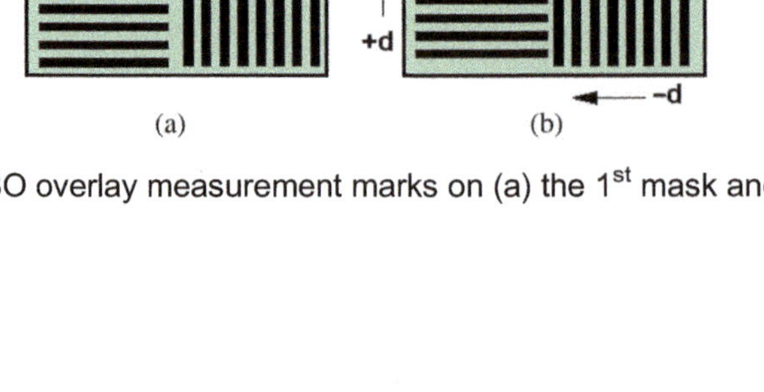

Figure 7.150 DBO overlay measurement marks on (a) the 1st mask and (b) the 2nd mask.

the directions indicated in Fig. 7.150(b). The overlay error is evaluated by measuring the asymmetry of the diffracted light from the overlay of these two marks and substituting this into a pre-derived formula.

7.7.2 Using measured data for alignment

Overlay inaccuracy occurs when the wafer is not correctly positioned to receive the image from the mask. There can be translations, rotations, magnifications of low and high orders due to misplacement of the wafer, thermal effects, chucking imperfections, unwanted particulates, process-induced stress, and many other causes. One needs to identify the misplacement and correct it accordingly.

After the wafer is loaded on the wafer chuck, the positions of a number of alignment marks are measured to determine the position, magnification, and rotation of the wafer. For the time being, just assume a first-order relationship of the wafer dislocation (ΔX, ΔY) at the wafer location (X, Y) and assume a full-wafer field system, as in proximity printing. The governing equations are

$$\Delta X = T_X + X \times M_X - Y \times R_X \tag{7.78}$$

and

$$\Delta Y = T_Y + Y \times M_Y + X \times R_Y, \tag{7.79}$$

where T_X and T_Y are translations from the correction positions in X and Y, respectively, M_X and M_Y are magnifications of the wafer in X and Y, respectively, and R_X and R_Y are rotations in X and Y, respectively. There are six unknown quantities to evaluate. ΔX and ΔY can be measured at three locations that are substituted into the above two equations, resulting in the following matrix:

$$\begin{bmatrix} \Delta X_1 \\ \Delta X_2 \\ \Delta X_3 \end{bmatrix} = \begin{bmatrix} 1 \, X_1 - Y_1 \\ 1 \, X_2 - Y_2 \\ 1 \, X_3 - Y_3 \end{bmatrix} \begin{bmatrix} T_X \\ M_X \\ R_X \end{bmatrix}. \quad \begin{bmatrix} \Delta Y_1 \\ \Delta Y_2 \\ \Delta Y_3 \end{bmatrix} = \begin{bmatrix} 1 \, X_1 \, Y_1 \\ 1 \, X_2 \, Y_2 \\ 1 \, X_3 \, Y_3 \end{bmatrix} \begin{bmatrix} T_Y \\ M_Y \\ R_Y \end{bmatrix}. \tag{7.80}$$

In reality, measurement errors occur in (ΔX_i, ΔY_i). Higher accuracy can be achieved by measuring more sites. Let there be n sites. Then there will be $2n$ equations to solve for only six variables. The least-mean-square method is used to construct six equations for solving the six variables. This takes advantage of the extra measuring sites for averaging out the measurement errors. Let R be the total number of errors in (ΔX_i, ΔY_i):

$$R^2 = \sum_{i=1}^{n} [(\Delta X_i - T_X - X_i M_X + Y_i R_X)^2 + (\Delta Y_i - T_Y - X_i M_Y + Y_i R_Y)^2]. \tag{7.81}$$

The condition that minimizes the change of R^2 from uncertainty in T_X is

$$\frac{\partial R^2}{\partial T_X} = -2\sum_{i=1}^{n}(\Delta X_i - T_X - X_i M_X + Y_i R_X) = 0. \tag{7.82}$$

Similarly, set $\frac{\partial R^2}{\partial T_Y} = 0,\ \frac{\partial R^2}{\partial M_X} = 0,\ \frac{\partial R^2}{\partial M_Y} = 0,\ \frac{\partial R^2}{\partial R_X} = 0,\ \frac{\partial R^2}{\partial R_Y} = 0.$

There are now six equations to solve for six variables, using $2n$ data points:

$$\begin{bmatrix} \sum_{i=1}^n \Delta X_i \\ \sum_{i=1}^n X_i \Delta X_i \\ \sum_{i=1}^n Y_i \Delta X_i \end{bmatrix} = \begin{bmatrix} n & \sum_{i=1}^n X_i & -\sum_{i=1}^n Y_i \\ \sum_{i=1}^n X_i & \sum_{i=1}^n X_i^2 & -\sum_{i=1}^n X_i Y_i \\ \sum_{i=1}^n Y_i & \sum_{i=1}^n X_i Y_i & -\sum_{i=1}^n Y_i^2 \end{bmatrix} \begin{bmatrix} T_X \\ M_X \\ R_X \end{bmatrix}.$$
$$\begin{bmatrix} \sum_{i=1}^n \Delta Y_i \\ \sum_{i=1}^n Y_i \Delta Y_i \\ \sum_{i=1}^n X_i \Delta Y_i \end{bmatrix} = \begin{bmatrix} n & \sum_{i=1}^n Y_i & \sum_{i=1}^n X_i \\ \sum_{i=1}^n Y_i & \sum_{i=1}^n Y_i^2 & \sum_{i=1}^n X_i Y_i \\ \sum_{i=1}^n X_i & \sum_{i=1}^n X_i Y_i & \sum_{i=1}^n Y_i^2 \end{bmatrix} \begin{bmatrix} T_Y \\ M_Y \\ R_Y \end{bmatrix}. \tag{7.83}$$

The values of T_X, T_Y, M_X, M_Y, R_X, and R_Y can be used to adjust the stepping of the wafer during exposure. It is preferred to adjust the wafer until R_X and R_Y are negligible. Otherwise, the rotation error will be repeated on every field as the wafer is stepped for exposure.

7.7.3 Evaluation of interfield and intrafield overlay error components

On a step-and-repeat or a step-and-scan system, there are interfield and intrafield overlay errors. The interfield overlay errors are denoted by T_X, T_Y, M_X, M_Y, R_X, and R_Y, while the intrafield overlay errors are denoted by t_x, t_y, m_x, m_y, r_x, and r_y. Interfield and intrafield areas are illustrated in Fig. 7.151. The field referring to the wafer coordinates (X, Y) is interfield, while the field referring to the mask coordinates (x, y) is intrafield.

The interfield and intrafield coordinates are related by Eq. (7.84), given below. The derivation is as follows. Assuming that $F_x + s \equiv G$, $F_y + s \equiv H$, we find that the leftmost edge of the mask fields is at X_L, the bottom edge of the mask fields is at Y_B, and

$$Integer\left[\frac{X - X_L}{G}\right] \equiv m,\ Integer\left[\frac{Y - Y_B}{H}\right] \equiv n, \tag{7.84}$$

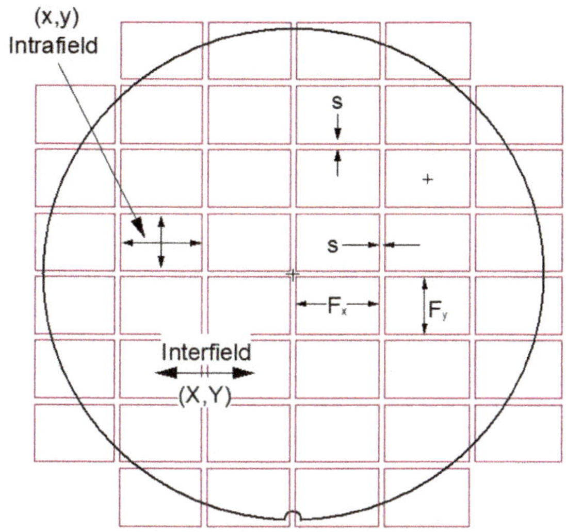

Figure 7.151 Interfield and intrafield coordinates.

where

$$x = X - (0.5 + m)G, \quad \text{and } y = Y - (0.5 + n)H.$$

Figure 7.152 is used to help the derivation of the next equation, Eq. (7.85). Here, a mask with two dies, D1 and D2, is stepped to cover the wafer in 21 exposure fields. Scribe lines are shown on the mask and on the wafer. Their width is exaggerated for illustrative purposes. The coordinates on the mask and on the wafer are also shown.

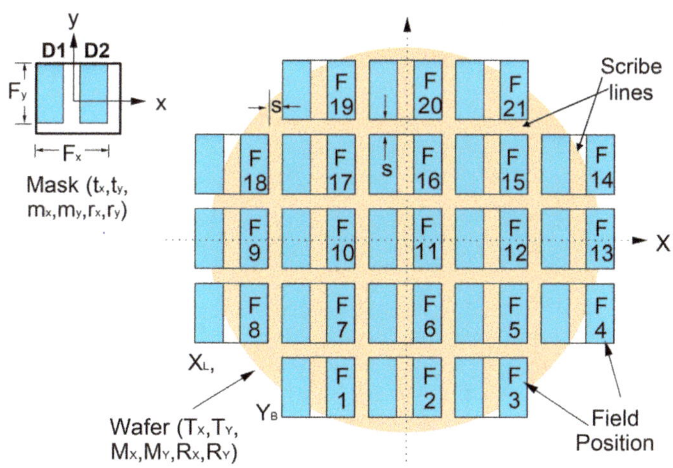

Figure 7.152 Interfield and intrafield coordinates.

After aligned exposure, the total overlay error may still contain some residual wafer-level alignment errors as well as intrafield errors. It is preferred to evaluate these variables together. Regardless of whether the overlay error is based on the interfield or the intrafield coordinates, eventually they all contribute to the error in the wafer coordinate, which is the only coordinate system that counts for the overlay of the IC element. Hence,

$$\begin{aligned} \Delta X &= T_X + X \times M_X - Y \times R_x, \\ \Delta Y &= T_Y + Y \times M_Y + X \times R_Y, \\ \Delta x &= t_x + x \times m_x - y \times r_x, \\ \Delta y &= t_y + y \times m_y + x \times r_y \end{aligned} \tag{7.85}$$

T_X, T_Y, M_X, M_Y, R_X, and R_Y will be different after exposure because most of them were reduced after alignment and before the exposure started. By the end, the (x, y) coordinates must be converted to (X, Y) coordinates because the overlay measurement coordinates are only the latter. Therefore, the conversion from (X, Y) to (x, y) is

$$\begin{aligned} \Delta X &= T_X + X \times M_X - Y \times R_x + t_x + x \times m_x - y \times r_x, \\ \Delta Y &= T_Y + Y \times M_Y + X \times R_Y + t_y + y \times m_y + x \times r_y. \end{aligned} \tag{7.86}$$

Applying the same least-mean-square technique as applied for a full-wafer field system, T_X, T_Y, M_X, M_Y, R_X, R_Y, t_x, t_y, m_x, m_y, r_x, and r_y can be evaluated. First the deterministic equations are given:

$$\begin{aligned} \begin{bmatrix} \Delta X_1 \\ \Delta X_2 \\ \Delta X_3 \\ \Delta X_4 \\ \Delta X_5 \\ \Delta X_6 \end{bmatrix} &= \begin{bmatrix} 1 & X_1 & -Y_1 & 1 & x_1 & -y_1 \\ 1 & X_2 & -Y_2 & 1 & x_2 & -y_2 \\ 1 & X_3 & -Y_3 & 1 & x_3 & -y_3 \\ 1 & X_4 & -Y_4 & 1 & x_4 & -y_4 \\ 1 & X_5 & -Y_5 & 1 & x_5 & -y_5 \\ 1 & X_6 & -Y_6 & 1 & x_6 & -y_6 \end{bmatrix} \begin{bmatrix} T_X \\ M_X \\ R_X \\ t_x \\ m_x \\ r_x \end{bmatrix}. \\ \begin{bmatrix} \Delta Y_1 \\ \Delta Y_2 \\ \Delta Y_3 \\ \Delta Y_4 \\ \Delta Y_5 \\ \Delta y_6 \end{bmatrix} &= \begin{bmatrix} 1 & Y_1 & X_1 & 1 & y_1 & x_1 \\ 1 & Y_2 & X_2 & 1 & y_2 & x_2 \\ 1 & Y_3 & X_3 & 1 & y_3 & x_3 \\ 1 & Y_4 & X_4 & 1 & y_4 & x_4 \\ 1 & Y_5 & X_5 & 1 & y_5 & x_5 \\ 1 & Y_6 & X_6 & 1 & y_6 & x_6 \end{bmatrix} \begin{bmatrix} T_X \\ M_X \\ R_X \\ t_x \\ m_x \\ r_x \end{bmatrix}. \end{aligned} \tag{7.87}$$

The equations for least-mean-square evaluation of the overlay errors follow the same argument as was used for Eq. (7.81):

$$R^2 = \sum_{i=1}^{n} \begin{bmatrix} (\Delta X_i - T_X - X_i M_X + Y_i R_X - t_x - x_i m_x + y_i r_x)^2 \\ +(\Delta Y_i - T_Y - X_i M_Y + Y_i R_Y - t_y - x_i m_y + y_i r_y)^2 \end{bmatrix}. \tag{7.88}$$

The partial derivative of one of the six criteria is shown explicitly here as an example for the derivation of all six of them:

$$\begin{aligned} \frac{\partial R^2}{\partial m_x} &= -2\sum_{i=1}^{n}(\Delta X_i - T_X - X_i M_X + Y_i R_X - t_x - x_i m_x + y_i r_x)\, x_i = 0, \\ \sum_{i=1}^{n} x_i \Delta X_i &= T_X \sum_{i=1}^{n} x_i + M_X \sum_{i=1}^{n} x_i * X_i - R_X \sum_{i=1}^{n} x_i Y_i \\ &\quad + t_x \sum_{i=1}^{n} x_i + m_x \sum_{i=1}^{n} x_i^2 - r_x \sum_{i=1}^{n} x_i y_i. \end{aligned} \tag{7.89}$$

The resultant equation in matrix form needed to evaluate the overlay error components using the least-mean-square technique is

$$\begin{aligned} &\begin{bmatrix} \sum_{i=1}^n \Delta X_i \\ \sum_{i=1}^n X_i \Delta X_i \\ \sum_{i=1}^n Y_i \Delta X_i \\ \sum_{i=1}^n \Delta X_i \\ \sum_{i=1}^n x_i \Delta X_i \\ \sum_{i=1}^n y_i \Delta X_i \end{bmatrix} = \begin{bmatrix} n & \sum_{i=1}^n X_i & -\sum_{i=1}^n Y_i & n & \sum_{i=1}^n x_i & -\sum_{i=1}^n y_i \\ \sum_{i=1}^n X_i & \sum_{i=1}^n X_i^2 & -\sum_{i=1}^n X_i Y_i & \sum_{i=1}^n X_i & \sum_{i=1}^n X_i x_i & -\sum_{i=1}^n X_i y_i \\ \sum_{i=1}^n Y_i & \sum_{i=1}^n X_i Y_i & -\sum_{i=1}^n Y_i^2 & \sum_{i=1}^n Y_i & \sum_{i=1}^n x_i Y_i & -\sum_{i=1}^n Y_i y_i \\ n & \sum_{i=1}^n X_i & -\sum_{i=1}^n Y_i & n & \sum_{i=1}^n x_i & -\sum_{i=1}^n y_i \\ \sum_{i=1}^n x_i & \sum_{i=1}^n X_i x_i & -\sum_{i=1}^n x_i Y_i & \sum_{i=1}^n x_i & \sum_{i=1}^n x_i^2 & -\sum_{i=1}^n x_i y_i \\ \sum_{i=1}^n y_i & \sum_{i=1}^n X_i y_i & -\sum_{i=1}^n Y_i y_i & \sum_{i=1}^n y_i & \sum_{i=1}^n x_i y_i & -\sum_{i=1}^n y_i^2 \end{bmatrix} \begin{bmatrix} T_X \\ M_X \\ R_X \\ t_x \\ m_x \\ r_x \end{bmatrix}. \\ &\begin{bmatrix} \sum_{i=1}^n \Delta Y_i \\ \sum_{i=1}^n Y_i \Delta Y_i \\ \sum_{i=1}^n X_i \Delta Y_i \\ \sum_{i=1}^n \Delta Y_i \\ \sum_{i=1}^n y_i \Delta Y_i \\ \sum_{i=1}^n x_i \Delta Y_i \end{bmatrix} = \begin{bmatrix} n & \sum_{i=1}^n Y_i & \sum_{i=1}^n X_i & n & \sum_{i=1}^n y_i & \sum_{i=1}^n x_i \\ \sum_{i=1}^n Y_i & \sum_{i=1}^n Y_i^2 & \sum_{i=1}^n X_i Y_i & \sum_{i=1}^n Y_i & \sum_{i=1}^n Y_i y_i & \sum_{i=1}^n Y_i x_i \\ \sum_{i=1}^n X_i & \sum_{i=1}^n X_i Y_i & \sum_{i=1}^n X_i^2 & \sum_{i=1}^n X_i & \sum_{i=1}^n X_i y_i & \sum_{i=1}^n X_i x_i \\ n & \sum_{i=1}^n Y_i & \sum_{i=1}^n X_i & n & \sum_{i=1}^n y_i & \sum_{i=1}^n x_i \\ \sum_{i=1}^n y_i & \sum_{i=1}^n Y_i y_i & \sum_{i=1}^n X_i y_i & \sum_{i=1}^n y_i & \sum_{i=1}^n y_i^2 & \sum_{i=1}^n x_i y_i \\ \sum_{i=1}^n x_i & \sum_{i=1}^n Y_i x_i & \sum_{i=1}^n X_i x_i & \sum_{i=1}^n x_i & \sum_{i=1}^n x_i y_i & \sum_{i=1}^n x_i^2 \end{bmatrix} \begin{bmatrix} T_Y \\ M_Y \\ R_Y \\ t_y \\ m_y \\ r_y \end{bmatrix}. \end{aligned} \tag{7.90}$$

We just demonstrated evaluation of low-order overlay correction parameters. In reality, there are higher-order errors that need to be evaluated and corrected. Figure 7.153 shows the improvement of overlay accuracy using 12 high-order terms in 9 wafers. The correction information can be used for feedback and wafer reworking to improve the overlay accuracy. This information can also be used to feed forward for better overlay of the subsequent batches of wafers.

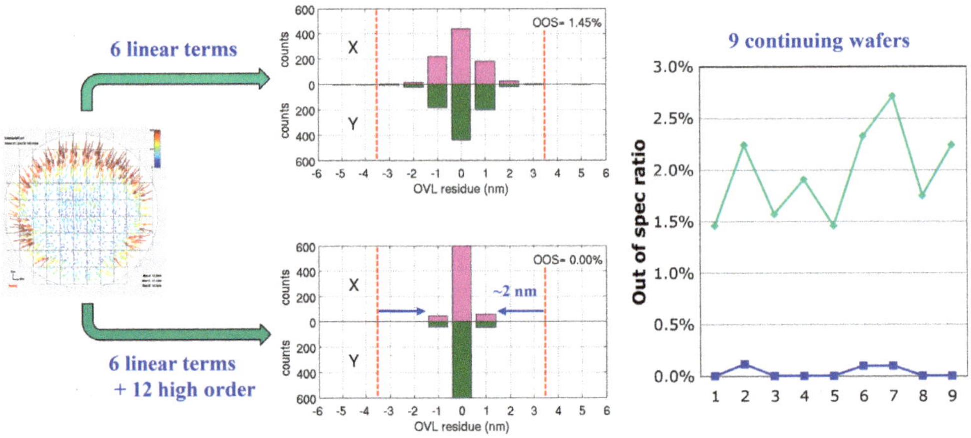

Figure 7.153 Comparison of overlay accuracy with and without high-order correction (images courtesy of S.Y. Lin).

References

1. B. J. Lin, "Optimum numerical aperture for optical projection microlithography," *Proc. SPIE* **1463**, p. 42 (1991) [doi: 10.1117/12.44773].
2. B. J. Lin "The optimum numerical aperture for attenuated phase-shifting masks," *Microelectronic Engineering* **17**, p. 79 (1992).
3. S. W. Choi, S. J. Lee, J. H. Shin, S.-G. Woo, H. K. Cho, and J.-T. Moon, "Influence of material on process focus budget and process window of 80 nm DRAM devices," *Proc. SPIE* **5752**, p. 570 (2005) [doi: 10.1117/12.600166].
4. Private communications with an ASML field engineer.
5. J. W. Bossung, "Projection printing characterization," *Proc. SPIE* **100**, p. 80 (1977) [doi: 10.1117/12.955357].
6. C. P. Ausschnitt, "Distinguishing dose from defocus for in-line lithography control," *Proc. SPIE* **3677**, p. 140 (1999) [doi: 10.1117/12.350800].
7. Commercialized by KLA.
8. T. A. Brunner, A. L. Martin, R. M. Martino, C. P. Ausschnitt, T. H. Newman, and M. S. Hibbs, "Quantitative stepper metrology using the focus monitor test mask," *Proc. SPIE* **2197**, p. 541 (1994) [doi: 10.1117/12.174449].
9. H. Nomura, "New phase-shift gratings for measuring aberrations," *Proc. SPIE* **4346**, p. 25 (2001) [doi: 100.1117/12.435738].
10. Y. Shiode, H. Morohoshi, A. Takagi, K. Fujimaru, K. Mizumoto, and Y. Takahashi, "A novel focus monitoring method using double side chrome mask," *Proc. SPIE* **5754**, p. 303 (2005) [doi: 10.1117/12.600170].
11. S. Nakao, J. Sakai, S. Maejima, A. Ueno, A. Nakae, S. Yanashita, K. Itano, A. Tokui Kozawa, and K. Tsujita, "Simple and highly sensitive

focus monitoring utilizing an aperture on backside of photomask," *Proc. SPIE* **5040**, p. 582 (2003) [doi: 10.1117/12.485324].
12. ASML scanner manual (2004).
13. R. A. McEachem, M. S. Lucas, and C. R. Simpson, "Lithography system using dual substrate stages" U.S. Patent 5,677,758 (1997).
14. B. J. Lin, "Step and repeat apparatus having enhanced accuracy and increased throughput," U.S. Patent 5,715,064 (1998).
15. W. M. Moreau, *Semiconductor Lithography Principles, Practices, and Materials*, Plenum Press, Section 13-2-2, p. 654 (1998).
16. A. Barraud, "Supermolecular engineering by the Langmuir–Blodgett method," *Thin Solid Films* **175**(Aug.), 73–80 (1989).
17. Reference 15, Section 6-6-1, p. 304.
18. L. F. Thompson, C. G. Willson, and M. J. Bowden, *Introduction to Microlithography, ACS Professional Reference Book, American Chemical Society*, Chapter 4, p. 311 (1994).
19. C. P. Auschnitt and D. A. Huchital, "Coating apparatus," U.S. Patent 4,290,384 (1981).
20. B. J. Lin, "In situ resist control during spray and spin in vapor," U.S. Patent 5,366,757 (1992).
21. T. J. Cardinali and B. J. Lin, "Material-saving resist spinner and process," U.S. Patent 5,378,511 (1994).
22. T. J. Cardinali, M.S. thesis, Rochester Institute of Technology (1990).
23. T. J. Cardinali and B. J. Lin, "Material-saving resist spinner and process," U.S. Patent 5,449,405 (1993).
24. F. H. Dill and J. M. Shaw, "Thermal effects on the photoresist AZ1350J," *IBM J. Res. Dev.* **21**, p. 210 (1977).
25. B. J. Lin, "Off-axis illumination—working principles and comparison with alternating phase-shifting masks," *Proc. SPIE* **1927**, p. 94 (1993) [doi: 10.1117/12.150417].
26. T.-S. Gau, C.-M. Wang, and C.-M. Dai, "Strategy for manipulating the optical proximity effect by postexposure bake processing," *Proc. SPIE* **3334**, p. 885 (1998) [doi: 10.1117/12.310823].
27. J. C. Matthews, M. G. Ury, A. D. Birch, and M. A. Lashman, "Microlithography techniques using a microwave powered deep UV source," *Proc. SPIE* **394**, p. 172 (1983) [doi: 10.1117/12.935136].
28. W. H.-L. Ma, "Plasma resist image stabilization (PRIST)," *1980 Proc. IEEE International Electron Devices Meeting*, p. 574, 8–10 Dec., Washington, D.C. (1980).
29. F. S. Lai, B. J. Lin, and Y. Vladimirsky, "Resist hardening using a conformable mold," *J. Vac. Sci. Technol. B* **4**(1), p. 426 (1986).
30. B. J. Lin, "Applications of the mold-controlled profile (MCP) technique for resist processing," *Proc. SPIE* **771**, p. 180 (1987) [doi: 10.1117/12.940324].

31. V. Marriott, "High resolution positive resist developers: A technique for functional evaluation and process optimization," *Proc. SPIE* **394**, p. 144 (1983) [doi: 10.1117/12.935132].
32. K. Deguchi, K. Miyoshi, T. Ishii, and T. Matsuda, "Patterning characteristics of a chemically amplified negative resist in synchrotron radiation lithography," *Jpn. J. Appl. Phys.* **31**, pp. 2954–2958 (1992).
33. T. Tanaka, M. Morigami, and N. Atoda, "Mechanism of resist pattern collapse during development process," *Jpn. J. Appl. Phys.* **32**, pp. 6059–6064 (1993).
34. S. A. MacDonald, W. D. Hinsberg, H. R. Wendt, C. G. Willson, and C. D. Snyder, "Airborne contamination of a chemically amplified resist. 1. Identification of problem," *Chem. Mater.* **5**, p. 348 (1993).
35. B. J. Lin, "Where is the lost resolution?" *Proc. SPIE* **633**, p. 44 (1986) [doi: 10.117/12.963701].
36. B. J. Lin, "Vibration tolerance in optical imaging," *Optical Engineering* **32**(3), p. 527 (1993) [doi: 10.1117/12.61215].
37. Y.-T. Wang and Y. C. Pati, "Phase shifting circuit manufacture method and apparatus," U.S. Patent, 5,858,580 (1999).
38. B. J. Lin, "The optimum numerical aperture for attenuated phase-shifting masks," *Microelectronic Engineering* **17**(1–4), 79–86 (1992).
39. A. Nitayama, T. Sato, K. Hasimoto, F. Shigemitsu, and M. Nakase, "New phase shifting mask with self-aligned phase shifters for quarter micron photolithography," *IEDM Technical Digest*, p. 57 (1989).
40. H. I. Smith, E. H. Anderson, and M. L. Shattenburg, "Lithography mask with a π-phase shifting attenuator," U.S. Patent 4,890,309 (1989).
41. B. J. Lin, "The attenuated phase-shifting mask," *Solid State Technology*, p. 43 (Jan. 1992).
42. N. Shiraishi, S. Hirukawa, Y. Takeuchi, and N. Magome, "New imaging technique for 64M-DRAM," *Proc. SPIE* **1674**, p. 741 (1992) [doi: 10.1117/12.130365].
43. M. Noguchi, M. Muraki, Y. Iwasaki, and A. Suzuki, "Subhalf-micron lithography system with phase-shifting effect," *Proc. SPIE* **1674**, p. 92 (1992) [doi: 10.1117/12.130312].
44. B. J. Lin, "Electromagnetic near-field diffraction of a medium slit," *J. Opt. Soc. Am.* **62**, pp. 977–981 (1972).
45. S. D. F. Hsu, N. Corcoran, and J. F. Chen, "Orientation dependent shielding for use with dipole illumination techniques," U.S. Patent 7,246,342 (2007).
46. B. J. Lin, "Optical projection printing threshold leveling arrangement," U.S. Patent 4,456,371 (1982).
47. B. J. Lin, A. M. Pelella, and A. E. Rosenbluth, "Lithographic process having improved image quality," U.S. Patent 4,902,899 (1990).

48. B. J. Lin, "Limits of optical lithography and status of EUV," *IEDM Short Course* (2009).
49. B. J. Lin, "Optical lithography with and without NGL for single-digit-nanometer nodes," *Proc. SPIE* **9426**, 942601 (2015) [doi: 10.1117/12.2087008].
50. B. J. Lin, "Making lithography work for the 7-nm node and beyond," *Microelectronic Engineering* **143**, pp. 91–101 (2015).
51. D. C. Owe-Yang, S. S. Yu, H. Chen, C. Y. Chang, B. C. Ho, C. H. Lin, and B. J. Lin, "Double exposure for the contact layer of the 65-nm node," *Proc. SPIE* **5753**, pp. 171–180, (2005) [doi: 10.1117/12.599651].
52. B. J. Lin, S. S. Yu, and B. C. Ho, "Improvement of hole printing by packing and unpacking (PAU) using alternating phase-shifting masks," ROC Patent 174, 380, U.S. Patent 666,4011 (2003).
53. B. J. Lin, "Multiple-mask step-and-scan aligner," U.S. Patent 6,777,143 (2005).
54. B. J. Lin, "Mask superposition for multiple exposures," U.S. Patent 7,394,080 (2008).
55. A. K. Wong, A. F. Molless, T. A. Brunner, E. Coker, R. H. Fair, G. L. Mack, and S. M. Mansfield, "Linewidth variation characterization by spatial decomposition," *J. Micro/Nanolith. MEMS, and MOEMS* **1**(2), p. 106 (2002) [doi: 10.1117/1.1488159].
56. A. K. Wong, A. F. Molless, T. A. Brunner, E. Coker, R. H. Fair, G. L. Mack, and S. M. Mansfield, "Characterization of linewidth variation," *Proc. SPIE* **4000**, p. 184 (2000) [doi: 10.1117/12.388994].
57. P. Vanoppen, O. Noordman, J. Baselmans, and J. B. P. van Schoot, "Analysis of full-wafer/full-batch CD uniformity using electrical line-width measurements," *Proc. SPIE* **4404**, p. 33 (2001) [doi: 10.1117/12.425226].
58. S. Wittekoek, J. van der Werf, and R. A. George, "Phase gratings as waferstepper alignment marks for all process layers," *Proc. SPIE* **538**, pp. 24–31 (1985) [doi: 10.1117/12.947743].
59. N. T. Sullivan, "Semiconductor pattern overlay," *Proc. SPIE* **10274**, *Handbook of Critical Dimension Metrology and Process Control: A Critical Review*, 102740C (1994) [doi: 10.1117/12.187454].
60. P. Leray, S. Cheng, D. Kandel, M. Adel, A. Marchelli, I. Vakshtein, M. Vasconi, and B. Salski, "Diffraction based overlay metrology: accuracy and performance on front end stack," *Proc. SPIE* **6922**, 69220O (2008) [doi: 10.1117/12.772516].
61. F. Dettoni, R. Bouyssou, C. Dezauzier, J. Depre, S. Meyer, and C. Prentice, "Enhanced 28nm FD-SOI diffraction-based overlay metrology based on holistic metrology qualification," *Proc. SPIE* **10145**, 101452B (2017) [doi: 10.1117/12.2258206].
62. E. Hecht, *Optics*, *5th Edition*, Pearson Education, pp. 498–499 (2017).

Chapter 8
Immersion Lithography

8.1 Introduction

Immersion lithography is a dark horse. The International Technology Roadmap for Semiconductors (ITRS)—documents adopted by all equipment and material suppliers—together with semiconductor manufacturing companies reduced the actinic wavelength from 436 nm to 365 nm, 248 nm, 193 nm, 157 nm, then 13.4 nm. The 436- and 365-nm wavelengths are two prominent lines produced by the mercury arc lamp. The 248-, 193-, and 157-nm wavelengths are generated by XeCl, KrF, and F_2 excimer lasers, respectively, followed by 13.4-nm EUV light, initially generated by Xe plasma, and then enhanced by switching to Sn plasma.

The ITRS continued to be followed over the years. Just before the year 2000, 193-nm lithography was projected to reach its limit with the 0.93-NA imaging optics, which can handle the 65-nm node and stretched to the 55-nm half node. However, following the roadmap, the industry started working on developing 157-nm lithography much earlier than year 2000.

Unfortunately, only CaF_2 is suitable for lens material at such a short wavelength. This material has been used in 193-nm applications, but the specification is much more stringent for 157 nm. A heavy investment was made to grow a large, perfect CaF_2 crystal for the imaging lens, but this attempt was not successful. A major reason for the failure is the 90-day turnaround time required to grow the large crystal. As many as 600 crystal-growing furnaces have been employed to develop the CaF_2 production process, but to no avail.

Developing a resist for 157 nm also faced tough challenges. Until the worldwide termination of the 157-nm program, no satisfactory resist had been demonstrated. The pellicle material was also wanting. No suitable membrane material was available. The only option was to use quartz or CaF_2 as a thick pellicle. Then, mounting and demounting the pellicle without affecting the optical performance of the imaging system became a difficult problem. Lastly, oxygen absorbs 157-nm light, so the optical path must be in nitrogen, making the system inconvenient and hazardous to operate. Even though nitrogen is not poisonous, it is not easy to detect. If a leak produces a nitrogen-dominant atmosphere, suffocation follows. Thus, flushing a system with nitrogen can be life threatening.

EUV lithography, the successor to 157-nm lithography, was quite immature in the year 2000. It would take another 20 years before EUV lithography could gain a chance at high volume manufacturing, as can be seen in the next chapter.

The concept of immersion lithography was first published[1] in 1987 as a final step in this author's roadmap. Fifteen years later, in 2002, he pointed out the advantage of water-immersion 193-nm lithography,[2] which offers an attractive alternative to 157-nm and EUV lithography. For example, the optical path before the last element of the imaging lens and after the immersion fluid is still at the 193-nm wavelength. All the difficulties with the 157-nm wavelength can be bypassed. Better yet, the 134-nm wavelength in the water immersion medium couples spatial frequencies that are 44% higher than those of the dry 193-nm system, which is 19% better than the dry 157-nm system. Of course, immersion lithography has its own challenges. However, it is easier to meet these challenges with water immersion, as is explained in the following sections.

Due to the tremendous obstacles associated with 157-nm and EUV lithography, immersion lithography has become the final technique for the incremental extension of dioptric/catadioptric optical lithography. Immersion lithography has carried semiconductor manufacturing from the 45-nm node through the 40-, 32-, 28-, 20-, 16-, 14-, 10-, and 7-nm nodes. Owa and Nagasaki[3] provide a review article on the history, status, and future prospects.

Although EUV lithography has made great strides in progress during the last decade since the first edition of this book was published, it still needs to establish itself as an economical workhorse for manufacturing.

8.2 Overview of Immersion Lithography

The basic idea of immersion lithography is to fill the air space between the last element of the imaging lens and the surface of the photoresist on the wafer, as shown in Fig. 8.1. With a higher-index medium filling the air space, light of higher spatial frequencies can be coupled to the resist, thus improving the resolution. Liquid coupling also minimizes the reflection loss at the two affected interfaces. The classical interpretation of the improvement of resolution is that $NA \equiv n \cdot \sin\theta$ is increased with the change of refractive index n in the coupling medium. But how does one visualize this? The wavelength in the fluid is reduced by the wavelength in vacuum according to the formula

$$\lambda = \lambda_0/n. \tag{8.1}$$

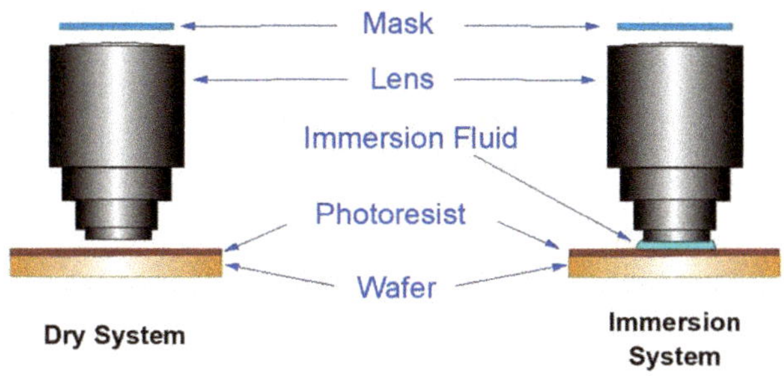

Figure 8.1 Schematic comparing a dry system to an immersion system.

From Eq. (5.3) (repeated here), the resolution[4] of a projection imaging system is given by

$$W = k_1 \frac{\lambda}{\sin\theta}, \tag{8.2}$$

where W is the resolution given as the half-pitch of a periodic object or the size of an isolated feature, λ is the wavelength in the medium of concern, and θ is the aperture angle, which is the angle sustained by the ray of the largest spatial frequency permitted by the optical system. When λ is reduced, W is also reduced. Folding this interpretation of resolution into the classical interpretation, where λ_0 is the wavelength of the actinic light in vacuum, we have

$$W = k_1 \frac{\lambda_0}{n \sin\theta} = k_1 \frac{\lambda_0}{NA}. \tag{8.3}$$

Basically, the NA acts as the threshold of a low-pass spatial frequency filter anywhere along the optical path. Therefore, when the NA in the gating part of the optical train is increased, the NA in other parts of the optical train also increase until the NA in a particular part becomes a gating. For example, when the NA is gated by the coupling medium, even though the lens or the resist can support a larger NA, the resolution is dictated by this coupling medium. When the coupling medium is switched from air to a fluid, higher spatial frequencies are coupled from the lens to the resist to improve the resolution. Note that the wavelength in the resist is also reduced by its refractive index, which is usually the highest in microlithography imaging systems, ensuring that it is not a gating factor.

Wavelength reduction in the medium caused by a change in refractive index is exhibited in the change of light velocity in the medium. The frequency of the light wave is not changed, as given by the well-known equation describing photon energy,[i]

[i]Consult any textbook in modern physics.

$$E = h\nu, \tag{8.4}$$

where E is the energy of a photon, h is Planck's constant, and ν is the frequency of the light wave. The energy of the light wave and therefore the characteristics of interactions among the materials are unchanged in any transparent medium. The mask, lens, and resist materials are illuminated with the same light frequency as that emitted by the light source. The optical characteristics of these materials do not change until the vacuum wavelength changes, implying a change in the light energy and thus in the frequency of the light. Because the optical characteristics of the materials change when the vacuum wavelength changes, new materials that work in the new frequency may need to be developed.

An immersion system is preferable to a system with a shorter vacuum wavelength because using the former one does not need to be concerned with developing new mask, lens, and resist materials. Purging or evacuating the ambient is not necessary either. The tradeoff is lower optical performance of the immersion system compared to a system using a dry wavelength identical to the immersed wavelength, and using the same NA. This is explained in more details in Sections 8.4 and 8.5.

Hence, it is possible to extend existing λ and NA combinations by two to three nodes, e.g., from 65 nm to 45, 32, and 28 nm using a 193-nm vacuum wavelength and infrastructure to meet both technology and economy requirements. However, there are many foreseeable challenges, including producing a fluid that has the required optical, mechanical, and chemical properties, designing and building immersion exposure systems, modifying resists for immersion, etc. Even before considering the materials and systems requirements, the immersion system is quite different from the conventional dry system because of the high-index coupling medium and extremely high spatial frequencies in the resist. The latter makes optical imaging very sensitive to the polarization of the imaging light, as is reported in Section 8.5.

Immersion microscopy has been well known and commercially available for decades. While the wavelength reduction principle of the coupling medium and its function are similar to those in immersion lithography, the former is much easier. Once an inert, transparent, high-index fluid is found and an immersion objective is fabricated, the challenge is over. For immersion lithography, meeting the first two requirements is much more difficult. The fluid must have a uniform and consistent refractive index, and the wafer must move at a high speed in the fluid for step-and-scan exposure, which has replaced step-and-repeat exposure as the frontrunner of the feature-size reduction trend. Making an immersion imaging lens is also not trivial. Even though $\sin\theta$ at the imaging side does not change, $\sin\theta$ at the mask side is increased by $(n - 1)$; i.e., the NA at the mask side is increased. For $\sin\theta = 0.9 \sim 0.95$, the conventional dioptric system must be replaced by the catadioptric system. In addition to these two classical requirements, many other challenges must be met.

Starting in the 1980s, there were proposals and demonstrations of immersion lithography. In 1989 Kawata et al.[5] demonstrated NA = 1.4 optical lithography using the 453-nm wavelength and a high-index oil on a commercial optical microscope. The smallest smooth resist line they could produce was a 230-nm isolated line. In 1999, Hoffnagle et al.[6] extended the limit of interferometric lithography using an index-matching fluid, cyclo-octane, which has a refractive index of 1.51 at the 257-nm wavelength. This study shows that a commercial chemically amplified resist designed for 248 nm works in an immersion mode. Switkes and Rothschild[7,8] have been reporting their findings on immersion fluids for 157- and 193-nm light since 2001. In 1987, in addition to discussing the advantages and concerns for such systems, Lin[1] proposed extending the lithography field size by scanning with an on-axis slot and extending the DOF with liquid immersion. The former concept points to modern step-and-scan exposure tools. The latter can extend scanners by several additional nodes.

Several publications dealing primarily with polarization effects in high-NA imaging have contributed to the understanding of immersion lithography, especially when it is not used to simply extend DOF but, in a hyper-NA mode, to increase resolution. Flagello and Rosenbluth[9] and (later) Flagello, Milster, and Rosenbluth[10] treat vectorial diffraction theory and high-NA imaging. Brunner et al.[11] provide valuable insight into polarization effects and hyper-NA lithography. Lin[4] defines a nonparaxial scaling coefficient k_3 for high-NA lithography and extends the scaling equations to include immersion lithography applications. Much of Section 8.3. is taken from his 2004 work.[12]

8.3 Resolution and DOF

The resolution and DOF of an immersion system require different considerations than a dry system. Due to the high-index coupling medium between the lens and the wafer, the wavelength in the medium reduces the incident and exiting angles of the imaging light. Higher spatial frequencies are accepted and propagated. Resolution and/or DOF are improved.

8.3.1 Wavelength reduction and spatial frequencies

The wavelength changes as a function of the refractive index of the propagation medium, as indicated by Eq. (8.1). In a dry system, the wavelength changes inside each piece of the lens element, the resist, and the underlying thin films on the wafer. In an immersion system, an additional wavelength change takes place in the coupling medium between the last element of the imaging lens and the resist, as depicted in Fig. 8.2. When an incident beam makes an angle to the optical axis, wavelength reduction induces a change in the beam angle according to the Snell's law:

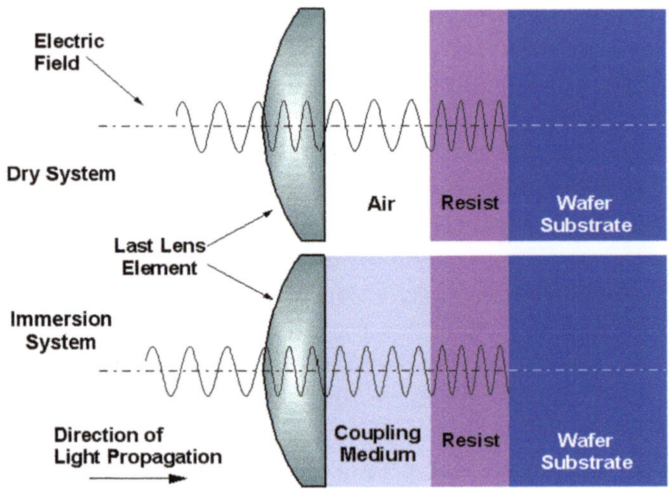

Figure 8.2 Wavelength reduction in an immersion system.

$$n_1 \sin\theta_1 = n_2 \sin\theta_2. \tag{8.5}$$

The change in the incident angle in the lens element, the coupling medium, and the resist is illustrated in Fig. 8.3. Larger spatial frequencies for higher resolutions sustain a larger θ. We emphasize here that the NA in the recording medium is what matters. The coupling medium performs the function of coupling the large angles, which a medium with a lower refractive index cannot sustain, to the recording medium. For example, with $NA = 1$ in vacuum, i.e., $\theta_2 = \pi/2$, and a refractive index $n_1 = 1.75$ for the resist, let 1 denote the lens medium, 2 denote the coupling medium, and 3 denote the resist. Equation (8.5) sets $\theta_3 = 35$ deg. There is ample room to increase θ_3, but

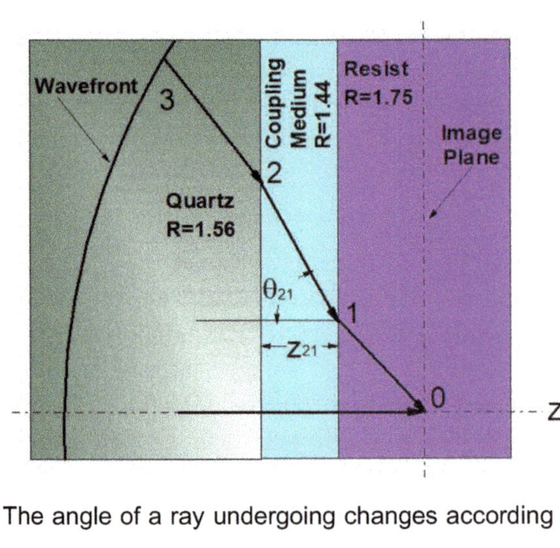

Figure 8.3 The angle of a ray undergoing changes according to Snell's law.

θ_2 is already saturated to 90 deg. However, with n_2 = 1.44, θ_3 increases to 55 deg. The NA is now 1.44 in the resist instead of just 1. The NA can be further increased by a coupling medium with a higher refractive index.

8.3.2 Resolution-scaling and DOF-scaling equations

The resolution scaling equation is given in Eq. (8.2), and the equation for DOF scaling is given in Eq. (5.4) (and repeated here):

$$DOF = k_3 \frac{\lambda}{\sin^2(\theta/2)}, \tag{8.6}$$

where k_3 is the DOF scaling coefficient for high-NA imaging, and the other two parameters were defined with Eq. (8.2). Representing Eq. (8.6) in terms of NA is less straightforward. One of several ways to present the DOF scaling equation is as follows:

$$DOF = k_3 \frac{n\lambda_0}{NHA^2} = \frac{\lambda_0}{n(1-\cos\theta)} = \frac{\lambda_0}{n\left(1-\sqrt{1-\frac{NA^2}{n^2}}\right)}, \tag{8.7}$$

where NHA is the numerical half aperture $n\sin(\theta/2)$. The first expression is taken from Ref. 3, and the latter two expressions are from Ref. 8. The former uses the form most similar to its paraxial counterpart and is easier to remember. The last form is expressed in terms of the invariant NA. Note that NHA is not an invariant.

8.3.3 Improving resolution and DOF with an immersion system

One application of immersion systems improves the DOF of a system in which the resolution is adequate but the DOF is marginal. In this case, the image-forming angle in the resist does not have to increase; ditto for the angle in the lens. The angle in the coupling medium is reduced, making focusing less sensitive to any physical longitudinal displacement of the wafer. Figure 8.4 shows an example using sinθ_{resist} = 0.51 and refractive indices 1.56, 1.44, and 1.75 for the lens material, water, and resist, respectively. Simple calculations of Snell's law set sinθ_{quartz} = 0.57, sinθ_{air} = 0.89, and sinθ_{water} = 0.62. If the image were to be recorded in the coupling medium, Eq. (8.6) predicts a gain in

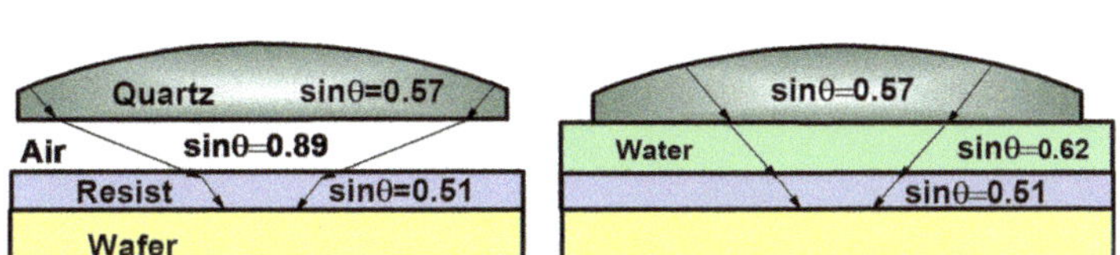

Figure 8.4 Preserving the incident angle in a resist to improve the DOF. Assumptions: n_{quartz} = 1.56, n_{water} = 1.44, n_{resist} = 1.75.

DOF of 77%. Remember that λ in the equation is reduced in the immersion fluid; it is much larger than the DOF obtained by just multiplying by the refractive index because of the leverage of $\theta/2$ in the high-NA DOF scaling equation. Since the image is not recorded in the coupling medium, the DOF should be expressed in terms of $\text{DOF}_{\text{avail}}$, which will be presented in Section 8.4. Nevertheless, the gain in DOF in such a scheme is appreciable. Note that if the paraxial DOF scaling equation is used, the gain in DOF is just by the refractive index, as shown in the following analytic derivation:

$$\frac{DOF_{\text{water}}}{DOF_{\text{air}}} = \frac{\lambda_{\text{water}}}{\sin^2\theta_{\text{water}}}\frac{\sin^2\theta_{\text{air}}}{\lambda_{\text{air}}} = \frac{n_{\text{water}}}{n_{\text{air}}}. \tag{8.8}$$

Immersion systems also improve resolution beyond what a dry system using the same light frequency can provide. Figure 8.5 shows an immersion system that preserves the physical angle in the coupling medium, using the distribution of refractive indices given in Fig. 8.4. If the image were to be recorded in the fluid, the resolution would be improved by 44%, while the DOF would be reduced by 44%, according to Eqs. (8.2) and (8.6). Similar to the statement on Fig. 8.4, since the image is not recorded in the coupling medium, the DOF should be expressed in terms of $\text{DOF}_{\text{avail}}$ (see Section 8.4). The actual imaging performance of the immersion system is slightly worse than that of a system using an equivalent reduced-vacuum wavelength due to polarization-dependent stray light, as will be discussed in Section 8.5.

8.3.4 NA in immersion systems

With different media on the mask and the wafer sides, the situation is complex. The lens designer makes it manageable by dictating the NA on both sides of the compound lens in terms of the image reduction ratio M:

$$NA_{\text{i}} = M(NA_{\text{o}}) \tag{8.9}$$

and

$$\sin\theta_{\text{i}} = M(n_{\text{o}}/n_{\text{i}})\sin\theta_{\text{o}}. \tag{8.10}$$

Thus, $\sin\theta_{\text{i}}$ is related to $\sin\theta_{\text{o}}$ by the reduction ratio and the refractive index

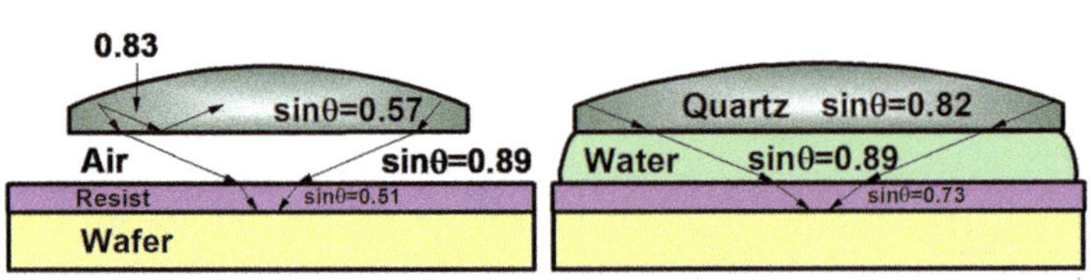

Figure 8.5 Preserving the physical angle in the coupling medium to improve resolution. Assumptions: n_{quartz} = 1.56, n_{water} = 1.44, n_{resist} = 1.75.

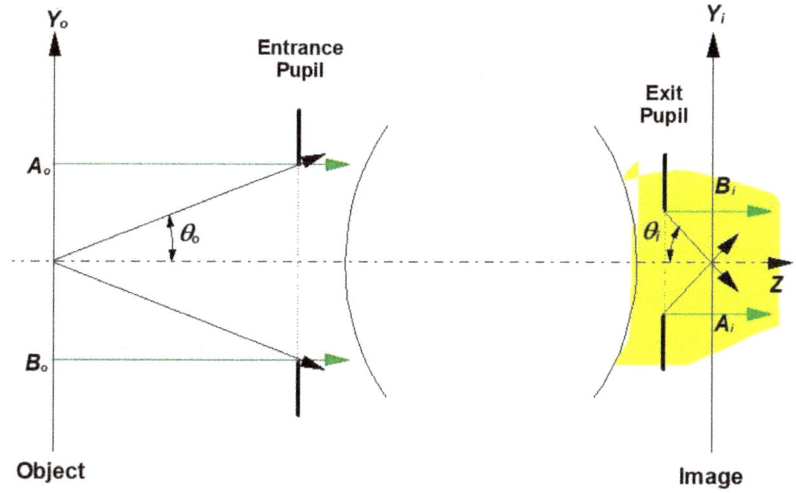

Figure 8.6 A 2X reduction dual-telecentric immersion system depicting NA at the mask and the wafer sides. The reduction ratio and the NA ratio are identical.

ratio. Figure 8.6 shows a dual-telecentric immersion system with 2X reduction. Because it is an immersion system, $n_i > n_o$.

8.4 DOF in Multilayered Media[13]

DOF is often taken from many different perspectives. There is the DOF dictated by diffraction, which supports all of the components in the DOF budget, such as the topography on the wafer, resist thickness, wafer and mask flatness, tilts, focusing error, image plane deviation, film nonuniformity, and scanning imperfections. The required DOF is a function of the budget components. The available DOF is the diffraction DOF with the resist thickness removed. This is the DOF available to the tool designer and process engineer for a given technology node after the resist thickness requirement is locked in.

8.4.1 Transmission and reflection in multilayered media

The situation we are analyzing is shown in Fig. 8.7, which illustrates the last lens element in a projection lens; the coupling medium can be gas, fluid, solid, or just vacuum. The figure also shows the resist contacting the coupling medium and the wafer substrate. We are accustomed to only examining the diffracted image in air. Immersion systems alert us to the need to consider the diffracted image in the resist as well as in the coupling medium between the lens and the resist.

Let us first look at the coupling medium, which has a refractive index close to 1. The image space is split into two types. Figure 8.8 shows a simulated diffraction image completely in air, then completely in a thick resist

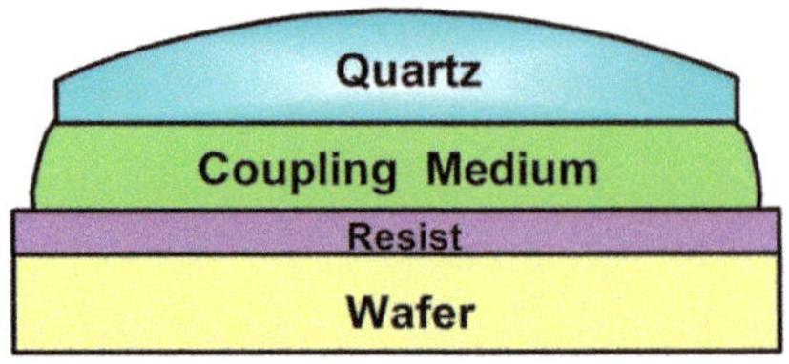

Figure 8.7 Lens, coupling medium, resist layer, and wafer.

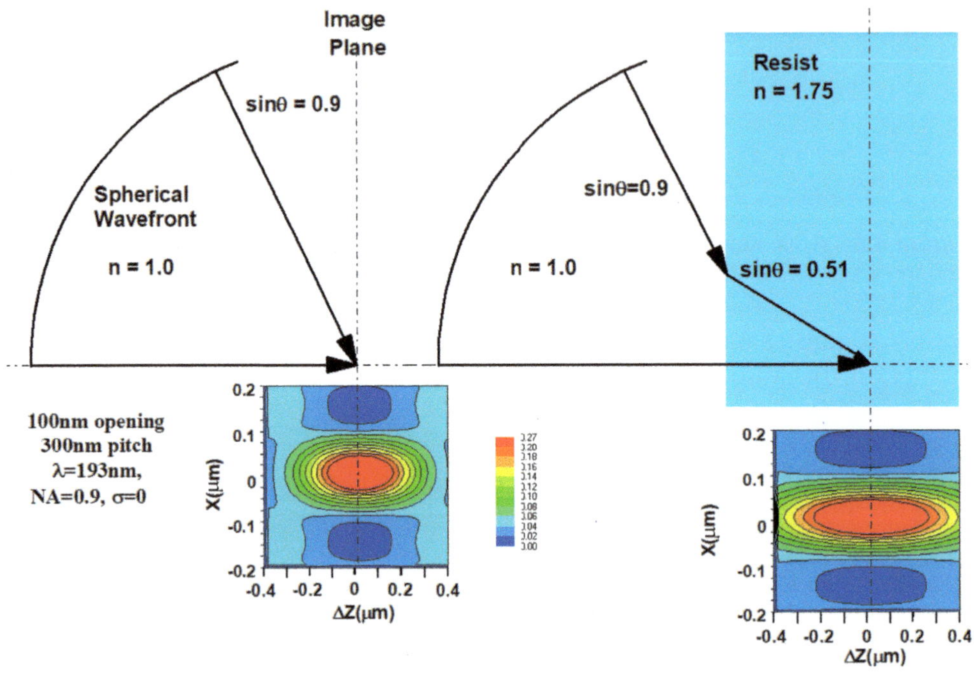

Figure 8.8 Simulated diffracted image near focus (right) with and (left) without a resist medium. The resist thickness is 800 nm.

system. Here, 100-nm line openings 200 nm apart are projected through an NA = 0.9, λ = 193-nm imaging lens using axial illumination at σ = 0.8. The optical system is assumed to be diffraction limited, so the wavefront in air is spherical. If we keep the spherical wavefront in air as shown, spherical aberration would automatically be incorporated as soon as the thick resist layer is introduced.[14] However, in the simulation, we assume that the lens is corrected for spherical aberration induced by the resist. This is equivalent to using a spherical wavefront in the resist. The diffracted images are shown in constant-intensity contours in the x-z space, where x represents the lateral direction and z is the longitudinal direction. Figure 8.8 shows that the diffracted images in air (left) and in thick resist (right) have the same lateral resolution because the NA in the resist remains unchanged. However, the DOF in the resist is improved because of a smaller θ in the resist.

When making an exposure of the resist on the wafer, the resist thickness must be kept within a fraction of the DOF derived from diffraction considerations using the NA, illumination condition, feature size, feature shape, and feature combinations. The remaining DOF, after deducting the resist thickness, is needed for focusing error, tilt, wafer and mask nonflatness, field curvature, scanning-induced errors, etc.

Figure 8.9 shows the situation when the resist thickness is 200 nm. Here, the diffracted image has two parts, one dictated by the wavelength in the coupling medium and the other by the wavelength in the resist. As the incident wave travels through the coupling medium, it is identical to a wave in a homogenous medium, shown in Fig. 8.8, until it reaches the resist surface. Then, it splits into two waves: the transmitted wave and the reflected wave. The latter is a mirror image of the transmitted wave adjusted for the refractive index in the coupling medium. The former can be thought of as a part of the diffracted wave in the resist shown in Fig. 8.8, taken within the depth of 200 nm because we assume perfect antireflection coatings above and below the resist, as well as an absorption-free coupling medium and resist. Hence, transmitted light with no change in intensity across the interface is plotted in Fig. 8.9.

Multiple reflections between the resist–wafer interface and the resist–coupling-medium interface are very important for DOF. When these reflections are present, one should not neglect them. Fortunately, they are

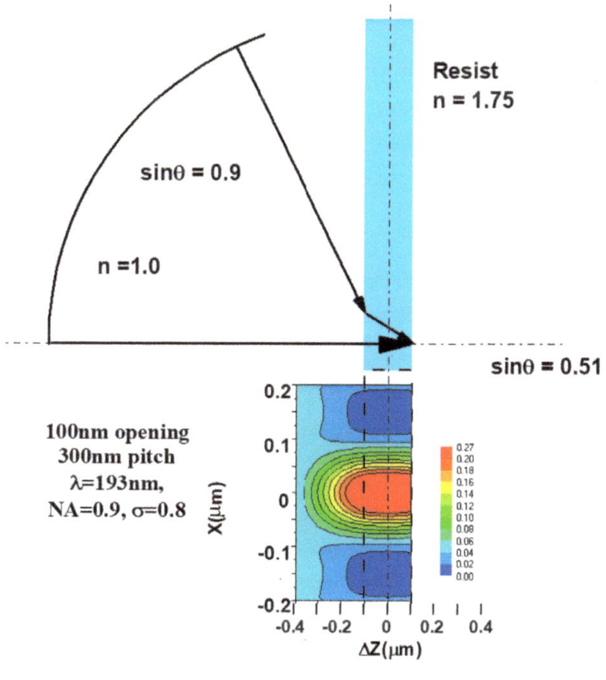

Figure 8.9 Same diffracted image as in Fig. 8.8, except using 200-nm resist thickness.

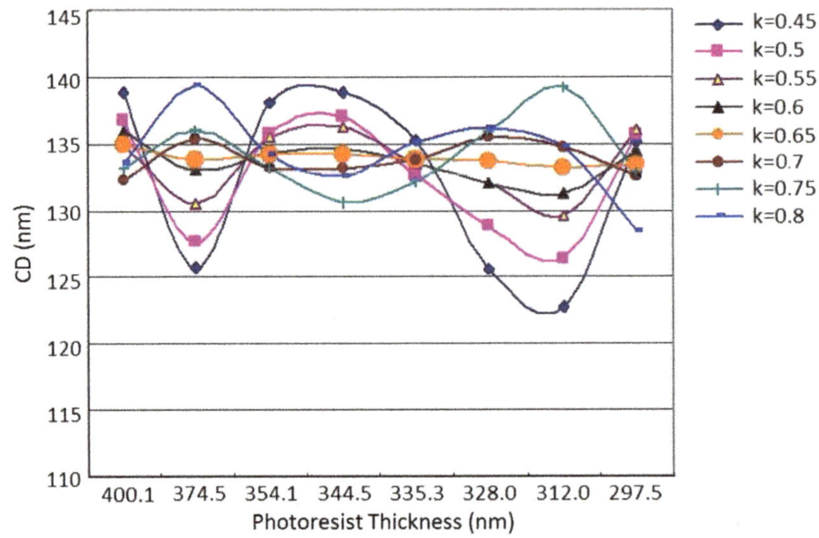

Figure 8.10 CD swing curve as a function of resist thickness on 26 nm of SiON.

often reduced to a minimum for control of the delineated image size, i.e., CD control. Figure 8.10 shows CD as a function of the resist thickness over a SiON bottom antireflection coating at different k values over an oxide layer. When k = 0.65, the CD variation is less than 1.5%. The reflectivity swing corresponding to this CD variation is about 1%, which is typically what can be achieved on the manufacturing floor. Consequently, we may assume very good antireflection properties at both interfaces. Figure 8.9 is plotted with these assumptions. With all the above simplifications, one can literally cut out the intensity contours in air between $Z = -400$ nm and -100 nm in Fig. 8.8 and paste them to another cutout from the thick resist space between $Z = -100$ nm and 100 nm to construct the diffraction pattern in Fig. 8.9, illustrating the diffracted intensity distribution in two media. In our case, a computer program was written for diffracting in the composite media and pasting. Strictly speaking, spherical aberration can be corrected for only one of the two media. For our purpose of illustrating the diffracted image in two media, neglecting spherical aberration is acceptable.

8.4.2 Effects of wafer defocus movements

Because the resist thickness does not change, when the wafer moves in the longitudinal direction the resist surfaces move together. The sensitivity of the longitudinal movement of the wafer is therefore mostly governed by the refractive index of the coupling medium instead of that of the resist. This point is illustrated in Figs. 8.11 and 8.12, using the imaging example in Fig. 8.8. The indices used for the resist and for the coupling medium are 1.75 and 1.44, respectively. The former shows the constant-intensity contours in air and in

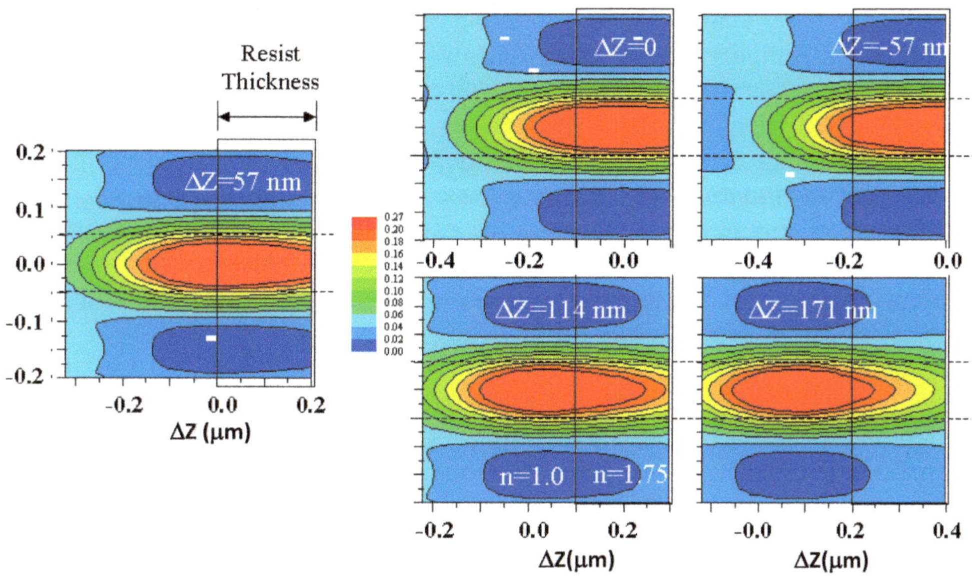

Figure 8.11 Resist layer in air at various ΔZ settings.

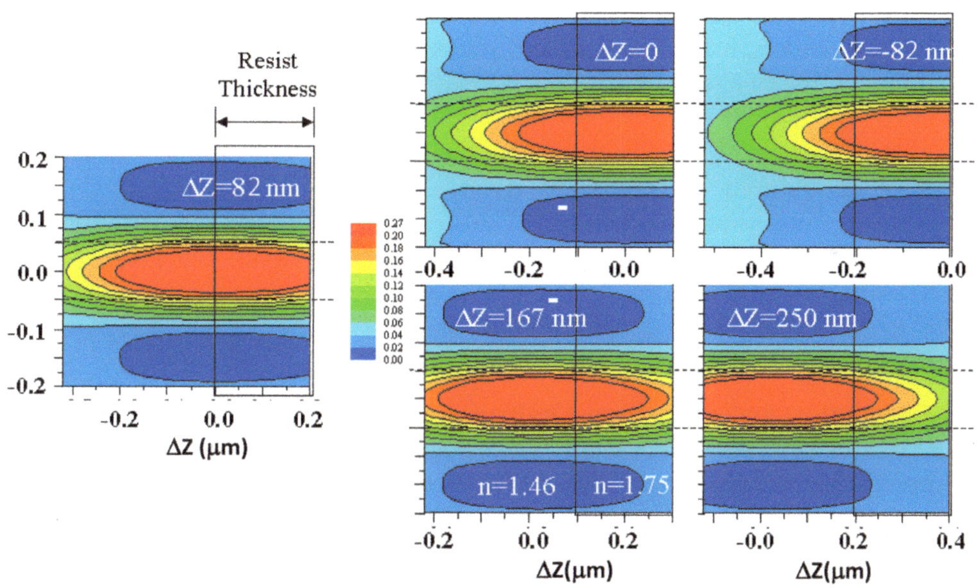

Figure 8.12 Resist layer in fluid at various ΔZ settings.

resist as the air–resist interface moves from –200 to 200 nm in 100-nm steps. ΔZ = 0 is the focal plane; i.e., the resist layer is perfectly centered when the air–resist interface is at –100 nm. The moving distance in 100-nm steps refers to movements in the resist imaging space. For an outside observer concerned with controlling the distance between the lens and the wafer, the steps should be adjusted for the difference in refractive index, i.e., 100 nm/1.75, resulting in 57-nm steps. Similarly, in the case of immersion lithography, the 100-nm steps in the resist become $100 \times 1.44/1.75 = 82$ nm to control the distance between the lens and the wafer. Using the refractive indices to adjust the defocusing steps will become clear after understanding Eq. (8.12), given in Section 8.4.4. When using simulation, one needs to be careful with the DOF given by the simulator. If this DOF is given in the resist space, the above adjustments should be performed. If it is given in air, the steps in a dry medium need no correction; however, the steps in a fluid coupling medium should be multiplied by the refractive index of the fluid.

When there is low light absorption and high development contrast in the resist, constant-intensity contours are good indicators of the resist profile at different exposure dosages. From Fig. 8.11, one sees that the resist profile is symmetrical when it is centered at the focal plane. There is more exposure at the top of the resist and less at the bottom as the wafer moves away from the lens, and the opposite occurs as the wafer moves closer to the lens. The span in the resist profile is indeed stretched in the z direction because of the higher refractive index in the resist. However, the contours in the coupling medium dictate the starting point of the contours in the resist. The former is crucial for controlling the distance between the lens and the wafer. Figure 8.12 shows the same trend as Fig. 8.11. The only difference here is that the coupling medium now has an index $n_{CM} = 1.44$, thus extending the contours in the z direction and producing more tolerance for the distance between the lens and the wafer. Note that at $\Delta Z = -82$ nm in the immersion case, the contours are still slightly better than without immersion at $\Delta Z = -57$ nm. The trend is even more obvious at the other ΔZ settings. This means that immersion lithography tolerates larger longitudinal physical movements between the lens and the wafer than just n_{CM}:1.

8.4.3 Diffraction DOF

The resolution and diffraction DOF in a single medium are represented by Eqs. (8.2) and (8.6), respectively. In the resist, the reductions of λ and $\sin\theta$ compensate for each other to keep the resolution unchanged. The effect of $\sin^2(\theta/2)$ is stronger than the effect of wavelength reduction, resulting in a larger $\mathrm{DOF}_{diffrac}$. Converting to dependency on NA, Eqs. (8.2) and (8.6) become Eqs. (8.3) and (8.7), respectively. Equations (8.6) and (8.7) describe the situation of Fig. 8.11 when the entire diffraction DOF is contained within one single medium. They are different from Eq. (8.3) in terms of being independent of the medium. With these equations, Snell's law makes NA an invariant in any medium in the optical imaging train, so W remains constant. Just as $\mathrm{DOF}_{diffrac}$ is not an invariant, neither is NHA.

Determining the correct k_3 in Eqs. (8.6) and (8.7) is very important. One takes the properly illuminated features of interest to either simulate or project the image through the lens and the coupling medium to the surface of the photoresist, then evaluates the range of defocus that keeps CD within specification for all features, and adjusts this range with the refractive index of the coupling medium. This sets the k_3 values to be used in these equations. One should bear in mind that, at a high NA, light polarization makes the scaling equations less effective, as discussed in Section 8.5.

During wafer printing, the diffraction DOF must support the required DOF of the imaging system. The requirement contains focus error, wafer and mask flatness within the exposure field, wafer tilt with respect to the optical axis, scanning imperfections, image plane deviation from the optical system, topography on the substrate, film non-uniformity, and resist thickness. Hence, for a given set of patterns on the mask, its diffraction DOF must be at least as large as the required DOF. These two types of DOF apply to the left and right sides of Eq. (8.11). In a way, the diffraction DOF is like a DOF reservoir for the required DOF to consume. Hence, a usable imaging system requires that

$$DOF_{\text{diffrac}} \geq DOF_{\text{required}}. \tag{8.11}$$

To determine $\text{DOF}_{\text{diffrac}}$, one takes all of the representative feature types and sizes under the predetermined imaging condition, such as λ, NA, illumination setting, and lens aberrations, to evaluate the CD at a given range of exposure dosage and defocus. The E-D tree of each feature type and size is evaluated from the CD data. The common E-D window from the common E-D tree is used to set the DOF of a given exposure latitude (EL). Substituting the DOF, λ, and θ into Eq. (8.6) or (8.7) yields k_3.

The foregoing procedure takes care of evaluating the DOF in vacuum and has been documented in various publications.[15,16,17] Since the refractive index of air is only slightly higher than that of vacuum, not much error is introduced. In a coupling medium having a refractive index much higher than that of vacuum, $\text{DOF}_{\text{diffrac}}$ is increased by n_{CM}. Similarly, in a thick-resist medium whose thickness is larger than $\text{DOF}_{\text{diffrac}}$, the DOF is increased by n_{resist}, as demonstrated in Fig. 8.11.

The diffraction DOF obeys the high-NA scaling law[4] as long as polarization effects are omitted. Polarization effects are treated in Section 8.5.

8.4.4 Required DOF

To determine $\text{DOF}_{\text{required}}$, all contributing components are combined either linearly or quadratically, depending on whether a particular component is systematic or statistical.[1,18,19,20] The following equation is our current understanding of $\text{DOF}_{\text{required}}$. While the wavelength is reduced in the resist medium that covers the wafer substrate and wafer topography, thus increasing the focal range of the acceptable diffraction image, many other contributing components are based on the resist surface moving in the coupling medium. As illustrated in the Fig. 8.12, the DOF contribution from these other components should use the refractive index of the coupling medium:

$$\frac{1}{n_{\mathrm{CM}}} DOF_{\mathrm{required}} = \frac{1}{n_{\mathrm{resist}}} (THK_{\mathrm{resist}} + Z_{\mathrm{topo}} + \Delta Z_{\mathrm{IPD}}) + \left[\begin{array}{l} \frac{1}{n^2_{\mathrm{resist}}} (\Delta Z^2_{\mathrm{resist}} + \Delta Z^2_{\mathrm{topo}} + \Delta Z^2_{\mathrm{FU}}) + \frac{1}{n^2_{\mathrm{air}}} \left(\Delta Z^2_{\mathrm{focus}} + \Delta Z^2_{\mathrm{SI}} + \frac{\Delta Z^2_{\mathrm{MF}} + \Delta Z^2_{\mathrm{tilts\ at\ mask}}}{M^2}\right) \\ + \frac{1}{n^2_{\mathrm{CM}}} (\Delta Z^2_{\mathrm{tilts\ at\ wafer}} + \Delta Z^2_{\mathrm{WF}}), \end{array} \right]^{\frac{1}{2}} \tag{8.12}$$

where IPD denotes image plane deviation, FU is film uniformity, CM is coupling medium, SI is scanning imperfections, WF is wafer flatness, MF is mask flatness, and M is the mask reduction ratio. All contributing components, except resist thickness $\mathrm{THK}_{\mathrm{resist}}$, are random, either within each exposure field or between exposure fields. By associating topography and film uniformity with n_{resist}, we assume that the resist is perfectly planar over the topography and the substrate film. We also assume that the image plane deviation is completely contained in the resist thickness. $\Delta Z_{\mathrm{tilts}}$ contains tilts at the wafer side as well as the mask side. Tilts at the mask side have to be divided by M^2. Film uniformity refers to local variations in the substrate films. The films are primarily planarized by the resist. Therefore, their DOF contributions are governed by the resist refractive index. Wafer flatness is spread out in a larger lateral range; therefore, the resist follows the wafer flatness conformably. As a result, its DOF contribution is governed by the index of the coupling medium.

The difference between a dry imaging system and an immersion system is in n_{CM}. Because of the larger n_{CM} with immersion systems, the DOF requirement is increased. Therefore, when the vacuum wavelength is identical to the immersed wavelength, a dry system requires less DOF. However, with the same light frequency, the immersed wavelength is shorter, resulting in a larger $\mathrm{DOF}_{\mathrm{diffrac}}$. This more than compensates for the loss in a higher required DOF.

8.4.5 Available DOF

$\mathrm{DOF}_{\mathrm{avail}}$ for device fabrication is $\mathrm{DOF}_{\mathrm{diffrac}}$ minus the contribution from resist thickness. It provides the "room" to contain the other defocus-contributing components. In an oversimplified view, $\mathrm{DOF}_{\mathrm{avail}}$ is the DOF measured in the resist image at the desired resist thickness through an exposure-defocus matrix measurement. The measurement must be carefully performed to remove any measurement error and any contribution from wafer nonflatness, tilting error, etc. Analytically, $\mathrm{DOF}_{\mathrm{avail}}$ is given as

$$DOF_{\text{avail}} = k_3 \frac{\lambda_{\text{CM}}}{\sin^2(\theta/2)} - \frac{n_{\text{CM}}}{n_{\text{resist}}} THK_{\text{resist}}, \tag{8.13}$$

and

$$\frac{1}{n_{\text{CM}}} DOF_{\text{avail}} = \frac{1}{n_{\text{resist}}} (Z_{\text{topo}} + \Delta Z_{\text{IPD}}) + \begin{bmatrix} \frac{1}{n^2_{\text{resist}}} (\Delta Z^2_{\text{resist}} + \Delta Z^2_{\text{topo}} + \Delta Z^2_{\text{FU}}) + \frac{1}{n^2_{\text{air}}} (\Delta Z^2_{\text{focus}} + \Delta Z^2_{\text{SI}} \ + \frac{\Delta Z^2_{\text{MF}} + \Delta Z^2_{\text{tilts at mask}}}{M^2}) \\ + \frac{1}{n^2_{\text{CM}}} (\Delta Z^2_{\text{tilts at wafer}} + \Delta Z^2_{\text{WF}}). \end{bmatrix}^{\frac{1}{2}} \tag{8.14}$$

DOF_{avail} from an immersion system is larger than that from a dry system by a factor that is greater than the refractive index. Let us continue with the example given in Figs. 8.11 and 8.12. With a resist thickness of 200 nm, n_{resist} = 1.75 and n_{CM} = 1.44. $DOF_{diffrac}$ and DOF_{avail} are listed in Table 8.1.

Here, $DOF_{diffrac}$ is evaluated with n_{CM} and n_{resist} equal to 1 according to Eq. (8.3), and with DOF_{avail} according to Eq. (8.14). The DOF_{avail} of this particular immersion system (top row of Table 8.1) is almost three times that of the dry system (bottom row). The slightly higher k_3 for the immersion system is due to less polarization-dependent stray light, which is discussed in more detail in Section 8.5.

DOF improvements are confirmed experimentally. Figure 8.13 shows that the DOF of the etched poly image (poly is the short form of polysilicon) exposed over the active layer is for a 90-nm SRAM circuit. The DOF determined by ±10% CD is 350 and 700 nm from dry and immersion imaging, respectively. The after-develop image has DOFs of 314 nm and 622 nm.

Figure 8.14 shows the DOF of 130-nm contact holes determined with an E-D window. Axial illumination at $\sigma = 0.6$, 0.75 NA, and a 6% AttPSM mask were used. Individual and common E-D windows at 260-nm, 400-nm, and 430-nm pitches were plotted. The DOF improvement for each individual window is between 30% and 50% and for the common window is 51%.

8.4.6 The preferred refractive index in the coupling medium

What is the optimum refractive index of the coupling medium? From the standpoint of index matching to reduce reflection, n_{CM} should be identical to the index of either the resist or the lens material that the fluid is in contact

Table 8.1 $DOF_{diffrac}$ and DOF_{avail} of a dry system and an immersion system.

λ (nm)	sinθ	k_1	k_3	$DOF_{diffrac}$	DOF_{avail}
193	0.9	0.466	0.35	240	126
134	0.625	0.466	0.367	451	289

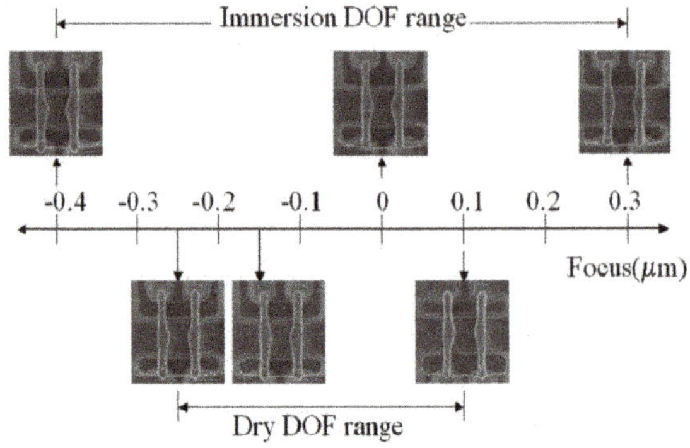

Figure 8.13 DOF of poly image on an active layer (reprinted from Ref. 48).

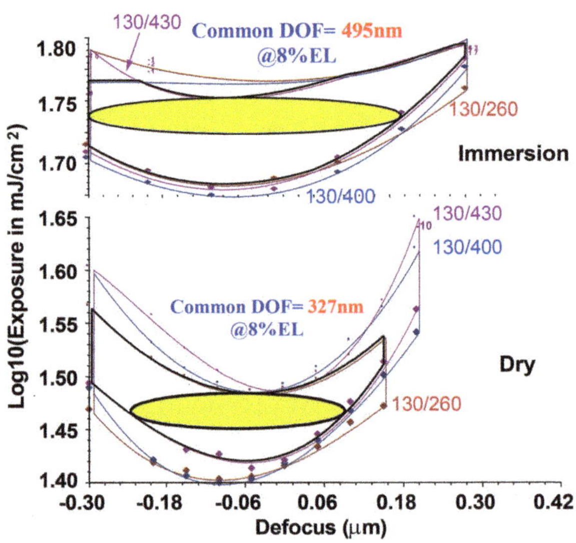

Figure 8.14 DOF of contact holes determined with an E-D window from immersion and dry exposures (reprinted from Ref. 48).

with, or it should be an optimum value between the indices of the two contacting materials. However, with the impressive progress made in antireflection coatings, reflection reduction with the coupling medium is not that important. One should make the index as high as possible to increase the DOF. If such a medium is available, one should not hesitate to use an index higher than that of the resist. Granted, total internal reflection will prevent high spatial frequencies from entering the resist, but the DOF is still increased as a function of its high index of refraction. The rejected frequencies are not available without the coupling medium, anyway. Therefore, the refractive index of the coupling medium should be as high as possible.

8.4.7 Tradeoff between resolution and $DOF_{diffrac}$

The tradeoff between resolution and $DOF_{diffrac}$ in an immersion system is similar to that of a wavelength reduction system. Equations (8.2) and (8.3) can be used to demonstrate the tradeoff, neglecting polarization-dependent stray light for the time being. There are two extremes in the tradeoff: maximum resolution and maximum $DOF_{diffrac}$. To maximize resolution, the angle θ is maintained as high as dry lithography can achieve. For example, $\sin\theta$ is kept at 0.9 in the presence of a high-index medium. This way, resolution is improved by $1/n_{CM}$, while $DOF_{diffrac}$ is reduced by the same factor. The situation is equivalent to simply reducing the vacuum wavelength. $DOF_{diffrac}$ reduction follows resolution improvement linearly instead of quadratically, as in the case of increasing NA. The other extreme, maximum $DOF_{diffract}$, is achieved by keeping the resolution constant by reducing θ such that $\sin\theta = \sin\theta_0/n_{CM}$. $DOF_{diffrac}$ is now improved by the following equation:

$$DOF_{diffrac} = k_3\frac{\lambda}{\sin^2(\theta/2)} = k_3\frac{\lambda_0}{n}\frac{1}{\sin^2\left[\frac{1}{2}\sin^{-1}\left(\frac{1}{n}\sin\theta_0\right)\right]}. \tag{8.15}$$

The tradeoff is illustrated in Fig. 8.15. Here, Eqs. (8.2) and (8.3) are plotted in the $DOF_{diffrac}$-versus-resolution space in the log scale, according to $k_1 = 0.315$ and $k_3 = 0.223$. λ and $\sin\theta$ are now parameters determining the resolution and attainable $DOF_{diffrac}$. This set of scaling factors is evaluated using the E-D forest methodology, 6% AttPSM, 8% EL, and ±10% CD tolerance. The common E-D window of three pitches corresponding to 1:1.5, 1:2, and 1:3 line:space ratios is used. In this figure, the three $DOF_{diffrac}$ curves

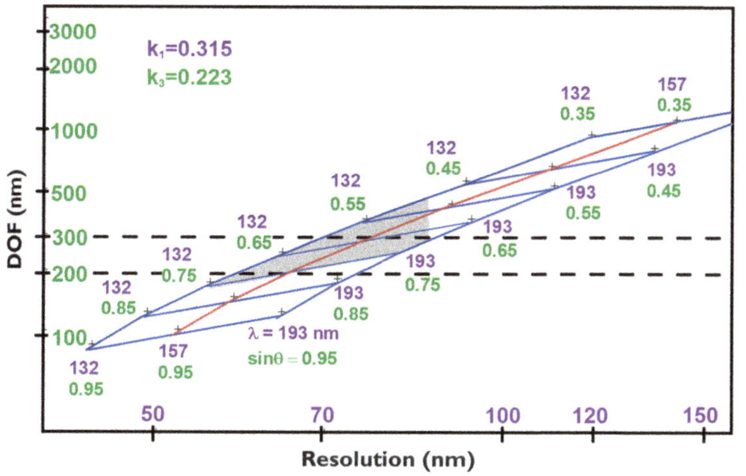

Figure 8.15 Tradeoff between resolution and $DOF_{diffrac}$ using annular illumination and AttPSM.

corresponding to 193, 157, and 134 nm are plotted, showing that 193-nm immersion is indeed a generation better than 157-nm dry systems. At 193 nm and 0.7 NA, the resolution is 87 nm and $DOF_{diffrac}$ is 300 nm. A feasible range of improvement using water immersion is shown in the shaded area between the 193- and 134-nm curves. Preserving the resolution improves $DOF_{diffrac}$ to 480 nm with $\sin\theta = 0.48$. Preserving the 300-nm $DOF_{diffrac}$ by moving the immersed $\sin\theta$ to 0.59 produces a resolution of 70 nm. Using the same $\sin\theta$ at 0.7 but increasing the wavelength to 134 nm with immersion produces a resolution of 59 nm while reducing $DOF_{diffrac}$ to 206 nm.

8.5 Polarization in Optical Imaging

Light, being an electromagnetic (EM) wave, has vectorial electric and magnetic fields; therefore, it has a polarity. The orientation of these fields, i.e., the polarity, affects the transmission and reflection characteristics at interfaces. Furthermore, the imaging process itself is also polarization dependent. The contrast of high spatial frequencies at the wrong polarization can be reduced. At low NA, the polarization effects are not pronounced and therefore have been ignored so far. Now, when NA is 0.6 or higher, especially when it is dramatically increased by immersion, the polarization effects can no longer be ignored.

In the last two decades, vectorial illumination and imaging began to gain attention from lithographers. Flagello and Rosenbluth[9] explored lithographic tolerances based on vector diffraction theory at high NA in the aerial image and in the resist. Brunner et al.[11] explained TM imaging and film reflection at high angles. They also produced various experimental demonstrations. Smith and Cashmore[21] showed polarization-dependent imaging in the resist and explored the impact of resist parameters.

This section analyses two prominent polarization effects: (1) loss of contrast in TM polarization when spatial frequencies are recombined to form the image and (2) stray light induced by specular reflection from the resist surface. Both effects can be accounted for quantitatively in terms of stray light. This polarization-dependent stray light (PDS)[22] is combined with the system stray light (SSL) normally existing in an optical imaging system in the intensity distribution of the optical image in order to generate E-D trees and windows that lead to the evaluation of DOF and EL of a set of given objects. We then compare the DOF loss due to PDS for line–space pairs, contact holes, disk illumination, annular illumination, and inclusion of more than the 1st-order spatial frequency in the object. Comparison is also made between 193-nm immersion and dry systems, as well as between 157-nm dry systems and 193-nm immersion systems.

8.5.1 Imaging with different polarizations

The imaging process can be viewed as the transmitting and filtering of spatial frequencies. As light leaves the mask, it carries all of the spatial frequencies contained in the mask towards the imaging lens, which assumes the role of recombining these spatial frequencies at the image plane to reproduce the mask pattern. Because of the finite NA, the lens behaves like a low-pass filter, cutting off the higher spatial frequencies that exceed the limit imposed by the NA and transmitting the lower frequencies for recombination. Figure 8.16 shows the angular distribution of the spatial frequencies generated from diffraction of the incident zero-frequency illumination beam by a large opening and a small opening in the mask. The lens pupil cuts off any spatial frequencies outside the $\pm\theta$ range. The large opening diffracts into a smaller angular spectrum and is reproduced better than the smaller opening.

8.5.1.1 Recombination of spatial frequency vectors in the resist

The recombination process is polarization dependent. Let's examine a single pitch combining two spatial-frequency vectors $\mathbf{S}_1$ and $\mathbf{S}_{-1}$ in the transverse electric field (TE) mode, making incident angles $\pm\alpha$ with respect to the optical axis of the lens, as shown in Fig. 8.17. The 0th-order beam $\mathbf{S}_0$ is shown in the same figure for reference, even though is not used for two-beam imaging. The vectors represent the direction of propagation of the positive and negative spatial frequencies. In the TE mode, i.e., when the electric field of the EM wave is perpendicular to the plane of incidence, the interference is straightforward. Let the electric field be $\mathbf{E}_1$ and $\mathbf{E}_{-1}$ for each vector:

$$\mathbf{E}_1 = \hat{y}A_1 e^{ik(-x\sin\alpha + \cos\alpha)} \tag{8.16}$$

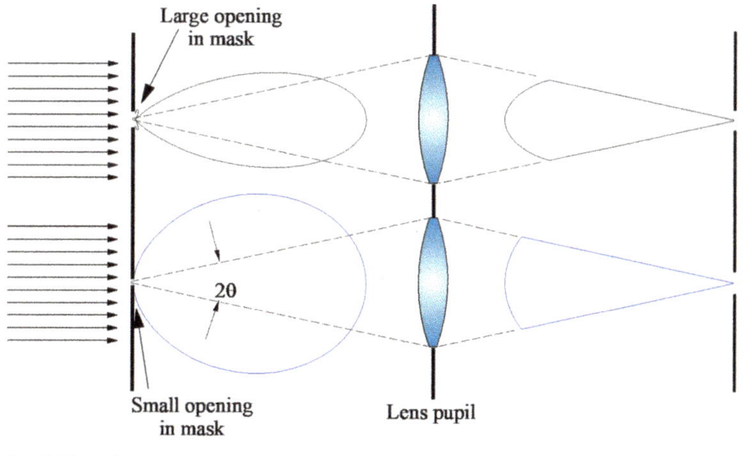

Figure 8.16 Diffraction from a large opening and a small opening. The lens cuts off the higher-spatial-frequency components, resulting in lost resolution. The diffraction patterns are normalized to the same intensity along the optical axis.

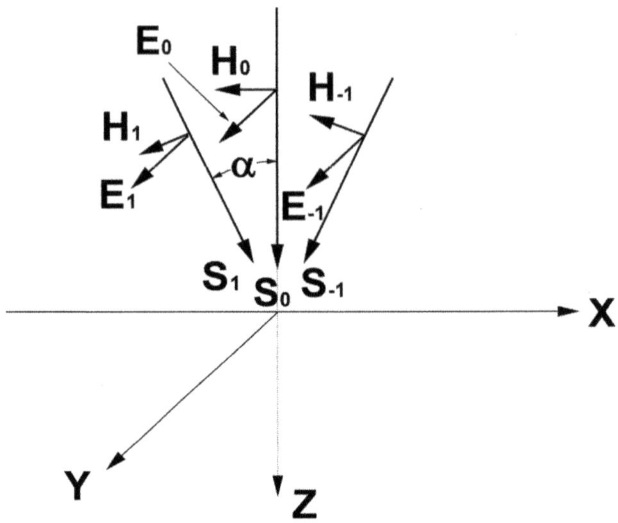

Figure 8.17 Imaging beams in TE mode.

and

$$\mathbf{E}_{-1} = \hat{y} A_{-1} e^{ik(x \sin\alpha + \cos\alpha)}, \tag{8.17}$$

where $k = 2\pi/\lambda$. The recombined intensity I is

$$I = [\mathbf{E}_1 + \mathbf{E}_{-1}]^*[\mathbf{E}_1 + \mathbf{E}_{-1}] = A_1^2 + A_{-1}^2 + 2A_1A_{-1}\cos(2kx\sin\alpha). \tag{8.18}$$

Assuming that $\alpha > 0$, when $2kx \sin\alpha = 0$, I is at its maximum, and when $2kx \sin\alpha = \pi$, I is at its minimum:

$$\text{Contrast}_{\text{TE,2beams}} = \frac{I_{\text{max}} - I_{\text{min}}}{I_{\text{max}} + I_{\text{min}}} = \frac{2A_1A_{-1}}{A_1^2 + A_{-1}^2}, \quad \alpha > 0. \tag{8.19}$$

Usually, $A_1 = A_{-1}$ and contrast = 1. Note that when $\alpha = 0$,

$$I_{\text{max}} = I_{\text{min}} = A_1^2 + A_{-1}^2; \tag{8.20}$$

that is,

$$\text{Contrast} = 0 \text{ when } \alpha = 0. \tag{8.21}$$

Therefore, when two beams making finite angles with respect to the optical axis are symmetrical to that axis and form an image, the image has a periodicity of $\lambda/(2\sin\alpha)$ and a contrast of 1, independent of the axial position; i.e., the DOF is infinite. The two-beam situation is exemplified by the alternating phase-shifting mask (AltPSM) or off-axis illumination (OAI). Note that we have used two coherent beams to illustrate imaging in the TE mode. When we are imaging through a lens using partially coherent illumination, the contrast is less than unity and the DOF is finite. In the case of off-axis illumination, one of the two interfering beams is actually the original 0th-order beam tilted by the off-axis light, and one of the 1st-order beams is thrown out of the coverage of the aperture angle θ. Hence, A_1 and A_{-1} are generally not identical to each other. In addition, when the tilting angle does not make the two beams symmetrical to the optical axis, many unwanted interference terms are produced.[23] Of course, one wants to optimize the tilting angle, but only one pitch can be optimized out of all of the pitches in the circuit layout.

In many situations, the 0th-order beam is included in imaging, for example, when the illumination is axial or when an alternating PSM is not used. Let's consider the case of three-beam imaging in the TE mode:

$$\mathbf{E}_0 = \hat{y} A_0 e^{ikz}. \tag{8.22}$$

$\mathbf{E}_1$ and $\mathbf{E}_{-1}$ were given in Eqs. (8.16) and (8.17), respectively. The recombined intensity is

$$I = [\mathbf{E}_0 + \mathbf{E}_1 + \mathbf{E}_{-1}]^* [\mathbf{E}_0 + \mathbf{E}_1 + \mathbf{E}_{-1}]. \tag{8.23}$$

In the three-beam case, the possibility of $A_1 = A_{-1}$ is high. Assuming this condition leads to

$$\begin{aligned} I = {} & A_0{}^2 + 2A_1{}^2 + 2A_1{}^2 \cos(2kx \sin\alpha) \\ & + 4A_0 A_1 \cos(kx \sin\alpha) \cos kz(\cos\alpha - 1), \end{aligned} \tag{8.24}$$

where the third term indicates a double-frequency component due to interference of A_1 and A_{-1}. The fourth term carries the fundamental frequency dictated by the pitch of the object. The double-frequency component is usually filtered out by the lens aperture angle θ for the critical dimensions in lithographic imaging. Assuming that $\alpha > 0$, when $kx \sin\alpha = 0$, I is at its maximum, and when $kx \sin\alpha = \pi$, I is at its minimum:

$$\text{Contrast}_{\text{TE,3beams}} = \frac{4r \cos kz(\cos\alpha - 1)}{1 + 4r^2}, \quad \alpha > 0, \tag{8.25}$$

where

$$r \equiv \frac{A_1}{A_0}.$$

Note that, even with coherent light, there is a Z-dependence of the image contrast in the three-beam case. Also, since the electric field vectors are combined as if they are scalars, the TE case is identical to image formation by scalar prediction as long as the imaging involves a single medium.

This is more complicated in the transverse magnetic field (TM) mode when the magnetic field vector of the imaging beams is perpendicular to the plane of incidence, as shown in Fig. 8.18. Let us first consider the two-beam case. Since the electric field is no longer perpendicular to the plane of incidence but resides in the x-z plane, it is decomposed into the x and z components for calculation:

$$\mathbf{E}_x = \mathbf{E}_{1x} + \mathbf{E}_{-1x} = (E_{1x} + E_{-1x})\hat{x}, \tag{8.26a}$$

$$\mathbf{E}_z = \mathbf{E}_{1z} + \mathbf{E}_{-1z} = (E_{1z} + E_{-1z})\hat{z}, \tag{8.26b}$$

and

$$E_{1x} = |\hat{E}_1 \cdot \hat{x}| A_1 e^{ik(-x\sin\alpha + z\cos\alpha)} = A_1 \cos\alpha\, e^{ik(-x\sin\alpha+z\cos\alpha)}\,. \tag{8.27}$$

Similarly,

$$E_{-1x} = |\hat{E}_{-1} \cdot \hat{x}| A_{-1} e^{ik(x\sin\alpha + z\cos\alpha)} = A_{-1} \cos\alpha\, e^{ik(x\sin\alpha+z\cos\alpha)}, \tag{8.28}$$

$$E_{1z} = |\overleftarrow{E}_1 \cdot \hat{z}| A_1 e^{ik(-x\sin\alpha + z\cos\alpha)} = A_1 \sin\alpha\, e^{ik(-x\sin\alpha+z\cos\alpha)}, \tag{8.29}$$

and

$$E_{-1z} = |\overleftarrow{E}_{-1} \cdot \hat{z}| A_{-1} e^{ik(x\sin\alpha + z\cos\alpha)} = -A_{-1} \sin\alpha\, e^{ik(x\sin\alpha+z\cos\alpha)}. \tag{8.30}$$

The recombined intensity is

$$\begin{aligned} I &= \mathbf{E}_x{}^*\mathbf{E}_x + \mathbf{E}_z{}^*\mathbf{E}_z = [\mathbf{E}_{1x} + \mathbf{E}_{-1x}]^*[\mathbf{E}_{1x} + \mathbf{E}_{-1x}] + [\mathbf{E}_{1z} + \mathbf{E}_{-1z}]^*[\mathbf{E}_{1z} + \mathbf{E}_{-1z}] \\ &= A_1{}^2 + A_{-1}{}^2 + 2A_1A_{-1}\cos(2kx\sin\alpha)\cos 2\alpha. \end{aligned} \tag{8.31}$$

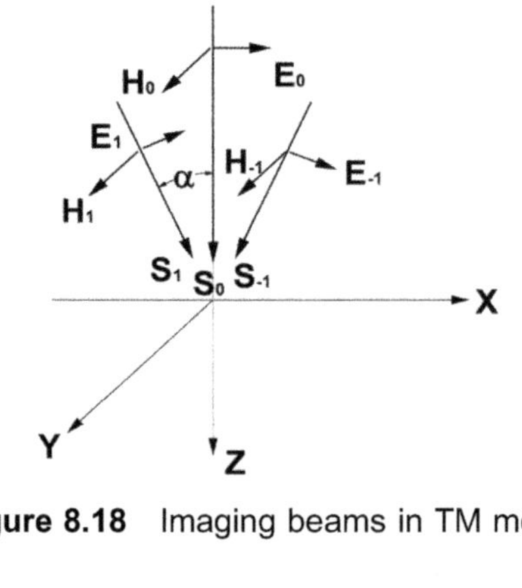

Figure 8.18 Imaging beams in TM mode.

Substituting $2kx\ \sin\alpha = 0$ and $2kx\ \sin\alpha = \pi$ for $I_{\max}$ and $I_{\min}$, we have

$$\text{Contrast}_{\text{TM,2beams}} = \frac{2A_1 A_{-1}\cos 2\alpha}{A_1{}^2 + A_{-1}{}^2},\quad \alpha > 0. \tag{8.32}$$

Equations (8.19) and (8.32) determine that the contrast in the TM mode is lower by the factor $\cos 2\alpha$ compared to the TE or scalar case. That is,

$$\frac{\text{Contrast}_{\text{TM,2beams}}}{\text{Contrast}_{\text{TE,2beams}}} = \cos 2\alpha. \tag{8.33}$$

For the three-beam case in the TM mode, we take the same E_1 and E_{-1} as in Eqs. (8.27) to (8.30) and add the 0th-order component E_0:

$$\mathbf{E}_0 = \hat{x}A_0 e^{ikz}. \tag{8.34}$$

With $A_1 = 4A_{-1}$,

$$\begin{aligned} I &= \mathbf{E}_x{}^*\mathbf{E}_x + \mathbf{E}_z{}^*\mathbf{E}_z \\ &= [\mathbf{E}_{0x} + \mathbf{E}_{1x} + \mathbf{E}_{-1x}]^*[\mathbf{E}_{0x} + \mathbf{E}_{1x} + \mathbf{E}_{-1x}] + [\mathbf{E}_{1z} + \mathbf{E}_{-1z}]^*[\mathbf{E}_{1z} + \mathbf{E}_{-1z}] \\ &= A_0{}^2 + 2A_1{}^2 + 2A_1{}^2\cos(2kx\sin\alpha)\cos 2\alpha \\ &\quad + 4A_0A_1\cos(kx\sin\alpha)\cos kz(\cos\alpha - 1)\cos\alpha. \end{aligned} \tag{8.35}$$

Assuming that $4\alpha > 0$, when $kx\ \sin\alpha = 0$, I is maximum, and when $kx\ \sin\alpha = \pi$, I is minimum:

$$\text{Contrast}_{\text{TM,3beams}} = \frac{4r\cos kz(\cos\alpha - 1)\cos\alpha}{1 + 2r^2(1 + \cos 2\alpha)},\quad \alpha > 0. \tag{8.36}$$

The comparison of Eqs. (8.25) and (8.36) to derive a contrast equation is less straightforward than the derivation of Eq. (8.33) in the TE case. Strictly speaking, the contrast in the TM case is

$$\frac{\text{Contrast}_{\text{TM,3beams}}}{\text{Contrast}_{\text{TE,3beams}}} = \frac{1 + 4r^2}{1 + 2r^2(1 + \cos 2\alpha)}\cos\alpha. \tag{8.37}$$

At a small angle α,

$$\frac{\text{Contrast}_{\text{TM,3beams}}}{\text{Contrast}_{\text{TE,3beams}}} = \cos\alpha. \tag{8.38}$$

The contrast in the TM case is now lower than in the TE or scalar case by $\cos\alpha$ at a small α. For a large α, the contrast in the TM case is slightly larger than what Eq. (8.38) predicts. With unpolarized (unpo) or circularly polarized light,

$$I = \frac{I_{\mathrm{TE}} + I_{\mathrm{TM}}}{2}, \tag{8.39}$$

$$\mathrm{Contrast}_{\mathrm{unpo,2beams}} = \frac{A_1 A_{-1}(\cos 2\alpha + 1)}{2(A_1{}^2 + A_{-1}{}^2)}, \tag{8.40}$$

and

$$\mathrm{Contrast}_{\mathrm{unpo,3beams}} = \frac{2r[\cos kz(\cos\alpha - 1)](\cos\alpha + 1)}{1 + 4r^2}, \text{ for small } \alpha \text{ only.} \tag{8.41}$$

Therefore, compared to the scalar results, the contrast differs by a factor of $(\cos\alpha+1)/2$ for three-beam recombination and by $(\cos 2\alpha+1)/2$ for two beams. This means that, at the largest possible α (i.e., $\pi/2$), the contrast drops to 0.5 for the former and to 0 for the latter. Fortunately, recombination imaging takes place inside the resist layer, which has a refractive index larger than that of the coupling medium. The angle α is reduced in the resist according to Snell's law and can be much less than π. For example, when light is directly incident on the resist from air, $\sin\alpha_{\mathrm{resist}} = (1/n_{\mathrm{resist}})\sin\alpha_{\mathrm{air}}$. In the case of immersion, $\sin\alpha_{\mathrm{resist}} = (n_{\mathrm{fluid}}/n_{\mathrm{resist}})\sin\alpha_{\mathrm{fluid}}$. The light that does not contribute to contrast is harmful.

Even though we have derived the $\cos 2\alpha$ and $\cos\alpha$ dependencies using coherent light, the relationship holds for partially coherent and incoherent light as well. These illumination conditions can be simulated with superposition of diffraction intensities from coherent point sources in the lens pupil.[24,25]

8.5.1.2 Polarized refraction and reflection at the resist surface

The introduction of the resist surface complicates the situation. The impact of this extra surface in terms of scalar diffraction and on the DOF budget was given in Section 8.4. Here, we examine the loss of the imaging light as a function of the polarization in the light beams. The Fresnel formulae[26] for reflection and refraction of polarized light from medium 2 to medium 3 in Fig. 8.3 are

$$R_{\mathrm{TE}} = \frac{n_2\cos\theta_2 - n_3\cos\theta_3}{n_2\cos\theta_2 + n_3\cos\theta_3} B_{\mathrm{TE}}, \tag{8.42}$$

$$T_{\mathrm{TE}} = 1 - R_{\mathrm{TE}}, \tag{8.43}$$

$$R_{\mathrm{TM}} = \frac{n_3\cos\theta_2 - n_2\cos\theta_3}{n_3\cos\theta_2 + n_2\cos\theta_3} B_{\mathrm{TM}}, \tag{8.44}$$

$$T_{\mathrm{TM}} = 1 - R_{\mathrm{TM}}, \tag{8.45}$$

and

$$n_2 \sin \alpha_2 = n_3 \sin \alpha_3, \tag{8.46}$$

where T and R are the transmittance and reflectance of the interface between media 2 and 3, and B_{TE} and B_{TM} are the amplitudes of the incident beam in the interface between media 2 and 3. Referring to Fig. 8.3 for designation of the media, medium 1 is the lens material, medium 2 is the coupling medium, and medium 3 is the resist.

We now plot the transmittance of light from the coupling medium to the resist in the TE and TM modes through an interface between $n_2 = 1$ and $n_3 = 1.75$ in Fig. 8.19. The differentiation in polarization is perceivable starting from $\sin\theta = 0.3$. However, it begins to make a difference at $\sin\theta = 0.6$. The loss of exposure energy is less important than the reflected light becoming stray light. At $\sin\theta = 0.8$, the total unpolarized transmission is about 90% and the reflection is 10%. If the reflected light is totally converted to stray light on the wafer, 10% reflection is significant, whereas, a loss of exposure affects only the throughput. From Eq. (8.44), $R_{TM} = 0$ when $n_3 \cos\alpha_2 - n_2 \cos\alpha_3 = 0$. This particular angle is Brewster's angle.[27] The impact of Brewster's angle on optical imaging is not as important as the loss of transmitted light in the TE mode. According to Section 8.5.1.1, TE is the preferred polarization in terms of reconstruction of spatial frequencies. Unfortunately, this preferred component is more reduced than the TM component at high NA.

This problem can be alleviated with a top antireflection coating (TARC), i.e., an ARC applied to the resist surface. A TARC couples more light into the resist, thus reducing the reflected light that can contribute to stray light. In the case of immersion lithography, the difference between the refractive indexes of the immersion fluid and the resist is lower than that between air and the

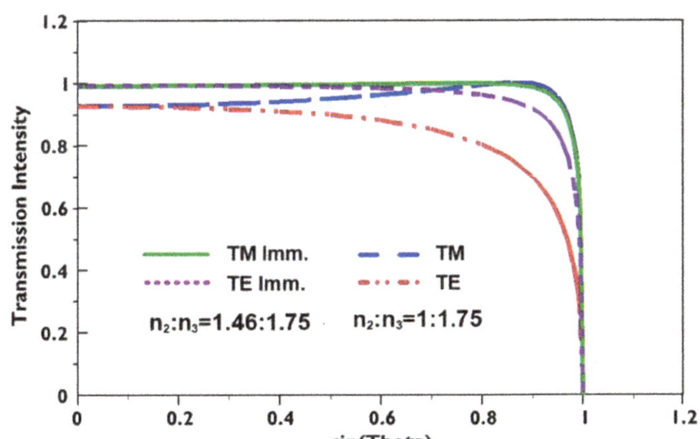

Figure 8.19 Polarization-dependent transmission at an interface. n_2:n_3 = 1:1.75 is used for the dry case and 1.44:1.75 is used for the immersion case.

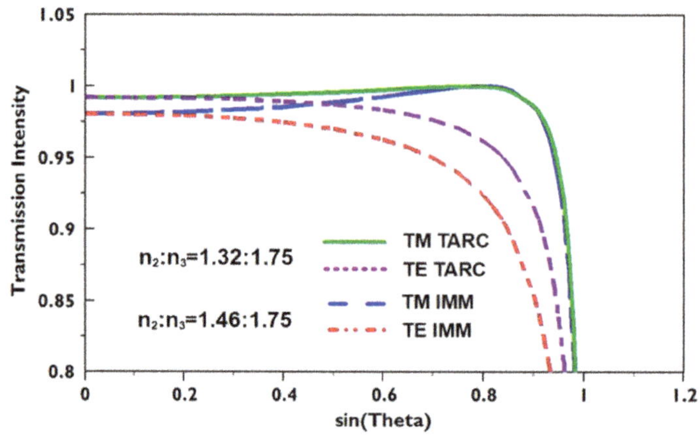

Figure 8.20 Polarization-dependent transmission at an interface. n_2:n_3 = 1.44:1.75 and 1.32:1.75. The former is the case with immersion lithography, and the latter is the best index matching to air with a TARC. Note that the transmittance scale is expanded compared to that in Fig. 8.19.

resist—the immersion fluid serves as a built-in TARC. With a fluid index of 1.44, the unpolarized transmitted light is more than 98%, leaving only less than 2% to be reflected at $\sin\theta = 0.8$. Even with the unfavorable TE polarization, the transmission and reflection are 96.2% and 3.8%, respectively. Figure 8.19 compares this scenario to the normal practice of using a wafer in air. The immersion case and the scenario using a TARC are compared in Fig. 8.20. Here, the refractive index is taken to be the optimum for matching the resist to air, i.e., $\sqrt{1.75} = 1.32$. Immersion shows better reflection/transmission characteristics than TARC. The difference between the curves is more subtle than the case of Fig. 8.19, so we have amplified the transmittance scale in Fig. 8.20.

8.5.1.3 Different effects of polarized illumination

In addition to the polarization dependence of spatial frequency recombination in the resist, pattern orientation in the mask also plays an important role in the final-image formation on the wafer. Figure 8.21 shows the diffracted waves from a 3λ slit in the mask.[28] Both TE and TM illuminations are used. Note that TE here is defined as the electric field vector parallel to the edge of the slit; Similarly, TM means that the magnetic field vector is parallel to the edge of the slit. These types of illumination are depicted in Fig. 8.21. Parts (a) and (b) of this figure are identical to the corresponding parts of Fig. 4.13. They are reproduced here to help illustrate the polarization-related discussion.

The TE-diffracted wave in Fig. 8.21(a) satisfies the boundary condition that the electric field in a conducting screen, i.e., the chrome part of the mask, is 0. The electric field has a structure in the mask plane. There are three peaks in the structure, separated by approximately one wavelength. The peak

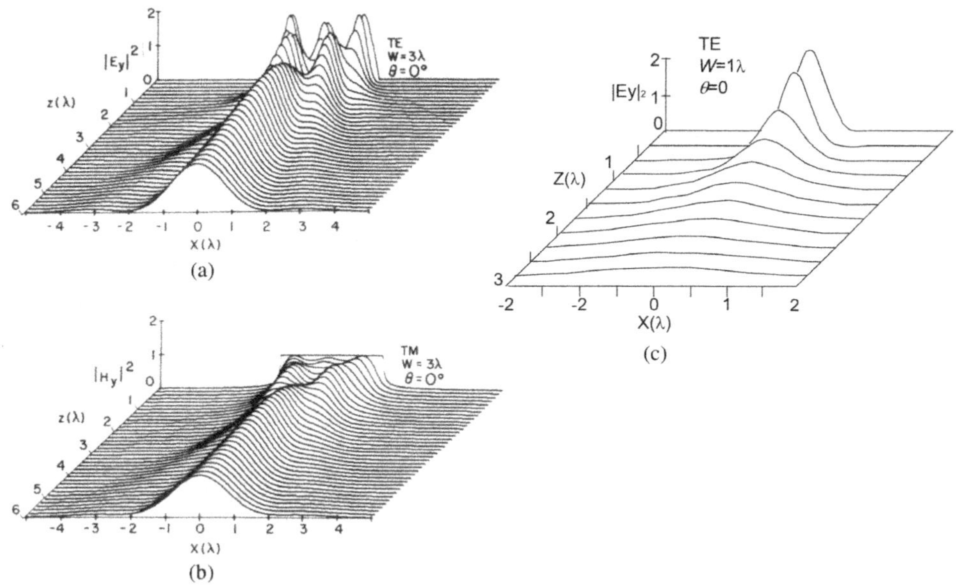

Figure 8.21 Polarization-dependent diffraction in three scenarios: (a) TE illumination on a 3λ slit, (b) TM illumination on the same slit, and (c) TE illumination on a 1λ slit (reprinted from Ref. 28).

intensity in the mask plane is about 2X higher than the incident intensity of the electric field at unity. There is a pronounced focus at about 3 wavelengths from the mask.

The TM-diffracted wave in Fig. 8.21(b) also satisfies the boundary condition that the intensity of the magnetic field in the slit is identical to the intensity of the incident magnetic field. The diffracted magnetic field exhibits a weak three-peaked structure slightly away from the mask plane. The three peaks become two, and finally one, just as in the case of TE illumination.

When the slit width becomes 1λ as in Fig. 8.21(c), the peak in the TE slit is even higher. The peak is most pronounced at a slit width of about 0.7λ. When the slit is smaller than 1λ, less light is transmitted through the TE slit, but this is not the case for the TM slit. The slit's different boundary conditions explain the discrepancy. In fact, subwavelength transmission gratings can be used as a TM filter to reject TE light. Therefore, TE slits transmit and focus more light when they are wider than 1λ, while they reject more light, otherwise.

Among the three types of polarization illumination conditions presented in Fig. 8.22, the type depicted in column (A) is found in imaging tools with NA below 0.6, where the electric field vectors are randomly oriented. At a high NA, the light with an electric field component perpendicular to the resist surface suffers from polarization-dependent stray light (PDS). These effects are analyzed in Section 8.5.2. The features on the mask are restricted to either vertical or horizontal orientations in the mask plane.

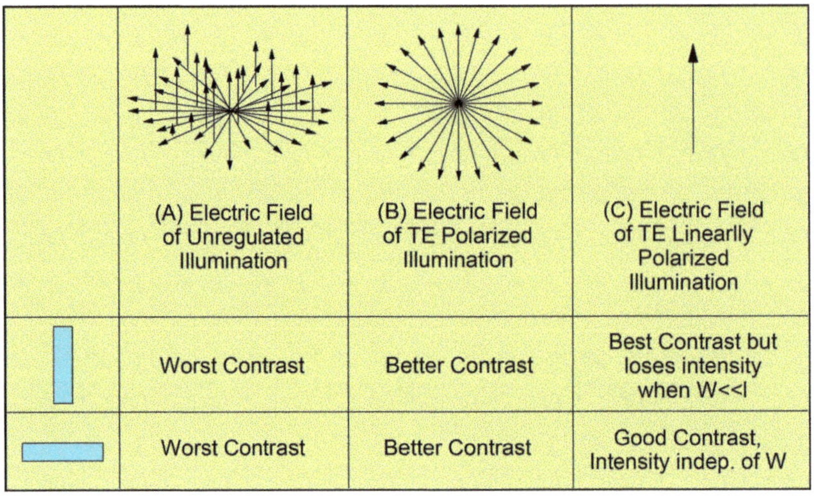

Figure 8.22 Effects of polarized illuminations on rectangular features.

The imaging lights combine to produce less PDS in Fig. 8.22 column (B), where the electric field vector is always perpendicular to the resist surface. However, the electric field vector can be randomly oriented with respect to the orientation of the features on the mask.

To get the most from polarized illumination, we use the scenario in Fig. 8.22 column (C), where the electric field vector is parallel to the edge of the mask patterns. Hence, optimization of polarization is yet another resolution-enhancement technique (RET). Polarization optimization calls for separating any arbitrary patterns into two perpendicular orientations and using two dipole illuminations in two separate exposures. This complements the type of illumination in Fig. 8.22 column (C). The only tradeoff is the throughput of the exposure tool and therefore the productivity. It is well known that aligning the gates in one direction helps to control the uniformity of the gate CD. This technique also works to align the gates to the electric field orientation of Fig. 8.22 column (C).

Figure 8.23 illustrates a method to produce the illumination of Fig. 8.22 column (B). (The illuminations in Fig. 8.22 are referred to here as A-type, B-type, and C-type illumination for convenience.) Light can be manipulated to be azimuthal in the lens pupil to result in elimination of the TM component while maintaining the symmetry in the orientation of the tangential electric-field vector. The orientation effect of B-type illumination is similar to that of annular illumination. Both are nondiscriminating to the orientation of the patterns on the mask, but the resolution-enhancement effect is not as strong as their discriminating counterpart in C-type illumination. Note that transmission and reflection at each optical surface between the light source and the resist have an effect on the polarization state of the actinic light. It is not

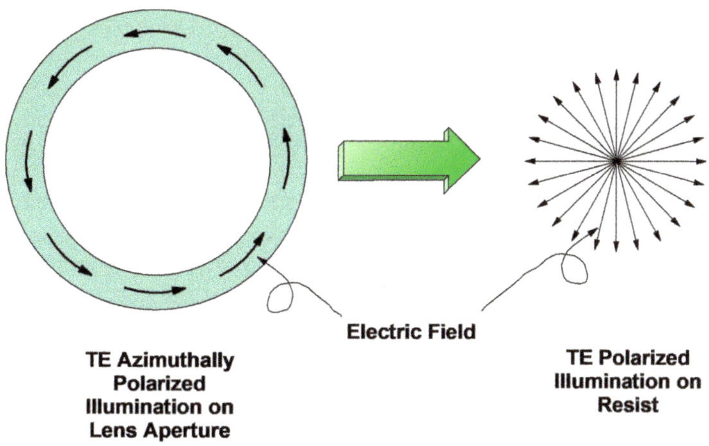

Figure 8.23 Azimuthally polarized illumination.

straightforward to produce pure polarization states such as those in B- or C-type illumination. However, residual unwanted polarization can be tolerated.

8.5.2 Stray light

The main equations for stray light were introduced in Section 8.5.1, and the physics related to them have been explained, so we now start to consider three types of stray light: system stray light (SSL), recombined polarization-dependent stray light (PDS_{recom}), i.e., PDS from spatial frequency recombination in the resist, and $PDS_{reflect}$, i.e., PDS from reflection off the resist and other surfaces on the wafer film stack.

Figure 8.24 shows an axial beam illuminating the mask pattern, creating the spatial frequencies in different diffraction orders. The 0th-, 1st-, and 2nd-order beams are illustrated. This angular spectrum is strictly a function of the mask pattern and the illumination beam; it is independent of the imaging lens so far. In this illustration, the 2nd-order beams are filtered by the lens. Only the 0th- and ± 1-order beams are incident on the resist surface. The transmitted light undergoes refractions according to Snell's law. The intensity resulting from the recombination of the electric field vectors exposes the resist. Specular reflection takes place at the resist surface, as shown. Although not plotted, specular reflection also takes place at all of the reflection surfaces on the wafer film stack. Most likely, there is a bottom antireflection coating (BARC) underneath the resist. Hence, the two most significant reflections are those from the resist surface and from the BARC.

8.5.2.1 System stray light

The transmitted light that does not recombine in the image becomes stray light. The lens itself induces SSL regardless of whether the imaging light is polarized. Basically, when a light beam strikes the lens surfaces, each lens

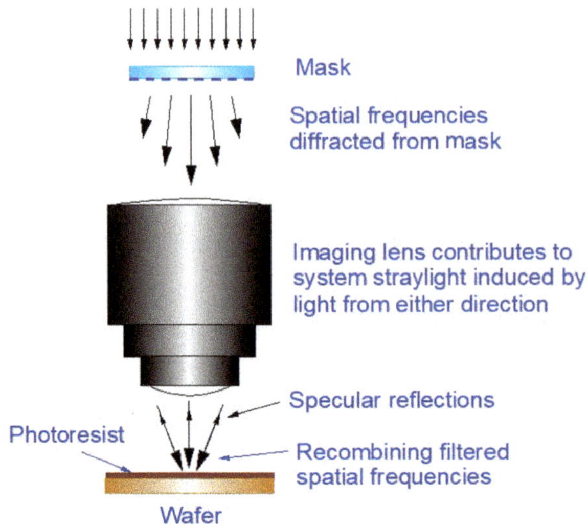

Figure 8.24 Light in a schematic projection system exhibiting illumination, diffraction, imaging through a lens, spatial frequency filtering, reflection, and exposure of the photoresist.

surface contributes to stray light from residual reflections from the lens ARC, surface roughness, surfaces blemishes, particulates on the lens surfaces, etc. This light beam can be either the normal imaging light moving forward or the reflected light from the wafer surfaces, moving in the opposite direction. Strictly speaking, the SSL is a complicated function of its location, which is determined by the lens configuration, including surface curvature, refractive indexes, ARC characteristics, surface roughness, position of lens elements, position of apertures, reflectivity of walls and slopes, distribution of surface contamination, etc. To simplify the problem, the SSL is approximated by a uniform percentage of the incident light whose level is determined by the ratio of transparent areas to opaque areas in the mask.

The reflected light from the wafer surfaces also produces SSL when it strikes the lens surfaces. After leaving the lens, it is reflected by the mask absorber. More information on this type of reflection will be presented in Section 8.5.2.3.

8.5.2.2 Stray light from recombination of spatial frequency vectors inside the resist

The normal lens-imaging process reduces contrast for images containing higher spatial frequencies. This takes place when the mask-diffracted light passes through the lens. Such diffraction contrast loss is handled with existing projection imaging simulations. After the lens, when spatial frequencies are recombining in the resist, the additional reduction of contrast in the recombination process shows up as stray light in its entirety—this is PDS_{recom}.

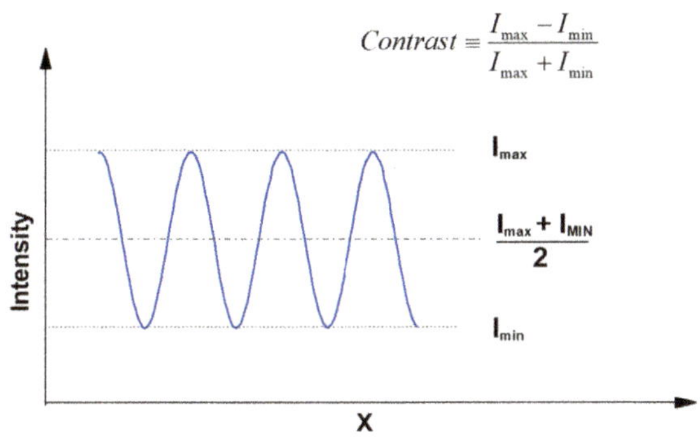

Figure 8.25 Contrast- and stray-light-related parameters.

As shown in Fig. 8.25, $I_{\min}$ is the background light in the resist, making up the stray light. Its portion with respect to the total light is normalized by $0.5{\cdot}(I_{\max} + I_{\min})$. Therefore,

$$PDS_{\text{recom}} = \frac{2I_{\min}}{I_{\max} + I_{\min}} = 1 - \text{Contrast}_{\text{recom}}. \tag{8.47}$$

The contrast of linearly polarized light making an arbitrary angle ϕ with respect to the electric vector of the TE mode is a function of ϕ and the additional recombination contrast of the TE and TM modes:

$$\text{Contrast}_{\text{recom}}(\phi) = \sin^2\phi \cdot \text{Contrast}_{\text{TM recom}} + \cos^2\phi \cdot \text{Contrast}_{\text{TE recom}}. \tag{8.48}$$

$\text{Contrast}_{\text{TErecom}} = 1$ and $\text{Contrast}_{\text{TMrecom}} = \cos\alpha_u$, where $\alpha_u = 2\alpha$ for two-beam interference, and $\alpha_u = \alpha$ for three-beam interference. The angle ϕ is depicted in Fig. 8.26. Equation (8.47) now becomes

$$PDS_{\text{recom}} = (1 - \cos\alpha_u)\sin^2\phi. \tag{8.49}$$

The most common illumination up to NA = 0.6 can be treated as having equal contributions from the TE and TM modes. That is,

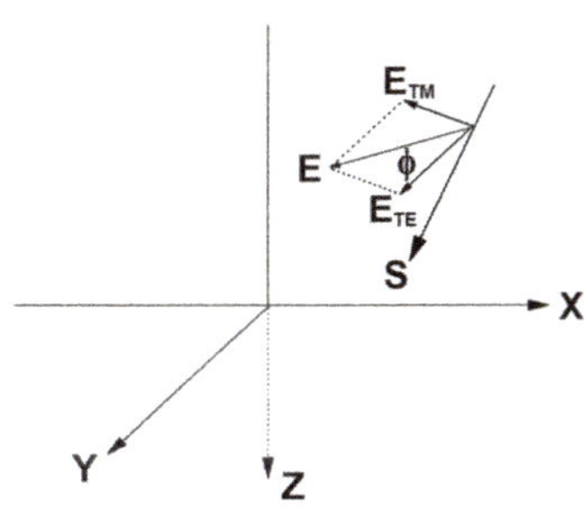

Figure 8.26 Electric field of an arbitrary linearly polarized beam, forming an angle ϕ with respect to the direction of the electric field in the TE mode.

$$\text{Contrast}_{\text{CP recom}} = 0.5\,(\cos\alpha_u + 1), \tag{8.50}$$

where CP stands for circular polarization. The equation can be used for circular polarization, linear polarization with $\phi = \pi/4$, or completely random polarization.

Note that the contrast used to calculate stray light is strictly the contrast produced by the interference of two coherent imaging beams whose incident angles are dictated by the spatial frequency of the patterns in the mask, i.e., $\sin\alpha = \lambda/p$, where p is the periodicity of the line–space pairs. Equations (8.40) and (8.41) are applied using this angle. The aperture angle θ of the imaging lens should not be confused with α. $\sin\theta$ sets the upper limit of α that is not blocked by the lens aperture. When $\sin\theta < \sin\alpha$, only the 0th-order beam passes through the lens; no image is formed. When $\sin\theta \geq \sin(n\alpha)$, the spatial frequencies λ/p, $2\lambda/p$, $\ldots$, $n\lambda/p$ recombine at the imaging plane to form the lens image. In lithography, because of economy, the imaging lens is extended to its full resolution potential. As a result, most imaging involves only the 1st-order spatial frequencies. Occasionally, some larger pitches are blended into the circuit design, leading to 2nd-order spatial frequencies. In extreme cases for pushing super-resolution, when $n\lambda/p > 1$, these frequencies turn into evanescent waves whose treatment is beyond the scope of this book.

When a two-beam RET system is used, $\alpha = \sin^{-1}(\lambda/2p)$, instead of $\sin^{-1}(\lambda/p)$, as shown in Fig. 8.27. In the case of off-axis illumination, using one of the illumination beams as an example, the beam is tilted by an angle $\sin^{-1}(\lambda/2p)$; as a result, the 1st-, 0th-, and -1st-order beams now make angles $-\sin^{-1}(\lambda/2p)$, $\sin^{-1}(\lambda/2p)$, and $\sin^{-1}(3\lambda/2p)$ with respect to the optical axis of the lens, respectively. The angle α for calculation of contrast is now $\sin^{-1}(\lambda/2p)$ instead of $\sin^{-1}(\lambda/p)$ in an axially illuminated BIM, which is equivalent to reducing the spatial frequency by half. In the case of an AltPSM, because of the phase shift in every other object, the 0th-order beam becomes zero and the two diffracted beams make angles $\pm\sin^{-1}(\lambda/2p)$ with respect to the optical axis, also reducing the spatial frequency by half.

The higher harmonics are separated by $\sin^{-1}(\lambda/2p)$. Naturally, the contrast in the two- and three-beam situations is calculated according to Eqs. (8.33) and (8.38), respectively.

With the same object, the $\text{PDS}_{\text{recom}}$ is similar in magnitude, independent of the described RET because, in the case of an axially illuminated BIM, $\cos\alpha = \cos[\sin^{-1}(\lambda/p)]$, whereas in the other two cases, $\cos 2\alpha = \cos[2\sin^{-1}(\lambda/2p)]$. Their magnitudes are almost identical with a small α and become significantly different only when $\sin^{-1}(\alpha/2p) \geq \pi/16$, as shown in Fig. 8.28. Since the range of α in both cases is between 0 and $\pi/2$, the main difference between them is that the contrast in the three-beam case cannot become negative. The tradeoff of reducing the spatial frequency to resolve an unresolvable object is negative

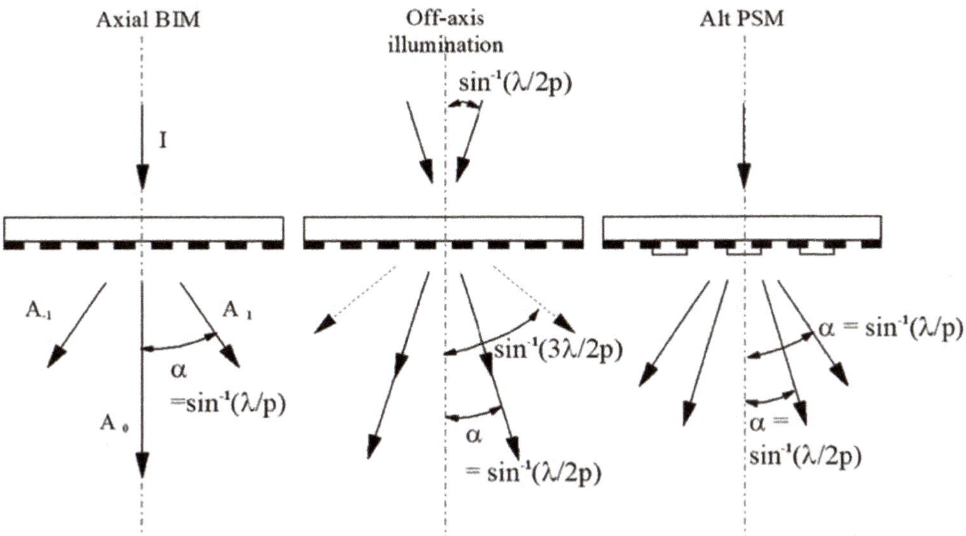

Figure 8.27 Angle α in an axially illuminated BIM, off-axis illumination, and an AltPSM.

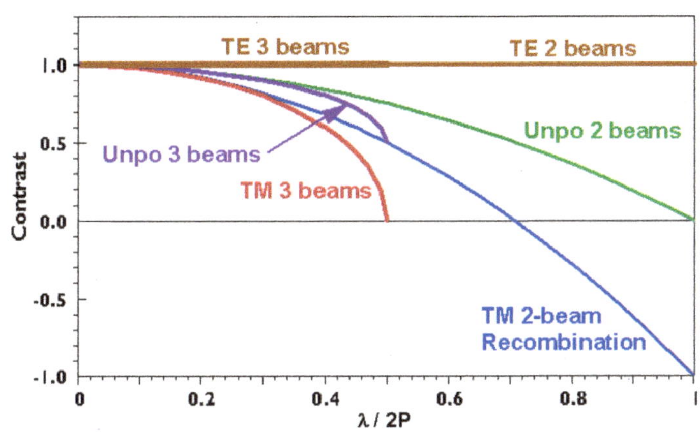

Figure 8.28 Contrast from a single object but with two- and three-beam illuminations.

contrast in the TM mode. Negative contrast means the image is reversed. It does not start at $\alpha/p = 1$ as usually speculated, but rather at $\sin^{-1}(\alpha/2p) = \pi/8$. Therefore, two-beam interference has better contrast than three-beam interference, even beyond the resolution limit of the latter.

8.5.2.3 Stray light from reflection off the resist surface

Only part of the specularly reflected light from the resist surface (as described in Section 8.5.1.2) contributes to $PDS_{reflect}$. After reflecting from the resist surface, light traces back to the mask as a conjugate of the image formed on the resist surface. Referring to Section 6.3.3.4 and Fig. 6.16, any light reflected off the conjugate of the mask plane contributes to stray light that reduces the contrast of the resist image. The reflectivity of the absorber ranges between 10% and 20%. In view of the above argument, it is preferable to use a low-reflectivity mask absorber unless all reflections from the wafer are suppressed.

With an actual imaging lens consisting of more than 30 elements—thus, about 70 glass–air interfaces—even the best ARC for each lens surface can add up to a few percent points of the total effective reflection. The reflected light, even under a perfectly conjugated reflection relationship, still produces stray light because of imperfect ARC at each lens surface and residual surface roughness.

8.5.2.4 Incorporating polarization effects into E-D windows

The E-D window methodology given in Chapter 5 is used in this section to study the effects of polarization on imaging. Rectangular E-D windows are used consistently throughout this section. The rectangular shape intersects the E-D trees more precisely than other shapes, especially in cases of small aperture angles θ, where the tree branches have a small curvature and are extended in the defocus direction. Throughout the rest of Section 8.5, the working DOF using elliptical windows is expected to be larger (by 30–150 nm) than that using rectangular windows.

The Signamizer software[29] is modified with the equations in this section to generate the results. To incorporate the SSL into the evaluation of E-D windows, a given level of constant intensity is superimposed on the aerial image based on the optical parameters such as λ, NA, illumination setting, and lens aberrations. For example, a 5%–10% constant intensity is added to the normalized intensity distribution of the aerial image of concern. This SSL level is typically found in modern diffraction-limited imaging lenses in scanners. Though it is a complicated spatial function of the lens ARC, surface roughness, optical configurations, baffles, etc., the SSL level is usually treated as constant for a 1st-order estimate of stray light effects.

PDS is clearly dependent on the incident angle, even though we still distribute it evenly in the resist. The level of PDS is calculated as a function of the incident angle according to the formulae given in this section. For PDS from recombination of spatial frequency vectors, its dependence on α can be determined from Eqs. (8.19), (8.25), (8.32), and (8.36), together with Eq. (8.47).

The PDS from reflection off the resist surface is treated by multiplying the reflectivity given in Eqs. (8.44) and (8.45) by the reflectivity of the absorber and by the lens stray light, which is induced by imperfect lens ARC, surface roughness, optical configurations, baffles, etc. Here, the incident beam is not from the illumination on the mask but from the reflection from the resist surface. Therefore, in the TM case,

$$PDS_{\text{reflect}} = \frac{R_{\text{TM}}^2}{B_{\text{TM}}^2}(\rho R_{\text{MA TM}} + SSL), \tag{8.51}$$

where ρ is an adjustment factor that takes care of the partial contribution to stray light discussed in Section 8.5.2.3. We have set adjustment factor ρ to 0.5 based on our experience. A rigorous setting awaits simulation with reflections from the multilayer stack on the wafer and the subsequent reflection from the mask absorber. For PDS due to reflection in the TE mode, one simply substitutes R_{TM} and B_{TM} with the corresponding B, R, and R_{MA} in Eq. (8.51). R_{MA} corresponds to reflection from the mask absorber. It is also polarization dependent.

Hence, the total modification of the intensity distribution $I(x, z)$ calculated from a regular simulation program is as follows:

$$I_{\text{modified}}(x, y) = [I(x, z) + I_{\text{avg}} \cdot (PDS_{\text{recom}} + PDS_{\text{reflect}}) + PD \cdot SSL] \cdot C, \tag{8.52}$$

where I_{avg} is $I(x, z)$ integrated over x at each z plane, PD is pattern density, and C is an adjustment constant that keeps the total intensity integrated over the mask unchanged before and after inducing stray light. This way, the exposure levels of different patterns are aligned to each other for evaluation of the common E-D window. The loss in transmitted light due to reflection is also incorporated through C:

$$C = \frac{I_{\text{avg}}}{I_{\text{avg}}(1 + PDS_{\text{in resist}} + PDS_{\text{reflection}}) + PD \cdot SSL}. \tag{8.53}$$

Both PDSs are multiplied by I_{avg} because the light incident on the resist and transmitted into the resist has already passed the imaging lens and is diffracting from there. The SSL is multiplied by PD because it is induced by the light that is diffracted by the mask but has not yet gone through the lens. The very action of the mask-diffracted light going through the lens induces the SSL.

8.5.2.5 Simulation results with PDS

Line–space pairs having L:S ratios of 1:1, 1:2, and 1:3 on a BIM are used to demonstrate simulation with PDS. Figure 8.29 shows the DOF of these features at 8% EL under 193-nm axial disk illumination at $\sigma = 0.82$. A 10% stray light from the optical system is included. The refractive index of the resist layer is assumed to be 1.75, and that of the immersion fluid is 1.46.

Let's first compare the PDS and No PDS effects within one feature type. Take the 1:1 pairs. At a very small aperture angle θ, there is no DOF because of lack of resolution. As the resolution builds up at larger aperture angles, the PDS-free curve shows an increase in DOF before it peaks. With PDS, DOF

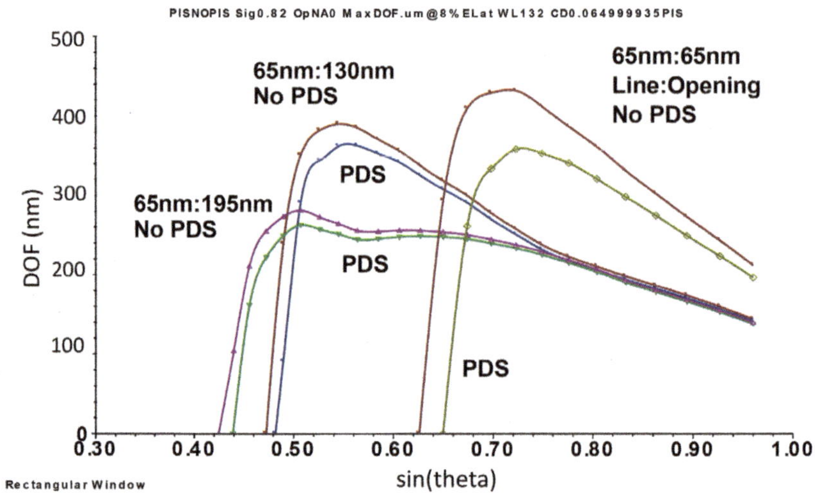

Figure 8.29 DOF as a function of sinθ, comparing PDS and No PDS with 193-nm immersion in water using a BIM, 65-nm lines, 1:1, 1:2, 1:3, $\sigma = 0.82$, $EL = 8\%$, $n_{\text{water}} = 1.44$, $n_{\text{resist}} = 1.75$, CD tolerance = ±10%, $SSL = 10\%$.

starts at a larger θ because the DOF becomes available only after the contrast in the optical image overcomes the stray light. Note that the PDS stays constant as θ increases because it is a function of α, which is determined by the spatial frequency component of the mask pattern, as discussed in Section 8.5.2.4. At a larger θ, the contrast increases while the PDS stays constant, explaining the smaller difference between PDS and No PDS at larger aperture angles. Among the three features, the 1:3 pairs show the least difference between PDS and No PDS because this group has the largest pitch, and the contrast with a larger pitch is higher. In this particular situation, the recombination PDS values for the 1:3, 1:2, and 1:1 features are 4.71%, 8.74%, and 23.43%, respectively; the reflection PDS values for these features are 0.13%, 0.17%, and 14.5%; and the values for the reflectivity on the resist surface for these features are 0.88%, 1.13%, and 100%. In the last case, the extremely high reflectivity is caused by the high incident angle in the coupling medium. The 130-nm pitch produces $\sin\alpha_u = 0.85$ in the resist at a refractive index of 1.75; however, in the reflected beam, $\lambda_{CM}/p > 1$. Light is completely reflected. The PDS and reflectivity values are summarized in Table 8.2.

The PDS stays constant with respect to θ only between each order of spatial frequency; i.e., when $n\lambda/p \leq \sin\theta < (n + 1)\lambda/p$. Take the 1:5 pairs shown in Fig. 8.30 as an example. At a pitch of 390 nm and a fictitious dry wavelength of 134 nm, the 1st-order spatial frequency is $\sin\alpha = \lambda/p = 0.338$. It passes through a lens with $\sin\theta \geq 0.338$. As soon as $\sin\theta \geq 0.676$, the 2nd-order spatial frequency is also admitted, contributing more to stray light. Here, the DOF curve shows a drop to 246 nm from 253 nm at $\sin\theta = 0.68$ due to the increase in recombination PDS from 0.94% to 3.86%. The PDS by reflection

Table 8.2 SSL, recombination PDS, reflection PDS, and reflectivity values of the coupling-medium–resist interface according to the scenario in Figs. 8.29 and 8.30. The second number in the last there columns includes both the 1st and 2nd harmonics.

System	L:S	PDS	PD·SSL	PDS_{recom}	PDS_{reflec}	Reflectivity
193-nm immersion	1:3	No	7.50%	—	—	—
		Yes	7.50%	4.70%	0.13%	0.88%
	1:2	No	6.67%	—	—	—
		Yes	6.60%	8.74%	0.17%	1.13%
	1:1	No	5.00%	—	—	—
		Yes	5.00%	23.43%	14.51%	100%
134-nm dry or 193-nm immersion	1:5	No	8.30%	—	—	—
134-nm dry	1:5	Yes	8.30%	0.94%, 3.86%	1.12%, 2.06%	7.46%, 15.8%
193-nm immersion	1:5	Yes	8.30%	2.03%	0.12%	0.83%

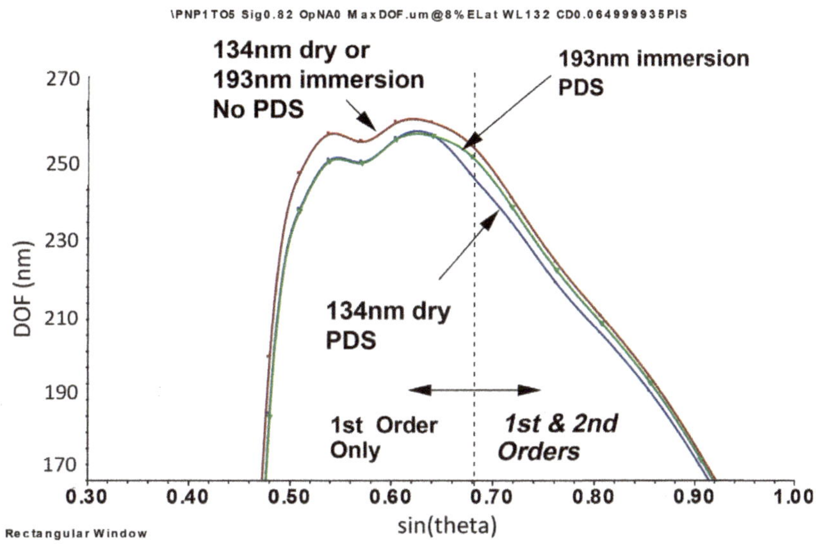

Figure 8.30 Same as Fig. 8.29, except the object consists of 65-nm lines at 1:5 line:space. Note the inclusion of the 2nd-order diffraction beam. The fictitious cases of a 134-nm dry system (No PDS) and a 193-nm immersion system (PDS) are compared.

off the resist surface also increases from 1.12% to 2.06%. Note that the change induced by the 2nd harmonic is purely due to the increase in stray light from an additional order of light. The cosα relationship in Eq. (8.38) is no longer exact, but we did not use a better formula in this simulation. A better formula should only slightly improve the simulation accuracy.

Figure 8.30 contains an additional curve from a 193-nm water-immersion system. There is no 2nd-order beam because at the mX mask side the wavelength is still 193 nm. Hence, sinθ = 0.497/m. The factor m is the reduction ratio of the imaging system. The wavelength only changes to 134 nm in the coupling medium. The 2nd harmonic at $\sin\theta = 0.994/m$ is smaller than $NA = 0.92/m$ and is rejected by the imaging lens within the θ range of our plot. It is ironic that rejecting the higher-order harmonic produces a larger DOF—this is due to PDS.

From Table 8.2, the recombination PDS is 2.03%, which is larger than that of the 134-nm dry case, as expected. However, after the 2nd harmonic is included in the image of the dry system at $\sin\theta \geq 0.676$, the recombination PDS of the dry system is larger. The reflection PDS at 0.12% is expected to be much smaller than its counterparts in the dry system, even with only the 1st harmonic, because of better index matching, as was discussed in Section 8.5.1.2.

Let's turn our attention to a two-beam interference system produced by annular illumination. The features used in Fig. 8.31 are identical to those in Fig. 8.29, except that AttPSM with 6% transmission and annular illumination with $\sigma_{in} = 0.42$ and $\sigma_{out} = 0.84$ are used. The DOFs without PDS are much higher than those in the three-beam case, exemplified by BIM with disk illumination, as expected from this RET. The two-beam leverage discussed in Section 8.5.2.2 makes PDS_{recom} smaller than that in the three-beam case, especially for the 1:1 feature. The SSL, PDSs, and reflectivity of this case are summarized in Table 8.3.

Does a dry system do better than the immersion system? Figure 8.32 shows the DOF of the 65-nm 1:1 feature versus $\sin\theta$ for three systems: 157-nm dry, 134-nm dry, and 193-nm immersion. The 193-nm dry system is not included because of lack of resolution. In this figure, we also include the DOF

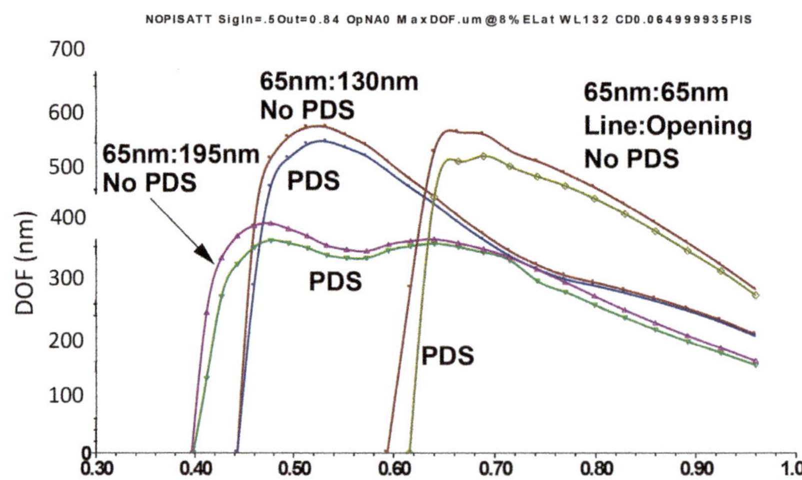

Figure 8.31 Same as Fig. 8.29, except using an attPSM and annular illumination σ_{in} = 0.42, σ_{out} = 0.84. The loss of DOF with PDS is less severe.

Table 8.3 SSL, PDS, and reflectivity values used in Fig. 8.31.

System	L:S	PDS	PD·SSL	PDS_{recom}	PDS_{reflec}	Reflectivity
193-nm immersion	1:3	No	7.50%	—	—	—
		Yes	7.50%	4.49%	0.12%	0.82%
	1:2	No	6.67%	—	—	—
		Yes	6.67%	7.97%	0.12%	0.83%
	1:1	No	5.00%	—	—	—
		Yes	5.00%	17.94%	0.13%	0.88%

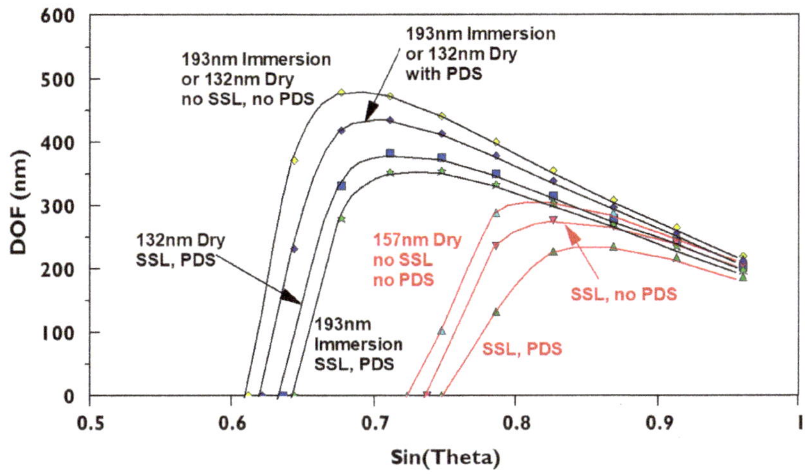

Figure 8.32 DOF of the 1:1 group in Fig. 8.29 for three systems: 157-nm dry, 134-nm dry, and 193-nm immersion. The curves with neither SSL nor PDS are included. The result from the 193-dry system is not included because there is no DOF in that case.

simulated with neither SSL nor PDS. The systems with complete absence of any stray light clearly outperform the others in the same wavelength group. The contributions from SSL and PDS with the 134-nm dry system are about the same, whereas, despite a smaller reflection PDS, the 193-nm immersion system shows less DOF than the 134-nm counterpart because PDS_{recom} is more significant here. When compared with the 157-nm dry systems, all 134-nm systems, including the 193-nm immersion system, are substantially better.

For the 1:2 group shown in Fig. 8.33, the trend is similar, but the margins are smaller. Here the 193-nm dry system is included. In this figure, we plotted the DOF against NA instead of sinθ. The 193-nm immersion system now moves to the same NA regime as the 193-nm dry system. This means that the complexity of the 193-nm immersion lens is similar to that of the 193-nm dry system, but the DOF is 2 to 3 times larger. The complexity of the 193-nm lenses is higher than that of the 157-nm system. However, because of reduced

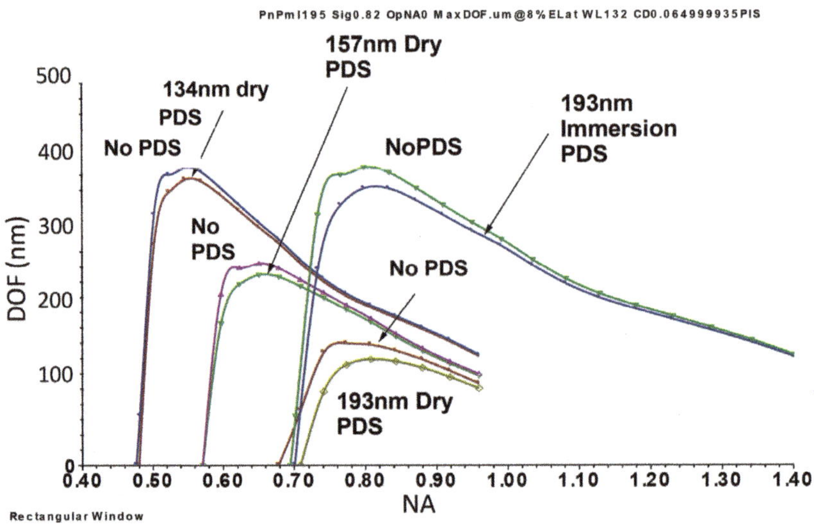

Figure 8.33 Same as Fig. 8.32, except that the object is the 1:2 group. The 193-nm dry system is also included in the comparison. Note that the horizontal axis now displays NA values instead of sinθ values.

material quality, it is not obvious that there is any advantage with 157-nm lenses.

Figure 8.34 shows the DOF of 65-nm contact holes, with hole:space ratios of 1:1, 1:2, and 1:3, using AttPSM, disk illumination at $\sigma = 0.82$. This is now a 2D object pattern. The stray light levels are evaluated in a manner similar to the 1D case because the spatial frequencies in x and y are independent. However, the pattern density is now lower, reducing the relative impact of SSL. The SSL values in Table 8.4, are 50% to 25% lower than those in Tables 8.2 and 8.3. The recombination PDS values are about three times smaller. Despite these smaller contributions, the DOF of contacts is smaller than that of line-space patterns due to the smaller PDS-free DOF to start with. The difference between PDS and No PDS is larger than that of line–space patterns because the proportion of stray light compared to the image contrast is larger.

We have been characterizing PDS with DOF at a fixed EL. Now the criterion is changed to EL with a fixed DOF. The scenarios in Figs. 8.29 and 8.34 are chosen for the comparison. The former examines 65-nm lines in line: space ratios of 1:1, 1:2, and 1:3. The latter covers 65-nm holes at hole:space ratios also of 1:1, 1:2, and 1:3.

Figure 8.35 is the EL counterpart of Fig. 8.29. The trend of smaller impacts from PDS for the larger pitches is similar to that from the DOF criteria. One noted difference is that the difference between PDS and No PDS decreases monotonically with sinθ according to the DOF criteria. In the EL

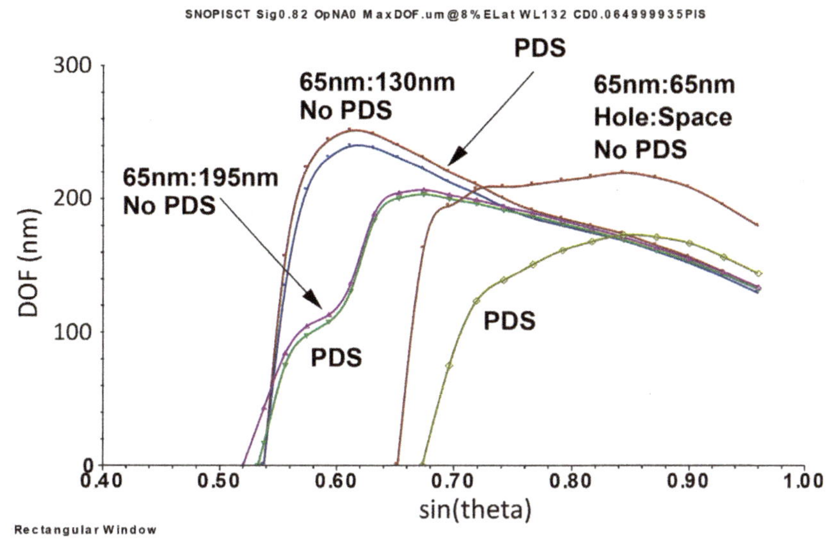

Figure 8.34 DOF of 65-nm holes, 1:1, 1:2, 1:3, with PDS and No PDS, using 193-nm immersion in water and AttPSM, $\sigma = 0.84$, 8% EL, $n_{\text{water}} = 1.46$, $n_{\text{resist}} = 1.75$, CD tolerance = ±10%, SSL = 10%.

Table 8.4 SSL, PDS, and reflectivity values used in Fig. 8.34.

System	H:S	PDS	PD·SSL	PDS_{recom}	PDS_{reflec}	Reflectivity
193-nm immersion	1:3	No	1.25%	—	—	—
		Yes	1.25%	4.71%	0.17%	0.88%
	1:2	No	1.67%	—	—	—
		Yes	1.67%	8.74%	0.13%	1.13%
	1:1	No	2.50%	—	—	—
		Yes	2.50%	23.43%	14.51%	100%

case, this difference is zero at both ends of sinθ, and it peaks at the optimum EL. This means that the gain in EL from polarized illumination is most significant at the sinθ for the optimum EL.

When one chooses to operate at the sinθ for the optimum DOF, the gain in EL is less. To explain this difference, the E-D windows used to determine the EL in Fig. 8.35 and the DOF in Fig. 8.29 are shown in Figs. 8.36 and 8.37, respectively. The E-D trees in these two figures are actually identical. Note that they stretch out in defocus at low sinθ then lose defocus but gain EL as sinθ becomes larger. Therefore, when E-D windows of the same EL are fitted within these trees, the No PDS E-D tree reaching the acceptable EL first shows a large DOF, while the DOF from the PDS E-D tree is still zero. As sinθ continue to increase, this difference becomes less significant, explaining the monotonic decrease in the difference in DOF between PDS and No PDS.

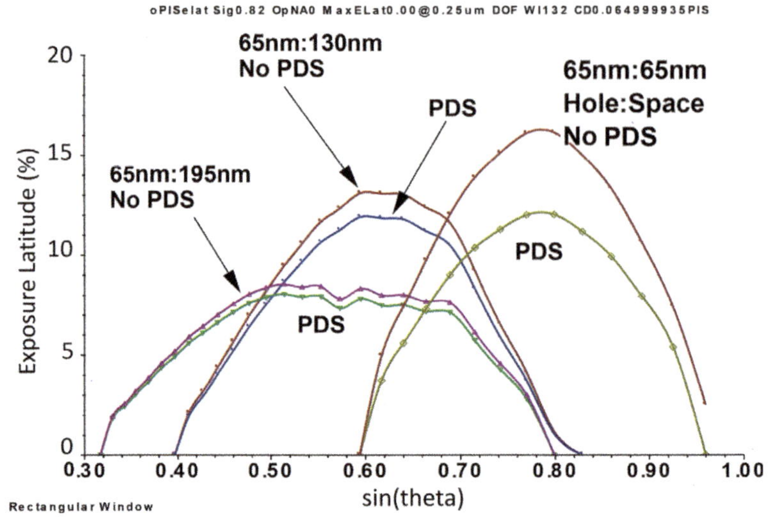

Figure 8.35 Same as Fig. 8.29, except that the EL is plotted instead of the DOF, which is set at 250 nm.

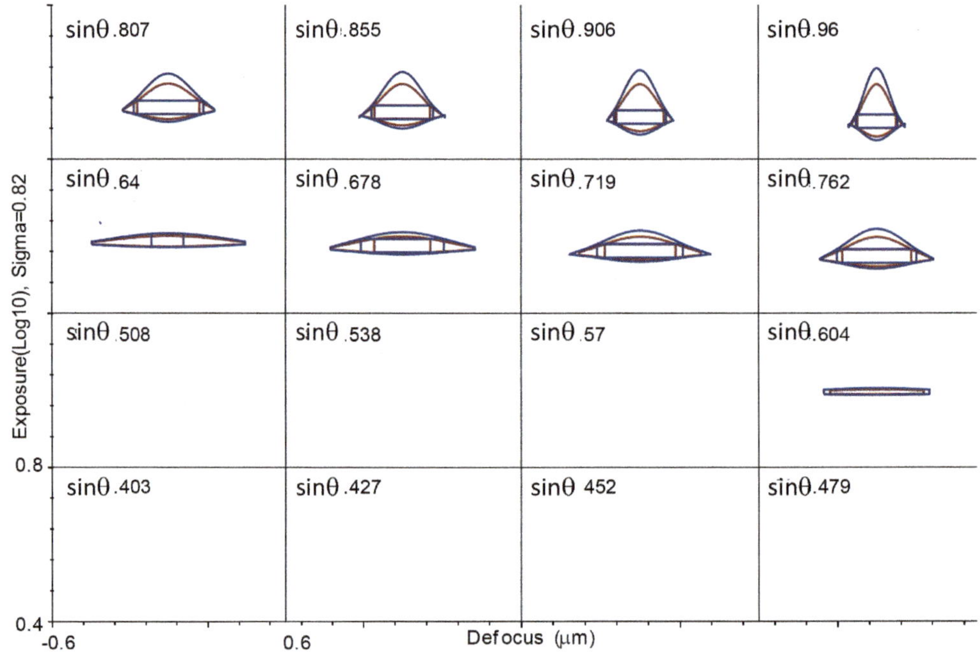

Figure 8.36 E-D windows defining the DOF of the 1:1 curves in Fig. 8.35. The E-D coordinates are 0.4 to 0.8; –0.6 to 0.6 repeats at each sinθ block.

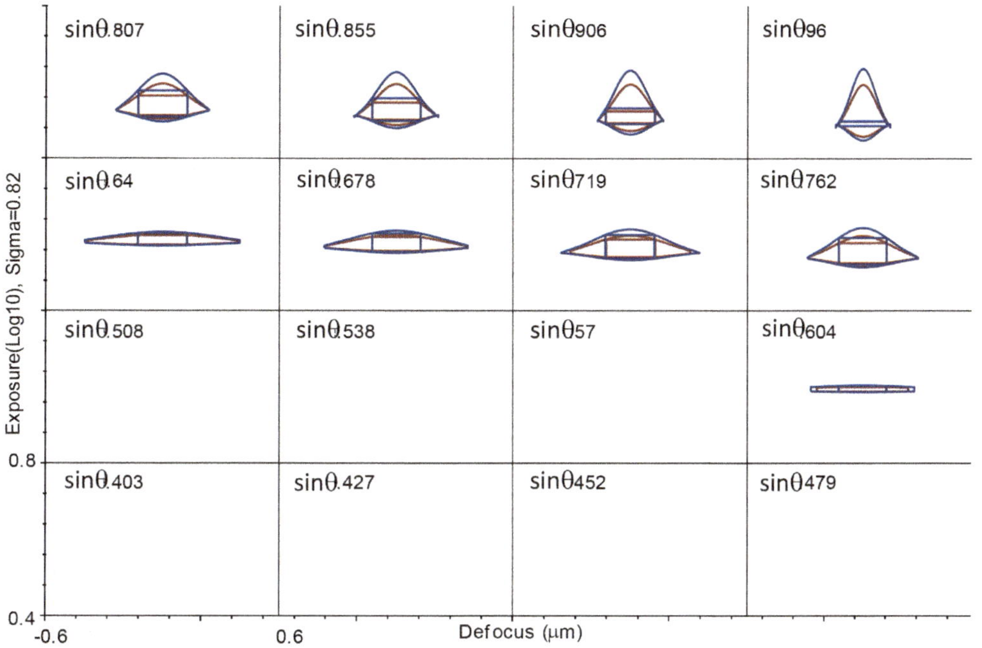

Figure 8.37 E-D windows defining the EL of the 1:1 curves in Fig. 8.29. The E-D coordinates are 0.4 to 0.8; –0.6 to 0.6 repeats at each sinθ block.

On the other hand, the difference in EL in the E-D trees at low sinθ is negligible when the fixed DOF is set at a usable medium value. The EL peaks at intermediate sinθ where the E-D windows are not stretching out near the extreme defocuses of the E-D trees.

EL as a function of sinθ for 65-nm holes using the same condition as Fig. 8.35 is plotted in Fig. 8.38. Here, the difference between PDS and No PDS for the tight pitch is more significant than in the case of DOF. For example, at sinθ = 0.732, the EL drops from 5.7% to 2.88%, almost by half. The EL does not close out at high sinθ because of the smile shape of the E-D trees, as shown in Fig. 8.39.

Polarized illumination helps DOF and EL. Its leverage is not uniform and is a strong function of the imaging conditions. As shown in the previous figures, its leverage is in tight pitches. Intuitively, one tends to think that polarized illumination helps EL more than DOF. However, the actual situation is quite complicated. In evaluating these gains, one should bear in mind that DOF_{avail} is the relevant criteria instead of $DOF_{diffrac}$. Table 8.5 shows a comparison of these gains for the common window of 32-nm lines at 90- and 113-nm pitches using AttPSM and annular illumination.

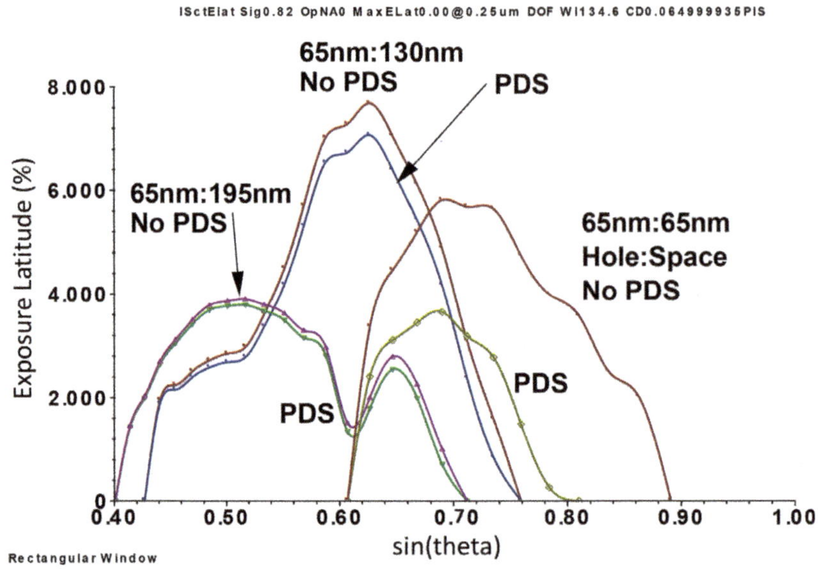

Figure 8.38 Same parameters as in Fig. 8.34, except that EL is plotted instead of DOF, which is set to 250 nm.

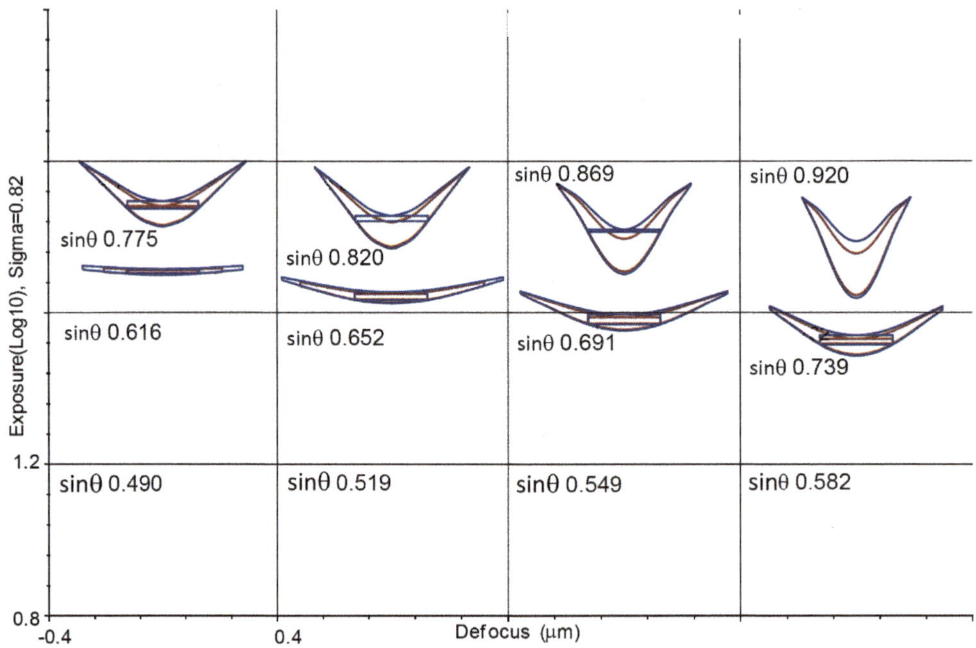

Figure 8.39 E-D windows defining the EL of the 1:1 curves in Fig. 8.35. The E-D coordinates are 0.8 to 1.2; –0.4 to 0.4 repeats at each sinθ block. The E-D trees at the sinθ = 0.616 line go off scale beyond EL = 1.2.

Table 8.5 Gain in EL and DOF with polarized illumination. The simulation was based on the common window of 90- and 113-nm pitches for the 32-nm poly line.

AttPSM OAI	NA	Sigma	EL	$DOF_{diffrac}$	DOF_{avail}	Improvement
Unpolarized	1.35	0.96:0.48	5%	137	55	
TE	1.3	0.96:0.48	5%	169	87	23/58%
Unpolarized	1.25	0.96:0.48	2.43%	200	118	
TE	1.25	0.96:0.48	3.57%	200	118	47%

8.6 Immersion Systems and Components

This section addresses the hardware side of immersion systems. Starting from the configuration of an immersion system, the coverage includes the immersion fluid, immersion lens, bubbles, particulates, mask, and photoresist.

8.6.1 Configuration of an immersion system

The lens-based configuration (LBC) is also known as the shower configuration.[30,31] As shown in Fig. 8.40(a), immersion fluid is applied to and extracted from a small region covering the len; this small, fluid-confined region remains stationary with respect to the lens as the wafer is stepped or scanned. A schematic plot of a commercial immersion hood is shown in Fig. 8.40(b).

There are many motivating advantages to adopting LBC:

1. The wafer stage is essentially identical to that of a dry system. This saves development cost and time.
2. It is possible to keep the alignment, focusing, and leveling setup unchanged.
3. The volume of fluid is small. Filling the cavity can be very quick, maintaining wafer throughput.

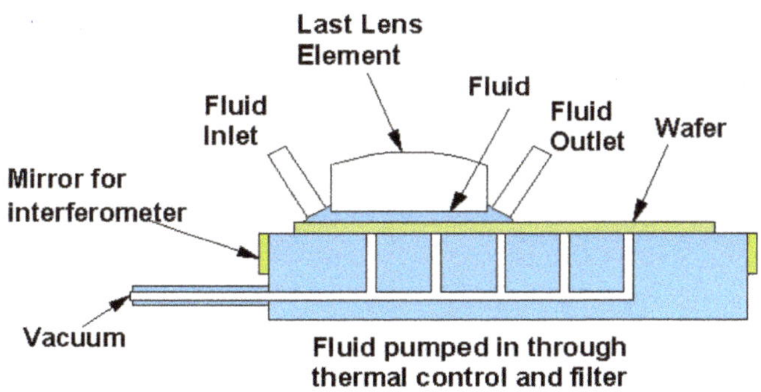

(a)

Figure 8.40(a) Lens-based configuration for immersion.

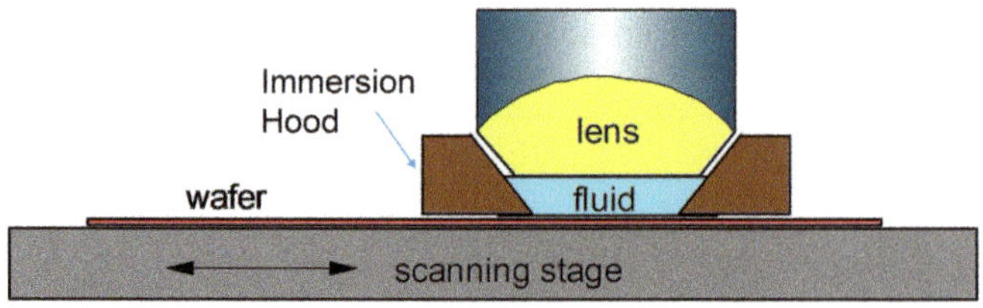

(b)

Figure 8.40(b) Schematic of a commercial immersion hood (image courtesy of ASML).

Foreseeable problems with LBC include the following:

1. Near the wafer edge, the immersion region includes the wafer and areas outside the wafer. There is discontinuity between the wafer and the chuck as well as between the wafer chuck and the area outside the chuck. Maintaining the hydrodynamics and the water cavity, and managing water extraction can be difficult. Particles at the back side of the wafer tend to be washed to the surface.
2. The immersion hood for the LBC system [shown in Fig. 8.40(b)] tends to leave trace amounts of water or moisture behind, which is the root cause of water stains.
3. The photoresist will have inconsistent water-contacting history at different locations. As the wafer is stepped from field to field, the neighboring fields (or parts of them) are covered by water. This can occur in a field several times and not necessarily in sequence nor the same number of times for each field. The inconsistency depends on the stepping pattern and the location of the field.

The wafer-based configuration (WBC) is also known as the bath configuration.[32] In this case, the wafer is completely immersed in water in a circulating tank in the wafer stage, as shown in Fig. 8.41. Water continuously circulates under and above the stage. It is filtered and temperature regulated as it moves in and out of the wafer surface area. The water can be completely drained for wafer loading and unloading. The cover prevents water from spilling over and stops foreign particles from falling in.

The advantages of WBC are the following:

1. Exposure at the edge of the wafer is no different from that at the center.
2. Each field contacts the wafer for the same amount of time.
3. There is no concern about water stains left outside the WBC head, unlike LBC.
4. Bubble generation from poor hydrodynamics near the edge of the wafer is not a problem.

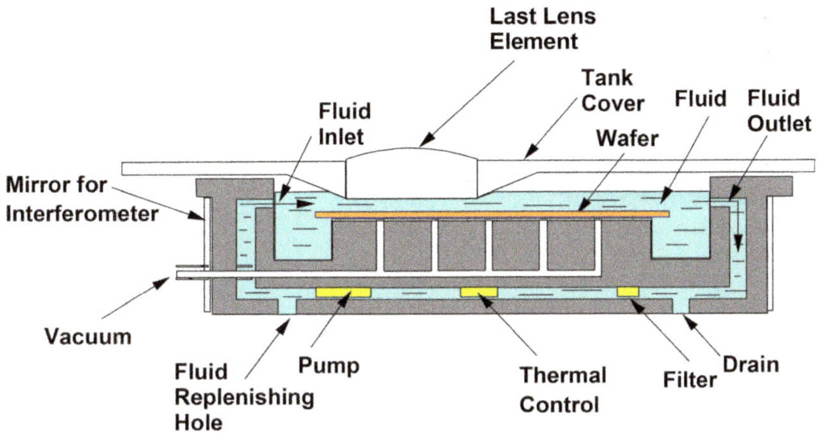

Figure 8.41 Wafer-based configuration for immersion.

The disadvantages include the following:

1. The pre-exposure and post-exposure soaking times of each exposure field are different.
2. WBC takes more effort or more time to fill and drain the water than LBC.
3. Focusing, tilting, and alignment must be performed in the immersion mode if a twin-stage setup is not used. However, from Eq. (8.12), $DOF_{required}$ is reduced if focusing and tilting are performed in the fluid—if practical difficulties can be overcome.
4. Substantial redesign of the wafer stage is necessary.

The WBC requires a lot more effort to build and will require quite a different scanner compared to the dry configurations. As a result, only the LBC has been commercialized, despite the need to constantly improve the immersion hood.

8.6.2 The immersion medium

The optical requirements of the immersion fluid are high refractive index and high transmission. Its refractive index should be consistent and uniform—unaffected by thermal variations during exposure. Table 8.6 lists various immersion fluids[8] for 193-nm and 157-nm exposures. The absorption coefficient of the perfluoropolyether (PFPE) material is too high. In order to limit light loss to 10% through the fluid, its thickness needs to be kept below 75 μm. The absorption coefficient is still too high at 193 nm. On the other hand, water appears to be the perfect medium for immersion at 193 nm. Not only is its absorption coefficient low, the refractive index 1.44 is surprisingly

Table 8.6 193-nm and 157-nm immersion fluids. The PFPE data is based on Ref. 8, and the water data is based on Ref. 33.

Fluid		λ_0 (nm)	n	α, cm^{-1} base 10	λ_1 (nm)
Perfluoropolyether	PFPE	157	1.37	6–19	115
		193	—	0.1–0.6	—
Water	H_2O	193	1.43664	0.036	134

high at 193 nm. In most other wavelengths, the refractive index of water is in the vicinity of 1.33.

The refractive index of water, as well as its thermal and spectral sensitivities at 193 nm, have been rigorously measured by Burnett and Kaplan.[33] They used two independent methods to calibrate one against the other. The first method fills a prism fluid cell with water. By carefully controlling the angle of the prism and measuring the deviation angle of the incident and refracted light, and by using exactly parallel, very flat and smooth, fused silica etalon plates as the windows for the fluid cell, the three optical constants can be determined to better than five digits. The second method measures the interference fringe positions in the transmittance spectrum from an etalon cell filled with water. This method is also accurate to better than five digits. At 21.5 °C and 193.39 nm, the refractive index and the thermal and spectral sensitivities of water are 1.43662, -1.00×10^{-4}/°C, and -0.002109 nm^{-1}, respectively. The immersed wavelength is 193.39 nm, which was used for all critical calculations. Some less critical calculations use the 1.46 index value given in an earlier publication.[8]

A fluid with a higher index can support a higher NA. However, this is not without limit. Table 8.7 lists $\sin\theta$ in the lens, coupling, and resist media, as well as the attainable NA. Assuming that (1) the last element in the lens is a flat piece of fused silica whose refractive index is 1.56, (2) 1.75 is the refractive index of the resist layer, and (3) $n_{\text{fluid}} = 1.44$ and $\sin\theta_{\text{fluid}} = 0.95$, the attainable NA is 1.37. Raising the index of the fluid to 1.56 leads to $NA = 1.48$. Further increasing the fluid index to 1.66 runs into the problem of a physically impossible $\sin\theta_{\text{lens}}$ of 1.01. The angle in the fluid has to be cut back to $\sin\theta_{\text{fluid}} = 0.93$ to keep $\sin\theta_{\text{lens}} < 1$. This is illustrated in Fig. 8.42(a), where the

Table 8.7 Maximum attainable NA as a function of n_{fluid}.

	$n_{\text{fused silica}} = 1.56$		$n_{\text{resist}} = 1.75$	
n_{fluid}	$\sin\theta_{\text{lens}}$	$\sin\theta_{\text{fluid}}$	$\sin\theta_{\text{resist}}$	NA
1.44	0.88	0.95	0.78	1.37
1.56	0.95	0.95	0.85	1.48
1.66	0.99	0.93	0.88	1.54
1.66	1.01	0.95	0.9	n/a

$\sin\theta_{fluid}$ = 0.95 line cannot be plotted in the lens. The NA is now limited to 1.54. Any further increase in the fluid index cannot support a larger NA unless the last lens element can have a concave surface, as shown in Fig. 8.42(b). The curvature makes the thickness of the fluid uneven, resulting in the requirement of a fluid transmission that is higher than that of water at 193 nm. This makes development of fluid with a refractive index higher than 1.66 even more difficult.

In addition to the optical properties of the immersion fluid, many material characteristics of the fluid must be considered. First, the fluid cannot cause any physical or chemical change to the surfaces it contacts, namely, the resist surface, the lens, and its antireflection coating. Also, it cannot deposit contaminants on these surfaces. The fluid is required to have a large dissolution capacity for gas so that any bubbles that are formed are readily dissolved. The viscosity should be low so that it can flow freely and does not

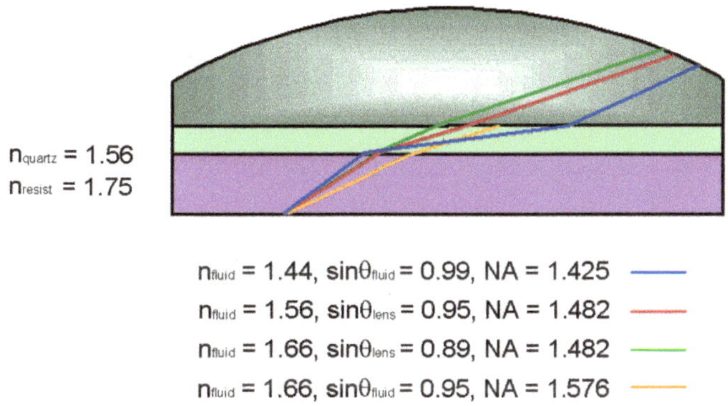

(a)

Figure 8.42(a) Impact of the fluid refractive index. When the fluid index is 1.66, $\sin\theta_{lens}$ is limited to only 0.89, not the 66% increase to NA = 1.64 because the corresponding $\sin\theta_{lens}$ exceeds 1.

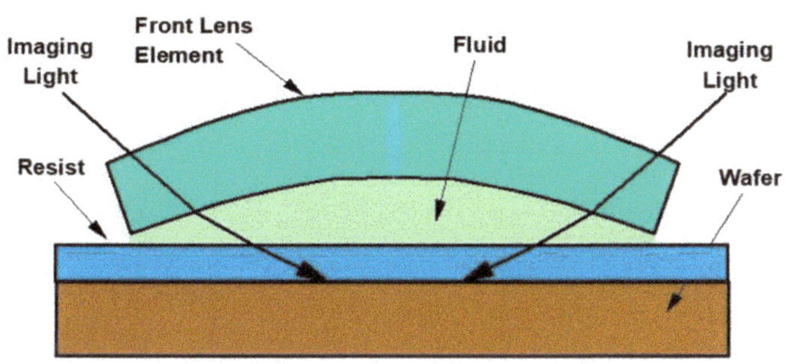

(b)

Figure 8.42(b) Curved lens–fluid interface to support hyper-NAs.

slow down the wafer scanning. The contact angle between the fluid and the resist surface (or the top-coating material on the resist, if used) should be carefully optimized. Too small an angle can lead to water residue in the area outside the lens immersion head. With too large an angle, air bubbles may be generated during scanning. During development, a small contact angle is preferred for better wetting of the developer.

The thickness of the immersion medium should be determined using optical and hydrodynamic considerations. From the optical point of view, in addition to an acceptable transmission, this thickness directly influences aberration sensitivity to refractive index changes. Suzuki[34] used the following formula to determine the relationship between the fluid thickness D and the angle θ of the image-forming beam in the fluid:

$$D \leq \frac{\cos\theta}{\Delta n} m\lambda, \tag{8.54}$$

where m is a fraction of the wavelength from which the imaging wavefront cannot deviate. Suzuki assumed that the fluid can be controlled to 0.01 °C. Using Burnett's number for dn/dT, 1.00×10^{-4}/°C, $\Delta n = 1 \times 10^{-6}$. When $\sin\theta = 0.95$, $D \leq 3.1 \times 10^5\, m\lambda$. When $m = 0.02$ and $\lambda = 193$ nm, $D \leq 1.2$ mm. D as a function of $\sin\theta$ is plotted in Fig. 8.43.

The fluid thickness also has to satisfy hydrodynamic requirements. Mulkens et al.[30] took the limits from the requirement of a laminar flow and an acceptable shear force on the projection lens to set the relationship between the fluid thickness and the scan speed using the density and viscosity of the fluid. They arrived at a fluid thickness between 1 and 2 mm to support a scan speed of almost 500 mm/sec for disturbance forces of only a few millinewtons (mN).

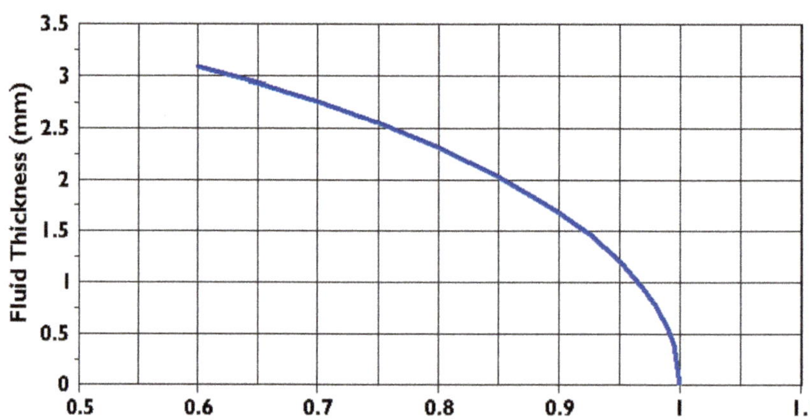

Figure 8.43 Upper limit of fluid thickness D for a wavefront error of less than λ/50.

8.6.3 The immersion lens

The challenge with immersion lenses varies greatly depending on the required NA. When the NA of a dry system is to be used for immersion, as in the case of Fig. 8.4, it is not difficult to design and fabricate the immersion lens. The major change involved in moving from dry to immersion is the reduction of the working distance to utilize the short optical distance required to minimize the effect of inhomogeneity. However, when $NA > 1$ is needed, the bending angles in the lens are harder to manage.[30,31] At $n_{CM} = 1.44$, the design limit appears to be 1.3 ~ 1.35 NA for some lens designers.[35] Above 1.2 NA, the field size of the conventionally configured lenses may need to be reduced to keep the size of the projection optics within physical and economic viability. It is well known that field size reduction substantially lowers the productivity of the exposure tool. Fortunately, the NAs of modern immersion lenses without a reduction in field size[36,37] are as large as 1.35. These are catadioptric systems, not simply dioptric. It is important that the catadioptric system does not introduce a mirror image of the mask. This can confuse the circuit designer, leading to unnecessary manufacturing errors. Therefore, the number of reflecting surfaces should not be an odd number. Figure 8.44 shows Zeiss high-NA immersion lenses[38] that have been installed in commercial scanners. Figure 8.45 shows four high-NA Nikon lenses[39] that were under development in 2007.

One needs to distinguish between the use of immersion lithography for attaining DOF extremes and for attaining resolution extremes. For the former purpose, immersion lithography extends the life of an existing imaging system to near-term new technology nodes (e.g., from the 65-nm node to the 45-nm

Figure 8.44 Zeiss high-NA immersion lenses that have been installed in commercial scanners (reprinted from Ref. 38).

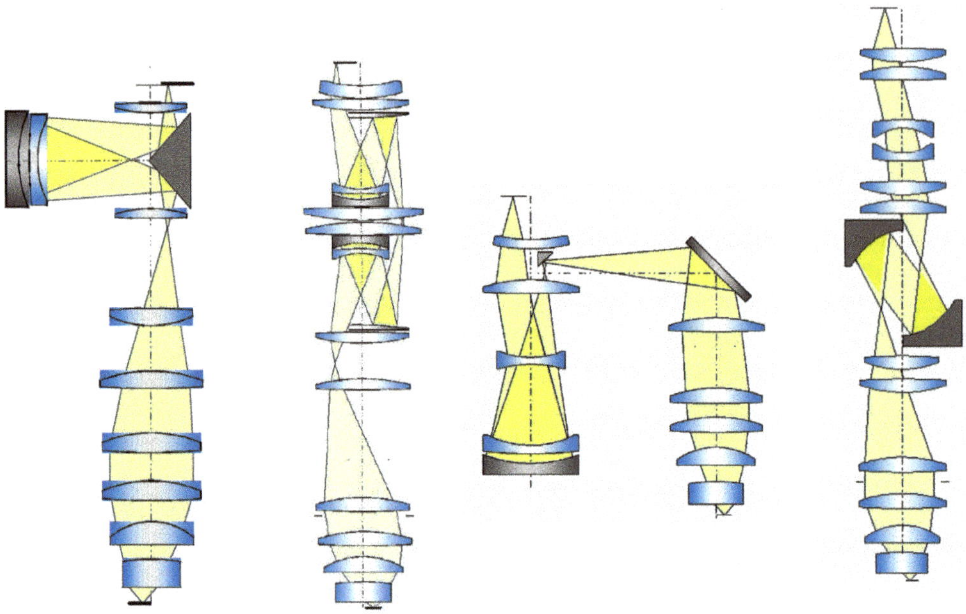

Figure 8.45 Nikon high-NA immersion scanners under development in 2007 (reprinted from Ref. 39).

node) and broadens the manufacturing margin of these systems. So far, the extra DOF capability of immersion lithography has not been needed. For the latter purpose, there is no dry system that can rival these applications. The technology nodes include 45 nm, 40 nm, 32 nm, 28 nm, 20 nm, 16 nm, 14 nm, 10 nm, and 7 nm.

8.6.4 Bubbles in the immersion medium

The effect of bubbles in the immersion medium is primarily to redirect the imaging light in the form of scattering, resulting in reduced image contrast. The scattering effect is a strong function of the size and quantity of the bubbles, as well as their distance from the resist surface.

Marston[40] investigated light scattering from bubbles in seawater. He analyzed the cases of simple bubbles and those having a coating of foreign material on their surface, using Mie theory and physical and geometrical optics approximations. The aspects of his investigation that are relevant to immersion lithography include total reflection from bubbles, angle-averaged scattering at the critical angle, enhanced backscattering, and optical radiation pressure on bubbles. Interested readers may find many more studies in the references of his article. Marston's explanation of the lifetimes of simple bubbles and coated bubbles is applicable to immersion lithography. Simple bubbles are readily dissolved in the suspending liquid. The extra coating is the

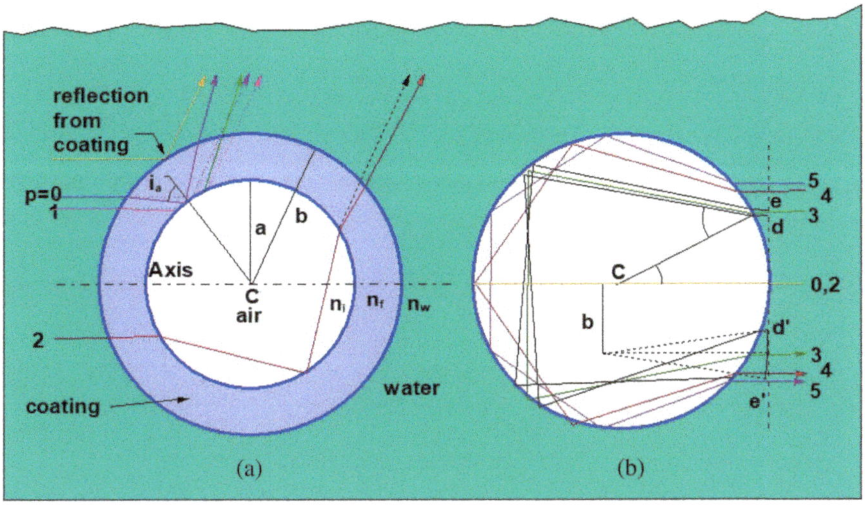

Figure 8.46 Scattering from bubbles as reported by Marston.[40] (a) Reflection and refraction of rays by a simple and a coated bubble. The rays from the former type are drawn in dashed lines. (b) Backward-directed rays for a simple bubble (adapted from Ref. 40).

major reason for undissolved bubbles. We have to make sure that there is no detrimental impurity in the water that protects the bubbles. Fortunately, Marston showed that the magnitude of the scattering from either type of bubble is similar.

The situations investigated by Marston are shown in Fig. 8.46. The normalized scattered irradiance I_j for a bubble of radius $a = 4.3$ μm is plotted in Fig. 8.47 as a function of the scattering angle, i.e., the angle the scattered light makes with the optical axis. The physical irradiance of the scattered i_j and incident i_{inc} light are related to I_j as follows:

$$i_j/i_{\text{inc}} = I_j a^2/4R^2, \tag{8.55}$$

where

$$I_j = |S_j|^2(\lambda/\pi a)^2, \tag{8.56}$$

where S_j are the complex scattering amplitudes from the Mie series. TE polarization is denoted by $j = 1$ and TM polarization by $j = 2$. R is the distance from the bubble to the observation point. R must be at the far field for the results to be exact. The originating point of R is not specified by Marston because he is mostly concerned with the far field. Let's assume that R originates in the center of the bubble at a reasonably far field. Take $R = 0.5$ mm, which is halfway into a proposed immersion fluid thickness of 1 mm. i_j / $i_{\text{inc}} = I_j \cdot 1.8 \times 10^{-5}$. With the 4.3-μm bubble, the scattering irradiance is quite small, except for a narrow range of small scattering angles. But when the bubble is closer to the

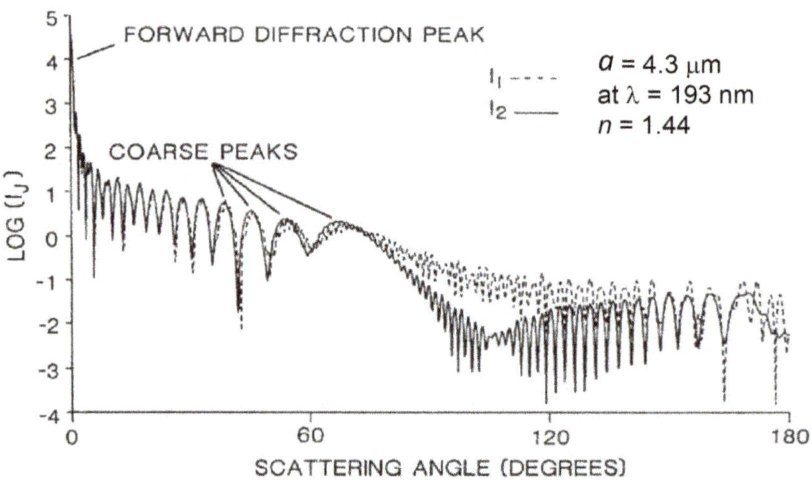

Figure 8.47 Normalized scattering irradiance evaluated from Mie theory as a function of the scattering angle for bubbles of radius $a = 100\lambda_{water}/\pi = 4.3$ μm (adapted from Ref. 40).

resist surface, i_j / i_{inc} can be much larger. Hence, the distance from the bubble to the resist surface is a very important factor.

Even though I_j is plotted for only one bubble size, we have used it to gain insight into the significance of bubbles of this size range in optical imaging. It is also useful to check the result of our work using Mie series.[41] Figure 8.48 shows the scattering spectra of bubbles with sizes of 100 nm, 1 μm, and 10 μm. The Marston result with a 4.3-μm bubble fits within the ranges of our 1-μm and 10-μm spectra. The shape of the Marston spectrum resembles that of our 10-μm case. Note that we used $n_{water} = 1.44$ at 193 nm, while Marston used $n_{water} = 1.33$ at the visible wavelengths.

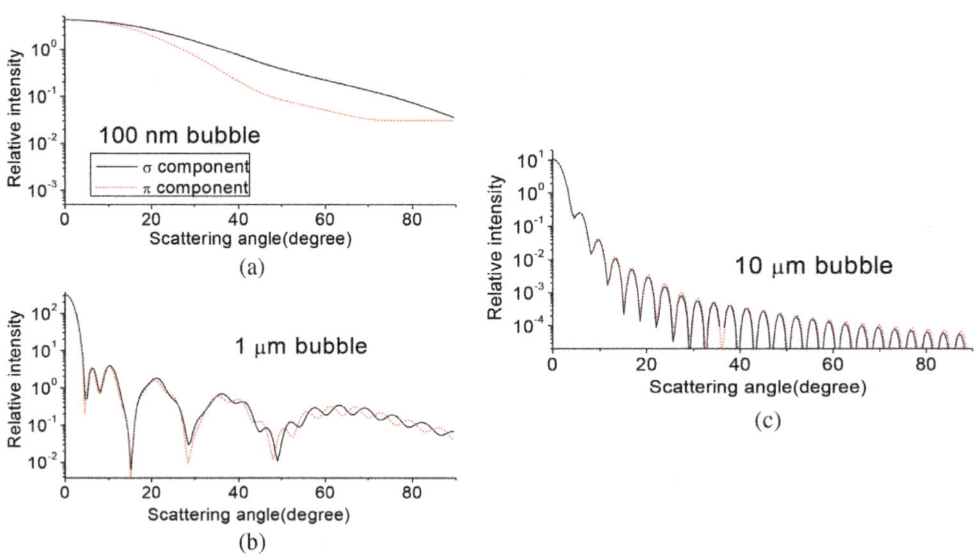

Figure 8.48 Scattering spectra for bubbles of sizes (a) 100 nm, (b) 1 μm, and (c) 10 μm (reprinted from Ref. 41).

Gau et al.[41] use a statistical approach and the scattering cross sections evaluated with Mie theory to investigate the impact of bubbles on imaging. In this scenario, the energy redirected by a group of bubbles is

$$P = \left[\frac{Na^3}{V}\right]^{\frac{2}{3}} \frac{\sigma_s}{\sigma_g}, \tag{8.57}$$

where N is the number of bubbles, V is the volume of liquid where the bubbles are equally distributed, a is the radius of the bubble, and σ_s/σ_g is the normalized scattering cross section. Taking all of the possible combinations of each diffraction order, this energy becomes stray light that reduces the contrast of the optical image.

Bubbles do deteriorate the optical image. It is best not to allow any bubble in the immersion fluid during exposure, which is not difficult. There are three possible causes for bubbles being produced in the fluid: (1) release of dissolved gas in the fluid, (2) entrapped air due to turbulent flow, and (3) outgassing of the resist.

The release of dissolved gas is mostly due to saturation of the gas already dissolved in the fluid. When temperature or pressure lowers the saturation point, gas is released. Since degassing tools are commercially available, there is no reason not to pretreat the immersion fluid before dispensing it to the coupling space. Doing so also facilitates the dissolution of any bubbles created by a turbulent flow or outgassing. Nevertheless, the turbulent flow should be avoided by hydrodynamic design of the immersion enclosure.

Bubbles from resist outgassing are also quickly dissolved in the immersion fluid.[34] Even if they do not dissolve, it is dubious that resist outgassing spoils the recorded image. Most resists outgas in a dry system. When these bubbles are released into the immersion fluid, the size and density near the resist–coupling-medium interface in the coupling medium must be similar to those inside the resist near the interface, maintaining the same level of perturbation to the image. Thus, if this type of perturbation does not affect dry systems, it should not affect immersion systems either.

As the bubbles move away from the resist surface, their sizes may increase due to multiple bubbles combining. The bubbles may also decrease in size due to dissolution into the fluid. When the net effect is a decrease in bubble size, bubbles are less of a problem compared to bubbles near the interface. Even if the net effect is an increase in bubble size, the density of the bubbles will be reduced accordingly, and the scattering effect decreases when the bubbles are farther from the interface, according to Eq. (8.57). Now let us stop and think.

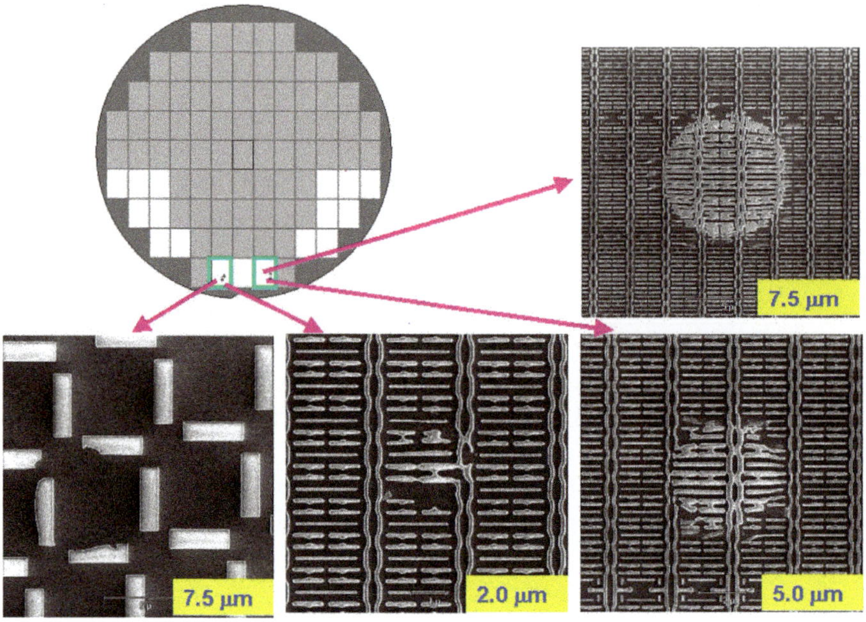

Figure 8.49 Resist image affected by bubbles (figure courtesy of C.G. Chen).

The size and density of the bubbles in an acceptable resist for dry exposure systems must not affect image recording in a way that can be detected. Otherwise, we would have discovered this effect with the innumerous dry exposures taken in worldwide manufacturing.

Following the foregoing argument, the perturbation of the bubbles inside the resist and in the fluid is of similar significance. Hence, outgassing from these resists does not affect the immersion image in a detectable way.

Using degassed water eliminates most concerns about bubbles. However, it is possible that a coating of foreign materials at the surface of the bubbles would prevent their dissolution, as described by Marston. Should this happen, certain materials need to be excluded or included in the immersion system to prevent the formation of the bubble coating. In the immersion-enabled technology nodes from 45 nm to 7 nm, coating of foreign materials has not been an issue.

Concluding from the above discussions, bubbles that can affect imaging are larger than a few micrometers. They are most likely generated by lack of laminar flow or by trapping of air during high-speed scanning. Figure 8.49 shows the resist image affected by air bubbles in the fluid during exposure. The sizes of these bubbles range from 2 μm to more than 7 μm.

8.6.5 The mask

To the first order, the mask for immersion lithography is the least impacted of the components. The illumination wavelength on the mask is unchanged. All the materials used in a dry system, such as the blank, absorber, and pellicles, as well as the fabrication process, remain unchanged. Is there a need to immerse the mask? In a 4X reduction system, the NA on the mask side is only 0.3375, even when the NA on the wafer side is 1.35. The mask NA is still far from unity. There is no need to immerse the mask to increase the NA beyond unity.

8.6.6 Subwavelength 3D masks

Despite the impossibility of having a large NA on the mask side, immersion lithography enables a given light frequency to extend its resolution limit to the point that even a 4X enlargement of the feature size makes the mask pattern subwavelength. For example, 32 nm on the wafer side becomes 128 nm on the mask side. This is only 0.66λ. Subresolution assist features can be as small as 0.15λ on the mask. There are other subwavelength features on the mask, such as jigs and jugs induced by optical proximity correction (OPC). In this case, the 3D nature of mask patterns and polarization effects must be taken into account. There is an anticipated dependence of mask diffraction on the feature orientation, absorber thickness, and absorber profile. OPC must be able to take care of this effect. Otherwise, other measures need to be taken. In addition, there are difficulties in making masks containing very small features. From the wafer imaging side, as the NA becomes very large, the size of the lens, its material, and its fabrication costs can be formidable.

These issues can be alleviated by reducing the field size of the lens from the well-established standard of $26 \times 33\ \mathrm{mm}^2$ and increasing the reduction ratio from 4X to 8X. Lens design and fabrication for smaller field sizes are becoming more forgiving. Mask making is also becoming easier with the requirement of covering much larger features with a lower feature count. Additionally, CD control and cycle time are inherently improved. Therefore, field-size reduction and increase of the reduction ratio complement each other and can be cost effective. Both of these methods have attracted the attention of potential beneficiaries.

There are two concerns with field-size reduction. The first is a loss of productivity of the exposure tools due to a smaller field size. The time spent in stepping through smaller fields increases, thus dropping the wafer throughput of exposure tools. A detailed discussion of this is in Section 7.1.5. A previous work[42] on productivity loss from 4X to 5X masks shows that the difference in cost between 4X and 5X is about 10%. The cost increase from 4X to 8X would be significantly higher. The second concern with field-size reduction is that, usually, many identical dies are placed within the $26 \times 33\ \mathrm{mm}^2$ exposure field. Having more dies per field raises the production yield. When the number of dies is reduced by half, the yield will be impacted. Worst of all, there are already ICs that need the entire $26 \times 33\ \mathrm{mm}^2$ area for a single die. Halving the field size necessitates field stitching, which is not trivial to implement and execute.

Note that, in the entire semiconductor manufacturing process consisting of 30 ~ 40 masking layers, if only one layer has a tool that requires a smaller field size, all of the other tools need to reset their field size unless extremely good field-stitching can be implemented. Cost savings on a tool may apply to one layer, but productivity loss affects all layers.

Perfect stitching is natural in the scanning direction on scanners. To increase the field size this way, one needs to use larger masks of, for example, 9-inches instead of 6 inches. Similar to the case of changing the wafer size, a large portion of the community in the industry will need to support this move before it is implemented.

Hence, productivity loss is by far the dominating factor. The mask reduction ratio should not increase unless the existing reduction ratio is prohibited by the laws of physics. Before taking the drastic measure of increasing the reduction ratio and shrinking the field, the limitations placed by the laws of physics can be avoided by using a solid immersion mask,[43] as discussed in Section 8.9.2.

8.6.7 The photoresist

As is the case in any type of lithography, the photoresist is an indispensable component in immersion lithography. Without it, the mask image cannot be successfully replicated and transferred. In the early days of immersion lithography, the first successful image was produced from experimental setups using commercially available resist designed for dry exposure.[44] It was not that resists do not have to be optimized for immersion exposure, but rather that resist development for 193-nm water-immersion lithography looked more feasible than developing 157-nm dry resists. Since the time shortly after the introduction of immersion lithography, no serious problems have been encountered when exposing resists in an immersion environment.[45] Some immersed resists suffer from pattern collapse during development, but many other resists are not affected.

Besides the optical requirements of a high refractive index, low optical absorption, and minimal outgassing, the resist must withstand a moment of contact with water. The longest possible time of contact is the time required to expose a wafer in WBC. With 100 wafers per hour, the water contact time per wafer is on the order of one minute. With LBC, the contact time is reduced by about 1/50, but some areas may be revisited, depending on the step-and-scan scheme. During contact with water, materials such as the photoacid, the photoacid generator, or the resin can leech into water. Water also soaks into the resist. The latter can affect the resist characteristics in the development process, while the former can become a source of contamination of the lens or a skin coating for bubbles, preventing the bubble's rapid dissolution. If these material exchanges are exposure dependent, then exposure and development uniformity may suffer.

In addition to water contact time, the water contact angle of the resist is also important.[46] Instead of the usual method of measuring the contact angle, it is now measured with a water droplet on an inclined surface to simulate the situation of water being dragged by the immersion head over the wafer during scanning and stepping in a LBC system. Figure 8.50 shows the water droplet on an inclined wafer surface, the receding and advancing contact angles, as well as the sliding and static contact angles. The advancing angle indicates the wetting property of the advancing immersion head, similar to the case of the receding immersion head with the receding angle. The receding contact angle influences water stains, and the advancing contact angle affects bubble formation.

To more effectively control the contact angles and isolate the resist material from the immersion fluid, a topcoat is often used. This topcoat works like a TARC in processing. For process simplicity, the topcoat is preferably removable during the development process. There are topcoats that need a separate solvent removal step. Initially, the latter is more effective. However, because of the need for simplicity, developer-removable topcoats have rapidly progressed. Ultimately, a single resist layer without a topcoat is the most preferable process for cost. The need for a topcoat is not obvious. Some resists[47] work perfectly without a topcoat. Instead of specifying the leeching specification too conservatively, thus leading to a limited choice of resist materials and necessitating topcoats, a better way to prevent lens contamination is to conduct an effective periodic cleaning.

The refractive index of the resist determines the angle of the spatial frequency recombination which, in turn, affects polarization-induced stray light. Hence, a high refractive index is always preferred for photoresists unless there are tradeoffs in light absorption, imaging performance, or other pertinent resist-processing characteristics.

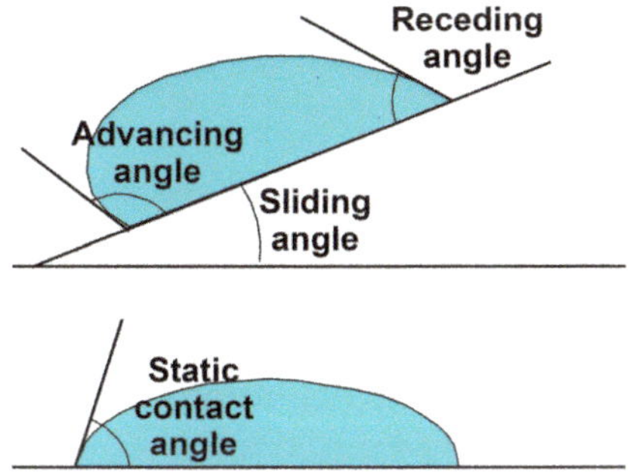

Figure 8.50 Receding, advancing, sliding, and static contact angles of a water droplet on an inclined wafer surface.

8.7 The Impact of Immersion Lithography on Technology

Having taken in all of the considerations of the immersion lithography system, the key question is: What technology nodes can be used to manufacture semiconductor products? Here, we evaluate DOF_{diff} and DOF_{avail} by simulation for the poly layer, contact layer, and metal layer of the 65-, 45-, and 32-nm nodes, using 193-nm and 157-nm, dry and immersion lithography systems. We also briefly explain the imaging feasibility of 157-nm immersion for the 22-nm node.

8.7.1 Simulation of immersion lithography

The simulation of immersion lithography is based on the Signamizer scheme,[29] which incorporates k_3 high-NA scaling,[4] PDS, and SSL to evaluate DOF_{diff} and DOF_{avail}. The technology requirement is based on a minimum half pitch, CD, and the other relevant pitches. No isolated features are considered because the use of dummy features is assumed. We have systematically set the poly photo CD to the technology node, i.e., 65-nm CD for the 65-nm node, etc. The CD tolerance is simplistically set at ±10% CD. For the poly layer, the half pitches are taken to be the CD of the last node, i.e., 65 nm for the 45-nm node, etc. For the contact layer, the CD is 1.5 times the half pitch, rounded to a close integer. Pitches of $2x$, $2.5x$, $3x$, $4x$, and $5x$ CD are included for the common E-D window. The metal CD is taken to be about 1.4 times the half pitch. The ratios of line openings to opaque spaces are 1:1, 1:1.5, 1:2, 1.5:1, and 2:1. The resist thicknesses are 200, 150, and 100 nm for the poly layer of the 65-, 45-, and 32-nm nodes, respectively. These thicknesses become 250, 175, and 100 nm for the contact and metal layers. The resist has to thin down to maintain a 3:1 height:width ratio and to consume less of the required DOF. SSL is set at 10% for the 65 nm node, then 5% for the other two nodes. This is on the pessimistic side; however, the stray-light level is adjusted for the pattern density. For example, it is multiplied by 0.5 for equal lines and spaces, and by 0.25 for equally spaced contact holes with a 1:1 hole:space ratio. The half pitches, CD, and features assumed for the three technology nodes for the poly, contact, and metal layers are shown in Table 8.8.

The results are plotted and tabulated in the following sections. They are taken from DOF-sinθ curves based on the Signamizer. DOF-NA curves can be readily obtained by multiplying sinθ by the refractive index of the coupling medium. Sample plots of the 45-nm node using 193-nm immersion and the corresponding E-D windows at the sinθ values are shown in Figs. 8.51 and 8.52, respectively. All of these calculations use 134.6 nm as the immersed wavelength and n = 1.4366 as the refractive index. The immersion wavelength is based on a 193.39-nm vacuum wavelength and the five-digit refractive index. Here, resolving the 130-nm minimum pitch requires the largest θ.

Table 8.8 Poly-layer, contact-layer, and metal-layer half pitches, CDs, and features assumed for the 65-, 45-, and 32-nm nodes.

Poly node	Half pitch 1 is the last node name	CD line	Pitch 1 2	Pitch 2 2.5	Pitch 3 3		
65	90	65	180	225	270		
45	65	45	130	162.5	195		
32	45	32	90	112.5	135		
Contact node	**Half pitch 1**	**CD hole**	**Pitch 1 2**	**Pitch 2 2.5**	**Pitch 3 3**	**Pitch 4 4**	**Pitch 5 5**
65	100	100	200	250	300	400	500
45	70	70	140	175	210	280	350
32	50	50	100	125	150	200	250
Metal node	**Half pitch 1 is the last node name**	**CD line**	**Pitch 1 2 (1:1)**	**Pitch 2 2.5 (1:1.5)**	**Pitch 3 3 (1:2)**	**Pitch 4 2.5 (1.5:1)**	**Pitch 5 3 (2:1)**
65	90	90	180	225	270	225	270
45	65	65	130	162.5	195	162.5	195
32	45	45	90	112.5	135	112.5	135

Nevertheless, the DOF is gated by the 195-nm pitch. The individual biases on the features optimize the common E-D window. They are very well optimized between $\sin\theta = 0.71$ and $\sin\theta = 0.82$. The common DOF coinciding with the 195-nm curve is the evidence of this optimization.

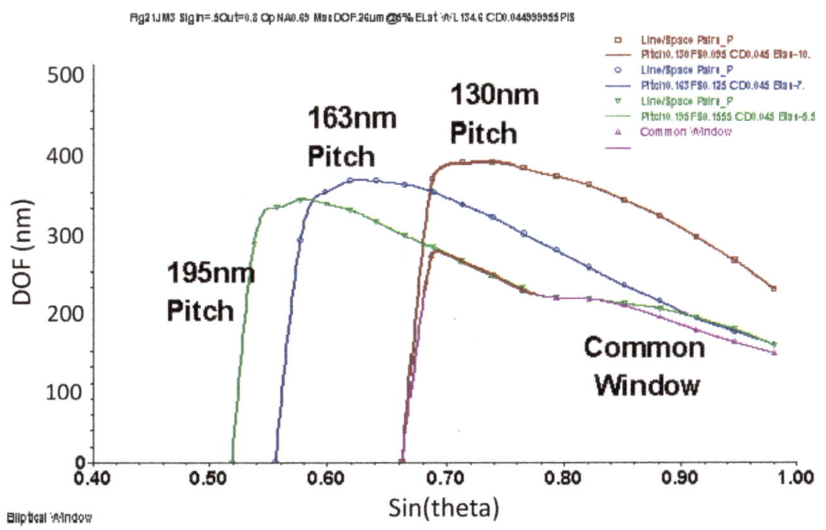

Figure 8.51 DOF versus $\sin\theta$ with 193-nm immersion, 6% AttPSM, σ_{in} 0.42, σ_{out} 0.84, 45-nm node, CD 45 nm ± 10%, pitches 130, 163, 195 nm, EL 5%, SSL 10%. Biases are –10, –7, and –5.5 nm.

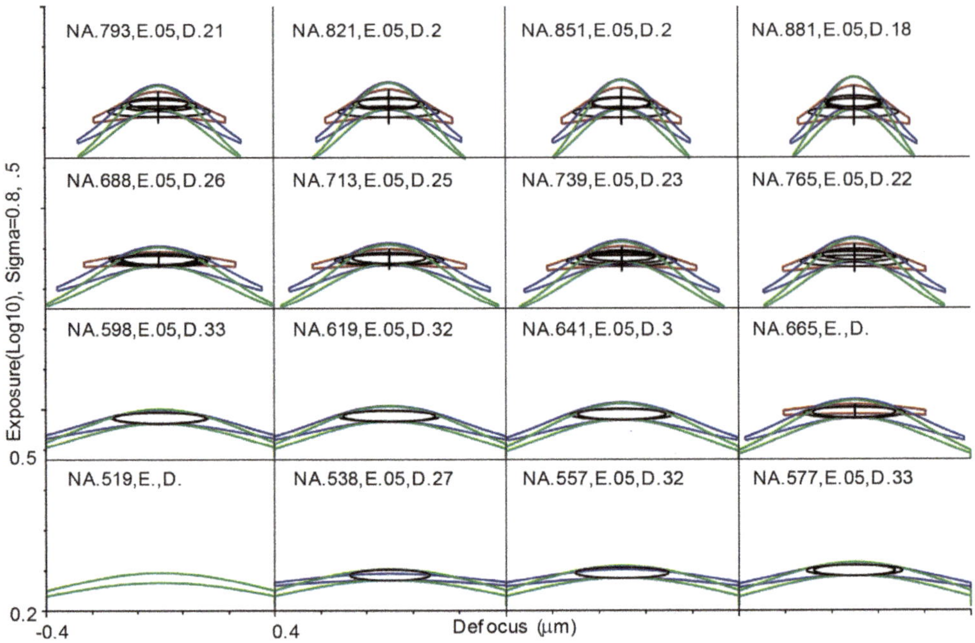

Figure 8.52 E-D windows leading to Fig. 8.51. The E-D coordinates are 0.2 to 0.5; –0.4 to 0.4 repeats at each sinθ block.

Both DOF_{diff} and DOF_{avail} are listed in the three tables that follow. Even though it is preferable to compare DOF_{avail}, providing DOF_{diff} makes it more straightforward to compute DOF_{avail} at a new resist thickness. However only DOF_{avail} values are plotted. We expect the exposure tools to support a DOF_{avail} of 250 nm for the 65-nm node, and 150 and 100 nm for the 45- and 32-nm nodes, respectively, based on discussions with exposure tool suppliers. Historically, these numbers have been 400 nm for the 130-nm node and 350 nm for the 90-nm node.

8.7.2 Poly layer

The simulation results from the poly layer for the three technology nodes are shown in Table 8.9 and Fig. 8.53. At the 65-nm node, the DOF_{diff} is 270 nm and the DOF_{avail} is 156 nm for 193-nm dry systems using annular off-axis illumination and 6% AttPSM. The required sinθ or NA is 0.888. The k_1 value based on the half pitch is 0.414. Changing to a 157-nm dry system relaxes the NA to 0.739, while it increases DOF_{avail} to 277 nm. The 193-nm immersion system provides a 315-nm DOF_{avail} at NA = 0.925. Switching to a lower NA of 0.839, a 286-nm DOF_{avail} is obtained. Note that, unlike the case in dry systems, the NA is no longer identical to sinθ in the immersion case. Only the 157-nm dry and 193-nm immersion systems support the 250-nm DOF_{avail} requirement. The NA = 0.839 requirement is important because this type of

Table 8.9 DOF of the poly layer at the three technology nodes.

			n_{resist}=1.75			$n_{193nmCM}$=1.4366			$n_{157nmCM}$=1.37	
		65 Poly	**90**			***CD*=65 nm, *P1*=180 nm, *P2*=225 nm, *P3*=270 nm**			***THK*$_{resist}$ (nm) =200**	**10%SSL, 8%Elat**
Line	No.	λ (nm)	HP(k_1)	sinθ	NA	Sigma	DOF$_{diff}$	DOF$_{avail}$	RET	Feature Bias (nm)
65P	1	193	0.414	0.888	0.888	0.84:0.42	270	156	AttPSM OAI	B-26,-23,-20
65P	2	157	0.424	0.739	0.739	0.84:0.42	391	277	AttPSM OAI	B-25,-22,-18.5
65P	3	134.6	0.431	0.644	0.925	0.84:0.42	479	315	AttPSM OAI	B-25,-22,-18.5
65P	4	134.6	0.390	0.584	0.839	0.84:0.42	450	286	AttPSM OAI	B-25,-22,-20
		45 Poly	**65**			***CD*=45 nm, *P1*=130 nm, *P2*=163 nm, *P3*=195 nm**			***THK*$_{resist}$ (nm) =150**	**5%SSL, 5%Elat**
Line	No.	λ (nm)	*HP*(k_1)	sinθ	NA	Sigma	DOF$_{diff}$	DOF$_{avail}$	RET	Feature Bias (nm)
45P	1	157	0.325	0.784	0.784	0.8:0.4	218	132	AttPSM OAI	B-9,-6.5,-5
45P	2	134.6	0.325	0.674	0.968	0.8:0.4	278	155	AttPSM OAI	B-8,-5,-3.8
45P	3	157	0.325	0.784	0.784	0.8:0.4	234	148	AttPSM OAI TE	B-9,-8,-5
45P	4	134.6	0.325	0.674	0.968	0.8:0.4	297	174	AttPSM OAI TE	B-11.5,-9.5,-8
45P	5	193	0.280	0.831	0.831	0.3	251	165	AltPSM	-7,-5,-4
45P	6	157	0.283	0.684	0.684	0.3	356	270	AltPSM	-5,-4,-6
45P	7	134.6	0.275	0.569	0.817	0.3	450	327	AltPSM	0,2,-5
						Pitches 130, 163 nm				
45P	8	157	0.337	0.814	0.814	0.8:0.4	277	191	AttPSM OAI	B-12.5,-10
45P	9	134.6	0.340	0.704	1.011	0.8:0.4	351	228	AttPSM OAI	B-12.8,-10
45P	10	134.6	0.339	0.701	1.007	0.8:0.4	347	224	BIM OAI	B-12.8,-10.2

(continued)

		32 Poly	45			*CD*=32 nm, *P*1=90 nm, *P*2=113 nm, *P*3=135 nm			THK_{resist} (nm) =100	5%SSL, 5%Elat
Line No.		λ (nm)	HP(k_1)	sinθ	NA	Sigma	DOF_{diff}	DOF_{avail}	RET	Feature Bias (nm)
32P	1	134.6	0.296	0.885	1.271	0.96:0.48	114	32	AttPSM OAI	B-5,-2.5,-1
32P	2	111	0.310	0.765	1.117	0.96:0.48	149	71	AttPSM OAI	B-10,-8,-6.5
32P	3	134.6	0.294	0.879	1.263	0.96:0.48	118	36	AttPSM OAI TE	B-5,-2.5,-1
32P	4	111	0.308	0.759	1.040	0.96:0.48	160	82	AttPSM OAI TE	B-10,-8,-7
32P	5	134.6	0.265	0.793	1.139	0.3	219	137	Alt PSM	B-8,-2,-6
32P	6	111	0.269	0.663	0.908	0.3	308	230	Alt PSM	B-10,-4,-9
32P	7	134.6	0.265	0.793	1.139	0.3	233	151	Alt PSM TE	B-7,-4,-9
						Pitches: 90, 113 nm				
32P	8	134.6	0.314	0.940	1.350	0.96:0.48	137	55	AttPSM OAI	B-8.7.-6
32P	8a	134.6	0.291	0.871	1.251	0.96:0.48	200	118	AttPSM OAI	B-9.-5.5
32P	9	111	0.321	0.793	1.086	0.96:0.48	202	124	AttPSM OAI	B-10,-7.7
32P	10	134.6	0.303	0.905	1.300	0.96:0.48	169	87	AttPSM OAI TE	B-9,-6.6
32P	10a	134.6	0.291	0.871	1.251	0.96:0.48	200	118	AttPSM OAI TE	B-9,-6.3
32P	11	111	0.313	0.771	1.056	0.96:0.48	226	148	AttPSM OAI TE	B-10,-7.7
32P	12	134.6	0.265	0.793	1.139	0.3	447	365	Alt PSM	B-17,-7

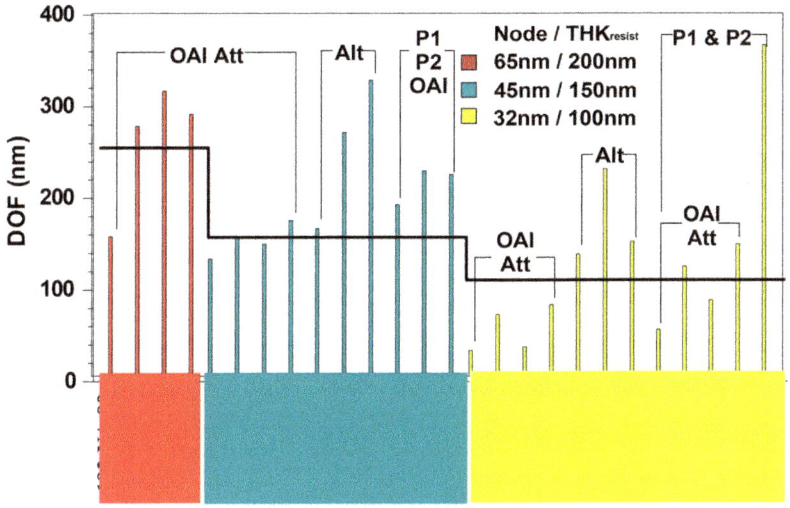

P1 = 2CD, P2 = 2.5CD, P3 = 3CD

Figure 8.53 DOF_{avail} of the poly layer at three technology nodes.

immersion tool is readily available. Even at the non-optimum lower NA, the immersion tool DOF_{avail} is still higher than that of either the 193-nm or the 157-nm dry system.

In 2008, the 65-nm node was in manufacturing in almost all semiconductor factories. Due to cost and timing issues, dry exposure tools were preferred. No immersion tool was needed. This was accomplished with a combination of using a stronger RET, relaxing the design rules, increasing the minimum pitch, forbidding the pitches that gate the DOF, and using a better $DOF_{required}$ from the exposure tools. Some of the anticipated performance enhancement techniques for the 45- and 32-nm nodes could also have been applied to the 65-nm node.

Turning to the 45-nm node, there is no chance of using a 193-nm dry system with annular illumination and 6% AttPSM, regardless of the NA. The 157-nm system provides a 134-nm DOF_{avail}, and the 193-nm immersion system provides a 155-nm DOF_{avail}, which is acceptable. However, if necessary, there are several ways to produce a larger margin:

1. Tighten the DOF control of the exposure tool.
2. Use a stronger RET, such as AltPSM or quadrupole illumination. The former can produce 165-, 270-, and 327-nm DOF_{avail} for 193-nm dry, 157-nm dry, and 193-nm immersion, respectively.
3. Remove the most restrictive pitch. Removing the 195-nm pitch produces a DOF_{avail} of 191 nm for the 157-nm dry system and 228 nm for the 193-nm immersion system with annular illumination

and 6% AttPSM. Note that, in this situation, DOF_{avail} with a BIM is only 4 nm smaller than that with an AttPSM.

4. Use the TE polarization. A DOF_{avail} of 148 nm can be obtained with 157-nm dry systems and 174 nm with 193-nm immersion.

At the 32-nm node, the 157-nm dry system, using 6% AttPSM and annular illumination, no longer supports a common window. 193-nm immersion produces 32 nm of DOF_{avail}, and 157-nm immersion produces 71 nm; resorting to the TE mode yields 36 nm and 82 nm, respectively. The relief is insignificant. DOF_{avail} of the previous four situations is insufficient to cover the 100-nm tool DOF control. Switching to a strong RET such as AltPSM suffices. Staying with 6% AttPSM/annular illumination and restricting the pitch to 90 and 113 nm produces a sufficient DOF_{avail} for 157-nm immersion, but not for 193-nm immersion. Tightening the $DOF_{required}$ needs to be pursued.

8.7.3 Contact layer

Table 8.10 and Fig. 8.54 show the DOF of the contact layer at the three nodes. At the 65-nm node, the 272- and 286-nm DOF_{avail} for 157-nm dry and 193-nm immersion systems are acceptable, while the 193-nm dry system is not. However, 65-nm manufacturing of the contact layer was still managed with dry systems by similar measures used for improving the DOF of the poly layer.

At the 45-nm node, DOF_{avail} for 157-nm dry and 193-nm immersion systems are insufficient, if staying with 6% AttPSM. However, the DOF_{avail} from the latter system is much closer to the tool requirement. There is no obvious reason that the focus control of the exposure tool cannot be tightened slightly to meet the requirement. Alternately, one may resort to restricted pitches, reduced resist thickness, slight change of design rule, or a stronger RET to bridge the gap. For example, keeping the pitch range from 140 to 210 nm facilitates a DOF_{avail} of 172 nm, as shown in Table 8.10. Changing the resist thickness to 150 nm results in DOF_{avail} = 163 nm. Relaxing the CD tolerance to ±6 nm between the 280- and 350-nm pitches extends DOF_{avail} to 166 nm.

For the 32-nm node, DOF_{avail} with 193-nm or 157-nm immersion using 6% AttPSM is again insufficient. To stay with 193-nm immersion, several approaches can be taken: (1) use an AltPSM and the pack-and-seal technique[32] to remove the 150-, 200-, and 250-nm pitches or (2) use annular illumination on a 6% AttPSM and remove the 200- and 250-nm pitches. If one wants to use 157-nm immersion instead, one must remove three pitches with disk illumination or two pitches with annular illumination.

Table 8.10 DOF of the contact layer at three technology nodes.

			n_{resist} = 1.75			$n_{193nmCM}$ = 1.4366				$n_{157nmCM}$ = 1.37
		65C T	**100**			***CD*=100 nm, Pitches=200, 250, 300, 400, 500 nm**			***THK*$_{resist}$ (nm) = 250**	**10%SSL, 8%Elat**
Line No.		λ (nm)	HP(k_1)	sinθ	NA	Sigma	DOF$_{diff}$	DOF$_{avail}$	RET	Feature Bias (nm)
65C	1	193	0.369	0.713	0.713	0.78	334	191	AttPSM	0,5,7,9.5,11
65C	2	157	0.385	0.604	0.604	0.78	415	272	AttPSM	0,4,6.5,9,9.5
65C	3	134.6	0.383	0.516	0.741	0.7	491	286	AttPSM	0,3,4.5,7,7
		45C T	**70**			***CD*=50 nm, Pitches=140, 175, 210, 280, 350 nm**			***THK*$_{resist}$ (nm) = 175**	**5%SSL, 6%Elat**
Line No.	λ (nm)	HP(k_1)	sinθ	NA	Sigma	DOF$_{diff}$	DOF$_{avail}$	RET	Feature Bias (nm)	
45C	1	157	0.339	0.760	0.760	0.78	223	123	AttPSM	0,4,5,7,8
45C	2	134.6	0.328	0.631	0.906	0.78	286	142	AttPSM	0,4.5,5,7.5,8
45C	3	134.6	0.328	0.631	0.906	0.78	310	166	AttPSM, ±6nm P4,P5	0,4.5,5.4,7.8,8.3
						Pitches: 140, 175, 210 nm				
45C	4	134.6	0.330	0.635	0.912	0.78	316	172	AttPSM	0,4.5,5.4
		32C T	**50**			***CD*=50 nm, Pitches=100, 125, 150, 200, 250 nm**			***THK*$_{resist}$ (nm) = 100**	**5%SSL, 5%Elat**
Line No.	λ (nm)	HP(k_1)	sinθ	NA	Sigma	DOF$_{diff}$	DOF$_{avail}$	RET	Feature Bias (nm)	
32C	1	134.6	0.320	0.862	1.238	0.86	122	40	AttPSM	0,3.8,4.5,6.5,7
32C	1	134.6	0.319	0.858	1.233	0.86	130	48	AttPSM TE	0,3.8,4.5,6.5,7
32C	2	111	0.316	0.702	0.962	0.86	161	83	AttPSM	0,3.8,4.5,6.5,7
32C	3	134.6	0.342	0.920	1.322	0.3	101	19	Alt PSM	0,–3.5,5 P1,P2,P5
32C	4	134.6	0.338	0.910	1.307	0.3	113	31	Alt PSM TE	5,-3.5,0
						Pitches: 100, 125 nm				
32C	3	134.6	0.321	0.863	1.240	0.86	150	68	AttPSM	0,3.9
32C	4	134.6	0.309	0.833	1.197	0.86	172	90	AttPSM TE	0,3.9
32C	5	134.6	0.329	0.885	1.271	0.3	266	184	Alt PSM	10,1
32C	6	111	0.323	0.716	0.981	0.86	221	143	AttPSM	0,3.8
32C	7	111	0.310	0.688	0.943	0.86	241	163	AttPSM TE	0,3.9
						Pitches: 100, 125, 150 nm				
32C	8	134.6	0.296	0.798	1.146	0.88:0.44	202	120	AttPSM OAI	–4,–1.2,0.4
32C	9	134.6	0.289	0.779	1.119	0.88:0.44	230	148	AttPSM OAI TE	–4,–1.6,0.1
32C	10	111	0.303	0.672	0.921	0.88:0.44	264	186	AttPSM OAI	–4,–1,0.3
32C	11	111	0.298	0.661	0.906	0.92:0.46	283	205	AttPSM OAI TE	–4,–1,0.3

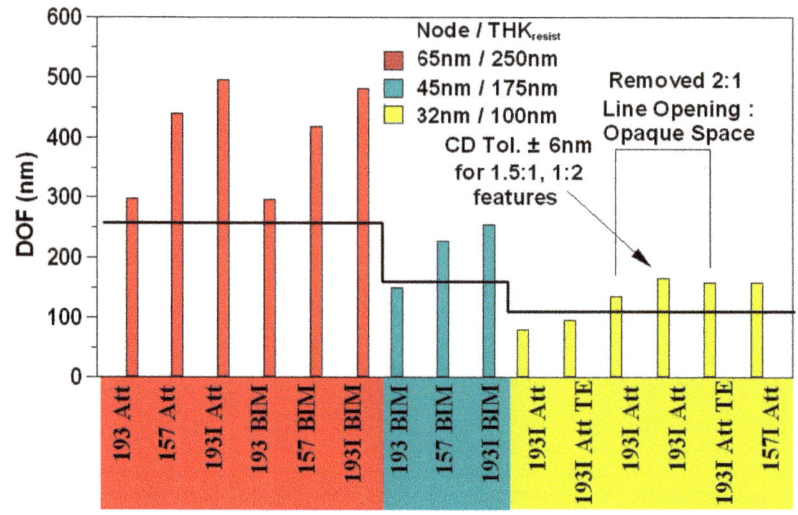

Figure 8.54 DOF$_{avail}$ of the contact layer at three technology nodes.

8.7.4 Metal layer

The simulation results for the metal layer are shown in Table 8.11 and Fig. 8.55. Note that there are three pitches but five types of feature because the line-opening:opaque-space ratios are 1:1, 1:1.5, 1:2, 1.5:1, and 2:1. At the 65-nm node, the DOF$_{avail}$ values of 298, 440, and 496 nm are abundant. There is even room to replace an AttPSM with a BIM. At the 45-nm node, the

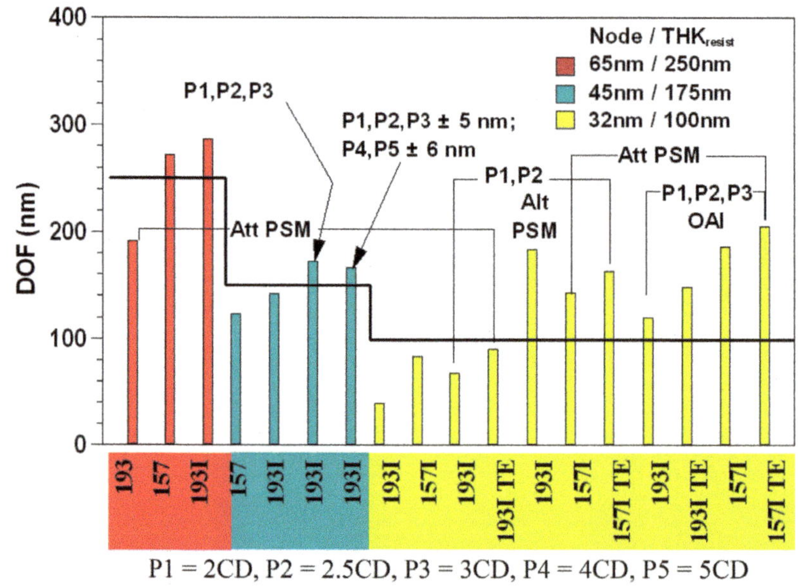

Figure 8.55 DOF$_{avail}$ of the metal layer at three technology nodes, using OAI throughout.

Table 8.11 DOF of the metal layer at three technology nodes.

		n_{resist} = 1.75				$n_{193nmCM}$ = 1.4366				$n_{157nmCM}$ = 1.37
		65 M1	**90**			***CD*=90, *P1*=180, *P2*=225, *P3*=270 nm**			***THK*$_{resist}$ (nm) = 250**	**10%SSL, 8%Elat**
Line	No.	λ (nm)	HP(k_1)	sinθ	NA	Sigma	DOF$_{diff}$	DOF$_{avail}$	RET	Feature Bias (nm)
65M	1	193	0.332	0.712	0.712	0.76:0.38	441	298	AttPSM OAI	5,0,5,19,27
65M	2	157	0.328	0.573	0.573	0.76:0.38	583	440	AttPSM OAI	5,0,5,19,27
65M	3	134.6	0.330	0.493	0.708	0.76:0.38	701	496	AttPSM OAI	5,0,5,19,27
65M	4	193	0.332	0.712	0.712	0.76:0.38	439	296	BIM OAI	5,0,5,19,27
65M	5	157	0.333	0.581	0.581	0.76:0.38	562	419	BIM OAI	5,0,5,19,27
65M	6	134.6	0.330	0.493	0.708	0.76:0.38	687	482	BIM OAI	5,0,5,19,27
		45 M1	**65**			***CD*=65, *P1*=130, *P2*=162.5, *P3*=195 nm**			***THK*$_{resist}$ (nm) = 175**	**5%SSL, 6%Elat**
Line	No.	λ (nm)	HP(k_1)	sinθ	NA	Sigma	DOF$_{diff}$	DOF$_{avail}$	RET	Feature Bias (nm)
45M	1	193	0.299	0.888	0.888	0.84:0.42	250	150	BIM OAI	0,5,8,11,17
45M	2	157	0.319	0.771	0.771	0.84:0.42	327	227	BIM OAI	0,5,8,10,14
45M	3	134.6	0.321	0.665	0.955	0.84:0.42	399	255	BIM OAI	0,5,8,10,14
		32 M1	**45**			***CD*=45, *P1*=90, *P2*=112.5, *P3*=135 nm**			***THK*$_{resist}$ (nm) = 100**	**5%SSL, 6%Elat**
Line	No.	λ (nm)	HP(k_1)	sinθ	NA	Sigma	DOF$_{diff}$	DOF$_{avail}$	RET	Feature Bias (nm)
32M	1	134.6	0.292	0.874	1.256	0.92:0.46	161	79	AttPSM OAI	0,4,6,8.5,13.1
32M	2	134.6	0.292	0.874	1.256	0.92:0.46	177	95	AttPSM OAI TE	0,4,6,7.5,11.5
32M	3	111	0.289	0.714	0.978	0.92:0.46	237	159	AttPSM OAI	0,4,6,7.5,13
						Pitches: 90, 112.5, 135(1.5:1 only) nm				
32M	4	134.6	0.296	0.884	1.270	0.92:0.46	217	135	AttPSM OAI	0,4,6.5,9.2
32M	5	134.6	0.287	0.857	1.231	0.92:0.46	240	158	Same as above but with TE	0,4,5.8,8.5,11.5
32M	6	134.6	0.297	0.888	1.276	0.92:0.46	248	166	AttPSM OAI, ±6nm P2(1.5:1), P3(1:2)	0,4,7.5,9.5

DOF_{avail} for 193-nm dry, 157-nm dry, and 193-nm immersion systems are also usable with a BIM.

At the 32-nm node, the DOF_{avail} for 193-nm immersion using 6% AttPSM and annular illumination is insufficient. Adding polarized illumination brings it close to the tool focus control limit, whereas 157-nm immersion has ample DOF_{avail} without resorting to TE illumination. AltPSM is not that useful for the metal layer because of the requirement to establish a common E-D window for lines and spaces at different pitches. To stay with 193-nm immersion, the 2:1 feature at the 135-nm pitch can be dropped to yield a DOF_{avail} of 135 nm. This can be extended to 158 nm with polarized illumination. Further relaxing the CD control to ±6 nm for the 1:2 feature at the 135-nm pitch and the 1.5:1 feature at the 112.5-nm pitch can extend the DOF_{avail} to 166 nm without resorting to polarized illumination.

8.7.5 Recommendation for the three technology nodes

For the 65-nm node, either the 157-nm dry system or the 193-nm immersion system is acceptable. The latter has a larger margin of DOF_{avail}. The 193-nm dry system requires more enhancements, such as tightening the tool focus control, using a stronger RET, or dropping some pitches. Even though the 157-nm dry system can support the DOF requirement, the technology is not yet ready because of a lack of photoresist, soft pellicle, and CaF_2 in quality and quantity. Replacing the soft pellicle with quartz pellicles has unsurmountable problems of its own. The considerations on the contact layer are similar. For the metal layer, DOF_{avail} for all three systems is acceptable.

For the 45-nm node at the poly level, the 193-nm immersion system and the 157-nm dry system are incidentally above and below the cutoff DOF_{avail} requirement of 150 nm, respectively, regardless of whether polarized illumination is used. This, of course, can be changed by further tightening the focus control of the exposure tool. Going to AltPSM or annular illumination with a designated pitch range moves both systems over the cutoff point. AltPSM can even move the 193-nm dry system over the cutoff point. The selection of AltPSM or BIM with annular illumination needs to take both performance and cost into consideration. At the contact layer, 193-nm immersion can be made to work more easily. At the metal layer, the DOF_{avail} values for all three systems are acceptable.

For the 32-nm node, from pure simulation results, 157-nm immersion provides more DOF_{avail} than 193-nm immersion. However, 157-nm immersion lithography is unlikely. There are combinations of RET and design rule changes that make it possible to use 193-nm immersion lithography.

Polarized illumination can improve the DOF of a given setup. A BIM works slightly worse than an AttPSM for this purpose. However, a BIM may be adopted in many cases to save mask cost.

Table 8.12 NA required for 193-nm water immersion at the three technology nodes.

Node	System	Poly	Contact	Metal	Tool
65 nm	dry	—	—	0.712 (65M1)	—
	immersion	0.839 (65P4)	0.741 (65C3)	0.708 (65M6)	0.84
45 nm	dry	—	—	0.888 (45M1)	—
	immersion	0.968 (45P2)	0.912 (45C4)	0.955 (45M3)	1.1
32 nm	immersion	1.14 (32P5)	1.15 (32C8)	1.28 (32M6)	1.3

The optimum-NA requirement of the three technology nodes with 193-nm water immersion is summarized in Table 8.12. The NA to support the tool focus control is listed under each layer in each node. The poly-node NA is taken from Table 8.9, lines 65C3, 45C4, and 32C8; the metal-node NA is taken from Table 8.11, lines 65M1, 65M6, 45M1, 45M3, 32M6. Under the Tool column is the NA suggested for the tool. For the 65-nm node, the required tool NA is taken to be a two-digit round-off of the highest NA that processing needs, namely 0.84. For the 45-nm node, tools of 0.97 NA suffice. However, the increment from the 65-nm tools is too small. Making the maximum NA 1.1 on the 45-nm tools comfortably covers the need of all the critical layers, plus some subnodes, before reaching the 32-nm node. Thus, the 45-nm tool might turn out to be quite a cross-generational workhorse. At the 32-nm node, the maximum tool NA is again selected to handle all of the 32-nm critical layers.

8.8 Practicing Immersion Lithography

When water-immersion 193-nm lithography was initially proposed, the industry had already substantially invested in 157-nm scanner development and the infrastructure to support it. There was strong momentum to remain on the 157-nm course until it was realized that producing a large quantity of high-quality CaF_2 is extremely difficult. Overcoming the high absorption in a 157-nm resist had a similar problem. Developing a durable soft pellicle was also very difficult. Therefore, the industry switched to water-immersion 193-nm lithography. In a relatively short time of two years, exposure tools had become available. Fervent efforts in resist development quickly followed. The mask infrastructure did not need much change. The burden has switched to semiconductor manufacturers to make immersion lithography a viable high-volume manufacturing technology. Here are some sample approaches that have been taken to make this happen.

8.8.1 Printing results

193-nm water-immersion has been applied to logic devices and circuits to explore its potential and shake out its problems. Figure 8.56 shows polysilicon images of a 90-nm-node static RAM (SRAM) chip overlaid on the active layer.[48] The active layer was exposed on a 0.75-NA 193-nm prototype immersion scanner at ASML, and the poly layer was exposed on an equivalent dry system at TSMC. This study demonstrated full-field imaging, a usable resist system, acceptable overlay, and large DOF. Turning to a 0.85-NA 193-nm immersion scanner, 55-nm-node SRAM chips were delineated. Figure 8.57 shows the metal layer image of this chip at different field locations. Figure 8.58 compares the number of good dies to the number of bad dies in a dry-immersion that is split at the contact layer.[49] The ratio is 72:70 versus 62:80 for immersion versus dry. The number of good immersion dies could increase by 25 (attaining 97:45) if nonphoto mis-operations were discounted.

8.8.2 Defect reduction

Immersion lithography has come a long way. The wafer throughput and overlay performance have been steadily improved. There is no longer a concern about these issues for 45- to 7-nm manufacturing, except for the ever-persisting demand on productivity and packing density. Perhaps the most imminent challenge is to reduce the defect level to a single digit, just as in dry lithography. By 2010, many semiconductor factories were able to control defects to an acceptable level. The major differentiation lies in whether a topcoat above the resist is needed. The company that does not need a topcoat enjoys cost savings and better imaging performance from innovation.

Figure 8.59 shows three major defect types from immersion exposure, namely, water stain, bubbles, and particles. Water stains are due to residual

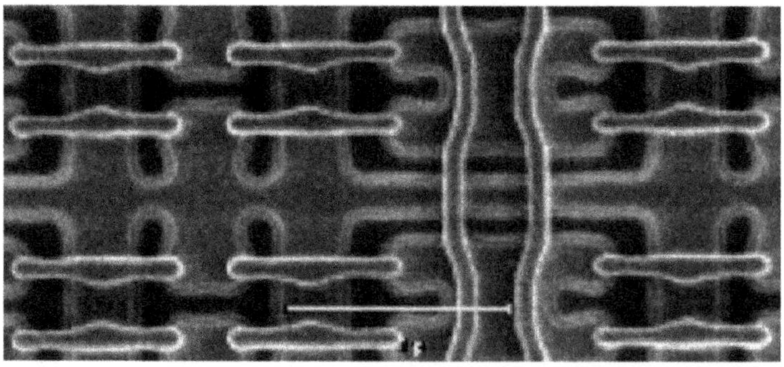

Figure 8.56 Immersion chip using 90-nm-node SRAM, showing the poly layer on top of the active layer. The poly layer was exposed with a 0.75-NA 193-nm prototype immersion scanner at ASML; the active layer was exposed with a dry scanner at TSMC (reprinted from Ref. 48).

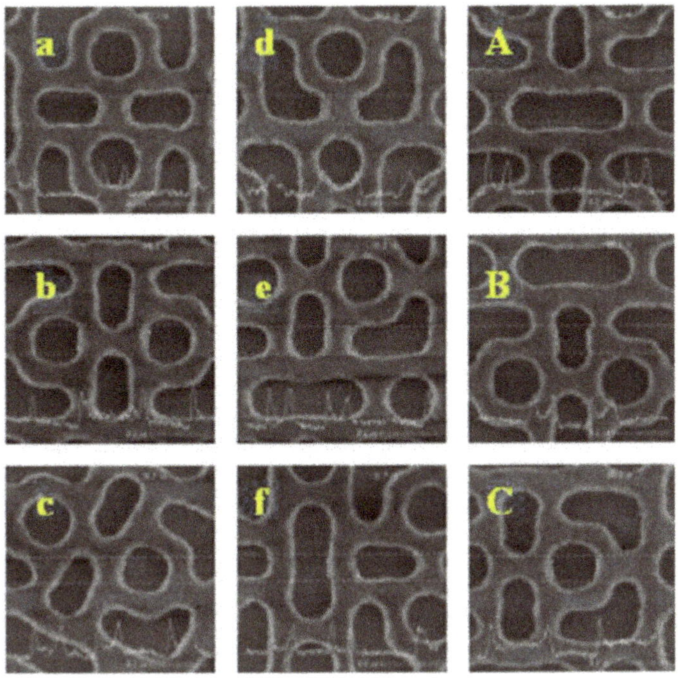

Figure 8.57 A 0.4-μm^2 55-nm-node SRAM metal layer exposed with a 0.85-NA 193-nm immersion scanner at different field locations (figure courtesy of K.S. Chen).

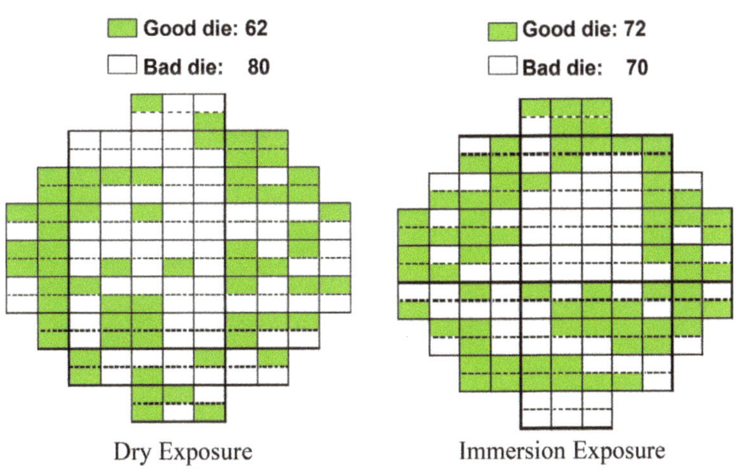

Figure 8.58 65-nm SRAM immersion yield from an R&D lot. One specific test chip was split for dry (left) and immersion (right) exposures at the contact layer. The yield would be higher if the center zone of the immersion wafer were not damaged by a nonphoto operation (figure courtesy of K.S. Chen).

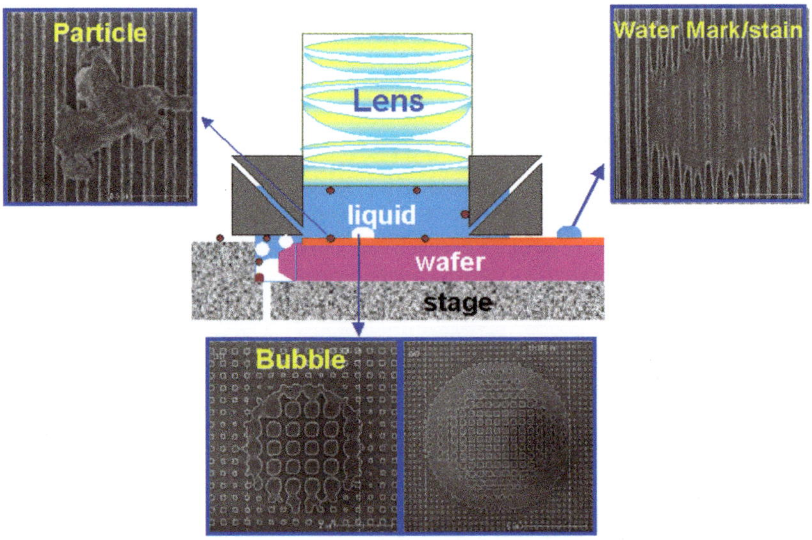

Figure 8.59 Major defects in immersion lithography: particles, water stains, and bubbles (figure courtesy of C.Y. Chang).

wetting of the resist surface after the immersion head passes through. They can be controlled by preventing residual wetting, material treatment, and/or special processing. Bubbles in water during exposure modify the resist image as if a magnifying glass were inserted during exposure. Bubbles are mostly induced by disruption of hydrodynamics during stepping and scanning. Particles can be in the water supply or can be carried to the wafer from any surface that water contacts. One primary suspect source is from under the wafer near its edge.

Figure 8.60 shows an effective post-exposure soaking method to remove water stains produced in a given range of conditions. New immersion hood designs have significantly reduced the number of bubbles.

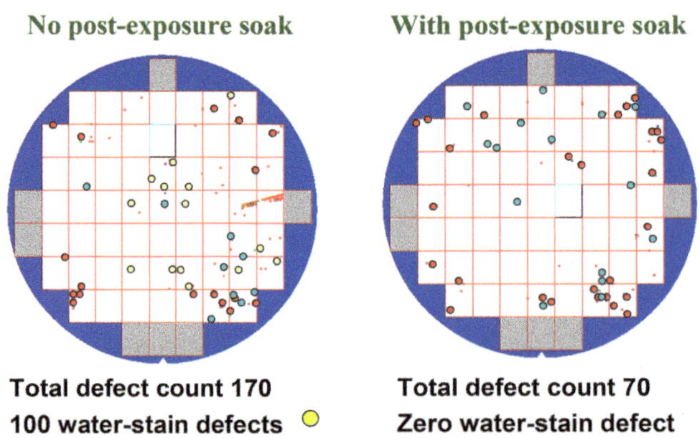

Figure 8.60 Post-exposure soak (right) reduces water-stain defects compared to no post-exposure soak (left) (figure courtesy of S.C. Wang).

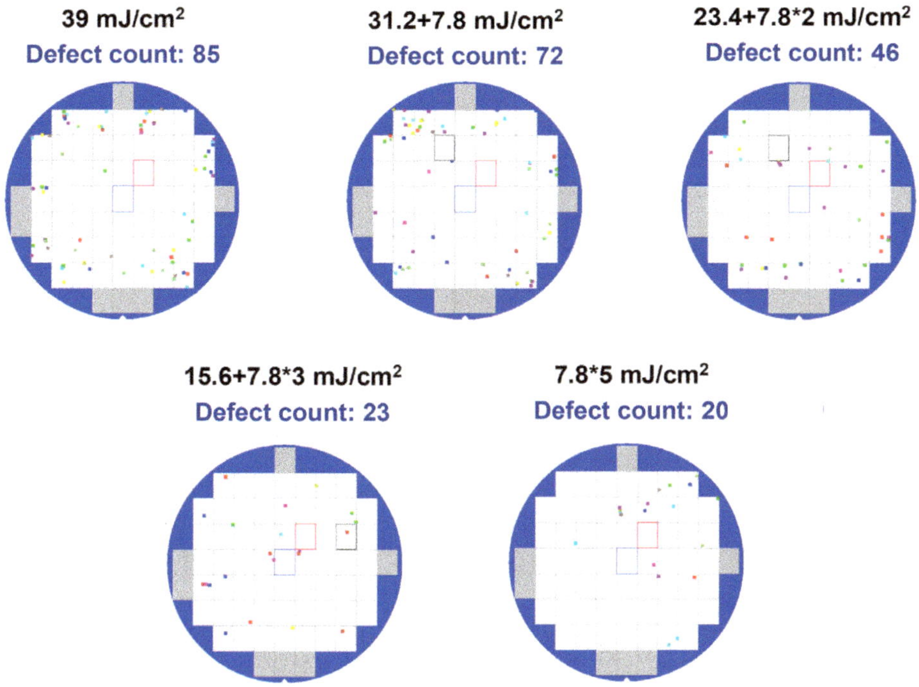

Figure 8.61 Immersion defect study by multiple exposures (figure courtesy of S.C. Wang).

Figure 8.61 shows the results of a multiple-exposure study on a plain wafer, aiming at differentiating static and dynamic defects. Out of the 85 defects found on the wafer, only about 1/4 survived after 5 mini-exposures. The other defects that were at different locations during each exposure were voted away. The experiment was conducted with a topcoat on the photoresist and with post-exposure bake to simplify the types of generated defects. We can attribute most of the dynamically located defects to bubbles and the static ones to particles.

8.8.3 Monitoring the immersion hood and special routing

Well-designed immersion systems together with hydrophilic materials can prevent production of bubbles. A post-exposure soak or a carefully selected resist/topcoat can prevent watermarks. Even just using a topcoat can greatly reduce the number of harmful particles present after development. Unfortunately, none of the above methods can prevent printing of particles deposited before exposure. A unique, special routing technique[50] can reduce the number of particle-printing defects to a single digit per wafer. This special routing methodology is the follow-up to a special technique for monitoring the immersion hood.

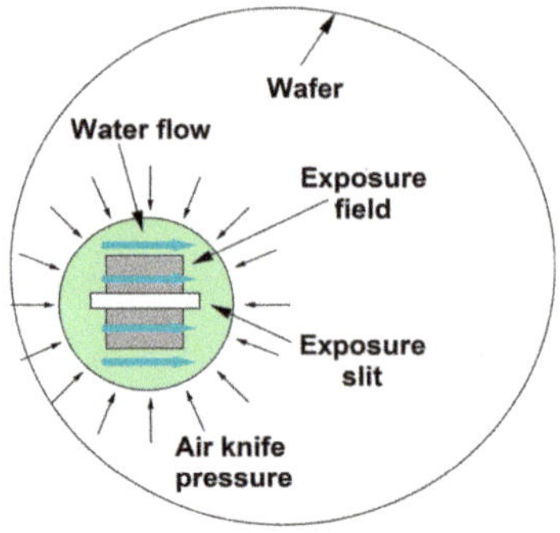

Figure 8.62 Conceptual configuration of the immersion hood.

The immersion hood (IH) is an important component of immersion scanners for supplying, confining, and draining the water during exposure. It is circular in shape and placed under the bottom lens element. Its edge is outside the exposure field; that is, the coverage area of the immersion fluid is larger than the exposure field. To confine the immersion fluid, an air knife surrounds the IH with inward-oriented force in the radial direction, preventing water leakage to the outside of the immersed area during wafer movement. Figure 8.62 conceptually depicts the IH configuration, showing the direction of the air and water flows, the positions of the exposure slit, and the exposure field.

Since particle printing is an important cause of defects in immersion lithography, it is necessary to identify where the particles come from and how they are printed.

The first possible scenario considered for particle printing is that the particles suspended in water near the resist surface are printed. Assuming a 500 mm/sec wafer scan speed and a 1.5 liter/min water flow, the particles in water are relatively static during exposure while the wafer passes by rapidly underneath. The resist image would consist of stripes if the particles in suspension were printed. Stripe-like defects have never been observed.

The second possible scenario is depicted in Fig. 8.63. During exposure, it is possible for particles to be transported through water from the IH to attach

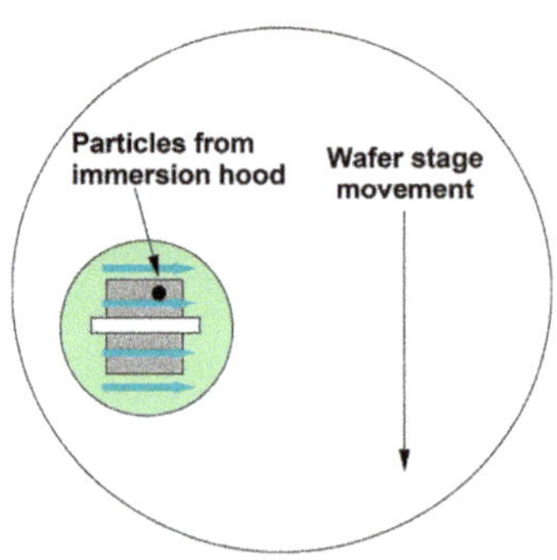

Figure 8.63 Particles transported through water from the immersion hood to the wafer during exposure.

to the wafer surface, and those inside the exposure slot are printed. There is no direct evidence for or against the existence of this type of defect. Nevertheless, there is no way to prevent particles from printing, except to entirely eliminate them in the water.

The third possible scenario is that particles left from the previous exposure fields are printed. They may be left there because the water droplet leaks out of the IH during the wafer scan. There are indeed water leaks during the wafer scan. Water leaks take place if the IH is not designed properly, the scan speed is too high, or the wafer surface is too hydrophilic. Figure 8.64(a) shows the wafer defect distribution accumulating on the defect map of 20 bare-silicon test wafers. Straight-line and arc trajectories are clearly observed. Water can only leak along the tangential directions at the edge of the IH along its path of travel since the water at the IH edge is constrained only in the radial direction. In addition to water leakage in tangential directions at the edge of the IH, the IH also leaks due to the inertial force during stage acceleration, especially at the rear edge while the stage starts to move, leading to an arc-shaped defect distribution. Hence, the defect map shown in Fig. 8.64(a) consists of straight lines and arcs from vertical and horizontal movements, respectively. The vertical direction is the wafer scanning direction.

Figure 8.64(b) shows the result of fitting the defect distribution to a special model used by the authors of Ref. 50 to describe the phenomenon of the third possibility. The good fitting shown affirms this possibility. If an air curtain containing circularly laid-out air holes was used, Fig. 8.64 would contain radial streaks of defects due to low-air-pressure regions between the holes around the edge of the IH, enhancing leakage during stage acceleration. The defect maps from Fig. 8.64 were taken from an IH using an air knife instead of an air curtain. The former supports a continuous radial airflow and does not produce radial streaks, explaining the absence of such streaks in Fig. 8.64.

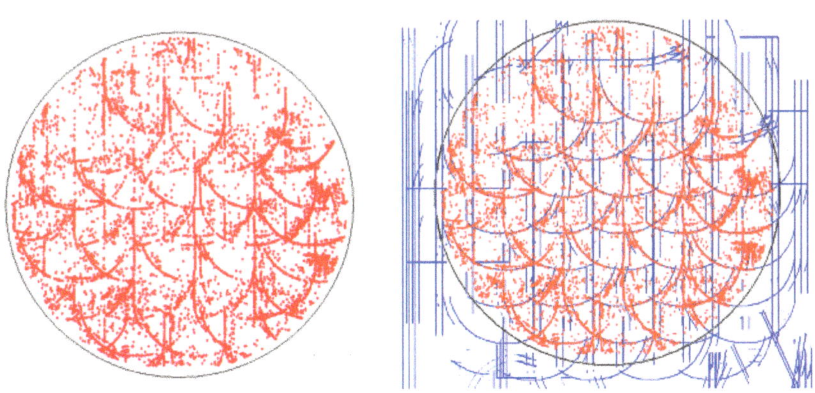

(a) (b)

Figure 8.64 (a) Accumulated defect distribution from 20 bare-silicon wafers. (b) Model fitting after considering wafer stage trajectories during exposure (adapted from Ref. 50).

Accumulating defect distributions from many wafers and fitting them to the predetermined model has become a powerful tool for defect analysis, from which a special routing technique was developed. To explain special routing, let us first look at Fig. 8.65, which illustrates which particles are left as residuals on the wafer after scanning. We classify these particles as effective and ineffective residuals according to whether they are printed. The residuals that are left on fields already exposed do not contribute to printed defects. These are ineffective residuals, as indicated in the figure. The effective residuals in the figure are particles deposited in the unexposed fields. These are printed subsequently if they remain on the wafer without being stirred up by the moving water. Since the hood is larger than the exposure field, it is possible that residuals are left from more than one neighboring field.

Special routings can overcome this problem. Figure 8.66 shows the normal exposure routing and two special routings. In the normal routing, the fields are exposed row by row and the scan directions alternate between

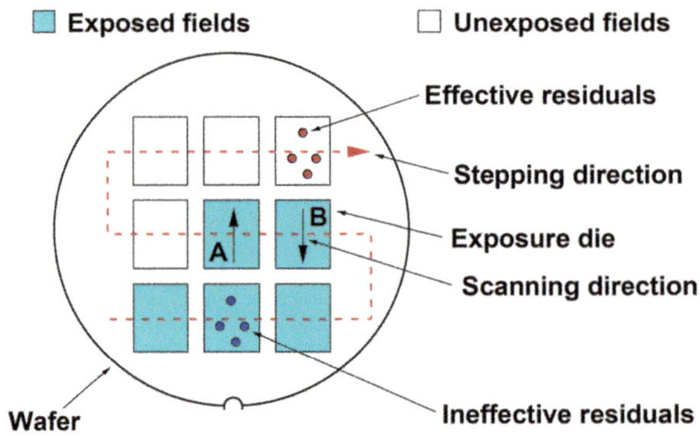

Figure 8.65 Effective and ineffective residuals. The effective residuals are left by field scan B, while the ineffective residuals are left by field scan A.

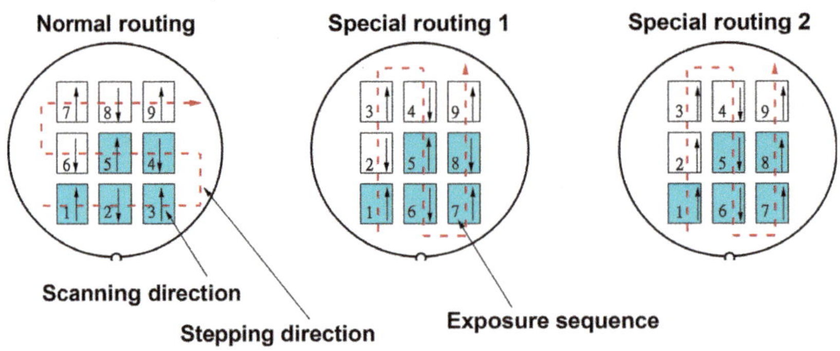

Figure 8.66 Normal routing and two defect-reducing special routings. The normal routing is used for dry scanner production runs, while special routings 1 and 2 are designated for immersion defect reduction.

upward and downward so that the time used to move the reticle back can also be used to expose the wafer for better productivity. The scan directions are perpendicular to the field-stepping directions. Special routing exposes the fields column-wise with the scan directions parallel to the field-stepping directions. Special routing 1 has opposite scan directions alternately assigned in a column, while the scan directions are all parallel to the field-stepping directions in special routing 2.

Figure 8.67 depicts the effective and ineffective residuals generated from normal routing. Right after scanning field 1, residuals are left outside the fields—these are the ineffective residuals. The wafer is now stepped to field 2 and is then scanned in the opposite direction. The residuals are left in their area for future exposures—these are the effective residuals. After all nine fields are scanned, there are three clusters of effective residuals.

Figure 8.68 depicts the effective and ineffective residuals generated from special routing 1. The locations of the effective residual cluster differ as shown. There are also three clusters of effective residuals after scanning all nine fields. The nine fields are used for clarity of illustration. In reality, the number of fields on a 300-mm wafer is on the order of 100. Let us examine whether special routing is less affected by effective residuals, even though Figs. 8.67 and 8.68 show an identical number of clusters. If we add two more rows above fields 7, 8, and 9 in the case of normal routing, i.e., fields 10 to 15, the distribution of effective and ineffective clusters will be identical to that of fields 4 to 9. There will be three additional clusters. Similarly, we can add two

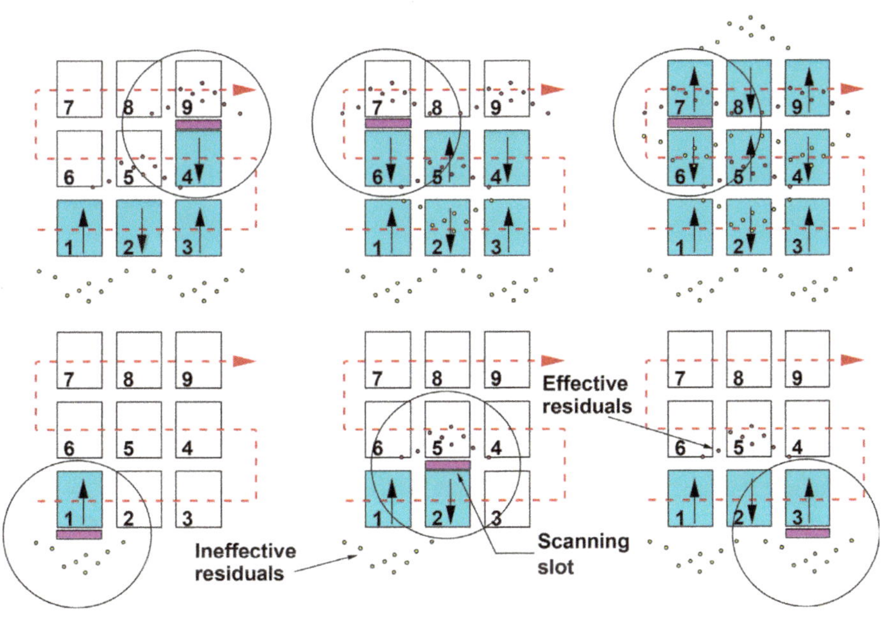

Figure 8.67 Effective and ineffective residuals generated from normal routing.

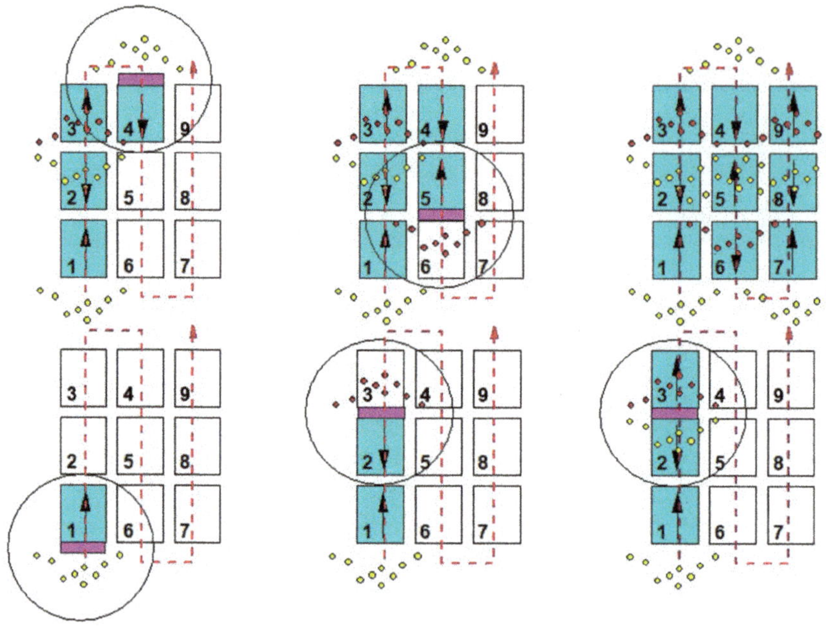

Figure 8.68 Effective and ineffective residuals generated from special routing 1.

more rows above fields, 3, 4, and 9 for special routing 1. There will be fields 3a, 3b, 3c, and 3d that will be scanned between fields 3 and 4. Then, two additional fields will be scanned after field 9. There will still be three additional effective clusters for the two additional rows. The defect-reduction capability of special routing 1 is not obvious. So far, we have only considered the arc trajectory of the defect map. Adding consideration of the tangential trajectory may differentiate the defect characteristic of the two routing schemes.

Figure 8.69 shows the effective and ineffective residuals generated from special routing 2. All residuals are ineffective. We expect this special routing to significantly reduce defects. Comparing two wafer lots of 25 wafers each, the wafers using special routing 2 average 4.8 defects per wafer against 19.7 defects per wafer for the lot without special routing.

To conclude, although special routing confirms the impact of water leakage from the IH, this method reduces the wafer throughput and should not be used in high-volume production.

8.8.4 Other defect-reduction schemes

The schemes presented in the following subsections are feasible for high-volume production.

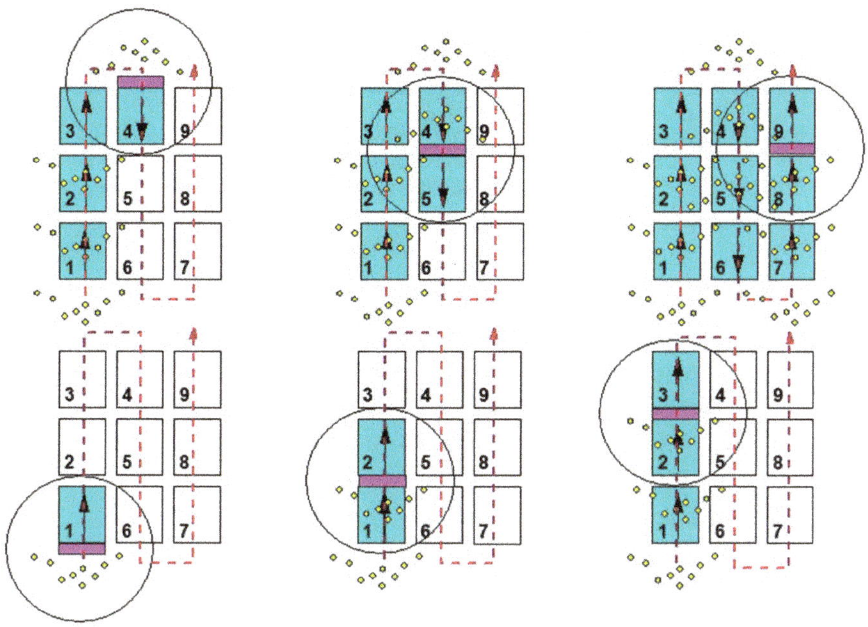

Figure 8.69 Effective and ineffective residuals generated from special routing 2.

8.8.4.1 Wafer and equipment cleanliness

Flowing water can stir up particles that are otherwise dormant in the exposure process. It can carry particles generated by or attached to any surface along its path. Water itself can contain particles from its source. All causes must be painstakingly eliminated. The supplied water must be carefully filtered. All parts in contact with water must be thoroughly cleaned in the assembly process. After the exposure tool is put to service, these surfaces must be periodically cleaned. All semiconductor-processing equipment having water passage requires periodic cleaning to remove particles and other contaminants. Immersion lithographic equipment is no exception. The wafer edge is susceptible to many nonlithographic processes that must be properly addressed to maintain its cleanliness. The wafer back side is usually not kept as meticulously clean as the front side so is particle prone.

8.8.4.2 Wafer seal ring

There is an easy way to isolate back-side particles to prevent them from reaching the front side. A thin, soft material can cover the gap between the wafer edge and the wafer chuck, as shown in Fig. 8.70. The soft material with a thickness of less than 0.1 mm is attached to a thicker frame for support and transportation. The wafer chuck is recessed to allow the seal ring to lie flat at the wafer edge. A vacuum through the gap between the wafer edge and the frame tightens the seal ring against the contacted surfaces. When a

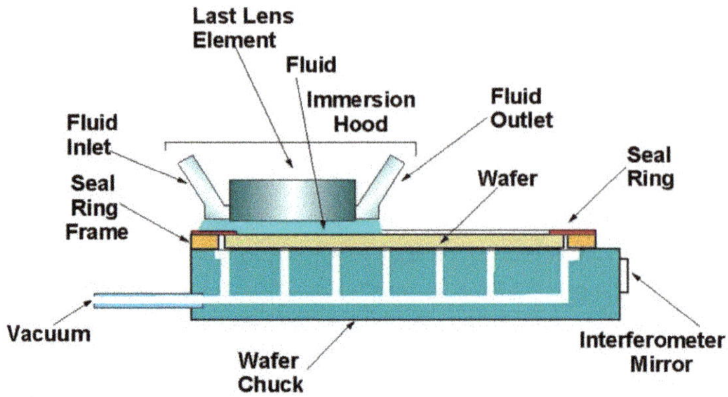

Figure 8.70 Wafer-chuck-supporting edge seal ring on a seal-ring frame placed at the edge of the wafer. The frame and wafer are coplanar.

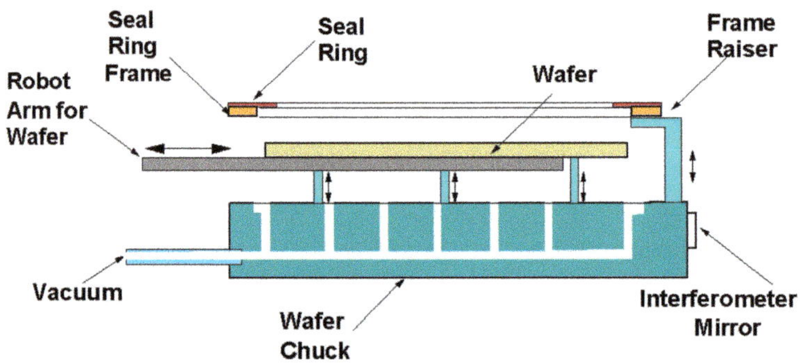

Figure 8.71 Wafer-edge seal ring and seal ring-frame raised above wafer, which is lifted by pins to allow a robot arm to load/unload the wafer to/from the wafer chuck.

lens-based-configuration (LBC) immersion head is at the wafer edge, water does not leak through it, bubble-induced turbulence is eliminated, and particles from the back of the wafer are sealed to prevent them from being stirred up. Figure 8.71 shows the loading and unloading of the wafer with the seal ring and its frame installed. A frame raiser lifts the seal ring and the frame above the wafer loading/unloading position so that normal wafer loading/unloading is unperturbed. In their raised positions, the frame and seal ring can be conveniently replaced.

8.8.5 Results

A defect monitor chart for an ASML 1700i immersion scanner at TSMC is shown in Fig. 8.72. Two resist systems were monitored. The defect characteristics of the two resist systems are quite different. Photoresist type 2 is clearly much preferable to photoresist type 1 in terms of defect count per wafer. Figure 8.73 is a fall-on particle trend chart for the poly layer of the

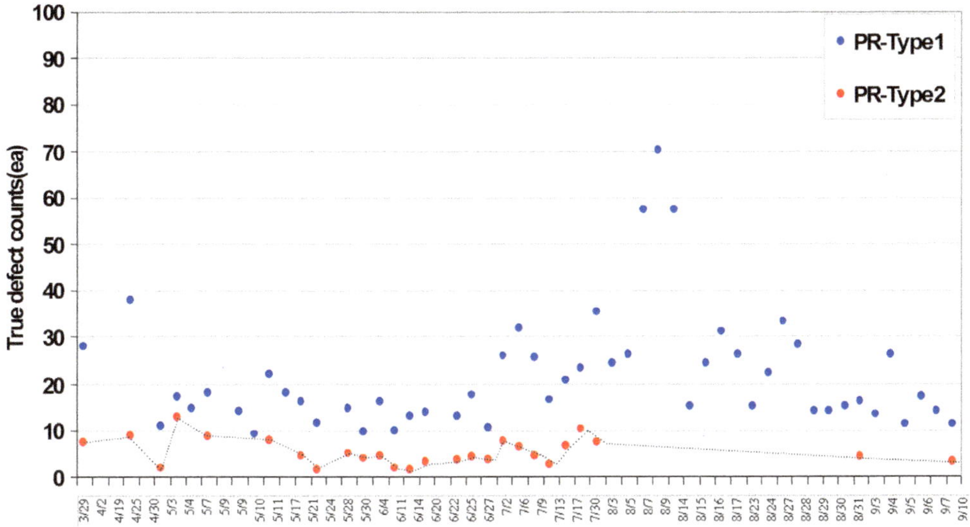

Figure 8.72 Defect monitor chart for the ASML 1700i immersion scanner.

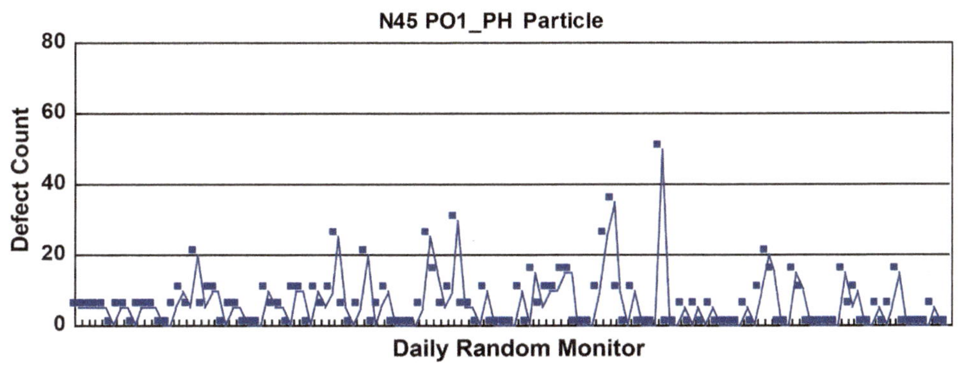

Figure 8.73 Fall-on particle trend chart for the poly layer.

45-nm logic node. Except for some excursions, the defect level is typically in the single digits and is often zero.

When the CMOS circuit is in a quiescent state, the supply current is called IDDQ. The yield and distribution of the quiescent supply current (IDDQ) of the 65-nm product integrated with immersion-imaged active, poly, contact, and metal-1 layers successfully achieved the dry-based production level, using the ASML XT-1400E and 1400I scanners with 0.93-NA dry and immersion lenses, respectively. A comparison of IDDQ distributions is shown in Fig. 8.74. Although immersion lithography is not needed for mass production of 65-nm products, the 65-nm node is a unique node for which dry and immersion scanners are available for comparison of yield and electrical performance. IDDQ testing is a method to test CMOS circuits for

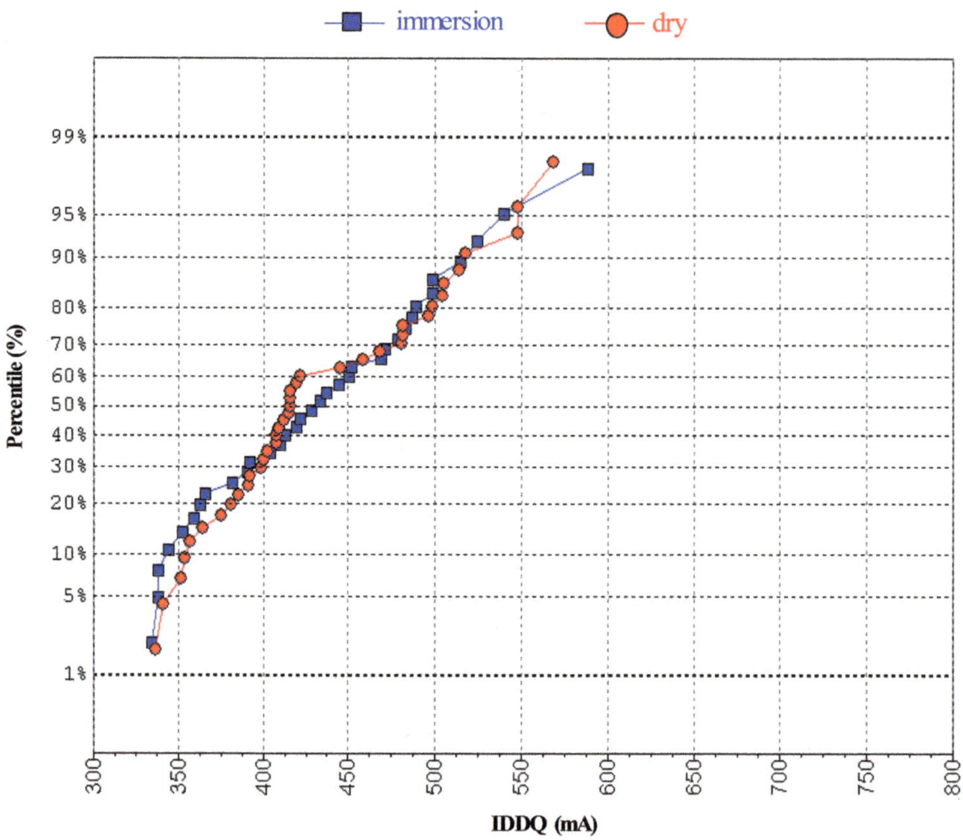

Figure 8.74 IDDQ distributions comparing products using all dry layers to those built with active, poly, contact, and metal-1 layers with immersion lithography (figure courtesy of K.S. Chen).

manufacturing faults. A higher IDDQ forces more shorts; i.e., a lower IDDQ percentile is preferable. In this comparison, the percentiles alternate between dry and immersion fabrication with neither type being overwhelmingly better or worse than the other.

8.9 Extension of Immersion Lithography

A few more techniques can be used with 193-nm water-immersion lithography to extend it beyond the 28-nm half pitch. These extension techniques use higher-refractive-index materials, solid-immersion masks, polarized illumination, and multiple patterning.

8.9.1 High-refractive-index materials

Air with a refractive index close to 1 has been limiting the spatial frequency coupling from the lens to the resist. Water with a refractive index of 1.44 at 193 nm is much better. The theoretical coupling limit sets the maximum NA of water-immersion 193-nm systems to 1.44. However, using a realistic upper limit of $\sin\theta = 0.95$ for the imaging lens, the realistic NA limit is 1.368. To break through this limit, a higher-refractive-index coupling fluid must be used. Figure 8.42(a) depicts the possibilities. Staying at $\sin\theta = 0.95$, a fluid with $n = 1.56$ moves the limit to $NA = 1.482$. Because the refractive index of fused quartz is about 1.56 and that of CaF_2 is lower, an immersion fluid with a refractive index higher than 1.56 cannot support a higher NA with a flat lens bottom. For example, at $n = 1.66$, the maximum realistic NA is still 1.482, but the DOF is larger due to a smaller angle in the fluid. Staying with $n < 1.56$ in the lens material, the only way to increase the NA beyond 1.482 is to bend the last lens surface, as shown in Fig. 8.42(b). Having a nonflat bottom lens surface requires extremely low absorption in the coupling medium. Otherwise, light would be absorbed unevenly in the coupling medium, leading to imaging difficulties. Hydrodynamics also suffers. In order to keep the bottom lens surface flat, higher-refractive-index optical materials are needed.

Although still not ready for practical applications, 193-nm immersion fluids with refractive indices ranging from 1.5 to 1.8 and lens materials with indices ranging from 1.6 to 2.0 were presented at the 2005 Second International Symposium on Immersion Lithography.[51,52,53] These attempts and other later attempts never succeeded. Although increasing the refractive index of the immersion fluid is quite difficult, it is still easier than increasing the refractive index of the lens material. Producing high-index materials of a quality suitable for nanometer imaging requires an investment similar in size to the investment in CaF_2 for 157-nm lithography, except that the required quantity of the high-index is much smaller. Only a thin piece of high-index material is needed on the last element of the imaging lens in comparison to almost the entire lens column needing this material in the former case. Thus, the return on investment is prohibitive. A second reason for not extending ArF H_2O immersion lithography is the DOF. Even though the NA can be increased, DOF reduction inevitably follows.

8.9.2 Solid-immersion masks

CD at 32 nm is subresolution on the mask. Even at 4X magnification, this is only 0.663λ for the 193-nm ArF wavelength. When subresolution assist features are used, they can be as small as 0.166λ on the mask. There are also subresolution jigs and jugs introduced by optical proximity correction (OPC). In addition, the absorber thickness is not negligible. A 70-nm thickness is 0.36λ. These 3D subwavelength features react to the illumination nonlinearly and can lead to imaging difficulties, not to mention challenges in making masks that contain very small features.

Instead of using a larger magnification at the mask, a solid immersion mask can be used to delay the onset of the 3D subwavelength effects. The solid immersion mask is depicted in Fig. 8.75. A high-refractive-index

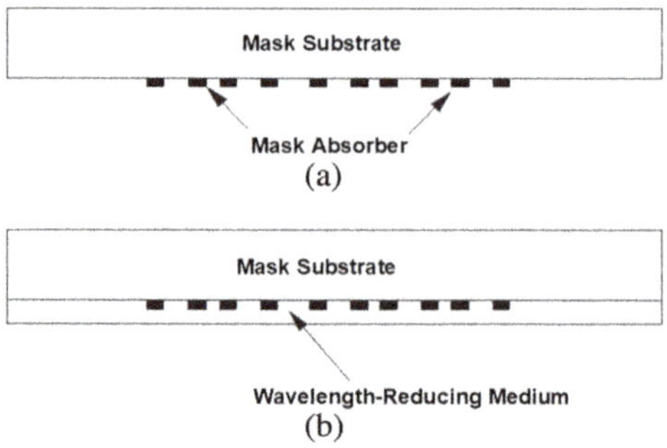

Figure 8.75 (a) A typical binary-intensity mask (BIM) and (b) a mask immersed with a wavelength-reducing solid medium.

transparent material planarizes over the 3D absorber structure on the mask. This way, the illuminating wavelength is reduced according to the refractive index of the planarizer. Reduction up to 70 ~ 80% is possible. Even though the wavelength is restored after light leaves the planarizer, the subwavelength diffraction zone is already taken care of.

In addition to having a high refractive index, the solid immersion material must be transparent to the imaging light. Its optical property should be consistent after repeated exposures. The material must be coated using a planarizing technique. If coated otherwise, such as using a conformal coating method, the material needs to be polished for planarization. The coating process should not trap any particulates or bubbles that could imbed defects on the mask. The coating should be strippable without damaging the mask. A resist resin would be a good starting candidate. Figure 8.76 shows the improvement in the measured mask error enhancement factor (MEEF) of an uncoated mask that was consecutively coated with 200 nm and 500 nm of a high-index material. The CD is 65 nm over a pitch range of 180 nm to 1 μm. Here MEEF is improved from 3.2 to 2.5 and DOF is improved from 220 to 270 nm for the 180-nm-pitch poly test patterns using the 500-nm-thick planarizer. The 200-nm-thick planarizer helps slightly less.

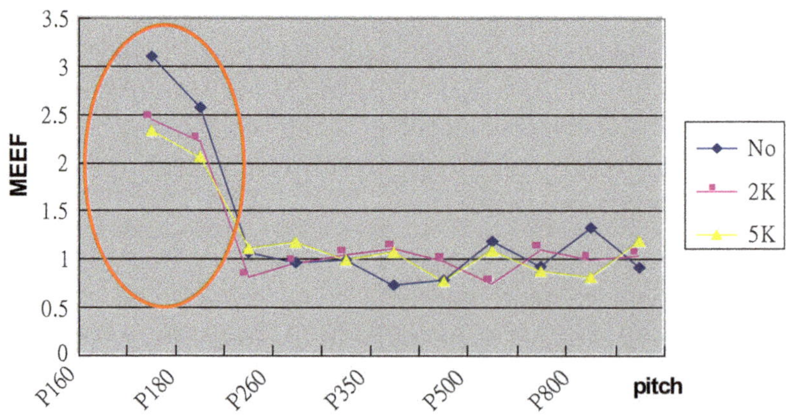

Figure 8.76 MEEF from an uncoated mask, a mask coated with 200 nm of high-index material, and a mask coated with 500 nm of high-index material (figure courtesy of C.H. Lin).

Table 8.13 Poly-layer exposure latitude (ELAT) and DOF gain with polarization, a 32-nm line, and 90-nm and 113-nm pitches.

AttPSM						
OAI	**NA**	**Sigma**	**ELAT**	**$DOF_{diffrac}$**	**DOF_{avail}**	**Improvement**
Unpolarized	1.35	0.96:0.48	5%	137	55	—
TE	1.3	0.96:0.48	5%	169	87	23/58%
Unpolarized	1.25	0.96:0.48	2.43%	200	118	—
TE	1.25	0.96:0.48	3.57%	200	118	47%

8.9.3 Polarized illumination

Polarization and its impact on optical imaging in high-NA dry and immersion systems is discussed in detail in Section 8.5. Its impact on high-NA systems is significant. As an example, Table 8.13 lists the gain in DOF_{diff} and DOF_{avail}[13] with TE-polarized illumination at 5% EL for the common exposure-defocus (E-D) window of 32-nm lines and 90-nm and 113-nm pitches, using an attenuated phase-shifting mask (AttPSM) and off-axis annular illumination (OAI). The gain in EL (or ELAT) with DOF_{diff} kept constant at 200 nm is also shown. The gains are 23% in DOF_{diff}, 58% in DOF_{avail}, and 47% in ELAT. All immersion systems should be equipped with polarization illuminations that can be customized to as many nanometer IC designs as possible.

8.9.4 Multiple patterning

The topic of multiple patterning is addressed in detail in Section 7.5. This method consists of pitch splitting by splitting the mask pattern, pitch splitting by sidewall imaging, end cutting, and pack-and-unpack. These are powerful techniques for extending lithography that can be used to extend immersion lithography as well.

8.10 Conclusion

Water immersion for 193-nm lithography was proposed by this author in 2002 to avoid the material difficulties of the well-supported 157-nm dry lithography. Not only did this prevent all of the technical difficulties and financial waste involved in developing 157-nm lithography, but the resolution was also substantially better than 157-nm lithography. At that time, the industry was at the 130-nm node, looking to manufacture the 90-nm node in about two years. After the 65-nm node and the 55-nm half-node, the remaining nodes up to the 7-nm node were manufactured with immersion lithography.

Table 8.14 shows the nodes that can be handled with each wavelength. Initially, at the 5000-nm node, IC fabrication uses proximity printing, which is

Table 8.14 Nodes per wavelength.

Nodes per wavelength						
Wavelength	**Node (nm)**					
Proximity	5000	3000				
All-reflective	5000	3000	2000	1500		
436 nm	3000	2000	1500	1000	700	
365 nm	700	500	350	250		
248 nm	250	180	130	90		
193 nm	90	65				
193 nm imm.	40	28	20	16	10	7
13.5 nm	7					

polychromatic. At around the 3000-nm node, projection printing using all-reflective optics replaced proximity printing, but it was still polychromatic. Using the g-line of the mercury arc lamp, 436-nm step-and-repeat projection printing becomes desirable at 1500 nm, then at 1000 nm. It cannot quite handle the 700-nm node. Thus, projection printing has two heavily used nodes, 1500 nm and 1000 nm. 365-nm systems are most effective for 700, 500, and 350 nm. A drastic change of the light source and resist was required to switch to 248 nm, which was used for the 250-, 180-, 130-, and 90-nm nodes. The 248-nm wavelength sustained an impressive 4 nodes, although the 90-nm node was a stretch for the most critical masking layers. The 193-nm dry system was used for the 90- and 65-nm nodes. Then it was quickly switched to 193-nm immersion, which carried semiconductor manufacturing through the 45-, 40-, 32-, 28-, 20-, 16-, 14-, 10-, and 7-nm nodes. In total, six full nodes and three subnodes can be handled with 193-nm immersion. This is more nodes than any other wavelengths could handle in the past and very likely also in the future. Even with MPT, immersion lithography continues to be the cost champion when compared with its challengers.

Regardless of Moore's scaling law, semiconductor production volume will continue to increase. The value of wafers manufactured with immersion lithography will always stand out. For example, the number of 12-inch wafers produced with immersion lithography was estimated to be 150 million in 2017 at an estimated value of US$75B. Immersion lithography, along with all of the RETs that make it work for the 7-nm node, needs to be mastered, if not further extended.

The next chapter explores a potential successor to immersion lithography and examines the cost and the infrastructure needed for its success.

References

1. B. J. Lin, "The future of subhalf-micrometer optical lithography," *Microelectronic Engineering* **6**(1–4), pp. 31–51, (1987).

2. B. J. Lin, "Drivers, prospects, and challenges for immersion lithography," Third International Symposium on 157 nm Lithography, held in Antwerp by IMEC, International SEMATECH, and Selete (2002).
3. S. Owa and H. Nagasaka, "Immersion lithography: its history, current status and future prospects," *Proc. SPIE* **7140**, 714015 (2008) doi: [10.1117/12.804709].
4. B. J. Lin, "The k_3 coefficient in nonparaxial λ/NA scaling equations for resolution, depth-of-focus, and immersion lithography," *J. Micro/Nanolith., MEMS, and MOEMS* **1**(1), pp. 7–12 (2002) [doi: 10.1117/1.1445798].
5. H. Kawata, J. M. Carter, A. Yen, and H. I. Smith, "Optical projection lithography using lenses with numerical apertures greater than unity," *Microelectronic Engineering* **9**(1–4), pp. 31–36 (1989).
6. J. A. Hoffnagle, W. D. Hinsberg, M. Sanchez, and F. A. Houle, "Liquid immersion deep-ultraviolet interferometric lithography," *J. Vac. Sci. Technol. B* **17**(6), pp. 3306–3309 (1999).
7. M. Switkes and M. Rothschild, "Immersion lithography at 157 nm," *J. Vac. Sci. Technol. B* **19**, pp. 2353–2356 (2001).
8. M. Switkes and M. Rothschild, "Resolution enhancement of 157 nm lithography by liquid immersion," *J. Micro/Nanolith., MEMS, and MOEMS* **1**(3), pp. 225–228 (2002) [doi: 10.1117/12/474568].
9. D. G. Flagello and A. E. Rosenbluth, "Lithographic tolerances based on vector diffraction theory," *J. Vac. Sci. Technol. B* **10**, pp. 2997–3003 (1992).
10. D. Flagello, T. Milster, and A. Rosenbluth, "Theory of high-NA imaging in homogeneous thin films," *J. Opt. Soc. Am. A* **13**, pp. 53–64 (1996).
11. T. A. Brunner, N. Seong, W. D. Hinsberg, J. A. Hoffnagle, F. A. Houle, and M. I. Sanchez, "High-NA lithographic imagery at Brewster's angle," *J. Micro/Nanolith., MEMS, and MOEMS* **1**(3), pp. 188–196 (2002) [doi: 10.1117/12.474473].
12. B. J. Lin, "Immersion lithography and its impact on semiconductor manufacturing," *J. Micro/Nanolith., MEMS, and MOEMS* **3**(3), pp. 377–395 (2004) [doi: 10.1117/12.534507].
13. B. J. Lin, "Depth of focus in multilayered media—a long-neglected phenomenon aroused by immersion lithography," *J. Micro/Nanolith., MEMS, and MOEMS* **3**(1), pp. 21–27 (2004) [doi: 10.1117/1.1637591].
14. V. N. Mahajan, *Optical Imaging and Aberrations, Part 1, Ray Geometrical Optics*, SPIE Press, p. 318 (1998) [doi: 10.1117/3.265735].
15. B. J. Lin, "Exposure-defocus forest," *Jap. J. Appl. Phys.* **33**, p. 6756 (1994).
16. B. J. Lin, "Partially coherent imaging in two dimensions and the theoretical limits of projection printing in microfabrication," *IEEE Trans. Electron. Dev.* **ED-27**, pp. 931–938 (1980).

17. C. A. Mack, "Focus effects in submicron optical lithography, part 4: metrics for depth of focus," *Proc. SPIE* **2440**, pp. 458–471 (1995) [doi: 10.1117/12.209276].
18. D. G. Flagello, H. van der Lann, J. B. P. van Schoot, I. Bouchoms, and B. Geh, "Understanding systematic and random CD variations using predictive modeling techniques," *Proc. SPIE* **3679**, p. 162 (1999) [doi: 10.1117/12.354328].
19. S. Inoue, M. Itoh, M. Asano, and K. Okumura, "Desirable reticle flatness form focus deviation standpoint in optical lithography," *J. Micro/Nanolith., MEMS, and MOEMS* **1**(3), p. 307 (2002) [doi: 10.1117/12.474600].
20. S. S. Sethi, A. Flores, P. McHale, R. Booth, S. W. Graca, S. Frezon, and C. Fruga, "Improving the backend focus budget for 0.5μm lithography," *Proc. SPIE* **2440**, p. 633 (1995) [doi:10.1117/.209290].
21. B. W. Smith and J. S. Cashmore, "Challenges in high NA, polarization, and photoresists," *Proc. SPIE* **4691**, p. 11 (2002) [doi: 10.1117/12.474562].
22. B. J. Lin, "Simulation of optical projection with polarization-dependent stray light to explore the difference between dry and immersion lithography," *J. Micro/Nanolith., MEMS, and MOEMS* **3**(1), pp. 9–20 (2004) [doi: 10.1117/1.1636769].
23. B. J. Lin, "Off-axis illumination—working principles and comparison with alternating phase-shifting masks," *Proc. SPIE* **1927**, p. 89 (1993) [doi: 10.1117/12.150417].
24. D. Gabor, "Light and information" in *Proc. Symposium on Astronomical Optics and Related Subjects*, Z. Kopal, Ed., North-Holland Publ., p. 17 (1956).
25. D. Gabor, "Optical Transmission," in *Information Theory*, C. Cherry, Ed., Butterworths Scientific Publ., p. 26 (1956).
26. M. Born and E. Wolf, *Principles of Optics, 6th Edition*, Cambridge University Press, p. 40 (1998).
27. Page 43 of Ref. 26.
28. B. J. Lin, "Electromagnetic near-field diffraction of a medium slit," *J. Opt. Soc. Am.* **62**(8), pp. 977–981 (1972).
29. B. J. Lin, "Signamization," *Proc. SPIE* **2726**, p. 71 (1996) [doi: 10.1117/12.240965].
30. J. Mulkens, D. Flagello, B. Streefkerk, and P. Graeupner, "Benefits and limitations of immersion lithography," *J. Micro/Nanolith., MEMS, and MOEMS* **3**(1), pp. 104–114 (2004) [doi: 10.1117/1.636768].
31. S. Owa and H. Nagasaka, "Advantage and feasibility of immersion lithography," *J. Micro/Nanolith., MEMS, and MOEMS* **3**(1), pp. 97–103 (2004) [doi: 10.1117/1.1637593].
32. B. J. Lin, "Semiconductor foundry, lithography, and partners," *Proc. SPIE* **4688**, 11 (2002) [doi: 10.1117/12.472292].

33. J. H. Burnett and S. G. Kaplan, "Measurement of the refractive index and thermo-optic coefficient of water near 193 nm," *J. Micro/Nanolith, MEMS, and MOEMS* **3**(1), pp. 68–72 (2004) [doi: 10.1117/1.1632501].
34. A. Suzuki, "Immersion lithography update," in the Second International Sematech Immersion Lithography Workshop, July 2003.
35. Private communications with Akiyoshi Suzuki.
36. J. Ishikawa, T. Fujiwara, K. Shiraishi, Y. Ishii, and M. Nei, "Latest results from the hyper-NA immersion scanners S609B and S610C," *Proc. SPIE* **6520**, 65201W (2007) [doi: 10.1117/12.712042].
37. J. de Klerk, C. Wagner, R. Droste, L. Levasier, L. Jorritsma, E. van Setten, H. Kattouw, J. Jacobs, and T. Heil, "Performance of a 1.35NA ArF immersion lithography system for 40-nm applications," *Proc. SPIE* **6520**, 65201Y (2007) [doi: 10.1117/12.712094].
38. R. Garreis, B. Kneer, P. Graupner, H. Feldmann, W. Kaiser, and T. Heil, "Catadioptric optics enabling ultra-high NA lithography," Third International Symposium on Immersion Lithography, Kyoto, 4 October (2006).
39. T. Matsuyama, Y. Ohmura, and D. M. Williamson, "The lithographic lens: its history and evolution," *Proc. SPIE* **6154**, 615403 (2006) [doi: 10.1117/12.656163].
40. P. L. Marston, "Light scattering from bubbles in water," *Proc. Oceans '89* **4**, pp. 1186–1193 (1989).
41. T. S. Gau, C. K. Chen, and B. J. Lin, "Image characterization of bubbles in water for 193-nm immersion lithography," *J. Micro/Nanolith., MEMS, and MOEMS* **3**(1), pp. 61–67 (2004) [doi: 10.1117/1.1630602].
42. B. J. Lin, "4X / 5X mask considerations for the future," ASML 157nm Users' Forum, September (2000).
43. B. J. Lin, H. T. Lin, and H. C. Hsieh, "A device and method for providing wavelength reduction with a photomask," U.S. Patent 20110244378A1 (2011).
44. The first demonstration of immersion exposure on a wafer was by J. Mulkens of ASML in 2003.
45. T. Hirayama, "Resist and cover material investigation for immersion lithography," Second International Sematech Immersion Workshop, Almaden, California, July (2003).
46. A. Otoguro, J. Santillan, T. Itani, K. Fujii, A. Kagayama, T. Nakano, N. Nakayama, H. Tamatani, and S. Fukuda, "Development of high refractive index fluids for 193nm immersion lithography," Second International Immersion Symposium, Bruges, September (2005).
47. Private communication with C.Y. Chang.
48. J. H. Chen, L. J. Chen, T. Y. Fang, T. C. Fu, L. H. Shiu, Y. T. Huang, N. Chen, D. C. Oweyang, M. C. Wu, S. C. Wang, J. C. H. Lin, C. K. Chen, W. M. Chen, T. S. Gau, B. J. Lin, R. Moerman, W. Gehoel-van Ansem, E. van der Heijden, F. de Johng, D. Oorschot, H. Boom,

M. Hoogendorp, C. Wagner, and B. Koek, "Characterization of ArF immersion process for production," *Proc. SPIE* **5754**, pp. 13–22 (2005) [doi: 10.1117/12/602025].

49. C. H. Lin, S. C. Wang, K. S. Chen, C. Y. Chang, T. C. Wu, M. C. Wu, M. T. Lee, J. H. Chen, S. W. Chang, Y. S. Yen, Y. H. Chang, T. C. Fu, T. S. Gau, and B. J. Lin, "193-nm immersion lithography for 65-nm and below," Second International Symposium on Immersion Lithography, Bruges, September (2005).
50. F. J. Liang, H. Chang, L. H. Shiu, C. K. Chen, L. J. Chen, T. S. Gau, and B. J. Lin, "Immersion defect reduction, part 1: analysis of water leaks in an immersion scanner," *Proc. SPIE* **6520**, 652012 (2007) [doi: 10.1117/12.712531].
51. S. Peng et al., "New developments in second generation 193nm immersion fluids for lithography with 1.5 numerical aperture," Second International Symposium on Immersion Lithography, Bruges, September (2005).
52. Y. Wang et al., "Material design for highly transparent fluids of the next generation ArF immersion lithography," Second International Symposium on Immersion Lithography, Bruges, September (2005).
53. Y. Inui et al., "Fluoride single crystals grown by the CZ method," Second International Symposium on Immersion Lithography, Bruges, September (2005).

Chapter 9
EUV Lithography

9.1 Introduction

Optical lithography in the regime of $k_1 = 0.28$ is very complicated. The image contrast is low and, except for 1D single-pitch patterns with special illuminations, the mask error enhancement factor (MEEF) is high. For contact hole patterning, the MEEF can be more than 4, which negates the gain of a 4X reduction system. Line ends can have MEEF as high as 10. In addition, the shape of patterns at low k_1 is rounded out and requires a significant amount of optical proximity correction to keep the patterns usable. Extreme-UV lithography[1,2] (EUVL) previously used the 13.4-nm wavelength. Recently, this technology migrated to 13.5 nm. Either wavelength offers an order of magnitude reduction in wavelength from the water-immersion ArF wavelength of 134 nm. This presents an opportunity to bring k_1 back to above 0.5. It is no wonder that EUV lithography has attracted large research and development efforts worldwide. The size of the EUV development effort has dwarfed two other well-known lithography developments, namely, 157-nm lithography and x-ray proximity printing. Take the 32-nm half pitch. At the water-immersion ArF wavelength of 134 nm and $\sin\theta = 0.95$, k_1 is 0.227. Turning to EUV at $\sin\theta = 0.25$, k_1 becomes 0.597. Initially, the term soft x-ray[3] was used to identify this lithography. However, the use of UV light has had an impressive history of success compared to x-ray lithography. It makes sense to associate the technology with the winning wavelength range, thus it is called EUV lithography. Since the publication of Ref. 1 in 1989, more than three decades have passed and many challenges have been met. EUVL for production of the 7-nm node has been demonstrated.[4] High-volume manufacturing is expected to follow. It will be worthwhile to track the number of nodes by which the three decades of development efforts can extend semiconductor scaling.

With such a dramatic drop in wavelength, the imaging system is substantially different from existing systems. First, EUV light is heavily absorbed by any substance, including gases. The optical path must be in vacuum. Second, because of EUV light's strong absorption characteristics, there is no transmitting material—EUV optical imaging depends on reflection.

However, the reflectivity is low, on the order of about 1%. Reflectivity is built up with a multilayer stack consisting of 40- to 50-layer pairs of Mo and Si, to achieve 65 ~ 70% reflectivity, enabling an all-reflective optical system that includes the illumination optics, the mask, and the imaging optics. Figure 9.1 shows an EUV imaging system developed at the Sandia National Laboratories and Lawrence Livermore National Laboratory.[5,6] A YAG laser beam hits a Xe-cluster target to generate the 13.4-nm light. The light is collected by the nearby collector and goes through the pupil optics, imaging optics, filter, and relay optics to illuminate the mask. Due to the lack of a beam splitter, the incident and reflection beams are oblique, making an angle on the order of 6 deg with respect to the direction perpendicular to the mask. The mask pattern undergoes four reflections in the imaging lens. The imaging light passes through a circular blocking slit to expose the resist on the wafer. The NA of this system is 0.1.

Zeiss developed a 0.25-NA system for the ASML EUV alpha and prototype tools. Later the 0.33-NA system was developed and has become the primary tool intended for high-volume manufacturing. Figure 9.2 shows the EUV source for the ASML NXE3100 0.25-NA prototype system.[7] Kilowatts of IR light from the CO_2 drive laser go through the beam delivery tube into the vacuum vessel, in which there is a droplet generator that sends liquid Sn droplets at a rate of tens of kilohertz. These droplets are hit by CO_2 laser pulses to generate a plasma for production of EUV light. A collector in the vacuum vessel collects the dispersing EUV light and directs it to the intermediate focus

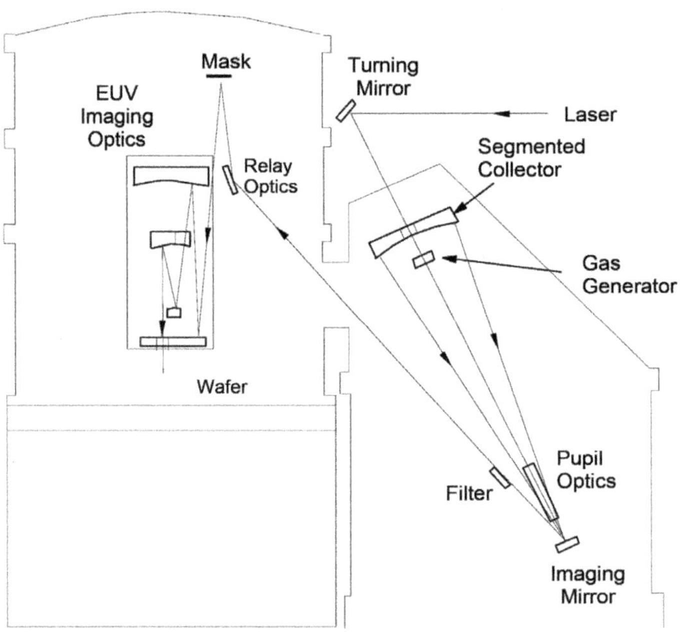

Figure 9.1 EUVL system developed at the Sandia National Laboratories and Lawrence Livermore National Laboratory (reprinted from Ref. 6).

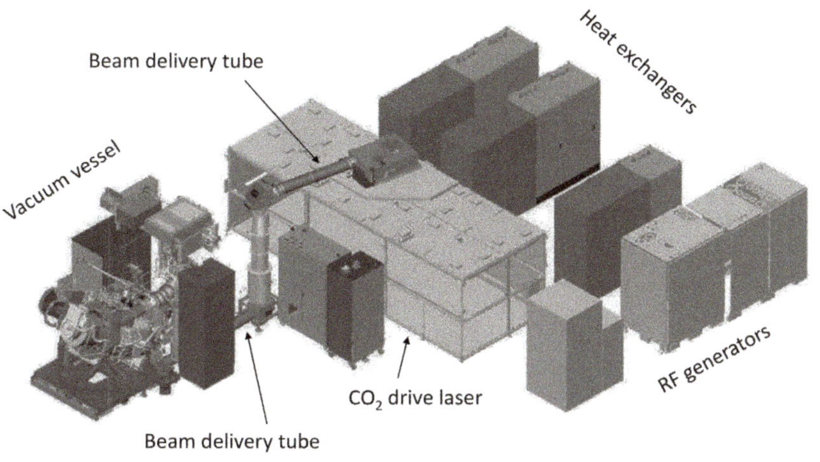

Figure 9.2 The EUV source for the ASML NXE3100 prototype system (adapted from Ref. 53).

(IF) such that it enters the scanner module shown in Fig. 9.3. The EUV light in Fig. 9.2 exits the vacuum vessel to the left to enter the scanner module seen on the left-hand side of Fig. 9.3. Figure 9.2 also shows the heat exchangers and RF generators that support the CO_2 drive laser.

Both the 0.25-NA and the 0.33-NA lenses consist of six reflecting surfaces. The optical layout of the latter[8] is shown in Fig. 9.4. Together with the two mirrors in the illuminator shown in Fig. 9.5, the reflective mask, and a light-collecting mirror, there are 10 normally incident reflecting surfaces.

Like an optical scanner, the imaging field takes the form of a slit that has a length of 26 mm. The entire 26 × 33-mm^2 field is filled by scanning. Unlike the straight slot of an optical scanner, a curved slit is used. A circular slit needs more travel distance for the beginning and ending of the scanning compared

Figure 9.3 An ASML EUV scanner module (reprinted with permission from ASML).

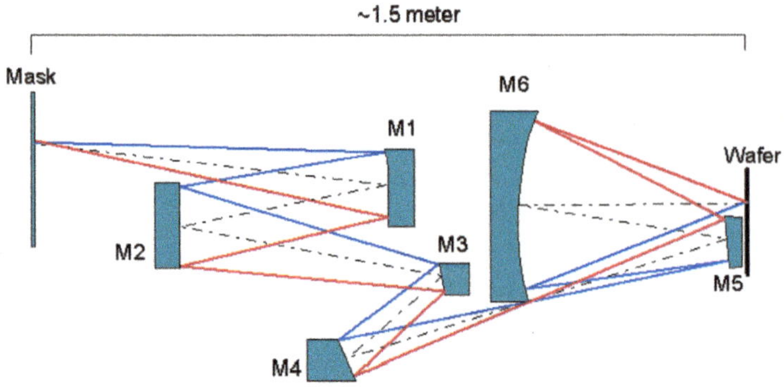

Figure 9.4 EUVL 0.33-NA projection optics developed by Zeiss for ASML (adapted from Ref. 7).

to a straight slot. A more important consideration is that the curved slit fields are taken from a ring field that has the best imaging only at an infinitesimal ring width. Imaging at other ring positions outside or inside of the ring is progressively worse as a function of the distance from the ring. The straight

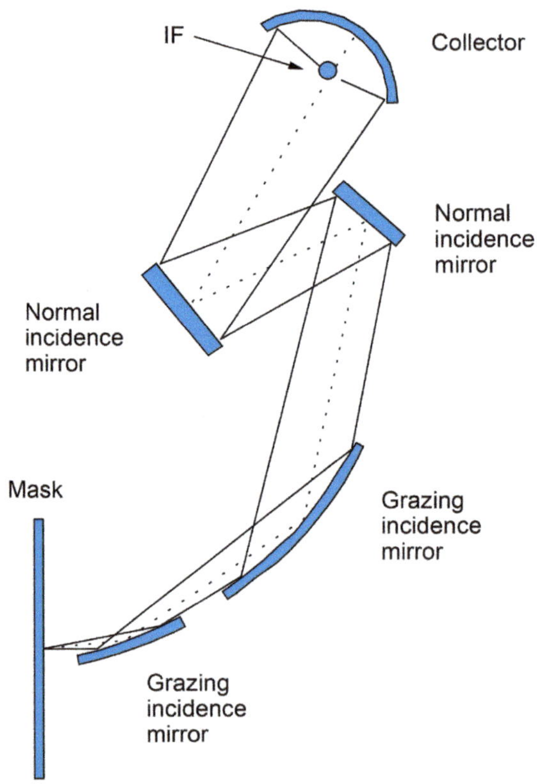

Figure 9.5 Example of an EUV illuminator for the ASML alpha tool. Grazing-incidence mirrors are used for their high reflectivity (adapted from Ref. 54).

slot field is taken from a large circular field, as in a scanner lens. The image quality should be uniformly good. Even after taking aberrations into consideration, the uniformity of the imaging quality of such a slot field should still be much better than that from a slit field.*

The illuminator used in the ASML EUV alpha tool is shown in Fig. 9.5. The collector reflects the EUV light at the IF such that the light goes through the illuminator optics (consisting of two normal-incidence mirrors and two grazing-incidence mirrors) to reach the mask. This simplified illuminator is used here to show a typical number of mirrors in an illumination system. Production-type illuminators that need to enable OAI, source optimization, and source mask optimization, as well as to control the partial coherence, are much more complicated, just like the DUV illuminator. Examples of some illuminators can be found in Section 5.5.3 of Ref. 19.

9.2 EUV Source

The EUV source has been a key bottleneck of EUVL for high-volume manufacturing; the power level has been insufficient to support a viable throughput. After an enormous investment of billions of dollars by suppliers and potential users that involved the research of thousands of individuals, engineering prowess combining talents in many disciplines, management and marketing skills, will power, and vision, the EUV source can now support an acceptable but not exciting throughput. The path to further increasing the source power is now visible, but this high power brings additional challenging problems concerning power grid loading, heat removal, and environmental impact. Before discussing the working principle of the source, let's first consider the source power requirements and impacts. Because there is no optical layout of the illuminator, we use the number of collectors and mirrors in the illumination module of the alpha tool.[54]

9.2.1 Source power requirement

An EUV scanner needs to be economical for manufacturing semiconductors. It is anticipated that the scanner itself will cost many times more than a 193-nm immersion scanner. The space required for the EUV source is also many times larger. The raw power required and the consumables such as water and gases are in much larger quantities. Therefore, to avoid cost escalation, it is very important to make the throughput of EUV scanners much higher than that of conventional scanners. This means economically increasing the source power and transmission of the optical system, overcoming the limits in scan speeds, or increasing the resist sensitivity.

*In a step-and-scan exposure tool, we use "slit" to denote a curved field and "slot" for a straight field.

Let us assume that S mJ/cm^2 is needed to expose the resist, where S is a variable to be substituted for the actual sensitivity of the resist in mJ/cm^2. There is a 700-cm^2 area to expose on a 300-mm wafer, and 65% of the time each wafer spends in the scanner is for exposure. Each wafer will spend 36 sec attaining a 100-wph throughput (where wph is wafers per hour); thus, 700 cm^2/65%/36 sec $\times$ S mJ/cm^2 = 30 S mW/cm^2 of in-band EUV power is needed per wafer.

The EUV powers incident on and absorbed by each EUV optical component are shown in Table 9.1. These values are marked in Fig. 9.6 for key components in a composite optical layout from Figs. 9.4 and 9.5.

According to Table 9.1, to reach 0.03S W, the EUV in-band light generated at the source and spreading out to 2π steradian (sr) is 12.87S W. The collector mirror sustains a solid angle that is 70% of 2π. Its reflectivity near the optical axis is assumed to be 67%, identical to the other normally incident multilayer coated mirrors. Farther away from the axis, the reflectivity is less because of a less favorable incident angle. This loss is assumed to be 20%. There are two normally incident mirrors for the illuminator and six for the projection optics. The mask reflectivity is also assumed to be 67%. The reflectivity of each grazing-incidence mirror[9] is taken to be 80%. Although not specified in Ref. 9, according to Ref. 10, the integrator in the illuminator (the integrator is not shown in Fig. 9.5 but should be included in any scanner

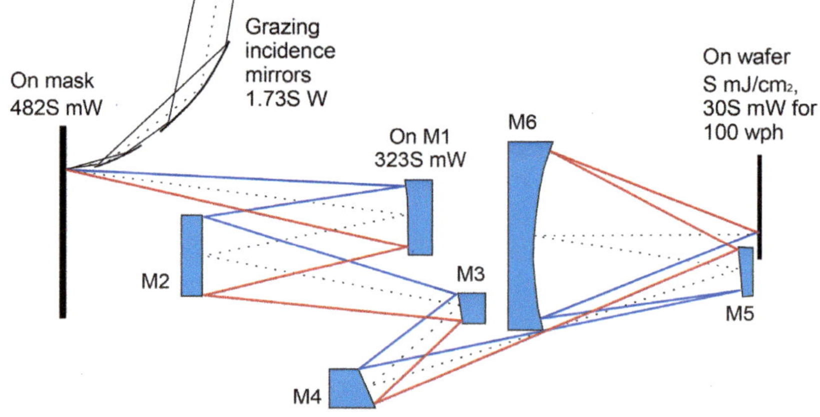

Figure 9.6 EUV powers indicated for the components in an optical layout combining the illuminator of Fig. 9.5 and the projection optics of Fig. 9.4.

Table 9.1 Power incident on and absorbed by each EUV optical component (NI is normal incidence, GI is grazing incidence, and BW is bandwidth).

Optical component	% of reflection or transmission	Power incident on (W)	Power absorbed (W)	Power survived (W)	Power absorbed (W)	Power survived (W)	Power absorbed (W)	Power survived (W)	Power absorbed (W)	Power survived (W)	Power absorbed (W)
Resist sensitivity (mJ/cm^2)		1		20		30		50		80	
EUV power from the Sn plasma (W)	12.866	12.866		257.320		385.980		643.300		1029.280	
Collection efficiency by area	70%	9.006	3.860	180.124	77.196	270.186	115.794	450.310	192.990	720.496	308.784
Collector reflectivity	54%	4.827	4.179	96.546	83.578	144.820	125.366	241.366	208.944	386.186	334.310
Incident on 1st NI mirror	67%	3.234	1.593	64.686	31.860	97.029	47.790	161.715	79.651	258.745	127.441
Incident on 2nd NI mirror	67%	2.167	1.067	43.340	21.346	65.010	32.020	108.349	53.366	173.359	85.386
Incident on 1st GI mirror	80%	1.734	0.433	34.672	8.668	52.008	13.002	86.679	21.670	138.687	34.672
Incident on 2nd GI mirror	80%	1.387	0.347	27.737	6.934	41.606	10.402	69.344	17.336	110.950	27.737
Incident on light integrator	64%	0.888	0.499	17.752	9.985	26.628	14.978	44.380	24.964	71.008	39.942
Before path attenuation in illuminating path	90%	0.799	0.089	15.977	1.775	23.965	2.663	39.942	4.438	63.907	7.101
Before BW mismatch in illuminating path	90%	0.719	0.080	14.379	1.598	21.569	2.397	35.948	3.994	57.516	6.391
Incident on mask	67%	0.482	0.237	9.634	4.745	14.451	7.118	24.085	11.863	38.536	18.980
Incident on 1st mirror	67%	0.323	0.159	6.455	3.179	9.682	4.769	16.137	7.948	25.819	12.717
Incident on 2nd mirror	67%	0.216	0.107	4.325	2.130	6.487	3.195	10.812	5.325	17.299	8.520
Incident on 3rd mirror	67%	0.145	0.071	2.898	1.427	4.346	2.141	7.244	3.568	11.590	5.709
Incident on 4th mirror	67%	0.097	0.048	1.941	0.956	2.912	1.434	4.853	2.390	7.765	3.825
Incident on 5th mirror	67%	0.065	0.032	1.301	0.641	1.951	0.961	3.252	1.602	5.203	2.563
Incident on 6th mirror	67%	0.044	0.021	0.871	0.429	1.307	0.644	2.179	1.073	3.486	1.717
Before path attenuation in imaging path	90%	0.039	0.004	0.784	0.087	1.176	0.131	1.961	0.218	3.137	0.349
Before BW mismatch in imaging path	90%	0.035	0.004	0.706	0.078	1.059	0.118	1.765	0.196	2.824	0.314
Through membrane	85%	0.030	0.005	0.600	0.106	0.900	0.159	1.500	0.265	2.400	0.424
On resist	1	0.030	0.030	0.600	0.600	0.900	0.900	1.500	1.500	2.400	2.400
Total power absorbed (W)		12.866		257.320		385.980		643.300		1029.280	
Total power absorbed other than resist, i.e., in optical components (W)		12.836		256.720		385.080		641.800		1026.880	

illuminator) most likely consists of two multifaceted mirrors with 80% reflectivity each. We also incorporate 10% loss for path length attenuation and another 10% loss for bandwidth mismatch in the illuminator as well as in the projection optics, as estimated in Ref. 10. A membrane is inserted between the last lens surface and the resist to filter out IR light that is simultaneously produced by the EUV-light-generating plasma, and to block any outgassing from the resist that can contaminate the optical elements. The membrane's transmissivity is estimated to be 85%.

Using an EUV resist sensitivity of 20 mJ/cm^2 leads to 257 W at the IF, according to the foregoing calculations. This is close to the 250 W given by Fomenkov et al.[11] Initially, 5 mJ/cm^2 was set by the developers of EUVL technology to project an appealing throughput. The target resist sensitivity was relaxed to 10 mJ/cm^2, but this is still too optimistic. The new goal is 20 mJ/cm^2 for the 7-nm node. It appears that at least 30 mJ/cm^2 is needed to satisfy the requirement of a sufficient number of photons to overcome the stochastic effect. More will be said on this in Section 9.6 on the EUV resist. As the technology node further advances, a lower sensitivity is required because a smaller feature size needs a higher photon density to maintain the number of photons hitting the target feature.

Resist sensitivities at 1, 20, 30, 50, and 80 mJ/cm^2 are listed in Table 9.1. The absorptivity is simply assumed to be $1 - R$, the reflectivity of each component. The power not collected by the collector mirror in the illuminator spreads out elsewhere in the illuminator and the imaging chamber. There is also the power lost in the optical path and bandwidth mismatch. Finally, the resist and the wafer absorb all incident power.

The absorbed power heats its respective optical element or the walls of the chambers, showing how much each element needs be cooled. The sum of these power values represents the total power that needs to be managed in an EUV optical system. The total power values are on the order of from several hundred W to more than 1 kW, starting at a resist sensitivity of 30 mJ/cm^2.

Only the power radiating in the 2% bandwidth is usable. The light outside this band is harmful. Because of the peaking of the spectral reflectivity of the multilayer reflector, reflectivity drops sharply outside the usable bandwidth; the light not reflected is absorbed and turns into heat, further complicating heat management of the EUV optics.

Reference 11 states that 200 W of EUV power at IF requires a laser power of 20 kW. This means that 20 kW of IR output from the CO_2 lasers converts to 200 W of EUV light. For a resist sensitivity of 30 mJ/cm^2, 386 W of EUV power is needed at the IF and a 39-kW IR output is required. For 50 and 80 mJ/cm^2, the EUV power is 643 W and 1029 W, and the IR power is 65 kW and 100 kW, respectively. The wall power needed for the CO_2 lasers would be about four times the IR output. The power required to power only the source is 260 ~ 400 kW.

The power absorbed by each optical element heats up that element, so each element must be cooled. Fortunately, the back sides of mirrors can always be used for cooling. There is more heat in the illuminator area, while the temperature must be more precisely controlled in the imaging optics area. The tolerance is tight. Because of reflective optics, any surface error in the direction of the incident light affects the imaging beam twice so the usual requirement of 0.02λ means that the surface accuracy must be $0.01\lambda = 0.014$ nm, which is on the order of a fraction of the dimension of an atom. It is impressive that this kind of precision can be attempted.

The IR light also heats up the area in the vacuum vessel where the IR beam hits the Sn droplets. This means that 40 to 100 kW of heat needs to be removed.

9.2.2 The adopted LPP source

EUV light is generated primarily by discharge-produced plasma (DPP)[12,13] or laser-produced plasma (LPP)[10,14] During the three decades of EUVL development, the LPP source emerged as the winner. Hence, we will focus on the LPP source, namely, the LPP from Cymer, an ASML company. As shown in Fig. 9.7, the LPP is in the vacuum vessel. The driving laser pulses come from the sub-fab floor through the beam transport tube to the main floor where the vacuum vessel and the scanner module are located. The sub-fab floor houses the seed unit and several stages of power amplification of the IR light. The sub-fab floor also houses the RF generators and heat exchangers shown in Fig. 9.2 but omitted here. In the vacuum vessel, there is a Sn droplet generator that keeps the Sn in molten liquid form and disperses Sn droplets (like an inkjet) at a frequency of between 40 and 100 kHz. A double pulse

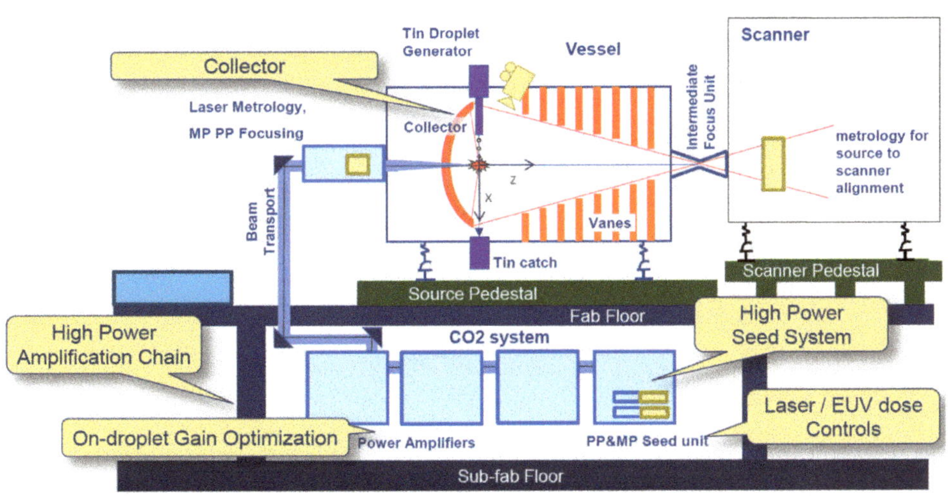

Figure 9.7 The Sn LPP MOPA (master oscillator power amplifier) architecture for the ASML NXE3300 (reprinted with permission from ASML).

consisting of a pre-pulse (PP) and a main pulse (MP) increases the efficiency of the conversion from IR to EUV light. The EUV light from the droplets is gathered by the collector and sent to the intermediate focus. Then it enters the scanner module that contains the illuminator and the imaging optics. The vanes are used to trap the Sn debris resulting from the laser-beam bombardment of the droplets.

We have already discussed the CO_2 drive laser and its challenges. The droplet generator also presents its own challenges. Reference 11 explains innovative ways to use gas pressure to efficiently produce uniform droplets of the right size and the desired frequency for the laser pulses. This is important because, if the laser beam misses the droplet, the powerful beam will hit the sidewall of the vacuum vessel and can pierce the wall, bursting the cooling water. Measures have been taken to protect the areas that can be accidentally hit by the laser beam. It is desirable to have a continuous supply of Sn instead of using cartridges. Heating up the droplet generator cartridge, switching out the empty cartridge, and calibrating all take time. This time can be reduced to increase the EUV scanner throughput.

The pre-pulse technique[15] significantly increases the EUV conversion efficiency. The idea, shown in Fig. 9.8, is to use a pre-pulse to (1) flatten the Sn droplet to increase the target area without needing to use more Sn (meaning less Sn debris) and to (2) pre-energize the Sn droplet.

The production of two pulses from the same laser beam utilizes a delay route for the main pulse, as shown in Fig. 9.9. Dichroic beam splitters are key elements that differentiate between the slightly different wavelengths from the pre-pulse seed laser and the main-pulse seed laser. These beam splitters let the pre-pulse beam pass through and redirect the main-pulse beam to a longer path length to produce the time delay.

The collector mirror presents a significant challenge. It is large and expensive. Because it is close to intensive radiation, damage can be severe.

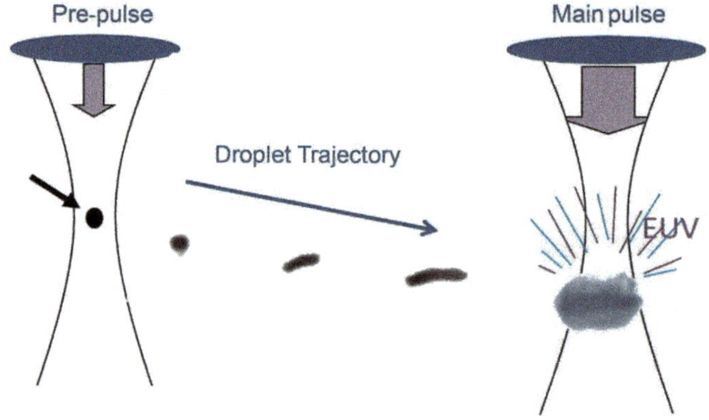

Figure 9.8 The pre-pulse technique for increasing the EUV conversion efficiency (reprinted from Ref. 50).

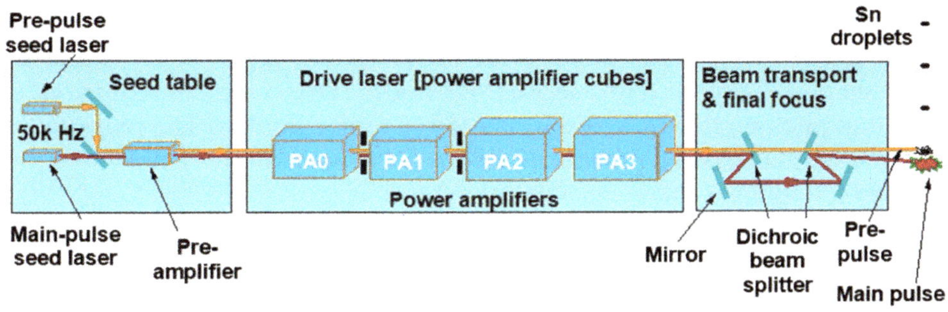

Figure 9.9 Generation of pre-pulses and main pulses (adapted with permission from ASML).

Also, replacing the collector mirror requires venting the vacuum vessel, which implies a loss of productivity due to venting and pump down. In addition to radiation damage, Sn debris can accumulate on the surface of the collector mirror, causing a loss of reflectivity and misdirection of the collected beam.

9.2.3 Wall-power requirement of EUV systems

In the past, power consumption of exposure tools was not an issue. However, with an exotic light source, two-stage conversion, and a surge in raw-energy costs, power consumption warrants careful examination. The calculation[16] of the raw input power required to run 59 100-layer-per-hour tools to sustain a 130,000 wafer-per-month capacity on 20 critical layers per wafer has been determined. The throughput is adjusted for typical field-size mix, availability, and utilization. The 150 MW required to run a 12-inch fab for such a capacity is used as the reference for the power needed for the exposure tools for the critical layers. The power for immersion scanners supporting double patterning assumes a 200-layer-per-hour single-exposure throughput. Each of these immersion tools consumes 115 kW to run the light source and 50 kW to run the exposure unit. The power required to run the tracks is not included. To run 59 immersion scanners this way requires 6.49% of the total power for the fab. To run 59 EUV scanners with an LPP source adds 57% of the power required to run the entire fab.

9.3 EUV Masks

The EUV mask is a reflective mask. Due to the lack of a beam splitter, the illumination needs to make an angle with respect to the optical axis of the mask. This angle is typically 6 deg. Similar to the mirrors in the EUV optics, reflectivity is built up with a multilayer stack with each layer pair contributing 1.5 ~ 2% of the total reflectivity of the stack. EUV masks have a shape and size that can be made similar to those of UV mask such that the former can physically fit into conventional mask writers and inspection and repair tools.

Not much change is expected for the EUV mask writer, except that with double patterning of immersion tools, the minimum pitch on the masks has remained at 320 nm for many technology nodes. Without pitch reduction, the minimum pitch on the EUV mask is now 160 nm or smaller. The resolution of the mask writer must be drastically improved. Furthermore, mask inspection requires actinic light. New mask inspection tools need to be developed. Mask repair is expected to be nontrivial, especially for the multilayer stack. Hence, the mask repair yield is expected to be lower than that of conventional masks.

9.3.1 Configuration of EUV masks

An EUV mask consists of a substrate, a multilayer stack, a capping layer, a buffer layer, and an absorber layer, as shown in Fig. 9.10. Because of the reflective optics, the mask substrate need not be transparent to the actinic light. The requirements are low thermal expansion, rigidity, stability, ease in smoothing and flattening, good adhesion in the multilayer stack, and compatibility with the electrostatic chuck.

Like all reflectors in the EUV optical train, EUV masks achieve their reflectivity with a multilayer thin-film stack. The stack consists of 40- or 50-layer pairs of Mo and Si[17,18] with a combined thickness of 6.94 nm per layer pair. The Mo thickness is about 40% of the thickness of the layer pair. To protect the multilayer stack, 2–3 nm of Ru[17] and 11 nm of Si[18] have been used as the capping layer. Buffer layers include 50 nm of SiO_2 and 10 to 50 nm of CrN.[17] The absorber layer is generally 70 to 100 nm in thickness and includes Cr,[18] TaN,[17] TaBN,[17] and W. There is a conductive layer at the back of the mask substrate for which Cr is a good choice because of its familiarity within the mask industry. Since the EUV scanner operates in vacuum, mask and wafer chucking by vacuum is impossible. Both the mask and the wafer are fastened to their respective supports by electrostatic force, which is not the best means to prevent defect particles from being attracted to the mask and the wafer.

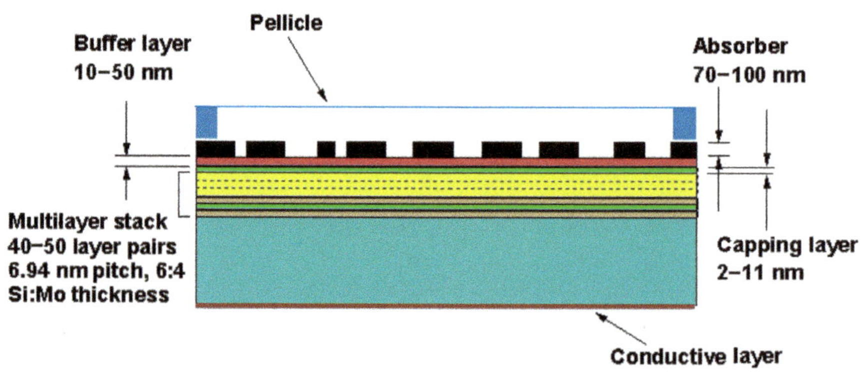

Figure 9.10 Configuration of the EUV mask.

The absorber thickness is 5.2X ~ 7.4X the 13.5-nm wavelength. In comparison, the 300-nm absorber thickness of the mask for 193 nm is 1.56X the actinic wavelength. As a result, the diffracted image from the mask must be considered as a 3D electromagnetic (EM) effect.

9.3.2 Effects of oblique incidence on the mask

9.3.2.1 Pattern shadowing

Due to the 6-deg chief ray at object (CRAO), the absorber thickness casts a shadow that induces an apparent change in the feature size, as shown in Fig. 9.11. There are two shadow effects: one due to the absorber thickness and the other due to the multilayer (ML) thickness. S_D is the width of the shadow cast by the CRAO:

$$S_D = D \tan\theta, \tag{9.1}$$

where D is the thickness of the absorber, and θ is the CRAO angle. S_d is the width of the shadow cast by the multilayers:

$$S_d = nd \cdot \tan\theta, \tag{9.2}$$

where n is the nth number of the multilayer pair, and d is the thickness of a pair of the layers in the multilayer. Note that S_d is independent of S_D. Even when the absorber thickness is zero, there is still a shadow effect at the reflective mask using a ML to reflect light. Substituting $D = 100$ nm, $n = 40$, and

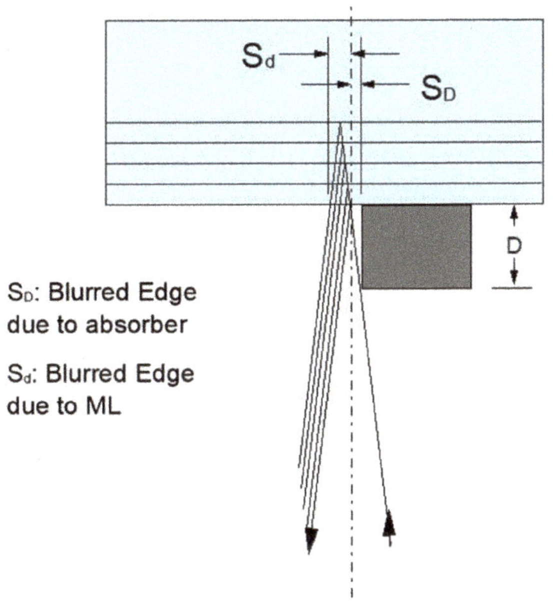

Figure 9.11 Shadow of edges due to a finite CRAO θ.

$d = 9.63$ nm, $S_D = 11$ nm and $S_d = 29$ nm. The latter is not as bad as 29 nm because it is taken from the 40th-layer pair. It is less after considering the averaging effect of all of the 40-layer pairs. Nevertheless, this discussion makes us aware that the shadow of a nanometer-sized feature can be on the order of tens of nanometers. With off-axis illumination, the situation can be more complicated.[19] Here we only consider geometric effects. For a more rigorous calculation, 3D EM simulation should be used.

The shadow can be compensated by proper proximity correction. However, this is not easy, for several reasons. First, 3D EM simulation of the situation is difficult and time consuming. Second, the shadow effect is a function of the feature orientation. Lorusso and Kim provide two figures to illustrate the edge-shadow effects.[20] Figure 9.11 shows the shadow of an absorber feature at different orientations in the exposing slit. Figure 9.12 shows that, even at the same feature orientation, there are still differences in the shadow at different positions in the slit. Note that the line feature is assumed to be infinitely long. Therefore, the shadow is drawn for its edges, but not for its ends. Figure 9.13 is a top view of the line features oriented in four different directions. For the 45- and 135-deg lines, the shadow is not

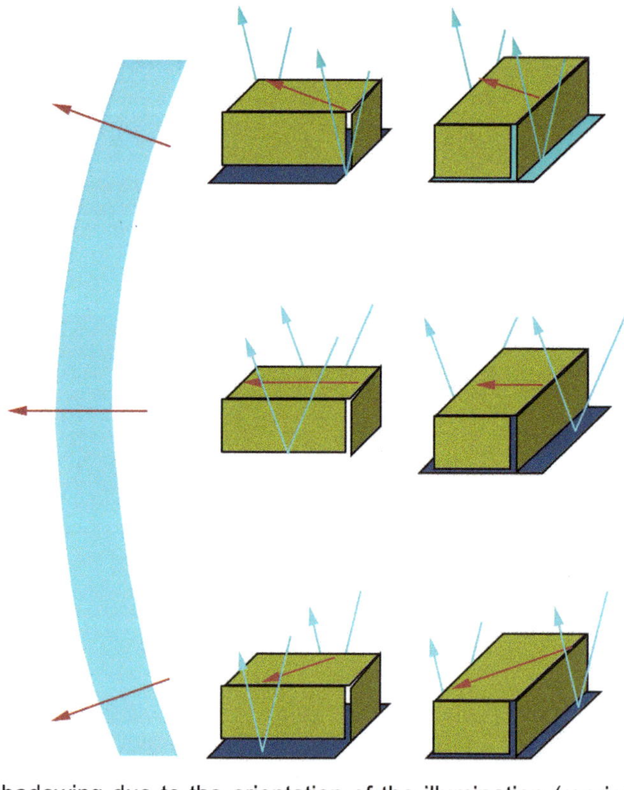

Figure 9.12 Shadowing due to the orientation of the illumination (reprinted from Ref. 20 with permission from IMEC).

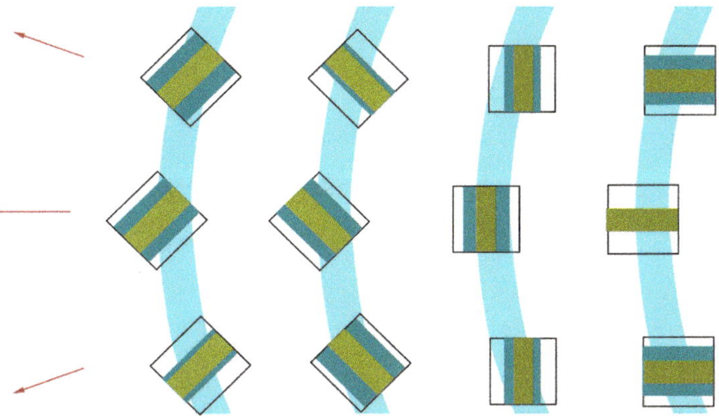

Figure 9.13 Shadow of edges as a function of the feature position in the illuminated slit (reprinted from Ref. 20 with permission from IMEC).

symmetrical with respect to the center of the feature because the relative orientations of the illuminating beam and the line are different at the ends of the slit.

This means that OPC needs to consider the feature orientation with respect to the circular scanning slit even though this orientation is identical from die to die. Identical cells in the circuit most likely need to be proximity corrected differently according to the exposure process.

9.3.2.2 Overlay and focusing errors from mask flatness

Another undesirable effect comes from the use of a reflective mask with a nonzero CRAO. When there is insufficient flatness on the mask, either from local thickness perturbation or local tilt, as shown in Fig. 9.14, a lateral displacement of the image and a focusing error take place, which will be analyzed in this section.

Figure 9.15(a) shows the ray tracing on the EUV imaging optics shown in Fig. 9.4 to analyze the impact of insufficient mask flatness. Here, we show the rays from the mask to M1, then from M6 to the wafer, and we hide the rays

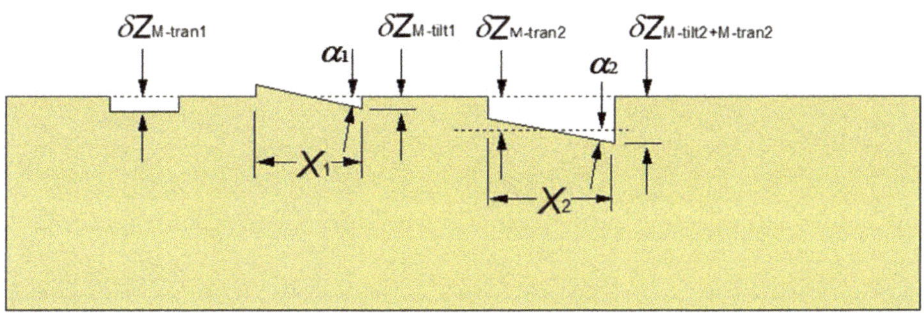

Figure 9.14 Mask roughness components consisting of local translations and local tilts.

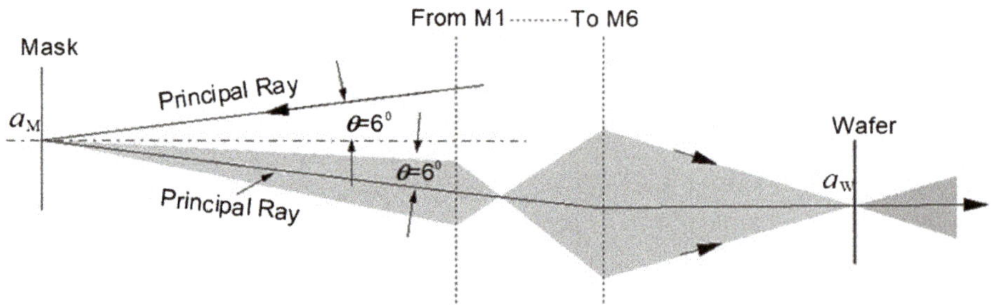

Figure 9.15(a) Ray tracing from the mask to the wafer, based on the system in Fig. 9.4. The rays from M1 to M6 are hidden. The point a_M on the mask and the point a_W on the wafer are conjugate points. The principal ray makes an angle θ with respect to the normal line to the mask but is normally incident on the wafer, making the image telecentric at the wafer side.

from M1 to M6. There is a conjugate relationship between points a_M on the mask and a_W on the wafer. All rays from a_M, regardless of the angle of tilt, converge to a_W. The principal ray propagating toward the wafer is normal to it such that the wafer image is telecentric; that is, the wafer image does not shift laterally away from the focal plane.

The most significant lateral image shift in this obliquely illuminated reflective system is caused by a translational displacement of the mask—in other words, a shift of the longitudinal direction from focus. This is illustrated in Fig. 9.15(b).

When the mask is translated in the z direction by $\delta Z_{\text{M-tran}}$, the translation directly produces a translation in the wafer image space:

$$\delta Z_W = \delta Z_{\text{M-tran}}/M^2 \tag{9.3}$$

when $\delta Z_{\text{M-tran}} = 50$ nm, and $\delta Z_W = 50/16$ nm $= 3.125$ nm. This is valid for any projection imaging system, whether it be dioptric, catadioptric, or reflective, 1X or MX, or illuminated perpendicularly or obliquely. For an obliquely illuminated reflective mask, a lateral shift δX_W is produced by a mask shift in

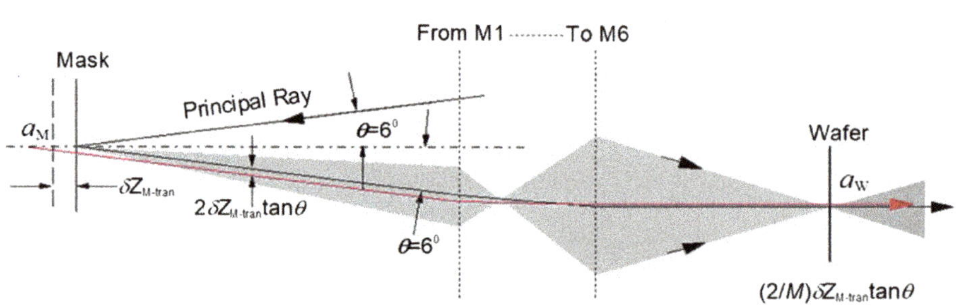

Figure 9.15(b) Mask longitudinal misplacement produces a lateral displacement in the image.

the longitudinal direction $\delta Z_{\text{M-tran}}$:

$$\delta X_W = \frac{2}{M}\delta Z_{\text{M-tran}} \tan\theta. \tag{9.4}$$

With $\theta = 6$ deg, $m = 4$, and $\delta Z_{\text{M-tran}} = 50$ nm, an image shift of 2.63 nm takes place.

Figure 9.15(c) shows a mask tilted by an angle α. The angle changes to 2α after reflecting off the mask, then to $2M\alpha$ after going through the reduction lens. A lateral shift of any out-of-focus position in the wafer image is introduced. As a result, the telecentricity is lost. The shift in x is

$$\delta X_W = \delta Z_W \tan(2M\alpha). \tag{9.5}$$

If $\alpha = 1$ deg, $M = 4$, and $\delta Z_W = 80$ nm, then $\delta X_W = 11$ nm. This image lateral shift is too great. Either the DOF or the mask mini-tilt needs to be restricted.

The overlay limit is used for breaking down the specification of each component as follows. The total lateral image shift from Eqs. (9.4) and (9.5) is

$$\Delta X_{\text{W-total}} = \sqrt{\left(\frac{2}{M}\delta Z_{\text{M-tran}} \tan\theta\right)^2 + [\delta Z_W \tan(2M\alpha)]^2}. \tag{9.6}$$

Usually one-third of the minimum feature is used as the overlay tolerance and one-third of this overlay tolerance can be allocated to $\Delta X_{\text{W-total}}$, assuming

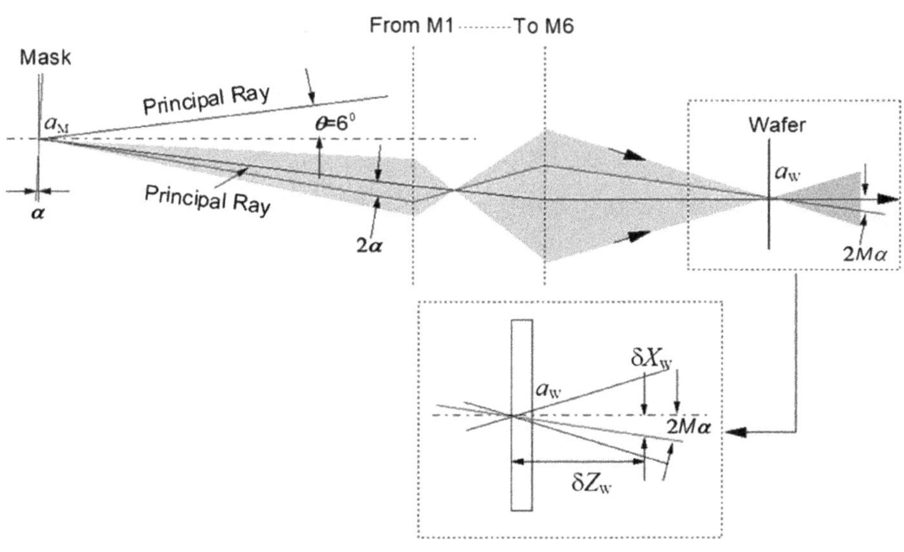

Figure 9.15(c) A tilt of α becomes 2α due to reflection, then it becomes $2M\alpha$ due to reduction ratio M. When the wafer is in focus, no lateral shift is induced. When the wafer is out of focus by δZ_W, there is a lateral displacement of $\delta X_w = \delta Z_W \tan(2M\alpha)$.

Table 9.2 Parameters controlling the overlay error contribution from the reflective-mask-induced component.

M	θ (deg)	α (deg)	$\delta Z_{\text{M-tran}}$ (nm)	$\delta X_{\text{W(from M-tran)}}$ (nm)	δZ_{W} (nm)	$\delta X_{\text{W(from M-tilt)}}$ (nm)	ΔX_{W} (nm)
4	6	1	50	2.63	80	11.17	11.48
4	6	0.05	10	0.53	50	0.35	0.63

that the other traditional overlay components can be squeezed to accommodate this extra reflective-mask-induced component. For a 16-nm half pitch, the minimum feature size after etch is assumed to be 5 nm. One third of this is 1.67 nm, and the allowance of the reflective-mask-induced component is 0.56 nm. We just showed that with $\theta = 6$ deg, $M = 4$, and $\delta Z_{\text{M-tran}} = 50$ nm, the image shift is 2.63 nm for the first term of Eq. (9.6); if $\alpha = 1$ deg, $M = 4$, and $\delta Z_{\text{W}} = 80$ nm, then $\delta X_{\text{W}} = 11$ nm. $\Delta X_{\text{W-total}} = 11.47$ nm. To reduce the contributions to bring $\Delta X_{\text{W-total}}$ to the 0.56-nm regime, we set $\alpha = 0.05$ deg, $\delta Z_{\text{M-tran}} = 10$ nm, and $\delta Z_{\text{W}} = 80$ nm. These parameters are summarized in Table 9.2 and indicate that the mask flatness must be held to within 10 nm and that the deviation from telecentricity limits the DOF to 50 nm and the local mask tilt to less than 0.05 deg. These requirements on the mask blank are extremely stringent.

9.3.3 EUV mask fabrication

The patterning of EUV masks should be like UV mask patterning in terms of the mask writing. The mask writer probably needs the least modification. The only concern is that, because of the stringent requirement of reflective mask roughness and flatness, the chucking in the mask writer and in the exposure tool should be as close to each other as possible. Electrostatic chucking for the mask writers is a new venture. Other than mask writing, subsequent processes in inspection, repair, and defect mitigation are not trivial. When it comes to actual practice, even mask writing is not as simple as originally perceived. One reason for this is that mask writing for immersion masks does not require many technology advances because of double and triple patterning. The minimum pitch does not have to shrink. Four technology nodes—20 nm, 16 nm, 10 nm, and 7 nm—had passed without much progress in reducing the minimum pitch. Now that the minimum pitch has decreased by at least a factor of 2 with EUVL, advancement in our understanding of the mask writer is lacking.

Starting with the mask substrate, this material must have a thermal expansion coefficient lower than quartz. The transmission property is not important, but the substrate must be flat to within 10 nm, as discussed in Section 9.3.2.2. It cannot have any blemishes or particles on it before the 40-layer pairs of multilayer reflecting coating are deposited onto it. These two requirements have not been completely mastered. As a result, defect-free EUV mask blanks have been extremely rare. Blanks with a defect number in the single digits fetch a high premium.

In terms of delineating contact holes, defects on the mask blank are less of an issue because the probability of single-digit defects being at the position of a contact hole is not high. Unfortunately, contact holes are difficult to image because the DOF is shallow. Shadowing effects are severe. Low transmission and insufficient area to collect photons make it necessary to increase the EUV power, thus reducing wafer throughput and increasing cost. There are suggestions to use negative resists or negative developing for contact holes, which would mean trading throughput for higher defect counts. The worthiness of this approach remains to be proven.

For other layers such as the gate layer and the metal layer, the pattern density can be close to 50%, making defects less likely to be at opaque locations. The mask blank's defect map must be matched to the pattern distribution to hide all of the defects. Four possibilities are created by rotating the defect map 90 deg four times. If there is still no match, the blank is put back into storage for the next matching opportunity. Figure 9.16 shows an example[21] of defect mitigation by matching the mask opaque areas.

To delineate the absorber, one must protect the multilayer from being damaged by etching and any related processing—hence, the need for a buffer layer. The capping layer is meant to protect the multilayer from environmental corrosion. Etch protection is generally left to the buffer layer. The buffer layer also protects the multilayer during the repair process, which typically uses a Ga focused ion beam. However, Ga stain, which is already present in the repair of UV masks, is more severe with EUV masks using a multilayer stack. Mechanical removal[22,17] of the absorber defect with atomic-force-microscope (AFM)-type equipment tends to leave debris that must be carefully removed. Mechanical removal has been shown to stop within the buffer layer while not scratching the capping layer and the multilayer. E-beam repair is new but holds promise.[23] With recent process improvements, the need for a buffer layer has decreased.

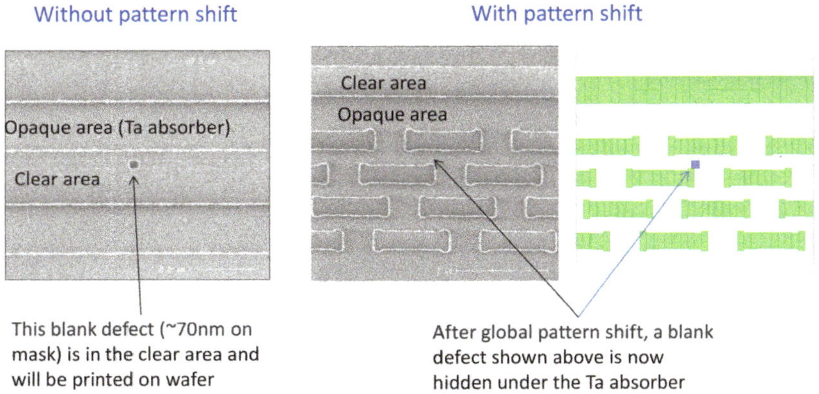

Figure 9.16 Mitigation of EUV mask blank defects by manipulating the placement of the mask pattern to cover the defect with opaque features (reprinted from Ref. 51).

The mask must be inspected after patterning to locate the repair sites. Inspection at the actinic wavelength calls for a completely new inspection technology. The general thinking is to inspect at a convenient UV wavelength, e.g., 257 nm, in continuous wave (CW). To facilitate such off-wavelength inspection, there must be sufficient contrast between the absorber and the reflecting layer stack at the inspection wavelength, and the EUV defects must be detectable. To date, many EUV printable defects have escaped inspection at off-wavelengths. Actinic inspection may be inevitable. A bright EUV source and reflective inspection optics are required.

9.3.4 EUV pellicles

Figure 9.10 includes a pellicle that keeps particles away from the DOF zone of the mask so that they do not image on the wafer to constitute defects. All optical masks enjoy the protection of a pellicle. Any particle accidentally falling on the mask is kept at a safe distance from focus. A sharp image of the particle is turned into an enlarged blurred image. The EUV mask is kept under an opaque cover during its storage and transportation but is without protection when it is at the exposure position in the scanner. The EUV optics is designed with the mask facing down at the exposure position to minimize the chance of particles falling on it. However, electrostatic chucking may unexpectedly attract particles. Figure 9.17 shows a micrometer-sized particle attached to the mask[24] at the 17th step of EUV exposure in an EUV scanner. Such an occurrence takes place at a frequency of every few hundred wafers.

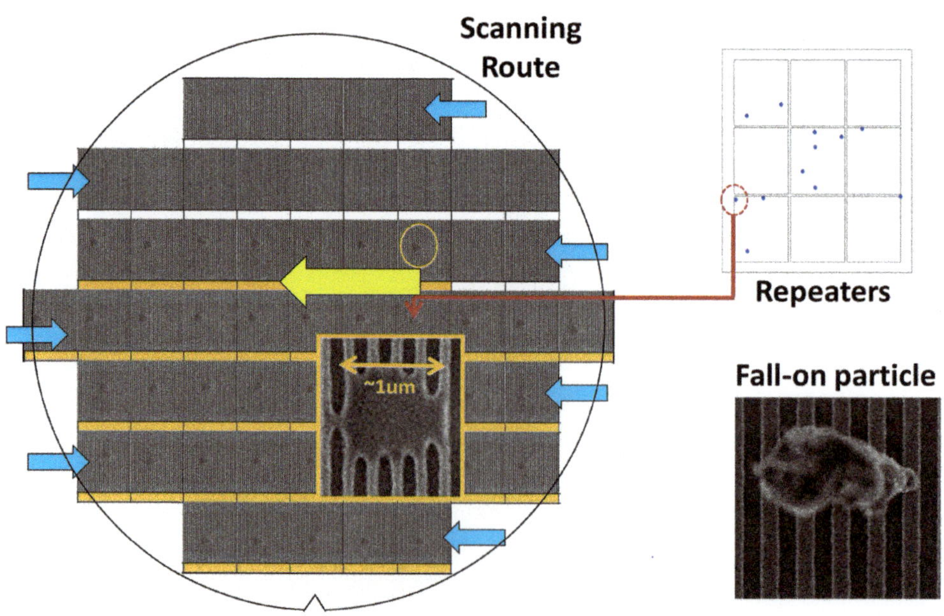

Figure 9.17 Micrometer-sized particle attached to the mask during exposure (reprinted from Ref. 51).

At a rated throughput of 100 wph, the mask must be cleaned after every production shift.

Because of low transmissivity, only extremely thin poly-Si capped with silicon nitride[25] on the order of several tens of nanometers can be used for the EUV pellicle. The transmission achievable by a pellicle is about 85%. Light goes through the pellicle in two passes due to the reflective mask. Therefore, about 28% of the imaging light is lost to the pellicle. If this loss were incorporated into Table 9.1, the power at the IF based on $S = 30$ mJ/cm^2 would increase from 386 W to 534 W. The power incident on the pellicle would be 20 W. The pellicle is a free-standing thin membrane and cannot be cooled by a contacting heat sink. There is no air cooling because it is in a vacuum environment. With no other way of cooling except by radiation, the temperature at the pellicle[26] can reach as high as 944 °C. Hence, in addition to needing high EUV transmission, the pellicle must be able to (1) withstand high heat without breaking, (2) withstand the mask cycling in and out of the vacuum environment, and (3) endure 10^4 exposure cycles per hour.

The most severe concern regarding the EUV pellicle is the requirement of uniformity of the membrane in thickness and stress. Areas that have weak points tend to initiate pellicle breakage. Unfortunately, no matter how homogeneous and uniform the pellicle is, when a particle attaches to the pellicle instead of the mask (as it is intended to work), the particle can become a hot spot where the pellicle can break. When the pellicle breaks, it can fly off to unexpected places. The broken pellicle pieces are mostly harmless, but it can be disastrous if a piece lands on the optical elements or the mask.

Earlier concerns over mounting and demounting the pellicle for mask inspection have been overcome with a clever tool presented by Brouns et al.[27]

This author's opinion of the EUV pellicle is that it is best not to use it. Increasing the cleanliness of the exposure tool to reduce the frequency of particle attachment to the mask is a more achievable way to reduce repeating defects.

9.4 Resolution-Enhancement Techniques for EUVL

Resolution-enhancement techniques (RETs) are effective ways to extend the resolution of existing exposure tools. For the 20-nm half pitch at 0.33 NA in the EUV, k_1 is already 0.488, which is well within RET territory. In addition, because EUVL is inherently expensive, RETs should be used wherever possible. For DUV lithography, illumination optimization, mask optimization, source-mask optimization, and multiple patterning have been effective. Illumination optimization includes off-axis illumination, free-form illumination, and polarized illumination. Implementing polarization is not trivial but, since the NA of EUVL has not exceeded 0.33 and may not exceed 0.55 in the future, polarization effects are not significant. Hence, only flexible illumination, which can also provide off-axis illuminations, will be discussed. Mask optimization in EUVL, including optical proximity correction and the phase-shifting mask, will also be discussed.

9.4.1 EUV flexible illumination

The principle of EUV OAI is like that of DUV. One difference between them is that strictly reflective optics must be used for the former. Initially, it was easier to implement OAI with transmission optics. The mating-cones setup in Fig. 6.12 is an example. Eventually, the need for much more complicated OAI lead to the invention of diffractive optical elements (DOEs) and Flexray®, the programmable array shown in Fig. 6.13. Because mirrors are required in Flexray, the programmable array concept can readily be transferred to EUVL. Figure 9.18 shows EUV flexible illumination with a programmable field-facet-mirror array.[28] The mirrors can be tilted to project the intermediate focus to a pupil facet mirror according to the program for the desired illumination. The left-most figure shows the field facets forming a vertical dipole illumination by illuminating the red-colored pupil facets. The right-most figure shows the field facets reprogrammed for a horizontal dipole. The center figure shows the switching in action: eight of the pupil facets are already switched, and two remain to be switched.

Figure 9.19(a) shows that four of the illuminated pupil facets in the annular setting (identified by the arrows) are moved to form an x dipole, as shown on the left. Instead of these four facets, four other facets can be moved to form a y dipole, as shown on the right.[29] Figure 9.19(b) shows that the more-complex illuminations 1, 2, and 3 can be set with the flexible illuminations on manufacturing tools containing more micro-mirrors.

Another difference between EUV OAI and DUV OAI is the 6-deg CRAO requirement for the former. As shown in Fig. 9.20, the off-axis beam drawn at the angle $\sin^{-1}[\sin 6° + \lambda/2P/4]$ further extends the edge shadow zone from a tall absorber. Here λ = 13.5 nm is the EUV wavelength. P is the pitch of the grating on the mask. $\lambda/2P$ determines the optimum angle for the oblique illumination. The number 4 comes from the 4X reduction system.

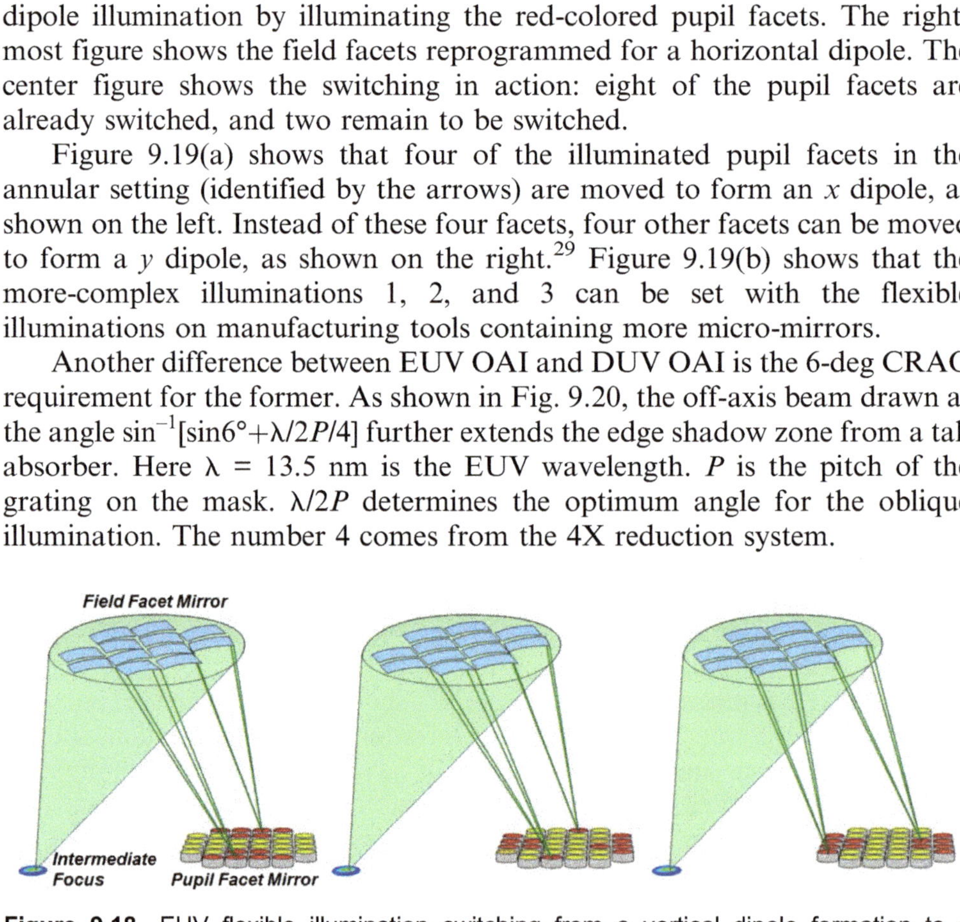

Figure 9.18 EUV flexible illumination switching from a vertical dipole formation to a horizontal dipole formation (reprinted from Ref. 28).

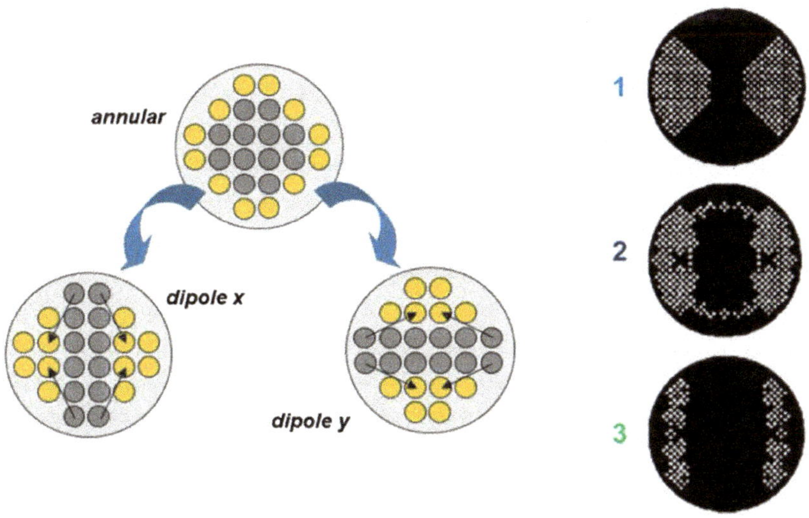

(a) (b)

Figure 9.19 Programmed pupil intensity distribution. (a) Annular setting reprogrammed to form dipole *x* and dipole *y* (reprinted from Ref. 29). (b) More-complex distributions made possible by using flexible illuminations (adapted from Ref. 52 with permission).

In summary, this EUV flexible illumination is an important contributor to k_1 reduction (RET) in EUVL and can be implemented with an understanding that there will be a larger edge shadow.

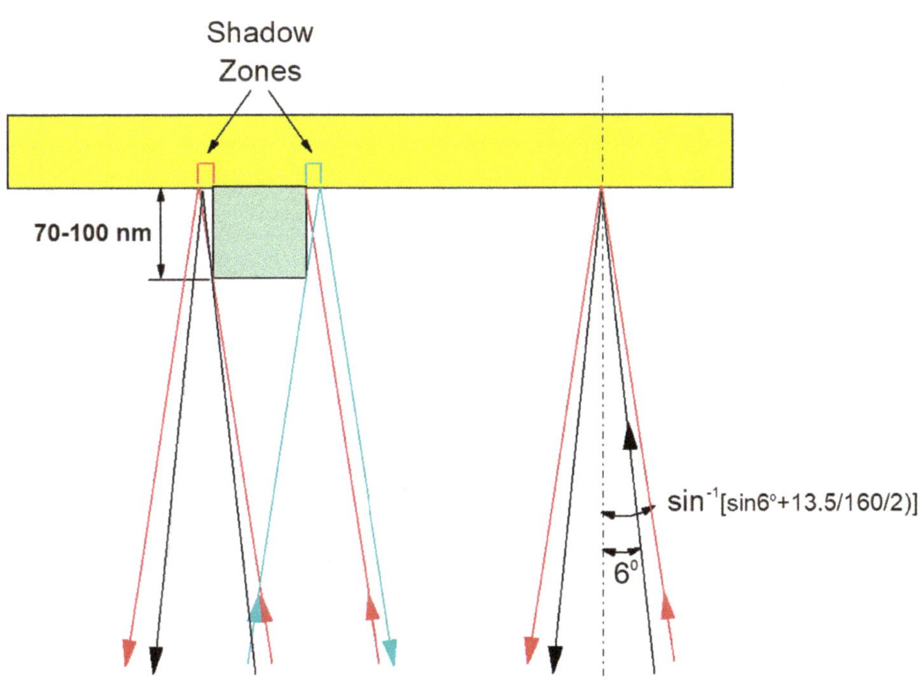

Figure 9.20 Edge shadow zone caused by OAI and the CRAO.

9.4.2 EUV proximity correction

Initially, because of the relatively high k_1, OPC for EUVL was expected to be simple. However, two new phenomena posed unconventional challenges to OPC for EUVL. Stray light in the EUV image is not only an order of magnitude higher than that of conventional optics, but it is also field dependent; i.e., the level of stray light varies at different intrafield locations. Hence, its contrast and exposure levels are field dependent. They must be compensated according to the location of stray light on the mask. It is also highly likely that the stray light distribution varies from tool to tool, making OPC tool dependent. The stray-light distribution may not remain constant as the optics become aged and contaminated. Therefore, OPC can also be a function of time. One must develop an algorithm to adjust OPC according to a given stray-light distribution. A fundamental solution is to reduce the stray light to about the same level as that in DUVL. Stray-light reduction is an ongoing effort of the tool supplier for which continuous progress has been made.

The second phenomenon, shadow effects, is also field dependent. It originates from the pattern shadowing discussed in Sections 9.3.2.1 and 9.4.1 and shown in Figs. 9.12 and 9.13.

9.4.3 EUV multiple patterning

Multiple patterning has been used for high-volume manufacturing for at least four generations since the 20-nm node. EUVL has a similar potential for adopting multiple-patterning techniques (MPTs). For DUVL, k_1 has been pushed to 0.28. Due to many issues considered here and by many other authors, it is not easy to reach 0.4 k_1 for EUVL; that is, with $NA = 0.33$, when the minimum pitch reaches 32 nm, pitch splitting is needed for further reductions. Another alternative is to use a higher NA such as 0.55, which was announced in 2017.[30] This alternative method is covered in Section 9.7.2.

As far as multiple patterning in EUVL is concerned, both the pitch splitting and the SADP techniques can work. The G-rule governing pattern splitability also applies. Line cutting with multiple patterning and pack-and-unpack with double patterning discussed in Section 7.5 can both be used. The hesitation to using EUV multiple patterning is due to the extra cost and an awkwardness factor. The latter is a concern because EUVL is usually advertised as a single-patterning technology, unlike DUV lithography.

9.4.4 EUV phase-shifting masks

When performing EUV phase shifting, one must be aware of the difference between reflection phase shifting and transmission phase shifting, as shown in Fig. 9.21. In reflection phase shifting, the phase difference ϕ is created by a step with t as its step height. Whereas in the case of transmission phase

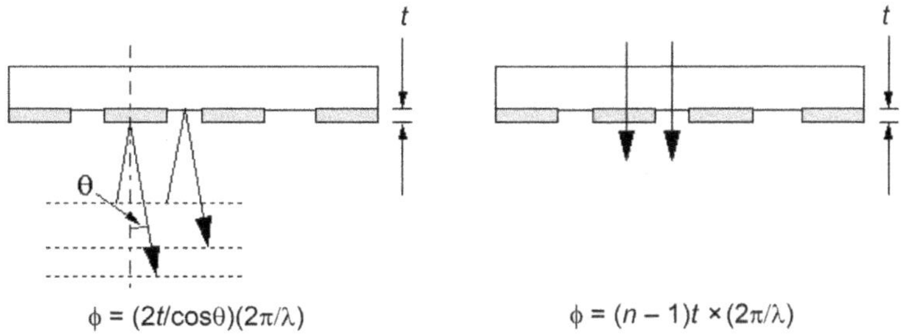

Figure 9.21 Thickness *t* comparison of (left) reflection and (right) transmission phase-shifting masks.

shifting, the phase difference comes from two beams going through two different media. For reflection phase shifting,

$$\phi = \left(\frac{2t}{\cos\theta}\right)\left(\frac{2\pi}{\lambda}\right) = \frac{4\pi t}{\lambda\cos\theta}. \tag{9.7}$$

For transmission phase shifting,

$$\phi = (n-1)t\left(\frac{2\pi}{\lambda}\right) = (n-1)\frac{2\pi t}{\lambda}, \tag{9.8}$$

where the refractive index of one medium is n and that of the other is 1 (as in vacuum) or close to 1 (as in air). For the common UV case of $n = 1.5$, the phase change as a function of t is 4 times greater in the case of reflection phase shifting. To shift the phase by π at the EUV wavelength of 13.5 nm, $t = (\lambda/4)\cos\theta = 3.36$ nm. To maintain the phase-shifting error to within 6 deg, Δt must be smaller than 0.11 nm.

If the thickness of the absorber or of a phase-shifting layer is not controlled to the required specification, random phase shifting takes place. This reduces the common E-D window, as shown in Fig. 9.22 (top). Here, the DOF based on four absorber thicknesses producing phase shifts of 0, 45, 90, and 180 deg is evaluated using the E-D window methodology. The absorber has a reflectivity of 2%. The feature is a 22-nm opaque line at 55-nm pitch. On-axis illumination at $\sigma = 0.66$, $\lambda = 13.5$ nm is used based on 8% EL and ±10% CD tolerance. The stray-light level is assumed to be 15%. The DOF versus $\sin\theta$ curves in Fig. 9.22 (bottom) from the four absorber thicknesses are all very similar, having a peak DOF above 300 nm. However, the common E-D window only supports a DOF of about 200 nm.

With this basic consideration, let us study two of the most popular phase-shifting masks—one strong and the other weak, namely, the alternating PSM and the attenuated PSM.

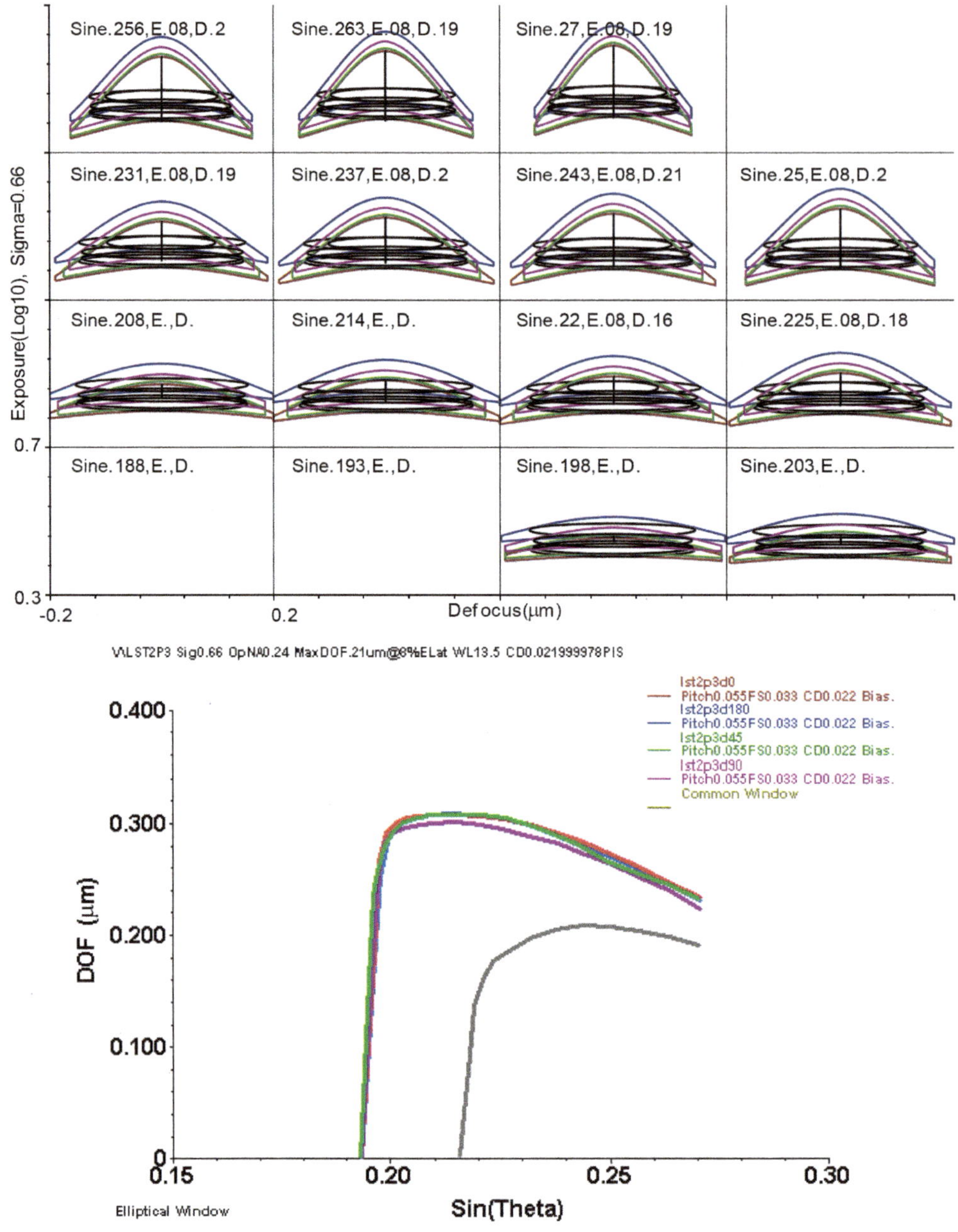

Figure 9.22 (top) Individual and common E-D windows. The four individual windows are evaluated with phases of 0, 45, 90, and 180 deg, respectively, for a 22-nm opaque line at 55-nm pitch, based on $\sigma = 0.66$, $CD_{\text{tol}} = \pm 10\%$, $EL = 8\%$, $\lambda = 13.5$ nm, and 15% stray light. The absorber has a reflectivity of 2%. (bottom) DOF versus $\sin\theta$ curves from the individual and common E-D windows in the top part.

9.4.4.1 EUV AltPSM

A straightforward way[31] to make an EUV AltPSM is to deposit the multilayer reflector on steps in the substrate with a step height h that produces a π phase shift. Figure 9.23 shows the mask substrate etched into steps of such a height. The bottom of the steps is designated at 0 deg, and the top of the steps at 180 deg. The multilayer is deposited, followed by deposition of the absorber. For a production mask instead of an experimental mask, a capping layer to protect the multilayer is necessary. A buffer layer between the capping and the multilayers is often used. Finally, the absorber is patterned, resulting in the top and side views shown at the right side of the figure. The step height h that can produce the 180-deg shift is governed by the following equation, where λ is the wavelength and n is a positive integer:

$$h = \left(\frac{1}{4} + \frac{n}{2}\right)\lambda, \tag{9.9}$$

where $n = 0, 1, \ldots$. Hence, h can be 3.375 nm, 10.125 nm, 16.88 nm, etc.

The advantage of this method is that the deposition of the multilayer, buffer layer, capping layer, and absorber layer are identical to the normal BIM fabrication process, even though the buffer and capping layers were not used in the experimental BIM. The shortcoming of this method is that the patterned mask must be made by a blank manufacturer that has the expertise and equipment to make low-defect EUV mask blanks. On the other hand, this can be a way for a rich semiconductor company to exclude others from using EUV AltPSMs. A fundamental disadvantage of this method is the tight thickness tolerance for EUV phase shifting that has just been discussed with Eq. (9.7).

Yan et al. also compared the performance[31] of AltPSMs to BIMs, as shown in Fig. 9.24. The comparison is on 15-nm lines with a 50-nm pitch and $\Delta CD = \pm 10\%$. The E-D window with the largest E-D area was used. The EUV AltPSM window shown is clearly much larger than that for the BIM. These E-D plots are similar to those in Figs. 7.52 and 7.53, where the larger E-D window of the AltPSM becomes smaller than that of the BIM, after

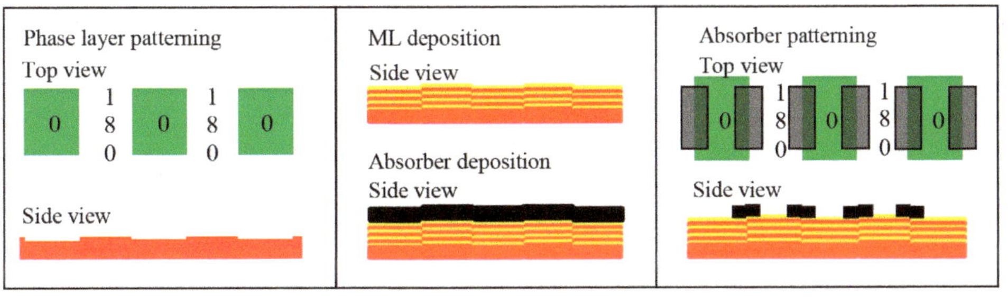

Figure 9.23 Fabrication of an AltPSM (reprinted from Ref. 31).

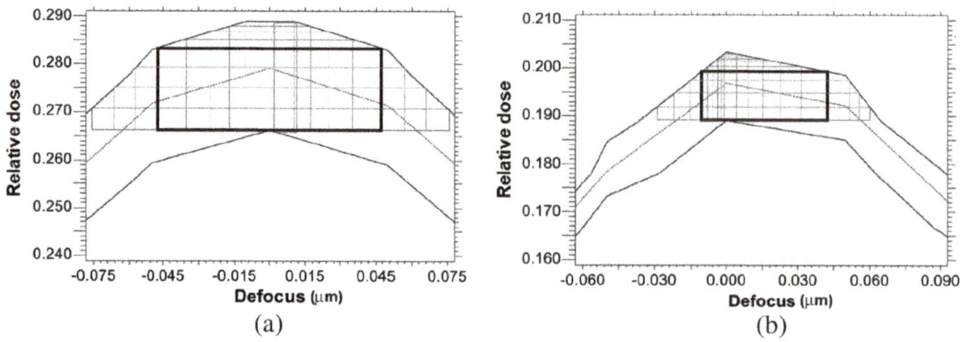

Figure 9.24 Performance of an AltPSM versus a BIM (adapted from Ref. 56).

considering the fabrication tolerance of an AltPSM. For an EUV AltPSM, a more rigorous comparison should also consider the fabrication tolerance.

There are other ways to make an AltPSM. Deng et al.[32] simulated the imaging of four different EUV mask configurations, as shown in Fig. 9.25. There are two configurations of AltPSMs: one with no absorber and the other with absorber patterns. These masks use multilayers of different numbers of layered pairs to produce the π phase shift. An interesting aspect of these configurations is the low sensitivity of their thickness variations to phase changes. Deng et al. only provided their thickness sensitivity results for the AttPSM configuration, which will be covered in the next section. This reduced sensitivity is due to the smaller impact of the entire multilayer on the accumulated phase stack compared to the impact of a single reflecting layer.

To make the etched multilayer work in high-volume manufacturing, the layers need to be protected by a capping layer. Special mask blanks with capping and etch stops can be prefabricated according to the AltPSM or AttPSM application. The exposed sidewalls of the multilayer after the etch must be capped with atomic layer deposition of the capping material. Needless to say, the configuration must be rigorously characterized both theoretically and experimentally before applying it to high-volume manufacturing.

In conclusion, the substrate method using an EUV AltPSM suffers from sensitivity to step height variations and multilayer deposition after mask

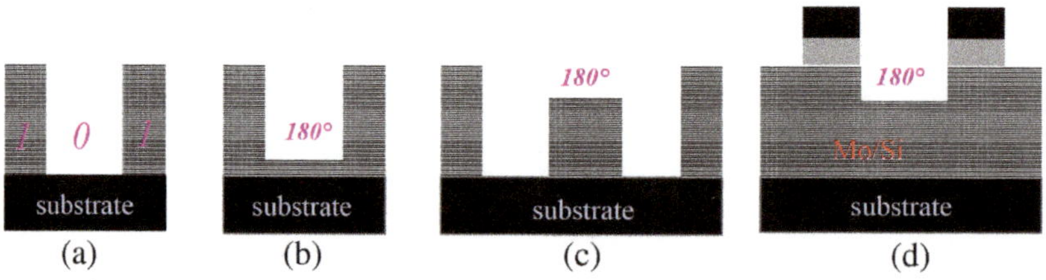

Figure 9.25 Various PSM configurations using an etched multilayer: (a) a BIM, (b) an AttPSM, (c) an AltPSM, and (d) an AltPSM with an absorber pattern on top (reprinted from Ref. 32).

patterning on the steps. The etched multilayer configuration has a lower sensitivity to etch tolerance, which can be further improved with an imbedded etch-stop layer. Imbedding two capping layers and using atomic layer deposition of the etched sidewalls (if fully developed) can improve the originally proposed configuration, moving it closer to readiness for high-volume manufacturing applications. Unfortunately, just like DUVL, the inherent problem of phase conflicts prevents AltPSMs from being used for arbitrary patterns. Only special patterns such as fully populated single-pitch line pairs or contact holes can take advantage of an AltPSM. Hence, the EUV AltPSM will have limited applications, even if all of its EUV problems are overcome.

9.4.4.2 EUV AttPSM

The AttPSM has been a workhorse for DUV lithography. Will the same occur in EUV lithography? There are several configurations of EUV AttPSMs. Deng et al.[32] studied the etched multilayer AttPSM shown in Fig. 9.26. The etched thickness tolerance has been shown to be 2.47 nm to maintain the 6-deg phase tolerance from etching. This is more forgiving than depending on a step to produce the phase shift. However, the fabrication process is quite complicated.

A simpler way to fabricate EUV AttPSMs is shown by Yan et al.[33] The scheme is shown in Fig. 9.27. Any commercially available multilayer with a capping layer is coated with an absorber layer pair (TaN/TiN) of 28.5 nm in thickness for each layer. This absorber reflects 6% and produces the required π-phase shift, whereas the same capped multilayer substrate coated with a total thickness of 85 nm of TaON and TaN reflects only less than 0.5% and qualifies as a BIM. With a blank coated with TaN/TiN, the user simply patterns the absorber as if patterning an EUV BIM. Hence, the EUV AttPSM can be

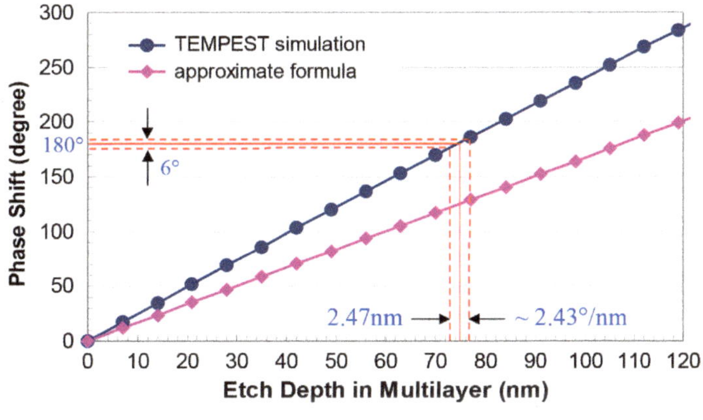

Figure 9.26 Tolerance in multilayer etch depth for an AttPSM from an etched multilayer (reprinted from Ref. 32).

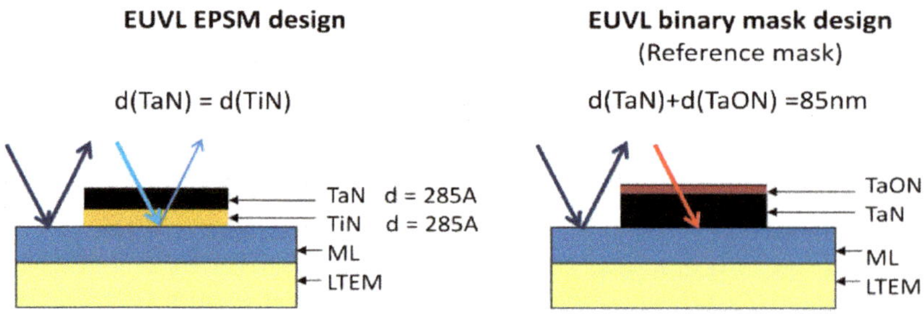

Figure 9.27 (left) AttPSM design (EPSM is embedded phase-shift mask; LTEM is low-thermal-expansion material) and (right) BIM design (reprinted from Ref. 33).

combined with OAI in the same that a DUV AttPSM can be combined with OAI. This combination has the potential for high-volume manufacturing.

9.5 EUV Projection Optics

Typical layouts of EUV optical components are shown in Figs. 9.1 to 9.3. The smoothness of these components is critical to reduce stray light. Note that reflection doubles the optical path difference from longitudinal surface changes in the optical element. Reflective elements are much more sensitive than refractive elements, as illustrated in Fig. 9.21. To maintain the same level of stray light, the reflective surface roughness and surface configuration must be four times better than the refractive surface roughness and configuration. The requirement of $\lambda/50$ in a 193-nm imaging wavefront translates to $13.5/200 = 0.067$ nm, which is a fraction of the dimension of an atom. Figure 9.28 shows the reduction of wavefront RMS error from >1 nm to about 0.2 nm in ASML EUV scanners.[34] Impressive progress has been made. The flare associated with these levels of RMS error can be seen in the same reference. This flare dropped from 16% to between 3 and 4%, as seen in Fig. 9.29.

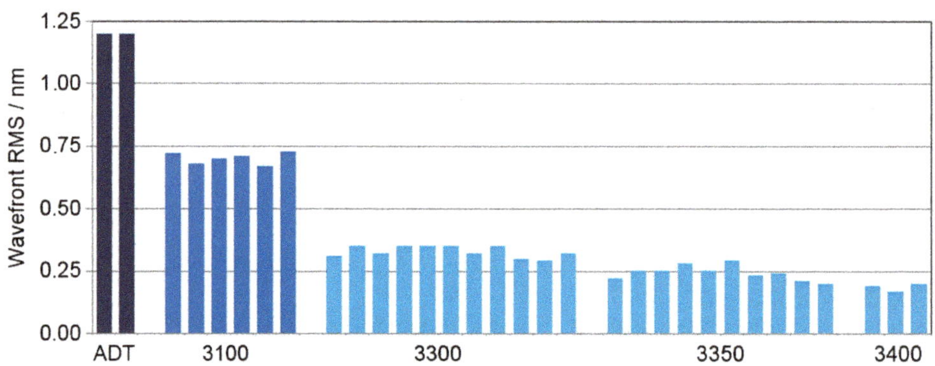

Figure 9.28 Progress in wavefront RMS error reduction in EUV imaging optics (ADT is alpha demo tool) (reprinted from Ref. 34).

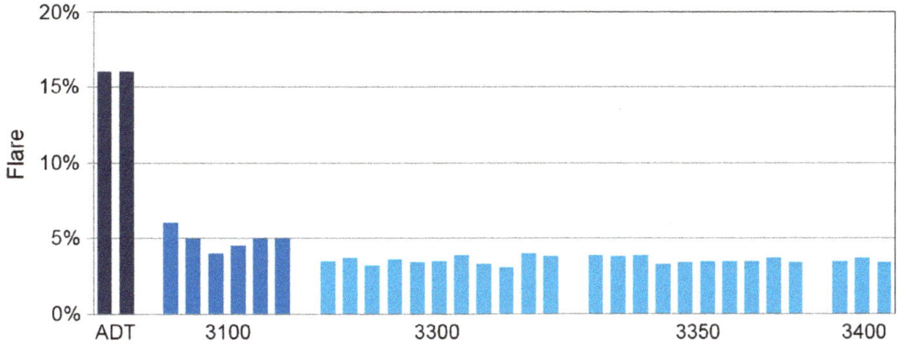

Figure 9.29 Progress in flare reduction in EUV imaging optics (reprinted from Ref. 34).

At $NA = 0.25$, a mid-spatial-frequency roughness of 0.14 nm can keep the system flare[11,35] at 10%. This precision must be maintained during fabrication, mounting, and assembly in the lens and scanner factories, as well as during the alignment and re-alignment in vacuum on the wafer-manufacturing floor. Contamination built up on the optical surfaces can also increase roughness above specification.

Contamination of the optical surface can deteriorate the ultrahigh smoothness and precision of the optical elements. Oxidation is irreversible. The capping layer on the multilayer structure is designed to prevent oxidation. Carbon buildup can presumably be cleaned. The number of cleaning cycles and the cleaning time and frequency should be kept at an acceptable level to minimize disruption of production and cost of ownership. The specification for a mirror lifetime should exceed 30,000 hours, or about 3.5 years.

Reflective optics has the inherent advantage of a polychromatic response and thus a broad bandwidth. Instead of the fraction-of-a-picometer bandwidth required by 193-nm lithography, the bandwidth requirement of EUVL is 2%. Most multilayer coatings can meet this requirement, as exemplified[11] by Fig. 9.30. A key requirement is to center all coatings to 13.5 nm. Spectral

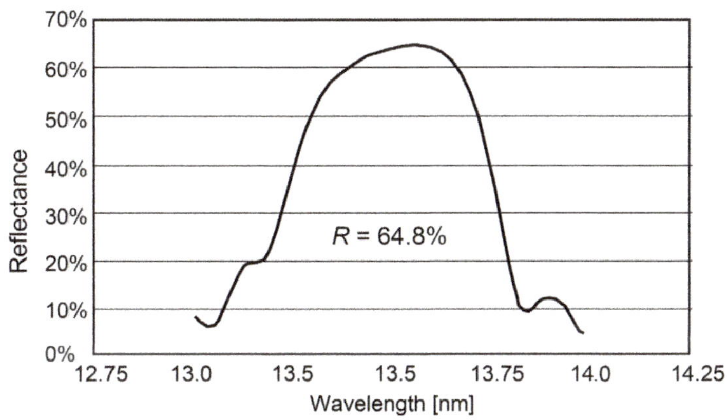

Figure 9.30 Spectral reflectivity of an EUV multilayer coating (adapted from Ref. 10).

mismatch can waste valuable EUV power. Power loss due to multilayer mismatching is about 10% according to Ref. 10.

Heat on the optical elements should be carefully managed. Figure 9.6 and Table 9.1 give an idea of the heat load the optical elements need to bear. The heat must be quickly and uniformly removed to maintain the subnanometer accuracy of the optical surfaces. From the viewpoint of heat management, it is desirable to use a fast resist. This is also true for maintaining the electric power in check for better power management, as discussed in Section 9.2.3.

9.6 EUV Resists

The resist issues for EUV light have changed rapidly over time. When soft-x-ray projection lithography was proposed, the main concern for resists was their high absorption. The absorption coefficient α was estimated[36] to be larger than 3 μm^{-1}. EUV light penetrates to 200 nm, but the effective imaging thickness is about 60 nm. At that time, 0.25 μm was considered a formidable resolution for optical lithography.[37] Only this author predicted 0.13-μm resolution.[38] EUV lithography was proposed in view of the pessimistic outlook of optical lithography. Therefore, resist absorption was clearly an issue. Bilayer and trilayer resist systems were considered necessary to support EUVL.

However, optical lithography marched on undeterred. The opportunity for a successor would barely arrive at the 32-nm half pitch, which was projected to emerge at least 20 years after the initial EUV proposal. At 20 ~ 30-nm feature sizes, resist thickness is no longer limited by absorption. Due to capillary force and other factors during resist development/rinse, as discussed in Section 7.2.4, resist images with height:width aspect ratios larger than 3:1 tend to collapse. The resist thickness is limited to less than 100 nm for the 30-nm feature size, 33 nm for the 10-nm feature size, and so on.

On the other hand, the throughput of optical scanners is on the order of 250 wph, while the field size remains 26×33 mm^2 at 4X reduction. Any successor will have to measure up to the cost of ownership of a state-of-the-art scanner. With the cost of equipment projected to be 1.5–2X that of an optical scanner, and with throughput of 100 wph, an EUV scanner is already inferior in terms of cost and productivity. Hence, high resist sensitivity is the only hope for making EUVL economical. If the resist sensitivity remains at 30–50 mJ/cm^2 for LER control, then the source power requirement will be extremely difficult to meet, as can be seen in Table 9.1. Even if physicists researching high energy succeed in producing an extremely intense EUV source, the incident power on the optical elements and the resultant heat will be difficult to manage. For throughput and heat management, it is desirable to have resist sensitivity between 1 and 2 mJ/cm^2. Of course, EUV shot noise at such a resist sensitivity must be considered, in addition to LER.

9.6.1 Mechanism of EUV resist exposure

The relationship between photon energy in electronvolts and wavelength is given by Brodie and Muray.[39] Table 9.3 shows the photon wavelengths and their equivalent energies in electronvolts according to

$$\lambda_{\text{photon}} = 1239.9\, E_0^{-1}\, \text{nm}. \tag{9.10}$$

From this table, it can be seen that the EUV energy is 14.3 times more than that for the 193-nm wavelength. With this much difference in energy, the exposure mechanism of EUV resists is quite different from that in the UV and DUV regimes. In the latter two cases, the incident light generates photoacid when it is absorbed by the photoacid generator (PAG). Its absorption by the polymer resin can be minimized. In the former case, the incident light has abundant energy to generate several orders of secondary electrons, each with energy loss in the single digit of electronvolts, to generate photoacid from the PAG and from the other polymer molecules in the resist. Therefore, the exposure mechanism of EUV resists is rather like exposing electrons of a similar electronvolt to the EUV light. As a result, EUV resists have a significant electron scattering cross section of several nanometers, just like electron resists. This electron blur reduces the resolution from the aerial image.

Because of the difference in energy, the number of EUV photons in a given volume is 14.3 times less than the number of ArF UV photons. With the same resist thickness, there are 14.3 more ArF UV photons with the same mJ/cm^2 dosage. The distribution of the photons can be shown[40] to be Poissonian. For consideration of the dosage variability of the Poisson distribution, the EUV variability is $\sqrt{14.3} = 3.8$ times higher than the ArF UV variability, as shown in Fig. 9.31. Too few photons make the CD and LER stochastic.

Hence, for CD control and LER, the EUV resist sensitivity is not allowed to be too high. On the other hand, low resist sensitivity reduces wafer throughput on the exposure tool, thus raising the exposure cost. An optimum sensitivity should be determined.

9.6.2 CAR EUV resists

Currently, there are two major types of EUV resists. Riding on the momentum of chemically amplified resists (CARs), there are CARs developed for EUV. Like CARs for DUV applications, EUV CARs consist of the

Table 9.3 Photon wavelengths and their equivalent energies in electronvolts (eV).

	g-line	i-line	KrF	ArF	F_2	EUV	X-ray
λ (nm)	436	365	248	193	157	13.5	1
eV	2.8	3.4	5.0	6.4	7.9	91.8	1240

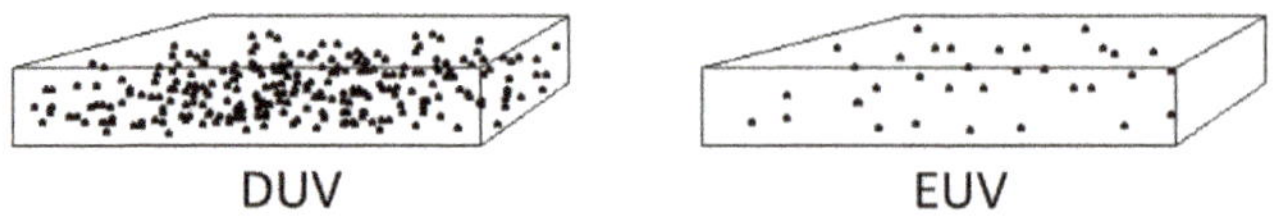

$$\frac{\text{No. of 13.5nm photons per volume}}{\text{No. of 193nm photons per volume}} = \frac{91.8\text{ eV}}{6.4\text{ eV}} = 14.3$$

$$\frac{\text{Variability of EUV photons}}{\text{Variability of ArF photons}} = \sqrt{\frac{91.8\text{ eV}}{6.4\text{ eV}}} = 3.8$$

Figure 9.31 Variability of photons for EUV and ArF UV.

polymer resin and a PAG. However, the exposure mechanism is much more complicated, as discussed in Section 9.6.1.

At least four resist companies have tried to develop an EUV resist with low LER and high sensitivity. No less than 48 samples have been evaluated.[41] Note that the LER values were converted to linewidth roughness (LWR) by multiplying by $\sqrt{2}$. The required 1-mJ/cm^2 sensitivity and the 2-nm LWR for the 32-nm node are located at the intersection of the horizontal and vertical dotted lines in Fig. 9.32. This author took the liberty to draw an asymptotic curve on the chart, showing how resist sensitivity statistically trades off with LWR.

A breakaway from the curve is needed to meet the sensitivity–LWR requirement. Several attempts have been made using acid amplifiers and photosensitized CARs. Brainard et al.[42] provide a comprehensive coverage of these two techniques. The author of this book agrees with the increase in sensitivity but is uncertain about whether they can overcome the

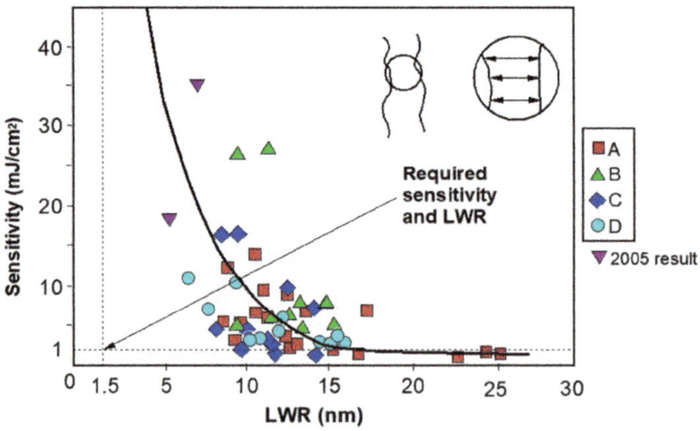

Figure 9.32 Tradeoff between sensitivity and LWR for resists from vendors A to D (adapted from Ref. 55). Two 2005 data points have been added. The asymptotic curve was drawn by the author of this book.

sensitivity–LWR tradeoff. There are indeed more photons to expose the resist. But these photons are generated from seed photons that are stochastic. Also, the acid amplification process and the photosensitization process contribute their own stochastic variabilities, just as Naulleau has demonstrated.[43] The net effect of reducing CD variability and LER must be understood, demonstrated, and optimized.

CAR-based resists have several weak points, thickness being one of them. The resist thickness must be reduced when the feature size shrinks because the resist would collapse under capillary force from removal of the developing solution, as discussed in Section 7.2.4. A 10-nm feature can only use up to 30 nm of resist thickness. With a small resist thickness, there is insufficient etch resistance to delineate the circuit patterns on the wafer. In addition, the DOF of the imaging system is consumed more by the resist in EUV systems than in DUV systems because the resist thickness is divided by its refractive index when calculating the consumption of the DOF budget. The refractive index of resists in the DUV regime is usually around 1.7, while it is usually close to 1 in the EUV regime.

Figure 9.33 shows a SEM image[44] of a 13-nm line–space pattern in an exposure-defocus matrix from a CAR. The dark-blue boundary lines define the usable E-D region of this optical setup and the CAR process. In order to quantify the E-D window, this author added three ellipses. The flat, green ellipse shows a DOF of 98 nm with an EL of 3.6%. The tall, red ellipse has a DOF of 23 nm with $EL = 10.3\%$. The third ellipse (brown) addresses the rugged shape of the blue E-D region. The abrupt loss of DOF in the middle of the EL might be due to resist collapse or other unexpected issues. Here, this author smoothed out the ruggedness to construct a generous process window. The DOF is 86 nm and the EL is 6.7%. This repaired process window has a barely usable EL and DOF. The required exposure dosage is 58 mJ/cm^2 with $LWR = 4.4$ nm and α (the absorption coefficient) $= 4\ \mu\text{m}^{-1}$. The sensitivity is far too low. Let's consider the 86-nm DOF. Since the authors of Ref. 44 did not explicitly specify the resist thickness, it can be 50 nm according to their Table 2, or 25 nm according to their processing information on page 3. For DUV, the 50-nm or 25-nm resist thicknesses would optically be 29 nm or 14.5 nm, respectively. Here, the entire 50 nm or 25 nm is counted for the DOF budget without reduction.

Another weak point of CAR-based resists is amplification. Chemical amplification and proton diffusion from the post-exposure bake introduce resist blur that ranges from 5 to 10 nm, limiting the resolution even when the aerial image can support a higher resolution. So, resist blur, sensitivity, LWR, etch resistance, and resist thickness must be improved for CARs.

9.6.3 Non-CAR EUV resists

The solution to these problems comes from introducing metal oxide such as HfO_2, ZrO_2, or SnO_x into the resist.[45,46,47] According to Yildirim et al.,[44]

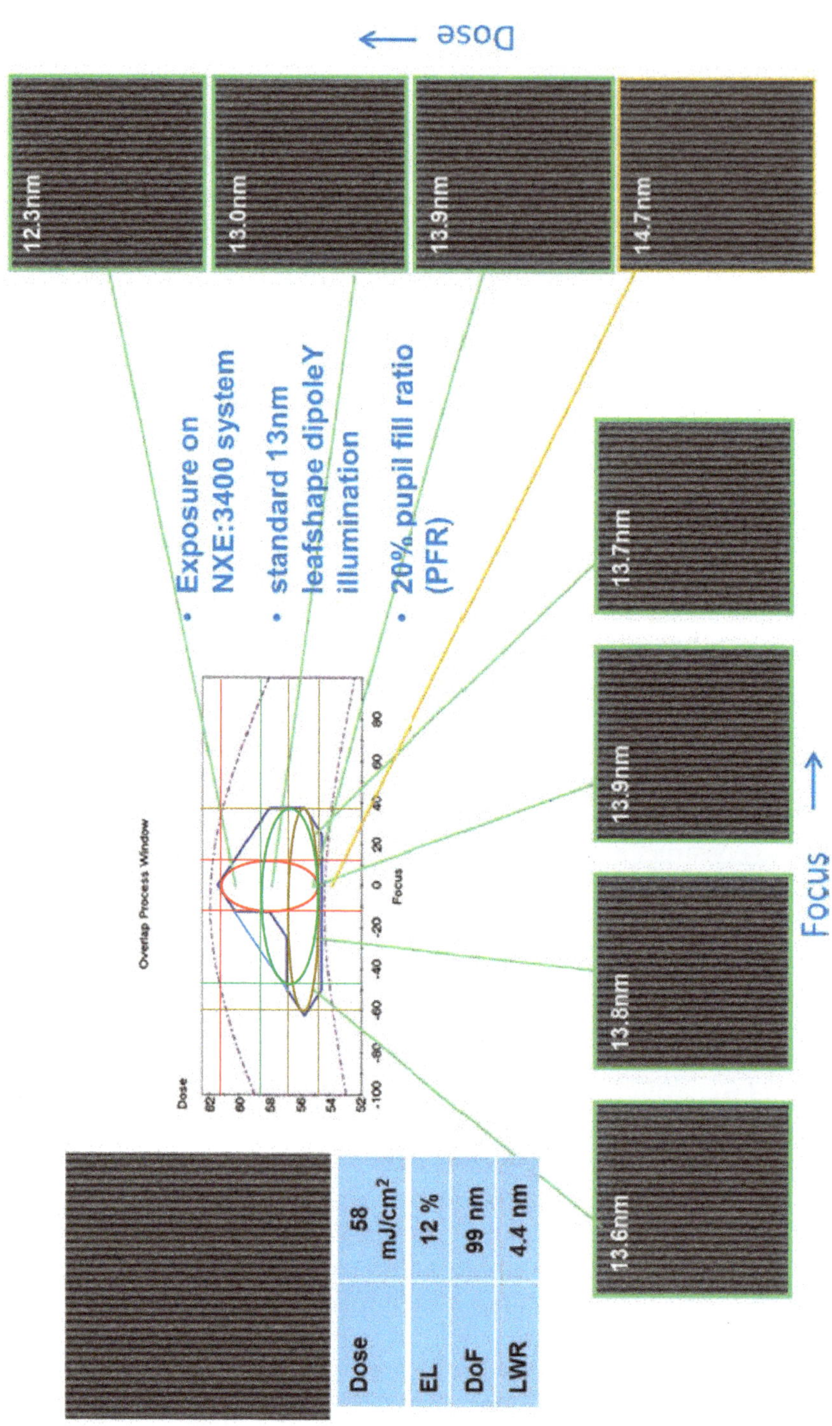

Figure 9.33 CAR image of a 13-nm line–space pattern (adapted from Ref. 44).

these oxide clusters are surrounded by organic ligands. Upon EUV exposure, the metal oxide crosslinks and becomes insoluble in organic solvents, rendering a negative polarity to the resist. According to Hinsberg and Meyers,[48] the differential condensation of oxo-networks upon EUV-patterned exposure induces different degrees of crosslinking, rendering the dissolution rate contrast in the latent image. These oxo-network formations can be driven to completion with PEB. Due to a higher etch resistance, the resist thickness can be substantially reduced. The resist blur, sensitivity, and LWR are also improved.

Figure 9.34 shows the result of imaging the same 13-nm line–space pattern in an E-D matrix but from a non-CAR. The usable E-D region is better behaved than that of the CAR. The exposure dosage is 34 mJ/cm^e, which is better but still has room for improvement. As indicated by the ellipse in the usable E-D region, there is a DOF of 131 nm at $EL = 10\%$.

For lack of the required NA and a mask for smaller pitches, Yildirim et al.[44] resorted to the EUV interference imaging tool at the Paul Scherrer Institute (PSI) to demonstrate the high resolution from the Inpria resist. Figure 9.35 shows the resist image from 13-nm line–space patterns, as well as from 12-, 11-, and 10-nm line–space patterns.

The dosages that delineate these features are 37, 55, 69, and 70 mJ/cm^2, respectively. The reason that a higher dosage is needed for smaller features is not clear. Granted, the dosage must increase to keep the same minimum number of photons in the exposed feature to maintain a similar LWR. If that criterion is used, since only the width of the resist line changes while the length and height remain constant, the dosage would be 37, 40, 44, and 48 mJ/cm^2, which are simply inversely proportional to the half pitch. The increase in exposure dosage as a function of half pitch is much higher than that caused by photon stochastics. There might be other additional stochastic sources such as the resist clusters, the ligands, and other sources that have not yet been considered. The LWR has improved from 4.4 nm using the CAR in Fig. 9.33 to 3.8 nm using the non-CAR in Fig. 9.34. With the interference imaging used for Fig. 9.35, the dosage increased to 37 ~ 55 mJ/cm^2, but the LWR improved slightly to 3.2 ~ 3.4 nm for the 13- and 12-nm half pitches. This large increase in dosage from the 13-nm half pitch to the 12-nm half pitch while the LWR deteriorates needs further investigation.

The absorption coefficient of a CAR is typically 4 μm^{-1}, and that for the Inpria resist can be as high as 20 μm^{-1}. During the i-line era, the use of resists possessing an absorption coefficient of 3 μm^{-1} was considered surface imaging, as shown in Fig. 4.39. Now, with the resist thickness reduced to double-digit nanometers, 4 μm^{-1} is not high enough. Table 9.4 shows the intensity of light as it is absorbed into a resist as a function of the penetration depth into the resist and the absorption coefficient α. It shows that, with $\alpha = 4$ μm^{-1}, only 21.7% is absorbed at a depth of 60 nm, whereas, at

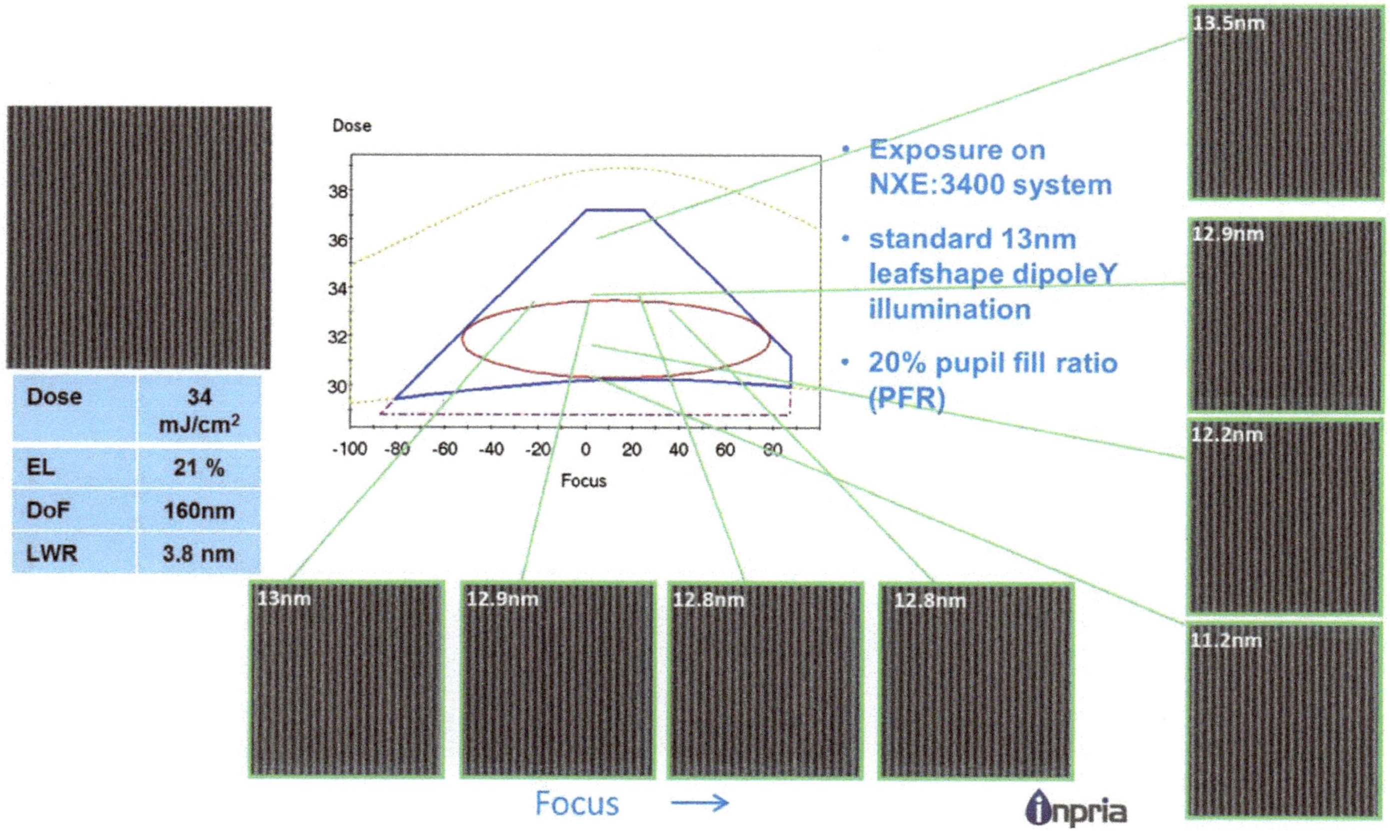

Figure 9.34 Non-CAR (Inpria YA-Series) image of a metal-containing resist for 13-nm line–space patterns. (reprinted from Ref. 44).

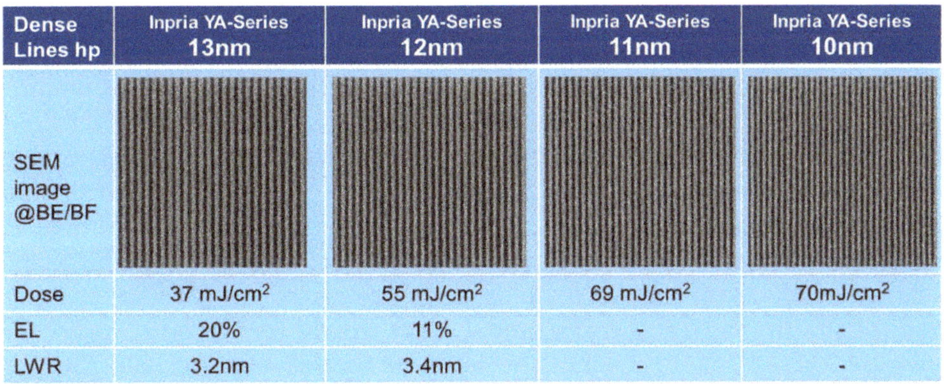

Figure 9.35 Metal-containing resist sensitivity versus resolution using interference imaging (hp is half pitch; BE is best energy, or dose-to-size value; BF is best focus) (reprinted from Ref. 44).

Table 9.4 Intensity in a resist as a function of α and the penetration depth in the resist.

	Intensity										
α (nm^{-1})	z (nm) 0	z (nm) 10	z (nm) 20	z (nm) 30	z (nm) 40	z (nm) 50	z (nm) 60	z (nm) 70	z (nm) 80	z (nm) 90	z (nm) 100
0.004	1.000	0.961	0.923	0.887	0.852	0.819	0.787	0.756	0.726	0.698	0.670
0.020	1.000	0.819	0.670	0.549	0.449	0.368	0.301	0.247	0.202	0.165	0.135
0.050	1.000	0.607	0.368	0.223	0.135	0.082	0.050	0.030	0.018	0.011	0.007

$\alpha = 20\ \mu m^{-1}$, 32% is absorbed at a depth of 20 nm. This is similar to the amount of absorbed light in the i-line era when the resist thickness is 1.3 μm and $\alpha = 0.3\ \mu m^{-1}$. Hence, the former requirement of 30% resist absorption still stands. Table 9.4 also includes an extremely high α at 50 nm^{-1}, which may not be realizable with real materials. Such a high absorption is not a problem because the intensity at $z = 20$ nm is still 36.8% of the incident intensity. For a negative resist, there should be sufficient exposure at the bottom; otherwise, the exposed area remaining on the substrate after development would be too small to keep the resist image adhered to the substrate.

9.7 Extendibility of EUVL

9.7.1 Resist sensitivity, throughput, and power at each technology node

It is now clear that the sensitivity of EUV resists must be high to support high wafer throughput, yet the resist must capture a minimum number of photons to overcome the stochastic effect. As the technology node advances, the feature size decreases and the thickness of the resist also decreases to maintain a 3:1 height:width ratio to prevent resist collapse during the drying process of resist wet development. The resist sensitivity must be suppressed to maintain the photon number threshold in the shrinking volume.

Table 9.5 shows the minimum features size (MFS) of the etched image in each node and the corresponding MFS of the resist image, the resist thickness, the required sensitivity, the wph, and the power at the IF. Here, the MFS of the etched image is used to define the node. This is a generic definition that can differ from the commercial definition of technology nodes. The MFS of the lithography image is typically larger than that of the etched image because etching is used to help trim the image to the required size. An etch bias of 2 nm is used for the 10- and 7-nm nodes, and 1.5 nm is used for the other nodes. The resist thickness is set to 3 times the MFS of the resist image, and the resist volume is the resist area multiplied by the thickness. Here, the resist image is an infinitely long straight line so the volume is taken to be per unit of feature length.

The sensitivity of the resist needs to decrease to maintain the same number of photons as in the 10-nm node. The 34 mJ/cm^2 sensitivity for the 14-nm node is based on the Inpria resist used to delineate 13-nm lines and spaces shown in Fig. 9.34. The sensitivities for the other nodes are calculated by keeping the product of the resist volume and the sensitivity constant through the nodes. The required 437 W of EUV power at IF for the 10-nm node is obtained by multiplying the resist sensitivity by 12.866 in Table 9.1, assuming that the EUV exposure tool is like the one used for that table. The EUV power required for the other nodes is evaluated the same way, according to Table 9.1.

To maintain the 100-wph throughput at the reduced sensitivity, the EUV power has to be increased for each node. This power is not linearly proportional to the mJ/cm^2 value because the wafer throughput depends on the exposure time, the stepping time, and the wafer load and unload times. The latter two t values are independent of the resist sensitivity. Hence, the throughput Q is evaluated with the following equations:

$$t_w = s \times t_{expo} + t_{step} + t_{load}, \tag{9.11}$$

Table 9.5 EUV power required for the 14-nm node to the 3.5-nm node. The calculation is based on maintaining the same number of photons for the shrinking resist volume as the node advances and maintaining a constant throughput of 100 wph for each technology node. Etch bias is used to increase the volume of the resist image for less required EUV power.

Node (nm)	14	10	7	5	3.5
Etched MFS (nm)	10	7	5	3.5	2.5
Resist MFS (nm)	12	9	6.5	5	4
Max. thickness before resist collapses (nm)	36	27	20	15	12
Resist volume at MFS / unit length (nm^2)	432	243	127	75	48
Resist sensitivity at constant shot noise	34	60	116	196	306
wph based on 437-W EUV power at IF	100.0	64.7	37.2	23.1	15.2
EUV power (W) at IF for 100 wph	437	675	1174	1893	2884

where s is the sensitivity ratio $s \equiv s_{new}/s_{orig}$, t_w is the time required to expose, step, and load and unload each wafer, t_{step} is the time to step the wafer, and t_{load} is the time to load and unload the wafer. Let

$$t_{step} + t_{load} \equiv t_c,$$

where t_c is the constant throughput with respect to resist sensitivity; then,

$$t_w = s \times t_{expo} + t_c.$$

The throughput Q is given as

$$Q = \frac{A}{t_w} = \frac{A}{s \times t_{expo} + t_c}, \tag{9.12}$$

where A is a constant of proportionality for determining Q when the t values are in dimensionless units.

For the 10-nm node, let $s = 1$, $t_c = 0.3$, $t_{expo} = 0.7$, and $A = 100$ wph. This means that 70% of the time is used for the exposure and 30% is used for stepping, loading, and unloading:

$$Q_{10} = \frac{100}{0.7s + 0.3} = 100\,\text{wph}.$$

Substituting s from the other nodes leads to Q of the corresponding node. The result is not at all promising for extending EUV to future nodes.

Five different methods might be attempted to improve the situation: (1) increase the etch bias to enlarge the resist MFS for collecting more photons; (2) improve the resist collapse threshold; (3) shorten t_c until the mechanical limits are reached; (4) make the IC design less sensitive to LER; or (5) improve the resist sensitivity without increasing LER.

Table 9.6 increases the etch bias to 2 nm per edge for the 10- and 7-nm nodes, then increases it to 1.5 nm for the other nodes. The resist collapse

Table 9.6 EUV power required for the 14-nm node to the 3.5-nm node when the etch bias is increased to 2 nm per edge for the 10- and 7-nm nodes, the etch bias is increased to 1.5 nm for the other nodes, the resist collapse threshold is stretched to 4:1 aspect ratio, and I_c is reduced to 10% of the total wafer time at the 10-nm node. IC design and resist improvement should further help the situation.

Node (nm)	14	10	7	5	3.5
Etched MFS (nm)	10	7	5	3.5	2.5
Resist MFS (nm)	14	11	8	6.5	5.5
Max. thickness before resist collapses (nm)	56	44	32	26	22
Resist volume at MFS / unit length (nm^2)	784	484	256	169	121
Resist sensitivity at constant shot noise	34	55	104	158	220
wph based on 437 W EUV power at IF (I_c = 0.3)	100.0	69.7	40.9	28.2	20.7
EUV power (W) at IF for 100 wph (I_c = 0.3)	437	626.61	1067.9	1550.2	2113.1
wph based on 437 W EUV power at IF (I_c = 0.1)	125.0	81.0	44.6	29.9	21.6
EUV power (W) at IF for 100 wph (I_c = 0.1)	349.6	539.21	980.52	1462.8	2025.7

threshold is improved to a 4:1 height:width aspect ratio, and the stepping time is reduced to 10% of the total wafer time at the 10-nm node. The IC design and resist improvement have not been incorporated. None of the five ways to improve the throughput is easy, but these methods should be pursued to contain the cost of EUVL.

The foregoing analysis of the resist sensitivity required for each node is based on 1D features such as lines and spaces. For 2D features, the resist volume calculation is based on the length, width, and height of the resist image. The sensitivity requirement is even less promising.

In terms of keeping the number of photons constant, we assume that the 34 mJ/cm^2 dosage for the 13-nm half pitch is adequate and use it as the starting point to maintain the same number of photons for each reduced volume. However, when tracing the nodes backwards, the constant-number-of-photons-per-MFS argument cannot be used. Otherwise, the density of photons would become too low. When going from the 14-nm resist MFS backwards, one needs to use the constant photon density as the criterion.

9.7.2 Increasing NA

It is difficult to reduce the EUV half pitch to 0.4 k_1. At $NA = 0.33$, the half pitch is 16 nm, which is not too much smaller than the 20-nm half pitch from double-patterning immersion lithography. To avoid multiple patterning, a $NA = 0.55$ tool has been announced[30] that would bring the half pitch to 9.8 nm. However, there are many problems with increasing the NA by brute force.

When the options of increasing the mask size and increasing the CRAO are both removed from consideration, a clever solution that involves a compromise is offered. The NA in the scan direction is increased to 0.55 to take advantage of the zero CRAO in that direction. The reduction ratio must be 8X to keep the mask NA at 0.1375; otherwise, the 3D mask absorber effect would worsen. Without increasing the mask size, the image field size in the y direction is reduced from 33 mm to 16.5 mm. In the x direction, the NA remains at 0.33 and the reduction ratio stays at 4X. Thus, the 26-mm field size in x is retained. This type of optics is called anamorphic.

This field size is now half of the customary field size, which is not as bad as being reduced to a quarter of its size when the NA is increased by brute force. In terms of productivity, more steps are required to cover the entire wafer. Throughput is thus reduced. A remedy is to increase the stepping and scanning speeds, which is possible for both speeds, but is not easy. If feasible, the same technique can also be used for the 0.33-NA scanners to improve the highly desired productivity.

A smaller field size is less forgiving in terms of defects and reduces yield. For an IC circuit utilizing fields larger than 26×16.5 mm^2, stitching is required. There is no way to allow for perfect stitching. Therefore, the IC circuit should be designed with loosened specifications at the stitching area.

Stitching also requires an additional mask, the cost of which would be comparable to double patterning with the 0.33-NA tool. But the latter, having a full-field lens, provides higher productivity than the former.

An obvious tradeoff for better resolution is DOF in the y direction. Applying the scaling equation, the DOF at 0.55 NA is $\sin^2(0.165) / \sin^2(0.275) =$ 2.94 times smaller. For a series of 12-nm spaces, van Schoot et al.[49] report a DOF of 150 nm. This translates to 7.2-nm spaces with a 50-nm DOF. For contact holes, the DOF would be even smaller.

The DOF using 0.55-NA optics is much larger, as is the DOF using a high-NA EUV scanner. The increase in cost and footprint is measured in multiples instead of percentages.

Hence, whether or not to employ high-NA EUV scanners depends on what the 0.33-NA scanners can do. Double patterning, triple patterning, and quadruple patterning have been proven in immersion lithography. Pattern-splitting techniques have been developed. The question is now whether EUVL can take advantage of these techniques. The challenge is in improving the overlay accuracy for the dimensions in EUVL. Let us assume that $k_1 = 0.35$ is possible for EUVL and that (at least) triple patterning can be achieved. This is a resolution of 4.77-nm half pitch, which is not much larger than the 4.3-nm half pitch using 0.55-NA and double patterning. With the advantages of a large field size, large DOF, no dimension shrinkage for mask making, smaller footprint, lower tool cost, and higher productivity for the 0.33-NA tool, it is difficult to carve out a niche for the 0.55-NA tool.

9.8 Summary of EUVL

EUVL offers an opportunity to extend optical lithography by at least two generations. Tremendous progress has been made in EUV optics, multilayer coating, source, resist, mask, and mask making. As EUVL is a lucratively funded project with a high-return commercial cause, much knowledge of its physics, chemistry, materials, processes, and metrology has been gained, and the related technology has been established. Regardless of whether EUVL will become an industrial workhorse, many highly talented scientists have been gainfully employed performing its research. Volumes of reports on EUVL have been produced, advancing its science and engineering.

The need for investment has not ceased. To spend the funds wisely, further research to increase the source power needs to consider heat-induced damage and defects. Resists with a small resist blur and that absorb more photons for better LER without trading off CD control and resist adhesion need to be developed. Mitigation of debris and extension of the life of the collector are also important, not to mention the need to reduce the probability of particles adhering to the mask during exposure.

The other challenges to be faced include off-axis optics; stray light; atomic-level configuration and smoothness of the optical components; optical components' shortened lifetime from debris, contamination and high-energy radiation; mask defects, mask inspection and repair, mask cost, mask pellicles, and mask flatness; random phase shifting; field-dependent OPC; system uptime, throughput, and footprint; energy consumption; and cost.

9.9 Outlook of Lithography

UV ~ DUV lithography with all of the resolution-enhancement techniques including multiple patterning has taken the semiconductor technology from the 5-μm node through the 3.5-, 2.5-, 1.8-, 1.25-, 1-, 0.7-, 0.5-, 0.35-, 0.25-, 0.18-, and 0.13- μm nodes, then through the 90-, 65-, 40-, 28-, 20-, 1-6, 10-, and 7-nm nodes. Altogether, lithography has gone through 19 nodes in 4 decades. The resolution has been reduced by almost 3 orders of magnitude. This has helped to sustain Moore's law with acceptable costs.

Figure 9.36 shows the progression of optical lithography starting from g-line tools using $NA = 0.15$, delineating $k_1 = 0.8$ features at 2.3 μm. Without improving any other parameters, the NA increased by several steps to 0.45, producing 0.78-μm features with k_1 still at 0.8. At that time RETs were not yet available and NA was considered too difficult to increase. The wavelength was shortened to 365 nm, while the NA was relaxed to 0.35, producing 0.83-μm features. It was during the 365-nm wavelength era that RETs came into play. With $k_1 = 0.6$, a barely subwavelength geometry of 357 nm was supported.

The wavelength line is marked in red. During the g-line era, all features were above the red line. At the i-line, the bottom of the node touched the red line.

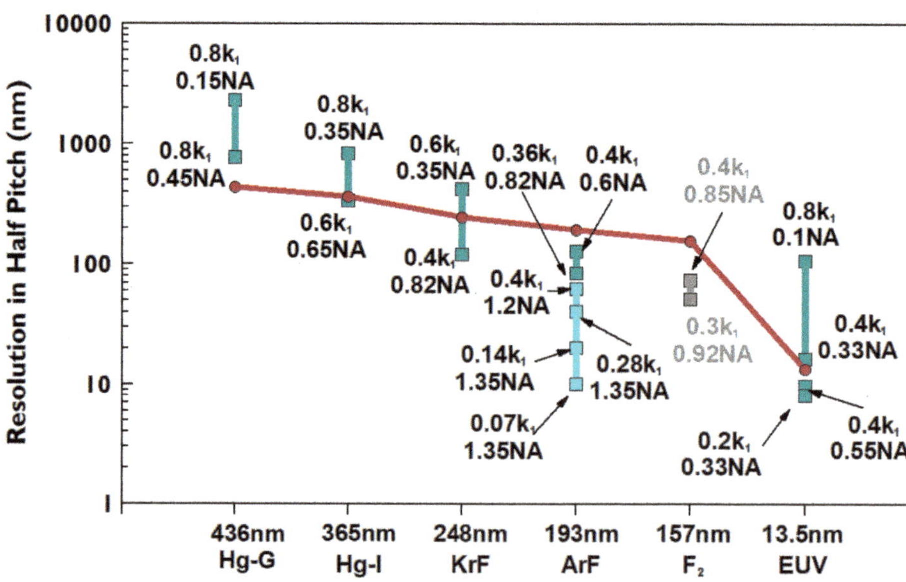

Figure 9.36 Evolution of lithography including immersion and EUV lithography.

A quantum jump happened at 248 nm. The wavelength was significantly reduced by switching from an Hg arc lamp light source to excimer lasers. Chemically amplified resists were introduced. Not only was the resolution improved, but productivity was improved as well. This was due to the much higher brightness from the new light source and the substantial increase in sensitivity with a CAR. Because of the narrow bandwidth and much less chromatic aberration to handle, the NA went from 0.35 to 0.82, and k_1 went from 0.6 to 0.4. The resolution went from 425 nm to 121 nm, which is substantially below the wavelength.

The ArF laser, the 193-nm resist, and the ArF optics took time and effort to develop. ArF tools were substantially used at the 65-nm node, while the KrF tools covered the 250-, 180-, 130-, and 90-nm nodes. By the 45-nm node, 193-nm water immersion was needed. Double patterning started at the 20-nm node through the 7-nm node with multiple patterning. Single-patterning k_1 started at 0.4 and finally reached 0.28; k_1 was extended by multiple patterning to 0.07.

EUVL was originally planned to be phased in at the 90-nm node. Its incorporation into production was stalled until the 7-nm node, during which time it faced the last few nodes of Moore's law of scaling. Whether it can cash in on its potential of a 4.77-nm half pitch remains to be seen. One should be aware that the definition of a technology node is no longer based on the MFS but rather depends on commercial designations.

Many other aspects of the semiconductor technology such as etching, deposition, polishing, and metrology are approaching their respective brick walls. Lithography may still last longer than many of the other technologies.

When scaling finally comes to an end, the semiconductor industry will still be thriving due to the never-ceasing need for more circuits, more functions, and more applications. By that time, lithography for manufacturing might fall into several distinct groups: the i-line non-CAR group up to the 350-nm node, the KrF-CAR group up to the 80-nm node, the ArF-CAR group (dry and immersion) up to the 7-nm node, and the EUV group—if the aforementioned issues can be overcome. EUVL is like a supersonic transport with a short commercial span due to its development and running cost. It would be cost effective for the institutions that cannot afford CARs to apply all of the RETs to the i-line group.

Breaking away from the inertia of scaling, more innovations, such as new devices with lower power, higher density, better performance, and lower cost, are needed to compete in the IC marketplace. New devices will be invented. Better IC designs and architectures can speed up calculations, save power, reduce cost, and improve yield by making ICs less prone to defects, noise, LER, and high physical accuracy dependency. More advanced packaging with reduced cost and improved performance is in demand. There will be new applications from the new capabilities. Competition in manufacturing cost will continue to play its indispensable role.

References

1. H. Kinoshita, K. Kurihara, Y. Ishii, and Y. Torii, "Soft x-ray reduction lithography using multilayer mirrors," *J. Vac. Sci. Technol. B* **7**, p. 1648 (1989).
2. J. E. Bjorkholm, J. Bokor, L. Eichner, R. R. Freeman, T. E. Jewell, W. M. Mansfield, A. A. Macdowell, E. L. Raab, W. T. Silfvast, L. H. Szeto, D. M. Tennant, W. K. Waskiewicz, D. L. White, D. L. Windt, O. R. Wood II, and J. H. Brunning, "Reduction imaging at 14 nm using multilayer-coated optics: Printing of features smaller than 0.1 μm," *J. Vac. Sci. Technol. B* **8**, p. 1509 (1990).
3. D. Attwood, *Soft X-Rays and Extreme Ultraviolet Radiation*, Cambridge University Press, p. 404 (1999).
4. G. Yeap, S. S. Lin, Y. M. Chen, H. L. Shang, P. W. Wang, H. C. Lin, Y. C. Peng, J. Y. Sheu, M. Wang, X. Chen, B. R. Yang, C. P. Lin, F. C. Yang, Y. K. Leung, D. W. Lin, C. P. Chen, K. F. Yu, D. H. Chen, C. Y. Chang, H. K. Chen, P. Hung, C. S. Hou, Y. K. Cheng, J. Chang, L. Yuan, C. K. Lin, C. C. Chen, Y. C. Yeo, M. H. Tsai, H. T. Lin, C. O. Chui, K. B. Huang, W. Chang, H. J. Lin, K. W. Chen, R. Chen, S. H. Sun, Q. Fu, H. T. Yang, H. T. Chiang, C. C. Yeh, T. L. Lee, C. H. Wang, S. L. Shue, C. W. Wu, R. Lu, W. R. Lin, J. Wu, F. Lai, Y. H. Wu, B. Z. Tien, Y. C. Huang, L. C. Lu, J. He, Y. Ku, J. Lin, M. Cao, T. S. Chang, and S. M. Jang, "5nm CMOS production technology platform featuring full-fledged EUV, and high mobility channel FinFETs with densest 0.021 μm^2 SRAM cells for mobile SoC and high-performance computing applications," *2019 IEEE International Electron Devices Meeting (IEDM)*, San Francisco, pp. 36.7.1–36.7.4 (2019).
5. D. A. Tichenor, G. D. Kubiak, W. C. Replogle, L. E. Klebanoff, J. B. Wronosky, L. C. Hale, H. N. Chapman, J. S. Taylor, J. A. Folta, C. Montcalm, R. M. Hudyma, K. A. Goldberg, and P. P. Naulleau, "EUV engineering test stand," *Proc. SPIE* **3997**, pp. 48–69 (2000) [doi: 10.1117/12.390083].
6. D. A. Tichenor, G. D. Kubiak, S. J. Haney, and D. W. Sweeney, "Extreme ultraviolet lithography machine," U.S. Patent 6,031,598 (2000).
7. H. Meiling, J. P. H. Benschop, U. Dinger, and P. Kuerz, "Progress of the EUVL alpha tool," *Proc. SPIE* **4343**, pp. 38–50 (2001) [doi: 10.1117/12.436675].
8. J. B. P. van Schoot and J. C. M. Jasper, "Fundamentals of EUVL Scanners," in *EUV Lithography, 2nd Edition*, V. Bakshi, Ed., SPIE Press, Chapter 9, Fig. 9.18 (2018) [doi: 10.1117/3.2305675.ch9].
9. K. Ota, K. Murakami, H. Kondo, T. Oshino, K. Sugisaki, and H. Komatsuda, "Feasibility study of EUV scanners," *Proc. SPIE* **4343**, pp. 60–69 (2001) [doi: 10.1117/12.436704].

10. H. Meiling, V. Banine, N. Harned, B. Blum, P. Kurz, and H. Meijer, "Development of the ASML EUV alpha demo tool," *Proc. SPIE* **5751**, pp. 90–101 (2005) [doi: 10.1117/12.600725].
11. I. V. Fomenkov, D. C. Brandt, A. I. Ershov, A. A. Schafgans, Y. Tao, and G. O. Vaschenko, "EUV Sources for High-Volume Manufacturing," in *EUV Lithography, 2nd Edition*, V. Bakshi, Editor, SPIE Press, Chapter 3A, p. 116 (2018) [doi: 10.1117/3.2305675.ch3A].
12. G. Schriever, M. Rahe, W. Neff, K. Bergmann, R. Lebert, H. Lauth, and D. Basting, "Extreme ultraviolet light generation based on laser produced plasmas (LPP) and gas discharge based pinch plasmas: A comparison of different concepts," *Proc. SPIE* **3997**, pp. 162–168 (2000) [doi: 10.1117/12.390051].
13. J. Pankert, R. Apetz, K. Bergmann, G. Derra, M. Janssen, J. Jonkers, J. Klein, T. Krucken, A. List, M. Locken, C. Metzmacher, W. Neft, S. Probst, R. Prummer, O. Rosier, S. Seiwet, G. Siemons, D. Vaudrevange, D. Wagemann, A. Weber, P. Zink, P. Zink, and O. Zitzen, "Integrating Philips' extreme UV source in the alpha-tools," *Proc. SPIE* **5751**, pp. 260–271 (2005) [doi: 10.1117/12.598650].
14. I. V. Fomenkov, D. C. Brandt, A. I. Ershov, A. A. Schafgans, Y. Tao, and G. O. Vaschenko, "EUV Sources for High-Volume Manufacturing," in *EUV Lithography, 2nd Edition*, V. Bakshi, Editor, SPIE Press, Chapter 3A, Fig. 3A.4 (2018) [doi: 10.1117/3.2305675.ch3A].
15. I. V. Fomenkov, D. C. Brandt, A. I. Ershov, A. A. Schafgans, Y. Tao, and G. O. Vaschenko, "EUV Sources for High-Volume Manufacturing," in *EUV Lithography, 2nd Edition*, V. Bakshi, Editor, SPIE Press, Chapter 3A, Fig. 3A.27 (2018) [doi: 10.1117/3.2305675.ch3A].
16. B. J. Lin, "NGL comparable to 193-nm lithography in cost, footprint, and power consumption," *Microelectronic Engineering* **86**(4–6), pp. 442–227 (2008).
17. S. Bajt, Z. R. Dai, E. J. Nelson, M. A. Wall, J. Almeda, N. Nguyen, S. Baker, J. C. Robinson, J. S. Taylor, M. Clift, A. Aquila, E. M. Gullikson, and N. V. G. Edwards, "Oxidation resistance of Ru-capped EUV multilayers," *Proc. SPIE* **5751**, pp. 118–127 (2005) [doi: 10.1117/12.597443].
18. M. Kureishi, R. Ohkubo, M. Hosoya, T. Shoki, N. Sakaya, H. Kobayashi, O. Nozawa, Y. Usui, and O. Nagarekawa, "Development of low damage mask making process on EUV mask with thin CrN buffer layer," *Proc. SPIE* **5751**, pp. 158–167 (2005) [doi: 10.1117/12.598613].
19. S. Migura, W. Kaiser, J. T. Neumann, H. Enkisch, and D. Hellweg, "Optical Systems for EUVL," in *EUV Lithography, 2nd Edition*, V. Bakshi, Editor, SPIE Press, Chapter 5, Fig. 5.19 (2018) [doi: 10.1117/3.2305675.ch5].
20. G. F. Lorusso and I. S. Kim, "Shadowing effect characterization and compensation," IMEC Technical Program week **H1**, L111 (2007).

21. A. Yen, "Semiconductor manufacturing using EUV lithography: progress and remaining challenges," 2013 EUV Symposium, Toyama, Japan, October 6 (2013).
22. H. Seitz, F. Sobel, M. Renno, T. Leutbeecher, N. Olschewski, T. Relchardt, R. Walter, H. Becker, U. Buttgereit, G. Hess, K. Knapp, C. Wies, and R. Lebert, "Recent results on EUV mask blank multilayers and absorbers," *Proc. SPIE* **5751**, pp. 190–199 (2005) [doi: 10.1117/12.600538].
23. T. Liang, E. Frendberg, D. Bald, M. Penn, and A. Stiveres, "E-beam repair: fundamental capability and applications," *Proc. SPIE* **5567**, pp. 456–466 (2004) [doi: 10.1117/12.569210].
24. M. Dusa, A. Yen, M. Phillips, I. Corp, Y. Ekinci, M. Vockenhuber, N. M. Mojarad, D. Fan, T. Kowaza, J. Joseph, S. Santillan, and T. Itani, "Progress and challenges of EUV lithography for high-volume manufacturing," Semicon Taiwan, Taipei, September 5 (2014).
25. D. Smith, "NXE pellicle progress update," EUV Mask Pellicle TWG, 2015 EUVL Symposium, 4–7 October, Maastricht, The Netherlands (2015).
26. L. Scaccaborozzi, D. Smith, P. R. Diago, E. Casimiri, N. Dzioomkina, and H. Meijer, "Investigation of EUV pellicle feasibility," *Proc. SPIE* **8679**, 867904 (2013) [doi: 10.1117/12.2015833].
27. D. Brouns, A. Bendiksen, P. Broman, E. Casimiri, P. Colsters, P. Delmastro, D. de Graaf, P. Janssen, M. van de Kerkhof, R. Kramer, M. Kruizinga, H. Huntzel, F. van der Meulen, D. Ockwell, M. Peter, D. Smith, B. Verbrugge, D. van de Weg, J. Wiley, N. Wojewoda, C. Zoldesi, and P. van Zwol, "NXE pellicle: offering an EUV pellicle solution to the industry," *Proc. SPIE* **9776**, 97761Y (2016) [doi: 10.1117/12.2221909].
28. B. Kneer, S. Migura, W. Kaiser, J. T. Neumann, and J. van Schoot, "EUV lithography optics for sub-9nm resolution," *Proc. SPIE* **9422**, 94221G (2015) [doi: 10.1117/12.2175488].
29. R. Peeters, S. Lok, E. van Alphen, N. Harned, P. Kuerz, M. Lowisch, H. Meijer, D. Ockwell, E. van Setten, G. Schiffelers, J.-W. van der Horst, J. Stoeldraijer, R. Kazinczi, R. Droste, H. Meiling, and R. Kool, "ASML's NXE platform performance and volume introduction," *Proc. SPIE* **8679**, 86791F (2013) [doi: 10.1117/12.2010932].
30. J. van Schoot, E. van Setten, G. Rispens, K. Z. Troost, B. Kneer, S. Migura, J. T. Neumann, and W. Kaiser, "High-numerical aperture extreme ultraviolet scanner for 8-nm lithography and beyond," *J. Micro/Nanolith., MEMS, and MOEMS* **16**(4), 041010 (2017) [doi: 10.1117/1.JMM.16.4.041010].
31. P.-Y. Yan, A. Myers, Y. Shroff, M. Chandhok, G. Zhang, E. Gullikson, and F. Salmassi, "EUVL alternating phase shift mask," *Proc. SPIE* **7969**, 79690G (2011) [doi: 10.1117/12.884790].

32. Y. Deng, B. M. La Fontaine, H. J. Levinson, and A. R. Neureuther, "Rigorous EM simulation of the influence of the structure of mask patterns on EUVL imaging," *Proc. SPIE* **5037**, pp. 302–313 (2003) [doi: 10.1117/12.484986].
33. P.-Y. Yan, I. Mochi, and K. Goldberg, "EUV actinic imaging tool aerial image evaluation of EUVL embedded phase shift mask performance," *Proc. SPIE* **8322**, 83221P (2012) [doi: 10.1117/12.919710].
34. S. Migura, W. Kaiser, J. T Neumann, H. Enkisch, and D. Hellweg, "Optical Systems for EUVL," in *EUV Lithography, 2nd Edition*, V. Bakshi, Editor, SPIE Press, Chapter 5, Fig. 5.11 (2018) [doi: 10.1117/3.2305675.ch5].
35. M. Lowisch, U. Dingeer, U. Mickan, and T. Heil, "EUV imaging: an aerial image study," *Proc. SPIE* **5374**, pp. 53–63 (2004) [doi: 10.1117/12.537338].
36. G. N. Taylor, R. S. Hutton, S. M. Stein, C. H. Boyce, O. R. Wood II, and D. R. Wheeler, "Extreme UV resist technology: the limits of silylated resist resolution," *Proc. SPIE* **2437**, pp. 308–330 (1995) [doi: 10.1117/12.209168].
37. T. E. Jewell, M. M. Becker, J. E. Bjorkholm, J. Bokor, L. Eishner, R. R. Freeman, W. M. Mansfield, A. A. Macdowell, M. L. O'Malley, E. L. Raab, W. T. Silfvast, L. H. Szeto, D. M. Tennant, W. K. Waskiewicz, D. L. White, D. L. Windt, O. R. Wood II, and J. H. Bruning, "20:1 projection soft x-ray lithography using tri-level resist," *Proc. SPIE* **1263**, pp. 90–98 (1990) [doi: 10.1117/12.20173].
38. B. J. Lin, "Quarter- and sub-quarter micrometer optical lithography," *Proc. 1991 Int. Symposium on VLSI Technology, Systems, and Applications (VTSA)* **1**, pp. 16–21, Taipei, 22–24 May (1991).
39. I. Brodie and J. J. Muray, *The Physics of Microfabrication*, Plenum Press, p. 80 (1982).
40. G. M. Gallatin and P. P. Naulleau "EUVL System Patterning Performance," in *EUV Lithography, 2nd Edition*, V. Bakshi, Editor, SPIE Press, Chapter 10, p. 677 (2018) [doi: 10.1117/3.2305675.ch10]; G. M. Gallatin, "Resist blur and line edge roughness," *Proc. SPIE* 5754, pp. 38–52 (2005) [doi: 10.1117/12.607233].
41. H. B. Cao, J. M. Roberts, J. Dalin, M. Chandhok, R. P. Meagley, E. M. Panning, M. K. Shell, and B. J. Rice, "Intel's EUV resist development," *Proc. SPIE* **5039**, pp. 484–491 (2003) [doi: 10.1117/12.485095].
42. R. L. Brainard, M. Neisser, G. Gallatin, and A. Narasimhan, "Photoresists for EUV Lithography," in *EUV Lithography, 2nd Edition*, V. Bakshi, Editor, SPIE Press, Section 8.7 (2018) [doi: 10.1117/3.2305675.ch8].
43. G. M. Gallatin and P. P. Naulleau "EUVL System Patterning Performance," in *EUV Lithography, 2nd Edition*, V. Bakshi, Editor, SPIE Press, Chapter 10, Table 10.1 (2018) [doi: 10.1117/3.2305675.ch10].

44. O. Yildirim, E. Buitrago, R. Hoefnagels, M. Meeuwissen, S. Wuister, G. Rispens, A. van Oosten, P. Derks, J. Finders, M. Vockenhuber, and Y. Ekinci, "Improvements in resist performance towards EUV HVM," *Proc. SPIE* **10143**, 101430Q (2017) [doi: 10.1117/12.2257415].
45. M. Trikeriotis, W. J. Bae, E. Schwartz, M. Krysak, N. Lafferty, P. Xie, B. Smith, P. A. Zimmerman, C. K. Ober, and E. Giannelis, "Development of an inorganic photoresist for DUV, EUV, and electron beam imaging," *Proc. SPIE* **7639**, 76390E (2010) [doi: 10.1117/12.846672].
46. C. Y. Ouyang, Y. S. Chung, L. Li, M. Neisser, K. Cho, E. P. Giannelis, and C. K. Ober, "Non-aqueous negative-tone development of inorganic metal oxide nanoparticle photoresists for next generation lithography," *Proc. SPIE* **8682**, 86820R (2013) [doi: 10.1117/12.2011282].
47. A. Grenville, J. T. Anderson, B. L. Clark, P. De Schepper, J. Edson, M. Greer, K. Jiang, M. Kocsis, S. T. Meyers, J. K. Stowers, A. J. Telecky, D. De Simone, and G. Vandenberghe, "Integrated fab process for metal oxide EUV photoresist," *Proc. SPIE* **9425**, 94250S (2015) [doi: 10.1117/12.2086006].
48. W. D. Hinsberg and S. Meyers, "A numeric model for the imaging mechanism of metal oxide EUV resists," *Proc. SPIE* **10146**, 1014604 (2017) [doi: 10.1117/12.2260265].
49. J. van Schoot, K. van Ingen Schenau, G. Bottiglieri, K. Troost, J. Zimmerman, S. Migura, B. Kneer, J. T. Neumann, and W. Kaiser, "EUV high-NA scanner and mask optimization for sub-8nm resolution," *Proc. SPIE* **9776**, 97761I (2016) [doi: 10.1117/12.2220150].
50. I. Fomenkov, D. Brandt, A. Ershov, A. Schafgans, Y. Tao, G. Vaschenko, S. Rokitski, M. Kats, M. Vargas, M. Purvis, R. Rafac, B. La Fontaine, S. De Dea, A. LaForge, J. Stewart, S. Chang, M. Graham, D. Riggs, T. Taylor, M. Abraham, and D. Brown, "Light sources for high-volume manufacturing EUV lithography: technology, performance, and power scaling," *Advanced Optical Topics* **6**(3–4), pp. 173–186 (2017).
51. B. J. Lin, "Optical lithography with and without NGL for single-digit nanometer nodes," *Proc. SPIE* **9426**, 942602 (2015) [doi: 10.1117/12.2087008].
52. R. Peeters' slide presentation, "ASML's NXE platform performance and volume," given at the SPIE Advanced Lithography 2013 Symposium.
53. D. C. Brandt, I. V. Fomenkov, N. R. Farrar, B. La Fontaine, D. W. Myers, D. J. Brown, A. I. Ershov, R. L. Sandstrom, G. O. Vaschenko, N. R. Böwering, P. Das, V. Fleurov, K. Zhang, S. N. Srivastava, I. Ahmad, C. Rajyaguru, S. De Dea, W. J. Dunstan, P. Baumgart, T. Ishihara, R. Simmons, R. Jacques, R. Bergstedt, P. Porshnev, C. Wittak, R. Rafac, J. Grava, A. Schafgans, Y. Tao, K. Hoffman, T. Ishikawa, D. Evans, and S. Rich, "CO_2/Sn LPP EUV sources for device development and HVM, *Proc. SPIE* **8679**, 86791G (2013) [doi: 10.1117/12.2011212].

54. M. Antoni, W. Singer, J. Schultz, J. Wangler, I. Escudero-Sanz, and B. Kruizinga, "Illumination optics design for EUV lithography," *Proc SPIE* **4146**, pp. 25–34 (2000) [doi: 10.1117/12.406673].
55. H. Meiling, V. Banine, P. Kuerz, and N. Harned, "Progress in the ASML EUV program," *Proc. SPIE* **5374**, pp. 31–42 (2004) [doi: 10.1117/12.534784].
56. P.-Y. Yan, "EUVL alternating phase-shift mask imaging evaluation," *Proc. SPIE* **4889**, pp. 1099–1105 (2002) [doi: 10.1117/12.468103].

Appendix A
Methods to Evaluate the Region of Validity Based on Lithography Applications

By Yen Hui Hsieh, Ming Xiang Hsieh, and Burn Lin

A.1 Motivation

Section 2.3 compares the various methods of approximation to evaluate their closeness to the exact solutions. An infinite slit opening is used for the comparison because the solutions for the exact TE and TM illuminations from such a slit are available. The criterion used is based on root mean square (RMS) error. Yen Hui Hsieh and Ming Xiang Hsieh, the two contributors to Figs. 2.12 to 2.21, and this author have been pondering whether there are better criteria for evaluating the validity of the approximation methods—criteria that are based on lithography applications instead of absolute fitness to the exact curves.

We went through several different criteria consisting of (1) the Pearson correlation coefficient for defining the similarity of two functions, (2) the log-slope criterion for determining the effectiveness of critical dimension (CD) control, (3) the CD at mid-waist of the diffraction pattern, as already covered in Section 2.3, and (4) the same RMS criterion as was given in Section 2.3, except that it was evaluated polychromatically using the five dominant mercury lines in the spectral region between 313 and 436 nm.

The above research is published here as an appendix to serve curious readers.

A.2 Similarity of the Approximation Methods According to the Pearson Correlation Coefficient

In Section 2.3, the RMS error is exploited as an indicator of the validity of the diffraction approximation methods. Another possible way to describe the correctness of the shape of two objects is by similarity measurement. Similarity measurement is often used in data analysis, image processing, and bioinformatics. For example, in bioinformatics, it can be used to quantify the degree of co-localization among fluorophores.[1] In image processing, it is used in image registration.[2] To quantitatively measure the similarity between the approximate and the exact patterns, the Pearson correlation coefficient is often used. The coefficient can be written as

$$C = \frac{1}{x_2 - x_1} \int_{x_1}^{x_2} \frac{F_{\text{approx}} - \overline{F}_{\text{approx}}}{\sigma_{\text{approx}}} \frac{F_{\text{exact}} - \overline{F}_{\text{exact}}}{\sigma_{\text{exact}}}, \tag{A.1}$$

which shows the sample covariance over the product of sample standard deviations. Here, $\overline{F}$ and σ are the means and the standard deviations of the diffraction patterns, respectively. The similarity is normalized, and it ranges from –1 to 1. The case for $C=1$ is called perfect correlation, and the case for $C=-1$ is called perfect negative correlation. It is interesting to compare the RMS error to the similarity error $\delta = 1 - |C|$ for some specific patterns. Table A.1 shows four conditions of correlation, while Fig. A.1 presents the corresponding intensity plots. Even if the similarity is 1, the RMS error can be quite large. However, the similarity error can also be larger than the RMS error. The case for $F_{\text{approx}} = -F_{\text{exact}}$ suffers the worst RMS error, while the similarity is unity. For the case of $F_{\text{approx}}(x) = F_{\text{exact}}(-x)$, just the opposite is true.

We set $\delta = 1.25\%$ as the threshold for the usable region determined with RMS error, as shown in Figs. 2.12 and 2.17. Figure A.2 shows the region of validity for Fraunhofer approximation with respect to the TE solution. Each pixel in the W-z space is represented by a color. The scale of the color level is the same as that of Figs. 2.12 to 2.17. The white lines plotted with W-z coordinates correspond to the Fresnel numbers, $\nu = 1.126$, 11, and 51. The red line $\nu = 2.59$ in Fig. A.2 marks the $\delta = 1.25\%$ border. It can be seen that the region of $\nu < 2.59$ is acceptable; the similarity error dramatically decreases as ν becomes less than 2.59. Using the diffraction patterns at $\nu = 11$ and 1.126 as

Table A.1 Four cases of similarity and their corresponding RMS errors.

	Exact Function	**Proportionally reduced by *n***	**Flipped upside down**	**Flipped left to right**
Function value	F_{exact}	$F_{\text{approx}} = nF_{\text{exact}}$	$F_{\text{approx}} = -F_{\text{exact}}$	$F_{\text{approx}}(x) = F_{\text{exact}}(-x)$
Mean	F_{exact}	$F_{\text{apx.avg}} = nF_{\text{exct,avg}}$	$F_{\text{apx.avg}} = -F_{\text{exct,avg}}$	$F_{\text{apx.avg}}(x) = F_{\text{exct,avg}}(-x)$
Standard deviation	σ_{exact}	$\sigma_{\text{approx}} = n\sigma_{\text{exact}}$	$\sigma_{\text{approx}} = \sigma_{\text{exact}}$	$\sigma_{\text{approx}} = \sigma_{\text{exact}}$
Similarity coefficient C	1	1	−1	0.38
Similarity error $\delta = 1 - \text{abs}(C)$	0	0	0	0.62
RMS error (δ_{RMS}) referring to exact function	0	0.38	0.58	0.29

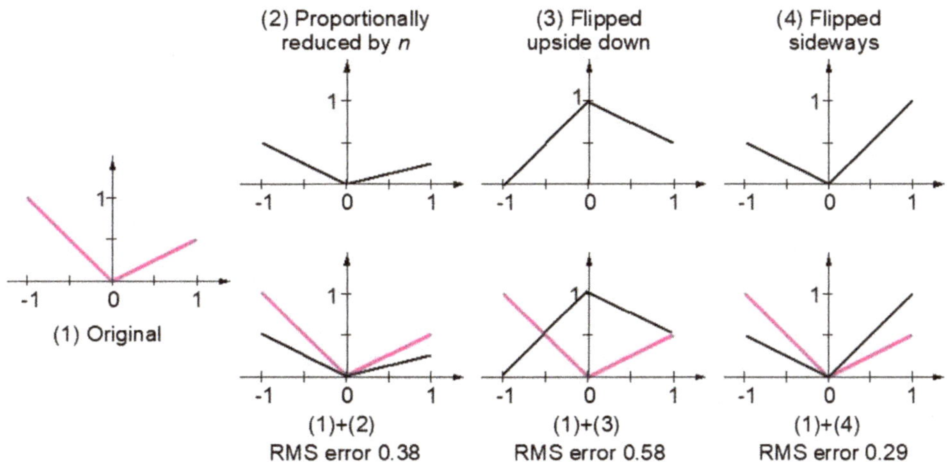

Figure A.1 Intensity plots of the four cases of similarity in Table A.1.

shown in Figs 2.10 and 2.11, respectively, we are convinced that the decrease in similarity error is attributed to the transition from multipeak to single-peak diffraction patterns. In other words, the similarity is higher when the diffraction patterns for F_{approx} and F_{exact} are smoother. This is the reason that the similarity error at $(W, z) = (5, 22.2)$ is smaller than the RMS error, and the region of validity indicated by using the similarity error is somewhat larger than the region indicated by the RMS error. In addition, the intensity contour of the similarity error is smoother than the contour of the RMS error since the nonlinear correlation that arises from the boundary condition difference

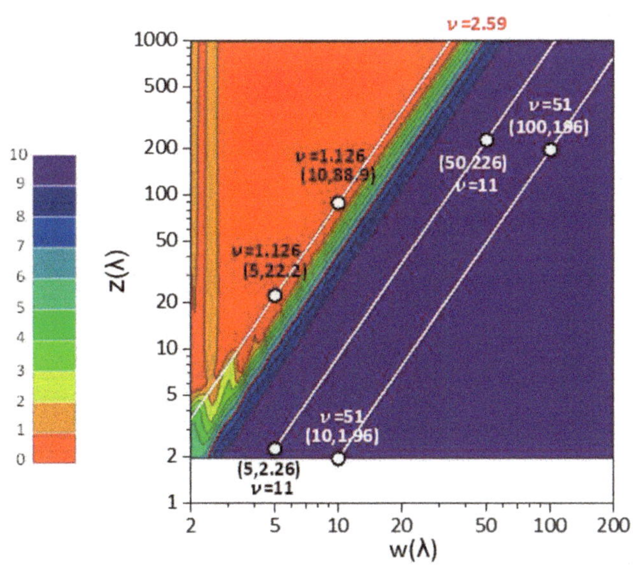

Figure A.2 Region of validity for Fraunhofer–TE.

between the TE case and physical optics approximation (POA) cannot be readily detected by the Pearson correlation coefficient.

Figure A.3 shows the region of validity for Fraunhofer–TM. The $\delta = 1.25\%$ zone boundary can also be regarded as $\nu = 2.59$, with the lower part of the red line moving up slightly compared to the line in Fig. A.2.

Figures A.4 and A.5 show the region of validity for Fresnel–TE and Fresnel–TM, respectively. In comparison with the Fraunhofer case, the valid

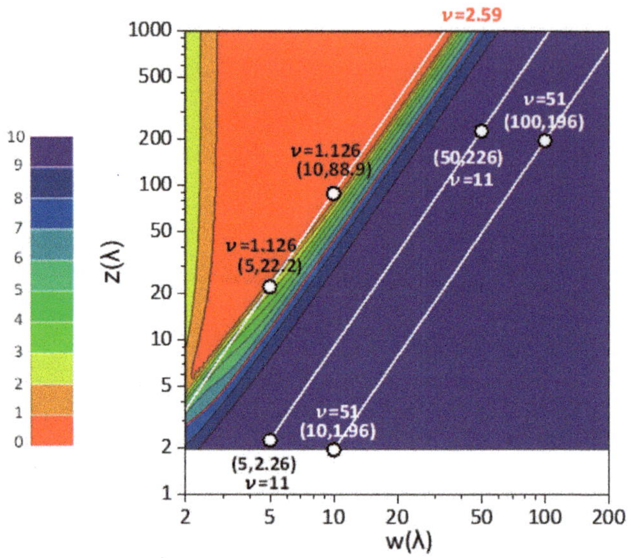

Figure A.3 Region of validity for Fraunhofer–TM.

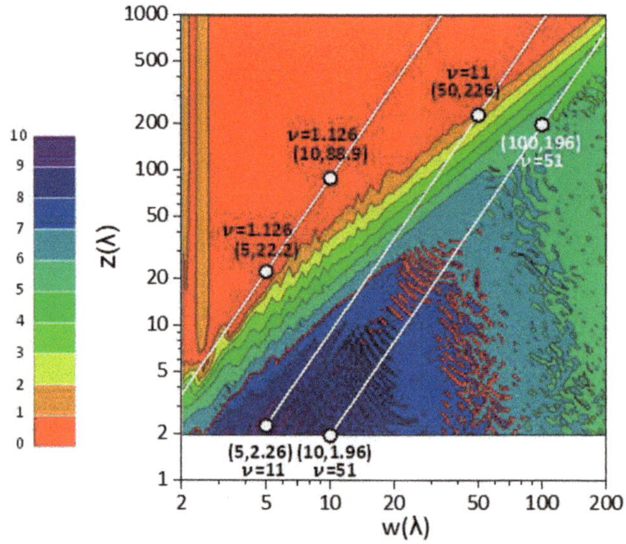

Figure A.4 Region of validity for Fresnel–TE.

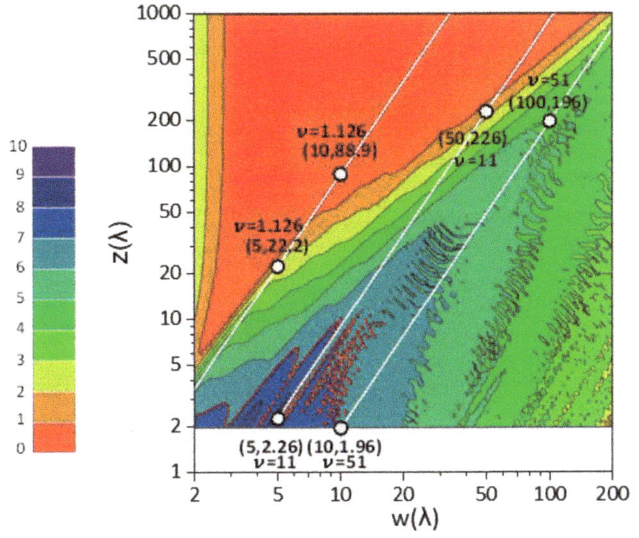

Figure A.5 Region of validity for Fresnel–TM.

W-z region is larger due to Fresnel being a better approximation. Again, the intensity contour of the similarity error in the acceptable region for the Fresnel–TE case is smoother than that of RMS error.

Figures A.6 and A.7 depict the region of validity for POA–TE and POA–TM, respectively. The region boundary is near the very bottom of the figure—similar to the boundary in the RMS case. Again, the ragged intensity contour in the POA–TE RMS case is not apparent in the similarity plot.

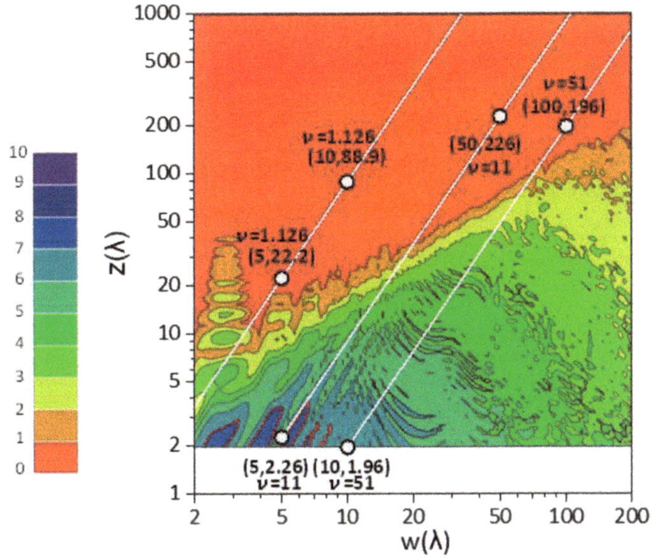

Figure A.6 Region of validity for POA–TE.

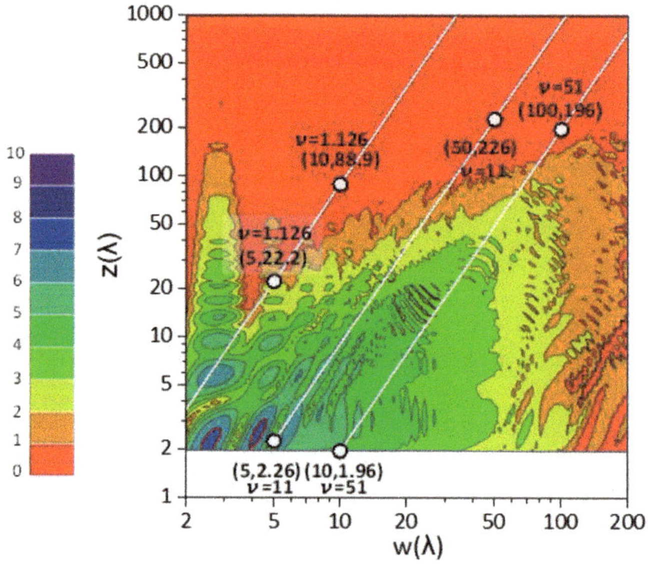

Figure A.7 Region of validity for POA–TM.

A.3 Critical Dimension

CD is indeed an important criterion for lithography. We made the comparison of validity based on CD at the mid-waist of a given diffraction curve, i.e., evaluating the CD at half of the peak intensity. This works when the mask-to-wafer gap is not too large. Further out, the peak intensity is too small, and the diffraction curve is too widely spread, making the resulting CD severely larger than any reasonable target CD. Hence, the log slope that determines the effectiveness of CD control is a better indicator.

A.4 Log-slope–CD Control

Because of the extremely small peak intensity and the severely flattened slope at large z (as discussed in Section A.3), we evaluate the log-slope error with X_{CD} located at the nominal edge of the slit, i.e., $X_{CD} = 0.5W$. The CD control aspect of the log-slope is discussed in Chapter 5.

The log-slope at X_{CD} is given by

$$S = \left. \frac{\partial \ln(F)}{\partial x} \right|_{x=X_{CD}}, \tag{A.2}$$

where F is the diffraction intensity of the Fresnel approximation, the TE solution, or the TM solution. Then, the slope error is defined as

$$\delta_s = \left| \frac{S_{approx} - S_{exact}}{S_{approx}} \right| \times 100\%. \tag{A.3}$$

The contour plots of log-slope error at the slit edge for the TE and TM cases with respect to the Fresnel approximation are shown in Figs. A.8 and A.9, respectively. The log-slope error is ≥10% at small *W*. This is quite different from the RMS-defined region of validity, as shown in Fig. 2.15.

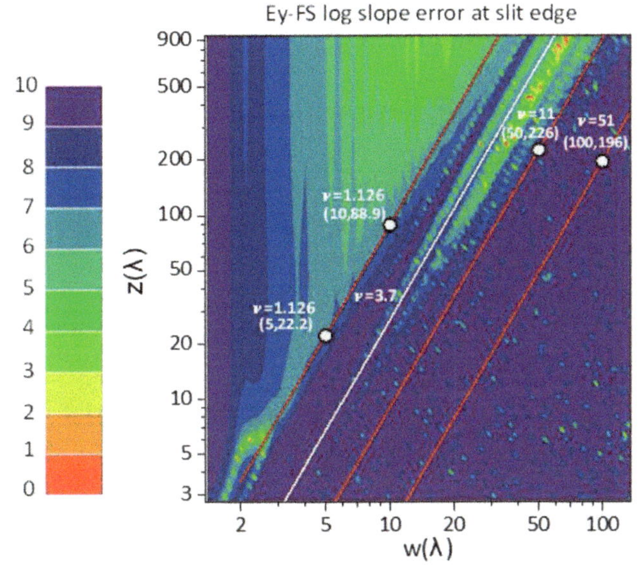

Figure A.8 Log-slope error contour plots for TE–Fresnel.

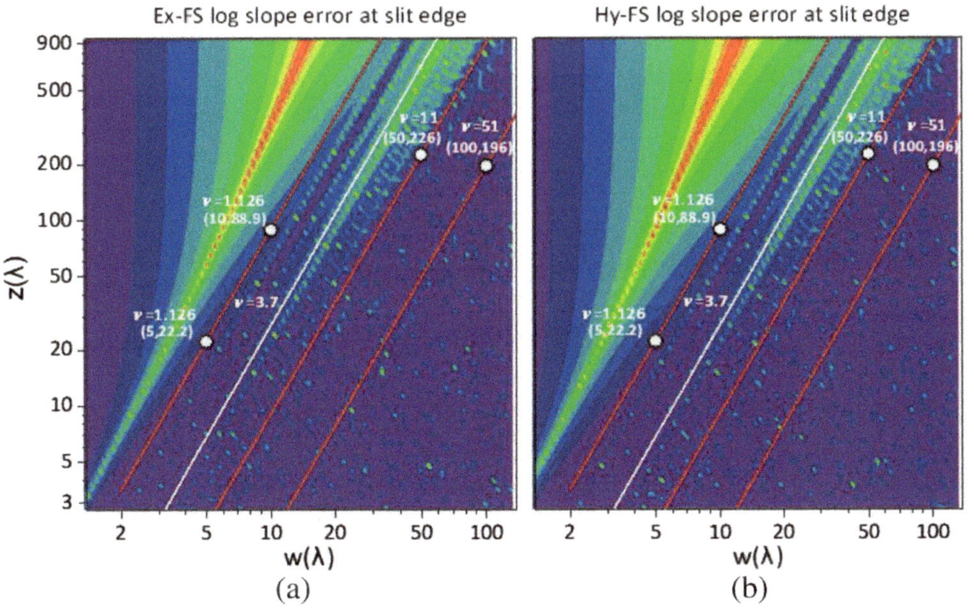

Figure A.9 Log-slope error contour plots for TM–Fresnel (a) using the electric field in the TM case for comparison and (b) using the magnetic field instead.

There, the RMS error is the smallest at the upper left corner of the W-z space. We are surprised and do not have a good explanation for the behavior of the log-slope error in that corner. Less surprising and more explainable is the rapidly changing and large slope error at the region between $\nu = 1.126$ and $\nu = 3.7$. The diffraction patterns take the shape of two steep slopes at both edges of the slit, and the intensity between them rapidly changes, as exemplified by Fig. 2.9. These middle peaks consolidate into a smaller and smaller number of peaks until there is no longer any middle peak, as exemplified by the shape in Fig. 2.10. The slopes at the two edges of the slit follow the consolidation of the middle peaks but are not fully synchronized among the different approximations and the exact solution, which explains the large and rapidly changing slope differences between $\nu = 1.126$ and $\nu = 3.7$. It is worth noting that a similar structure can be seen in the CD error plots shown in Figs. 2.20 and 2.21—for the same reason.

A.5 Polychromatic Illumination

In Section 2.3, the region of validity and CD error for approximation methods under the assumption of a single spectral line are evaluated. In reality, when proximity printing is practiced, a broad spectrum is more desirable. This not only provides a higher intensity for better throughput, but also smooths out the diffraction curves for less erratic behavior in the diffraction image. Here, the influence of polychromatic light on the region of validity is investigated. Fresnel approximation is selected as the paradigm because it can be effectively evaluated from well-tabulated Fresnel integrals. All five spectral lines from the Hg arc lamp are given by 313 nm, 334 nm, 365 nm, 405 nm, and 436 nm, with the relative intensities being 84, 60, 192, 136, and 169, respectively.[3] Accordingly, the polychromatic intensity is expressed as

$$I(x) = \frac{\sum_{i=1}^{5} C_i E_i^*(x, \lambda) \cdot E_i(x, \lambda)}{\sum_{i=1}^{5} C_i}, \tag{A.4}$$

where C_i is the relative intensity. Figures A.10 to A.12 are diffraction patterns of five spectral lines from the Hg arc lamp and the polychromatic diffraction pattern for Fresnel approximation and TE and TM illuminations. Both the slit width W and the propagation distance z are set to 7.5 μm (≈20λ) for convenience. Overall, the diffraction patterns of polychromatic intensity are smoother than those of any single wavelength. Furthermore, the polychromatic diffraction patterns of the three situations being compared are similar to each other, implying a good region of validity.

The RMS error is plotted in the W-z space in Figs A.13 and A.15. Each parameter is normalized to the central wavelength of 365 nm. It is seen that

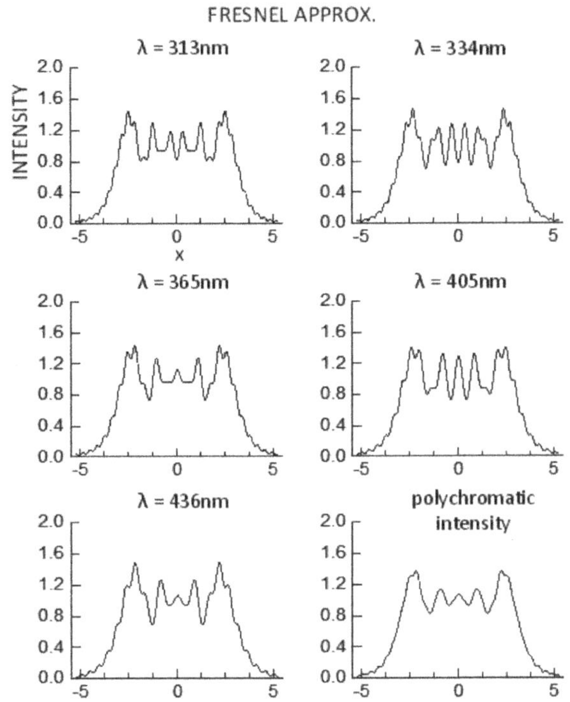

Figure A.10 Fresnel diffraction patterns of five spectral lines from the Hg arc lamp and their polychromatic counterparts.

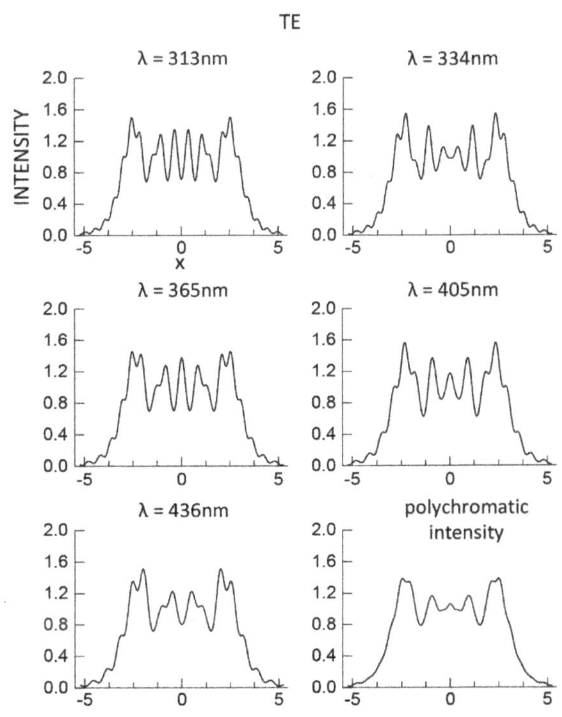

Figure A.11 TE diffraction patterns of five spectral lines from the Hg arc lamp and their polychromatic counterparts.

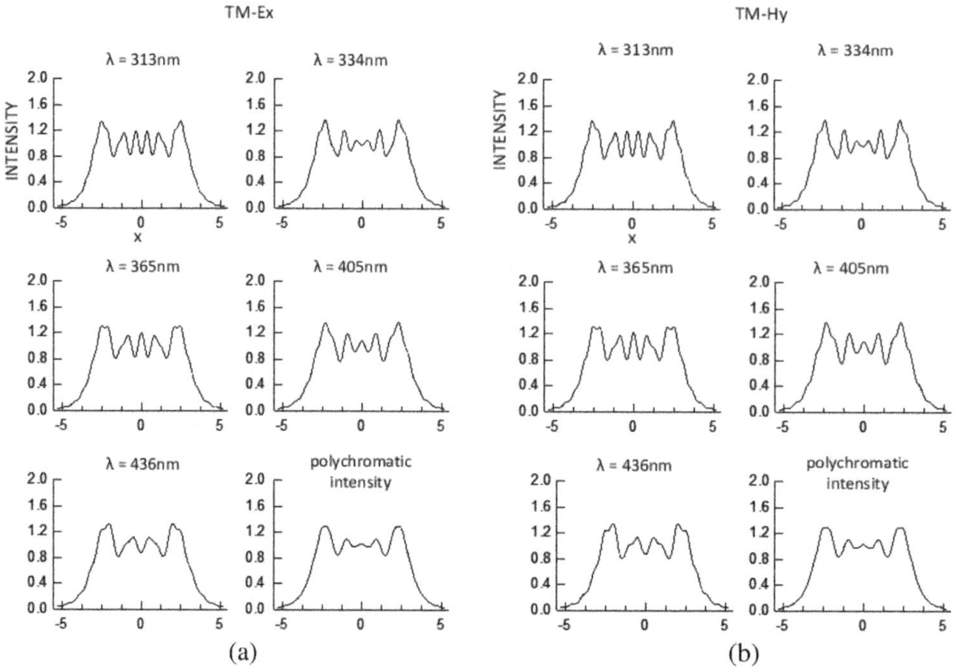

Figure A.12 TM diffraction patterns of five spectral lines from the Hg arc lamp and their polychromatic counterparts for (a) the electric field and (b) the magnetic field.

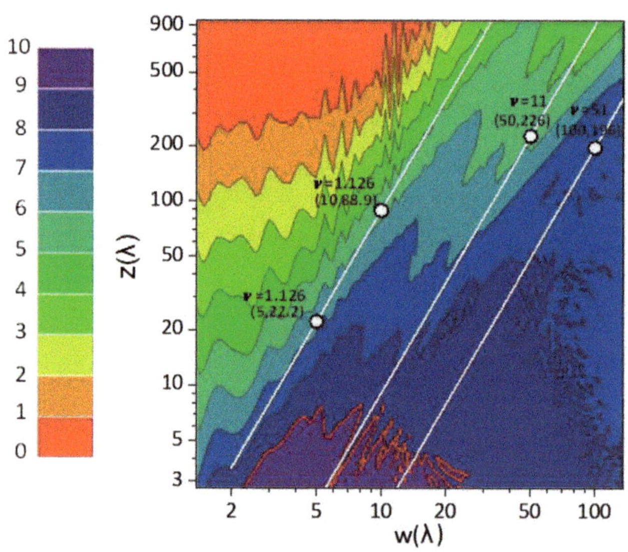

Figure A.13 Polychromatically improved region of validity for Fresnel–TE.

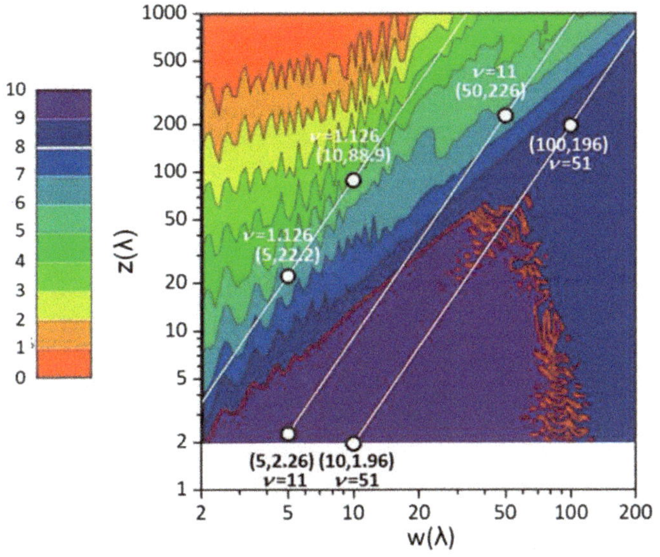

Figure A.14 Monochromatic region of validity for Fresnel–TE (same as Fig. 2.14).

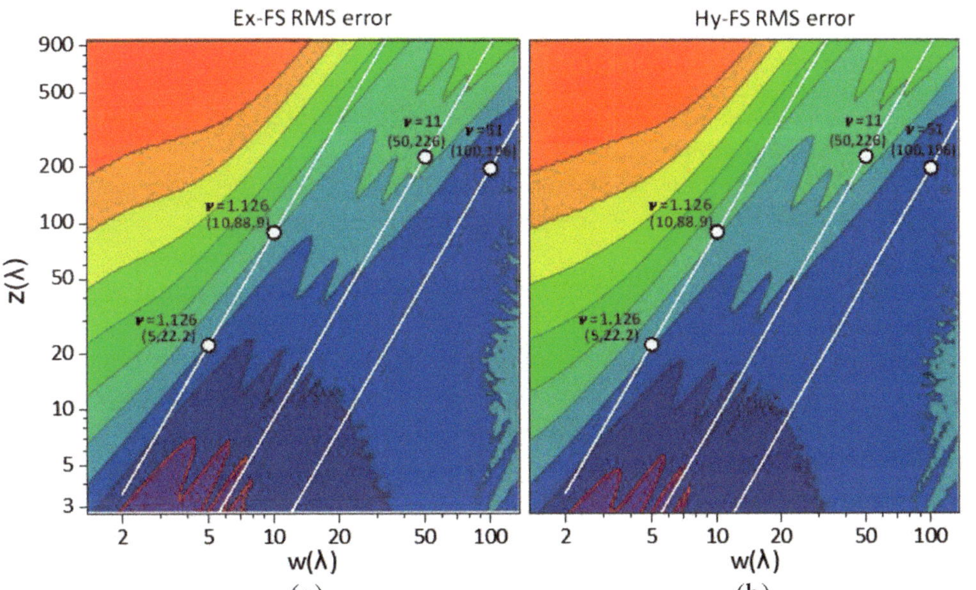

Figure A.15 Polychromatically improved region of validity for Fresnel–TM in (a) the electric field and (b) the magnetic field.

the averaging effect of the five wavelengths leads to an improved region of validity with respect to the regions in Figs. 2.14 and 2.15, replotted here as Figs. A.14 and A.16. Furthermore, the polychromatic Fresnel–TE case has less jitter than its single-wavelength counterpart.

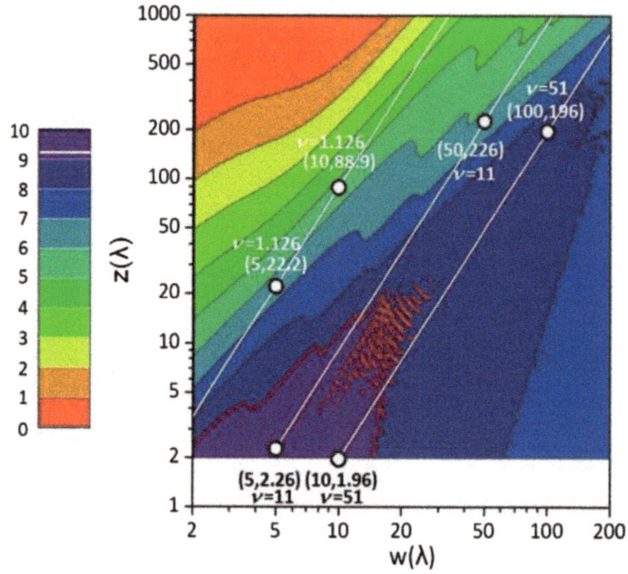

Figure A.16 Monochromatic region of validity for Fresnel–TM (same as Fig. 2.15). Here, only the magnetic field is used for comparison.

A.6 Summary and Conclusion

We have discussed several criteria for evaluating the correctness of approximation methods compared to the exact solutions. Our conclusions are as follow. (1) The similarity evaluated with the Pearson correlation coefficient was exploited to display generally more optimistic regions of validity than those based on RMS error. (2) The CD control ability was used to indicate the region of validity for a CD defined by the nominal edge. This criterion is less promising but is more relevant for lithography applications. (3) Finally, polychromatic illumination was shown to be helpful for improving the region of validity based on RMS error. We also expect it to improve the similarity. The log-slope from polychromatic illumination is expected to be slightly worse.

References

1. J. Adler and I. Parmryd, "Quantifying colocalization by correlation: The Pearson correlation coefficient is superior to the Mander's overlap coefficient," *Cytometry* **77A**(8), pp. 733–742 (2010).
2. W. K. Pratt, "Correlation techniques of image registration," *IEEE Transactions on Aerospace and Electronic Systems* **AES-10**(3), pp. 353–358 (1974).
3. R. Newman, Ed., *Fine Line Lithography*, North-Holland Publishing Co., Chapter 2, p. 137 (1980).

Index

Burn J. Lin is a Distinguished Research Chair Professor at the National Tsing Hua University in Taiwan and the Director of the Tsing Hua-TSMC Joint Research Center. From 2011 to 2015 he was Vice President and a Distinguished Fellow at TSMC, Ltd., which he joined in 2000 as a Senior Director. Earlier he held various technical and managerial positions at IBM in the U.S., after joining the IBM T.J. Watson Research Center in 1970. He has been extending the limit of optical lithography for half a century. In 1991, he founded Linnovation, Inc. and is still the CEO of the company.

Dr. Lin is the inaugurating Editor in Chief of the *Journal of Micro/Nanolithography, MEMS, and MOEMS*, a life member of the U.S. National Academy of Engineering, Academician of Academia Sinica, Laureate of the Industrial Technology Research Institute in Taiwan, IEEE and SPIE Life Fellow, and a joint distinguished professor at the National Chiao Tung University and the National Taiwan University. He is also Distinguished Alumni of the Ohio State University and the National Taiwan University.

Dr. Lin is the recipient of the 2018 Future Science Prize in Mathematics and Computer Science, a 2017 SPIE Award for inaugurating the Optical Microlithography Conference in 1987, the 2013 IEEE Jun-Ichi Nishizawa Medal, the first Semi IC Outstanding Achievement Award in 2010, the 2009 IEEE Cledo Brunetti Award, the 2009 Benjamin G. Lamme Meritorious Achievement Medal, the 2007 Industrial Technology Advancement Award, the 2006 Distinguished Optical Engineering Award from the Optical Engineering Society of R.O.C., the 2005 Most Valuable Player in VLSI Research Inc.'s All-Stars of the Chip Making Industry, the 2005 Two Best R&D Managers in Taiwan, the 2004 Outstanding Research Award from PWY Foundation, the first SPIE Fritz Zernike Award in 2004, the 2003 Outstanding Scientific and Technological Worker Award, and the 2002 Ten Best Engineers in Taiwan award. Throughout his career, he has received 2 TSMC Innovation Awards, 10 IBM Invention Awards, and an IBM Outstanding Technical Contribution Award.

Dr. Lin has been pioneering deep-UV lithography since 1975, multilayer resist systems since 1979, simulation of 2D partially coherent imaging since 1980, exposure-defocus methodology since 1980, scaling equations of resolution and depth of focus since 1986, k_1 reduction since 1987, proofs of 1X mask limitations since 1987, vibration in optical imaging since 1989, electrical measurement of contact holes since 1989, E-G tree for x-ray proximity printing since 1990, experimental demonstration of the impact of mask reflectivity on imaging since 1990, optimum lens NA since 1990, attenuated phase-shifting mask since 1991, signamization since 1996, LWD-η and LWD-β since 1999, nonparaxial scaling equations of resolution and depth of focus since 2000, 193-nm immersion lithography since 2002, and polarization-induced stray light since 2004. His innovations and research efforts have spanned 21 lithography generations, starting at the 5000-nm node and extending to the 5-nm node.

Dr. Lin has written 2 books, 3 book chapters, and more than 132 articles, 71 of which he is the sole or the first author. He also holds 88 U.S. patents.